Advances in Questionnaire Design,
Development, Evaluation, and Testing

Advances in Questionnaire Design, Development, Evaluation, and Testing

Edited by

Paul C. Beatty
Debbie Collins
Lyn Kaye
Jose-Luis Padilla
Gordon B. Willis
Amanda Wilmot

WILEY

This edition first published 2020

Registered Office
John Wiley & Sons, Inc., 111 River Street, Hoboken, NJ 07030, USA

Editorial Office
111 River Street, Hoboken, NJ 07030, USA

For details of our global editorial offices, customer services, and more information about Wiley products visit us at www.wiley.com.

Wiley also publishes its books in a variety of electronic formats and by print-on-demand. Some content that appears in standard print versions of this book may not be available in other formats.

Library of Congress Cataloging-in-Publication Data is applied for

Paperback: 9781119263623

Cover Design: Wiley
Cover Image: © Tuomas Lehtinen/Getty Images

Set in 10/12pt WarnockPro by SPi Global, Chennai, India

Printed in the United States of America

V10014773_101619

We dedicate this book to some of our most ardent supporters:

Paul: to my mother, Mary Beatty

Debbie: to Marc, Lilly, and George

Lyn: to Deb Potter, my mentor and my good friend, who first started me on this path

Jose-Luis: to my wife, Carmen; my son, Andres; and my mother, Josefina

Gordon: to the memory of my mentor, Monroe Sirken

Amanda: to my husband Lawrence LeVasseur for his love and support, and in memory of my mother Christine Wilmot

And also to the memory of our colleague Scott Fricker, a valued contributor to this book and to our field, who was taken from us far too soon.

Contents

List of Contributors

Antuane Allen
Health Analytics, LLC
Silver Spring, MD, USA

Jo d'Ardenne
The National Centre for Social Research
London, UK

Dorothée Behr
Survey Design and Methodology, GESIS
Leibniz Institute for the Social Sciences
Mannheim, Germany

Michael Braun
Survey Design and Methodology, GESIS
Leibniz Institute for the Social Sciences
Mannheim, Germany

J. Michael Brick
Westat
Rockville, MD, USA

Sarah Brockhaus
Institute for Statistics, Ludwig Maximilian University of Munich
Munich, Germany
and
Mannheim Centre for European Social Research (MZES)
University of Mannheim
Mannheim, Germany

Dynesha Brooks
Montgomery County Public Schools
Rockville, MD, USA

Ian Brunton-Smith
Department of Sociology
University of Surrey
Guildford, UK

Debbie Collins
The National Centre for Social Research
London, UK

James M. Dahlhamer
Division of Health Interview Statistics
National Center for Health Statistics
Hyattsville, MD, USA

Trine Dale
Department of Research and Development
Kantar TNS
Oslo, Norway

Mary C. Davis
Office of Survey and Census Analytics
US Census Bureau
Washington, DC, USA

Don A. Dillman
Department of Sociology
Washington State University
Pullman, WA, USA

Jennifer Dykema
University of Wisconsin Survey Center
University of Wisconsin-Madison
Madison, WI, USA

Stephanie Eckman
RTI International
Washington, DC, USA

W. Sherman Edwards
Westat
Rockville, MD, USA

Rory Fitzgerald
ESS ERIC HQ. City
University of London
London, UK

Stephanie Fowler
Office of the Director, All of Us Research Program
National Institutes of Health
Bethesda, MD, USA

Scott Fricker
Office of Survey Methods Research
Bureau of Labor Statistics
Washington, DC, USA

Dana Garbarski
Department of Sociology
Loyola University
Chicago, IL, USA

Emily Geisen
RTI International
Research Triangle Park, NC, USA

Pamela Giambo
Westat
Rockville, MD, USA

David Grant
RAND Corporation
Santa Monica, CA, USA

Felix Henninger
Mannheim Centre for European Social Research (MZES)
University of Mannheim
Mannheim, Germany

and

Department of Psychology
University of Koblenz-Landau
Landau, Germany

Jody L. Herman
The Williams Institute
UCLA School of Law
Los Angeles, CA, USA

Temika Holland
Data Collection Methodology & Research Branch, Economic Statistical Methods Division
US Census Bureau
Washington, DC, USA

Sue Holtby
Public Health Institute
Oakland, CA

Rachel Horwitz
US Census Bureau
Washington, DC, USA

Michael Hout
Department of Sociology
New York University
New York, NY, USA

Jonathan Jackson
Department of Methodology
London School of Economics and Political Science
London, UK

Matt Jans
ICF
Rockville, MD, USA

Lars Kaczmirek
Library and Archive Services / AUSSDA
University of Vienna
Vienna, Austria

and

ANU Centre for Social Research and Methods
College of Arts and Social Sciences
Australian National University
Australia

and

GESIS – Leibniz Institute for the Social Sciences, Monitoring Society and Social Change
Mannheim, Germany

Robin L. Kaplan
Office of Survey Methods Research
Bureau of Labor Statistics
Washington, DC, USA

Florian Keusch
Department of Sociology, School of Social Sciences
University of Mannheim
Mannheim, Germany

Pascal J. Kieslich
Mannheim Centre for European Social Research (MZES)
University of Mannheim
Mannheim, Germany

and

Department of Psychology, School of Social Sciences
University of Mannheim
Mannheim, Germany

Brandon Kopp
Office of Survey Methods Research
Bureau of Labor Statistics
Washington, DC, USA

Frauke Kreuter
Joint Program in Survey Methodology
University of Maryland
College Park, MD, US;

School of Social Sciences
University of Mannheim
Mannheim, Germany;

Institute for Employment Research
Nuremberg, Germany

Nicole Lordi
Public Health Institute
Oakland, CA

Aaron Maitland
Westat
Rockville, MD, USA

Jaki S. McCarthy
National Agricultural Statistics Service
US Department of Agriculture
Washington, DC, USA

Katharina Meitinger
Department of Methodology and Statistics
Utrecht University
Utrecht, The Netherlands

Joe Murphy
RTI International
Chicago, IL, USA

Elizabeth Nichols
Center for Behavioral Science Methods
US Census Bureau
Washington, DC, USA

Erica Olmsted-Hawala
Center for Behavioral Science Methods
US Census Bureau
Washington, DC, USA

Kristen Olson
Department of Sociology
University of Nebraska-Lincoln
Lincoln, NE, USA

Yfke Ongena
Department of Communication studies, Centre for Language and Cognition
University of Groningen, Groningen
The Netherlands

José-Luis Padilla
Faculty of Psychology
University of Granada
Granada, Spain

Royce Park
UCLA Center for Health Policy Research
Los Angeles, CA USA

Joanne Pascale
Center for Behavioral Science Methods
US Census Bureau
Suitland, MD, USA

Polly Phipps
Office of Survey Methods Research
Bureau of Labor Statistics
Washington, DC, USA

Stanley Presser
Sociology Department and Joint Program in Survey Methodology
University of Maryland
College Park, MD, USA

Melanie Revilla
RECSM University Pompeu Fabra
Barcelona, Spain

Amy Anderson Riemer
Data Collection Methodology & Research Branch, Economic Statistical Methods Division
US Census Bureau
Washington, DC, USA

Heather Ridolfo
Research and Development Division
National Agricultural Statistics Service
Washington, DC, USA

Paul Scanlon
Collaborating Center for Questionnaire Design and Evaluation Research
National Center for Health Statistics
Centers for Disease Control and Prevention
United States Department of Health and Human Services
Hyattsville, MD, USA

Nora Cate Schaeffer
Department of Sociology
University of Wisconsin-Madison
Madison, WI, USA

and

University of Wisconsin Survey Center
University of Wisconsin-Madison
Madison,WI, USA

Malte Schierholz
Institute for Employment Research
Nuremberg, Germany

Tom W. Smith
NORC, University of Chicago
Chicago, IL, USA

Jolene D. Smyth
Department of Sociology
University of Nebraska-Lincoln
Lincoln, NE, USA

Martha Stapleton
Instrument Design, Evaluation, and Analysis (IDEA) Services
Westat, Rockville, MD, USA

Darby Steiger
Instrument Design, Evaluation, and Analysis (IDEA) Services
Westat, Rockville, MD, USA

Patrick Sturgis
Department of Methodology
London School of Economics and Political Science
London, UK

Jane Tom
Independent scholar
USA

Daniele Toninelli
Department of Management Economics and Quantitative Methods
University of Bergamo
Bergamo, Italy

Roger Tourangeau
Westat
Rockville, MD, USA

Shirley Tsai
Office of Technology Survey Processing
Bureau of Labor Statistics
Washington, DC, USA

Sanne Unger
College of Arts and Sciences
Lynn University
Boca Raton, FL, USA

Joseph Viana
Los Angeles County Department of Public Health
Los Angeles, CA, USA

Heidi Walsoe
Department of Research and Development
Kantar TNS
Oslo, Norway

Douglas Williams
Westat
Rockville, MD, USA

Gordon B. Willis
Behavioral Research Program
Division of Cancer Control and Population Sciences, Cancer Institute National
National Institutes of Health
Rockville, MD, USA

Amanda Wilmot
Westat
Rockville, MD, USA

Bianca D.M. Wilson
The Williams Institute, UCLA School of Law
Los Angeles, CA, USA

Ting Yan
Statistics and Evaluation Sciences Unit
Westat
Rockville, MD, USA

Diana Zavala-Rojas
Department of Political and Social Sciences
Universitat Pompeu Fabra
RECSM, Barcelona, Spain

Bruno D. Zumbo
Measurement, Evaluation, and Research Methodology Program, and The Institute of Applied Mathematics
University of British Columbia
Vancouver, Canada

Preface

The chapters in this volume were all invited papers at the Second International Conference on Questionnaire Design, Development, Evaluation, and Testing (QDET2), which was held in Miami, Florida, November 9–13, 2016. Over the course of four days, QDET2 brought together 345 attendees from 29 countries, united by an interest in improving the art and science of producing survey instruments that collect high-quality data.

Planning for QDET2 began in 2013, when participants of an international working group known as QUEST (Question Evaluation Standards) discussed the possibility of such a conference and invited Amanda Wilmot to serve as Chair. Since the late 1990s, QUEST has consisted of a small group of researchers, mainly from statistical agencies and academic institutions, who meet about every two years to discuss guidelines and potential advances to question evaluation practices. Many of the regular QUEST participants had also participated in the first QDET conference, held in 2002. QDET, as the first major conference on its topic, played an important role in creating a broad understanding of the tools available for questionnaire development, evaluation, and testing. Many of the common methods had evolved somewhat differently in various institutions, not necessarily based on the same assumptions or using the same terminology, and without a shared understanding of the strengths and weaknesses of alternative approaches. The QDET conference helped to build those understandings, and the book resulting from the conference (Presser et al. 2004) was the first to compile the collective methodological guidance into one volume.

More than a decade had passed since this pivotal conference, and it seemed likely that the field had sufficiently matured since then to justify a second one. Whereas the first QDET had helped to establish the key elements of the toolkit, the second would document the continuing evolution of these methods, as well as new additions. It would also foster more critical assessments of the relative contributions of the various methods. Also, the scope of the conference was expanded to include questionnaire design; after all, questionnaires need to be created before they can be evaluated and improved.

At the same time, some of the most important aspects of the first QDET were retained or even enhanced. Most critically, this included a call for international participation, to generate the widest possible incorporation and dissemination of new ideas. The conference was designed to share knowledge through a combination of presentations – keynote addresses, formal papers, electronic posters – as well as workshops, demonstrations, exhibits, courses, and less-formal networking opportunities. Calls for participation began in 2015 and were encouraged across the academic, government, and private sectors.

At first glance, paper proposals seemed to span much of the same methodological territory as the original QDET – for example, including numerous papers about the use of probes to understand interpretations, and experiments to identify the effect of particular design variations. Closer inspection revealed that a great deal had changed, not necessarily in terms of the introduction of completely novel methods, but through new applications of them in response to major changes in the survey landscape.

One of the most significant of these changes, described by Dillman in Chapter 2 of this volume, was a substantial move away from traditional interviewer-based surveys toward self-administered modes. Such movement increases the burden on questions, and the challenge for questionnaire designers, as questions must be understandable without the presence of an interview to help navigate uncertainties of meaning. These challenges are compounded by the proliferation of mobile devices, with small screen sizes that significantly constrain question length and complexity. Most researchers would not choose such devices as the primary response mode if they could avoid it – but increasingly, obtaining respondent cooperation requires accepting their preferred mode for interacting with the larger world. The increase in self-administration in general, and of small screens in particular, has required methodologists to pay greater attention to visual design principles when developing questionnaires.

These changes have also required methodologists to adjust their approaches for questionnaire evaluation. One example is an increased prominence of direct observation of self-response, and the incorporation of more standardized probes into self-administered instruments, relative to traditional interviewer-administered testing protocols. Another is an increase in the prominence of usability testing, which focuses on respondent interaction with the survey instrument, and is an important complement to the cognitive testing that focuses more on matters of question wording. Electronic devices also create paradata – information about the response process itself, such as how respondents navigate through the instrument – which may have useful implications for evaluating questionnaires.

Another significant change since the first QDET conference is that rates of survey participation have continued to decline precipitously. In fact, in the early stages of QDET2 planning, we considered the possibility that traditional questionnaires may play a significantly reduced role in a future dominated by Big Data and other alternatives to surveys. Although we concluded that questionnaires are unlikely to be completely replaced anytime soon, even as other data sources rise in prominence, the challenges of maximizing respondent cooperation are impossible to ignore. Consequently, survey burden was a prominent topic at QDET2. Several chapters in the book deal with this topic explicitly – defining it and understanding its role in survey participation. However, the topic is also in the background of many other chapters, as contributors consider the relationship between design features and various forms of satisficing, as well as questionnaire evaluation strategies that require minimal additional effort from respondents.

Survey questionnaires continue to play a unique role in producing data on attitudes and behaviors for which there is no viable alternative. Our need for such data has never been greater – for distinguishing actual public opinion from speculative commentary, for producing demographic and behavioral data to guide policy planning and resource allocation, and for developing comparative measures of well-being on a global scale – among others. It is our hope that the chapters presented here will help practitioners address recent changes and challenges, and enhance the viability of survey questionnaires for years to come. With those goals in mind, we have organized this book into five major sections.

The first section, "Assessing the Current Methodology for Questionnaire Design, Development, Testing, and Evaluation," begins with two broad overviews. Willis's chapter, based on his keynote address at QDET2, offers analysis of the "trends, development, and challenges" facing the field from the perspective of a researcher-practitioner immersed in questionnaire evaluation for several decades. Dillman's chapter offers the somewhat different perspective of a methodologist whose work centers more on design and survey participation than testing methods per se. Both point to important future developments. For example, Willis considers the application of current methods to emerging data sources, while Dillman calls for the field to expand its focus to respondent motivation, visual aspects of questionnaires, and multimode evaluation. The remaining chapters in this section offer broad assessments of the various methods currently available while addressing several practical questions: how do we decide among current methods, what can different methods tell us about response validity, and how do we make sense of apparently contradictory findings?

Chapters in the second section, "Question Characteristics, Response Burden, and Data Quality," center more on matters of design than evaluation methodology. Both the Dykema, Schaeffer, et al. and Dahlhamer, Maitland, et al. chapters broadly consider the relationship between question characteristics and data quality – collectively providing insight into how various design decisions affect different indicators, and providing an extensive framework to guide future investigations. The next two chapters consider the relationship between questionnaire attributes and perceptions of burden (Yan, Fricker, and Tsai) or respondent behaviors that seem to indicate burden (Kreuter, Eckman, and Tourangeau), while also considering the overarching issue of what burden actually is. The final two chapters in this section focus on how particular questionnaire characteristics affect particular response behaviors. Findings in these chapters may be interesting not only for their specific design implications, but also as models for additional research on related topics.

Chapters in the third section, "Improving Questionnaires on the Web and Mobile Devices," address methodological issues arising from increased use of self-response through electronic modes of data collection. As with other sections of the book, this begins with relatively broad overviews, including Geisen and Murphy's review of the overall toolkit for web and mobile survey pretesting, and Nichols et al.'s overview of usability testing methods for online instruments. Empirical studies of more specific topics follow, respectively addressing the challenges of designing instruments with various screen sizes; an application of paradata from mouse movements to identify measurement issues; and the use of web probing, a response to both proliferation of web surveys and the need for larger pretesting samples.

As its title suggests, the fourth section, "Cross-Cultural and Cross-National Questionnaire Design and Evaluation," takes on the particular challenges of measurement across nations and cultures. The chapters within address conceptual considerations of design (Smith), pretesting methodology (Fitzgerald and Zavala-Rojas; Behr, Meitinger et al.), and both, in practical application (Wilmot). Developing measures that function so broadly is surely one of the biggest measurement challenges methodologists currently face.

Finally, the fifth section, "Extensions and Applications," brings together chapters offering methodological innovation, and examples of novel design and evaluation methodology in practice. Offerings in this section are quite wide-ranging, including new analytic applications to familiar verbal protocols (Sturgis, Brunton-Smith, and Jackson), an effort to bridge concepts with the sister discipline of psychometrics (Zumbo and Padilla), developing management standards for question evaluation projects (Stapleton, Steiger, and Davis), in addition to a series of studies that combine, enhance, and apply methods in different ways to address the continually evolving challenges of collecting high-quality data through survey questions.

In assembling this book, our objective was to create not simply a proceedings volume of the QDET2 conference, but rather an integrated volume reflecting the current state of the field that points to promising new directions. We selected chapters for inclusion and designed our editorial process with that end in mind. Each chapter was reviewed by at least two editors, one primary and one secondary. Editors were chosen purposively based on key interests, but also mixed as much as possible given those interests, such that very few chapters were reviewed by the same pair. We believe that this resulted in a high degree of editorial collaboration and intermingling of ideas, such that the book as a whole reflects our collective guidance. We are hopeful that the results are useful and interesting, but of course defer the final evaluation to the reader.

For us, the completion of this book marks the end of five years of continuous involvement with the QDET2 conference. We are grateful to many for supporting us throughout this effort that consumed much time and energy, including our employing institutions, colleagues, and families. We also gratefully acknowledge the patience of the contributing authors through the extensive editorial process. QDET2 might not have happened at all without the financial and logistical support of the American Statistical Association; we are all grateful for their commitment and confidence in the effort. Finally, we thank all who contributed to the QDET conference – too many to name individually, but certainly including all who served on the organizing and program committees, in addition to the sponsors, volunteers, presenters, and of course the attendees.

Paul C. Beatty
Debbie Collins
Lyn Kaye
Jose-Luis Padilla
Gordon B. Willis
Amanda Wilmot

Disclaimer

Any views expressed here are the authors' and do not necessarily reflect those of our institutions, in particular the US Census Bureau.

Reference

Presser, S., Rothgeb, J.M., Couper, M.P. et al. (eds.) (2004). *Methods for Testing and Evaluating Survey Questionnaires*. Hoboken, NJ: Wiley.

Part I

Assessing the Current Methodology for Questionnaire Design, Development, Testing, and Evaluation

1

Questionnaire Design, Development, Evaluation, and Testing: Where Are We, and Where Are We Headed?

Gordon B. Willis

Division of Cancer Control and Population Sciences, National Cancer Institute, National Institutes of Health, Rockville, Maryland, USA

Although questionnaire development, evaluation, and testing (QDET) could be considered a fairly specific area of survey methods, it spans a wide range of activities and perspectives that the chapters in this book endeavor to describe and to further develop. This opening chapter – itself based on my keynote address delivered at the 2016 International Conference on Questionnaire Design, Development, Evaluation, and Testing (QDET 2) – outlines the contours of the QDET field and delineates the major trends, developments, and challenges that confront it. Further, I attempt to project into the future, both in terms of promising areas to pursue, as well as the types of methodological research that will further this goal. To this end, I consider three basic questions:

1) What is the current state of the art, and of the science, of QDET?
2) Whatever that may be, how important is methodological attention to the questionnaire, in the present world of information collection?
3) As information collection needs evolve, how can the QDET field anticipate new trends and directions?
   In addressing these issues, I make reference to chapters within this volume that relate to each of them.

1.1 Current State of the Art and Science of QDET

A half-century ago, Oppenheim (1966, p. vii) stated that "The world is full of well-meaning people who believe that anyone who can write plain English and has a modicum of common sense can produce a good questionnaire." Although that viewpoint may to some extent persist, there has been considerable evolution in the status of questionnaire design within survey methods,

Advances in Questionnaire Design, Development, Evaluation and Testing, First Edition.
Edited by Paul C. Beatty, Debbie Collins, Lyn Kaye, Jose-Luis Padilla, Gordon B. Willis, and Amanda Wilmot.

and researchers in the social and health sciences especially have increasingly focused on question design as a scientifically backed endeavor, as opposed to a skill attained simply through familiarity with the use of language (Fowler 1995; Schaeffer and Dykema 2011b). Further, researchers and practitioners assert that many pitfalls in survey measurement derive from characteristics of survey questions that produce error – especially in the form of response error, e.g. the departure of a reported value from the true score on the measure of interest (Sudman and Bradburn 1982). Finally, as a vital component of the Questionnaire Design, Development, Evaluation, and Testing (QDET) approach, evaluation and testing – the "E" and "T" within the acronym – have become pronounced as means for assessing the performance of survey items in minimizing such error.

1.1.1 Milestones in QDET History: The Quest to Minimize Error

It was the challenge of controlling response error that largely motivated the organizers of the QDET2 conference, as this has been a consistent theme throughout the history of questionnaire design. To a considerable extent, efforts in this area have focused on understanding sources of error through modeling the survey response process, based on the supposition that humans need to think in order to answer survey questions, and on the observation that they can experience difficulty in doing so. These efforts have a relatively longer history than is sometimes acknowledged. Some highlights of this development are summarized in Table 1.1.

Table 1.1 Selected milestones in the history of questionnaire design, evaluation, and pretesting as a scientific endeavor.

- Description of cognitive probing: Cantril and Fried (1944)
- A systematic set of rules of question design: Payne (1951)
- A psychological model of the survey response process: Lansing et al. (1961)
- The 1978 Royal Statistical Society and Social Science Research Council (U.K.) seminar on retrospective and recall data in social surveys (Moss and Goldstein 1979)
- A cognitive/motivational model of survey response: Cannell et al. (1981)
- The 1983 CASM Seminar (CNSTAT Advanced Research Seminar on Cognitive Aspects of Survey Methodology; Jabine et al. 1984)
- The four-stage model of the survey response process: Tourangeau (1984)
- The Zentrum fur Umfragen, Methoden und Analysen (ZUMA)-sponsored international conference on social information processing and survey methodology (Hippler et al. 1987)
- The Second Advanced Research Seminar in the Cognitive Aspects of Survey Methodology (CASM II, 1997)
- The International Conference on Question Development, Evaluation, and Testing (QDET 1, 2002)
- The International Conference on Question Design, Development, Evaluation, and Testing (QDET 2, 2016)

Within this series of events, a key turning point in QDET history, associated with the 1983 conference on the Cognitive Aspects of Survey Methodology (now referred to as CASM 1), was the introduction by Tourangeau (1984) of the Four-Stage model of survey response, which infused survey methodology with key concepts of cognitive psychology by specifying the importance of Comprehension, Recall, Judgment/Estimation, and Response processes. This model was elegant in its simplicity and has served to capture much of what goes wrong with survey questions at each of these stages. The model was operationalized at the National Center for Health Statistics (NCHS) though the inception of the NCHS Cognitive Lab in the 1980s (Royston et al. 1986; Sirken and Fuchsberg 1984; Sirken et al. 1988); as well as by the development of the cognitive laboratories at the US Census Bureau (DeMaio and Rothgeb 1996) and Bureau of Labor Statistics; and also at Statistics Canada and several European survey organizations (Prüfer et al. 2003). The embedding of the Tourangeau Four-Stage model into the larger rubric of CASM – along with the advent of the cognitive labs – in conjunction worked to propel the science of QDET further over the following decades (Jobe and Mingay 1991; Willis 2005).

1.1.2 Influences from Outside of Cognitive Psychology

There have, however, always been indications that the Four-Stage model, within the CASM rubric, did not tell the entire story concerning question function. A complementary, if not alternative, strand that has existed throughout QDET history concerns not how to ask questions, but rather what to ask – *are we asking the right question* in the first place? This notion has persisted, under different guises – and can be thought of as a non-statistical analog of *specification error* (Lazarsfeld 1959) – in simple form, where the investigator desires one thing, but measures another. In this vein, there was the associated recognition that the Four-Stage cognitive model did not capture all problem types revealed through cognitive testing. In response, Willis et al. (1991) added Logical/Structural errors, as a fifth element, and where problems are more "in the question" than "in the respondent" – that is, in the interaction between the two, in a conceptual sense.

As a simple example, item testing revealed that physical activity questionnaires devised for university graduates were found to fail to function adequately for low-income Hispanic women, but not because of failures that could be cleanly tied to the four proposed cognitive stages. Rather, the questions failed because they focused on leisure-time activities of types that failed to characterize the ways in which this subdomain of the population experienced physical activity in their everyday lives (Willis 2005). Such problems required additional conceptualization of error that took into account the appropriateness of the question's logical foundations (see also Fowler 1995).

1.1.3 Development of a Socio-Linguistic Model

Despite the general recognition that the cognitive model was limited, what was still lacking was a theoretical underpinning or disciplinary connection that more explicitly recognized additional influences on the survey response process. To address this need, CASM has subsequently been augmented to incorporate additional academic fields that provided a context for thinking about influences other than respondent cognition, such as socio-linguistics, sociology, and anthropology, into the emerging interdiscipline. To this end, QDET theorists have endeavored to incorporate perspectives that had existed in earlier literature sources (e.g. Cannell et al. 1977), or developed novel multidisciplinary orientations (Gerber and Wellens 1997; Miller et al. 2014; Schaeffer 1991). Further, Schaeffer and Dykema, (2011a) have distinguished broadly between cognitive and interactional models, where the latter largely subsume a sociological model of the survey response process. According to this view, survey errors involve interaction between the question and the respondent – and also the interviewer, for interviewer-administered questionnaires (IAQs) – such that QDET must focus on how interviewer and respondent characteristics, as opposed to only characteristics of the survey questions, enter into the mix. The newer, multidisciplinary viewpoint has become especially relevant as surveys increasingly target cross-cultural and multinational populations, and in ways that incorporate disadvantaged groups (Miller et al. 2011; see also Chapter 19 by Smith and Chapter 20 by Fitzgerald and Zavala-Rojas).

1.1.4 A Comprehensive Model of Survey Response

Returning to my initial question concerning where the QDET field resides: it is a science, within the more general field of survey research methods, that focuses on a range of error types, and it has taken ownership of response error as its focus. The QDET field is supported by some modicum of theory that is largely psychological in nature, but that has become increasingly elaborate, diverse, and rich over time. A depiction of a generalized model of the survey response process that takes into account both intra- and interpersonal factors, as well as characteristics of the survey question, is presented in Figure 1.1.

Current questionnaire evaluation and testing procedures, in particular, focus on different aspects of the total model, as they variably take into account (i) the investigators (in terms of their measurement objectives), (ii) the survey questionnaire, (iii) the interviewer, and (iv) the respondent; as well as (v) the overall environmental and cultural context in which the interview is conducted. The development of alternative QDET procedures naturally relates to the fact that they focus on "different parts of the elephant." For example, expert review focuses mainly on the questionnaire, and sometimes on the researcher, as it judges the match between measurement objectives and question selection, format, and wording. Behavior coding, in classic form, targets the trio consisting

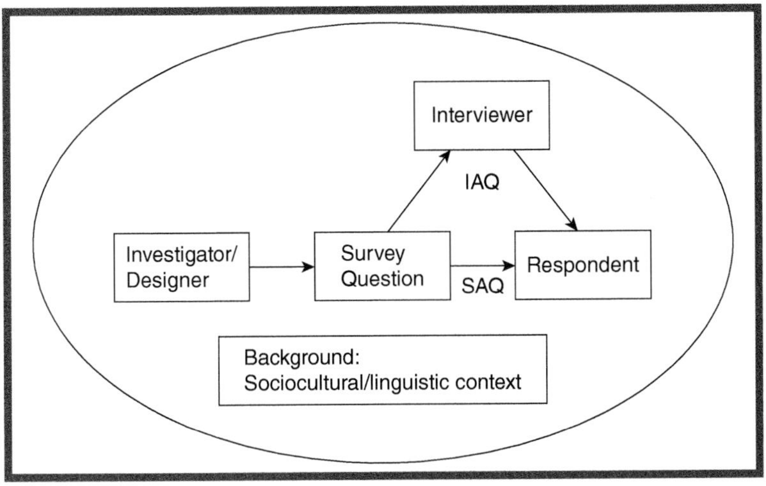

Figure 1.1 Model of the survey response process that incorporates interviewer, respondent, question characteristics, and contextual factors (adapted from Willis (2016) QDET2 Keynote Address). Note: IAQ, Interviewer-Administered Questionnaire; SAQ, Self-Administered Questionnaire.

of the interviewer, respondent, and questionnaire, as it overtly considers the interaction between these entities (see Fowler and Cannell 1996; Chapter 11 by Ongena and Unger; and Chapter 27 in this volume by Kaplan, Kopp, and Phipps). Cognitive interviewing focuses heavily on the respondent, as he/she relates to the survey questionnaire (Beatty and Willis 2007; Collins 2015).

Looking at the overall process from this perspective, one conclusion is that no method serves as a direct substitute for another, due to this differential targeting. However, methods do overlap, both in terms of goals and in the results they produce, to the extent that they would not be expected to be independent (e.g. problems for the interviewer in reading a question may in turn create difficulties for respondents in answering it). A natural conclusion, given this perspective, is that a key challenge is optimizing practices for selecting and integrating these methods – and a significant subset of chapters in the current book endeavor to tackle this (e.g. Tourangeau, Maitland, Steiger, and Yan [Chapter 3]; Maitland and Presser [Chapter 4]; d'Ardenne and Collins [Chapter 5]; Wilmot [Chapter 22]; Pascale [Chapter 26]; and Jans, Herman, Viana, et al. [Chapter 30]). I consider the solution to this ongoing challenge to be a persistent theme for QDET researchers.

1.1.5 Design Rules vs. Empirical Testing

Although Figure 1.1 focuses on a number of distinctions, I believe that the most critical distinction reduces to a choice between two major alternative types: (i)

QDET methods that rely on *design rules* or established sources of guidance, and (ii) those that are dependent on *empirical testing* of survey questions. Chapter 4 by Maitland and Presser further addresses this distinction, which underlies much of the variation in current QDET practice. Further, I propose that the selection of rule-based versus empirical methods is largely (although by no means completely) influenced by the investigator's viewpoint with respect to an unresolved and somewhat sublimated debate in the field, concerning our notions of whether question quality should be thought of as absolute and static, on the one hand; versus relative and dynamic on the other.

An *absolutist* viewpoint, first, holds that questions are, in essence, either good or bad, and that this quality of the question persists – somewhat like a personality trait – across time and context. Further, survey questions incorporate particular features that determine their ultimate quality or validity as measures of a particular construct – e.g. length, clarity, and so on (see Chapter 7 by Dahlhamer, Ridolfo, Maitland, Allen, and Brooks). This philosophy of design leads to the notion that a database or compendium of good versus poor questions can be assembled, such as through a computer-driven analysis of the characteristics of the items. The Survey Quality Predictor (SQP) approach, described by Saris et al. (2003), represents this approach.

As an alternative to an absolutist perspective, a *relativistic* orientation considers question function to be considerably more variable, in that it is largely contextually determined and is strongly dependent on both the measurement objectives it is used to satisfy and the socio-cultural context in which the item is administered. One of the purported purposes of cognitive testing – perhaps the major empirical QDET approach currently in use – is to assess the functioning of the item as it connects to the lives of the particular respondents to whom the item will be administered. This basic view has been expressed through varied terminology, by a number of authors (Gerber and Wellens 1997; Miller et al. 2014; Willis, et al. 1991). As stated by Maitland and Presser (Chapter 4), the assumption of the empirically oriented philosophy is that: "The response process is set within a sociocultural context, and some approaches, such as cognitive interviews, allow the researcher to observe the process in that context." Further, because context may vary over time or space, the use of an item in a new context is often regarded as a trigger for additional empirical testing, as the functioning of the item requires assessment in this new situation. As a consequence, an item's "goodness" is not an inherent feature, but varies with these circumstances – and is a somewhat slippery entity that must constantly be reevaluated.

In reviewing the current state of questionnaire design, it is instructive to consider the viewpoints and dominant approaches adopted by a variety of QDET researchers concerning absolutism versus relativism. There is clearly no consensus, even though this inherent conflict in worldviews has seldom been directly discussed, debated, and evaluated. I believe that one reason that pretesting – and cognitive testing in particular – is often assumed to be a basic, ongoing requirement for virtually all surveys is that it is commonly felt that questions are unreliable, in the sense that a particular item may function in one study, yet fail in another. Put another way, as Oppenheim (1966) suggested decades ago, questions may have a range of validities, depending on the uses to which they are put.

As an alternative, the absolutist's focus on question features that lead those items to be either effective or ineffective leads to a more inwardly focused analysis of a question's specific design features, through use of either established question design rules or else experimental research on the effects of these features on data quality. For example, Payne (1951) introduced a compendium of questionnaire design rules that were subsequently widely adopted; a more recent analog is the Question Appraisal System (QAS) (Willis and Lessler 1999), which posits that questions may either "pass" or "fail" an evaluation of 24 characteristics – such as Length, Vagueness, and Response Category problems – that can be identified through review by an expert, assisted by the QAS problem checklist.

Design-oriented approaches to assessing item quality are often less structured and algorithmic than rule-based or QAS-type checklist-based systems, and may be completely informal in nature. In any form, they are commonplace. In a pre-QDET 2 key informant interview, Dykema (personal communication, October 24, 2016) suggested that the use of empirical methods such as cognitive testing are not as widespread or ubiquitous as is sometimes claimed by their proponents. Rather, a common sequence is to develop or select survey questions (perhaps informed by the survey methods literature), to appraise these through expert review methods, and to consider this to represent an adequate application of QDET methodology. Although expert review, as a design-based approach, can certainly attempt to incorporate contextual elements (i.e. how the item will be expected to function in a novel environment or for a special population), and therefore incorporate a relativistic orientation, it often is divorced from that context. As an extreme case, computer-based approaches such as the Question Understanding Aid (QUAID) (Graesser et al. 2000) tend to exhibit a "one-size-fits-all" approach in which question features, rather than features of respondents or of the environment, dominate the assessment of item quality.

1.1.6 Practical Drivers of Design-Based vs. Empirical Approaches

In her capacity as expert informant (as described previously), Dykema suggested that the underlying philosophy concerning absolutist versus relativistic approach is certainly not the sole driver of choice of empirical versus expert methods. Rather, practical and organizational factors often dominate the selection of QDET methods. In particular, the US government sector emphasizes, and has been a major driver of the adoption of, empirical cognitive testing as the dominant pretesting procedure of the past 30 years for development of federal surveys. Arguably, cognitive interviewing has become, at least to some extent, associated with federal staff and contract research organizations that serve government agencies, consisting of the establishment of permanent cognitive laboratories featuring dedicated personnel and physical spaces, and requiring the contribution of significant ongoing financial and staff resources.

On the other hand, some survey organizations and centers may not generally make use of cognitive laboratories, partly due the lack of the same mandate to do so that characterizes the US federal statistical establishment, but also due to resource constraints. In such cases, relatively more emphasis may be placed on the grooming of questionnaire-design professionals who are adept at conducting expert reviews, or through making use of a copious questionnaire-design literature that produces general design rules: for example, that response scales function better when labeled than when left unlabeled (Alwin and Krosnick 1991). A telling distinction is that reflected by researchers with opposing views concerning what it means to be a proficient QDET practitioner. For example, Miller (personal communications, September 20, 2016) lamented the paucity of training in *empirical question evaluation and testing*; whereas Dykema (personal communication, October 24, 2016) expressed a similar opinion but with respect to *training in questionnaire design*. The key distinction – between empiricism and reliance on design rules – again permeates this debate within the QDET field.

1.1.7 Resolving Design-Based vs. Empirical QDET Methods

One approach to resolving the tension between design-based and empirical methods might be to combine these approaches, as through expert review along with empirical testing. I have suggested previously (Willis 2005) that expert review can be conducted initially, in order to both identify potential problems and assist with the creation of probes for cognitive testing that pointedly investigate these problems. As Beatty stated (personal communication, September 16, 2016), rules alone are unlikely to be wholly sufficient, but purely empirical testing – especially when unguided by prior knowledge of how survey questions function – is also not likely to be effective; a carefully planned combination may therefore obviate the intrinsic limitations of both.

Another way to combine notions of design rules and empirical testing would be to reduce the strong "dustbowl empiricist" nature of cognitive testing or other empirical techniques: that is, the tendency to treat each project as a new tabula rasa endeavor by instead working to incorporate the results of testing into the questionnaire design literature as the basis for guidance to design, if not outright rules. To this end, Kristen Miller and colleagues have developed the Q-Bank database of cognitive testing results (http://wwwn.cdc.gov/qbank/

home.aspx), which attempts to chronicle the outcome of empirical testing in such a way that future studies need not constantly reinvent the same wheel. Of course, the novel context of new uses for items may dictate a need for new testing and evaluation, but the provision of prior empirical results can in principle be extremely helpful in narrowing the search for effective question variants.

1.1.8 How Should Each QDET Method Be Practiced?

The previous discussion concerns the challenge of selecting and combining QDET methods. Beyond this general level, QDET practitioners are also confronted with the more specific question of how each method should be carried out. Cognitive testing, behavior coding, expert review, or any other method can be conducted in a multitude of ways – and at this point there is no established set of "best practices" that has been agreed upon, with respect to any of these (although Chapter 25 by Stapleton, Steiger, and Davis suggests some basic principles). Cognitive testing in particular appears to have given rise to varied practice, with respect to key elements such as necessary sample size, use of think-aloud versus verbal probing techniques, and analysis methods. In fact, it is not necessarily clear from cognitive testing reports and publications what procedures have been used, as these are often unspecified. In response, Boeije and Willis (2013) have introduced the Cognitive Interviewing Reporting Format (CIRF), which specifies 10 major elements that should be described, and which promotes the widespread reporting of these in any report that documents the conduct and outcome of a cognitive interviewing project. Adherence to such standards would, at the least, make clear what the state of the science actually is, in terms of current practices. The same logic of course applies to all QDET methods. To summarize, I propose that further development of any method is dependent on enhanced attention to the documentation of seemingly mundane procedural details, as these are critical to understanding how each method is applied, and perhaps to elucidating the effectiveness of critical variants.

1.2 Relevance of QDET in the Evolving World of Surveys

To this point, the current chapter has chronicled the history of QDET methods and assessed their status within the current world of survey methodology. However, a clear question that emerges concerns the relevance of these methods, as the underlying landscape shifts. A perusal of the literature concerning the place of surveys immediately suggests, if not a crisis, a critical point that demands a reassessment of the role of self-report surveys generally – and by extension, that of QDET. Most pressing are the societal shifts that have greatly influenced the attractiveness, and even acceptability, of survey participation to members of targeted populations, and resultant major drops in cooperation and response rates. Simultaneously, technological shifts, and most pointedly the diminishing use of landline telephones, and concomitant increase in web-based and mobile devices, have resulted in a sea change in choice of administration mode.

Especially telling is the comment by Robert Groves, previous director of the US Census Bureau and prominent survey methodologist, that – at least with respect to some data-collection challenges – "I don't have much hope for surveys" (Habermann et al. 2017, p. 4). This somewhat downbeat assessment might lead one to wonder whether, despite the developments and progress in relevant theory and methods described earlier, there is much point to the further consideration of QDET – or of the utility of questionnaires, generally. That is, how important is it to further study the self-report questionnaire in a world where the underlying foundation of survey research has been fundamentally shaken, and could even be in danger of collapse? (See Couper (2013)). Most pressing is the potential trend toward data sources other than respondent self-report (e.g. social media, "big data") that do not necessarily involve asking questions of people and accruing individual-level self-reports. Are QDET proponents, in effect, perfecting the proverbial buggy whip?

1.2.1 Should We Even Ask Questions?

The most radical, yet intriguing possibility that has gathered some steam in recent years is that information seekers can make extensive use of alternatives to asking survey questions – that is, to supplant self-report-based data collection with another data source – or what I refer to as the movement from Q(uestion) sources to non-Q sources. In particular, it has been proposed that the information needs traditionally served through the conduct of self-report surveys can instead be satisfied though reliance on administrative data, social media, activity monitoring, or even biological-marker-based tests. For example, it has been proposed that DNA changes, in the form of methylation alterations, could serve as markers of a person's smoking history that have advantages over standard self-report methods (see Joehanes et al. 2016). An implication of this view is that there is no longer the need to rely upon generally unreliable, unmotivated survey respondents who increasingly produce high levels of response error, even when we can induce them to answer our questions – no matter how well-designed. Rather, the argument holds, we increasingly can make use of more straightforward, objective data that obviate the need for the traditional survey, somewhat as the diesel locomotive eventually eclipsed the use of steam power in the evolution of rail power.

I believe that there are indications that a trend toward objective, non-Q sources is definitely upon us. However, there are severe limits to the degree to which a complete conversion in data sources is possible, or desirable. First, note that such trends may be even countered by movements in the reverse direction. A case in point is the field of clinical care research and practice, where emphasis is increasingly placed on patient-reported outcomes (PROs), which have been developed in the interest of improving upon the traditional practice of relying on physician reports, chart records, laboratory tests, and other "objective" measures of phenomena such as pain, symptoms, and medical adverse events (Deshpande et al. 2011). PROs focus on the experience of the patient, by asking him/her to provide information through self-report based on carefully developed standardized items. Ironically, the use of self-reported Q measures is here viewed as representing an evolutionary improvement away from more traditional non-Q measures. As such, we see subfields of research moving in opposite directions simultaneously, and the most apt observation may be that the cliché "the grass is always greener on the other side" applies to information collection. In any event, it is far from clear that self-report is headed the way of the buggy whip or the steam engine.

1.2.2 Which Is Better? Q vs. Non-Q Information

If we envision a world in which we retain self-report but might have the luxury of choice between self-report survey questions versus supposedly objective sources, then selecting between these requires that we can assess the quality of each approach. This report-versus-records debate is not new. Despite the well-known and acknowledged limitations of self-report question-based measures, it has also been well documented that record data, as a major form of non-Q information, frequently contain non-ignorable levels of error. In the QDET realm, Willson (2009) found, within a cognitive interviewing study of birth certificates, that such official records are subject to considerable error, ultimately due to the fact that they are completed by a human respondent. An earlier investigation involving responses to death certificates by Smith et al. (1992) obtained parallel results, ironically suggesting that error in records in fact persists across the entirely of the human lifespan. Any form of data, whether Q or non-Q based, may contain unacceptable levels of an array of error types. As such, a better question than "Which is better?" is to ask when to incorporate either, and perhaps when to attempt to make use of both in some blended combination.

1.2.3 How Can Non-Q Data Sources Be Selected and Evaluated?

To reiterate, new potential approaches could tip the balance of obtaining data in favor of non-Q sources, but these may not fall into our laps in useful form. The prospective non-Q source that has perhaps received the most attention is the use of social media – currently dominated by Facebook and Twitter (although note that in a decade or two, either or both may have vanished in place of

Q versus Non-Q sources

Representation category / Measurement category	Q; survey self-report: ask questions	Non-Q; social media, device, records...
Probability Probability sampling	Standard model: garden variety probability survey Probability-based Web panels	Sampling based on demographics, etc.; processing of obtained (social media/record/device) data, as are survey data
Non-Probability Not sampled via probability mechanisms	Non-probability surveys - e.g., (some) Web panels	Big Data sources - Mega Focus Group? - "Pseudo-survey?" - Other novel developments?

Figure 1.2 Embedding Q(uestion) and non-Q sources of information into a total survey error model.

whatever emerges as the next development in this area). However, the scope of potential uses of such social media in the systematic measurement of social phenomena, or for official statistics, is not clear (Japec et al. 2015). In order to systematically organize the potential contributions of non-Q data sources, I propose that these – as well as Q-based sources – can be conceptually organized into a framework that extends a well-known model of the survey response process proposed by Groves et al. (2004), which distinguishes between errors of *measurement* and errors of *representation.*

I propose that any data source – whether Q or non-Q – needs to be evaluated with respect to both its measurement properties (whether it produces accurate information) and its effects on representation (whether it allows us to collect information from the right individuals). The model presented here (Figure 1.2) therefore distinguishes both of these critical dimensions, for both Q and non-Q data. The element of measurement is portrayed by the Q and non-Q columns at the top of the 2 × 2 table. The second relevant dimension, involving representation, is indicated by the major rows. At its endpoints, representativeness is expressed as either (i) random and representative, or else (ii) nonrandom and presumably far less representative of the targeted population.

Each of the four cells within Figure 1.2 represents a unique case. The cell in the upper left concerns the traditional manner of conducting a survey: a set of self-report survey questions is posed to representatively sampled respondents.

Next, the lower left reflects the case in which Q sources are administered under non-representative conditions (whether or not the intent of the researchers is to faithfully represent any particular population). This case is not novel – as many questionnaires have, for better or worse, been administered under conditions that fail to involve statistical principles that ensure representativeness (e.g. convenience samples, surveys with high levels of nonresponse bias).

The third cell, at the upper right of Figure 1.2, encompasses non-Q sources of information other than those obtained through question-based self-report: administrative records, sensor data, and so on, under conditions of representation in which these records are sampled on a probability basis. Again, this approach features an established history: NCHS, for example, has conducted such sampled studies of hospital records for many years (e.g. https://www.cdc.gov/nchs/nhds/about_nhds.htm). New forms of non-Q data, such as social media sources, could be viewed as filling this cell, to the extent that these records are selected on a random or probability-based basis. For example, it may be possible to make use of social media information (e.g. Facebook posts or tweets) through first sampling the individuals producing them, as opposed to simply "sampling tweets" without regard to the producer. Again, the underlying logic is not new, but a further example of "old wine in new bottles."

Finally, the lower-right cell of Figure 1.2 represents the truly novel, "Wild West" of information data collection, in which non-Q data – especially in novel form – are collected through means that lack a probability basis (e.g. the analysis of 1 million tweets or Facebook posts, which may not constitute a random selection according to any statistical model).

It is this case that is often directly contrasted with the standard approach depicted in the first cell. However, as a major thesis, I suggest that such a comparison is confounded in a way that Figure 1.2 makes clear. Arguments concerning uses of Q versus non-Q sources must take into account both measurement and representation, when researchers purport to assess data quality, especially relative to cost. As such, in evaluating the novel, non-probability/non-Q paradigm, we must ascertain its effects with respect to each of these major dimensions. In summary, by partitioning our conceptualization of Q and non-Q sources such that representation and measurement are clearly distinguished, QDET researchers can more effectively design investigations that consider the potential contributions of non-Q sources to the future of surveys.

More broadly, it does seem evident that a strategy of remaining within the upper-left quadrant of Figure 1.2, and considering only question-based probability surveys, may be too limited as the world of information collection moves to the approaches of the other three cells. QDET professionals, in practice, must broaden their focus beyond a narrow perspective on question function, to the point where the term *questionnaire designer* might best be relabeled as *Information System Design Specialist*. Such individuals are adept at making use of multiple forms of information that are suited to the research question or information need at hand, without necessarily involving the knee-jerk tendency to "put together a questionnaire and field a survey" as the necessary solution. This expansive view of information-collection challenges is likely to help QDET practitioners to remain relevant through the course of an uncertain, but certainly changing, future.

1.3 Looking Ahead: Further Developments in QDET

If one accepts that QDET is likely to remain relevant within the future world of survey data collection, by adapting to key changes in the world of data collection, there are still several potential developments that may serve to enhance its effectiveness and impact. To close the chapter, three of these are identified:

1) The development of better *measures of survey question quality*
2) The development of novel *empirical pretesting and evaluation techniques*
3) The development of more specific and useful *questionnaire design rules* and guidelines

1.3.1 Developing Measures of Survey Question Quality

The assessment of QDET methods ultimately depends on a determination of whether they have improved the quality of survey questions. The notion of how to define and measure "quality" is elusive, however. Chapter 16 by Horwitz, Brockhaus, Henninger, et al.; Chapter 23 by Sturgis, Brunton-Smith, and Jackson; and Chapter 24 by Zumbo and Padilla. examine the issue of quality from different perspectives. From the perspective of a psychometrician, quality implies notions of validity and reliability; whereas for those steeped in the administration of surveys "on the ground," this may also bring to mind notions of feasibility, acceptability to respondents, degree of burden imposed, and other elements that cannot necessarily be measured in terms of question-level error effects. Further, some notions of cognitive testing in particular focus less on error, and more on concerns such whether an item produces information that is useful to the researcher, given the key objectives of the investigation: that is, whether it is "fit for use" within the current investigation (Beatty 2004; Miller et al. 2014). Reiterating the initial discussion of specification error, a survey measure that is free of response error may in fact be useless if it does not measure an attribute or construct that is relevant to the research question being targeted. To date, there has been insufficient attention to definition of error in such a way that QDET practitioners can assess the effectiveness of their techniques in alleviating it.

Even with respect to the more straightforward notions of error that emphasize item validity, as expressed by a difference on a behavioral measure between a true score and that reported by the respondent, there has unfortunately been too little research directly focused on determining whether QDET methods result in reduction in such error. An important caveat that underlies all QDET work, although not commonly stated, is that we do not have direct and compelling evidence that these methods decrease error in the manner expected or claimed. A common logical pitfall reflected within manuscripts evaluating cognitive testing is to argue that cognitive testing was effective because it was found to improve the evaluated survey questions; and on the other hand, that the survey questions are known to have been improved solely because cognitive testing is known to be effective. Such purely internal analyses are clearly prone to circularity and are therefore unsatisfactory. As an escape from this logical loop, we require an external and independent measure of question "goodness" or quality. I am left, at this point, asking a question that is the title of a German text on questionnaire design (Faulbaum et al. 2009): "Was ist Eine Gute Frage?" (What Is a Good Question?) – one that I think we still have to answer more clearly, before we can successfully ascertain whether QDET methods are truly on the right track.

1.3.2 Novel Techniques for Pretesting and Evaluation

Beyond establishing best practices for the tools within the existing toolbox of QDET methods, another direction for the future of QDET involves the development of novel evaluation and testing approaches (again, the "E" and "T" within the acronym). As the world of surveys has turned, especially in terms of dominant technologies, several new potential applications have arisen. One imaginative development has been the incorporation of internet-based platforms that enable survey pretesting in ways that are substantially quicker, easier, and more cost-effective than those traditionally used in the past. For either single-instrument pretesting or comparison between alternate versions, Mechanical Turk (mTurk), an inexpensive online labor platform, has been advocated (Murphy et al. 2014). Further, concerning developments in cognitive testing specifically, it may be possible to conduct face-to-face interviews effectively using internet-based platforms that project both visual and auditory information (e.g. Skype or future equivalents).

1.3.2.1 Web Probing

Beyond developments related to survey platform, there a few truly novel methodological possibilities. The most intriguing at this point may be the updating of an older procedure – embedded probing – but in the form of *web probing*, in which cognitive probe questions are embedded within a questionnaire and are self-administered by respondents (that is, there is no human cognitive interviewer) (see Behr et al. 2013, 2014); and chapters within this book by Scanlon (Chapter 17); Fowler and Willis (Chapter 18); and Behr, Meitinger, Braun, and Kaczmirek (Chapter 21). In combination with the adoption of practical web-based platforms for item administration to relatively large groups, such as mTurk, web probing may hold considerable promise for QDET purposes. These potential benefits must be balanced against potentially serious limitations, however, especially the degree to which web probing involves testing "at arms length," as there is no way for the interviewer/investigator to improvise and directly determine the course of the investigation (Beatty, personal communication, Sept. 16, 2016).

1.3.2.2 Use of Social Media in Pretesting and Evaluation

As a further potential development in pretesting and evaluation, I again raise the prospect of making use of non-Q information as a contributor. Even if, as discussed, those sources are in fact found to be not especially useful as an alternative to the traditional survey questionnaires within the sample survey, they may nevertheless hold promise as a tool for helping us to evaluate instruments and measures. One possibility is that social media content could be seen as constituting, in effect, a very large naturalistic, "big data focus group" – in which qualitative trends, boundaries and contours of beliefs, and dominant conceptual themes could be ascertained – through mining existing "organic" information that conveys attitudes, beliefs, and behavior, as opposed to the formal convening of a series of focus groups. There has been, to date, little work in the QDET field that treats social media content as a source of such ethnographic information, but it could potentially be used to inform the early phases of instrument development in which we attempt to understand the phenomenon under study, and ultimately to ask the right questions in a way that minimizes specification errors. I therefore advocate a wide consideration of the role of non-Q data.

1.3.2.3 Usability Testing of Questionnaires

Finally, another recent development in the empirical study of how individuals attend to survey questions is the increased adoption of usability testing in contexts in which visual layout of information is key. In particular, this may be facilitated by the ongoing development of usability testing methods as applied to survey questionnaires (Geisen and Romano Bergstrom 2017; Nichols, Olmsted-Hawala, Holland, and Riemer, Chapter 13); and eye-tracking procedures (described by Geisen and Murphy, Chapter 12), which have developed to the point of being feasible for production use in evaluating self-administered questionnaires (SAQs), and for web surveys in particular.

1.3.3 Future Developments Involving Design Rules

Here, I have mainly discussed, and speculated on, directions relevant to empirical QDET techniques. In keeping with the fundamental distinction I have maintained through the chapter, there is additionally the potential for further development of design/rule-based approaches – e.g. a modern updating of Payne

(1951) or of the QAS (Willis and Lessler 1999). The best recent summary may be in Chapter 6, by Dykema, Schaeffer, Garbarski, and Hout, which first classifies a range of question features and then proposes research to determine how these features interact to impact item functioning. Many of these developments involve the influence of visual presentation, as for web surveys, especially as self-administration has increasingly become ascendant in the world of self-report surveys.

I suggest that the following areas may be vital directions for QDET work focusing on question and questionnaire features, whether as extensions of an existing body of work, or as novel forms of methods research (see Schaeffer and Dykema, 2011a, for detailed discussions of several of these):

- *The impact of visual aspects of question design*, such as information layout, on responses to self-administered instruments
- *Supra-question-level effects*: context and ordering (e.g. Kreuter, Eckman, and Tourangeau, Chapter 9, systematically assessed the effects of ordering of questions that follow a filter item)
- *Inclusion of psychological variables* and personality constructs, as these influence the survey response process generally
- *Greater attention to psychometrics* (see Chapter 24 by Zumbo and Padilla)
- *Adaptation of QDET methods for surveys of business establishments*
- *Research on factors that define and influence respondent perceptions of survey burden* (Yan, Fricker, and Tsai, Chapter 8)
- *Mixed method* (quantitative/qualitative) approaches (see chapter 26 by Pascale)
- *The expanded use of survey experiments* – especially based on the use of practical and low-cost platforms such as web panels and Mechanical Turk
- *The challenge of device screen size.* A particularly vexing design problem involves the use of screen territory to display questions in cases where small-screen devices such as smartphones are used: how can complex questions be displayed in a way that minimizes response error? (see Chapter 10 by Smyth and Olson; Chapter 14 by Toninelli and Revilla and Chapter 15 by Dale and Walsoe).

1.4 Conclusion

In this chapter, I have reviewed past, present, and potential future trends affecting QDET methods. Looking to the future, staying ahead of the developmental curve demands that we be proactive in seeing new trends develop, as opposed to always reactively adapting to the world as it passes us by. Anticipating the future is, however, not an easy task. When looking ahead, there is always the danger of focusing too narrowly and failing to keep pace with developments in the world of survey (and other) data collection. As a case in point, the final item within my list of projected design challenges, within the previous section, advocates the continued study of how to fit survey questions into vanishingly smaller device screens. As a cautionary note, Fitzgerald (personal communication, November 13, 2016) suggested at the close of the QDET2 conference that too much attention to "the small screen problem" could end up missing the boat, if in 15 years the dominant mechanism for displaying survey questions in fact does not involve small-screen devices, but something else entirely that has not yet even been conceptualized by QDET researchers. Given such uncertainties, we can only hope that our crystal ball nevertheless remains clear. In summary, I look forward to the possibility of a QDET 3 meeting, some years from now, if only to see which of the ideas contained within this book are predictive of the methodological world of the future.

References

Alwin, D. and Krosnick, J. (1991). The reliability of survey attitude measurement. *Sociological Methods and Research* 20 (1): 139–181.

Beatty, P. (2004). Paradigms of cognitive interviewing practice, and their implications for developing standards of best practice. In: *Proceedings of the 4th Conference on Questionnaire Evaluation Standards*, 8–25. Zentrum fur Umfragen, Methoden und Analysen (ZUMA).

Beatty, P.C. and Willis, G.B. (2007). Research synthesis: the practice of cognitive interviewing. *Public Opinion Quarterly* 71 (2): 287–311.

Behr, D., Braun, M., Kaczmirek, and Bandilla, W. (2013). Testing the validity of gender ideology items by implementing probing questions in Web surveys. *Field Methods* 25 (2): 124–141.

Behr, D., Braun, M., Kaczmirek, and Bandilla, W. (2014). Item comparability in cross-national surveys: results from asking probing questions in cross-national surveys about attitudes towards civil disobedience. *Quality and Quantity* 48: 127–148.

Boeije, H. and Willis, G. (2013). The cognitive interviewing reporting framework (CIRF): towards the harmonization of cognitive interviewing reports. *Methodology: European Journal of Research Methods for the Behavioral and Social Sciences.* 9 (3): 87–95. https://doi.org/10.1027/1614-2241/a000075.

Cannell, C.F., Marquis, K.H., and Laurent, A. (1977). A summary of studies of interviewing methodology. *Vital and Health Statistics* 2 (69) (DHEW Publication No. HRA 77-1343). Washington, DC: U.S. Government Printing Office.

Cannell, C.F., Miller, P.V., and Oksenberg, L. (1981). Research on interviewing techniques. In: *Sociological Methodology* (ed. S. Leinhardt), 389–437. San Francisco: Jossey-Bass.

Cantril, H. and Fried, E. (1944). The meaning of questions. In: *Gauging Public Opinion* (ed. H. Cantril). Princeton, N.J: Princeton University Press.

Collins, D. (2015). *Cognitive Interviewing Practice*. London: Sage.

Couper, M. (2013). Is the sky falling? New technology, changing media, and the future of surveys. *Survey Research Methods* 7 (3): 145–156.

DeMaio, T.J. and Rothgeb, J.M. (1996). Cognitive interviewing techniques: in the lab and in the field. In: *Answering Questions: Methodology for Determining Cognitive and Communicative Processes in Survey Research* (eds. N. Schwarz and S. Sudman), 177–195. San Francisco: Jossey-Bass.

Deshpande, P.R., Rajan, S., Sudeepthi, B.L., and Nazir, C.P.A. (2011). Patient-reported outcomes: a new era in clinical research. *Perspectives in Clinical Research* 2 (4): 137–144. https://doi.org/10.4103/2229-3485.86879.

Faulbaum, F., Prüfer, P., and Rexroth, M. (2009). *Was Ist Eine Gute Frage*. Heidelberg, Germany: Springer VS.

Fowler, F.J. (1995). *Improving Survey Questions*. Thousand Oaks, CA: Sage.

Fowler, F.J. and Cannell, C.F. (1996). Using behavioral coding to identify problems with survey questions. In: *Answering Questions: Methodology for Determining Cognitive and Communicative Processes in Survey Research* (eds. N. Schwarz and S. Sudman), 15–36. San Francisco: Jossey-Bass.

Geisen, E. and Romano Bergstrom, J. (2017). *Usability Testing for Survey Research*. Waltham, MA: Morgan Kaufman.

Gerber, E.R. and Wellens, T.R. (1997). Perspectives on pretesting: cognition in the cognitive interview? *Bulletin de Methodologie Sociologique* 55: 18–39.

Graesser, A.C., Wiemer-Hastings, K., Kreuz, R. et al. (2000). QUAID: a questionnaire evaluation aid for survey methodologists. *Behavior Research Methods, Instruments, and Computers* 32 (2): 254–262. https://doi.org/10.3758/BF03207792.

Groves, R., Lyberg, L., Fowler, F.J. et al. (2004). *Survey Methodology*. New York: Wiley.

Habermann, H., Kennedy, C., and Lahiri, P. (2017). A conversation with Robert Groves. *Statistical Science* 32 (1): 128–137. https://doi.org/10.1214/16-STS594.

Hippler, H.J., Schwarz, N., and Sudman, S. (1987). *Social Information Processing and Survey Methodology*. New York: Springer-Verlag.

Jabine, T.B., Straf, M.L., Tanur, J.M., and Tourangeau, R. (eds.) (1984). *Cognitive Aspects of Survey Methodology: Building a Bridge Between Disciplines*. Washington, DC: National Academy Press.

Japec, L., Kreuter, F., Berg, M. et al. (2015). Big data in survey research: AAPOR task force report. *Public Opinion Quarterly* 79 (4): 839–880. https://doi.org/10.1093/poq/nfv039.

Jobe, J.B. and Mingay, D.J. (1991). Cognition and survey measurement: history and overview. *Applied Cognitive Psychology* 5 (3): 175–192.

Joehanes, R., Just, A.C., Marioni, R.E. et al. (2016). Epigenetic signatures of cigarette smoking. *Cardiovascular Genetics* 9: 436–447. https://doi.org/10.1161/CIRCGENETICS.116.001506.

Lansing, J.B., Ginsburg, G.P., and Braaten, K. (1961). *An Investigation of Response Error*. Urbana: University of Illinois, Bureau of Economic and Business Research.

Lazarsfeld, P. (1959). Latent structure analysis. In: *Psychology: A Study of a Science*, vol. 3 (ed. S. Koch), 476–535. New York: McGraw-Hill.

Miller, K., Mont, D., Maitland, A. et al. (2011). Results of a cross-national structured cognitive interviewing protocol to test measures of disability. *Quality and Quantity* 45 (4): 801–815.

Miller, K., Willson, S., Chepp, V., and Padilla, J.L. (2014). *Cognitive Interviewing Methodology*. New York: Wiley.

Moss, L. and Goldstein, H. (eds.) (1979). *The Recall Method in Social Surveys*. London: NFER.

Murphy, J., Edgar, J., and Keating, M. (2014). Crowdsourcing in the cognitive interviewing process. Annual Meeting of the American Association for Public Opinion Research, Anaheim, CA, May.

Oppenheim, A.N. (1966). *Questionnaire Design and Attitude Measurement*. New York: Basic Books.

Payne, S.L. (1951). *The Art of Asking Questions*. Princeton, NJ: Princeton University Press.

Prüfer, P., Rexroth, M., and Fowler, F.J. Jr., (2003). *QUEST 2003: Proceedings of the 4th Conference on Questionnaire Evaluation Standards*. Mannheim, Germany: Zentrum für Umfragen, Methoden und Analysen (ZUMA) http://nbnresolving.de/urn:nbn:de:0168-ssoar-49673-6.

Royston, P.N., Bercini, D., Sirken, M., and Mingay, D. (1986). Questionnaire design research laboratory. In: *Proceedings of the Section on Survey Research Methods*, 703–707. American Statistical Association.

Saris, W.E., van der Veld, W., Galhofer, I.N., and Scherpenzeel, A. (2003). A scientific approach to questionnaire development. *Bulletin de Methodologie Sociologique* 77: 16. https://journals.openedition.org/bms/1425?lang=en.

Schaeffer, N.C. (1991). Conversation with a purpose–or conversation? Interaction in the standardized interview. In: *Measurement Errors in Surveys* (eds. P. Biemer, R.M. Groves, L. Lyberg, et al.), 367–391. New York: Wiley.

Schaeffer, N.C. and Dykema, J. (2011a). Questions for surveys: current trends and future directions. *Public Opinion Quarterly* 75 (5): 909–961. https://doi.org/10.1093/poq/nfr048.

Schaeffer, N.C. and Dykema, J. (2011b). Response 1 to Fowler's chapter: coding the behavior of interviewers and respondents to evaluate survey questions. In: *Question Evaluation Methods: Contributing to the Science of Data Quality* (eds. J. Madans, K.A. Miller and G. Willis), 23–39. Hoboken, NJ: Wiley.

Sirken, M.G. and Fuchsberg, R. (1984). Laboratory-based research on the cognitive aspects of survey methodology. In: *Cognitive Aspects of Survey Methodology: Building a Bridge Between Disciplines* (eds. T.B. Jabine, M.L. Straf, J.M. Tanur and R. Tourangeau), 26–34. Washington, DC: National Academies Press.

Sirken, M.G., Mingay, D.J., Royston, P. et al. (1988). Interdisciplinary research in cognition and survey measurement. In: *Practical Aspects of Memory: Current Research and Issues*, vol. 1: *Memory in everyday life* (eds. M.M. Gruneberg, P.E. Morris and R.N. Sykes), 531–536. New York: Wiley.

Smith, A.F., Mingay, D.J., Jobe, J.B. et al. (1992). A cognitive approach to mortality statistics. In: *Proceedings of the American Statistical Association Meeting, Survey Research Methods Section*, 812–817. Boston/Alexandria, VA: American Statistical Association.

Sudman, S. and Bradburn, N.N. (1982). *Asking Questions: A Practical Guide to Questionnaire Design*. San Francisco: Jossey Bass.

Tourangeau, R. (1984). Cognitive science and survey methods: a cognitive perspective. In: *Cognitive Aspects of Survey Design: Building a Bridge Between Disciplines* (eds. T. Jabine, M. Straf, J. Tanur and R. Tourangeau), 73–100. Washington, DC: National Academies Press https://www.ncbi.nlm.nih.gov/pmc/articles/PMC3713237.

Willis, G.B. (2005). *Cognitive Interviewing: A Tool for Improving Questionnaire Design*. Thousand Oaks, CA: Sage.

Willis, G.B. (2016). Questionnaire design, development, evaluation, and testing: where are we, and where are we headed? Keynote address presented at the Second International Conference on Questionnaire Design, Development, Evaluation, and Testing (QDET2), Miami, Florida.

Willis, G.B. and Lessler, J. (1999). *The BRFSS-QAS: A Guide for Systematically Evaluating Survey Question Wording*. Rockville, MD: Research Triangle Institute.

Willis, G.B., Royston, P., and Bercini, D. (1991). The use of verbal report methods in the development and testing of survey questionnaires. *Applied Cognitive Psychology* 5: 251–267.

Willson, S. (2009). Exploring the 2003 revision of the U.S. standard certificate of live births. https://wwwn.cdc.gov/QBANK/report/Willson_NCHS_2008_Birth%20Certificate%203.pdf.

2

Asking the Right Questions in the Right Way: Six Needed Changes in Questionnaire Evaluation and Testing Methods

Don A. Dillman

Department of Sociology, Washington State University, Pullman, Washington, USA

2.1 Personal Experiences with Cognitive Interviews and Focus Groups

In the mid-1990s, I was taught by US Census Bureau staff to evaluate questionnaires by individually doing cognitive interviews with volunteers from outside the Census Bureau. My memory of that experience was to sit in a small room with the test respondent and ask them to read aloud each question and give reactions to the content and wording of those questions. Probes were provided along the lines of: What does this question mean to you? What are you thinking right now? How did you come up with your answer to this question? The main goal of those interviews was to find out if the wording of each question worked as intended. Methodologically, we were trying to find out whether the words and phrases being used elicited a uniform understanding, thus assuring us that all respondents were being stimulated to answer the same question as other respondents.

The development of cognitive interviewing was a timely achievement. A breakthrough 1984 book edited by Jabine et al. (1984) had made a much-needed connection between cognitive psychology and the design of surveys. Dividing four major components of the survey response process into four tasks – comprehension of the question, retrieval of relevant information, using that information to make judgments, and then selecting or otherwise reporting an answer, described in a book by Tourangeau (1984) – was a key development that led to creation of the practice of cognitive interviewing.

My initial experience with cognitive interviewing was that the wording of questions seemed improved by the interviews. Nonetheless, I also found the outcomes unsettling for dealing with the issues most applicable to my work. I was attempting to design improved self-administered questionnaires that would be sent in the mail for household responses in the 2000 Decennial

Advances in Questionnaire Design, Development, Evaluation and Testing, First Edition.
Edited by Paul C. Beatty, Debbie Collins, Lyn Kaye, Jose-Luis Padilla, Gordon B. Willis, and Amanda Wilmot.

Census. I was concerned with whether people would *see and read* each question, and answer all of the questions on each page, as appeared not to happen as desired in the 1990 Census. Questions were often skipped entirely or answered out of order. In addition, I was concerned with whether people were truly motivated to return the questionnaire or were only filling it out because we had offered a modest reimbursement for coming to the Census Bureau to be interviewed.

The charge assigned to me by advisors to Census was to make the questionnaires "respondent-friendly." Knowing how people interpreted each question remained important. But there was a lot more that I needed to find out, and doing cognitive interviews in the way that I observed them being done by others seemed at best a partial solution. Occasionally, I left the interview room thinking that answers were not provided to a particular question because the respondent didn't realize an answer was needed, perhaps because no question number provided navigational guidance or for some other reason. In order for the comprehension described by Tourangeau (1984) to take place, it seemed that respondents first needed to perceive that it was a question needing an answer (Dillman and Redline 2004). At this time, I also realized through conversations with Norbert Schwartz, a key contributor to cognitive evaluation of the survey process, that to understand problems with the perception process, I needed to begin understanding more about psychological influences from a branch of psychology that focused on motivation rather than only cognition. And it seemed plausible that cognitive interviewing could be adapted to these concerns.

Those initial frustrations with cognitive interviews (at least as I observed them being used) soon extended to another commonly used qualitative evaluation tool, focus groups. Colleagues and I had developed a proposed addition to Census mailing envelopes informing respondents that a response from all households was mandatory. We placed, "US Census Form Enclosed: Your response is required by Law" in large print within a rectangular box on the address side of the outgoing envelope. We were trying to solve the problem of getting the Census envelope opened, a factor known to have reduced response to the 1990 Census (Fay et al. 1991). Since this did not require finding out how a question was interpreted, we were advised to conduct focus groups in several cities, showing people envelopes with and without the message, and asking them which envelope would most encourage them to answer the Census form it contained. An alternative message was also presented to the same focus groups that read: "It pays to be counted: US Census Form Enclosed." Results from these evaluations across cities were quite similar. It was reported that focus group members in all cities objected to the mandatory message and were described as being far more likely to answer the Census when the benefit message appeared on the envelope. The contractor's recommendation was not to use the mandatory message.

After observing each of those focus groups, I found myself wondering whether people were actually saying what they would do or not. It seemed plausible to me that people were being influenced by societal norms to respond with what they thought other people would do, instead of saying what they, themselves, would actually do when they received the questionnaire in the privacy of their homes. After much discussion we decided to ignore the report recommendations and instead continue with the large-scale experimental test of what would happen when actual Census forms were sent to a national sample of households, with different messages on the envelopes. The results were unequivocal. The mandatory message improved response rates by 8–10 percentage points over the inclusion of no message on the envelope, while the benefit message showed no significant effect compared to imprinting it with no message (Dillman et al. 1996b). In retrospect, had we followed the focus group findings, we would not have included the single most powerful influence on response rates from among the 16 factors tested in experiments leading up to the 2000 Census (Dillman 2000, Chapter 6).

A third memorable experience from my Census encounters with the use of qualitative methods was being informed that Census forms must be tested before any questions could be changed. I was introduced to this requirement when I overheard a Census researcher asked to "do a few" cognitive interviews or focus groups on a proposed change in question wording, to find out if it would change answers. The well-intentioned inquiry was asked in hopes of avoiding the need to run an experimental test. Deciding to ask a few volunteers to come to the Census Bureau to respond to a think-aloud cognitive interview, instead of asking a random sample of households to respond to two versions of a questionnaire, seemed to me a less adequate way of testing whether the proposed change in questions would improve results. The statistician who made the request explained that he was just being practical. Doing a few interviews was defended by pointing out that designing and implementing an experiment would cost more and take much longer to complete. A methodology colleague defended the proposal to do a few cognitive interviews, noting it was a significant improvement over the past when a few people sitting in a room would often decide on new question wording without any testing.

These three experiences created a personal dilemma for me. On the one hand, I felt that interacting with potential respondents in cognitive interviews and focus groups could provide useful information for evaluating draft questionnaires and data collection procedures. Such interviews seemed likely to identify potential explanations for why answers to the current and proposed question wordings might produce differences. However, it would be impossible to make quantitative estimates of effects, if any, that resulted.

Central to my concern was the small numbers of people who would be asked to participate in the testing. That number was often less than 10, and seldom more than 15 or 20. This practice was often defended by arguing that doing more interviews was unlikely to produce additional insights. Yet, the peril of small numbers of tested respondents remained. It also appeared that as the number of interviews increased and the evaluator was faced with any differences in viewpoints about responses to a particular question, it became exceedingly difficult to summarize those differences in a reliable manner.

This challenge was amplified by the frequent desire of methodologists to evaluate all aspects of a questionnaire at once, using a single set of interviews. For example, that set of interviews might be undertaken to simultaneously evaluate people's understanding of each question, potential barriers to response, whether branching questions were correctly followed, and why some answers to questions were not provided at all. In addition, the evaluator might be instructed to test these issues on a heterogeneous set of respondents, including people of different racial and ethnic groups, different age categories, and various levels of educational achievement. The more a questionnaire evaluator was asked to incorporate into a test, the more likely they would be faced with the eventual challenge of reconciling what one or two people said with different comments from other respondents.

In addition, when trying to interpret cognitive interview reports, it appeared likely that one or two issues would float to the top as headline findings focused on only a small part of what was ostensibly being evaluated. The temptation to focus reports on one or two major, and perhaps dramatic, findings seemed considerable, and as a frequent conductor of cognitive interviews, I found myself sometimes doing this. For example, while testing all aspects (from question wording to visual layout and navigation) of a complete mock-up of a proposed short form proposed for the 2000 Decennial Census, the response to one question from one respondent stood out. A strong objection was raised to the race question, which the respondent elaborated on at length. She reported being of one race and married to someone of a different race. She objected to the instruction to mark only one racial category for her children. Emphatically she stated it made no difference into what category she, herself, was placed. But she refused to fill out a form that did not allow placing the couple's children into both of their parents' racial categories. This interview provided a compelling example that became cited repeatedly in subsequent staff discussions about whether the single-category response should be changed to multiple categories.

Much of the angst I felt over the qualitative methods being used to test questionnaires was how much confidence could be placed in interview judgments about what was an important finding, compared to the use of statistical tests for large random samples of respondents. My personal solution involved trying to limit sharply the focus of cognitive interviews so that I was not placed in a situation of trying to look for everything wrong with a questionnaire. My mind seemed unable to simultaneously handle assessing all aspects of a questionnaire in a single interview. Sometimes the focus I chose was on wording, and sometimes it was on testing visual layout for compliance with a prescribed navigational path. I also began using retrospective methods whereby I simply watched, and sometimes video recorded, respondents answering questions, always following that process with detailed questions based on hesitations, missed questions, and other mistakes.

In addition, I became increasingly concerned with the need for selecting respondents to fit very precise eligibility categories, rather than trying to represent entire populations by choosing one or two people with each of several characteristics such a race, household size, age, etc. A memorable experience in this regard was trying to provide insight into why some parents seemed not to report on their Decennial Census form a baby as a "person" living in their household, a problem evidenced by the Census counts producing fewer people in the 0–1 age category than the number available later from birth records for that time period. In an attempt to answer this question through cognitive interviews, I searched diligently to find parents of children less than six months old. Most of the new parents I recruited reported their babies on the test form, but two respondents did not. Both of those respondents were from countries in quite different parts of the world, and each explained their omission by saying that in their native language, babies were not yet considered "persons." These results were informative, but we also realized that we were probably just touching the surface of reasons for this undercount.

Against this background of intriguing, but mixed, experiences with cognitive interviewing and focus groups, I welcomed the opportunity to attend the first International Conference on Questionnaire Design, Development, Evaluation, and Testing (QDET) in 2002. I had hoped to get a sense of how different kinds of issues could benefit from various approaches to assessing the measurement and response problems associated with commonly used survey methods.

2.2 My 2002 Experience at QDET

The first QDET conference was somewhat disappointing to me. I did not find it very helpful in a practical sense for solving many of the questionnaire evaluation questions I was attempting to answer. The intriguing aspect of the conference for me was how the papers made it abundantly clear that the specific cognitive interview procedures I had learned at Census were only one of many approaches being used by various organizations.

The conference included papers that described the value of traditional pretesting and simply surveying a small sample of the population to see what happened. Other papers described techniques such as behavior coding, reviews by experts, vignette evaluations, observations of item nonresponse, focus group procedures, usability testing, and respondent debriefings. Some papers were analytic in nature, extolling the virtues of experimental tests of alternative wordings and statistical modeling, such as the multitrait-multimethod

(MTMM) approach developed by Saris et al. (2004). Special population issues, ranging from translation across languages and cultures to studying children, were also covered by some of the papers.

I left the conference feeling unfulfilled. It was as though I had observed a sports event in which the excellent players on the field were doing their own thing, without taking into account other players, and I left the conference feeling it had done little to advance my knowledge of how to improve the testing methods I was then using. I had looked for connectedness among testing approaches, and instead found researchers who were individually trying to solve testing challenges in their own ways, much as I had been doing. Also I did not sense from listening to individual conference papers that the relative contributions of cognitive interviewing and focus groups versus experiments and other evaluation methods was being clearly articulated, or perhaps appreciated.

Some of the frustration I felt at QDET1 stemmed from the paper Cleo Redline and I had prepared for the conference (Dillman and Redline 2004). In it, we compared the results of cognitive interviews we had done with field experiments on the same issues for three different design challenges. In all three studies, the cognitive tests and sample survey experiments were implemented at the same times. For all three studies, the results supported making the same administrative decisions on whether to adopt or reject the new design features under investigation.

One of the three comparisons evaluated the effects of three proposed questionnaires and related communications proposed for the 2000 Decennial Census. A "marketing appeal" had been developed, resulting in two of the questionnaires and envelopes being printed in bright yellow. One of the treatment envelopes had the message that had improved response dramatically, i.e. "your response is required by law" placed on the front of the envelope in a bright blue circle, where it gave the cultural appearance of a marketing appeal that might be found on a department store advertisement. The other treatment envelope relegated the same message to the back side of the yellow envelope in a black bar across the bottom of the envelope with the mandatory message written in white print.

Other research we were doing at the time on visual design principles had convinced us that in neither case was the mandatory message likely to be seen (Jenkins and Dillman 1997). Thus we were *not* surprised by the experimental findings showing that these envelopes reduced response rates by 5–10 percentage points below that obtained by the official government business look with the large rectangle on the front of the envelope that presented the mandatory response message previously found so effective for improving response (Dillman 2000, pp. 306–310). This envelope message was the only factor among 16 different techniques hypothesized to improve response rates that showed such a dramatic increase in response rates.

Unfortunately, the marketing appeal had already been rolled out in a general announcement at the Department of Commerce headquarters as being the approach to be used in the 2000 Decennial Census. The concern we had was that we did not think that the results from the cognitive interviews, although negative toward the new forms, would be powerful enough to convince the director and others who had approved the yellow marketing appeal forms to revert to the previously tested "official business" forms (Dillman, 2000). However, the combined results of cognitive interviews and experimental test convinced them to abandon the yellow marketing forms envelopes and forms.

For purposes of the paper written for the first QDET conference, it was satisfying to see that for this investigation as well as the other two comparisons of cognitive interviews versus experiments, results were in the same direction and mutually supportive. However, this produced some consternation for us as authors over how the results might be used. It seemed plausible that this paper might end up providing strong citation support for agency officials to decide that it was okay to evaluate major changes in survey procedures based only upon a few cognitive interviews or focus groups. In our view that would entail considerable risk because the three cognitive interview/field test comparisons we presented were too few upon which to base a scientifically definitive conclusion.

We resolved this practical concern about how the results from these three tests might be used to justify not needing to do field experiments, through another conclusion in the paper. It was that while the experiments quantified the empirical effects very likely to occur when those procedures were used in samples of the same population, they provided little insight into why those effects happened. In contrast, while the cognitive interviews did not provide a basis for quantifying likely effects, they provided hypotheses, based upon interviewee comments as to why the marketing appeal produced lower response rates (Dillman and Redline 2004).

These personal experiences in testing explained only some of the angst I was experiencing during QDET as I found myself unable to organize take-home points from the conference for changing my approaches to questionnaire testing. Most of the conference was focused on qualitative evaluations of questions for use in interviews, yet telephone interviewing was experiencing sampling and response issues that seemed to be fueling a decline in use. The most important survey development occurring in 2002 was the rapid rise in the use of web surveys to collect survey information. Very few of the QDET papers dealt with web surveys, for which software and computer usability issues seemed as critical as question structure and wording.

The QDET papers were also focused mostly upon question wording for interviews but without attention to how the structure and wording of interview questions would likely need to be changed to make them work on the web. In addition, the conference focus was almost entirely on measurement effects of question wording, and not on whether people would respond to the survey.

Finally, the conference seemed to concentrate its attention on resolving cognition issues rather than on factors that motivate people to respond or not respond to surveys. And, although statistical evaluations of questionnaires were on the program, the tilt of the conference seemed mostly focused toward qualitative method evaluations often conducted by professionals with little background in sampling, coverage, and statistical methods for analyzing data.

Despite these concerns, I also experienced immediate benefits from the conference. It helped me put into context why I had changed some of my own interviewing from concurrent think-aloud cognitive interviewing to retrospective interviews in which I observed how people responded to self-administered questions by silently watching how people responded, followed by a systematic debriefing (Dillman 2000, pp. 141–146). Much of the concern Census staff had about Census forms and other questionnaires was that questions were left unanswered. By asking test respondents to read questions and interrupting their thought processes with queries about what the question meant to them, I concluded that they might be giving more attention to questions than would otherwise be the case. Thus, by listening silently and observing hesitations as people answered questions, seeing where mistakes were made, trying to follow people's eye movements, and observing when and where questions appeared not to be completely read or were completely skipped, I collected information that could be revisited in the extensive debriefing. Also, the question I wanted most to answer was how response rates might be affected – and what would cause people to abandon the answering process midstream. I was also trying to understand the impacts of different communication languages needed for conveying meaning as well as response. And I was trying to locate question structures that would work equally well across survey modes, even though they might not maximize benefit for a particular mode, as I sensed had been done for telephone interviews.

Two years after the conference, when the papers from the conference were published (Presser et al. 2004), I found myself reassessing the QDET conference in a somewhat different way. QDET was probably the right conference at the right time. Like many others, I had gone to the conference not knowing what to expect regarding issues that survey practitioners had begun to deal with in their own practically oriented ways. The impetus for creating this conference about cognitive interviewing and other qualitatively oriented investigation techniques was that important aspects of designing surveys had been mostly ignored by traditional survey methodologists. Thus, this conference had a been a first-of-its kind gathering of professionals who had been working on many different dimensions of measurement, often in quite different ways, in order to identify commonality as well as differences in approaches, with a view toward building a stronger and sustainable approach to testing. The QDET conference appeared to me in hindsight as what I believe the organizers had intended it to be – an effort to transition the disparate thinking that prevailed at the turn of the century into conceptually sound and systematic ways of thinking about the design and testing of questionnaires. It seemed in retrospect to signify the shift from art to science that might lead to increased resources being devoted to these efforts, as articulated by Beatty and Willis (2007) in their synthesis on the practice of cognitive interviewing that was produced a few years after the conference.

When I saw announcements for QDET2, the 15 year follow-up conference, I wondered what was likely to be different about this meeting than the preceding one. It also seemed like an excellent opportunity to identify ahead of time the changes happening in the survey environment that should be affecting our evaluation and testing methods. The next section of this chapter describes six major trends I hoped to see taken into account in the 2017 conference presentations and discussions.

2.3 Six Changes in Survey Research that Require New Perspectives on Questionnaire Evaluation and Testing

2.3.1 Self-administration Has Replaced Interviewing as the Dominant Way of Answering Surveys

The final 20 years of the twentieth century witnessed the rise of voice telephone interviewing as a replacement for face-to-face interviews, the result of random digit dialing (RDD) making it possible to access nearly all households on residential lines. Its development required the elimination of showcards that were an integral part of in-person interviews. Telephone interviewing also required shortening questions and answer categories to a length that could be remembered. It also encouraged changing the structures of scales from long and fully labeled to shorter with polar point labeling. Greater use of branching questions came about, partly in an effort to reduce the length and complexity of questions, and also because computers made the process transparent to respondents, thus making the implementation of surveys easier for interviewers. In addition, the design and implementation of telephone surveys became a survey specialization to the extent that some organizations were focused entirely on that mode of data collection. Thus, for many, telephone interviews became the "right" way to do surveys (Dillman 2017).

It is not surprising that the increased application of cognitive science to surveys in the 1990s resulted in the development of testing methods that mostly targeted the improvement of telephone and face-to-face interviews. Consequently, it is not surprising that cognitive interviewing was established as a testing method that focused almost entirely upon question wording and used an interactive process that emulated the same kind of interview process as the one engaged in by interviewer and respondent. There also existed a sort of disconnect between these methods and their intended use.

Cognitive interviews to improve interview questionnaires typically relied upon face-to-face communication, which was appropriate for face-to-face interviews, but less adequate for telephone evaluations where only verbal communication could be used without a physical presence or showcards.

By the end of the 1990s, telephone interviews were facing multiple problems. Residential landlines that provided the sample frame for probability household surveys were being disconnected in favor of mobile phones, which tended to be individually used rather than for all members of one's household. The most likely response to a ringing telephone became a no answer as answering machines became part of most phones, and people could see from what number they were being called. The decline in response rates that accompanied these developments accelerated to the point that response rates for many surveys are now in single digits and likely to stay there. Culturally, the use of voice telephone is a less significant part of most people's lives, making telephone surveys seem "old school" in nature. Interactions with business acquaintances and even casual friends are much more likely now to occur by email and texting instead of voice telephone.

The development of internet surveys, seen by most surveyors as the likely, and much-needed, replacement to telephone and face-to-face interviews, was well underway by the late 1990s, although it was hampered by people having multiple email addresses and there being no algorithm for probability-sampling households or individuals. Mail surveys began somewhat of a comeback as an alternative to telephone and were capable of producing much higher response rates. In addition, our best sample frame for drawing probability surveys became US Postal Service residential delivery addresses. This has led to the development of web-push methodologies whereby people are contacted by mail but asked to respond over the internet, with a paper questionnaire being sent later in the implementation process to achieve better household representation (Dillman 2017). Telephone interviewing has become a niche methodology, sometimes being used as a follow-up to the use of self-administered web and mail questionnaires. In addition, the use of face-to-face methods has continued to decline because of cost and access challenges.

A consequence of the shift toward self-administration is that the process of testing questionnaires in a way that was structured around interviewer and respondent interaction processes, seemed increasingly inadequate. In addition, this transition has ushered in new issues that must now be considered in such testing. Because of the absence of an interviewer to encourage, cajole, and/or correct survey responses, the burden on question structure and wording for motivating accurate responses has increased greatly. In addition, the cognitive processes associated with self-administration need to guide cognitive interviewing and other means of testing.

2.3.2 Motivation Effects Have Joined Cognition Effects in Being Crucial to Adequate Evaluations of Questionnaires

Understanding the role of cognition in determining how respondents answer survey questions was perhaps the most important development that accompanied the rise of the telephone in the 1980s and 1990s (Jabine et al. 1984). It played a major role in the development of cognitive interviewing and other testing methods. When cognitive interviewing and other testing methods were created, it is not surprising that the focus was mostly on individual measurement error with little attention being focused on how the construction of questions and questionnaires affected people's willingness to answer surveys. The role of the interviewer was seen as key to obtaining survey responses, but that motivation was viewed as coming mostly from introductory comments of interviewers and perhaps prior contact, rather than as part of the questioning process itself.

Survey methodology has a long history of attempting to understand what causes people to respond or not respond to surveys. When the interviewer and their controlling influence on response is left out of the response process, increased pressure is relegated to how all aspects of the survey implementation process, including communications requesting a response and the questionnaire itself, affect whether and how people respond to surveys. It remains essential to evaluate question wording effects, but obtaining high-quality responses also requires that questionnaires be created in ways that are easy to understand and motivational. In addition, instructions and potential reminders that interviewers could provide naturally when appropriate needed to be built into the questions themselves.

Thus, it is not surprising that attempts to estimate effects on response rates became a key part of the development and testing of questionnaires. Attempting to predict overall response and item nonresponse effects was a key aspect of the cognitive testing of Census questionnaires that I was focused on in the 1990s while preparing for the 2000 Census.

It is noteworthy that the initial development of testing for web surveys emphasized "usability." Responding to internet surveys was a new task that posed new challenges. Response errors, and in some cases the decision to quit, seemed likely to occur when it was unclear how to transform guidance on answering that had been previously left up to the interviewer into something usable by the respondent. Getting beyond the interviewer-respondent interaction model became a critical consideration because the guidance and persuasion to respond that had been left up to the interviewer had to be conveyed through a succession of words and computer screens. Motivation for responding has thus joined cognition as a key factor that needs to be evaluated in the questionnaire testing process.

2.3.3 Visual and Aural Communication Require Different Approaches for Evaluating and Testing Questionnaires

Fundamental differences exist between how aural (interviewer-respondent) and visual (web and mail) communication occurs (Dillman et al. 2014, Chapter 6). The shift toward visual self-administration requires significant modifications in how one approaches the testing and evaluation of questionnaires. In telephone surveys, words and the manner in which they are conveyed communicate question meaning, with tone, inflection, and other aspects of delivery adding nuance to the question-answer process.

Visual communication used in web and mail surveys relies on words as the primary means of communication, but that meaning is also influenced by three other communication languages: i.e. numbers, symbols, and graphics. Together these four means of communication create the meaning of the survey question, as well as instructions in how to navigate through the questionnaire. Many different concepts, which have been enumerated elsewhere (e.g. Jenkins and Dillman 1997; Christian and Dillman 2004) are helpful in understanding how visual layout influences answers. Information that is communicated in words *and* through visual display is more likely to be perceived, remembered, and used than is information that utilizes only words, especially if delivered aurally. The respondent also has more control over the speed at which that information is provided and can review it without needing to ask the interviewer to redeliver it.

At the time of the first QDET conference, my research team was facing the challenge of finding an improved visual design for a National Science Foundation survey in which a web survey posed a seemingly simple month-and-year problem. The respondent was required to put two digits in the first box and four digits in the second box, but as shown in the example at the top of Figure 2.1, many respondents did not meet the expectation indicated by the MM/YYYY symbols that followed the box.

We used multiple aspects of visual design theory to develop a succession of tests across three experiments, as shown in Figure 2.1, in which different visual layouts were used. For each design, the percent of who "correctly" answered the question with the requested two digits for month and four digits for year was recorded. It can be seen that the poorest compliance with this instruction occurred when two disconnected words "month" and "year" were placed below the answer boxes. Responses were improved when boxes were connected together, even though the words were unchanged, and especially when the month box was sized for fewer digits than was the year box. Gestalt psychology suggests that things seen together (as suggested by the connection of the boxes) encourages treating them in the same way. Consequently, fewer respondents used words for month, but instead used digits as needed to be used for reporting years in a conventional manner (Christian et al. 2007).

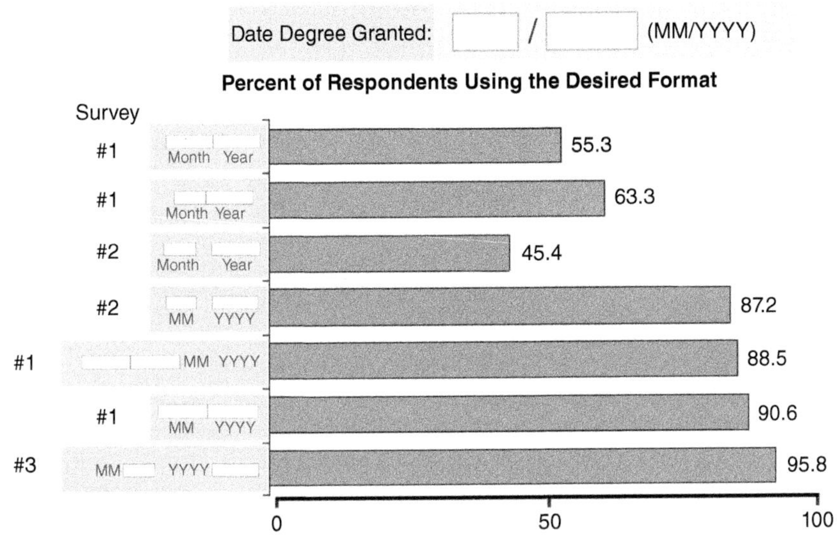

Figure 2.1 Responses provided in intended format based on the design of the request.

The remainder of Figure 2.1 shows that the percent of respondents using the desired two-digit month and four-digit year format was improved dramatically when symbols indicating the number of digits (MM/YYYY) were substituted for the month/year words. Compliance with the request in the desired manner was nearly doubled over the versions using the less-specific words. It is also significant that the best performance of all was received when the symbolic indicators preceded the appropriately sized boxes in natural reading order. Related research also showed that the positive effects of visual layout of the answer boxes with symbols instead of words occurred even when the question specifically instructed respondents to provide their answer using two digits for the month and four digits for the year. Carefully designed visual display clearly augmented the use of words.

This is one set of experiments from among many that have showed how many different visual design concepts can dramatically influence how people respond to visually presented questions (Dillman et al. 2014). Other research has shown that distances between response categories, the sizes of answer spaces provided for open-ended answers, and whether response categories are presented in multiple rows and columns, also influences the answers people give to visually delivered questions. In addition, question numbers, spacing on pages, consistency across question layouts, and a host of other visual features influence whether questions are skipped (a common problem in mail questionnaires) or correctly interpreted and the needed answers provided.

These studies have significant implications for questionnaire evaluation and testing. As the use of self-administered surveys increases, our methods of testing must also change. Focusing mostly on "what the wording of this question means to you" is insufficient. It becomes just as important to learn whether the visual layout of questions is working in the desired manner. We also need to learn whether the structure of questions, and the degree to which questions are liked or not liked by respondents, encourage answers to be provided. Mid-survey terminations are one of the largest problems we now face with web surveys.

2.3.4 Unified Mode Construction Has Emerged as an Important Criterion for Questionnaire Construction

The domination of telephone interviewing in the late twentieth century brought with it an emphasis on maximizing the use of question structures that were easier to administer by telephone. Specifically, surveyors attempted to use conceptually narrower questions that required fewer words, as surveyors adjusted to the difficulty of not being able to use showcards to assist the answering process. Also, it became common to withhold less-desired answer categories (e.g. "no opinion") from the offered choices unless the respondent was having difficulty choosing any of the categories. In the rare circumstances that mail methods were used for some respondents in the same study, this practice produced different rates of item nonresponse, as those respondents were either given the "no opinion" category up front or denied its use entirely, so that item nonresponse became more common (Dillman and Christian 2005).

The introduction of web surveys encouraged the use of a "check all that apply" question-answer format that was seldom used on the telephone, thus producing undesirable mode differences in measurement (Smyth et al. 2006). The web provided two types of closed question formats – the radio button that allowed only one of a series of answers to be marked versus the HTML box that allowed the selection of multiple answer choices. Although check-all formats had been commonly used on mail surveys, the format was awkward in interviews. Typically, it was replaced there by asking respondents for a yes/no answer accepted after each item was read. Extensive research has now shown that these two formats are not interchangeable, producing different answers on interview versus visual self-administered surveys (Smyth et al. 2008).

In the multi-mode environment that now exists, considerable effort is being made to reduce the use of quite different question structures across survey modes through the use of unified mode construction. The central idea underlying these proposals is that surveyors should attempt to optimize construction across modes while recognizing that some differences are necessary. An example is automatic branching of questions in telephone and web surveys, which for mail surveys can only be accomplished with written instructions and appropriate visual formatting (e.g. Redline et al. 2003). Achieving equivalency in the face of necessary differences provides an additional challenge for evaluating questionnaires for mixed-mode surveys that collect data by mail as well as web and/or telephone.

When multiple response modes are used for surveys, it is especially important that questions be evaluated from the perspective of how they are structured and will appear in all modes. A compelling case for this effort was made by Martin et al. (2007) upon the discovery of quite different wording of questions being promulgated by different mode-specific working groups for conducting the planned 2010 Census. Accepting such differences would have resulted in different answer distributions for each mode.

It is now evident the questionnaire evaluation and testing process must move beyond the idea of maximizing construction for individual modes. Instead it needs to identify common ground in order to minimize unnecessary differences in measurement. Unified mode construction that seeks to optimize measurement across all modes is an essential feature of the mixed-mode world in which we now live.

2.3.5 Mixed-Mode Surveys Are Replacing Single-Mode Surveys as the Dominant Way of Conducting Surveys

With few exceptions, single modes are no longer the best choice for doing surveys. Face-to-face interviews are extremely costly, face coverage problems from gated communities and locked apartment buildings, and are slow. Voice telephone surveys face severe coverage, sampling, and response problems, and no longer fit with societal culture as a way of normal communication without prior contact and agreement. US Postal Service residential lists now provide our most adequate general public sample frame, and mail response rates are better than telephone, but this response mode does not handle branching questions well. Web surveys, though now in use for 20 years, have no general public sample frame (as do RDD telephone and postal contact surveys), and significant portions of the general public with different demographic profiles are not likely to respond (Dillman 2017).

Common to all of these modes is the fact that some people in many survey populations do not use all contact methods evenly, and may ignore some altogether, setting up a situation that allows coverage to be increased if more than one contact mode can be used. Use of more than one mode of contact makes it possible to deliver more requests for response in ways that may increase the likelihood of response and offer different response modes.

An issue now facing survey methodologists is how to combine contact and response mode possibilities in ways that will improve response and data quality. Typically, this means trying to encourage use of cheaper modes of response early, and save the most costly modes until later in the data collection process.

One mixed-mode possibility that has emerged is the use of web-push methods, whereby requests to respond over the web are delivered by postal mail.

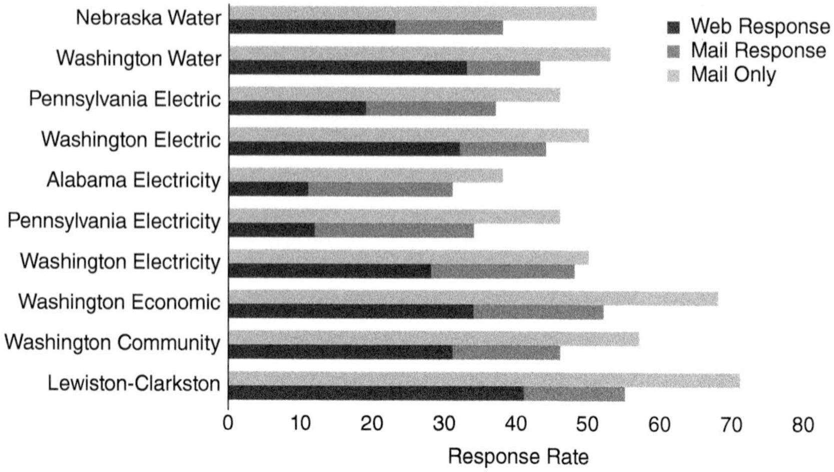

Figure 2.2 Response rates for 10 experiments comparing mail-only response designs with web-push designs that asked for internet responses in initial mail contacts, and provided a mail response option in later contacts.

Such a request can include an incentive, typically cash, and other response contact and response modes may be made later to obtain responses from people unable or unwilling to respond over the internet.

A series of tests of such a methodology was undertaken by a Washington State University team between 2007 and 2014, using regional and state-wide samples of US Postal Service residential addresses (Smyth et al. 2010). All tests involved questionnaires consisting (in mail form) of 12 pages containing 90–140 individual questions that would have taken 20–30 minutes to administer over the telephone. Ten comparisons were made of web-push responses, as shown in Figure 2.2.

The web-push response rates average 43% with a range of 30–54% (Dillman et al. 2014). The mail-only comparisons produced a mean household response rate of 53% ranging from 38–71%. Approximately 60% of the responses to the web-push treatments came over the internet, with the remainder coming by mail. The initial request included a $4–$5 cash incentive, which was found in an experiment to improve the web response rates dramatically, from 13–31% (Messer and Dillman 2011). Another key finding was that web respondents were quite different than the later responders on paper questionnaires, being younger, better educated, and more likely to have children at home. However, the demographics for the combination of web and mail responders to the web-push methodology were virtually the same as all respondents to the mail-only treatment groups.

The web-push methodology has been developed further for use in other survey situations, including periodic population censuses in Japan, Australia, and Canada. In the 2016 Canadian Census, 68% of the households responded over the internet, 20% by follow-up mail, and an additional 10% through enumerator visits, reaching 98% total response (Statistics Canada 2016). The web-push methodology is also being used for national surveys in European countries (Dillman 2017). When it is possible to use both postal and email contact with survey samples, response rates, and in particular response on the web, can be pushed to higher levels (Millar and Dillman 2011).

I think it is likely that web-push data collection procedures using multiple modes of data collection in sequence will replace large numbers of surveys that have relied mostly on interviewing methods. It will also require research on how to articulate modes of contact and response for greater effectiveness. This means that designing questionnaires with an emphasis on unified construction across modes, as discussed in the previous section, will increase.

2.3.6 Devices Have Replaced Interviewers as a Principal Consideration in Development of Questions and Questionnaires

People now have a variety of devices that may be used for responding to web surveys, including smartphones, tablets, laptops, and desktop computers. In particular, the use of smartphones as a way of responding to surveys appears to be increasing, and also creating new response problems, which might be described as the purse/pocket problem (Dillman 2017). Many people want a size of mobile phone that they can carry with them everywhere, so that size needs to fit in pants and coat pockets as well as small purses or hands-free carrying cases. The screen sizes available for displaying the types of questions asked in today's surveys, even on telephones, are hard to fit on small smartphone screens. As a consequence, many surveyors finding themselves trying to eliminate items in a series, whereby people are given a set of response categories to use in responding to a list of items. It also means that horizontal scales and long, fully labeled scales sometimes become hard to fit onto screens (Barlas and Thomas 2017).

We now face the necessity that surveyors must construct questions differently for smartphones than has often been done for all other modes. The need for unified question construction raises the prospect of having to design questions differently for other modes of surveying as well in order to get the same measurement as is feasible on smartphones. As a consequence, long-used survey formats, such as 7- or 10-point scales validated on traditional survey modes, and items in a series whereby a scale is presented for use in answering a series of a dozen or more items, may need to be changed in order to avoid obtaining different answers on this newest device for completing self-administered surveys.

It is also apparent that mobile phones obtain lower response rates, higher breakoff rates, and longer completion times than do web surveys on personal computers (Couper et al. 2017). In addition, the nature of smartphone use is that there is a great likelihood that emails arrive and are processed while prospective respondents are in the midst of other activities, so that the request is easily dismissed and deleted.

For questionnaire testing and evaluations that include the use of web responses, it is no longer acceptable to only test question structures and wordings on laptops and/or tablets. It is now essential that the consequences for smartphone administration also be evaluated. This need makes testing more complicated. Underlying this concern is the realization that we may be moving into a period of greater simplicity and perhaps survey brevity in order to achieve satisfactory mixed-mode survey results.

2.4 Conclusion

The six issues described in this chapter cover issues that were not well-represented at the 2002 QDET conference, or had not yet emerged, as is the case with smartphones:

- The need to emphasize survey self-administration as well as interview modes of data collection
- Evaluation of motivational qualities of questionnaires, as well as their cognitive aspects
- Evaluation of the complexities associated with visual communication that involve the use of numbers, symbols, and graphics for communication as well as wording used in aural communication
- Utilization of unified-mode question construction across survey modes to reduce measurement differences
- Evaluation of whether equivalent measurement is achieved when more than one survey mode is used for data collection, as now happens with far greater frequency
- How to design questions for use in the limited space provided by smartphones and carry those questionnaire formats over to other survey modes to avoid measurement differences

The implication I draw from these changes is that we are living and working in a more challenging time for questionnaire evaluation and testing than we have ever experienced. Much that was taken for granted in the late 1990s leading up to QDET1 is badly out of date. Interviewing, whether in person or by telephone, is transitioning to less use for data collection and is now more likely to be used in mixed-mode designs.

Question wording must now be evaluated on the basis of getting common measurement across modes. Evaluating the cognitive properties of questionnaires remains important, but we must add to it evaluations of motivational qualities to facilitate the loss of interviewers as the motivators in chief for obtaining survey responses. We have also had to learn new concepts and theories for how communication occurs in order to incorporate visual layout assessments for self-administered mail and web questionnaires, now that words are not completely responsible for communicating meaning and response tasks.

It was pleasing to see how well participants in QDET2 responded to these challenges. All of the needed changes described here received attention from multiple papers, many of which are included in this book. There was a heavy emphasis on experimental results, and away from simply describing how questionnaires and procedures had been subjectively evaluated. Mixed-mode studies dominated the reports, with most including internet data collection, while relatively few papers focused only on interview methods. Experiments on the effects of visual layout differences and eye-tracking studies, barely feasible at the time of the first QDET, were also reported. Dependent variables in many studies emphasized response rates and response quality, thus bringing motivational issues into consideration. QDET2 also drew participants from throughout the world, with many thought-provoking papers making clear how survey evaluation methods had become international in scope.

The 15 years that elapsed between QDET and QDET2 witnessed a much-needed process of maturation from somewhat isolated efforts to solve practical measurement problems to systematic efforts to develop and utilize the scientific foundations of evaluating and testing questionnaires. As evidenced by QDET2, methodologists throughout the world are utilizing these foundations to conduct far more sophisticated evaluations that not only promise improved measurement, but also address the rapid changes in survey research that require the use of such methods.

References

Barlas, F. and Thomas, R. (2017). Good questionnaire design: best practices in the mobile era. American Association for Public Opinion Research webinar, January 19.

Beatty, P.C. and Willis, G.B. (2007). Research synthesis: the practice of cognitive interviewing. *Public Opinion Quarterly* 71 (2): 287–311.

Christian, L.M., Dillman, D.A., and Smyth, J.D. (2007). Helping respondents get it right the first time: the influence of words, symbols, and graphics in web surveys. *Public Opinion Quarterly* 71 (1): 113–125.

Christian, L.M. and Dillman, D.A. (2004). The influence of symbolic and graphical manipulations on answers to paper self-administered questionnaires. *Public Opinion Quarterly* 68 (1): 57–80.

Couper, M.P., Antoun, C., and Mavletova, A. (2017). Mobile web surveys. In: *Total Survey Error in Practice* (eds. P.P. Biemer, E.D. de Leeuw, S. Eckman, et al.). Hoboken, New Jersey: Wiley.

Dillman, D.A. (2017). The promise and challenge of pushing respondents to the Web in mixed-mode surveys. *Survey Methodology* 43 (1) Statistics Canada, Catalogue No. 12-001-X, Paper available at http://www.statcan.gc.ca/pub/12-001-x/2017001/article/14836-eng.htm.

Dillman, D.A. (2000). *Mail and Internet Surveys: The Tailored Design Method*, 2e. Hoboken, NJ: Wiley.

Dillman, D.A. and Christian, L.M. (2005). Survey mode as a source of instability across surveys. *Field Methods* 17 (1): 30–52.

Dillman, D.A. and Redline, C.D. (2004). Testing paper self-administered questionnaires: cognitive interview and field test comparisons. In: *Methods for Testing and Evaluating Survey Questionnaires* (eds. S. Presser, J.M. Rothgeb, M.P. Couper, et al.), 299–317. New York: Wiley-Interscience.

Dillman, D.A., Singer, E., Clark, J.R., and Treat, J.B. (1996b). Effects of benefits appeals, mandatory appeals, and variations in confidentiality on completion rates for census questionnaires. *Public Opinion Quarterly* 60 (3): 376–389.

Dillman, D.A., Smyth, J.D., and Christian, L.M. (2014). *Internet, Phone, Mail and Mixed-Mode Surveys; the Tailored Design Method*, 4e. Hoboken, NJ: Wiley.

Fay, R.E., Bates, N., and Moore, J. (1991). Lower mail response in the 1990 census a preliminary interpretation. In: *Proceedings of the Bureau of the Census 1991 Annual Research Conference*, 3–22. Washington, DC: US Department of Commerce.

Jabine, T., Straf, M., Tanur, J.K., and Tourangeau, R. (eds.) (1984). *Cognitive Aspects of Survey Methodology: Building a Bridge between Disciplines.* Washington, DC: National Academy of Sciences.

Jenkins, C.R. and Dillman, D.A. (1997). Towards a theory of self-administered questionnaire design, Chapter 7. In: *Survey Measurement and Process Quality* (eds. L. Lyberg, P. Biemer, M. Collins, et al.), 165–196. New York: Wiley-Interscience.

Martin, E., Childs, J.H., DeMaio, T. et al. (2007). *Guidelines for Designing Questionnaires for Administration in Different Modes.* Suitland, MD: US Census Bureau.

Messer, B.L. and Dillman, D.A. (2011). Surveying the general public over the Internet using address-based sampling and mail contact procedures. *Public Opinion Quarterly* 75 (3): 429–457.

Millar, M.M. and Dillman, D.A. (2011). Improving response to Web and mixed-mode surveys. *Public Opinion Quarterly* 75 (2): 249–269.

Presser, S., Rothgeb, J.M., Couper, M.P. et al. (2004). *Methods for Testing and Evaluating Survey Questionnaires.* Hoboken, New Jersey: Wiley-Interscience.

Pullman, C. (2000). Sense and Census. *Critique: the Magazine of Graphic Design Thinking* 16 (Summer): 54–61.

Redline, C.D., Dillman, D.A., Dajani, A., and Scaggs, M.A. (2003). Improving navigational performance in US census 2000 by altering the visual languages of branching instructions. *Journal of Official Statistics* 19 (4): 403–420.

Saris, W.E., van der Veld, W., and Gallhofer, I. (2004). Development and improvement of questionnaires using predictions of reliability and validity. In: *Methods for Testing and Evaluating Survey Questionnaires* (eds. S. Presser, J.M. Rothgeb, M.P. Couper, et al.), 275–297. New York: Wiley-Interscience.

Smyth, J., Christian, L.M., and Dillman, D.A. (2008). Does "yes or no" on the telephone mean the same as check-all-that-apply on the web? *Public Opinion Quarterly* 72 (1): 103–111.

Smyth, J.D., Dillman, D.A., Christian, L.M., and O'Neill, A. (2010). Using the Internet to survey small towns and communities: limitations and possibilities in the early 21st century. *American Behavioral Scientist* 53: 1423–1448.

Smyth, J.D., Dillman, D.A., Christian, L.M., and Stern, M.J. (2006). Comparing check-all and forced-choice question formats in Web surveys. *Public Opinion Quarterly* 70 (1): 66–77.

Statistics Canada. (2016). 2016 Census of Population collection response rates. http://www12.statcan.gc.ca/census-recensement/2016/ref/response-rates-eng.cfm. Accessed October 24, 2016.

Tourangeau, R. (1984). Cognitive science and survey methods. In: *Cognitive Aspects of Survey Methodology: Building a Bridge between Disciplines* (eds. T. Jabine, M. Straf, J.K. Tanur and R. Tourangeau). Washington, DC: National Academy of Sciences.

3

A Framework for Making Decisions About Question Evaluation Methods

Roger Tourangeau, Aaron Maitland, Darby Steiger, and Ting Yan

Westat, Rockville, MD, USA

3.1 Introduction

This chapter provides an overview of methods available for questionnaire development and testing and recommendations about when to use each one. It is organized by the setting in which the methods are used (by experts, in the laboratory, or in the field), which often determines the cost, staffing, and resources needed to implement them. The budget and time available for questionnaire development and testing are generally the most important factor determining which methods can be used. Expert methods (which require no data collection from members of the target population) are the least expensive and require the least time; field-based methods are the most costly and time-consuming. Other considerations, such as the population of interest and the mode of data collection, also affect the choice of methods. For example, usability testing is most often used with methods of data collection in which respondents interact directly with a computer. Similarly, a focus group discussion is unlikely to be used if the target population consists of young children. As we review each method, we list its main advantages and disadvantages.

We begin by discussing expert methods for questionnaire development and testing (Section 3.2). In Section 3.3, we review laboratory methods, in which relatively small purposive samples of respondents are recruited as part of the question evaluation process. Then, we discuss field-based methods that recruit respondents to participate in a field test approximating realistic survey conditions (Section 3.4). Section 3.5 discusses statistical modeling techniques to assess data quality. Section 3.6 presents some recent studies that have compared convergence within and across various testing methods. The final section summarizes our recommendations and presents some fairly typical combinations of methods used.

Advances in Questionnaire Design, Development, Evaluation and Testing, First Edition.
Edited by Paul C. Beatty, Debbie Collins, Lyn Kaye, Jose-Luis Padilla, Gordon B. Willis, and Amanda Wilmot.

3.2 Expert Reviews

Expert methods for testing survey questions are often divided into those that involve human experts and those that involve computer systems. We begin our review with the methods involving human experts.

3.2.1 Unaided Expert Reviews

Expert review is a quick and efficient appraisal process conducted by survey methodologists. Based on accepted design principles and their own research experiences, expert reviewers identify possible issues with questions and suggest revisions. In some cases, a cognitive appraisal form is used that identifies specific areas (e.g. conflicting instructions, complex item syntax, overlapping response categories) to review for each question in the instrument. A typical outcome of an expert review is improved question wording, response formats, instructions, and questionnaire flow. The focus includes not only the content, but also extends to the ease and efficiency of completion, avoidance of errors, ability to correct mistakes, and presentation of the content (overall appearance, organization, and layout of the instrument).

An important goal for an expert review is that survey questions are understood by all respondents the way the question designers intended. Although there is general agreement about this principle, experts looking at the same question often disagree about whether a question is ambiguous. Apart from identifying clearly problematic questions, expert reviews often identify questions to be targeted for additional testing (for example, in cognitive interviews).

Several systems have been developed for assessing potential problems with questions. Lessler and Forsyth (1996) present a scheme that distinguishes 26 types of problems with questions; the problem types are based largely on a cognitive analysis of the response process. We discuss their scheme in Section 3.3. Graesser et al. (1996) distinguish 12 major problems with questions, most involving comprehension issues; they have also developed a computer program that analyzes draft questions in terms of a subset of these issues, serving as an automated expert appraisal. We discuss this automated tool – Question Understanding Aid (QUAID) – in Section 3.3.

Expert reviews are widely thought to have several virtues:

- They are fast and relatively inexpensive to do;
- They generally detect more problems than other methods (see Section 3.6); and
- Experts often make good suggestions about how to fix the problems.

Yan et al. (2012) found that expert reviews generally detect more problems than other methods do. Expert reviews also have their limitations: They show low levels of consistency across trials (e.g. Presser and Blair [1994] found that the correlation between the judgments of two expert panels as to whether an item had a problem was only 0.37). Given this low level of agreement, it is likely that many of the problems turned up by expert reviews are false alarms; Conrad and Blair (2004, pp. 67–87) make a similar point about cognitive interviews. In addition, expert are generally most helpful when the questionnaire is nearly finished.

3.2.2 Questionnaire Appraisal System (QAS)

Although expert reviews have been used for many years, forms appraisal methods offer a more structured method for evaluating survey questions. The primary goal for these methods is to provide a systematic tool for evaluating survey questions, one that can be employed by non-experts. A forms appraisal is conducted by evaluating each question for a specified set of problems. The Question Appraisal System, or QAS (Willis and Lessler 1999), is designed to focus attention on problems that are likely to affect the accuracy of the answers – that is, their agreement with records data or other criterion measures (Lessler and Forsyth 1996). The QAS requires the evaluator to check each question for seven classes of problems involving question reading, question instructions, question clarity, assumptions, knowledge or memory, sensitivity or bias, and response categories. In total, each question is evaluated for the presence of 26 potential problems.

Raters are encouraged to be liberal in assigning codes in order to identify as many potential problems as possible. Accordingly, the QAS tends to find more problems than other methods (Rothgeb et al. 2001). The QAS can be used spot problems and improve the questions or to identify questions that need additional testing. Willis (2005) recommends that researchers review draft questions using the QAS prior to cognitive testing to identify issues for further investigation; he recommends different cognitive probes be used in the cognitive testing depending on the type of problem found by the QAS.

Other forms appraisal schemes can be used to evaluate survey questions. Van der Zouwen and Dijkstra (2002) proposed the Task Difficulty Score (TDS). This appraisal form focuses on 11 potential elements of the response task that may cause problems with survey questions. The results from the TDS are highly correlated with the results from the QAS (Van der Zouwen and Smit 2004).

Some potential benefits of these questionnaire appraisal systems are:

- They systematically focus on the cognitive issues in the response process;
- They help experts provide more consistent feedback;
- They are low-cost enhancements to expert review; and
- They do not require experts with a survey methodology or substantive background.

A potential drawback with these systems is that they tend to identify problems with almost every question.

3.2.3 Question Understanding Aid (QUAID)

QUAID is a computer tool developed by Graesser et al. (2006). QUAID was inspired by computational models from computer science, computational linguistics, discourse processing, and cognitive science. The software identifies technical features of questions that may cause question comprehension problems. The current version of QUAID critiques each survey question on five classes of comprehension problems: unfamiliar technical terms, vague or imprecise predicate or relative terms, vague or imprecise noun phrases, complex syntax, and working memory overload.

QUAID generally identifies these problems by comparing the words in a question to several databases or data files (e.g. Coltheart's MRC Psycholinguistics Database). For example, QUAID identifies a word as unfamiliar if it falls below a threshold level of frequency or familiarity in several lexicons. Similarly, vague or imprecise predicate or relative terms (e.g. *frequently*) are identified by QUAID if they have multiple senses or exceed some abstractness threshold.

Expert ratings of a corpus of survey questions were critical in the development of QUAID. The corpus consisted of 505 questions on 11 surveys done by the US Census Bureau. This suggests that the program may have more relevance for factual than for attitudinal questions or may be limited in its usefulness for other countries. The threshold levels of the computer program were determined by identifying values that maximized the correlations with expert ratings of these questions.

QUAID is thought to have three major advantages:

- It is good (and consistent) at finding problems with the syntax of the question;
- It often produces different results from what experts provide; and
- There is some evidence that when experts use QUAID, they create better questions than when they don't use it.

And, of course, it can be very inexpensive to use. The potential drawbacks are that there is not much evidence that QUAID results are better than those from experts or other techniques, and the QUAID results sometimes disagree with those of other methods (Maitland and Presser 2016).

3.2.4 The Survey Quality Predictor (SQP)

The Survey Quality Predictor (SQP) was created by Saris and colleagues based on meta-analyses of a large number of multitrait-multimethod (MTMM) experiments (Saris and Gallhofer 2014). Saris and colleagues conducted regression analyses using the characteristics of questions in the MTMM experiments to predict the quality criteria (e.g. reliability, validity, and method effects). The parameters from their regression models are the basis for SQP, which researchers can use to predict the quality of their own survey questions. Users code each question on such features as the number and type of response categories. SQP also automatically codes certain characteristics, such as the number of words in the question. SQP then outputs predicted coefficients for reliability, validity, and method effects. Currently, the program requires the researcher to code variables ranging from fairly objective factors, such as the mode of administration and type of response options, to more subjective factors, such as degree of social desirability and centrality of the question. As a result, the predicted quality measures are subject to coding errors.

Like QUAID, SQP can be very fast and very inexpensive to use since no data collection is required. Two potential problems with it are (i) it may be difficult for researchers to code the characteristics of their questions reliably (since the most of the coding is not automated and requires considerable judgment); and (ii) the predictions are based largely on experiments with attitudinal rather than factual or behavioral items.

3.3 Laboratory Methods

The methods reviewed here typically involve small purposive samples of respondents who are brought into a laboratory or a similar setting where they can be observed closely and the environment can be controlled. In principle, many of these methods could also be used in field settings, but the laboratory has been the usual setting. Although these methods are typically used at an early stage of question development, they can also be helpful for understanding problems found when a question has been fielded. We discuss four such methods – cognitive interviews, focus groups, usability testing, and eye-tracking. Because they usually are done on a small scale, they tend to be relatively fast and inexpensive, but the participants are not necessarily representative of the target population for the survey.

3.3.1 Cognitive Interviews

Cognitive interviews are in-depth, semi-structured administrations of the survey that are designed to yield insights into cognitive problems that may lead to response errors (Beatty and Willis 2007; Willis 2005). Typically, these interviews present a survey item to the respondent, allow the respondent to answer, and probe to discover the basis of the response or the interpretation of the question. Much of the probing is developed prior to the pretesting, so as to have some standardization across interviews. But often a cognitive interviewer probes spontaneously in response to unanticipated difficulties in answering questions, including nonverbal cues suggesting respondent confusion or difficulty.

Information about a respondent's thought processes is useful because it can be used to identify problems such as

- Instructions that are insufficient, overlooked, misinterpreted, or difficult to understand;
- Wordings that are misunderstood or understood differently by different respondents;
- Vague or ambiguous definitions and instructions;
- Items that ask for information to which the respondent does not have; and
- Confusing response options or response formats.

The interviews are typically done in person (even if the actual survey is to be administered via mail or telephone) by researchers with expertise in questionnaire design or qualitative methodologies. In addition, interviews are typically done at a central location (e.g. a focus group facility) and limited to about an hour, and respondents are paid. The interviews are often tape-recorded so that they can be reviewed for careful analysis.

The advantages of cognitive interviews are:

- They gather data from members of the target population;
- They often reveal comprehension or memory problems that other methods wouldn't find; and
- They allow researchers to try out new versions of the questions, a feature referred to as *iterative testing* (Esposito 2004).

The method also has its specific drawbacks:

- Data reduction/summary can be difficult and subjective (some simple coding techniques have been explored, but none has been widely adopted), and it may not always be clear whether there is a problem with a question and, if so, what the problem is (Presser and Blair 1994; Yan et al. 2012);
- Cognitive interview practices are not standardized, and different techniques are used by different organizations and by different interviewers within an organization (DeMaio and Landreth 2004; Rothgeb et al. 2001); and
- There is no way to judge how common the problems found are within the target population.

3.3.2 Focus Groups

Focus groups are widely used to capture participants' perceptions, feelings, and suggestions about a survey topic. In the context of designing a population-based survey, focus groups can be used to elicit important areas that should be addressed by the survey or to ensure that the language of the survey questions matches the terminology of potential respondents. Comparative information may also be gathered from different segments of the target population.

Traditionally, a focus group is a discussion among about 8–10 people who share some common interest, trait, or circumstance. Focus group sessions are generally an hour to an hour-and-a-half long and are led by a trained moderator. The moderator follows a prepared script that lists questions and issues relevant to the topic. Preparation steps for focus groups include developing the protocol, recruiting group members, arranging for facilities, and conducting the groups. Focus groups can be conducted in a number of modalities, including a traditional in-person setting, by telephone, online using video-conferencing, online using a synchronous chat room feature, or asynchronously over a number of days. The traditional in-person focus group is still considered the "gold standard" (Hine 2005), because it promotes interactions among participants and encourages breadth and depth of responses (Gothberg et al. 2013).

Among the potential advantages of focus groups are:

- They can explore issues of terminology and understanding of key concepts before the survey designers attempt to draft the questions;
- They can elicit a detailed understanding of experiences or perceptions underlying survey responses; and
- Telephone and online groups allow for geographic diversity at relatively low cost.

And the potential drawbacks include:

- A few individuals can dominate the discussion or set the group off on unproductive paths, reducing the amount of information obtained;
- Data reduction/summary can be difficult and subjective; and
- Telephone focus groups produce lower levels of interaction between participants, and online focus groups may underrepresent lower-income populations.

3.3.3 Usability Testing

The goal of usability testing of an instrument is to uncover (and fix) problems that interviewers or respondents may encounter before the instrument is fielded (e.g. Couper 1999). Issues that may lead to reporting errors, nonresponse, and break-offs are of utmost importance. Typically, usability testing is conducted with one individual at a time, while an observer notes the following:

- General strategies that respondents use to complete the instrument;
- The path(s) they take (when multiple paths are possible);
- The points at which they become confused or frustrated (which can indicate a risk of break-offs in the field);
- The errors they make, how they try to correct them, and the success of their attempts; and
- Whether the respondent is able to complete the instrument.

Think-aloud testing is commonly done with a new or previously untested instrument. This approach provides a wealth of qualitative data that serves as the basis for recommendations for instrument improvements. Respondents may also be asked to comment on the need for and usefulness of instructions, ease of navigation, the overall appearance of the instrument, ease of completion, and their overall satisfaction.

The major advantages of usability testing are the relatively low costs and quick turnaround; in addition, usability testing often reveals instrumentation problems that other methods would not detect. Still, it has its particular problems as well. One major issue is that the conditions under which the testing takes place are not usually representative of those encountered in the field; and, as with cognitive testing and focus groups, the data are rich and detailed and may, as a result, be hard to summarize reliably. It may be possible to carry out usability testing in field settings, but we are unaware of any attempts to do this.

3.3.4 Eye Tracking

Survey researchers have begun to utilize technology that enables them to track the eye movements of survey respondents and to understand the visual aspects of question design more fully. The technology allows researchers to measure how long respondents fixate on particular areas of the question and the total number of fixations on the question. Galesic et al. (2008) used eye tracking to show that respondents attend more to response options that come earlier in the list of options than those that come at the end. This helps explain the phenomenon of response-order effects in surveys. In addition, they demonstrated that respondents pay less attention to response options that do not initially appear on the screen compared to those that are always visible. Eye tracking can thus provide valuable insights into the visual aspects of question design.

In addition, certain eye movements may signal comprehension or other difficulties with a question. For example, "regressive eye movements" (going back to reread the question or answer categories) are likely to reflect respondent confusion. "Real-world" eye tracking has recently become available, allowing researchers to track eye movements outside laboratory settings. In the past, eye-tracking studies required that respondents answer questions on a computer equipped with the necessary hardware and software.

An advantage of eye tracking is that it can reveal usability problems or other issues with questions in a relatively objective way (for example, the presence of a regressive eye movement could be taken as a sign of a problem), whereas cognitive and usability testing both depend to some extent on the respondent's ability to verbalize problems. The availability of real-world eye tracking has made it possible to test all types of questionnaires (e.g. paper questionnaires), not just questions administered by computer. Still, there are potential drawbacks to eye tracking:

- It requires new data collection;
- It requires special equipment and software;
- To date, most eye tracking has been done with computer-administered questionnaires; and
- Eye-tracking data can require considerable pre-processing before they are ready to analyze.

3.4 Field Methods

3.4.1 Pretests

Pretests, or pilot studies, are small-scale rehearsals of the data collection protocol before the main survey. Their purpose is to evaluate the survey questionnaire and detect any problems with the data collection procedures and software. Pretests are done with "small" samples (although for a large survey, the pretest sample may include more than 1000 cases), often by relatively few interviewers. Pretesting has been standard survey practice in survey research for many years.

In analyzing pretest data, survey designers examine one-way distributions of the answers, looking for items with high rates of missing or problematic responses (such as out-of-range values or values that are inconsistent with other answers) or items that produce little variation in the answers.

The key advantages of pretests are:

- They test the entire package of procedures and questionnaires;
- They test them in a realistic setting; and
- They produce more objective, quantitative data than most of the other methods.

The main drawbacks of pretests are that they are generally much more expensive and time-consuming than the expert or laboratory methods. In addition, many problems may go undetected. People will often answer questions that they misunderstand. Finally, pretests require extensive efforts to prepare, implement, and analyze, and most are cost- or time-prohibitive.

3.4.2 Randomized Experiments

Survey designers sometimes conduct studies that compare different methods of data collection, different field procedures, or different versions of the questions in randomized experiments. Such experiments may be done as part of a field pretest. When random portions of the sample get different questionnaires or procedures, as is typically the case, the experiment is called a "randomized" or "split-ballot" experiment. Tourangeau (2004) describes some of the design issues for such studies.

Experiments like these offer clear evidence of the impact on responses of methodological features – differences in question wording, question order, the mode of data collection, and so on. Unfortunately, although they can conclusively demonstrate that the different versions of the instruments or procedures produce different answers, many split-ballot experiments cannot resolve the question of which version produces *better* data, unless the study also collects some external validation data against which the survey responses can be checked. Still, experimental results are also often interpretable when there are strong theoretical reasons for deciding that one version of the questions is better than another. Fowler (2004) describes several split-ballot experiments in which, despite the absence of validation data, it seems clear on theoretical grounds which version of the questions produced better data.

A major advantage of randomized experiments is that they offer the potential to evaluate the impact of proposed wording changes or other variations in procedures during pretesting phases of a study. This comes with the added cost and complexity of producing multiple versions of the instrument, which may exceed the modest budgets of many pretests. Also, the results may not make it clear which version of the question should actually be adopted.

3.4.3 Interviewer and Respondent Debriefings

Interviewer briefings are often held after a pretest to get the interviewers' evaluations of the questions and the data collection procedures. These debriefings are often guided discussions (a bit like focus groups) in which the interviewers discuss their concerns about specific questions or other issues that emerged during the pretest. The interviewers may be asked to give their suggestions for streamlining the procedures or improving the questions. Interviewer debriefings are typically done by telephone since the interviewers may be scattered across a wide area. Although a single session is probably more common, interviewer debriefings can also be conducted in multiple sessions addressing different topics, such as gaining cooperation, administering the instrument, and administrative aspects of data collection. Additionally, interviewers may be invited to respond to a debriefing questionnaire in which they provide feedback on their experiences collecting the data.

Interviewer debriefings are a relatively inexpensive add-on to a pretest. A potential problem with them is that interviewers may favor strategies that make their jobs easier but run counter to other objectives (such as accurate reporting).

Respondents may also be asked for their opinions about the survey experience at the conclusion of a pretest interview, known as a "respondent debriefing." Respondents may be asked to comment on topics such as:

- The overall survey response experience;
- Whether any questions were difficult to understand, overly burdensome, or potentially embarrassing (Martin 2001);

- Their confidence in their answers to particular items (Campanelli et al. 1991);
- Difficulties in recalling information (Biderman et al. 1986);
- Events or facts that the respondent failed to report, or reported incorrectly during the survey (Esposito 2004); or
- Reactions to features of the survey design, such as the mailing of both English and Spanish language forms (Wobus and de la Puente 1995)

Respondent debriefings are generally done in the same mode as the questionnaire itself and at the conclusion of the main interview. However, respondent debriefings may also be done in a follow-up interview (Belson 1981). In addition, probes can be administered online, immediately after a question (Meitinger and Behr 2016).

Respondent debriefing items can be relatively inexpensive additions to a pretest questionnaire that help identify problematic items. On the other hand, respondents may be reluctant to admit they had difficulty with any of the questions.

3.4.4 Behavior Coding

Recording pretest interviews, then making systematic observations of how questions are read and answered, can provide additional useful information about the questions. Recently, automated recording on the interviewer's laptop has become more common (Hicks et al. 2010). Behavior coding is the systematic coding of interviewer–respondent interactions. Simple coding schemes are often used (Was the question read verbatim? Did the respondent ask for clarification?). The researchers might code every question in every interview or just a subset of them. For each question coded, the coder makes judgments about the interviewer's reading of the question and about whether the respondent exhibited behaviors indicating difficulty. The question designer then looks for problem items – those the interviewers often did not read as worded or ones where respondents frequently asked for clarification.

Computer audio recorded interviewing (CARI) can be used to assess the quality of survey data collection. CARI recording is relatively unobtrusive, since it is done directly through the interviewer's laptop. Programming in the survey instrument turns the recording on and off for each question designated for recording. The recordings are transmitted in the same manner as the actual survey data. The audio files can then be behavior coded. CARI can be used to validate interviews, monitor interviewer performance, or evaluate survey questions.

The advantages of behavior coding and CARI are that they are objective methods for picking up problems as they occur with real respondents in the field. The disadvantages are that (i) they can only be used in computer-assisted interviewer-administered surveys and when respondents consent to being recorded; (ii) they can produce a mass of data (depending on how many items are coded and how many codes are used), increasing the cost and time needed to process and analyze the data; (iii) the coding may be subjective and unreliable; and (iv) the process may help detect problems, but it may not suggest solutions for them.

3.4.5 Response Latency

Another approach to diagnosing problems with a question is to measure response latency (the amount of time it takes respondents to answer the question). Although response latencies can be measured in laboratory settings, it is now common to capture timing information in the field as part of the survey paradata. In general, the use of response latency to evaluate survey questions rests on two assumptions. The first is that response latency reflects the amount of information processing required to answer the question. This includes the amount of time it takes to comprehend the question, retrieve information from memory, integrate that information into a judgment, and select an answer. A second assumption is that problems with a question lead to slower response times, because resolving the problems requires processing time (Bassili and Scott 1996). Like behavior coding, response latencies provide a quantitative assessment of the amount of difficulty that respondents are having with a question. However, the second assumption has been challenged for behavioral frequency questions, because it is often difficult to tell whether longer response latencies are due to a problem with the question or to more careful processing. In addition, researchers have seen *shorter* response latencies as potential indicators of shortcutting the response process (Ehlen et al. 2007). It seems likely that both very fast and very slow reaction times are signs of trouble, though of different sorts.

Response latencies can thus provide a systematic way to measure how much processing respondents are doing as they read and answer the questions, with both very long and very short latencies signaling problems. On the negative side, response latencies are not always easy to interpret, because it can be tricky to determine the appropriate range of response times for a given question. For example, longer questions almost always take longer to answer than short ones so that simple comparisons of response times across items may be misleading.

3.4.6 Vignettes

Vignettes are brief hypothetical scenarios presented to survey respondents that allow them to respond to survey questions without necessarily having direct experience with the topic. Vignettes can be implemented at many stages of the questionnaire design process – during focus groups, in cognitive interviews, in pretests, and even in the final fielded instrument; however, most of the instances we are familiar with use vignettes in a field setting.

Vignettes can also be used as a methodological tool for designing, redesigning, and evaluating questionnaires (Biderman et al. 1986). As part of the redesign of the key labor force items on the Current Population Survey (Martin 2004), vignettes were used to determine whether respondents' interpretations of concepts were consistent with the intended definitions and to evaluate the effects of different question wordings on the interpretations of survey concepts.

Vignettes have a number of attractive features:

- If they can be added to an ongoing survey or pretest, they are a relatively inexpensive way to evaluate the performance of survey questions;
- The results can be generalized to the survey population if the vignettes are administered to a probability sample of the target population; and
- Many vignettes can be included in a single instrument.

Vignettes have their limitations as well. They only provide indirect evidence about item performance because they are based on hypothetical situations and respondents may process questions about hypothetical situations differently (e.g. more superficially) than questions about their own situations or experiences. In addition, vignettes may be subject to their own sources of error (such as misunderstanding of the vignettes).

3.5 Statistical Modeling for Data Quality

The main goal of question evaluation is to establish the validity and reliability of the survey questions. Many statistical methods are available to assess validity and reliability given an appropriate research design. Statistical modeling can be done using pretest data or existing survey data.

3.5.1 Assessing Validity

"Validity" refers to the extent that the survey measure accurately reflects the intended construct (Groves 1989; Groves et al. 2009, Chapter 8). According to classical test theory, the observed value (the answer given by a respondent) reflects both a true value and an error component (the deviation from the true value). In practice, validity is estimated using data external to the survey, using multiple indicators of the same construct in one survey, or examining relationships with other constructs.

Many survey items are attitudinal, and it is not clear how to decide whether respondents have answered accurately. Consider this question: *Do you prefer mathematics or history?* This item is intended to assess preferences across two subjects. Which classes respondents are currently taking (or have taken) is not necessarily relevant to the accuracy of the answer. In fact, it is difficult to say *what* facts are relevant to deciding whether a respondent's answer to this question is a good one. As a result, it is unclear what the true value is for this question for any respondent. Measuring validity is thus more complicated for subjective or attitudinal questions than for factual ones.

Because there is no clear external standard for answers to questions about subjective states, like preferences or attitudes, the evaluation of the validity generally rests on one of three kinds of analyses:

1) Correlation of the answers with answers to other survey questions with which, in theory, they ought to be highly related;
2) Comparison between groups whose answers ought to differ if the answers are measuring the intended construct; and
3) Comparisons of the answers from comparable samples of respondents to alternative question wordings or protocols for data collection (split-ballot studies).

The first approach is probably the most common for assessing validity. For example, if an answer is supposed to measure how healthy a person is, people who rate themselves at the high end of "healthy" should also report that they feel better, that they miss fewer days of work, that they can do more things, and that they have fewer health conditions. The results of such analyses are called *assessments of construct validity*. If researchers do not find the expected relationships, it casts doubt on the validity of the health measure.

Validity assessment can also be applied using sophisticated modeling approaches that evaluate the evidence for validity by looking at the patterns and strength of the correlations across several measures of the same construct (e.g. Andrews 1984; Saris and Andrews 1991).

Another approach to assessing validity is to compare the answers of groups of respondents who are expected to differ on the underlying construct of interest. For example, teachers might identify groups of students who are high or low in persistence. We would expect the students classified by their teachers as high on persistence to score higher, on average, on a battery of items designed to measure persistence than those identified as low on persistence. Such evaluations depend critically on our theory about the relationships among certain variables. In general, it is difficult or impossible to distinguish poor measurement from inaccurate theories in the assessment of construct validity.

Assessing validity may be relatively inexpensive if multiple items assessing each construct are incorporated into an existing questionnaire. On the other hand, it is usually impossible to find external gold standards with which to compare responses to factual items; similarly, it can be difficult to find converging measures for assessing the validity of attitudinal items. Finally, analyses of construct validity must make assumptions about which items *should* correlate, assumptions that may not hold in every population or setting.

3.5.2 Assessing Reliability

"Reliability" refers to the consistency of the answers over repeated trials (for example, Forsman and Schreiner 1991). There are two main methods that survey researchers use to assess the reliability of reporting – repeated interviews with the same respondents and administration of multiple indicators of the same construct in a single interview.

Repeated interviews can be used to assess simple response variance (a common measure for assessing reliability) if the following assumptions are met:

1) There are no changes in the underlying construct between the two interviews;
2) All the important aspects of the measurement protocol remain the same (such as the method of data collection); and
3) The first measurement has no impact on an answer given in the second interview (for example, respondents do not remember their first answer).

Of course, there can be problems with each of these assumptions. Still, reinterview studies, in which survey respondents answer the same questions twice, are probably the most common approach for obtaining estimates of simple response variance (e.g. Forsman and Schreiner 1991; O'Muircheartaigh 1991).

Another approach to assessing reliability is to ask multiple questions assessing the same underlying construct in a single interview. This approach is used most often to measure subjective states. It makes the following assumptions:

1) The questions are all indicators of the same construct;
2) All the questions have the same expected response deviations (i.e. their simple response variance or reliability is the same); and
3) The questions are independent (i.e. answers to one question do not influence how the respondent answers another).

The reliability of an index that is created by combining answers to multiple items is often measured by Cronbach's alpha (α), which depends on the number of items and their average intercorrelation. A high value of Cronbach's alpha implies high reliability or low error variance. Unfortunately, it can also indicate that the answers to one item affected responses to the others (e. g. respondents may remember their earlier answers and try to be consistent), inducing an artifactually high correlation. A low value can indicate low reliability or can indicate that the items do not really measure the same construct.

Assessments of reliability provide direct evidence as to how well items are performing. However, they have some potential drawbacks. Both reinterviews and use of multiple items to measure reliability require assumptions that can be difficult to meet in practice. For example, reinterviews assume that the errors in the two interviews are independent. Reinterview studies are particularly time- and labor-intensive, and they can be costly.

3.5.3 Multitrait-Multimethod (MTMM) Experiments

An MTMM study usually includes the measurement of three different traits (e.g. memory, comprehension, and vocabulary) using three different methods (e.g. child's report, parent's report, teacher's report). The usual model is that observed responses reflect three things: variance in the true scores (that is, the person's position on the underlying construct of interest), method variance (the systematic effect of a specific method on observed responses), and random error variance. Both true score variance and methods variance contribute to the reliability of the observed responses, whereas only the true score variance contributes to their validity. MTMM experiments produce a matrix of correlations, consisting of different combinations of traits measured by different methods. These data are used to produce estimates of the reliability and validity of each measure (e.g. Saris and Gallhofer 2014).

The advantage of MTMM studies is that they provide quantitative estimates of reliability and validity (when all the model assumptions are met). In addition, they provide estimates of the impact of formal characteristics of survey questions (such as the number and type of scale points). The drawbacks of this approach are the difficulty of coming up with multiple versions of a question that can be administered in a single questionnaire and the specialized statistical skills needed to analyze the results.

3.5.4 Latent Class Analysis (LCA)

Latent class analysis (LCA) is a statistical procedure that models the relationship between an unobserved latent variable (the "construct") and the multiple observed indicators of the construct (Biemer 2011). Both the latent variable and the observed indicators are categorical. The indicators are not assumed to be error-free. However, the errors associated with the indicators have to be independent, conditional on the latent variable. This assumption – the local independence assumption – is almost always made in applications of LCA models. When it is satisfied, LCA produces unbiased estimates of the unconditional probabilities of membership in each of the latent classes and estimates of the probability of each observed response conditional on membership in each latent class. For example, there may be two latent classes, and the model estimates how likely someone is to be in each of the latent classes given his or her responses to each "indicator" – that is, on each of the observed variables. Two of the conditional probabilities (the false positive probability and the false negative probabilities) represent error rates. A high false positive or false negative probability signals a problem with a particular item. The primary purpose of applying LCA to the evaluation of survey questions is to identify questions that elicit high rates of false positives or false negatives.

The appeal of LCA is that it produces estimates of the error rates associated with questions, and these reflect both biases and variances in the answers.

Still, these estimates are themselves biased when the model assumptions are not met (e.g. Kreuter et al. 2008). As with MTMM, fitting latent class models requires specialized statistical expertise.

3.5.5 Item Response Theory (IRT)

Item response theory (IRT) refers to a family of models that describe the relationship between a person's response to a survey question in a scale or battery and the person's standing on the latent construct that the scale measures (Embretson and Reise 2000). For example, the latent construct may be an ability (say, arithmetic ability) and the questions forming the scale may consist of problems assessing that ability. The responses are typically dichotomous (e.g. whether it represents a correct or incorrect answer), and the underlying model assumes that there is a monotonic, logistic (S-shaped) curve relating the person's value on the construct (their arithmetic ability) to the probability of a positive (in this case, correct) answer. The item slope (or discrimination parameter) represents the strength of the relationship between the item and the latent construct. The item difficulty or threshold parameter identifies the location along the construct's latent continuum where the probability of a given response (e.g. a correct answer) is 0.50. The parametric, unidimensional IRT models make three key assumptions – unidimensionality, local independence, and a good fit of the model to the data. Violations of these assumptions mean that the model can produce misleading results.

The most attractive feature of IRT models is that these models produce item and ability parameters that are independent of one another and independent of the particular sample used to estimate the parameters. In contrast, in methods relying on classical test theory, the parameter estimates do depend on the particular sample used. Question evaluators can use the item parameters reflecting various item characteristics, including the item difficulty and the strength of the relationship between item and construct, to draw conclusions about the quality of different items. Although IRT models were initially used with dichotomous items, polytomous models are being used increasingly. Differential item functioning (DIF) analyses can be done to assess whether the items work equally well in different populations. Some drawbacks are that the method can only be used with multi-item scales, and the parameter estimates will be biased when the model assumptions are not met. Finally, like MTMM and LCA, IRT models require staff with specialized statistical skills.

3.6 Comparing Different Methods

There is no clear consensus about best practices for question evaluation. Still, a growing literature compares different methods of evaluation and provides a good starting point for developing best practices.

3.6.1 Convergence Within Method

A few studies have examined repeated use of the same methods. Presser and Blair (1994) compared behavior coding, cognitive interviewing, expert review, and conventional pretesting. They tested five supplements from the National Health Interview Survey that covered a variety of topics. Each question evaluation method was repeated, and the authors assessed the extent to which the methods detected the same problems across trials and across methods. Unexpectedly, the between-method correlations were not much lower than the within-method correlations for both conventional pretests and expert reviews. Behavior coding was by far the most reliable method across trials, followed distantly by cognitive interviews and expert review.

A study by Willis et al. (1999) compared cognitive interviewing, behavior coding, and expert review. Their design was similar to Presser and Blair's; however, Willis and his colleagues included cognitive interviews from two different survey organizations. They again found that the correlation between trials of behavior coding was the highest (0.79). The correlation between trials of cognitive interviewing were somewhat lower, but was still quite high (0.68). Contrary to Presser and Blair's findings, the within-method correlations were higher than the correlations across methods for both behavior coding and cognitive interviewing.

A potential problem with cognitive interviewing is that there is not an agreed-upon set of procedures for conducting the interviews. The study by Willis et al. (1999) suggests that experience level of the interviewers may not lead to different results. Other studies are less clear about this conclusion. DeMaio and Landreth (2004) compared the results from cognitive testing across three different organizations that implemented cognitive interviewing in quite different ways. Although there were few differences in the number of problems and problematic questions found, the organizations differed on the types of problems they found.

These studies suggest that behavior coding is the most reliable of the methods that have been studied. In addition, there is some degree of consistency between cognitive interviews conducted across trials or organizations. However, this consistency seems to be higher at the question level (does the question have any problems?) than at the level of specific problems (does the problem involve the comprehension of the question?). Finally, these studies suggest that expert reviews have the lowest levels of consistency across trials.

3.6.2 Convergence Across Methods

How closely do the methods agree with each other about whether a survey question has problems? Or should we not expect the methods to agree but instead to yield different but complementary information about specific survey questions?

Several of the methods we have discussed provide either qualitative or quantitative evidence of problems with survey questions. One metric that has been used to compare methods is the number of problems found. The literature is quite clear that forms appraisal methods, in particular, and expert reviews to a slightly lesser extent, detect the highest number of problems with survey questions (Rothgeb et al. 2001). This is not surprising since the users of forms appraisal methods are usually encouraged to identify as many problems as possible.

Another way to compare the results from these methods is to categorize the problems that are found using some common coding scheme. Although these coding schemes can be quite elaborate (e.g. Rothgeb et al. 2001), most of them can be mapped to Tourangeau's (1984) four stages of cognitive processing: comprehension, retrieval, estimation, and mapping. One major finding from the literature is that a clear majority of the problems found by cognitive interviewing, expert review, and forms appraisal methods involve the interpretation of survey questions (Presser and Blair 1994; Rothgeb et al. 2001; Willis et al. 1999).

The picture becomes less clear when the analyses focus on which questions are problematic and the specific nature of the problem. In fact, the results are quite inconsistent. For example, some studies find that there are at least moderate levels of agreement between expert review and cognitive interviewing (Presser and Blair 1994; Willis et al. 1999). In contrast, other studies report weak or even negative relationships between expert review and cognitive interviewing (Yan et al. 2012). Clearly, there is a need for more research to understand the circumstances under which one should expect the methods to give converging or diverging results.

The literature also suggests that the newly developed expert systems detect different types of problems from the traditional methods. For example, QUAID's focus on technical issues of question structure leads to different results from those found by expert reviews or cognitive interviewing (Graesser et al. 2006). Likewise, the existing studies have found either weak or negative agreement between the SQP and traditional question evaluation methods. In contrast to the traditional methods, SQP tends to focus on issues with the response scale (Saris 2012). The authors of these new tools often highlight these differences to suggest that these methods may complement the traditional methods by finding different types of problems.

3.6.3 Other Evidence Regarding Effectiveness

There are relatively few studies that attempt to confirm the results of question evaluation methods in the field. Two studies used indirect indicators of data quality to confirm the results from expert or laboratory methods. Forsyth et al. (2004) tallied the number of problems found cumulatively by expert review, forms appraisal, and cognitive interviewing. They then conducted a field study to see whether questions with more problems according to these methods produced higher levels of item nonresponse, problematic behavior by the interviewers or respondents, and problems identified by the field interviewers. The questions with more problems according to expert review, forms appraisal, and cognitive interviewing did tend to have more problems in the field. Specifically, the questions with interviewer problems according to expert review, forms appraisal, and cognitive interviewing had interviewer problems in the field (identified by behavior coding and interviewer ratings). Similarly, questions with respondent problems according to expert review, forms appraisal, and cognitive interviewing had more respondent problems in the field (found by behavior coding). Finally, recall and item sensitivity problems were related to item nonresponse. Unfortunately, it is not possible to tell from their analysis which of the three methods (expert review, forms appraisal, and cognitive interviewing) was most predictive, since the analysis lumped the three together.

A study by Blair et al. (2007) also investigated the extent to which problems identified by cognitive interviewing show up in the field. Their study included 24 questions with problems identified by cognitive interviewing; they identified problems in the field using behavior coding techniques. Overall, they found that 47% of the problematic interviewer-respondent exchanges matched a problem found in cognitive testing.

Few studies have used the results from question evaluation to change the questions and then assess the results in the field. Willis and Schechter (1997) is an exception; they found support for four of the five hypotheses that they tested concerning how a repair to an item would affect responses to the item in the field.

Other studies have compared the results from the traditional evaluation methods to measures of reliability and validity. Yan et al. (2012) compared expert reviews, cognitive interviews, quantitative measures of reliability and validity, and error rates from latent class models. They generally found low consistency across the methods in how they rank ordered questions in terms of quality. There was, however, considerable agreement between the expert ratings and the latent class method and between the cognitive interviews and the validity estimates. They concluded that the methods yielded different, sometimes contradictory conclusions with regard to the 15 items they tested. Their overall conclusion regarding best practice in question evaluation echoed that of Presser et al. (2004): "...until we have a clearer sense of which methods yield the most valid results, it will be unwise to rely on any one method for evaluating survey questions" (Yan et al. 2012, pp. 523).

Reeve et al. (2011) compared cognitive interviewing and IRT and found some problems on which the methods converge and others where they diverge. Both methods were useful at identifying items that were redundant or needed clarification. The methods disagreed on the most appropriate reference period for questions. Cognitive interviewing favored shorter reference periods, but IRT found that longer reference periods had greater discriminatory power. In addition, IRT differential item functioning (DIF) analyses found that at least two of the items performed differently within the three ethnic groups tested, but cognitive interviewing did not discover these differences across groups, even though members of all three groups were interviewed. The small sample sizes of the cognitive interview group (a total of 30) may account for this difference in the results.

Behavior coding results have been found to have the strongest link with reliability and validity. Hess et al. (1999) behavior-coded 34 questions on food security for which they obtained test-retest measurements. They used the behavior codes to predict the reliability of the questions. Two respondent behavior codes were significantly related to reliability (specifically, the index of inconsistency). The percentage of adequate answers was negatively related to the index of inconsistency, and the percentage of qualified answers was positively related. Dykema et al. (1997) behavior-coded 10 medical history questions for which they obtained medical records to verify responses; they attempted to predict inaccurate responses with the behavior codes at the respondent level. They found no consistent relationship between interviewers' misreading of the questions and the accuracy of the answers. In contrast, respondent behavior codes (including qualified or don't-know answers and interruptions) were significant predictors of inaccuracy.

Draisma and Dijkstra (2004) conducted a study that included both behavior coding and response latency. The authors had access to records that provided the true values for several questions. In bivariate analyses, longer response latencies and linguistic indicators of doubt (such as "I think" or "I believe") were associated with incorrect answers. In addition, some paralinguistic indicators (e.g. answer switches) and the number of words used by the respondent to answer were also associated with incorrect answers. However, a multivariate logistic regression model showed that only response latencies and expressions of doubt were significant predictors of response error.

3.7 Recommendations

Questionnaire design is difficult and often requires tradeoffs. For example, a short recall period may improve the accuracy of the answers but produce fewer reports of the type sought, increasing the variance of estimates of characteristics of these events. By providing insights into various sources of error, different questionnaire testing techniques can help researchers make better design choices.

We believe that most researchers see certain low-cost methods of questionnaire evaluation as almost always valuable for testing questionnaires. These include expert review (particularly when guided by a formal review system, such as QAS) and cognitive testing. These methods produce reasonably reliable results and tend to predict problems found when the instrument is fielded. Yan et al. (2012) also found that cognitive testing results predicted estimates of construct validity. When the questions being developed are new, focus groups are almost certainly helpful to ensure that the researchers understand how members of the target population think and talk about the topic of the questions. Whenever possible, focus groups should be done face-to-face, though this may be impractical for geographically dispersed or very rare populations.

There is also consensus that, if the budget and schedule make this feasible, some form of field test is generally useful for testing new items. This may take the form of a "dress rehearsal" in which the questions are administered under realistic field conditions. The value of a field test may be enhanced in several ways:

1) Recording some or all of the interviews and coding interviewer-respondent interactions;
2) Debriefing the interviewers;
3) Including a reliability study in which some (or all) respondents complete the questionnaire twice;
4) Conducting "construct" validity analyses or comparing survey responses to external "gold standards" (such as records data);
5) Including one or more split-ballot experiments to compare different item formats or different sets of questions; and
6) Getting respondent feedback via debriefing questions.

The first two of these options do not apply when the items are self-administered; in addition, comparisons to external gold standards are feasible only rarely.

Two other considerations for the selection of evaluation methods are the mode of data collection and the population being surveyed. Many evaluation methods are suitable for all modes. For example, expert review, cognitive interviewing, and field testing can easily be adapted to any mode. On the other hand, methods such as eye tracking are more suitable for self-administered modes, and methods such as behavior coding and interviewer debriefing are only relevant for interviewer-administered modes. In general, the methods are relevant for surveys of all populations, but methods relying on respondent verbalizations (such as focus groups) may be unsuitable for children.

Table 3.1 shows some typical packages of pretesting methods, based on the budget (and time) available and whether the questionnaire consists mostly of new content or of existing items from prior surveys. Focus groups, expert review or forms appraisal, and cognitive interviewing are recommended for surveys with new content and a relatively low budget. If more money is available, the project should add a field test including interviewer and respondent debriefing, experiments, behavior coding, and, if possible, reliability

Table 3.1 Recommended methods based on content and budget.

	Low budget	High budget

New content	• Focus groups
	• Expert review or forms appraisal
	• Cognitive interviewing
		• Focus groups
		• Expert review or forms appraisal
		• Cognitive interviewing
		• Field test
		  ◦ Interviewer debriefing
		  ◦ Respondent debriefing
		  ◦ Experiments
		  ◦ Behavior coding
		• Reliability assessment

Existing content	• Expert review • Cognitive interviewing • Usability testing if switching modes or using new devices	• Expert review • Cognitive interviewing • Usability testing if switching modes or using new devices • Field test • Interviewer debriefing

assessment. Surveys with mostly existing content and a low budget should consider expert review, cognitive interviewing, and usability testing. Usability testing is particularly important if new data collection devices are being used or a mode switch is occurring. Surveys with existing content and a slightly higher budget should also consider a field test with interviewer debriefing.

References

Andrews, F. (1984). Construct validity and error components of survey measures: a structural modeling approach. *Public Opinion Quarterly* 48 (2): 409–442.

Bassili, J.N. and Scott, B.S. (1996). Response latency as a signal to question problems in survey research. *Public Opinion Quarterly* 60 (3): 390–399.

Beatty, P.C. and Willis, G.B. (2007). Research synthesis: the practice of cognitive interviewing. *Public Opinion Quarterly* 71 (2): 287–311.

Belson, W.A. (1981). *The Design and Understanding of Survey Questions*. London: Gower.

Biderman, A.D., Cantor, D., Lynch, J.P., and Martin, E.A. (1986). *Final Report of the National Crime Survey Redesign Program*. Washington DC: Bureau of Social Science Research.

Biemer, P. (2011). *Latent Class Analysis of Survey Error*. Hoboken, NJ: Wiley.

Blair, J., Ackerman, A., Piccinino, L., and Levenstein, R. (2007). The effect of sample size on cognitive interview findings. In: *Proceedings of the ASA Section on Survey Research Methods*. Alexandria, VA: American Statistical Association.

Campanelli, P., Martin, E.A., and Rothgeb, J.M. (1991). The use of respondent and interviewer debriefing studies as a way to study response error in survey data. *The Statistician* 40 (3): 253–264.

Conrad, F.G. and Blair, J. (2004). Data quality in cognitive interviews: the case of verbal reports. In: *Methods for Testing and Evaluating Survey Questionnaires* (eds. S. Presser et al.). New York: Wiley.

Couper, M.P. (1999). The application of cognitive science to computer assisted interviewing. In: *Cognition and Survey Research* (eds. M.G. Sirken, D.J. Herrmann, S. Schechter, et al.), 277–300. New York: Wiley.

DeMaio, T.J. and Landreth, A. (2004). Do different cognitive interviewing techniques produce different results? In: *Methods for Testing and Evaluating Survey Questionnaires* (eds. S. Presser, J.M. Rothgeb, M.P. Couper, et al.), 89–108. Hoboken, NJ: Wiley.

Draisma, S. and Dijkstra, W. (2004). Response latency and (para) linguistic expressions as indicators of response error. In: *Methods for Testing and Evaluating Survey Questionnaires* (eds. S. Presser, J.M. Rothgeb, M.P. Couper, et al.), 131–147. Hoboken, NJ: Wiley.

Dykema, J., Lepkowski, J., and Blixt, S. (1997). The effect of interviewer and respondent behavior on data quality: analysis of interaction coding in a validation study. In: *Survey Measurement and Process Quality* (eds. L. Lyberg, P. Biemer, M. Collins, et al.), 287–310. New York: Wiley.

Ehlen, P., Schober, M.F., and Conrad, F.G. (2007). Modeling speech disfluency to predict conceptual misalignment in speech survey interfaces. *Discourse Processes* 44 (3): 245–265.

Embretson, S.E. and Reise, S.P. (2000). *Item Response Theory for Psychologists.* Mahwah, NJ: Lawrence Erlbaum.

Esposito, J.L. (2004). Iterative multiple-method questionnaire evaluation research: a case study. *Journal of Official Statistics* 20 (2): 143–183.

Forsman, G. and Schreiner, I. (1991). The design and analysis of reinterview: an overview. In: *Measurement Error in Surveys* (eds. P. Biemer, R.M. Groves, L. Lyberg, et al.), 279–302. New York: Wiley.

Forsyth, B., Rothgeb, J.M., and Wills, G.B. (2004). Does pretesting make a difference? An experimental test. In: *Methods for Testing and Evaluating Survey Questionnaires* (eds. S. Presser, J.M. Rothgeb, M.P. Couper, et al.), 525–546. Hoboken, NJ: Wiley.

Fowler, F.J. (2004). The case for more split-sample experiments in developing survey instruments. In: *Methods for Testing and Evaluating Survey Questionnaires* (eds. S. Presser, J.M. Rothgeb, M.P. Couper, et al.), 173–188. Hoboken, NJ: Wiley.

Galesic, M., Tourangeau, R., Couper, M.P., and Conrad, F.G. (2008). Eye-tracking data: new insights on response order effects and other cognitive shortcuts in survey responding. *Public Opinion Quarterly* 72 (5): 892–913.

Gothberg, J., Reeves, P., Thurston, L. et al. (2013). Is the medium really the message? A comparison of face-to-face, telephone, and internet focus group venues. *Journal of Ethnographic and Qualitative Research* 7 (3): 108–127.

Graesser, A.C., Bommareddy, S., Swamer, S., and Golding, J.M. (1996). Integrating questionnaire design with a cognitive computational model of human question answering. In: *Answering Questions: Methodology for Determining Cognitive and Communicative Processes in Survey Research* (eds. N. Schwarz and S. Sudman), 143–174. San Francisco: Jossey-Bass.

Graesser, A.C., Cai, Z., Louwerse, M.M., and Daniel, F. (2006). Question understanding aid (QUAID): a web facility that tests question comprehensibility. *Public Opinion Quarterly* 70 (1): 3–22.

Groves, R.M. (1989). *Survey Errors and Survey Costs*. New York: Wiley.

Groves, R.M., Fowler, F.J., Couper, M.P. et al. (2009). *Survey Methodology*, 2e. Hoboken, NJ: Wiley.

Hess, J., Singer, E., and Bushery, J. (1999). Predicting test-retest reliability from behavior coding. *International Journal of Public Opinion Research* 11 (4): 346–360.

Hicks, W.D., Edwards, B., Tourangeau, K. et al. (2010). Using CARI tools to understand measurement error. *Public Opinion Quarterly* 74 (5): 985–1003.

Hine, C. (2005). Introduction. In: *Virtual Methods: Issues in Social Research on the Internet* (ed. C. Hine), 1–13. New York, NY: Berg.

Kreuter, F., Yan, T., and Tourangeau, R. (2008). Good item or bad – can latent class analysis tell? The utility of latent class analysis for the evaluation of survey questions. *Journal of the Royal Statistical Society, Series A* 171 (3): 723–738.

Lessler, J.T. and Forsyth, B.H. (1996). A coding system for appraising questionnaires. In: *Answering Questions: Methodology for Determining Cognitive and Communicative Processes in Survey Research* (eds. N. Schwarz and S. Sudman), 259–292. San Francisco: Jossey-Bass.

Maitland, A. and Presser, S. (2016). How accurately do different evaluation methods predict the reliability of survey questions? *Journal of Survey Statistics and Methodology* 4 (3): 362–381.

Martin, E.A. (2001). Privacy concerns and the census long form: some evidence from census 2000. In: *Proceedings of the Section on Survey Research Methods*. Washington, DC: American Statistical Association.

Martin, E.A. (2004). Vignettes and respondent debriefings for questionnaire design and evaluation. In: *Methods for Testing and Evaluating Survey Questionnaires* (eds. S. Presser, J.M. Rothgeb, M.P. Couper, et al.), 149–171. Hoboken, NJ: Wiley.

Meitinger, K. and Behr, D. (2016). Comparing cognitive interviewing and online probing: do they find similar results? *Field Methods* 28 (4): 363–380.

O'Muircheartaigh, C. (1991). Simple response variance: estimation and determinants. In: *Measurement Error in Surveys* (eds. P. Biemer, R.M. Groves, L. Lyberg, et al.), 551–574. New York: Wiley.

Presser, S. and Blair, J. (1994). Survey pretesting: do different methods produce different results? *Sociological Methodology* 24: 73–104.

Presser, S., Rothgeb, J.M., Couper, M.P. et al. (2004). *Methods for Testing and Evaluating Survey Questionnaires*. Hoboken, NJ: Wiley.

Reeve, B.R., Willis, G., Shariff-Marco, S. et al. (2011). Comparing cognitive interviewing and psychometric methods to evaluate a racial/ethnic discrimination scale. *Field Methods* 23 (4): 397–419.

Rothgeb, J.M., Willis, G.B., and Forsyth, B.H. (2001). Questionnaire pretesting methods: do different techniques and different organizations produce similar results? In: *Proceedings of the Section on Survey Research Methods*. Alexandria, VA: American Statistical Association.

Saris, W. (2012). Discussion: evaluation procedures for survey questions. *Journal of Official Statistics* 28 (4): 537–551.

Saris, W. and Andrews, F. (1991). Evaluation of measurement instruments using a structural modeling approach. In: *Measurement Error in Surveys* (eds. P. Biemer, R.M. Groves, L. Lyberg, et al.), 575–597. New York: Wiley.

Saris, W.E. and Gallhofer, I.N. (2014). *Design, Evaluation, and Analysis of Questionnaires for Survey Research*. New York: Wiley.

Tourangeau, R. (1984). Cognitive science and survey methods. In: *Cognitive Aspects of Survey Design: Building a Bridge between Disciplines* (eds. T. Jabine, M. Straf, J. Tanur and R. Tourangeau), 73–100. Washington, DC: National Academy Press.

Tourangeau, R. (2004). Experimental design considerations for testing and evaluating questionnaires. In: *Methods for Testing and Evaluating Survey Questionnaires* (eds. S. Presser, J.M. Rothgeb, M.P. Couper, et al.), 209–224. Hoboken, NJ: Wiley.

Van der Zouwen, J. and Dijkstra, W. (2002). Testing questionnaires using interaction coding. In: *Standardization and Tacit Knowledge: Interaction and Practice in the Survey Interview* (eds. D.W. Maynard, H. Houtkoop-Steenstra, N.C. Schaeffer and J. van der Zouwen), 427–447. New York: Wiley.

Van der Zouwen, J. and Smit, J.H. (2004). Evaluating survey questions by analyzing patterns of behavior c codes and question-answer sequences: a diagnostic approach. In: *Methods for Testing and Evaluating Survey Questionnaires* (eds. S. Presser, J.M. Rothgeb, M.P. Couper, et al.), 109–130. Hoboken, NJ: Wiley.

Willis, G.B. (2005). *Cognitive Interviewing: A Tool for Improving Questionnaire Design*. Thousand Oaks, CA: Sage.

Willis, G.B. and Lessler, J.T. (1999). *Questionnaire Appraisal System: QAS-99*. Rockville, MD: Research Triangle Institute.

Willis, G.B. and Schechter, S. (1997). Evaluation of cognitive interviewing techniques: do results generalize to the field? *Bulletin de Methodolgie Sociologique* 55 (1): 40–66.

Willis, G.B., Schechter, S., and Whitaker, K. (1999). A comparison of cognitive interviewing, expert review, and behavior coding: what do they tell us? In:

Proceedings of the Section on Survey Research Methods. Alexandria, VA: American Statistical Association.

Wobus, P. and de la Puente, M. (1995). Results from telephone debriefing interviews: the Census Bureau's Spanish Forms Availability Test. In: *Proceedings of the Section on Survey Research Methods*. Alexandria, VA: American Statistical Association.

Yan, T., Kreuter, F., and Tourangeau, R. (2012). Evaluating survey questions: a comparison of methods. *Journal of Official Statistics* 28 (4): 503–529.

4

A Comparison of Five Question Evaluation Methods in Predicting the Validity of Respondent Answers to Factual Items

Aaron Maitland[1] and Stanley Presser[2]

[1] *Westat, Rockville, Maryland, USA*
[2] *Sociology Department and Joint Program in Survey Methodology, University of Maryland, College Park, Maryland, USA*

4.1 Introduction

Researchers use various methods to understand the quality of survey questions prior to fielding them. These methods vary considerably in cost from those requiring no data collection from respondents, such as expert review, the Questionnaire Appraisal System (QAS), the Question Understanding Aid (QUAID), and the Survey Quality Predictor (SQP), to those requiring data collection either in the lab (e.g. cognitive interviews) or in the field (e.g. behavior coding and response latency). The well-established finding that different methods of evaluating a question often lead to different conclusions about the question (Presser and Blair 1994; Rothgeb et al. 2001; van der Zouwen and Smit 2004, Willis et al. 1999; Yan et al. 2012) suggests the possibility that evaluation methods vary in the degree to which they identify problems that need to be repaired.

In line with this possibility, we recently found differences across evaluation methods in their prediction of both item reliability (Maitland and Presser 2016) and four different problems (missing data, requests for clarification, inadequate initial answers, and response latency) produced in actual survey conditions (Maitland and Presser 2018). Especially with factual questions, however, probably the most useful way to define problems in need of repair is in terms of the validity of the answers produced by the questions. Yet there are only a handful of studies that examine how well question evaluation methods predict the validity of factual questions and even fewer studies that compare the methods on this dimension.

In a study of behavior coding, Dykema et al. (1997) found that qualified or don't know answers and interruptions were significant predictors of inaccuracy (as assessed with records) for the majority of 10 medical history questions. In

Advances in Questionnaire Design, Development, Evaluation and Testing, First Edition.
Edited by Paul C. Beatty, Debbie Collins, Lyn Kaye, Jose-Luis Padilla, Gordon B. Willis, and Amanda Wilmot.

a similar study of 22 questions, Draisma and Dijkstra (2004) found that longer response latencies, as well as linguistic indicators of doubt, such as "I think" or "I believe," and paralinguistic indicators, such as answer switches and the number of words used by the respondent to answer a question, were generally associated with incorrect answers to the questions. In a study of 17 questions from two surveys (one about divorce and the other about financial aid for college students), Olson (2010) found that expert ratings, but not those from QUAID, were related to response accuracy as measured by records. Finally, in a case study, Schaeffer and Dykema (2004) provided a qualitative assessment of the contribution of focus groups, behavior coding, and cognitive interviews to improving the measurement of child custody among divorced parents using court records as a criterion.

The very small number of questions on a restricted range of topics tested with only one or two methods means that much more research is needed on how the various evaluation methods perform in identifying item validity. In this chapter we contribute to this end by comparing the performance of five of the most widely used evaluation methods in predicting the validity of 44 factual questions that range across a wide range of topics. The five methods are expert review, the QAS, the QUAID, the SQP, and cognitive interviews.

Our analysis is guided by two hypotheses. The first hypothesis stems from the finding that evaluation methods often disagree in their diagnoses of problems. This is frequently interpreted to mean that the methods are complementary and therefore it is better to use multiple methods than to rely on a single one (Presser et al. 2004; Yan et al. 2012). We call this the "complementary methods hypothesis."

The second hypothesis – the "test environment hypothesis" – proposes that methods that more closely observe the response process have an advantage over those that observe the process less closely or not at all. The response process is set within a sociocultural context, and some approaches, such as cognitive interviews, allow the researcher to observe the process in that context (Gerber and Wellens, 1997; Miller 2011). Expert review does not directly observe the process, but rather draws on prior experience and research to predict problems. The QAS draws less on sociocultural context and computerized evaluations, such as QUAID and SQP, do so the least.

4.2 Methods

4.2.1 Measure of Validity

We searched the literature for survey questions whose answers were checked (at the individual level) against records and located a total of 44.[1] In the

---

1 We regret not having used the Wentland and Smith (1993) review of 33 record-check studies as a source in our search. (Wentland and Smith say they identified 37 studies, but list only 35, and 3

two cases where results were reported separately for different modes of administration, we used only the interviewer-administered mode as all the single mode studies used interviewers. Appendix 4.A shows the 44 questions and associated percent correct that is the basis for our dependent variable. Of course, the record check methodology is not perfect, as records can contain errors and the matching process between survey responses and records can yield still other errors. Partly for this reason, we present results using the exact percent correct as well as using a trichotomy – the 25% most accurate items (higher than 91% accuracy), the middle 50% (between 74% and 91% accurate), and the 25% least accurate items (lower than 74% accuracy).

4.2.2 Pretesting Methods

4.2.2.1 Computer-Based Systems

We used two computer-based systems. The first, QUAID,[2] is based on computational models developed in the fields of computer science, computational linguistics, discourse processing, and cognitive science. The software identifies technical features of questions that have the potential to cause comprehension problems. It rates questions on five classes of comprehension problems: unfamiliar technical terms, vague or imprecise predicate or relative terms, vague or imprecise noun phrases, complex syntax, and working memory overload. QUAID identifies these problems by comparing the words in a question to several databases (e.g. Coltheart's MRC Psycholinguistics Database).

The second automated system, SQP, which is based on a meta-analysis of multitrait-multimethod (MTMM) studies, predicts the reliability, validity, method effects, and total quality of questions (Saris and Gallhofer 2014). Total quality is the product of reliability and validity. To use SQP, each question is coded according to the variables from the MTMM studies. One of the authors coded the questions using the 2012 version, SQP 2.0.[3]

4.2.2.2 Expert Review

Three reviewers, each of whom had either a PhD in survey methodology or a related discipline or more than five years of experience as a survey researcher, were given the questions and the following instructions:

---

of those – Parry and Crossley, Crossley and Fink, and Cahalan – use the same data.) Although we would have excluded many of their studies because they involve very special populations (e.g. middle school students and formerly incarcerated drug addicts), very small samples (e.g. 15 and 46), or questionnaires not in English, their review indicates that our literature search was not exhaustive. Nonetheless, as shown in Appendix 4.A, our 44 items cover a broad range of subject areas, recall periods, and degrees of sensitivity.

2 See Graesser et al. (2006) and http://www.memphis.edu/iis/projects/quaid.php.

3 See http://sqp.upf.edu. We present results using the total quality score, but the results are identical using the validity score.

Question wordings, introductions associated with questions, and response categories are considered in scope for this evaluation. For each survey question, identify and briefly explain each specific problem you find. Please type a brief description of the problem immediately following the question in the attached document. You may observe multiple problems with a question. Please describe each one. You do not need to type anything after questions for which you do not observe a problem.

4.2.2.3 Questionnaire Appraisal System

Students from a Joint Program in Survey Methodology graduate-level course on questionnaire design were asked to evaluate the questions using the QAS.[4] Students were assigned different sets of questions, and each was asked to evaluate independently whether the questions had any of 26 potential problems. The form also called for a brief description of each problem found.

4.2.2.4 Cognitive Interviews

About a month later, the same students who did the QAS coding conducted cognitive interviews of the questions. Three students tested 22 questions, and four students tested the other 22. Three students used only questions they had not coded with the QAS, and the other four students did cognitive interviews in which approximately half of the questions were ones they had coded with the QAS.

In the language of Willis (2015), these constituted a single round (as opposed to an iterative round) of reparative (as opposed to interpretive) interviews. Each student was instructed to develop a cognitive protocol including think-aloud exercises and probes and then to interview four subjects, recruited from among their friends, neighbors, co-workers, or other convenient populations.[5] All interviews were recorded so that the students could review the recordings when preparing reports on their findings. Both the reports and audio tapes were turned in.

---

4 See Willis and Lessler (1999).

5 There are many ways to conduct cognitive interviews (e.g. Beatty and Willis 2007; Miller et al. 2014; Willis 2015). Our approach differs from a commonly recommended approach to use a single protocol for each round of interviews. We know of no way to estimate how often this approach is used as opposed to our approach (in which the protocols vary across interviews). But even when a single protocol is used there is apt to be significant variability in how it is implemented. Indeed, cognitive interviewing is inherently unstandardized because "the objective is not to produce cookie-cutter responses to standard stimuli, but to enable our participants to provide rich elaborated information" (Willis 2015). Based partly on the finding that two organizations using different protocols arrived at similar conclusions (Willis et al. 1999), we think it is likely our approach yielded a set of problems similar to those that would have been produced using a single protocol.

4.2.2.5 Problem Identification across the Methods

The problems identified from QUAID, QAS, expert review, and cognitive interviews were coded according to the scheme used by Presser and Blair (1994), which has three major categories of problems: respondent semantic, respondent task, and interviewer. Respondent semantic problems refer to respondents having difficulty understanding or remembering a question or having diverse understandings of the question's meaning. They are divided into two types: problems due to the structure of the question (for instance, item wordiness) and those due to the meaning of terms or concepts in the question. Respondent task problems are of three types: difficulty recalling information or formulating an answer; insufficient response categories; and question sensitivity. Interviewer problems refer to problems reading the question or difficulty understanding how to implement a question. The 26 problems identified by cognitive interviewing and the 279 problems identified by expert review were coded by the first author. In a preliminary sample of similar problems double-coded by two research assistants, the overall inter-coder agreement, as measured by Cohen's (1960) kappa, was 0.76; and by category, the kappas were 0.91 for respondent semantic, 0.73 for respondent task, and 0.68 for interviewer.

The crosswalk shown in Table 4.1 was used to code the QUAID and QAS problems into Presser-Blair categories. The first author then determined which problems matched across methods and assigned an identifier to each problem.

4.3 Results

4.3.1 Descriptive Statistics

Most of our variables involve the number of times that a method discovered a problem type across the 44 questions. Descriptive statistics for each method-problem type combination (as well as for the SQP score) and for the two versions of the accuracy variable are shown in Table 4.2. As can be seen, the items vary in estimated accuracy between 54% and 100% (with an average of 83%), and, in line with previous findings, the methods differ substantially in their identification of problems.

Table 4.3 shows the Pearson correlation coefficient for each pair of variables. Again consistent with previous findings, there is generally little association between the problems identified by the different methods (although there are a handful of notable exceptions).

Our main focus is how well the problems identified by the various methods predict validity. Surprisingly, QAS meaning problems are correlated with accuracy in the wrong direction – items identified as having such problems by QAS were more likely to produce accurate answers. An examination of these items sheds no light on this anomaly and – given the large number of associations in the table – we believe it is best treated as due to chance.

Table 4.1 Crosswalk between QUAID, QAS, and Presser-Blair codes.

Presser-Blair codes		QUAID	QAS
Semantic I: Structure of question or questionnaire	Information overload	Working memory overload	
	Structure/organization	Complex syntax	2a–b: Conflicting or inaccurate instructions, Complicated instructions 3a: Wording: complex syntax
	Transition problem		
Semantic II: Meaning of words	Boundary lines	Vague or imprecise relative or technical term, Vague or ambiguous noun phrase	3c–d: Vague, Reference period
	Technical term not understood	Unfamiliar technical term	3b and 7c. Technical term
	Common term not understood		
	Double-barreled		4c. Double barreled
Respondent Task I: Recall/Response Formation	Recall/response difficult		5d. Computation
	Recall/response impossible		5a–c: Knowledge, Attitude, Recall
	Recall/response redundant		
	Recall/response resisted		4a–b: Inappropriate assumptions, Assumes constant behavior
Respondent Task II: Response Categories	Overlapping	Vague or imprecise relative or technical term or noun phrase	7d–e: Vague, Overlapping
	Insufficient		7a and f: Open ended, Missing
	Too fine distinction		
	Inappropriate		7b. Mismatch
Respondent Task III: Sensitivity	Sensitivity		6a–c: Sensitive content, Sensitive wording, Socially acceptable
Interviewer	Procedural		
	Reading problem		1a–c: What to read, Missing information, How to read
	Coding answers to open		

Table 4.2 Means, standard deviations, and ranges for all variables ($n = 44$ questions).

Variable	Mean	Standard deviation	Minimum value	Maximum value
Measure of validity				
Accuracy – percent correct	83.14	11.31	54	100
Accuracy – collapsed into a trichotomy	1.05	0.71	0	2
Problem identification by method				
Semantic I: question structure				
QUAID	0.23	0.48	0	2
Expert review	0.98	0.95	0	3
QAS	0.39	0.87	0	4
Cognitive interviews	0.05	0.21	0	1
Semantic II: meaning				
QUAID	1.82	2.18	0	11
Expert review	3.80	1.94	0	8
QAS	3.32	2.76	0	11
Cognitive interviews	0.34	0.71	0	3
Respondent task I: recall				
QUAID	—	—	—	—
Expert review	0.68	1.07	0	3
QAS	1.75	1.06	0	4
Cognitive interviews	0.18	0.58	0	2
Respondent task II: response categories				
QUAID	0.02	0.15	0	1
Expert review	0.05	0.21	0	1
QAS	0.11	0.49	0	3
Cognitive interviews	0	0	0	0
Respondent task III: sensitivity				
QUAID	—	—	—	—
Expert review	0.80	0.85	0	2
QAS	0.98	0.88	0	3
Cognitive interviews	0.02	0.15	0	1
Interviewer problems				
QUAID	—	—	—	—
Expert review	0.05	0.30	0	2
QAS	0.05	0.30	0	2
Cognitive interviews	0	0	0	0
Other method				
SQP total quality score	0.62	0.05	0.5	0.68

Note: Entries with a "—" represent a problem that the method is not designed to identify.

Table 4.3 Pearson correlations between variables ($n = 44$ questions).

		Semantic I				Semantic II				Respondent task I			Respondent task II			Respondent task III			Interviewer		Survey	Percent correct	
		Quaid	Expert review	QAS	Cog. Int.	Quaid	Expert review	QAS	Cog. Int.	Expert review	QAS	Cog. Int.	Quaid	Expert review	QAS	Expert review	QAS	Cog. Int.	Expert Review	QAS	Quality Predictor	Numeric percent correct	Categorical percent correct
Semantic I	Quaid		0.11	0.51*	0.59*	0.80*	−0.18	−0.04	0.11	−0.22	−0.21	0.18	0.57*	0.13	0.58*	−0.40*	−0.32*	−0.07	0.25	0.25	−0.30*	0.16	0.11
	Expert review			−0.02	0.24	0.04	−0.14	0.08	−0.09	−0.03	−0.1	0.01	0	0.01	0.06	−0.09	−0.17	−0.16	−0.16	−0.16	−0.02	0.01	0
	QAS				0.66*	0.48*	−0.01	−0.07	0.16	−0.16	−0.04	−0.05	0.64*	−0.1	0.55*	−0.39*	−0.2	−0.07	0.11	0.11	0.1	0.13	0.05
	Cog. Int.					0.63*	−0.26	−0.15	0.05	−0.14	−0.16	−0.07	0.70*	−0.05	0.32*	−0.21	−0.25	−0.03	−0.03	−0.03	−0.02	0.05	−0.01
Semantic II	Quaid						−0.16	−0.27	0.18	−0.04	−0.29	−0.08	0.65*	−0.03	0.33*	−0.2	−0.26	−0.13	−0.06	−0.06	−0.27	0.04	−0.01
	Expert review							0.3	−0.07	−0.14	−0.05	−0.17	−0.30*	−0.26	−0.39*	−0.15	0.18	0.1	−0.06	−0.06	−0.14	−0.08	−0.11
	QAS								0.07	−0.49*	0	0.19	−0.19	0.21	−0.18	−0.49*	−0.28	−0.02	−0.13	−0.13	0.05	0.44*	0.43*
	Cog. Int.									−0.04	0.02	0.07	−0.07	0.05	0.32	0	−0.02	−0.07	−0.07	−0.07	−0.16	0.12	0.24
Respondent task I	Expert review										0.38*	−0.2	−0.1	−0.14	−0.15	0.49*	0.09	−0.1	−0.1	−0.1	0.34*	−0.30*	−0.25
	QAS											0.23	−0.11	0.05	−0.12	0.1	−0.03	−0.11	0.33*	0.33*	0.31*	−0.21	−0.2
	Cog. Int.												−0.05	0.69*	0.39	−0.2	−0.17	−0.05	0.48*	0.48*	0	0.14	0.09
Respondent task II	Quaid													−0.03	0.90*	−0.14	−0.17	−0.02	−0.02	−0.02	0.18	0.05	−0.01
	Expert review														0.17	−0.08	−0.12	−0.03	−0.03	−0.03	0.05	0.25	0.3
	QAS															−0.11	−0.21	−0.04	−0.04	−0.04	0.11	0.08	0.05
Respondent task III	Expert review																0.49*	0.04	−0.14	−0.14	0.11	−0.45*	−0.41*
	QAS																	0.36*	−0.17	−0.17	−0.24	−0.28	−0.3
	Cog. Int.																		−0.02	−0.02	−0.07	−0.40*	−0.23
Interviewer	Expert Review																			1.00*	−0.06	−0.15	−0.23
	QAS																				−0.06	−0.15	−0.23
Survey Quality Predictor																						0.01	0.06
Percent correct	Numeric percent correct																						0.90*
	Categorical percent																						

*p<.05

The other significant predictions of accuracy in Table 4.3 are in line with expectations. Sensitivity problems identified by expert review and such problems identified by cognitive interviews, as well as recall problems identified by expert review, are associated with higher levels of inaccuracy. These results, taken together with the finding that the evaluations made by the two computerized approaches (QUAID and SQP) are not associated with inaccuracy, provide support for the test environment hypothesis.[6] To examine the complementary methods hypothesis (and further examine the test environment hypothesis), we turn to multivariate analysis.

4.3.2 Multivariate Analyses

Our multivariate analysis began by including all the significant bivariate effects ($p < 0.05$) except for the presumably chance one for QAS meaning problems. Thus Model 1 of Table 4.4 predicts accuracy (the numeric percent) using expert review sensitivity problems, cognitive interview sensitivity problems, and expert review recall problems. The intercept in these models refers to the level of accuracy for a question with no problems identified by the methods. The coefficients in the model represent the reduction in accuracy for each problem detected by a method. The diagnoses explain almost half the variation in accuracy, though only two – sensitivity problems identified by expert review and cognitive interviewing – are statistically significant. Model 2 shows that the *R*-squared does not change by dropping the recall problem predictor. Models 3–5 show the extent to which each method predicts accuracy when used alone.

Reducing to only expert review or only cognitive interviewing leads to a significant reduction in explanatory power, lending partial support to the complementary methods hypothesis that a combination of methods is better than any single method. Similarly, there is partial support for the test

Table 4.4 Predicting percent accurate (as judged by record checks) for 44 questions.

Predictor	Model 1	Model 2	Model 3	Model 4	Model 5
Intercept	88.8*(2.0)	88.4*(2.0)	0.87.8*(2.1)	83.8*(1.6)	85.3*(2.0)
Expert review sensitivity	−4.7*(1.9)	−5.8*(1.7)	−6.0*(1.8)		
Cog. interview sensitivity	−30.1*(9.5)	−28.6*(9.5)		−29.8*(10.6)	
Expert review recall	−1.8(1.5)				−3.2*(1.6)
R-squared	0.37	0.34	0.20	0.16	0.09
Adjusted R-squared	0.32	0.31	0.18	0.14	0.07

*$p < 0.05$.

6 The finding that SQP is unrelated to measures of validity echoes the Maitland and Presser (2016) finding that SQP did not predict an indicator of the reliability of factual items (though it did predict the reliability of attitudinal items). As noted in that article, this may be because the SQP data base consists largely of attitudinal items.

environment hypothesis, which implies that cognitive interviewing should be the best predictor of accuracy, followed by expert review, QAS, and the machine-based approaches (QUAID and SQP). In fact, expert review was the best, though it was followed by cognitive interviews, with QUAID and SQP performing least well.

4.4 Discussion

We found that the accuracy of survey questions as assessed by record checks was predicted by three testing methods – expert review, cognitive interviews, and QAS; that a combination of expert review and cognitive interviews provided the best prediction of accuracy; and that two computer-based methods (SQP and QUAID) were not significant predictors of accuracy. These results provide partial support for both the complementary methods hypothesis and the test environment hypothesis.

The results for the complementary methods hypothesis are similar to those in our earlier studies – using a different set of items-of how well various question evaluation methods predicted item reliability (Maitland and Presser 2016) and problems observed in the field (Maitland and Presser 2018). The present study's results for the test environment hypothesis are also similar to those from the study of observable field problems, though less similar to those from our study of reliability, which found expert review to be particularly predictive but few differences among the remaining methods. This suggests the possibility that validity and observable field problems are more closely linked than either is to reliability.

Most of the limitations of our earlier studies apply to the present study as well. With the exception of the computer-based methods, there is no one correct way to implement evaluation methods. Earlier we cited references discussing the many different ways of conducting cognitive interviews. Although it has been much less discussed, there are also many different ways to implement the other methods. Expert reviews, for instance, can be done in groups or individually, by people with only methodological expertise or also by those with subject matter expertise, unstructured or structured by issues to address, and so on. Likewise, QAS can be carried out by coders with varying degrees of survey expertise. We know little about how such factors influence the outcome of the methods. Consequently we do not know how sensitive our results are to the particular ways we implemented the methods. This is particularly true for cognitive interviews and QAS since there was some overlap in the people who produced those results. Similarly, we don't know how sensitive our results are to the particular sample of items we used. Thus further work using other items and other variants of the testing methods is needed. Nonetheless, we see our contribution as an important step toward better understanding how well different question evaluation methods predict the accuracy of the answers to questions.

References

Beatty, P. and Willis, G. (2007). Research synthesis: the practice of cognitive interviewing. *Public Opinion Quarterly* 71: 287–311.

Belli, R.F., Traugott, M.W., Young, M., and McGonagle, K.A. (1999). Reducing vote over-reporting in surveys: social desirability, memory failure, and source monitoring. *Public Opinion Quarterly* 63: 90–108.

Blumberg, S.J. and Cynamon, M.L. (1999). Misreporting medicaid enrollment: results of three studies linking telephone surveys to state administrative records. In: *Proceedings of the Seventh Conference on Health Survey Research Methods, Sep 24–27; Williamsburg, VA* (eds. M.L. Cynamon and R.A. Kulka), 189–195. Hyattsville, MD: Department of Health and Human Services; 2001; Publication No. (PHS) 01-1013.

Cohen, J. (1960). A coefficient of agreement for nominal scales. *Educational and Psychological Measurement* 20: 37–46.

Draisma, S. and Dijkstra, W. (2004). Response latency and (para) linguistic expressions as indicators of response error. In: *Methods for Testing and Evaluating Survey Questionnaires* (eds. S. Presser, J.M. Rothgeb, M.P. Couper, et al.), 131–148. Hoboken, NJ: Wiley.

Duncan, G.J. and Hill, D.H. (1985). An investigation of the extent and consequences of measurement error in labor-economic survey data. *Journal of Labor Economics* 3: 508–532.

Dykema, J., Lepkowski, J., and Blixt, S. (1997). The effect of interviewer and respondent behavior on data quality: analysis of interaction coding in a validation study. In: *Survey Measurement and Process Quality* (eds. L. Lyberg, P. Biemer, M. Collins, et al.). New York: Wiley.

Gerber, E. and Wellens, T. (1997). Perspectives on pretesting: "Cognition" in the cognitive interview? *Bulletin de Methodologique Sociologique* 55: 18–39.

Graesser, A.C., Cai, Z., Louwerse, M.M., and Daniel, F. (2006). Question understanding aid (QUAID): a web facility that tests question comprehensibility. *Public Opinion Quarterly* 70: 3–22.

Jay, G, Belli, R., and Lepkowski, J. (1994). Quality of last doctor visit reports. In: *Proceedings of the ASA Section on Survey Research Methods*, 362–372.

Kreuter, F., Presser, S., and Tourangeau, R. (2008). Social desirability bias in CATI, IVR, and web surveys: the effects of mode and question sensitivity. *Public Opinion Quarterly* 72: 847–865.

Locander, W., Sudman, S., and Bradburn, N. (1976). An investigation of interview method, threat and response distortion. *Journal of the American Statistical Association* 71: 269–275.

Loftus, E.F., Smith, K.D., Klinger, M.R., and Fiedler, J. (1992). Memory and mismemory for health events. In: *Questions About Questions: Inquiries into the Cognitive Bases of Surveys* (ed. J.M. Tanur), 102–137. Thousand Oaks, CA: Sage.

Maitland, A. and Presser, S. (2016). How accurately do different evaluation methods predict the reliability of survey questions? *Journal of Survey Statistics and Methodology* 4: 362–381.

Maitland, A. and Presser, S. (2018). How do question evaluation methods compare in predicting problems observed in typical survey administration? *Journal of Survey Statistics and Methodology*.

Miller, K. (2011). Cognitive interviewing. In: *Question Evaluation Methods* (eds. J. Madans, K. Miller, A. Maitland and G. Willis), 51–76. New York: Wiley.

Miller, K., Chepp, V., Willson, S., and Padilla, J.L. (eds.) (2014). *Cognitive Interviewing Methodology*. New York: Wiley.

Olson, K. (2010). An examination of questionnaire evaluation by expert reviewers. *Field Methods* 22: 295–318.

Parry, H.J. and Crossley, H.M. (1950). Validity of responses to survey questions. *Public Opinion Quarterly* 14: 61–80.

Presser, S. and Blair, J. (1994). Survey pretesting: do different methods produce different results? *Sociological Methodology* 24: 73–104.

Presser, S., Rothgeb, J.M., Couper, M.P. et al. (eds.) (2004). *Methods for Testing and Evaluating Survey Questionnaires*. Hoboken, NJ: Wiley.

Rothgeb, J.M., Willis, G.B., and Forsyth, B.H. (2001). Questionnaire pretesting methods: do different techniques and different organizations produce similar results? In: *Proceedings of the ASA Section on Survey Research Methods*. Alexandria, VA: American Statistical Association.

Saris, W. and Gallhofer, I. (2014). *Design, Evaluation, and Analysis of Questionnaires for Survey Research*. Hoboken, NJ: Wiley.

Schaeffer, N.C. and Dykema, J. (2004). A multiple-method approach to improving the clarity of closely related concepts: distinguishing legal and physical custody of children. In: *Methods for Testing and Evaluating Survey Questionnaires* (eds. S. Presser, J.M. Rothgeb, M.P. Couper, et al.), 475–502. Hoboken, NJ: Wiley.

van der Zouwen, J. and Smit, J. (2004). Evaluating survey questions by analyzing patterns of behavior codes and question-answer sequences: A diagnostic approach. In: *Methods for Testing and Evaluating Survey Questionnaires* (eds. S. Presser, J.M. Rothgeb, M.P. Couper, et al.), 109–130. Hoboken, NJ: Wiley.

Wentland, E.J. and Smith, K.W. (1993). *Survey Responses: An Evaluation of the Validity*. San Diego, CA: Academic Press.

Willis, G.B. (2015). *Analysis of the Cognitive Interview in Questionnaire Design Understanding Qualitative Research*. New York: Oxford University Press.

Willis, G.B. and Lessler, J. (1999). *The BRFSS-QAS: A Guide for Systematically Evaluating Survey Question Wording*. Rockville, MD: Research Triangle Institute.

Willis, G.B., Schechter, S., and Whitaker, K. (1999). A comparison of cognitive interviewing, expert review, and behavior coding: what do they tell us? In: *Proceedings of the ASA Section on Survey Research Methods*. Alexandria, VA: American Statistical Association.

Yan, T., Kreuter, F., and Tourangeau, R. (2012). Evaluating survey questions: a comparison of methods. *Journal of Official Statistics* 28: 503–529.

4.A Record Check Questions and Their Accuracy[7]

Source	Question	Percent correct
Belli et al. (1999)	The next question is about the elections in November. In talking to people about elections, we often find that a lot of people were not able to vote because they weren't registered, they were sick, or they just didn't have time. How about you – did you vote in the elections this November?	80
	The next question is about the elections in November. In talking to people about elections, we often find that a lot of people were not able to vote because they weren't registered, they were sick, or they just didn't have time. We also sometimes find that people who thought they had voted actually did not vote. Also, people who usually vote may have trouble saying for sure whether they voted in a particular election. In a moment, I'm going to ask whether you voted on Tuesday, November 5th, which was ____ (time fill) ago. Before you answer, think of a number of different things that will likely come to mind if you actually did vote this past election day; things like whether you walked, drove, or were driven by another person to your polling place (pause), what the weather was like on the way (pause), the time of day that was (pause), and people you went with, saw, or met while there (pause). After thinking about it, you may realize that you did not vote in this particular election (pause). Now that you've thought about it, which of these statements best describes you? (INTERVIEWER:READ STATEMENTS IN BOXES 1–4 TO R) 1. I did not vote in the November 5th election. 2. I thought about voting this time but didn't. 3. I usually vote but didn't this time. 4. I am sure I voted in the November 5th election. 7. (VOLUNTEERED) I VOTED BY ABSENTEE BALLOT.	87

7 As some of these questions were originally part of surveys of populations restricted to residents of particular cities (Denver and Chicago) or graduates of a particular college or university (the University of Maryland), in our cognitive interviews we substituted the respondents' city/state and college or university (or skipped the latter questions if the respondent had not attended a college or university). In addition, we changed all the dates in the registration and voting questions to more recent ones that corresponded to the elapsed time between the original date and the original administration time. Likewise in the two Loftus et al questions we changed "GHC" to "health care coverage".

Blumberg and Cynamon (1999)	Is CHILD covered by Medicaid, a health insurance program for low income families?	90
Duncan and Hill (1985)	Is your current job covered by a Union Contract	99
	Do you belong to that union?	100
	Do you have medical, surgical, or hospital insurance that covers any illness or injury that might happen to you when you are not at work?	99
	Do you receive sick days with full pay?	91
	Are dental benefits provided to you on your main job?	95
	Do you have life insurance that would cover a death occurring for reasons not connected with your job?	90
	Do you get paid vacation days?	99
	Do you have (maternity/paternity) leave that will allow you to go back to your old job or one that pays the same as your old job?	64
	How about (maternity/paternity) leave with pay. Is that available to you on your main job?	84
	Now I need to get some information about any pension or retirement plan you (HEAD) may be eligible for at your place of work. Not including Social Security or Railroad Retirement, are you covered by a pension or retirement plan on your present job?	97
	Have you worked under the main or basic plan long enough to earn (the right of vesting?)	89
	If you wished to retire earlier (than time needed to receive full benefits), could you receive part but not full benefits from this plan?	72
Jay et al. (1994), Dykema et al. (1997)	I'm going to ask you a series of questions about different procedures you may have had done during your last visit to a medical doctor or assistant. This includes X-rays, lab tests, surgical procedures, and prescriptions. For each of these areas, I'll ask you whether or not it happened, and whether you paid any of your own money to cover the costs. First, during your last visit to a medical doctor or assistant, did you have an X-ray, CAT scan, MRI, or NMR?	84
	During your last visit to a medical doctor or assistant, did you have any lab tests done that required blood, urine, or other body fluids?	70

	During your last visit to a medical doctor or assistant, did you have any surgical procedures?	96
	I've asked you a number of questions about X-rays, lab tests, or surgical procedures that you have had done at your last visit. This is an important area for our research. Can you think of any other tests or procedures you had done at your last visit to a medical doctor or assistant that you have not already had a chance to tell me about?	79
Kreuter et al. (2008)	During the time you were an undergraduate at the University of Maryland, did you ever drop a class and receive a grade of "W"?	75
	Did you ever receive a grade of "D" or "F" for a class?	79
	Were you ever placed on academic warning or academic probation?	90
	Did you graduate with cum laude, magna cum laude, or summa cum laude?	95
	Are you a dues-paying member of the University of Maryland Alumni Association?	91
	Since you graduated, have you ever donated financially to the University of Maryland?	73
	Did you make a donation to the University of Maryland in calendar year 2004?	79
Locander et al. (1976)	Are you now a registered voter in the precinct where you live?	84
	Did you vote in the last primary election-the one that took place last March (1972)?	65
	Do you have your own Chicago Public Library card?	80
	Have you ever been involved in a case in any of the following courts? Bankruptcy court	69
	During the last 12 months, have you been charged by a policeman for driving under the influence of liquor?	54
Loftus et al. (1992)	During the past 2 months, since (date), have you had any of the following procedures done under your GHC coverage? Blood pressure reading	82
	During the past 6 months, since (date), have you had any of the following procedures done under your GHC coverage? Blood pressure reading	73
Parry and Crossley (1950)	Have you been registered to vote in Denver at any time since 1943?	82

	We know a lot of people aren't able to vote in every election. Do you know for certain whether or not you voted in any of these elections? November 1948 Presidential election	86
	September 1948 primary election	74
	November 1947 city charter election	67
	May 1947 Mayoralty election	71
	November 1946 Congressional election	77
	November 1944 Presidential election	74
	Do you have a library card for the Denver public library in your own name?	89
	Do you have a Colorado driver's license that is still good?	88
	Do you or your family rent, or own, the place where you live?	97
	Is there a telephone in your home in your family's name?	99

5

Combining Multiple Question Evaluation Methods: What Does It Mean When the Data Appear to Conflict?

Jo d'Ardenne and Debbie Collins

The National Centre for Social Research, London, UK

5.1 Introduction

The use of question testing (QT) methods in the development and evaluation of survey questionnaires and instruments has become well established in national statistics institutes and social research organizations over the past 30 years (Presser et al. 2004). Methods include desk appraisal (e.g. Willis and Lessler 1999), expert review (e.g. Campanelli 2008), in-depth interviewing and focus groups (e.g. Blake 2014), cognitive interviewing (e.g. Miller et al. 2014), usability testing (e.g. Geisen and Romano-Bergstrom 2017), field piloting and debriefing methods (e.g. DeMaio and Rothgeb 1996), behavior coding (e.g. Schaeffer and Dykema 2011), use of paradata (e.g. Couper and Kreuter 2013), split ballot experiments (e.g. Krosnick 2011), and statistical methods such as latent class analysis (e.g. Biemer 2004), structural equation modeling (e.g. Saris and Gallhofer 2007), and item response theory (e.g. Reeve 2011). Such a plethora of methods affords the question evaluator considerable choice, which can be helpful when time and money are in short supply; but it also raises questions about how evaluators are to decide which method or methods to use in which contexts.

Various studies have compared the findings from different question testing methods (e.g. DeMaio and Landreth 2004; Presser and Blair 1994; Rothgeb et al. 2007; Willis et al. 1999; Yan et al. 2012; Maitland and Presser, this volume). The findings from these studies indicate that different methods produce different results and that there are problems with the validity and reliability of findings from different methods. The results suggest that "until we have a clearer sense

Advances in Questionnaire Design, Development, Evaluation and Testing, First Edition.
Edited by Paul C. Beatty, Debbie Collins, Lyn Kaye, Jose-Luis Padilla, Gordon B. Willis, and Amanda Wilmot.

of which methods yield the most valid results, it will be unwise to rely on any one method for evaluating survey questions" (cf Presser et al. 2004, cited in Yan et al. 2012, p. 523).

There are several research design options for combining question testing methods (Benitez and Padilla 2014). Mono-method designs involve combining different qualitative methods or quantitative methods (but not mixing the two) to address specific research questions: for example, using both cognitive interviews and focus groups to explore the conceptual boundaries of a topic (e.g. Cortés et al. 2007). Mixed-model designs apply different analytical strategies to the same set of data: for example, analyzing cognitive interview data quantitatively and qualitatively to identify responses to patient satisfaction questions (Blanden and Rohr 2009). Mixed-methods designs typically involve quantitative and qualitative research methods, which are used in an integrated research design to address specific research questions and produce in-depth insights (Tashakkori and Creswell 2007). An example of a mixed-methods study would be where cognitive interview data are used to interpret the findings from a question wording experiment. However, even if a mixed-methods approach is adopted, this does not mean that findings from different methods will necessarily be consistent.

This chapter adds to the literature comparing questionnaire evaluation methods, looking in more detail at how question-testing aims, choice of question testing methods, and survey life-cycle stage may influence findings. This chapter presents four case studies that combined different question testing methods in different ways, at different points in the survey development lifecycle, and considers the circumstances in which different QT methods result in contradictory or complementary findings. A decision framework is proposed, to aid researchers in selecting question testing methods and in making sense of the data different methods produce.

5.2 Questionnaire Development Stages

The QT projects reviewed took place at different questionnaire development stages in the survey life cycle, involving the development and testing of new questions, or the evaluation and amendment of existing survey questions, in order to improve measurement in repeated, continuous, or longitudinal surveys. In the case of the latter, these QT projects could be described as "post-testing," as the development work occurred after the questions had been used in an established survey. Table 5.1 presents different questionnaire development stages in the survey lifecycle.[1]

1 Note: This framework includes some QT methods not used in the case studies. These methods are shown in gray.

Table 5.1 Questionnaire development stages.

1) **Scoping:** The scoping stage is the earliest phase of questionnaire development and occurs prior to a new questionnaire being composed. The scoping stage involves determining what the aims of data collection should be and what practical constraints exist regarding what information can and cannot be collected. The scoping stage may include discussions on the mode of data collection and on how key concepts should be described and explained to the survey's target audience.
2) **Questionnaire development:** This stage involves writing the first drafts of the new survey questions. Some form of desk-based pre-testing may occur concurrently with this stage. For example, quality criteria checklists, such as the QAS (Willis and Lessler 1999), can be referred to while questions are being drafted.
3) **Exploratory testing:** This stage involves collecting initial *qualitative feedback* on new questions. This stage may be iterative, in that it can involve the alteration and retesting of amended draft questions. Exploration can be completely open (i.e. simply asking for views on the questions) or can be shaped by existing theory (e.g. cognitive probing based on Tourangeau's four-stage question and answer model (Tourangeau 1984)).
4) **Quality assessment:** This phase involves the generation of *quantitative metrics* that indicate how well new questions perform in a survey context. This phase could also involve the collection of metrics related to cost-effective survey implementation (e.g. questionnaire length testing), which are of relevance to interviewer-administered surveys. It can also involve generating metrics to demonstrate whether a revised design has led to quantifiable differences in data quality or not.
5) **Explanatory/Data driven retesting:** This stage involves collecting *qualitative feedback* data to explain and redress known issues detected within the survey data. The key difference between this stage and exploratory testing is that explanatory testing aims to explain *specific phenomena* observed in existing survey data that could be a result of measurement error. Therefore, this type of testing can only occur post–survey data collection. This stage of questionnaire development may be most relevant for measures designed for repeated use.

5.3 Selection of Case Studies

As the aim of the review was to explore why different QT methods may yield different findings, we purposively selected prior QT studies that combined at *least two* different QT methods. We also purposively selected studies that varied in terms of the type of material being tested and the sorts of data used to evaluate question performance. Studies varied by:

- Substantive area (e.g. topics covered and whether questions were factual or attitudinal)
- Data collection method (retrospective questionnaires or diaries)
- Data collection mode (face-to-face or web instruments)
- Type of evaluation data collected: subjective or objective
- Sources of evidence (e.g. data collected from participants or from experts)

All QT studies were ones carried out by NatCen Social Research within the UK. QT projects were only selected for comparison where there appeared to be some overlap in test aims across the methods used in a study. Note that all focus groups and cognitive interviews were audio-record with participants' consent. Detailed notes on what was discussed were made with reference to the recording and analyzed using the Framework method (Ritchie and Spencer 1994).

Documentation related to each project, e.g. copies of the survey questions or material that were tested; copies of any protocols such as topic guides, probe sheets, and instructions to interviewers; and the final pre-testing report were collated. Note that is was not possible to return to the raw data as this had been destroyed once the projects were completed. Findings from each QT method were reviewed, with a particular focus on identifying conflicting and complementary findings between two or more of the methods used. Again, comparisons were only made when test aims between methods were similar enough to allow for direct comparisons. The final stage involved producing a case study summarizing the background of each study, the aims and methodology of the QT methods being compared, what conflicting and complementary findings were found (if any), and what wider conclusions might be drawn from these findings. Table 5.2 provides details on the four case studies included in this chapter. Two illustrate different types of conflict between different method

Table 5.2 Details of QT projects reviewed.

Case study	Description of project	Questionnaire development stage(s)	Case study illustrates
1. Testing questions on "extremism"	Testing new questions on attitudes toward extremism for use in the Citizenship survey using a mono-method design	Focus groups: **scoping** (exploring views on appropriateness of content and terminology for key concepts) and **exploratory testing** of draft questions Cognitive interviews: **exploratory testing** of draft questions	Conflicting findings between **focus groups** and **cognitive interviews**
2. Testing an Event History Calendar	Testing a new online Event History Calendar (EHC) developed for a longitudinal cohort study using a mono-method design, with hybrid interviews using a mixed model design	Hybrid interviews and pilot interviewer and respondent debriefing: **exploratory testing** of the prototype EHC format	Conflicting findings between **hybrid eye-tracking/cognitive interviews** and **pilot respondent and interviewer debriefing**

Table 5.2 (Continued)

Case study	Description of project	Questionnaire development stage(s)	Case study illustrates
3. Testing questions on "benefit receipt"	Testing questions on state benefit receipt already used in the Family Resources Survey (FRS). Exploring reasons for discrepancies between survey data and administrative data using a mono-method design.	Cognitive interviews and interviewer feedback: **explanatory/data driven retesting** of existing survey questions	Complementary findings between **cognitive interviews** and **interviewer feedback**
4. Testing a travel diary	Testing a travel diary used as part of the National Travel Survey (NTS). Designing and testing a revised version of the diary and evaluating the changes made using a mixed method design.	Cognitive interviews and pilot data on data quality: **quality assessment** of new diary design	Complementary findings between **cognitive interviews** and **pilot**

pairings, and two illustrate how certain combinations of pre-testing worked well in tandem.

5.4 Case Study 1: Conflicting Findings Between Focus Groups and Cognitive Interviews

Our first case study has been selected to illustrate how some question problems may go undetected in a focus group setting, yet be detected using cognitive interviewing. In this case study, the same questions were tested using a mono-method design involving both focus group and cognitive interviewing methods to explore understanding of a key phrase – "violent extremism." Sampling and recruitment criteria were similar between the two methods. However, despite the fact that more participants were exposed to the same test questions in the focus groups than the cognitive interviews, some key problems with understanding were only raised in the cognitive interviews.

5.4.1 Background

New questions on public attitudes toward extremism designed to be included in a UK government multitopic, continuous "Citizenship Survey" from 2009 were assessed using focus groups, cognitive interviews, an expert panel on question translation, and a survey pilot. This case study focuses on a conflicting finding found between the focus groups and the cognitive interviews.

5.4.2 Aims and Methods

Twelve focus groups were held in three different locations across England and Wales involving a total of 103 participants. Quotas were set to ensure participants varied in terms of their sex, age, religion, and ethnic origin.

The focus groups explored understanding of the term "violent extremism" and public attitudes toward extremism to gauge whether this line of questioning was considered appropriate in a voluntary government-funded survey. Six draft questions on violent extremism were tested to ascertain whether people would be willing and able to answer them, and whether the wording of the questions could be improved.

Thirty cognitive interviews were also conducted in four locations. Quotas were set to ensure participants varied in terms of their sex, age, religion, and ethnic origin. The aims of the cognitive interviews were to test the six draft questions developed for the survey in more detail, by looking at:

1. Whether participants were able to understand the questions and were able to provide an answer that matched their attitude using the answer response options available.
2. Whether there was any perceived sensitivity around answering the questions.

5.4.3 Conflicting Findings

The cognitive interviewing revealed issues with the comprehension of the questions that were not mentioned in the focus groups.

The focus groups found that the key term "violent extremism" used throughout the questionnaire was understood by all participants. The focus group participants provided various examples of extremism and discussed how the term was used by government and the media. For example, participants discussed how the term referred to violent actions or "terrorism" and cited examples such as acts of violence by Islamic fundamentalists, far-right nationalists, or animal rights activists.

In contrast, the cognitive interview participants did not always understand the term "violent extremism" as intended. A definition of violent extremism was included as a read-out at the start of the new questions, but this was considered by some as too long-winded and difficult to retain. The cognitive interviewing also revealed people experiencing difficulties answering the questions accurately due to misunderstanding terminology around violent extremism or simply not knowing what the term meant. Response mapping errors were also noted in the cognitive interviews due to respondent difficulties in processing the agree-disagree scale used. These errors resulted in some participants erroneously condoning violent extremism. As a result of these findings, it was recommended that a definition of violent extremism be displayed on a showcard and that the response options used should be reviewed. All questions used in later tests had response options ranging from "Always right" to "Always wrong" rather than using an agree-disagree scale.

5.4.4 Conclusions

There are two plausible explanations for the disparity found between the focus group and the cognitive interview findings. The first is that comprehension problems are easy for individuals to hide in a focus group setting. If some members of a group are able to give examples of what a word means to the moderator, people who are less articulate or less forthcoming can nod and provide similar interpretations to the person who spoke first.

The second is that participants may feel less comfortable admitting they don't understand something in a group environment compared to a more private one-to-one interview. Perceived sensitivity of answering comprehension probes may be exacerbated in a focus group setting with multiple bystanders present. Therefore, one-to-one cognitive interviews are likely to be more effective than focus groups when it comes to identifying comprehension issues with questions. Finally, issues with response mapping were not detected in the focus group setting as participants were asked to look at and comment on the questions, but were not asked to provide their answers. This meant that participants did not have to perform the cognitive task of response mapping in practice and so issues with this were not detected.

5.5 Case Study 2: Conflicting Findings Between Eye-Tracking, Respondent Debriefing Questions, and Interviewer Feedback

In our next case study, we look at conflicting findings on instrument usability, using a mono-method design involving hybrid interviews combining both eye-tracking and cognitive interviewing (Neuert and Lenzner 2016) with findings from respondent debriefing questions and interviewer feedback. Hybrid interviews detected issues with a computerized calendar interface that went undetected in a pilot. The hybrid interview data were analyzed in two ways, using a mixed-model approach: qualitatively and quantitatively. The qualitative data were participants' accounts; the quantitative data, fixation durations on defined areas of the screen.

5.5.1 Background

Next Steps (formerly known as the Longitudinal Study of Young People in England) is a longitudinal survey following the lives of a cohort of young people born in 1989–1990. As part of development of a mixed-mode questionnaire that took place ahead of the 2015 survey, the usability of an Event History Calendar (EHC) to be used in the computer-assisted personal interviewing (CAPI) and online questionnaires was assessed.

The EHC was a visual representation of the participants' life that displayed key life events (changes in relationship status, changes in employment or education status, address changes) on a timeline. The contents of the EHC were updated as information was entered into the questionnaire. The intended purpose of the EHC was to help participants improve accuracy and recall of key dates, increase consistency of information collected across life domains (relationship status, employment status, and housing), and enhance navigation across the life history module. This was considered to be especially important in the web mode as no interviewer would be present to assist in the recording of complex histories. The EHC itself could be used as a form of data entry. For example, if participants wanted to edit their historical information, they could do so by clicking on the relevant stripe of the calendar and dragging and dropping the entry. The calendar also had zoom and scroll features so that participants could look at month-by-month entries up close if required. A single page of instructions was provided to explain these functions prior to the data entry screen.

5.5.2 Aims and Methods

Eleven hybrid interviews, combining eye-tracking with cognitive interviewing, explored whether participants: (i) read and understood the EHC instructions; (ii) looked at and used the EHC when inputting historical data; (iii) found the EHC helpful (or not) in improving the accuracy of their historical reports; and (iv) how participants interacted with the EHC. For example, we wanted to observe whether participants ever clicked on the EHC in order to edit their responses and whether they used the zoom-in and zoom-out functions. These interviews were carried out face-to-face.

During the hybrid interviews, all participants were asked to complete a shortened version of the Next Steps web questionnaire while eye-tracking equipment recorded what they looked at. Participants were then shown a gaze replay video showing them the screen elements they had looked at and were asked to think aloud while reviewing the video to gain a deeper insight on what they did and did not look at and why.

Recruitment quotas were set to ensure participants varied in terms of their relationship, housing, and activity histories. All participants were aged 23–27 (roughly the same age group as the Next Steps cohort). Participants varied in terms of their highest qualification, and those with low-level education were included.

Participant feedback was collected using four respondent debriefing questions, asked at the end of the CAPI and online pilot questionnaire. Participants were first asked whether they had used the EHC calendar when answering any of the history questions (an image of the EHC was shown to participants in order to clarify what feature was being referred to). Participants who reported using it were then asked to rate the EHC in terms of its ease of usage and perceived usefulness. These participants were also asked to provide open, qualitative feedback on how the EHC could be improved. A total of 35 web participants and 28 CAPI participants fully participated in the pilot.

Interviewers who had worked on the CAPI pilot were also asked to provide qualitative feedback on the EHC. This information was collected via a single open question on an interviewer feedback form and during a face-to-face interviewer debriefing session. Interviewer feedback was collected after the pilot fieldwork was completed.

5.5.3 Conflicting Findings

The hybrid interviews, pilot interview, and respondent debriefing data identified a common set of issues, the main one being that the EHC function was not being used by all participants. However, there was some discrepancy between the hybrid interviews and the pilot interviews regarding the extent to which the calendar was being used. In the hybrid interviews, the EHC was not used by any of the participants. In contrast, 4 out of 28 face-to-face pilot participants and 8 out of 35 web pilot participants reported using the EHC. Interviewer feedback from the CAPI pilot was that the EHC feature was rarely or never used.

In this case, we believe the evidence from the hybrid interviews on usage is likely to be more accurate, as the eye-tracker provided an objective measure of how often the EHC was looked at, rather than relying on self-reported usage or interviewer observation. The eye-tracking data showed that the EHC was only ever looked at it in passing and not read, as participants scanned for other screen elements (i.e. the Next button). It was clear from the gaze replay recordings that the EHC was not being used in the intended way and that participants were not reviewing their timeline when answering the historical questions.

The different pre-testing methods drew different conclusions regarding why the EHC was not being used. The pilot interview feedback suggested that the calendar function was not being used because participants were able to easily recall dates without any visual prompts. This finding was not supported by the hybrid interviews, as the cognitive probes uncovered that some changes in address and working status were going unreported. Certain types of economic activity (short-term work, periods of job seeking, and unpaid work) were also being excluded by participants when the intention was for them to be included.

Examination of the eye-tracking data also revealed that the EHC instructions were not being read, e.g. some eye-tracking participants appeared to skim over the instruction page rather than fully read it. This was indicated by eye movements flitting quickly over the screen, with some text not being glanced at. The cognitive probes elicited feedback that people were put off by the volume of text, and also that they did not understand the purpose of the EHC. Findings from the hybrid interviews led to the EHC instruction page being replaced with a short tutorial video, explaining the purpose of the EHC and how it should be used in practice.

In contrast, the pilot respondent debriefing questions and interviewer feedback provided only brief comments on how the EHC could be improved. Comments typically focused on improvements to the visual design, e.g. by making the EHC larger, making scrolling easier, etc. These comments demonstrated that participants (and interviewers) were not aware of how to use the zoom and scroll functions explained in the initial instructions. Therefore, the pilot feedback in itself did not provide much information on how the EHC could be improved.

5.5.4 Conclusions

Eye-tracking data collected objective evidence on what screen features were actually being used by participants, in a way that was potentially more accurate than participant self-reports captured in the pilot debriefing questions. The eye-tracking data also demonstrated key instructions were not being read and that the functionality of the EHC was not being communicated effectively. The inaccuracies in reporting dates were only uncovered by cognitive probing used in the hybrid interviews. Issues with recall were not detected by interviewers in the pilot or mentioned spontaneously by participants in the open debriefing questions.

Although the pilot did reveal the EHC feature was being underused, it did not yield enough information on how this feature could be improved. The qualitative data collected from the pilot respondents and interviewers was insufficient in depth to generate helpful recommendations on design improvement. The hybrid interviews provided more useful information in this regard.

5.6 Case Study 3: Complementary Findings Between Cognitive Interviews and Interviewer Feedback

This case study illustrates when complementary findings are produced using different qualitative methods from a mono-method design involving interviewer workshops and cognitive interviews, and the findings that were unique to the cognitive interviews. In particular, we focus on the how the interviewer feedback and cognitive interviews identified similar types of data in the qualitative phase.

5.6.1 Background

The Family Resources Survey (FRS) collects information on the incomes and circumstances of around 20 000 UK households each year. As part of a face-to-face CAPI interview, individual household members are asked about whether they are currently receiving a variety of state-funded welfare benefits. The survey interviewer presents respondents with a series of showcards that list 30 state benefits. Respondents are asked to identify which, if any, of these benefits they receive. If respondents indicate they receive any of the benefits listed, further information is collected, including the amount received and how it is paid. Interviewers encourage respondents to consult documentation during the interview, to improve the accuracy of information collected. These questions are asked toward the end of a long interview on a variety of topics related to family finances.

The UK Department for Work and Pensions (DWP), which sponsors the FRS, carried out a comparison of individual responses to the 2009/10 FRS benefits questions with administrative data in cases where consent to data linkage was obtained. This was the quality assessment phase of testing. The validation analysis showed that there were a range of circumstances in which FRS survey respondents had either:

- Not reported a benefit that the administrative record showed they were in receipt of (an omitted benefit); or
- Reported a benefit that was not indicated in the corresponding administrative data (a misreported benefit/false positive).

Of the two problems, omission was the more common and was more prevalent among respondents aged over 80. Certain benefits were found to be more likely to be omitted (mainly Pension Credit and Attendance Allowance, though Industrial Injuries Disablement Benefit, Disability Living Allowance, Incapacity Benefit, and Retirement Pension were also affected), and certain types of benefit were more prone to being misreported/false positives (notably Incapacity Benefit and Income Support).

A qualitative retesting study was undertaken to explore the reasons for the discrepancies detected in the quality assessment phase and to suggest ways in which the FRS benefits questions could be improved.

5.6.2 Aims and Methods

The qualitative retesting involved both collecting interviewer feedback and cognitive interviews with FRS respondents whose individual survey responses were found to be different than their administrative record data for the same time period. The aim of both methods was to ascertain the reasons for errors detected in the quality assessment phase.

Interviewer feedback was collected during three interviewer debriefing sessions. These were conducted face-to-face, in a similar way to a focus group. In total, 29 FRS interviewers were consulted during these sessions. Each session was composed of interviewers working in different parts of the UK, with differing levels of overall survey and FRS-interviewing experience. The debriefing explored interviewers' views on the FRS interview generally and the benefits questions specifically, in terms of respondents' reactions to the questions, what kinds of difficulties respondents and interviewers experienced, and how interviewers dealt with these difficulties. The debriefing also garnered interviewers' suggestions about how the benefits questions could be improved.

Cognitive interviews were conducted with 31 respondents who had taken part in the FRS in 2009/10 and given their consent to be re-contacted. Sample quotas were set to ensure questions concerned with specific benefits, known to be at risk of being misreported or omitted, were tested. The cognitive interviews took place in 2011/12, and participants were asked the FRS benefits questions afresh. Interviews focused exclusively on participants' responses to the FRS benefits questions asked during this follow-up interview (i.e. they were not asked to provide feedback on how they had answered the questions in their original FRS survey interview).

Think-aloud followed by probing was used to explore respondents' understanding of the benefit questions and key terms, recall of the information being sought, and strategies for answering the questions. The cognitive interviews explored any sensitivity around asking about benefit receipt, and the use of documentation.

5.6.3 Complementary Findings Between Methods

There was a high degree of overlap between findings from the interviewer debriefings and the cognitive interviews in determining why response errors were occurring. For example, both methods demonstrated that participants were unaware of the terminology used to describe certain benefits they were receiving: e.g. FRS respondents get confused between Attendance Allowance and Disability Living Allowance, and between the different types of Disability Living Allowance available. Also, respondents were aware of receiving a "retirement-related benefit" but unaware that this is sometimes a "Pension credit."

The cognitive interviews also highlighted issues that the interviewer debriefing did not. For example, confusion between Child Tax Credit and Child Benefit was only detected in the cognitive interviews. Likewise, the broader finding that respondents do not think of "Child Benefit" or "State Retirement" as a benefit at all (but rather as an entitlement) was only detected in the cognitive interviews (Balarajan and Collins 2013).

5.6.4 Conclusions

First, this case study illustrates how qualitative retesting following a quantitative quality assessment of data collected can be targeted to investigate specific known data issues with key groups and explain why problems are occurring in practice. For example, the interviewer debriefing focused specifically on interviewers' reactions to known problematic questions and what issues they had observed, if any, when asking these questions in the field. The cognitive interviewing protocol included scripted probes on the types of welfare benefits and payments found to be under-reported by the quality assessment phase. The quality assessment phase was also used directly in the selection of cognitive interview participants. The individuals approached to take part in cognitive interviews were survey respondents for whom there was a mismatch between their survey answer and the administrative data held.

This case study also illustrates that interviewer feedback can provide insights similar to cognitive interviews when conducting explanatory or data driven retesting. Compared to case study 2 (Section 1.5), in which interviewer feedback collected as part of a pilot failed to detect key issues in questionnaire design, this case study shows that survey interviewers can be well-placed to provide accurate insights into measurement error in some circumstances. For surveys like the FRS that run continuously or are repeated frequently, interviewers may know why certain questions are not working well. This is because interviewers have regular exposure to survey respondents in a realistic setting and to a wider variety of respondents than in a pilot, with the likelihood of problem detection increasing with this wider exposure. Also, interviewers are likely to have had a greater exposure to respondents who naturally articulate problems when answering (even if actual survey respondents have not been asked to provide this type of feedback). During the mainstage survey, interviewers may not consistently pass on these issues for a variety of reasons, e.g. because there is not a mechanism for capturing this type of feedback at the time of interview and so it is forgotten, or because interviewers are not encouraged to provide feedback.

This case study also illustrates the point that cognitive interviewing post quality assessment can identify issues that are hidden from the experienced survey interviewer. These covert issues may only come to light through careful qualitative probing that exposes respondents' thought processes. In the survey interview, the interviewer asks the questions and the respondent is expected to answer them. In the cognitive interview, the interviewer gives explicit permission to the respondent to voice confusion or difficulties they have with the survey question. Contextual information about the participant's circumstances can also be collected in the cognitive interview that allows for an exploration of the validity of their survey response (Miller 2011, pp. 51–76).

5.7 Case Study 4: Combining Qualitative and Quantitative Data to Assess Changes to a Travel Diary

Our final case study illustrates how both qualitative and quantitative QT methods can be effectively combined to assess proposed changes to a travel diary. This was a simultaneous QUAN/QUAL mixed research design, in which neither method was dominant (Creswell 1995).

5.7.1 Background

The National Travel Survey (NTS) provides the Department for Transport (DfT) with its main measure of the personal travel behavior for the British public. Each year, in the region of 7500 household computer-assisted personal interviews are conducted. Subsequently, each household member is asked to complete a seven-day paper travel diary. Both adults and children complete a travel diary, the versions differing slightly. The diary collects detailed information about each journey made including its purpose and the mode of travel, as well as other details such as duration and ticketing details. Information on each journey is recorded in each row of the diary. The diary contains detailed instructions and an example page, and each day's travel is collected on a separate page. When interviewers pick up the diaries at the end of the seven-day recording period they check over the diary with the participant, correcting any recording errors and adding in any missing journeys or information using a unique color pen.

The seven-day travel diary method places a significant response burden on survey participants, seeking detailed information that participants may not possess, which can lead to missing or erroneous data. A mixed-methods evaluation of the existing NTS travel diary involving secondary analysis of the diary survey data, a focus group with diary coding staff, and cognitive interviews with members of the public found evidence of such problems, identifying specific problems with its visual layout (McGee et al. 2006). For example, the evaluation of the existing diary found that instructions on how to complete the diary, included on flaps at the front and back of the diary, were often missed, and that errors and omissions arose as a result. A redesign of the travel diary aimed to simplify its visual layout by improving navigation and the way in which information was organized and displayed, using principles proposed by Jenkins and Dillman (1997). Figure 5.1 shows the design of the original and redesigned travel diary recording pages.

The new design was evaluated, again using a mixed-methods approach, to assess whether the alterations made to the original diary design had addressed

Existing diary page

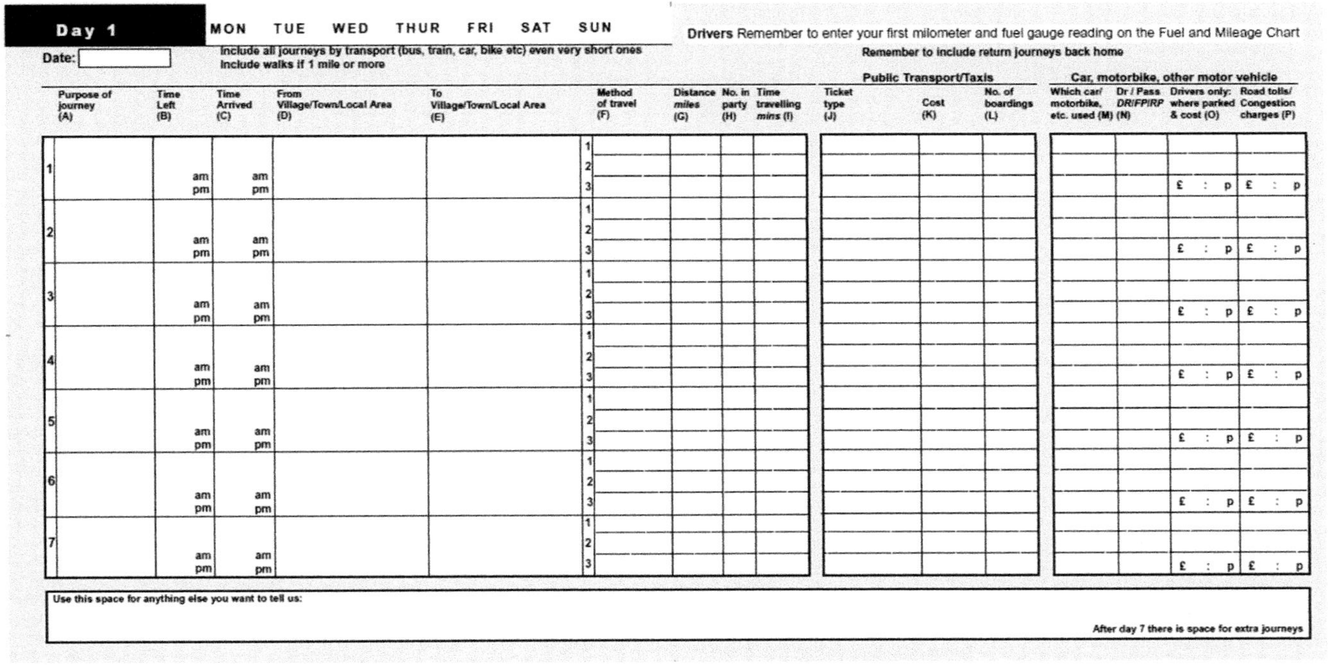

Figure 5.1 Existing and proposed new design of National Travel Survey diary recording page.

Proposed new diary page design

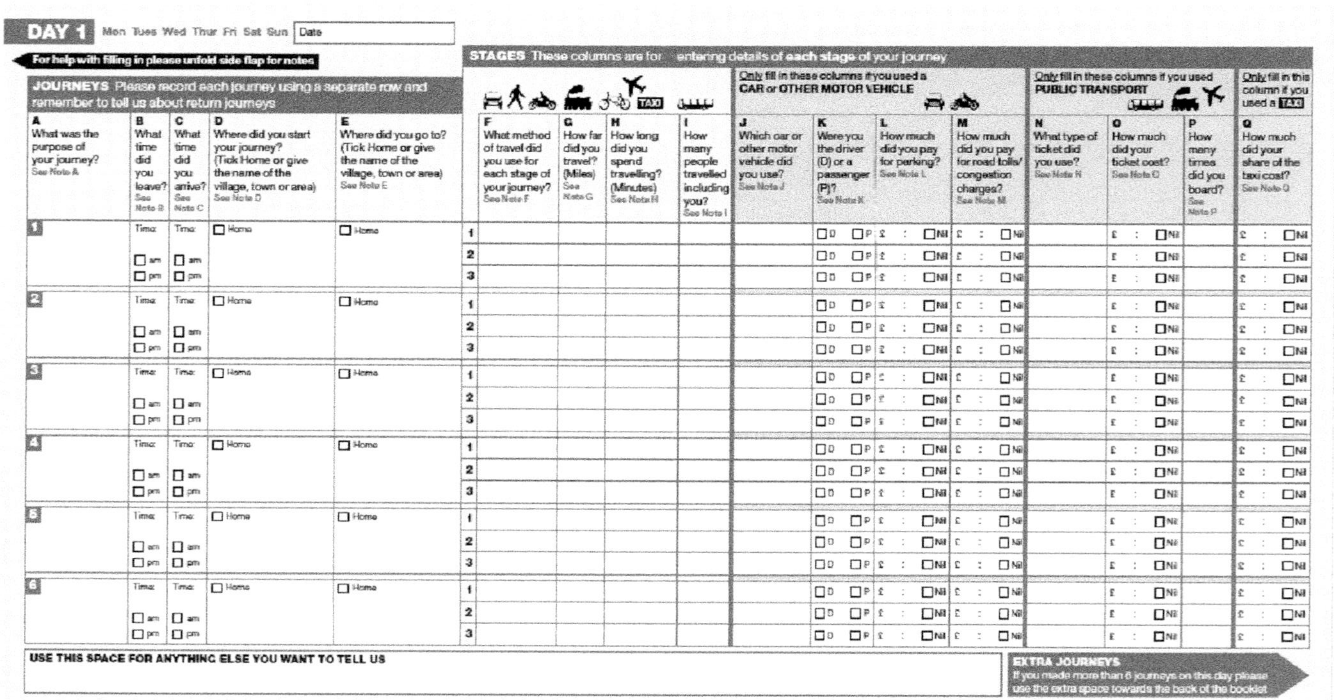

DAY 1 Mon Tues Wed Thur Fri Sat Sun Date

For help with filling in please unfold side flap for notes

JOURNEYS Please record each journey using a separate row and remember to tell us about return journeys

STAGES These columns are for entering details of each stage of your journey

Only fill in these columns if you used a CAR or OTHER MOTOR VEHICLE

Only fill in these columns if you used PUBLIC TRANSPORT

Only fill in this column if you used a TAXI

A What was the purpose of your journey? See Note A	B What time did you leave? See Note B	C What time did you arrive? See Note C	D Where did you start your journey? (Tick Home or give the name of the village, town or area) See Note D	E Where did you go to? (Tick Home or give the name of the village, town or area) See Note E		F What method of travel did you use for each stage of your journey? See Note F	G How far did you travel? (Miles) See Note G	H How long did you spend travelling? (Minutes) See Note H	I How many people travelled including you? See Note I	J Which car or other motor vehicle did you use? See Note J	K Were you the driver (D) or a passenger (P)? See Note K	L How much did you pay for parking? See Note L	M How much did you pay for road tolls/congestion charges? See Note M	N What type of ticket did you use? See Note N	O How much did your ticket cost? See Note O	P How many times did you board? See Note P	Q How much did your share of the taxi cost? See Note Q
1	Time: ☐ am ☐ pm	Time: ☐ am ☐ pm	☐ Home	☐ Home	1						☐ D ☐ P	£ : ☐ Nil	£ : ☐ Nil		£ : ☐ Nil		£ : ☐ Nil
					2						☐ D ☐ P	£ : ☐ Nil	£ : ☐ Nil		£ : ☐ Nil		£ : ☐ Nil
					3						☐ D ☐ P	£ : ☐ Nil	£ : ☐ Nil		£ : ☐ Nil		£ : ☐ Nil
2	Time: ☐ am ☐ pm	Time: ☐ am ☐ pm	☐ Home	☐ Home	1						☐ D ☐ P	£ : ☐ Nil	£ : ☐ Nil		£ : ☐ Nil		£ : ☐ Nil
					2						☐ D ☐ P	£ : ☐ Nil	£ : ☐ Nil		£ : ☐ Nil		£ : ☐ Nil
					3						☐ D ☐ P	£ : ☐ Nil	£ : ☐ Nil		£ : ☐ Nil		£ : ☐ Nil
3	Time: ☐ am ☐ pm	Time: ☐ am ☐ pm	☐ Home	☐ Home	1						☐ D ☐ P	£ : ☐ Nil	£ : ☐ Nil		£ : ☐ Nil		£ : ☐ Nil
					2						☐ D ☐ P	£ : ☐ Nil	£ : ☐ Nil		£ : ☐ Nil		£ : ☐ Nil
					3						☐ D ☐ P	£ : ☐ Nil	£ : ☐ Nil		£ : ☐ Nil		£ : ☐ Nil
4	Time: ☐ am ☐ pm	Time: ☐ am ☐ pm	☐ Home	☐ Home	1						☐ D ☐ P	£ : ☐ Nil	£ : ☐ Nil		£ : ☐ Nil		£ : ☐ Nil
					2						☐ D ☐ P	£ : ☐ Nil	£ : ☐ Nil		£ : ☐ Nil		£ : ☐ Nil
					3						☐ D ☐ P	£ : ☐ Nil	£ : ☐ Nil		£ : ☐ Nil		£ : ☐ Nil
5	Time: ☐ am ☐ pm	Time: ☐ am ☐ pm	☐ Home	☐ Home	1						☐ D ☐ P	£ : ☐ Nil	£ : ☐ Nil		£ : ☐ Nil		£ : ☐ Nil
					2						☐ D ☐ P	£ : ☐ Nil	£ : ☐ Nil		£ : ☐ Nil		£ : ☐ Nil
					3						☐ D ☐ P	£ : ☐ Nil	£ : ☐ Nil		£ : ☐ Nil		£ : ☐ Nil
6	Time: ☐ am ☐ pm	Time: ☐ am ☐ pm	☐ Home	☐ Home	1						☐ D ☐ P	£ : ☐ Nil	£ : ☐ Nil		£ : ☐ Nil		£ : ☐ Nil
					2						☐ D ☐ P	£ : ☐ Nil	£ : ☐ Nil		£ : ☐ Nil		£ : ☐ Nil
					3						☐ D ☐ P	£ : ☐ Nil	£ : ☐ Nil		£ : ☐ Nil		£ : ☐ Nil

USE THIS SPACE FOR ANYTHING ELSE YOU WANT TO TELL US

EXTRA JOURNEYS If you made more than 6 journeys on this day please use the extra space towards the back of the booklet

Figure 5.1 *(Continued)*

specific problems identified by the earlier evaluation. Three predictions were tested:

1. A narrower range of problems would be identified in cognitive testing the new diary as compared to the existing diary.
2. Problems relating to information organization and navigation would be diminished for the new diary compared to the existing diary, as evidenced from cognitive testing and respondent debriefing questions asked as part of a pilot.
3. The new diary design would produce fewer errors than the existing one, with evidence coming from an error recording sheet (ERS) as part of a pilot.

This case study describes the methods used in the evaluation of the new diary design.

5.7.2 Aims and Methods

The evaluation of the new diary involved a small-scale pilot involving 123 participants (102 adults and 21 children) and, due to limited development time, in parallel, 32 cognitive interviews.

The pilot aimed to test the usability of the new diary and to assess data quality. Evidence on usability came primarily from responses to respondent debriefing questions asked by the interviewer at the end of the diary recording period, which provided quantitative and qualitative data. These questions asked respondents to: (i) rate their ease or difficulty of completing the travel diary, (ii) indicate whether they read the instructions at various points in the diary, (iii) rate the helpfulness of instructions, (iv) indicate whether any parts of the diary were confusing, and (v) describe what they found confusing.

Data quality was assessed using an ERS, developed using findings from the evaluation of the existing diary. It recorded different types of respondent error that related to participants providing insufficient information, or not providing any information at particular places in the diary (e.g. whether in the Purpose column, insufficient information was recorded to allow coding or if a return journey was omitted). Evidence of errors came from a review of the completed diaries and the corrections made to them by the interviewer (in red ink) and/or data processing team (in green ink).

Pilot interviewers were issued preselected addresses and asked to collect at least 120 completed travel diaries: at least 100 from adults (16 years and over) and 20 from children. Up to three people per household could be selected to complete the diary. In addition, each interviewer was asked to obtain travel diaries from at least one household:

- That contained a child aged under 16 years
- That contained three or more adults

- Where the diary was explained by proxy

These requirements aimed to maximize the heterogeneity of the pilot sample and ensure that households at greater risk of producing poor-quality travel diaries were included. The usual NTS diary placement rules were followed.

Due to timetable and cost constraints, the use of a split-ballot experiment to compare the existing and new diary designs was not possible. Instead, the ERS was applied to 100 new (pilot) diaries and a matched sample of 100 existing NTS diaries (completed as part of the main NTS). This meant that error rates in old and new versions could be compared. Each diary was reviewed, and each time a pre-specified problem code was observed, it was systematically recorded on the ERS. Reviewing all the completed ERSs, the total number of errors was calculated across the two samples.

The cognitive interviews aimed to assess whether the changes to the existing diary improved participants' understanding of the diary completion task and navigation through it, and whether this encouraged (accurate) recording of travel in the diary.

At the start of the cognitive interview, participants were asked to complete two days of the travel diary, thinking about the previous two days. A mixture of think-aloud and cognitive probing techniques were used to explore how participants went about completing the diary and their understanding of the diary completion task. Cognitive participants were purposively recruited to ensure sample heterogeneity. Quotas reflected demographic variables known to affect travel behavior: age, sex, employment status, household composition, and car ownership.

5.7.3 Complementary Findings Between Methods

Data from the pilot and the cognitive interviews supported the three predictions we tested, indicating that the new diary design improved data quality. We also found consistency in the data obtained from the different evaluation methods for our predictions, and in particular for our second prediction that problems relating to information organization and navigation would be diminished for the new diary compared to the existing diary. In testing this prediction, we used multiple methods: respondent debriefing questions (as part of the pilot) and the ERS (as part of the pilot) and cognitive interviews. The new design, in which all instructions were in one place and were better signposted, had lower rates of errors and omissions.

The respondent debriefing questions indicated high levels of awareness of where the instructions could be found (80% of pilot respondents said it was clear that the side flap contained instructions) and high use of the instructions (76% said they had referred to them). Of those who said they used the instructions, 92% said they found them either "very helpful" or "quite helpful."

Evidence from the ERS indicated that fewer respondents left out stages of their journeys when filling in the new diary as opposed to the existing one (21% of respondents missed a stage at least once while filling in the current diary compared with 5% of respondents completing the new diary). We speculate that the use of the travel icons at the top of column F in the new diary design helped respondents identify which methods of travel they had used in their journeys and so reduced the likelihood of stages being missed.

The cognitive interview participants were observed looking at the instruction flap, and understood that the "See Note [letter]" at the end of each column label on the travel recording page referred to instructions for that column contained within the instruction flap. Cognitive interview findings also allowed us to assess whether problems with understanding of the column headings in the old design were improved by turning headings into questions. Findings suggested mixed success: while the cognitive testing indicated that changes made to the column collecting information about whether the diary keeper was the driver or passenger of the motor vehicle (column N of the old diary and column K of the new) improved participants' understanding of the information they were supposed to provide, changes to the wording of column A ("purpose of journey") were not successful in communicating that participants needed to record return journeys. The cognitive interview evidence supported the evidence from the ERS, providing an explanation of the quantitative findings (McGee et al. 2006).

5.7.4 Conclusions

The consistency of the findings across the different methods we used, we contend, results from two features of the design of this evaluation. First, we were able to draw on evidence from the earlier evaluation of the existing travel diary. This helped us to redesign the diary and develop a sharply focused evaluation of it.

Second, we purposively adopted a mixed-methods approach, using both quantitative and qualitative pretesting methods to assess different aspects of the new diary design. The question testing methods were carefully selected to complement each other and to provide evidence that would allow us to test each of our predictions. The pilot allowed us to assess the size and extent of errors and behaviors and compare these for the existing and new diary designs. Cognitive interviewing was used to look in detail at the interaction between diary design and diary completion behavior and assess whether problems identified with the existing diary were still present when testing the new diary or whether new problems arose from a new design. Putting the evidence together from each pretesting method allowed us to build a detailed picture of how successful the new diary design was in combatting the problems found with the existing design.

5.8 Framework of QT Methods

Based on our conceptualization of QT stages and lessons drawn from our cases studies, we have produced a framework to help research practitioners select an appropriate question testing methodology based on their objectives. This framework is shown in Table 5.3.

Deciding which QT methods are most appropriate will depend on the type of survey that is being conducted (one-off or repeated) and the stage in the survey life cycle at which testing takes place. For example, scoping occurs prior to writing questions and confirming measurement objectives, whereas quality assessment can only occur once some survey fieldwork has been conducted. Piloting is one way of collecting preliminary quality assessment metrics, but this assumes that a pilot has a sufficient sample size to draw meaningful conclusions. In cases where only a small-scale pilot is conducted, quality assessment metrics may only exist after a survey has been completed. Quality assessment may only be possible once a survey has been conducted, meaning this phase may be more appropriate for questions intended for use in repeat, continuous, or longitudinal surveys.

Post-survey analysis of data can encompass a range of quantitative quality assessment metrics not covered in our case studies. These metrics include item non-response rates, item response distributions, item response latencies (i.e. via paradata), statistical tests for reliability and validity, latent class analysis of scale items, etc.

Explanatory/data driven retesting can only be undertaken once quality assessment metrics have been reviewed and problems with data identified. This phase involves qualitative research to establish probable causes of data issues identified. The efficacy of any changes made can be evaluated by further quality assessment, i.e. collecting quantitative data to see if changes have remedied issues detected. In practice, this may mean piloting the revised questionnaire to establish if there are any improvements in the data collected or running the new measures in the next wave of the survey. Alternatively, the revised measures could be tested using a split-ballot experiment where the survey participants are randomly asked either the old questions or the new questions.

5.9 Summary and Discussion

Our case studies illustrate that different QT methods sometimes produce conflicting results even when test aims are similar enough to allow comparison. In our review, we found several factors that contributed to conflicting results.

Table 5.3 Framework for selecting a question testing methodology.

Stage	Objectives	QT method
Scoping phase	To clarify research objectives and priorities	Expert panel
	To clarify what methodology is feasible given costs and other logistical considerations	
	To explore how the survey's target audience understands the key concepts and ideas being researched	Focus groups
	To explore the experiences, examples, and language used by the survey target audience about the research subject	
Questionnaire design	To write new questions that avoid common pitfalls	Desk-based appraisals (e.g. QAS)
Exploratory testing	To check whether new questions are understood as intended, and whether participants are able to answer them accurately	Cognitive interviewing
	To check instruments in terms of "usability," e.g. whether participants are able to easily navigate self-completion questionnaires, whether they understand how to use computerized functions, and so on	Hybrid interviews Cognitive interviewing Eye-tracking
	To check instrument usability/navigation in interviewer-administered questionnaires	Interviewer feedback (post pilot)
	To obtain information on how to improve overall survey processes, e.g. how to sell the survey to the public	
Quality assessment metrics	To check questionnaire length	Piloting
	To conduct live testing of computerized questionnaire systems	
	To check for overt issues in data quality, e.g. high levels of item non-response/missing data/high levels of uncodeable responses	Piloting and post-survey quantitative analysis
	To check levels of willingness to additional requests, e.g. consent to re-contact, data-linkage, etc.	
	To check data collected by comparing it to other data sources and/or existing measures	Validation
	To demonstrate efficacy of changes made to questions; in practice, this means repeating quality assessment metrics after making changes to the questionnaire, so this stage will only apply to continuous or longitudinal studies, or questions that have been designed for inclusion in multiple surveys	Re-piloting changes Post-survey quantitative analysis after changes have been made Split-ballot experiments All these methods can involve collecting metrics to assess quality such as respondent debriefing questions or error-coding schemes.
Explanatory/ Data driven retesting	To establish why known data issues are occurring and explore how these issues can be fixed	Expert panel Desk-based appraisals Cognitive interviews Interviewer feedback

5.9.1 Test Aims

Methods may be used to explore similar or different research questions. If similar research questions are to be explored using different methods, then researchers should think about whether they anticipate findings will be similar across methods or whether different methods will produce different findings and why that might be. Using the mixed-methods paradigm may be helpful here (Benitez and Padilla 2014; Tashakkori and Teddlie 1998).

5.9.2 Development Stage

At the scoping and exploratory stages, the pre-testing aims may be broad, reflecting the fact that there is no evidence of how the questions perform. Often the aims of pre-testing will be based on the theoretical underpinnings of common types of issues (e.g. Tourangeau's four-stage model) and the choice of testing methods will be driven by more generic issues related to the topic rather than individual test questions themselves. In contrast, at the explanatory retesting stage, research aims may be focused on known errors and the QT methods selected and tailored to reflect the material being tested. Specific subgroups can be targeted for retesting rather than relying on general population testing (i.e. groups in which errors are known to be occurring). Quantitative and qualitative QT methods can be combined effectively to conduct explanatory or data driven retesting and to repeat quality assessment metrics to establish if any changes made to questions have resulted in any quantifiable improvements in data quality. This is an example of mixed-methods testing as typified by Benitez and Padilla (2014).

5.9.3 Context

Methods can be placed on a continuum in terms of how closely they replicate the actual survey context. Depending on the aims of the test, deviations from the proposed survey protocol may produce false findings. Mimicking the mode of survey administration may be important for detecting some types of problems, e.g. issues with question length or visual presentation. Mode mimicking can be more difficult using a method such as focus groups than in a pilot.

5.9.4 Method Sensitivity

Some methods have greater sensitivity to identifying certain types of potential measurement issues than others. This sensitivity reflects the underlying etiology of the method and its focus.

5.9.5 Resources and Time

The lack of sensitivity of a pretesting method in detecting problems may reflect constraints placed on the pretesting design. Our review suggests that conflicting results may sometimes reflect the competing demands placed on limited pretesting resources. In the survey production environment in which we work, we often have to navigate a design path that balances the needs and priorities of different survey stakeholders within limited resources.

Our findings suggest that some methods are more suitable than others for addressing particular research aims and at particular stages of questionnaire development.

Focus groups are useful in the scoping stage of questionnaire development to explore conceptual boundaries and language usage at the exploratory stage. However, focus groups may be less useful for exploratory testing of draft survey questions because comprehension of the question, or individuals' difficulties with the response task, with carrying out the task (e.g. calculating an answer), or with having the knowledge required to be able to answer may not be detected. In the group setting, individuals may not feel willing or comfortable or have the opportunity to voice their interpretations of the question.

Hybrid interviews (combining eye-tracking and cognitive interviewing) appear better at detecting issues with usability compared to using respondent debriefing questions and collecting interviewer feedback in a pilot. Eye-tracking data provide objective evidence of what visual features are actually looked at by participants. When paired with cognitive interviews (to explain *why* some areas are looked at and some are overlooked), eye-tracking data provide more detailed feedback on how to improve visual design.

Interviewer feedback may be more useful in explanatory retesting, particularly in the case of interviewers who are working on continuous or longitudinal surveys.

Given the tension of "depth" versus "breadth" of aims, it should not be surprising that some piloting techniques are less successful in detecting questionnaire design issues than others. In our experience, where pilots have failed to detect certain types of problem, this failure has arisen as a consequence of the pilot having multiple test aims, with the collection of feedback on specific questions being only a small part of the overall remit. Piloting is the only QT method appropriate for testing key assumptions prior to mainstage fieldwork, such as average length of interview and whether systems work in practice. Therefore, it is not our intention to suggest that piloting is ineffective. Rather, it is our recommendation that an integrated mixed-methods approach to testing is optimal. Qualitative methods such as cognitive interviewing may be more sensitive to detecting and fully mapping the types of problems participants encounter when answering questions and exploring the reasons for them, and as a result provide more detailed evidence on how to fix problems identified. Pilots are essential for checking underlying survey processes but may be less helpful in terms of detecting specific question issues, especially when little is known about what the potential problems might be. Quantitative methods for quality assessment can be put in place for continuous or ongoing surveys and can be used to examine how often specific issues occur in practice, and whether issues are fixed if questions are altered.

Two case studies (case studies 3 and 4) illustrate where *different* methods result in complementary findings. In particular, these case studies illustrate how combining quantitative techniques (quality assessment) and qualitative techniques (to explain reasons for known data quality issues) can result in ongoing improvement in the case of repeated, continuous, or longitudinal surveys. These conclusions are based on our experience, and we encourage others working in the field to consider the utility of this approach in developing multi-method question evaluations and making sense of findings.

References

Balarajan, M., and Collins, D. (2013). A Review of Questions Asked About State Benefits on the Family Resources Survey. Department for Work and Pensions working paper. https://assets.publishing.service.gov.uk/government/uploads/system/uploads/attachment_data/file/199049/WP115.pdf.

Benitez, I. and Padilla, J. (2014). Cognitive interviewing in mixed research. In: *Cognitive Interviewing Methodology* (eds. K. Miller, S. Wilson, V. Chepp and J. Padilla), 133–152. Hoboken, NJ: Wiley.

Biemer, P. (2004). Modeling measurement error to identify flawed questions. In: *Methods for Testing and Evaluating Survey Questionnaires* (eds. S. Presser, J. Rothgeb, M. Couper, et al.), 225–246. Hoboken, NJ: Wiley.

Blake, M. (2014). Other pretesting methods. In: *Cognitive Interviewing Practice* (ed. D. Collins), 28–56. London: SAGE Publications.

Blanden, A.R. and Rohr, R.E. (2009). Cognitive interview techniques reveal specific behaviors and issues that could affect patient satisfaction relative to hospitalists. *Journal of Hospital Medicine* 4 (9): E1–E6.

Campanelli, P. (2008). Testing survey questions. In: *International Handbook of Survey Methodology* (eds. E. de Leeuw, J. Joop and D. Dillman), 176–200. New York: Taylor & Francis Group.

Cortés, D.E., Gerena, M., Canino, G. et al. (2007). Translation and cultural adaptation of a mental health outcome measure: the basis-R? *Culture, Medicine and Psychiatry* 31 (1): 25–49.

Couper, M. and Kreuter, F. (2013). Using paradata to explore item level response times in surveys. *Journal of the Royal Statistical Society: Series A (Statistics in Society)* 176 (1): 271–286.

Creswell, J. (1995). *Research Design: Qualitative and Quantitative Approaches.* Thousand Oaks, CA: Sage.

DeMaio, T. and Landreth, A. (2004). Do different cognitive interview techniques produce different results? In: *Methods for Testing and Evaluating Survey*

Questionnaires (eds. S. Presser, J. Rothgeb, M.P. Couper, et al.), 89–108. Hoboken, NJ: Wiley.

DeMaio, T. and Rothgeb, J. (1996). Cognitive interviewing techniques: in the lab and in the field. In: *Answering Questions: Methodology for Determining Cognitive and Communicative Processes in Survey Research* (eds. N. Schwarz and S. Sudman), 177–195. San Francisco: Jossey-Bass.

Geisen, E. and Romano-Bergstrom, J. (2017). *Usability Testing for Survey Research*. Cambridge, MA: Elsevier Ltd.

Jenkins, C.R. and Dillman, D.A. (1997). Towards a theory of self-administered questionnaire design. In: *Survey Measurement and Process Quality* (eds. L. Lyberg, P. Biemer, M. Collins, et al.), 165–196. New York: Wiley.

Krosnick, J. (2011). Experiments for evaluating survey questions. In: *Question Evaluation Methods: Contributing to the Science of Data Quality* (eds. J. Madans, K. Miller, A. Maitland and G. Willis), 215–238. Hoboken, NJ: Wiley.

McGee, A., Gray, M., Andrews, F. et al. (2006). National Travel Survey Travel Record Review: Stage 2. Department for Transport working paper. https://webarchive.nationalarchives.gov.uk/+/http://www.dft.gov.uk/pgr/statistics/datatablespublications/personal/methodology/ntsrecords/ntstravelrecord2.pdf.

McGee, A., Gray, M., and Collins, D. (2006). National Travel Survey Travel Record Review: Stage 1. Department for Transport working paper. https://webarchive.nationalarchives.gov.uk/+/http://www.dft.gov.uk/pgr/statistics/datatablespublications/personal/methodology/ntsrecords/ntstravelrecord1.pdf.

Miller, K. (2011). Cognitive interviewing. In: *Question Evaluation Methods: Contributing to the Science of Data Quality* (eds. J. Madans, M. Miller, A. Maitland and G. Willis). Hoboken, NJ: Wiley.

Miller, K., Chepp, V., Willson, S., and Padilla, J. (eds.) (2014). *Cognitive Interviewing Methodology*. Hoboken, NJ: Wiley.

Neuert, C.E. and Lenzner, T. (2016). Incorporating eye tracking into cognitive interviewing to pretest survey questions. *International Journal of Social Research Methodology* 19 (5): 501–519.

Presser, S. and Blair, J. (1994). Survey pretesting: do different methods produce different results. *Sociological Methodology* 24: 73–104. Retrieved from http://www.jstor.org/stable/pdf/270979.pdf.

Presser, S., Couper, M., Lessler, J. et al. (2004). Methods for testing and evaluating survey questions. In: *Methods for Testing and Evaluation Survey Questionnaires* (eds. S. Presser, J. Rothgeb, M. Couper, et al.), 1–22. Hoboken, NJ: Wiley.

Reeve, B. (2011). Applying item response theory for questionnaire evaluation. In: *Question Evaluation Methods: Contributing to the Science of Data Quality* (eds. J. Madans, K. Miller, A. Maitland and G. Willis), 105–124. Hoboken, NJ: Wiley.

Ritchie, J. and Spencer, L. (1994). Qualitative data analysis for applied policy research. In: *Analyzing Qualitative Data* (eds. A. Bryman and R. Burgess), 173–194. London: Routledge.

Rothgeb, J., Willis, G., and Forsyth, B. (2007). Questionnaire pretesting methods: do different techniques and different organizations produce similar results? *Bulletin de Méthodologie Sociologique* 96: 5–31.

Saris, W. and Gallhofer, I. (2007). *Design, Evaluation, and Analysis of Questionnaires for Survey Research*. Hoboken, NJ: Wiley.

Schaeffer, N.C. and Dykema, J. (2011). Response 1 to Fowler's chapter: coding the behavior of interviewers and respondents to evaluate survey questions. In: *Question Evaluation Methods: Contributing to the Science of Data Quality* (eds. J. Madans, K. Miller, A. Maitland and G. Willis), 23–40. Hoboken, NJ: Wiley.

Tashakkori, A. and Creswell, J. (2007). Editorial: exploring the nature of research questions in mixed methods research. *Journal of Mixed Methods Research* 1 (3): 207–211.

Tashakkori, A. and Teddlie, C. (1998). *Mixed Methodology: Combining Qualitative and Quantitative Approaches*. Thousand Oaks, CA: Sage.

Tourangeau, R. (1984). Cognitive science and survey methods: a cognitive perspective. In: *Cognitive Aspects of Survey Design: Building a Bridge between Disciplines* (eds. T. Jabine, M. Straf, J. Tanur and R. Tourangeau), 73–100. Washington DC: National Academy Press.

Willis, G., and Lessler, J. (1999). Question appraisal system QAS-99. Research Triangle Institute. http://appliedresearch.cancer.gov/areas/cognitive/qas99.pdf.

Willis, G., Schechter, S., and Whitaker, K. (1999). A comparison of cognitive interviewing, expert review, and behavior coding: what do they tell us? In: *Proceedings of the ASA Section on Survey Research Methods*. Alexandria, VA: American Statistical Association.

Yan, T., Kreuter, F., and Tourangeau, R. (2012). Evaluating survey questions: a comparison of methods. *Journal of Official Statistics* 28 (4): 503–529.

Part II

Question Characteristics, Response Burden, and Data Quality

6

The Role of Question Characteristics in Designing and Evaluating Survey Questions

Jennifer Dykema[1], Nora Cate Schaeffer[1,2], Dana Garbarski[3], and Michael Hout[4]

[1] *University of Wisconsin Survey Center, University of Wisconsin-Madison, Madison, WI, USA*
[2] *Department of Sociology, University of Wisconsin-Madison, Madison, WI, USA*
[3] *Department of Sociology, Loyola University, Chicago, IL, USA*
[4] *Department of Sociology, New York University, New York, NY, USA*

6.1 Introduction

When developing questions for standardized measurement, question writers often focus on the characteristics of the question. These characteristics include features like the question's length (e.g. number of words or clauses included in the question), the question's level of difficulty (e.g. question's readability level using measures such as the Flesch-Kincaid Grade Level score), or the format for the respondent's answer (e.g. whether the question is formatted for a yes-no response, selection of an ordered category from a rating scale, or a number for a discrete value question). Researchers have developed recommendations for writing questions that are formulated around these characteristics (e.g. data quality is higher for scales that use fully labeled categories in comparison to those that use only partially labeled categories) and that are based on research (or at times just beliefs) about the impact of characteristics on outcomes, including measures of validity or reliability or proxies for data quality, such as differences in response distributions or response times.

However, while questionnaire designers know a lot about the effects of some question characteristics on data quality (e.g. Alwin 2007; Bradburn et al. 2004; Dillman et al. 2014; Fowler 1995; Fowler and Cosenza 2009; Krosnick and Presser 2010; Presser et al. 2004; Saris and Gallhofer 2007; Schaeffer and Presser 2003; Schaeffer and Dykema 2011a, 2015; Schuman and Presser 1981; Tourangeau et al. 2000), the field of questionnaire design is still in the process of developing a comprehensive typology in which question characteristics are cataloged and organized; the effects of question characteristics on interviewers' and respondents' cognitive processing are understood; and the effects

Advances in Questionnaire Design, Development, Evaluation and Testing, First Edition.
Edited by Paul C. Beatty, Debbie Collins, Lyn Kaye, Jose-Luis Padilla, Gordon B. Willis, and Amanda Wilmot.

of question characteristics on data quality are documented. To further this discussion, we present a taxonomy of question characteristics that labels and reinforces potentially important distinctions and groupings among characteristics (Schaeffer and Dykema 2011a), and we briefly describe an interactional model of the response process that positions questions characteristics in the context of other variables and mechanisms that influence data quality.

We have several goals for this chapter. We begin by providing an overview of some of the approaches used to measure, code, and evaluate question characteristics. Work in this area has led us to propose that one way to identify and organize characteristics is to think about the decisions question writers make when writing questions. We refer to these classes of decisions as a *taxonomy of characteristics*. We provide an overview of our taxonomy of characteristics and illustrate how the taxonomy can be used to analyze the effects of two kinds of characteristics. Following that discussion, we explore the effects of some general question characteristics and coding systems in two case studies to examine how well summary measures of question characteristics – or "problems" associated with those characteristics – predict measures of data quality. In the first case study, which uses data from the Wisconsin Longitudinal Study (WLS), a health study of older adults, we examine the effects of question characteristics on indicators of processing difficulties for interviewers and respondents (e.g. question reading accuracy by interviewers and expressions of processing difficulty by respondents). In the second case study, we evaluate the effects of question characteristics on the reliability of individual items using the General Social Survey (GSS). Finally, we conclude by offering some summary comments, notes on limitations presented by this line of research, and some recommendations for future directions.

6.2 Overview of Some of the Approaches Used to Conceptualize, Measure, and Code Question Characteristics

6.2.1 Experimental vs. Observational Approaches

Most studies about how characteristics of survey questions affect data quality are "experimental" or "observational" (see Table 6.1 for an overview). In traditional experiments (see Krosnick 2011 for a review), researchers identify a limited of set of question characteristics to examine – often only one or two. They write alternative forms of questions that incorporate those characteristics, while holding all other characteristics of the question constant, and they administer the questions randomly so each respondent is only exposed to questions with specific characteristics. They then evaluate the effect of the characteristic or characteristics on an outcome or outcomes built into the experimental design. For example, an experiment might examine whether

Table 6.1 General approaches to identify and evaluate question characteristics.

Dimensions	Experimental approach	Observational approach
Scope	Limited characteristics considered	Many characteristics considered
Design	Write alternative question forms that incorporate the characteristic	Code questions along the dimensions of all characteristics
	Other characteristics held constant	Other characteristics vary across questions
Administration	Respondents randomly exposed to a subset of characteristics	Respondents exposed to all questions and characteristics
Evaluation	Assess effect on predetermined criterion	Assess impact on outcome available for all questions

measures of political efficacy using agree-disagree versus construct-specific response formats yield higher concurrent validity by examining which format better predicts self-reported voting (Dykema et al. 2012). Although the experimental approach has been enormously productive, it offers some limitations in generalizing effects found for a specific characteristic to questions that also vary on other characteristics. For example, can experimental findings that compare the results obtained using an odd versus an even number of categories for bipolar items be generalized to unipolar scales or to questions that have the same number of categories but different category labels?

In contrast to the experimental approach, with the observational approach, researchers often use an entire survey conducted for another purpose and so can identify a wider range of characteristics and questions to examine. They code the questions using the characteristics selected, but, because such studies are not experiments, the combinations of the characteristics vary in uncontrolled or haphazard ways across the questions. This allows researchers to examine naturally occurring interactions among question characteristics, but some possible combinations of characteristics may not be observed, or there may be too few observations to estimate a given combination of characteristics reliably (Dykema et al. 2016). In an observational approach, all respondents usually answer all questions and are exposed to multiple question characteristics. Finally, this approach assesses the impact of the characteristics on an outcome available across all questions, such as response times or item-missing data.

6.2.2 Observational Approaches: Ad hoc and System-Based

We identified two kinds of approaches used by observational studies, ad hoc and system-based (Schaeffer and Dykema 2011a). A common way to identify and evaluate question characteristics is to do so in an ad hoc manner to meet the needs of a specific analysis (Couper and Kreuter 2013; Holbrook et al. 2006; Johnson et al. 2015; Kasabian et al. 2014; Kleiner et al. 2015; Knauper et al. 1997; Olson and Smyth 2015; Yan and Tourangeau 2008). These "ad hoc" studies differ in which question characteristics they examine, in how those characteristics are operationalized, and in the outcomes selected for evaluation. For example, these studies have included general characteristics like question type, question length, response format, and instructions to respondents as well as more specific characteristics that are likely to be associated with data quality, such as the inclusion of ambiguous terms, whether scales are fully labeled versus only end-point labeled, and mismatches between the format of the response projected by the question and the actual response categories presented. Outcomes examined included response times (both response latencies and question administration times), various kinds of item-missing responses, and behaviors by respondents and interviewers that research has shown are associated with lower data quality, such as respondents providing answers that cannot be coded and interviewers misreading questions (Schaeffer and Dykema 2011b). (Often, the variation in outcomes examined across the studies makes it difficult to generalize about the effects a given characteristic might have.)

A second approach is to associate the characteristics identified in a systematic analysis of the characteristics of questions with "problems" that the characteristics might cause interviewers or respondents. These "system-based" approaches employ an established scheme to code multiple characteristics in order to identify specific problems the question might cause respondents or interviewers or in order to obtain a "problem" score that can be used to predict the level of data quality the question will yield. Thus, in contrast to ad hoc approaches, system-based approaches provide systematic compilations of individual question characteristics, developed to code characteristics of questions from any type of study.

The ways in which different coding systems have associated question characteristics with problems or data quality vary. These associations may be based on: observations of problems, for example from cognitive interviews; analyses that use criteria such as reliability or validity; or speculation about problems questions might have for interviewers or respondents – often guided by the four-stage response process model of comprehension, retrieval, judgment, and response (Tourangeau et al. 2000) – with particular emphasis on problems due to comprehension and retrieval.

System-based approaches, like ad hoc approaches, provide a method for identifying and aggregating many discrete and specific question characteristics. They are, typically, more comprehensive than ad hoc approaches, and potentially offer powerful methods for assessing and revising draft questions during the critical development phase when no evaluation criteria are available, provided the coding schemes can be applied reliably and the characteristics they identify are indeed associated with data quality. Three prominent system-based approaches include:

1. *Problem Classification Coding Scheme (CCS).* The CCS provides a system for coding 28 problems, such as "vague topic or term" or "question too long," that are grouped under the stages of answering a question (Forsyth et al. 2004).
2. *Question Appraisal System (QAS).* The QAS codes for 27 problems a question could present in the following categories: reading, instructions, clarity, assumptions, knowledge/memory, sensitivity bias, response categories, and other (Willis 2005; Willis and Lessler 1999). For example, a type of characteristic related to response categories coded in the QAS is whether there is a mismatch between the question and response categories.
3. *Survey Quality Predictor (SQP).* The SQP is a very comprehensive online tool for coding variables related to the language, structure, content, and administration of a question. Depending on the content and structure of the question, a human coder can code over 50 different question characteristics. After the item is coded, the online tool produces predicted reliability, validity, and quality. The predictions use coefficients that were previously estimated in empirical analyses of split-ballot multitrait multimethod (MTMM) experiments; the experiments used the same analysis of question characteristics, and the data structure allowed measures of data quality to be estimated (see analysis in Saris and Gallhofer 2007 and http://sqp.upf.edu).

The CCS and QAS, and to a lesser extent the SQP, also include categories for features that affect interviewers, such as whether the coder determines it would be difficult for interviewers to read the question uniformly to respondents or whether the question includes a showcard. These three approaches are comprehensive, are accessible to researchers, and have potential for improving question development, and so our case studies evaluate them, using observational data.

6.2.3 Question Characteristics in Context: How They Affect Data Quality

The characteristics of survey questions are embedded in a complex set of social and cognitive processes in the interview that results in answers (data) of varying quality. The interactional model of the question-answer sequence depicted in Figure 6.1 summarizes paths that link the practices of standardization, characteristics of questions, cognitive processing of survey participants, and the production of a survey answer (see earlier version in Schaeffer and Dykema 2011b).[1] This model is informed by a variety of sources: evidence

1 Objects in boxes are more or less directly observable, while objects in circles are inferred. Although Figure 6.1 focuses on data obtained through an interview, a similar figure could be developed for self-administered data (see, for example, Toepoel and Dillman 2011).

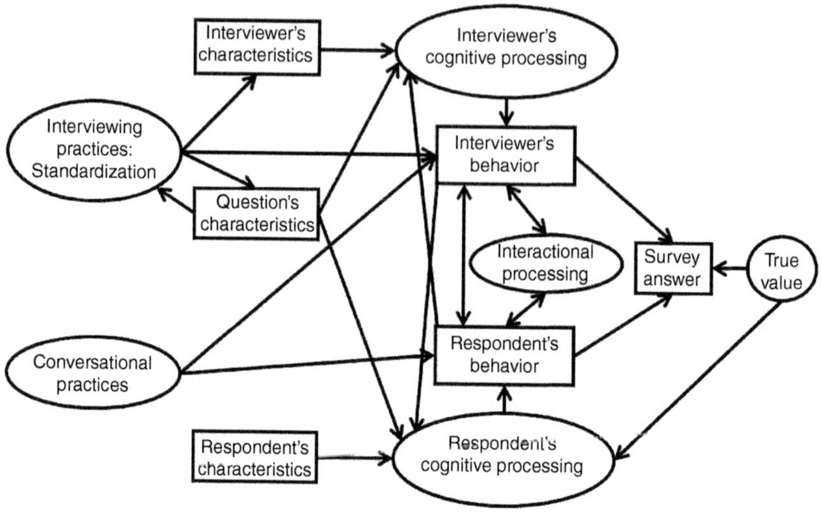

Figure 6.1 Interactional model of the question-answer sequence.

of interviewer variance, which motivates the practices of standardization; evidence that behaviors such as response latency are associated with cognitive processing; and the importance of conversational practices for measurement (Garbarski et al. 2011; Schaeffer et al. 2010; Schaeffer and Maynard 2002; Schwarz 1996).

The model takes as a starting point that there are external – and at times contradictory – influences of standardization and conversational practices. Standardization has an impact on interviewers through their role characteristics – their understanding of the rules of standardized interviewing – and on their behavior (see discussion of a "typology of interviewer characteristics" in Schaeffer et al. 2010). Standardization or standardized measurement also affects the characteristics of questions and how we write and structure survey questions. Conversely, we draw an arrow from question characteristics back to standardized interviewing practices because a limited set of interviewing rules, such as how to follow up when respondents provide answers that cannot be coded, are based on specific question characteristics. Conversational practices – practices of talk in a variety of social situations – also influence both the interviewer's and respondent's behavior (and may also influence behavior and processing in questions written for self-administration [see Schwarz 1996]).

Next, the model considers the impact of the interviewer's, question's, and respondent's characteristics on the interviewer's and respondent's cognitive processing. The interviewer's cognitive processing is affected by her characteristics – including her integration of knowledge about standardization, her understanding of her role, and her personal characteristics (e.g.

experience, competence) – and the characteristics of the question. Similarly, the respondent's cognitive processing is affected by the characteristics of the question and personal characteristics such as cognitive ability and motivational level (Krosnick 1991).

These sources influence the production of answers within the interaction between the interviewer and respondent. Processing occurs at three sites – within the interviewer, within the respondent, and through their interaction – all of which are directly (for the interviewer's and respondent's cognitive processing) and indirectly affected by characteristics of the question. For a given question, the interviewer's behavior before, during, and after the administration of the question – such as whether she reads the question exactly as worded and probes nondirectively – is influenced by features such as the number of words in the question. The stages of the respondent's cognitive processing (e.g. comprehensive, retrieval, judgment, and response; Cannell et al. 1981; Tourangeau et al. 2000) are affected directly by question characteristics and by the interviewer's actual interviewing practices and behaviors. The arrows from the interviewer's behavior to the respondent's behavior and back indicate that processing is interactional and each party may respond to behaviors produced by the other. The processing itself is internal and interactional: we cannot observe it, only its outcomes. Most frequently, an outcome of this interaction is the respondent's answer, which will vary in terms of its validity and reliability based in part on the influences and processes described. We also draw arrows from the true value of the construct being measured to the survey answer and the respondent's cognitive processing to emphasize that a survey response is also a product of its true value (see Groves 1991).

6.2.4 Measures of Data Quality and Their Proxies

Observational studies, like our case studies, require outcomes for evaluation that can be assessed for all individual items, items which can vary substantially based on their topic, question type, response format, and so on. One method for assessing the impact of question characteristics, which we employ in the first case study, incorporates interviewer-respondent interactional outcomes (i.e. outcomes obtained from coding the observable behaviors of interviewers and respondents, often referred to as *interaction* or *behavior coding*). Motivation for examining this interaction is provided by Figure 6.1 and supported by analyses of the relationship among the characteristics of survey questions, the behavior of interviewers, the behavior and cognitive processing of respondents, and the measurement properties of the resulting answers (Dykema et al. 1997; Fowler and Cannell 1996; Schaeffer and Dykema 2011a; see also Ongena and Dijkstra 2007). Some of the processes displayed in Figure 6.1 find expression in behavior, with prior research finding support for the association between specific behaviors and measures of both reliability and validity (summary in

Schaeffer and Dykema 2011b). As a criterion, behavior has the advantage that it can be observed and coded for all items from any interviewer-administered instrument and provides information about the performance of individual items. In addition, although recording, transcribing, and coding interactions is not cheap, it may be less expensive than other designs for assessing data quality. There are disadvantages, however: some problems may be internal to the participants and leave no trace in the interaction. In addition, behaviors may be influenced by factors that are not direct influences on data quality, and even if this measurement error is random, it may attenuate observed relationships.

Behaviors that have been examined in prior research include whether interviewers read the questions exactly as worded or engage in probing (Mangione et al. 1992), and whether respondents exhibit behaviors that could indicate problems in processing the question. Although the respondent's disfluencies and response latency have received particular attention (e.g. Bassili and Scott 1996; Schober et al. 2012; Maitland and Presser 2016), other behaviors, such as providing qualifications (e.g. "probably" or "about"), seeking clarification of terms, and expressing uncertainty (e.g. saying "don't know" or the equivalent in addition to a codable answer) have also been shown to be associated with data quality (Dykema et al. 1997; Schaeffer and Dykema 2004).

Estimates of an item's reliability and validity potentially offer more direct measures of data quality (e.g. Saris and Gallhofer 2007; Alwin 2007), but different measures of reliability and validity also vary in their quality. Estimates of the reliability of a set of items from a scale, such as Cronbach's alpha, include correlated error variance and do not provide values for individual items, and estimated test-retest reliabilities, over the short intervals that are commonly used, may be too compromised by memory or reliable method effects to provide a strong criterion (Alwin 2007). Two designs that provide estimates of quality for individual items are those based on the MTMM matrix, which compares measures of a set of constructs using different methods in a strong experimental design (Rodgers et al. 1992; Saris and Gallhofer 2007) and those based on a minimum of three waves of panel data (Alwin and Krosnick 1991; Alwin 2007; Heise 1969; Hout and Hastings 2016), although there is recurring debate about the advantages of each (Alwin 2011; Saris 2012). If the respondent's true score is stable, two measures can identify reliability (assuming the respondent is not influenced by their initial answer), and three measures can identify reliability even in the presence of change. For our second case study, we selected Alwin-Heise estimates, because that is the most used method in previous research (see the discussion in Hout and Hastings (2016)).[2]

2 Hout and Hastings also discuss reliability estimates based on within-person variance; those estimates are lower, on average, than the Alwin-Heise estimates (0.96), but they are highly correlated.

6.3 Taxonomy of Question Characteristics

6.3.1 Overview

There is a close relationship among the characteristics of survey questions and the decisions faced by question writers – in some sense, each question characteristic presents the question writer with a decision (e.g. how many response categories and whether to include a middle category with a rating scale). In the long run, improvements in data quality rely on an understanding of these characteristics and their inter-relationships. Based on an analysis of decisions that question writers have to make when drafting items, we proposed a *taxonomy of characteristics* (Schaeffer and Dykema 2011a).[3] The taxonomy, which we highlight in Table 6.2 and describe in more detail here, summarizes classes of characteristics that affect the task faced by respondents and interviewers and the reliability and validity of the data. As presented, the classes of characteristics are ordered roughly from most abstract to most specific:

- *Question topic* can refer to broad domains, like "health," or more refined domains, like "comparison of own health to others." Topics are important because they vary along dimensions that affect measurement, including how sensitive and difficult to report about they might be.
- *Question type* refers to groups of questions that require similar decisions, have a similar structure, or share response dimensions. Two important types of questions are questions about events and behaviors and questions that ask for evaluations or judgments (Schaeffer and Dykema 2015). Another common question type is a classification question, a question that requires respondents to classify themselves based on past behaviors or current statuses such as "What is your current marital status?"
- *Response dimension* (see too Fowler and Cosenza 2009) is the dimension or continuum that the question asks the respondent to think about in constructing an answer. For example, for questions about events and behaviors, some commonly used response dimensions include whether an event occurred (occurrence), how often it occurred (frequency), how long it lasted (duration), and when it occurred (date or timing). For questions about evaluations and judgments that use rating scales, response dimensions can

3 See Schaeffer and Dykema (2011a, 2015) for a detailed discussion regarding many of the decisions that a question writer must make when writing a question to measure an event or behavior versus an evaluation or judgment. For events and behaviors, for example, decisions have to be made about whether to include a filter question to establish relevance, how to name the target event or behavior and whether to include a definition of the event or behavior, how and where to specify the reference period, and what response dimension to use. If the response dimension asks about frequency, additional decisions need to be made about whether to ask about absolute or relative frequencies, whether to include an occurrence filter, and whether to format the question using categories or with a discrete value (open) response format.

establish whether the respondent's evaluation of the target object is positive or negative (valence or direction), the degree to which the respondent holds the evaluation (intensity), the certainty or strength of the evaluation, and how important the target object is (importance).

- Most survey questions are written about target objects – events, objects to be rated, policy positions, and so on – and *conceptualization and operationalization of the target object* refers to how to label the target object and the response dimension. For example, a question about "doctors' visits" could label the target object as "doctor," "medical doctor," or "health care provider," and the question could ask for the number of visits in the past year, a response dimension that captures frequency.
- *Question structure* concerns the relationship between questions. A common example is the decision about whether or not to use an occurrence filter followed by a frequency question versus a frequency question alone or about whether to precede a rating scale with a "no opinion filter." We discuss the decision about whether to use a filter question in Section 6.3.2. Another important type of question structure is a *battery* – a set of items that share an introduction and response categories.
- *Question specifications* capture decisions about the inclusion and wording of elements that question writers use to aid respondents in comprehension or recall. These include such elements as task instructions, definitions, examples, parenthetical statements, and reference periods.
- *Response format or question form* refers to the type of answer format projected by a question and decisions about which response format to use. Some common formats include the yes-no format, selection from ordered categories (such as a rating scale), and the discrete value format (such as a "number of times"). Three other common response formats include selection from unordered categories (such as from a list of choices); record-verbatim open questions (in which the interviewer records in the respondent's own words); and field-coded open questions (in which the respondent is asked an open question and an interviewer codes their response into one of a set of categories).
- Some important decisions about *response categories* concern the type, number, and labeling of the categories. However, the decisions about response categories for questions about events and behaviors are different from those for evaluations and judgments that use rating scales, an issue we return to in more depth in Section 6.3.2.
- A substantial group of question characteristics concern *question wording* and language, particularly language issues that affect comprehension (e.g. because the vocabulary or grammar is complex) or retrieval (e.g. because the question provides cues or time for retrieval). These characteristics include

Table 6.2 Overview of the taxonomy of characteristics.

Classes of characteristics	Examples of individual features
Question topic	Health, politics
Question type	Event or behavior, evaluation or judgment, classification
Response dimension	For events and behaviors: occurrence, frequency, duration, date, or timing For evaluations and judgments: valence or direction, intensity, certainty, strength, importance
Conceptualization and operationalization of the target object	Labels for target object (e.g. "doctor" versus "health care provider") and response dimension (e.g. number of visits in the past year)
Question structure	Filter and follow-up question, battery
Question specifications	Task instructions, definitions, examples, parenthetical statements, reference periods
Response format or question form	Yes-no, selection from ordered categories, selection from unordered categories, discrete value, record-verbatim open, field-coded open
Response categories	Type, number, and labeling
Question wording	Length, readability, vocabulary difficulty, term ambiguity, grammatical complexity
Question implementation	Orientation of response categories on page/screen, instructions to interviewers

features such as question length; reading difficulty of the vocabulary; number of words that are vague, ambiguous, or technical; and grammatical complexity (see, for example, Graesser et al. 2006).

- An item is ultimately tailored in some way to fit the mode in which it is presented. Features of implementation include, for example, orientation of response categories on the page or screen and the placement and presentation of instructions to interviewers. Understanding the impact of the way an item is *implemented* – that is, actually administered to a respondent – is increasingly important as we field more mixed-mode designs for which measurement needs to be comparable across different implementations (Dillman et al. 2014). Issues surrounding implementation also are important for studying question characteristics as research increasingly demonstrates that characteristics related to visual design for self-administered questionnaires – but also interviewer-administered questionnaires – affect the interpretability of survey questions and quality of survey responses (see Dillman et al. 2014; Toepoel and Dillman 2011).

6.3.2 Illustrations: Using the Taxonomy to Analyze Effects of Question Characteristics

Using information about how different classes of characteristics are related to one another, we offer two examples to illustrate (i) how the meaning of a question characteristic, such as a response category label, may depend on another characteristic, such as question type, and, as a result (ii) how the impact of one question characteristic on data quality may depend on other characteristics.

One of the first decisions faced in writing questions about events and behaviors or about evaluations and judgments is whether or not to use a filter question (see the decision flowcharts in Schaeffer and Dykema 2011a, 2015). A question that asks for absolute frequencies, that is, for the number of times an event occurred, may be preceded by an occurrence filter question, which asks if the event occurred, as illustrated in the left panel of Figure 6.2 for Question Structure 1 (see Knauper (1998) for an example). Questions that ask for evaluations sometimes also use filter questions; one sort of filter asks whether a quality, "concern" for example, is present, that is, whether the respondent's level of concern exceeds some threshold for saying "yes" the concern is present. This sort of "threshold filter" is illustrated in the right panel of Figure 6.2 for Question Structure 1 (see Sterngold et al. (1994) for an example). In each panel, Question Structure 1 uses a filter question and Question Structure 2 does not. In both of these cases, the writer decides about something that might be labeled a "filter question," but the conceptualization and operationalization of the target objects, the response dimensions, and the cognitive processes differ for the questions in each panel. As a result, we might expect that the meaning of the filter question and the impact of using a filter question on the quality of the resulting data might be different for the two types of questions – or at least we might not want to assume that the impact of using a filter question would be the same for these two questions by treating "filter question present" as a characteristic of a question without controlling for the type of question.

A second example, suggested by Figure 6.3, illustrates that the meaning of the decision about whether or not to label response categories depends on the type of question. Version 1 of the item in the left panel is a discrete-value question that asks for the absolute frequency or count of an event. The categories are "unlabeled" because the numbers are counts; unless the question is changed to a selection question and the frequencies are grouped in some way, "category labels" are not relevant. Version 1 of the item in the right panel is one of the core items in the GSS and presents what might be a bipolar scale – that is, the response scale has a sort of negative quality on one end and a positive quality on the other. There are verbal labels only for the extreme categories, or endpoints, so most of the response categories have only numeric labels. For Version 2 of the item in the right panel, we have sketched in a fully labeled version of the response scale (simply indicating that there would be labels, rather

Type of Question	
Events & Behaviors	**Evaluations & Judgments**
Question Structure 1: With Filter	**Question Structure 1: With Filter**
During the last twelve months, did you go to a museum? (Yes/No) **IF YES:** During the last twelve months, how many times did you go to amuseum?	Are you concerned about pesticides in your food? (Yes/No) **IF YES:** How concerned are you about pesticides in your food: a little bit, somewhat, very, or extremely?
Question Structure 2: Without filter	**Question Structure 2: Without filter**
During the last twelve months, how many times, if any, did you go to a museum?	How concerned are you about pesticides in your food: not at all, a little bit, somewhat, very, or extremely?

Figure 6.2 Decision: filter questions for two types of questions.

than trying to choose specific labels). Comparing Versions 1 and 2 for the two types of question in Figure 6.3 indicates that what it means to be an "unlabeled response category" differs for these two question types. We can imagine comparing Version 1 of the evaluation question on the right with a version that had verbal labels for every point on the scale. Note, however, that it is difficult to construct a comparable Version 2 for the discrete-value question event question on the left.

These examples illustrate that question characteristics – like the use of a filter question – potentially have different consequences for the quality of the resulting data depending on the question type. Another way of saying this is that there may be statistical interactions among question characteristics or classes of characteristics. In addition, some decisions – like decisions about category labels – are relevant only for some question types or some question forms. In other words, there are structural dependencies among question characteristics. We describe these challenges for three reasons: (i) These structural dependencies need to be taken into account during analysis. In the case of scale polarity, for example, if we model the impact of question characteristics on reliability, and we include in the analysis both questions about events and questions that ask for evaluations, all the bipolar questions in the analysis will be questions that ask for evaluations. The cell of "bipolar" questions about "events" is empty. (ii) Of course, whether or not we can construct powerful tests of these interactions depends on the structure of the data, that is, the number of questions and the combination of question characteristics in an analysis. In observational studies in particular,

Type of Question

Events & Behaviors	Evaluations & Judgments
Response Format: Discrete Value	**Response Format: Selection with Ordered Categories**
Version 1: Unlabeled Categories	**Version 1: Unlabeled Categories**[a]
In the last 12 months, how many times, if any, did you visit a museum? ____ NUMBER OF TIMES	Do people in these groups tend to be unintelligent or intelligent? HAND CARD A15 — Unintelligent 1 2 3 4 5 6 7 Intelligent; DON'T KNOW 8; RATING A. Where would you rate whites in general on this scale? B. Blacks?
Version 2: Labeled Categories	**Version 2: Labeled Categories**
---Null---	Do people in these groups tend to be unintelligent or intelligent? HAND CARD A15 — Label 1, Label 2, Label 3, Label 4, Label 5, Label 6, Label 7: 1 2 3 4 5 6 7; DON'T KNOW 8; RATING A. Where would you rate whites in general on this scale? B. Blacks?

[a]Illustrative questionnaire item from the General Social Survey.

Figure 6.3 Decision: category labels for two types of questions.

the joint distribution of question characteristics affects whether group sizes are sufficient to estimate main effects and interactions (Dykema et al. 2016). (iii) The applicability of the results of previous studies may depend on which combinations of characteristics were taken into account in those analyses, and so these structural dependencies and potential interactions among question characteristics are relevant when we try to apply results of previous studies.

6.4 Case Studies

Next we turn to our two case studies, which serve several purposes. First, they illustrate both the benefits and limitations of observational studies. They include, for example, a wide range of items in a realistic field setting without the benefit of randomly assigning questions that vary on selected characteristics to respondents. Second, the case studies incorporate two of the different approaches to evaluating data quality that we discussed in Section 6.2.4, the behavior of interviewers and respondents (Case 1) and item reliability (Case 2). Finally, the case studies include measures of question characteristics that are accessible to question writers who want to evaluate candidate questions: the three system-based approaches (e.g. CCS, QAS, and SQP) discussed in

Table 6.3 Overview of the design and methods employed in the case studies.

	Wisconsin Longitudinal Study (WLS)	General Social Survey (GSS)
Sample	1/3 random sample of Wisconsin high school class of 1957	National area probability sample, noninstitutionalized US population
Mode	Telephone	Face-to-face
Dates	2003–2005	2006–2010, 2008–2012, and 2010–2014
Question topics	Health	Varied, public opinion
Question types	Events and behaviors	Varied, 2/3 are evaluations
Outcomes	(1) Interviewers' question-reading accuracy (exact versus not exact) (2) Respondents' "problem" behaviors (any versus none)	Individual question reliabilities
Unit of analysis	Question-answer sequences (n = 8150)	Questions (n = 196–205)
Analytic method	Cross-classified random effects logistic models	Regression
Further details	Dykema et al. 2016	Schaeffer et al. 2015

Section 6.2.2 as well as other question characteristics (e.g. number of words, readability, and comprehension difficulty).

Table 6.3 provides an overview of the design and methods employed by the case studies. Our analysis presents bivariate relationships rather than multivariate relationships because a practitioner testing items with limited resources would want to know which of these methods (e.g. which discrete question characteristics or system-based approaches) to invest in to get the best evaluation of their questions or to set priorities about which questions to devote resources to improving. For the practitioner, this question – which method or methods are most effective at identifying questions that require revision or disposal – is best answered by examining the relationship between the individual method and the data quality outcome.

6.4.1 Case Study 1: Wisconsin Longitudinal Study (WLS)

6.4.1.1 Survey Data

Data for our first case study are provided by the 2003–2005 telephone administration of the WLS, a longitudinal study of a one-third random sample of the 1957 class of Wisconsin high school graduates (n = 10,317) (Sewell et al. 2003; AAPOR RR2 = 80%). Our analytic sample of 355 cases was obtained by randomly selecting interviewers and then randomly selecting respondents within interviewers, stratified by the respondent's cognitive ability measured in adolescence (see Dykema et al. (2016) and Garbarski et al. (2011) for further details). Comparisons between our analytic sample and the entire sample indicated they were similar across a range of sociodemographic characteristics.

We examine interviewer-respondent interaction during the administration of a series of health questions. These questions were the first substantive module in the survey and contained items about self-rated health, physical and mental health functioning, and diagnosed health conditions (http://www.ssc.wisc.edu/wlsresearch/documentation/flowcharts/Full_Instrumentation_1957_2010_vers8_Final.pdf). Although many of the questions are from standardized instruments, such as the Health Utilities Index, prior qualitative work demonstrated they were written in ways difficult for older respondents to process because the questions contained complicated or unfamiliar terms or because the questions were written using complex sentence structures or complex syntax. While 76 questions were included in this module, respondents received fewer questions because of skip patterns, and we limited analysis to the 23 questions administered to all respondents.

6.4.1.2 Interaction Coding

We identified over 100 behaviors for coding based on conversation analysis of a subset of the transcripts, detailed examination of the interviews used for the conversation analysis, and the literature on interaction in survey interviews (see Dykema et al. 2009). Some codes were specific to the interviewer (e.g. codes for question-reading accuracy and follow-up behaviors); some codes were specific to the respondent (e.g. codes for the adequacy of the respondent's answer to the survey question including whether the answer was codable using the response format provided; whether the answer contained other kinds of talk in addition to a codable answer, such as considerations; or whether the answer was an uncodable report); and some codes were relevant for both interviewers and respondents, such as whether they uttered tokens (e.g. "um" or "ah") while speaking. Coding was done from transcripts using the Sequence Viewer program (Wil Dijkstra, http://www.sequenceviewer.nl) by five former WLS interviewers, who received extensive training. To assess intercoder reliability, a sample of 30 cases was independently double-coded by five coders, and a measure of inter-rater agreement, Cohen's kappa, was produced. While kappa values varied across the behaviors coded (available upon request), the average overall kappa for all coded events in the health section was high at 0.861. Our unit of analysis is the question-answer sequence ($n = 8150$), which begins with the interviewer's question reading and ends with the last utterance spoken by the interviewer or respondent before the interviewer reads the next question.

6.4.1.3 Measures

Table 6.4 shows descriptive statistics for the variables in the analysis. Two binary outcomes serve as indicators of problems with cognitive processing

Table 6.4 Descriptive statistics for outcomes, question characteristics, and coding systems.

	Mean or proportion	Standard deviation	Min	Max	n
Wisconsin Longitudinal Study					
Dependent variable					
Interviewer exact question reading (vs. not)	71.03		0	1	8150
Any respondent problem behaviors (vs. none)	25.23		0	1	8150
Question characteristics and coding systems					
Response format					
Yes/No	82.61				19
Other	17.39				4
Question length (number of words)	18.47	10.24	1	43	23
Question readability (Flesch Grade Level)	8.74	4.66	0	20.20	23
Comprehension problems (QUAID)	1.91	1.41	0	6	23
Problem Classification Coding System (CCS)	3.96	1.68	0	7	23
Question Appraisal System (QAS)	1.78	0.83	1	3	23
Survey Quality Predictor (SQP)	0.53	0.93	0.48	0.58	23
General Social Survey					
Dependent variable					
Reliability	0.71	0.15	0.32	1	205
Question characteristics and coding systems					
Response format					205
Yes/No	0.18				
Other	0.82				
Question length (number of words)	21.91	20.22	1	120	205
Question readability (Flesch Grade Level)	7.74	5.38	0	32	205
Comprehension problems (QUAID)	10.64	11.01	0	51	205
Problem Classification Coding System (CCS)	0.95	0.85	0	3	205
Question Appraisal System (QAS)	1.01	0.97	0	4	205
Survey Quality Predictor (SQP)	0.67	0.07	0.49	0.82	195

and are associated with measurement error. First, we assess how accurately interviewers read the questions. We code readings as exact versus any change. Changes included slight change (i.e. diverges slightly from the script but does not change the meaning of the script), major change (i.e. diverges from the script in a way that changes the meaning), or verification (i.e. alters wording of initial question asking to take into account information provided earlier). Second, for respondents, we look at an index of behaviors indicative of potential problems answering the question, including providing reports, considerations, expressions of uncertainty, and other uncodable answers. For the analysis, the index is collapsed to a binary indicator of no problems versus one or more problems.

Questions are classified based on their values for several general characteristics and also for three system-based approaches (see Table 6.5 for a comparison of characteristic values for the question characteristics and system-based approaches for two questions analyzed in this study). The general characteristics we examine are response format and three variables that capture different aspects of question wording, including question length, readability, and comprehension difficulty. As discussed during our overview of the *taxonomy of characteristics*, response format refers to how the question is formatted for response. Two formats appear in our data: yes-no questions that provide "yes" or "no" as categories and selection questions that provide a set of predetermined categories. Based on past research, which indicates selection questions are harder to process than yes-no questions, we predicted interviewers would be more likely to make a reading error with a selection question and respondents would be more likely to display a problem behavior when answering a selection question. Question length is measured as the raw number of words in the question. Question readability – which indicates comprehension difficulty in a passage of text – in this case, the question – is measured using the Flesch-Kincaid Grade Level score. A final characteristic related to question wording that we examine is comprehension difficulty, which we assess using the Question Understanding Aid or QUAID (http://quaid.cohmetrix.com, Graesser et al. 2006). QUAID is an online tool in which the user enters their question text and the program returns a list of problems, such as whether the question contains unfamiliar technical terms, vague or imprecise relative terms, vague or ambiguous noun phrases, complex syntax, or working memory overload. We tallied the number of problems across categories in the analysis (see also Olson 2010). Based on past research, we expected that longer questions, questions that are harder to read, and questions with higher comprehension difficulty scores would decrease the likelihood that interviewers would read the question exactly as worded and increase the likelihood that respondents would display a problem behavior.

The systems we use to classify characteristics in a more comprehensive way include the CCS, the QAS, and the SQP discussed in Section 6.2.2. For the CCS and QAS, we adopted a straightforward sum of the problems identified by each system to provide a global evaluation of each system's ability to predict data quality. In addition, practitioners seeking to set priorities for question testing and revision are likely to benefit from such a straightforward operationalization of total problems – if it is useful in predicting data quality. For the SQP we use the value of the "quality estimate" generated by the online program. Based on past research, we expected that interviewers would be less likely to

Table 6.5 Relative values for general question characteristics and system-based approaches for two questions from the Wisconsin Longitudinal Study.

	Question length	Question readability: Flesch Grade Level score	Comprehension problem score: Question Understanding Aid (QUAID)	Problem score: Problem Classification Coding Scheme (CCS)	Problem score: Question Appraisal System (QAS)	Data quality prediction: Survey Quality Predictor (SQP)
Question 1						
Because of any impairment or health problem, do you need the help of other persons in handling your routine needs, such as everyday household chores, doing necessary business, shopping, or getting around for other purposes?						
	Long	Low	Medium	High	High	Low
Question 2						
Have you ever been diagnosed with a mental illness?						
	Short	High	Medium	Low	Low	Medium

read questions exactly as worded when problem tallies from the CCS and QAS were higher and when quality scores from the SQP were lower; similarly, we expected that respondents would be more likely to display a problem behavior with questions with higher CCS and QAS scores and lower SQP quality scores.

6.4.1.4 Analysis

To account for the complicated crossed and nested structure of the data, we implement a mixed-effects model with a variance structure that uses crossed random effects. Models include random effects for interviewers, questions, and respondents (nested within interviewers and crossed with question). Question characteristics are modeled as fixed effects that are nested within and crossed with the random effects. The response variables are binary; logit models were computed in R using the glmer function from the lme4 package.

6.4.1.5 Results

Next, we turn to the results from bivariate cross-classified random effects logistic models of interviewer and respondent outcomes on question characteristics. Results are presented in Table 6.6 and shown separately for the outcomes of exact question reading by interviewers and any problem behaviors by respondents. The odds ratio provides the proportional change in the odds of interviewers reading exactly or respondents exhibiting a problem behavior. For example, contrary to predictions, whether the question is formatted for a yes-no response or a different response – in the case of the WLS questions these other response formats were for selection questions – does not significantly affect interviewers' ability to deliver the question as worded. Looking at the results for the other question characteristics and coding systems, however, we find a relatively consistent pattern associated with interviewers' exact question reading. For the question characteristics and coding systems in which larger values predict interviewers will have more difficulty reading the question exactly as worded, increases are associated with lower odds of exact reading, and the effects are significant for question length, comprehension difficulty as measured with QUAID and the CCS. The SQP is coded so that higher values are associated with greater quality, and here we see that a unit increase in the SQP score is associated with increased odds of the interviewer reading the question exactly.

The second set of results is for respondents' problem behaviors (a dummy variable collapsed from an index of behaviors exhibited by the respondent that may be associated with response errors, such as providing an uncodable answer or qualifying an answer). Again, the results are quite consistent across the various approaches to coding question characteristics. The odds of respondents exhibiting any problem behaviors are significantly higher for selection questions, longer questions, questions with higher levels of

Table 6.6 Results from bivariate models regressing outcome on question characteristics.

	Outcome					
	Wisconsin Longitudinal Study[a)]				General Social Survey[b)]	
	Interviewer exact question reading		Any respondent problem behaviors		Item reliability	
Question characteristic and coding system	Odds ratio	95% Confidence interval	Odds ratio	95% Confidence interval	Coefficient	Standard error
Response format is other [vs. yes/no][c)]	1.63	[−0.17, 1.17]	3.38*	[0.16, 2.31]	−0.131***	0.026
Question length (number of words)	0.91***	[0.89, 0.94]	1.10***	[1.06, 1.15]	−0.001**	0.001
Question readability (Flesch Grade Level)	0.96	[0.87, 1.06]	1.05	[0.93, 1.19]	−0.003	0.002
Comprehension problems (QUAID)	0.71*	[0.53, 0.94]	1.80***	[1.20, 2.48]	−0.007***	0.001
Problem Classification Coding System (CCS)	0.71**	[0.57, 0.89]	1.40*	[1.02, 1.90]	−0.052***	0.012
Question Appraisal System (QAS)	0.74	[0.54, 1.02]	1.20	[0.78, 1.85]	−0.073***	0.010
Survey Quality Predictor (SQP)	1.71**	[1.17, 2.53]	0.74	[0.43, 1.30]	−0.391**	0.150

*$p < 0.05$, **$p < 0.01$, ***$p < 0.001$.

a) Models are cross-classified random effects logistic regressions. Cells show odds ratios. 23 items are included in the analysis.

b) Models are linear regressions. Cells present regression coefficients. The number of items varies between 196 and 205 depending on the outcome.

c) The omitted category is shown in square brackets.

comprehension difficulty, and questions with more problematic scores from the CCS. Flesch-Kincaid Grade Level, QAS, and SQP are not significantly associated with respondents' problem behaviors.

We provide a more in-depth summary of the results in the discussion, after we present the results from our second case study.

6.4.2 Case Study 2: General Social Survey (GSS)

6.4.2.1 Survey Data

For our second case study we analyze estimates of reliability for the core items in the GSS. The GSS uses an area probability sample of the non-institutionalized population of adults in the United States. Respondents are randomly selected within household and were interviewed in person in all three waves. Response rates ranged from 71% to 69% for the full sample at the first wave. (http://gss.norc.org/documents/codebook/GSS_Codebook_AppendixA.pdf). The estimates use three three-wave panels of the GSS that were conducted at two-year intervals, beginning in 2006, 2008, and 2010 (Hout and Hastings 2016). As summarized, because items were asked in three waves, it is possible to obtain estimates of reliabilities for the individual items. The Alwin-Heise reliabilities analyzed here were based on polychoric correlations (or Pearson correlations for items with 15 or more categories) and average estimates from each of the three waves. Because we focus on how questions affect answers, we omit from our analysis items answered by the interviewer (e.g. the respondent's sex), scales constructed from sets of items, and measures from the household enumeration form. For categorical items, we included reliability for one of the categories, choosing when possible the category we judged least socially desirable. Because SQP does not provide estimates for field-coded open questions, we excluded those questions from the SQP analysis.

6.4.2.2 Measures, Analysis, and Results

For the GSS we use measures of question characteristics that are comparable to those we used for the WLS, with modifications to take into account differences between the studies. The GSS includes a larger number of items than the WLS, and the items are more heterogeneous with respect to question type and response format. We relied on the GSS documentation of the instruments to assess characteristics of questions, but note that most available GSS documentation is based on earlier paper versions of the instrument and does not include the screens that the interviewer currently sees. The GSS also includes a number of batteries, sets of items that share a preamble and response categories. Because of the preamble, the first item in a battery typically includes more words than subsequent items. The use of batteries presumes that the respondent retains information from the preamble, but we use only the words that the available documentation suggests were actually read to the respondent at a given item.

In contrast to the WLS, the large majority of items in the GSS – both yes-no and selection questions – ask about attitudes rather than health conditions, and the items are heterogeneous in content. Because the GSS offers a larger number of items, the analysis affords more precision and the ability to detect smaller differences and patterns. Despite differences in the outcome and models – because the dependent variable is estimated reliability, most of the relationships are predicted to be negative – the results for the GSS in Table 6.6 are generally consistent with those from the WLS. In the GSS, response formats other than yes-no are associated with lowered reliability. The number of words read to the respondent, and the number of problematic question characteristics detected by QUAID, CCS, and QAS each have a significant negative relationship with reliability: as the number of problematic characteristics identified by the method increases, reliability decreases. As was the case for the WLS, the Flesch scores do not predict reliability. Surprisingly, the reliabilities predicted by SQP are negatively associated with the actual estimated reliabilities.

6.5 Discussion

A good survey question obtains data that are reliable and valid, and we address in this chapter issues researchers face in trying to write good questions. The taxonomy and conceptual framework help identify the decisions that arise when writing questions and guide the application of existing literature to those decisions. Researchers also need methods to set priorities for the limited resources available for question development and testing. Our comparative analysis of several prominent methods – QUAID, CCS, QAS, and SQP – suggest that these methods successfully identify questions whose characteristics would reduce the quality of measurement (see too Forsyth et al. 2004). Development and testing could then focus on modifying these problematic question characteristics.

Our two case studies were conducted with different populations; they use different modes; they contain different question types; and they used very different outcomes for evaluation – interviewer and respondent behaviors versus reliability estimates for individual questions. Our outcomes are complementary: the analysis of interviewers' and respondents' behavior illuminates the possible mechanisms for differences in reliability – decreased standardization by interviewers, and increased processing challenges for the respondents. The differences between our cases strengthen our examination of how well the methods succeed in identifying problematic question characteristics and also highlight some major limitations with observational studies. For example, the questions are not randomly sampled from a population of questions with many different characteristics, so that researchers typically have a limited pool of item characteristics available for study. Further, the characteristics may have idiosyncratic joint distributions, and so the conclusions may be limited to the comparisons tested in the study. For example, the majority of questions from the WLS asked about events and behaviors, while most of the questions from the GSS were about evaluations and judgments. In addition, while an observational study gives the advantage of working with items in an actual operational setting, a disadvantage is that the items themselves might not conform to current best practices, and so the analyses might not address the choices that informed question writers are currently making.

Nevertheless, the consistencies across our two case studies are striking and lead to useful and practical results. Reviewing the results from Table 6.6, we find that with only a few exceptions (e.g. the effect of the QAS and SQP in the GSS), the question characteristics and system-based approaches that are associated with inexact question reading by interviewers and the expression of problem behaviors by respondents in the WLS are also associated with lower item reliability in the GSS.

One of the strongest predictors across both studies was question length, which was associated with more reading errors by interviewers, more problem behaviors by respondents, and lower reliability (see also Alwin and Beattie 2016; Holbrook et al. 2006; Mangione et al. 1992; Presser and Zhao 1992). This finding does not, however, imply that longer questions will necessarily yield poor-quality data (see Blair et al. 1977). Questions that were long, for example, were also complex. A close look at the items in the case studies also makes clear that question type, response format, and the number of words have complicated relationships. As an example, selection questions, those that contain response categories that are read to the respondent, are probably more common for questions that ask about evaluations, and they are typically longer because of the response categories they include. But these results do provide support for writing questions as simply as possible.

Although word counts are easy to obtain, and they are likely to have little random error – which makes them attractive as predictors – in themselves they are not informative about what specific features of a question need revision. Because of QUAID's focus on issues of vocabulary and syntax, it is a plausible and accessible complement to word counts, and it does suggest specific language and syntax problems to correct. Questions flagged by QUAID as being problematic were associated with negative outcomes in both studies. QUAID identifies complex syntax, and a more detailed analysis of the GSS (Schaeffer et al. 2015) suggests that, once vocabulary issues and grammatical complexity are accounted for, the number of words loses some predictive power. Evidence from both studies indicates QUAID can be used to identify, and presumably correct, problems in the language of the question (Graesser et al. 2006), although the strongest evidence would be provided by a before-after comparison with an outcome. But again, we see the possible impact of the pool of questions (and possibly of mode): although the analysis clearly shows that for both studies the number of problems identified by QUAID predicts the quantitative measure of data quality, QUAID identified at least one problem with 92% of the GSS items; in contrast, in her comparison between QUAID and expert ratings (for an interviewer-administered and mixed-mode study with web as the primary mode), Olson (2010) noted that overall QUAID identified few problems with the items in her study.

In neither case study was the Flesch-Kincaid Grade Level associated with indicators of data quality. These findings are consistent with the work of Lenzner (2014), who discusses and demonstrates the inadequacies of readability formulas for assessing question difficulty. For example, many survey questions are not composed of well-formed sentences – they incorporate casual conversational practices (such as using a brief phrase for second and later items in a battery) and do not use consistent punctuation. Our examination of the Flesch scores for individual items suggests that such problems make this measure less useful for even quick assessment of most survey questions. An exception might be long text passages, in vignettes, for example – a task for which a question writer might use repeated applications of Flesch measurements to track on-the-fly their progress in simplifying the vocabulary and syntax of the text or to demonstrate to a client how demanding the text of the vignette might be for a respondent. In addition, both studies we examined were interviewer-administered, and the length of the question could have a different impact – both cognitive and motivational – in self-administered instruments, which lack the intonation, pacing, and emphasis of the interviewer as supplements to the respondent's literacy skills (Rampey et al. 2016).

We included several system-based approaches, which are important to an analysis of question characteristics, in our analyses. System-based approaches compile measures of individual question characteristics – some of which are isolated because they are believed to be associated with problems (like containing an term or phrase that might be unfamiliar to respondents) and some of which just identify features of questions (like the number of response categories). As noted in our earlier discussion, two of the systems we examine – the CCS and QAS – are oriented toward identifying possible cognitive and other problems that questions pose for interviewers and respondents, rather than being rooted in an analysis of question structure. The SQP, in contrast, relies on an analysis of question characteristics and has commonalities with our taxonomy.

In both of our case studies, the CCS was a strong predictor of the negative outcomes examined as was the QAS for the GSS, suggesting that these systems can usefully identify problems that lead to a reduction in data quality – although again, before-after comparisons targeted at specific revisions suggested by the methods are needed (see examples in Presser et al. 2004). Maitland and Presser's (2016) analysis suggests that some of the power of the CCS and QAS may come from their ability to identify recall problems (although most questions in the GSS do not demand difficult recall). Identifying the specific strengths of each of these tools – and ways in which they overlap – requires additional study.

In the analysis of the WLS, exact question reading by the interviewer was associated with higher quality predicted by SQP, but the reliability predicted by SQP was negatively associated with the estimated reliabilities for the GSS. Maitland and Presser (2016) found that overall the SQP was unrelated to their index of inconsistency (their measure of reliability), but, in analyses for which we do not have a comparison, the reliability predicted by SQP did predict discrepant answers for subjective questions. It is possible that differences in the distribution of the types of items or the complexity of coding items using the SQP coding scheme may account for differences between their results and our findings for the GSS.

Each of the systems in this analysis represents a potential tool for the question writer, and although our analyses suggest that most of these tools have the potential for identifying problems that, if corrected, could improve data quality, practice would benefit from a restricted set of accurate and reliable, accessible, and affordable assessments that locate where revisions are needed. Methods for analyzing question characteristics would need to be integrated with other methods, such as cognitive interviewing, in a sequence of steps that can be implemented in a realistic time frame for question development.

Our study, like many others that compare various methods of question assessment, considers the overall performance of the approach. Our understanding of where these methods succeed and fail could be strengthened with subsequent analyses of which specific question characteristics were associated with low reliability or high levels of interactional problems – and which questions with low reliability or high levels of interactional problems were not identified by the methods. A necessary complement to such analyses is examination of which question characteristics and which decisions by the question writer are critical to developing reliable and valid questions. We argue that this work should consider not just the presence (or absence) of a characteristic or problem caused by a characteristic, but also how the functioning or meaning of a characteristic may depend on other characteristics. The potential importance of these structural dependencies and the resulting interactions among question characteristics is illustrated in Table 6.7, which presents only a few common types of questions with two categories. Although all of these questions have two response categories, response category labels would usually be offered only for the bipolar questions, which ask the respondent to make some sort of judgment rather than to recall autobiographical facts. These structural dependencies and statistical interactions have largely been

Table 6.7 Selected decisions about response categories and implementation by question type, response dimension, and response format, for interviewer-administered instruments.

Question type		**Decisions about response categories**			
Response dimension	**Response format**	**How many?**	**What labels?**	**What order?**	**Example**
Events and *behaviors*					
Occurrence	Yes-no	2	NA	NA	**In the last 12 months have you seen a medical doctor about your own health?**
Occurrence	Yes-no with formal balance	2	NA	NA	In the last 12 months have you seen a medical doctor about your own health or not?
Evaluations and judgments					
Bipolar – valence	Selection – Forced choice	2	Positions	Yes	Do you think most people would try to take advantage of you if they got a chance, or would they try to be fair?
Bipolar – valence	Selection – Ordered categories	2	Evaluative dimensions	Yes	Do you find cleaning house boring or interesting?
Unipolar – threshold	Yes-no	2	NA	NA	Are you concerned about possible bacteria in your food? Do people understand you completely when you speak?
Unipolar – threshold	Yes-no with formal balance	2	NA	NA	Are you concerned about possible bacteria in your food or not? Do people understand you completely when you speak or not?
Event-based classification					
Occurrence or label	Yes-no	2	NA	NA	Are you currently married?
Occurrence or label	Yes-no with formal balance	2	NA	NA	Are you currently married or not?

Note: "NA" = not applicable. Question about preferences for a job based on Schuman and Presser (1981, p. 89). Questions about fairness are based on the General Social Survey. Questions about concern about bacteria in food are based on Sterngold et al. (1994). Other items are loosely based on existing survey questions or were composed as examples for the table.

ignored in the experimental studies that typically manipulate only one or two characteristics. We advocate for more research that takes these structural dependencies into account during analysis.

Organizing such efforts requires a taxonomy of question characteristics, whether explicit, as in Table 6.2, or implicit, that structures analysis and allows the question writer to organize the results of previous research and bring it to bear on specific decisions (Figures 6.2 and 6.3). Although the taxonomy in Figure 6.2 is only one possible approach (e.g. compare that in SQP), it, and our model of decisions, recognizes the importance of question type and response format in the sequence of possible decisions and their implication. With regard to the taxonomy, we believe that pursing the kinds of underlying distinctions we described can identify issues about survey questions and their characteristics that may have been overlooked, and provide a useful framework for organizing the results of past and future research, potentially helping to explain where and why there are discrepancies in past research. We believe that a systematic framework such as that provided by the taxonomy of features and the decision-making framework discussed in Section 6.3 could be particularly useful to aid in understanding how question characteristics are related to each other. For example, as previously discussed, some decisions, like those for response category labels, are relevant only for some question types or some question forms. Question writers need to keep these decisions in mind when writing questions, evaluating them, and interpreting findings from research studies. In addition, the conclusions one reaches about the importance of a question characteristic may depend on the range of other characteristics included in a given analysis, and we currently have little theory or practice to guide decisions about what range of characteristics should be included in an analysis.

All studies about the impact of question characteristics on data quality require a criterion. Ultimately, we want to be able to say which version of a question or characteristic is more valid or reliable, but we rarely have direct measures of validity or reliability. In fact, the availability of item-specific reliability measures for questions from the GSS motivated that study, even though we knew that the distribution of item characteristics would limit the conclusions we might draw. Instead of direct measures of validity and reliability, we often rely on proxy measures – measures that we assume are related to validity or reliability, such as our measures of interviewer and respondent behaviors from the WLS. Progress in the field requires consideration of how results may vary depending on the criterion selected, because criteria vary in their quality (for example, the extent to which they include stable method variance) and their sensitivity to particular problems (e.g. item nonresponse may be particularly sensitive to the motivation of the respondent in self-administered instruments, reinforcing the importance of the class of "implementation" issues in our taxonomy). In our case studies we examine proxy measures – behavior of interviewers and respondents – as well as high-quality measures of individual item reliability, and our use of different criteria in our case studies can be seen as a strength.

Some proxy measures now have accumulated substantial evidence of their relationship with more direct measures of data quality. These proxy measures include behaviors of interviewers and respondents such as question-reading accuracy, response latency, use of disfluencies, and "problem" behaviors like qualifications and expressions of uncertainty; problems identified in cognitive interviewing; and problems suggested by models of error (e.g. that threatening behaviors are under-reported). These data-quality proxy measures, however, have limitations, and each may be sensitive to some problems in item construction but not others. Item nonresponse and simple test-retest measures of reliability are more affordable and accessible (e.g. Maitland and Presser 2016; Olson 2010) than detailed interaction codes or three-wave estimates of individual item reliabilities.

We conclude by recommending that in order to further our understanding of which question characteristics affect the quality of survey data, when they do so, and how to use the available diagnostic tools to improve data quality, studies of several sorts are needed: (i) those that incorporate characteristics of questions that previous research has shown are associated with data quality; (ii) those that compare existing coding systems for evaluating the characteristics of questions and explore new coding systems; and (iii) those that examine these characteristics and coding systems across a wide range of questions, modes, and populations. We believe that future work should see more complementarity between experimental and observational approaches, with strong results from observational studies providing an agenda for subsequent experiments, particularly those with before-after comparisons that look at the success of the revisions indicated by a particular approach on improving data quality – which must then be replicated in production surveys.

Acknowledgments

This research was supported by the following at the University of Wisconsin-Madison: a National Institute on Aging grant to Schaeffer (under P01 AG 21079 to Robert M. Hauser); the Center for the Demography of Health and Aging (NIA Center Grant P30 AG017266); the University of Wisconsin Survey Center (UWSC); and the use of facilities of the Social Science Computing Cooperative and the Center for Demography and Ecology (NICHD core grant P2C HD047873). This research uses data from the Wisconsin Longitudinal Study (WLS) of the University of Wisconsin-Madison. Since 1991, the WLS has been supported principally by the National Institute on Aging (AG-9775, AG-21079, AG-033285, and AG-041868), with additional support from the

Vilas Estate Trust, the National Science Foundation, the Spencer Foundation, and the Graduate School of the University of Wisconsin-Madison. Since 1992, data have been collected by the University of Wisconsin Survey Center. A public use file of data from the Wisconsin Longitudinal Study is available from the Wisconsin Longitudinal Study, University of Wisconsin-Madison, 1180 Observatory Drive, Madison, Wisconsin 53706 and at http://www.ssc.wisc.edu/wlsresearch/data. The authors thank Nadia Assad, Minnie Chen, Curtiss Engstrom, Barbara Forsyth, and Gordon B. Willis for their help coding the questions. The opinions expressed herein are those of the authors.

References

Alwin, D.F. (2007). *Margins of Error: A Study of Reliability in Survey Measurement*. Hoboken, NJ: Wiley.

Alwin, D.F. (2011). Evaluating the reliability and validity of survey interview data using the MTMM approach. In: *Question Evaluation Methods: Contributing to the Science of Data Quality* (eds. J. Madans, K. Miller, A. Maitland and G. Willis), 263–293. Hoboken, NJ: Wiley.

Alwin, D.F. and Beattie, B.A. (2016). The KISS principle in survey design: question length and data quality. *Sociological Methodology* 46: 121–152.

Alwin, D.F. and Krosnick, J.A. (1991). The reliability of survey attitude measurement: the influence of question and respondent attributes. *Sociological Methods and Research* 20: 139–181.

Bassili, J.N. and Scott, B.S. (1996). Response latency as a signal to question problems in survey research. *Public Opinion Quarterly* 60: 390–399.

Blair, E., Sudman, S., Bradburn, N.M., and Stocking, C. (1977). How to ask questions about drinking and sex: response effects in measuring consumer behavior. *Journal of Marketing Research* 14: 316–321.

Bradburn, N.M., Sudman, S., and Wansink, B. (2004). *Asking Questions: The Definitive Guide to Questionnaire Design*. New York: Wiley.

Cannell, C.F., Miller, P.V., and Oksenberg, L. (1981). Research on interviewing techniques. In: *Sociological Methodology*, vol. 12 (ed. S. Leinhardt), 389–437. San Francisco: Jossey-Bass.

Couper, M.P. and Kreuter, F. (2013). Using paradata to explore item level response times in surveys. *Journal of the Royal Statistical Society: Series A (Statistics in Society)* 176: 271–286.

Dillman, D.A., Smyth, J.D., and Christian, L.M. (2014). *Internet, Phone, Mail, and Mixed-Mode Surveys: The Tailored Design Method*, 4e. Hoboken, NJ: Wiley.

Dykema, J., Garbarski, D., Schaeffer, N.C. et al. (2009). Code manual for codifying interviewer-respondent interaction in surveys of older adults. University of Wisconsin Survey Center, University of Wisconsin-Madison. https://dxs0oxbu2lwwr.cloudfront.net/wp-content/uploads/sites/18/2018/09/CodeManual_Master_V69_PostProduction.pdf

Dykema, J., Lepkowski, J.M., and Blixt, S. (1997). The effect of interviewer and respondent behavior on data quality: analysis of interaction coding in a validation study. In: *Survey Measurement and Process Quality* (eds. L. Lyberg, P. Biemer, M. Collins, et al.), 287–310. New York: Wiley-Interscience.

Dykema, J., Schaeffer, N.C., and Garbarski, D. (2012). Effects of agree-disagree versus construct-specific items on reliability, validity, and interviewer-respondent interaction. Annual meeting of the American Association for Public Opinion Research, May, Orlando, FL.

Dykema, J., Schaeffer, N.C., Garbarski, D. et al. (2016). The impact of parenthetical phrases on interviewers' and respondents' processing of survey questions. *Survey Practice* 9.

Forsyth, B., Rothgeb, J.M., and Willis, G.B. (2004). Does pretesting make a difference? An experimental test. In: *Methods for Testing and Evaluating Survey Questionnaires* (eds. S. Presser, J.M. Rothgeb, M.P. Couper, et al.), 525–546. Hoboken, NJ: Wiley.

Fowler, F.J. Jr., (1995). *Improving Survey Questions: Design and Evaluation*. Thousand Oaks, CA: Sage.

Fowler, F.J. Jr., and Cannell, C.F. (1996). Using behavioral coding to identify cognitive problems with survey questions. In: *Answering Questions: Methodology for Determining Cognitive and Communicative Processes in Survey Research* (eds. N. Schwarz and S. Sudman), 15–36. San Francisco, CA: Jossey-Bass.

Fowler, F.J. Jr., and Cosenza, C. (2009). Design and evaluation of survey questions. In: *The Sage Handbook of Applied Social Research Methods* (eds. L. Bickman and D.J. Rog), 375–412. Thousand Oaks, CA: Sage.

Garbarski, D., Schaeffer, N.C., and Dykema, J. (2011). Are interactional behaviors exhibited when the self-reported health question is asked associated with health status? *Social Science Research* 40: 1025–1036.

Graesser, A.C., Cai, Z., Louwerse, M.M., and Daniel, F. (2006). Question Understanding AID (QUAID): a web facility that tests question comprehensibility. *Public Opinion Quarterly* 70: 3–22.

Groves, R.M. (1991). Measurement error across disciplines. In: *Measurement Errors in Surveys* (eds. P.P. Biemer, R.M. Groves, L.E. Lyberg, et al.), 1–28. Hoboken, NJ: Wiley.

Heise, D.R. (1969). Separating reliability and stability in test-retest correlation. *American Sociological Review* 34: 93–101.

Holbrook, A., Cho, Y.I., and Johnson, T. (2006). The impact of question and respondent characteristics on comprehension and mapping difficulties. *Public Opinion Quarterly* 70: 565–595.

Hout, M. and Hastings, O.P. (2016). Reliability of the core items in the General Social Survey: estimates from the three-wave panels, 2006–2014. *Sociological Science* 3: 971–1002.

Johnson, T.P., Shariff-Marco, S., Willis, G. et al. (2015). Sources of interactional problems in a survey of racial/ethnic discrimination. *International Journal of Public Opinion Research* 27: 244–263.

Kasabian, A., Smyth, J., and Olson, K. (2014). The whole is more than the sum of its parts: understanding item nonresponse in self-administered surveys. Annual meeting of the American Association for Public Opinion Research, May, Anaheim, CA.

Kleiner, B., Lipps, O., and Ferrez, E. (2015). Language ability and motivation among foreigners in survey responding. *Journal of Survey Statistics and Methodology* 3: 339–360.

Knäuper, B. (1998). Filter questions and question interpretation: presuppositions at work. *Public Opinion Quarterly* 62: 70–78.

Knäuper, B., Belli, R.F., Hill, D.H., and Herzog, A.R. (1997). Question difficulty and respondents' cognitive ability: the effect on data quality. *Journal of Official Statistics* 13: 181–199.

Krosnick, J.A. (1991). Response strategies for coping with the cognitive demands of attitude measures in surveys. *Applied Cognitive Psychology* 5: 213–236.

Krosnick, J.A. (2011). Experiments for evaluating survey questions. In: *Question Evaluation Methods: Contributing to the Science of Data Quality* (eds. J. Madans, K. Miller, A. Maitland and G. Willis), 215–238. Hoboken, NJ: Wiley.

Krosnick, J.A. and Presser, S. (2010). Question and Questionnaire Design. In: *Handbook of Survey Research*, 2e (eds. P.V. Marsden and J.D. Wright), 263–313. Bingley, UK: Emerald Group Publishing Limited.

Lenzner, T. (2014). Are readability formulas valid tools for assessing survey question difficulty? *Sociological Methods & Research* 43: 677–698.

Maitland, A. and Presser, S. (2016). How accurately do different evaluation methods predict the reliability of survey questions? *Journal of Survey Statistics and Methodology* 4: 362–381.

Mangione, T.W., Fowler, F.J. Jr., and Louis, T.A. (1992). Question characteristics and interviewer effects. *Journal of Official Statistics* 8: 293–307.

Olson, K. (2010). An examination of questionnaire evaluation by expert reviewers. *Field Methods* 22: 295–318.

Olson, K. and Smyth, J.D. (2015). The effect of CATI questions, respondents, and interviewers on response time. *Journal of Survey Statistics and Methodology* 3: 361–396.

Ongena, Y.P. and Dijkstra, W. (2007). A model of cognitive processes and conversational principles in survey interview interaction. *Applied Cognitive Psychology* 21: 145–163.
Presser, S., Rothgeb, J.M., Couper, M.P. et al. (2004). Methods for testing and evaluating survey questions. *Public Opinion Quarterly* 68: 109–130.
Presser, S. and Zhao, S. (1992). Attributes of questions and interviewers as correlates of interviewing performance. *Public Opinion Quarterly* 56: 236–240.
Rampey, B.D., Finnegan, R., Goodman, M. et al. (2016). Skills of U.S. unemployed, young, and older adults in sharper focus: results from the program for the international assessment of adult competencies (PIAAC) 2012/2014: first look. NCES 2016-039rev. Washington, D.C.: U.S. Department of Education. National Center for Education Statistics. http://nces.ed.gov/pubs2016/2016039rev.pdf.
Rodgers, W.L., Andrews, F.M., and Herzog, A.R. (1992). Quality of survey measures: a structural modeling approach. *Journal of Official Statistics* 8: 251–275.
Saris, W.E. (2012). Discussion: evaluation procedures for survey questions. *Journal of Official Statistics* 28: 537–551.
Saris, W.E. and Gallhofer, I.N. (2007). *Design, Evaluation, and Analysis of Questionnaires for Survey Research*. Hoboken, NJ: Wiley.
Schaeffer, N.C. and Dykema, J. (2004). A multiple-method approach to improving the clarity of closely related concepts: distinguishing legal and physical custody of children. In: *Methods for Testing and Evaluating Survey Questionnaires* (eds. S. Presser, J.M. Rothgeb, M.P. Couper, et al.), 475–502. Hoboken, NJ: Wiley.
Schaeffer, N.C. and Dykema, J. (2011a). Questions for surveys: current trends and future directions. *Public Opinion Quarterly* 75: 909–961.
Schaeffer, N.C. and Dykema, J. (2011b). Response 1 to Fowler's chapter: coding the behavior of interviewers and respondents to evaluate survey questions. In: *Question Evaluation Methods: Contributing to the Science of Data Quality* (eds. J. Madans, K. Miller, A. Maitland and G. Willis), 23–39. Hoboken, NJ: Wiley.
Schaeffer, N.C. and Dykema, J. (2015). Question wording and response categories. In: *International Encyclopedia of the Social and Behavioral Sciences*, 2e, vol. 23 (ed. J.D. Wright), 764–770. Oxford, England: Elsevier.
Schaeffer, N.C., Dykema, J., and Maynard, D.W. (2010). Interviewers and interviewing. In: *Handbook of Survey Research*, 2e (eds. P.V. Marsden and J.D. Wright), 437–470. Bingley, UK: Emerald Group Publishing Limited.
Schaeffer, N.C. and Maynard, D.W. (2002). Occasions for intervention: interactional resources for comprehension in standardized survey interviews. In: *Standardization and Tacit Knowledge: Interaction and Practice in the Survey Interview* (eds. D.W. Maynard, H. Houtkoop-Steenstra, N.C. Schaeffer and J. van der Zouwen), 261–280. New York: Wiley.
Schaeffer, N.C. and Presser, S. (2003). The science of asking questions. *Annual Review of Sociology* 29: 65–88.

Schaeffer, N.C., Chen, M., Dykema, J., Garbarski, D. et al. (2015). Question characteristics and item reliability. Annual meeting of the Midwest Association for Public Opinion Research, November, Chicago, IL.

Schober, M.F., Conrad, F.G., Dijkstra, W., and Ongena, Y.P. (2012). Disfluencies and gaze aversion in unreliable responses to survey questions. *Journal of Official Statistics* 28: 555–582.

Schuman, H. and Presser, S. (1981). *Questions and Answers in Attitude Surveys: Experiments on Question Form, Wording and Context*. Orlando, FL: Academic Press.

Schwarz, N. (1996). *Cognition and Communication: Judgmental Biases, Research Methods, and the Logic of Conversation*. New York: Lawrence Erlbaum Associates.

Sewell, W.H., Hauser, R.M., Springer, K.W., and Hauser, T.S. (2003). As we age: a review of the Wisconsin Longitudinal Study, 1957–2001. *Research in Social Stratification and Mobility* 20: 3–111.

Sterngold, A., Warland, R.H., and Herrmann, R.O. (1994). Do surveys overstate public concerns? *Public Opinion Quarterly* 58: 255–263.

Toepoel, V. and Dillman, D.A. (2011). How visual design affects the interpretability of survey questions. In: *Social and Behavioral Research and the Internet: Advances in Applied Methods and Research Strategies* (eds. M. Das, P. Ester and L. Kaczmirek), 165–190. New York: Routledge.

Tourangeau, R., Rips, L.J., and Rasinski, K. (2000). *The Psychology of Survey Response*. Cambridge, England: Cambridge University Press.

Willis, G.B. (2005). *Cognitive Interviewing: A Tool for Improving Questionnaire Design*. Thousand Oaks, CA: Sage.

Willis, G. B., & Lessler, J. T. (1999). *Question Appraisal System: QAS-99*. Retrieved from Rockville, MD. https://www.researchgate.net/publication/259812768_Question_Appraisal_System_QAS_99_Manual

Yan, T. and Tourangeau, R. (2008). Fast times and easy questions: the effects of age, experience and question complexity on web survey response times. *Applied Cognitive Psychology* 22: 51–68.

7

Exploring the Associations Between Question Characteristics, Respondent Characteristics, Interviewer Performance Measures, and Survey Data Quality

James M. Dahlhamer[1], Aaron Maitland[2], Heather Ridolfo[3], Antuane Allen[4], and Dynesha Brooks[5]

[1] *Division of Health Interview Statistics, National Center for Health Statistics, Hyattsville, MD, USA*
[2] *Westat, Rockville, MD, USA*
[3] *Research and Development Division, National Agricultural Statistics Service, Washington, DC, USA*
[4] *Health Analytics, LLC, Silver Spring, MD, USA*
[5] *Montgomery County Public Schools, Rockville, MD, USA*

7.1 Introduction

In interviewer-administered surveys, the quality of data produced by survey questions is determined by a complex interaction between interviewers and respondents through the medium of a survey instrument (Sudman and Bradburn 1974). Questions on many ongoing surveys such as the National Health Interview Survey (NHIS) may have undergone some initial evaluation, via cognitive testing and/or small-scale field tests, prior to their introduction on the survey; however, further evaluation may be needed. Other questions may not have been evaluated at all due to time or budgetary constraints. Therefore, it may be prudent to examine the quality of the data produced by survey questions in the field. Analysis of field or production data may identify characteristics of questions or specifically problematic questions leading to further testing or better question design.

To identify potential problems with questions using production data, there needs to be one or more indicators of data quality that are available for all questions in the field. Most ongoing surveys do not typically incorporate research designs to directly measure the reliability or validity of survey questions. The valuable space available on survey instruments is best filled with questions that are used to create key estimates or address important research questions. This means that any indicator of quality is likely to be indirect. Probably the two most widely used indirect indicators of data quality are item nonresponse rates

Advances in Questionnaire Design, Development, Evaluation and Testing, First Edition.
Edited by Paul C. Beatty, Debbie Collins, Lyn Kaye, Jose-Luis Padilla, Gordon B. Willis, and Amanda Wilmot.

and item response times. We briefly describe literature that suggests these two indicators have an important relation to data quality.

The cognitive mechanisms that are used to explain item nonresponse suggest a link to data quality. For example, a model of the decision to respond to a survey item by Beatty and Hermann (2002) demonstrates that a respondent's cognitive state (i.e. how much the respondent knows) and communicative intent (i.e. what the respondent wants to reveal about herself) are primary factors influencing item nonresponse. Satisficing theory also suggests that respondents who have lower levels of motivation and questions that are more difficult have higher levels of item nonresponse (Krosnick 1991). Shoemaker et al. (2002) found that sensitive survey items had higher levels of refusals than non-sensitive items and items that required more cognitive effort had higher levels of don't know responses than items that required less cognitive effort. Therefore, item nonresponse is an indicator of data quality to the extent that it identifies questions that respondents do not possess the relevant information to answer or do not wish to disclose an answer.

There has also been research linking response times to data quality. In general, this literature assumes that response time is a measure of the amount of information processing undertaken by a respondent to answer a question. Hence, from this perspective longer response times are an indicator of questions that are more difficult for a respondent to answer. There is a long history of analyzing response latencies (duration of time between the delivery of a question and a response) in the context of attitude measurement. For example, response latencies have been used as a measure of the accessibility of an attitude with shorter response latencies indicating more-accessible attitudes and longer response latencies indicating less-accessible attitudes (Fazio et al. 1986). The longer response latencies are an indication that the respondent is constructing an attitude in response to a stimulus rather than retrieving an existing attitude (see Strack and Martin (1987) for a discussion of attitude construal). Both Bassili and Fletcher (1991) and Heerwegh (2003) present evidence in a survey context that respondents with less-stable attitudes take longer to answer attitude questions. In addition, Draisma and Dijkstra (2004) present evidence that response times are related to validity. They asked respondents about different aspects of their membership in an environmental organization and linked the survey responses to administrative records. They found that correct answers have the shortest response times followed by incorrect answers and nonsubstantive answers such as don't know and refuse.

Response times have also been used to identify problematic questions. Bassili (1996) found that questions with double negative or vague wording took relatively longer to answer than other questions. Bassili and Scott (1996) found that double-barreled questions and questions with superfluous phrases took longer to answer. There has also been an interest in studying short response times with regard to problematic questions; however, most of this research is focused on identifying respondents who are satisficing throughout a questionnaire rather than identifying problematic items (e.g. Malhotra 2008).

The literature demonstrates the relevance of item nonresponse and response times to data quality; however, as stated earlier, these two measures are indirect (and imperfect) measures of data quality. Therefore, we believe the best use of these measures is to identify questions that might be problematic and then follow up with other methods to further diagnose potential problems. Hence, these two measures provide commonly available data on all questions that can be used as initial indicators of questions that may need follow-up. Another desirable feature of analyzing production data to identify problematic questions is that the researcher can account for the fact that the survey interview is an interaction and that, in addition to the questions themselves, the interviewers and respondents also affect the quality of the data. We now review research that uses multilevel models to study indicators of data quality in the context of this interaction.

The use of multilevel models to study the effects of different components of the survey interaction on data quality has increased (Olson and Smyth 2015; Couper and Kreuter 2013; Yan and Tourangeau 2008; Pickery and Loosveldt 2001). These models are flexible enough to accommodate the nesting and cross-classified nature of interviewers, respondents, and questions, while still allowing one to control for characteristics that explain variability at each level of the model. Surprisingly, there is very little research using multilevel models to study item nonresponse. This dearth of literature may be due to the low prevalence of item nonresponse on most surveys (Olson and Smyth 2015). What studies that exist on the topic tend to focus on the contributions of interviewers and respondents, as opposed to questions, to variance in item nonresponse (e.g. Pickery and Loosveldt 2004, 2001). More research, however, has utilized multilevel models to study item response times in interviewer-administered surveys. One major finding from this research is that the majority of the variance in response times is due to the question (Olson and Smyth 2015; Couper and Kreuter 2013). Respondents tend to account for a substantially lower amount of the variance in item response times, and interviewers account for the least amount of variance.

In addition to respondent and interviewer characteristics, Olson and Smyth (2015) present a multilevel framework relating four different types of question features to item response times: necessary question features, features that affect respondent task complexity, features that affect interviewer task complexity, and features that affect respondent processing efficiency. We utilize this framework and highlight some of the key findings from studies using multilevel models to study item response times.

Some of the *necessary question features* that have been found to be positively related to response times include question length (Olson and Smyth 2015; Couper and Kreuter 2013; Yan and Tourangeau 2008) and number of response options (Olson and Smyth 2015; Yan and Tourangeau 2008). Olson and Smyth (2015) found that behavioral questions take longer to answer than attitudinal and demographic questions, while Yan and Tourangeau (2008) found that factual and attitudinal questions take longer to answer than demographic questions. Open-ended textual questions tend to take longer to answer than questions with most other types of answer choices (Olson and Smyth 2015; Couper and Kreuter 2013). Olson and Smyth (2015) found that open-ended textual questions had longer response times than questions with closed nominal response categories, but similar response times to questions with ordinal categories. Yan and Tourangeau (2008) found no difference between questions with response categories that formed a scale and questions with response categories that did not form a scale. Couper and Kreuter (2013) found that response times are faster in audio computer assisted interviewing (ACASI) compared to computer assisted personal interviewing (CAPI). There is some disagreement in the literature about the effect of position in the questionnaire on response times. Couper and Kreuter (2013) found that response times get slower, whereas Yan and Tourangeau (2008) found that response times get faster as the respondent moves through the questionnaire.

Question features that affect respondent task complexity include question reading level, number of clauses, number of words per clause, and question sensitivity (Olson and Smyth 2015; Yan and Tourangeau 2008). Using the Flesch-Kincaid Grade Level measure, Olson and Smyth (2015) found that the higher the reading level, the longer the response time. Similarly, Yan and Tourangeau (2008) found that more clauses in the question and more words per clause both significantly increased response times. Finally, question sensitivity has been found to be negatively related to response times (Olson and Smyth 2015).

Some *question features affecting interviewer task complexity* that are positively related to response times include showcards (Couper and Kreuter 2013) and help screens (Couper and Kreuter 2013). Couper and Kreuter (2013) found that question fills were positively related to response times for male respondents and negatively for female respondents. Interviewer instructions have been found to be negatively related to response times (Couper and Kreuter 2013). Only one variable affecting *respondent processing efficiency* has been found to be significantly related to response times. Olson and Smyth (2015) found that the presence of definitions in the question text is positively related to response times.

Respondent characteristics are also important predictors of response times. Response times increase with age (Olson and Smyth 2015; Couper and Kreuter 2013; Yan and Tourangeau 2008) and decrease with education (Couper and Kreuter 2013; Yan and Tourangeau 2008). Couper and Kreuter (2013) also found longer response times for respondents who are non-white compared to white; married, cohabitating, or formerly married compared to never married;

and those who completed the interview in Spanish. Olson and Smyth (2015) found shorter response times for respondents who are employed. *Interviewer characteristics* or experience can also influence response times. Olson and Smyth (2015) found that interviews conducted later in the interviewer's caseload have shorter response times. Couper and Kreuter (2013) found that Spanish-speaking interviewers have longer response times compared to English-speaking interviewers.

In this chapter, we build and expand on the existing literature using multilevel models to understand the joint effects of question and respondent characteristics, and interviewer performance measures on data quality using a large, nationally representative survey. We model the variability in both response time and item nonresponse. We follow the framework presented by Olson and Smyth (2015), identifying different question features that may affect either item response times or item nonresponse. We also include a set of paradata or process variables that measure any reluctance the respondent had about doing the interview to control for the potential difficulty of completing a case. We conclude with a discussion of the varying contributions of questions, respondents, and interviewers to data quality, and the implications of our findings for questionnaire design and evaluation.

7.2 Methods

7.2.1 Data

The data used in these analyses were drawn from 9336 sample adults aged 18 and over who participated in the 2014 NHIS. The NHIS is a multipurpose, nationally representative health survey of the civilian, noninstitutionalized US population conducted continuously by the National Center for Health Statistics. Interviewers with the US Census Bureau administer the questionnaire using CAPI. Telephone interviewing is permitted to complete missing portions of the interview (National Center for Health Statistics 2015).

Variables used in the analyses are based on data from multiple NHIS files. To ensure that the dependent variables most closely reflect how the question and answer process unfolded, we utilized initial, partially edited data files as opposed to final, fully edited data files. The response time and item nonresponse dependent variables were based on data collected with 270 questions in the Sample Adult Core module of the NHIS. The Sample Adult Core module, administered to one adult aged ≥18 years randomly selected from each family, collects information on adult sociodemographics, sexual orientation, health conditions, health status and limitations, health behaviors, and heath care access and use. The sample adult answers for himself/herself unless mentally or physically unable to do so, in which case a knowledgeable family member serves as a proxy respondent. The final sample adult response rate for 2014 was 58.9% (National Center for Health Statistics 2015).

The respondent (sample adult) characteristics are based on information collected with the Household Composition and Family Core modules. Data from both modules were collected via self and proxy reports. Several paradata measures included in the analysis are based on data collected with the Contact History Instrument (CHI), an automated instrument that is completed each time an interviewer makes a contact attempt on a household. CHI is used to collect basic contact history information such as outcome of the attempt, whether or not householders expressed reluctance, and any strategies interviewers used to overcome that reluctance or ensure contact at a future attempt. Additional paradata measures were collected at the end of the NHIS instrument via questions posed directly to the interviewer. Finally, measures capturing the social environment, such as region of residence, were taken from the NHIS sample frame file.

7.2.2 Measures

7.2.2.1 Dependent Variables

The first dependent variable is a continuous, log-transformed measure of response time. For this analysis, a response time corresponds to the time spent on the longest visit to the question, including the time taken by the interviewer to read the question, receive an answer from the respondent, and enter that answer. In contrast to an active timer approach, we use a latent timer approach in which the entire time spent on the question is considered the response time (Yan and Tourangeau 2008). In interviewer-administered surveys, active timer approaches require the interviewer to press a key to start the timer as soon as they finish reading a question and press the key again as soon as the respondent starts to answer. Response time is measured as the elapsed time between the two key presses. Conceptually, the primary difference between latent and active timer approaches is the assumption each makes about when the response process starts (Yan and Tourangeau 2008). Active timer approaches assume the process starts only after the question has been fully read to the respondent, while the latent timer approach assumes the process begins as soon as the interviewer starts reading the question. As Yan and Tourangeau (2008) note, it is common for respondents to interrupt with an answer before the interviewer finishes reading a question. In addition, active timer approaches are difficult to implement in face-to-face interview settings and therefore prone to error (Bassili 1996; Bassili and Fletcher 1991). Finally, empirical evidence shows a significant correlation between the times produced and considerable overlap in the results achieved by the two approaches (Mulligan et al. 2003).

We limited the measure of response time to the longest question visit so as to eliminate time that accrued from extraneous movement through the instrument (e.g. to go back to a previous question, reviewing responses, etc.).[1] By taking the longest visit to a question, we assume it is the visit in which the response process unfolded. As expected, the distribution of response times was highly right-skewed. Consistent with past studies of response times, we first trimmed the distribution of times by replacing all values below the 1st percentile and all values above the 99th percentiles with those percentile values, respectively. We then took the natural log transformation of the truncated response times (Olson and Smyth 2015; Yan and Tourangeau 2008). The average number of seconds per question before truncation was 8.61 (SD = 89.07). After truncation, the average was 7.77 (SD = 8.32) seconds per question. With the log transformation, the average number of logseconds per question was 1.65 (SD = 0.88).

Item nonresponse, the second dependent variable, is a dichotomous measure defined as 1 = "don't know" or refused response versus 0 = valid response to a question under analysis. Item nonresponse is low in the Sample Adult Core interview. Nineteen of the 270 questions included in the analysis had no item nonresponse. An additional 234 questions had an item nonresponse rate below 5.0%. The overall item nonresponse rate across all visits to the 270 questions was 1.1%.

7.2.2.2 Question Characteristics

The primary independent variables in the analysis of response time and item nonresponse are eight measures capturing characteristics of 270 questions. Table 7.1 provides detailed information on variable creation, and Table 7.2 presents descriptive statistics for each measure. Measures of respondent task complexity include a measure of question sensitivity and whether or not the question includes one or more unfamiliar technical terms. Whether or not the question includes clarifying text or definitions of key terms is the lone measure of respondent processing efficiency in the models. Two measures of interviewer task complexity were also included: whether or not the question is accompanied by a separate screen of help text and whether or not the question includes optional text to be read at the interviewer's discretion. Finally, we included three measures that are conceptually grouped as necessary features of questions: whether or not the question captures factual or demographic versus attitudinal or subjective information, type of response option format used with the question (yes/no, integer, pick one from a list, text, and enter all that apply), and length of the question measured in characters (<70 characters,

---

1 As an example, an interviewer enters and exits a question or field three times for a given respondent, with the resulting item times being 1 second, 10 seconds, and 1 second. The 10-second visit would be retained for analysis and the two 1-second visits would be dropped. The total time spent on this question for this respondent would be 10 seconds.

≥70 characters and <103 characters, ≥103 characters and <143 characters, and ≥143 characters).

7.2.2.3 Respondent/Case Characteristics

A set of respondent and case characteristics ($n = 9336$) is also included in the models (see Table 7.1 for coding and Table 7.2 for descriptive statistics). Sample adult sociodemographic measures include age (18–24, 25–44, 45–64, and 65 and over), sex, race/ethnicity (Hispanic, non-Hispanic white, non-Hispanic black, and non-Hispanic other race), education (less than a high school diploma, high school diploma or GED, some college or an AA degree, and bachelor's degree or higher), whether or not the adult was born in the United States, and total family income from the prior calendar year (less than $35 000, $35 000–$74 999, $75 000–$99 999, $100 000 or more, and unknown [don't know/refused responses]). Two health-related measures are included in the analysis: reported health status (poor or fair health, good health, and very good or excellent health) and whether or not the adult has cognitive difficulties. Two measures of the social environment are also included: region of residence (Northeast, Midwest, South, and West) and metropolitan statistical area (MSA) status[2] (MSA, central city; MSA, noncentral city; and non-MSA). Finally, four paradata measures are included in the analysis: whether or not the sample adult interview was conducted primarily by telephone, whether or not household members made refusal-like statements during one or more contacts, whether or not household members expressed privacy concerns and/or asked questions about the survey content during one or more contacts, and whether or not household members expressed time constraints during one or more contacts.

7.2.2.4 Interviewer Performance Measures

Traditional measures of interviewer characteristics such as sex, education, race/ethnicity, and survey experience were not available for this analysis. Instead, we included three measures of interviewer performance: interviewer item nonresponse rate, interviewer cooperation rate, and interviewer sample adult interview pace (measured in seconds per item). All three measures were recoded into categorical variables using quartiles of the respective distributions. The measures were lagged one calendar quarter to avoid direct overlap between the coding of the interviewer performance measures and the

2 A metropolitan statistical area (MSA) is defined as a county or group of contiguous counties that contain at least one urbanized area of 50 000 population or more. Adults were defined as living in the central or principal city of an MSA (MSA, central city), in an MSA but not in the central city (MSA, noncentral city), or not in a MSA. "Not in a MSA" indicates that the adults lives in a nonmetropolitan area, defined as an area that does not include a large urbanized area; these areas are generally thought of as more rural. See "2010 Standards for Delineating Metropolitan and Micropolitan Statistical Areas; Notice," 75 Federal Register 123 (28 June 2010), pp. 37246–37252.

Table 7.1 Description of variables included in the models of response time and item nonresponse.

Variable	Description and coding
Question characteristics: factors affecting respondent task complexity	
Unfamiliar technical term(s)	Based on Question Understanding AID (QUAID) software coding (see Graesser et al. 2000). The following definition of unfamiliar technical terms comes from the QUAID software online documentation: "There is a word or expression that may be unfamiliar to some respondents. The term may be rare in the English language. The term may involve an abbreviation or acronym (e.g. IRS, TVA) that is unfamiliar to some individuals and cultures. The term may contain a symbol that is not frequently used. The term may be misspelled. You need to make a decision whether the term is sufficiently unfamiliar that it will present a problem to your population of respondents." 1 = Yes 2 = No (reference category)
Sensitivity score	The authors rated each of the 270 questions using the following questions: Rating item #1: "This question is very personal." 1 = completely disagree to 5 = completely agree. Rating item #2: "I would be uncomfortable asking this question." 1 = completely disagree to 5 = completely agree. Each rater's score for each question was summed for each of the two rating items. For example, for a given question under analysis the following scores were assigned by each of the five raters using rating item #1: 1, 3, 3, 2, and 1. The total score assigned to that question using rating item #1 would be 10. The same procedure was followed using rating item #2. Since the summed ratings based on these two rating items were highly correlated (0.94), the two scores were summed to create an index of sensitivity. Actual scores ranged from 10 to 37, with higher scores indicating greater sensitivity. Due to a highly skewed distribution, the scores were recoded into four discrete categories (using quartiles as cut points) for analysis: 1 = Score of 10 2 = Scores of 11–12 3 = Scores of 13–17 4 = Scores of 18–37 Note: To assess inter-rater reliability for each rating item, we computed Kendall's coefficient of concordance (i.e. Kendall's W). Kendall's W ranges from 0 (no agreement) to 1 (complete agreement). For rating item #1, Kendall's W was 0.59, while for rating item #2, Kendall's W was 0.51. Both indicate a moderate level of concordance or agreement among the five raters.

Table 7.1 (Continued)

Variable	Description and coding
Question characteristics: factors affecting respondent processing efficiency	
Definition/clarifying text	This variable captures whether or not there are definitions of key terms or other clarifying text provided in the question itself. Example: Have you EVER had a pneumonia shot? This shot is usually given only once or twice in a person's lifetime and is different from the flu shot. It is also called the pneumococcal vaccine. 1 = Yes 2 = No (reference category)
Question characteristics: factors affecting interviewer task complexity	
Help screen	This variable captures whether or not a separate screen, accessed by pressing the F1 function key, with help text accompanies the question. If needed, the interviewer can read the help screen text to the respondent. 1 = Yes 2 = No (reference category)
Optional text	This variable captures whether or not there is text, placed in parentheses or an interviewer instruction, which can be read to the respondent at the interviewer's discretion. Example: What kind of business or industry was this? (For example: TV and radio mgt., retail shoe store, State Department of Labor) 1 = Yes 2 = No (reference category)
Question characteristics: necessary question features	
Type of question	This variable measures whether or not the question captures factual or demographic versus attitudinal or subjective information. Example of a factual/demographic question: Have you ever held a job or worked at a business? Example of an attitudinal/subjective question: How worried are you right now about not being able to pay medical costs for normal healthcare? 1 = Factual/demographic 2 = Attitudinal/subjective (reference category)
Response option format	Example of "yes/no" format: Have you EVER been told by a doctor or other health professional that you had coronary heart disease? Example of "integer" format: On the average, how many cigarettes do you now smoke a day? Example of "pick one answer from a list" format: Compared with 12 MONTHS AGO, would you say your health is better, worse, or about the same?

Table 7.1 (Continued)

Variable	Description and coding
	Example of "text entry" format: For whom did you work at your MAIN job or business? (Name of company, business, organization or employer) Example of "mark-all-that-apply" format: Why did you not buy the plan? 1. Turned down, 2. Cost, 3. Pre-existing condition, 4. Got health insurance from other source, 5. Other reason 1 = Integer 2 = Pick one answer from a list 3 = Text entry 4 = Mark all that apply 5 = Yes/no (reference category)
Question length	Number of characters in the question. Due to a highly skewed distribution, the variable was recoded into four discrete categories (based on quartile thresholds) for analysis: 1 = <70 characters 2 = ≥70 and <103 characters 3 = ≥103 and <143 characters 4 = ≥143 characters (reference category) Note: Optional text was included in the computation of question length.
Respondent/case characteristics	
Age	1 = 18–24 2 = 25–44 3 = 45–64 (reference category) 4 = 65+
Sex	1 = Male 2 = Female (reference category)
Race/ethnicity	1 = Hispanic 2 = Non-Hispanic black 3 = Non-Hispanic other 4 = Non-Hispanic white (reference category)
Education	1 = Less than high school/GED 2 = High school diploma/GED 3 = Some college/AA degree 4 = Bachelor's degree or higher (reference category)
Nativity	1 = Foreign born 2 = US born (reference category)

Table 7.1 (Continued)

Variable	Description and coding
Reported health status	1 = Poor/fair 2 = Good 3 = Very good/excellent (reference category)
Cognitive difficulties	Based on the following question: "{Is person} LIMITED IN ANY WAY because of difficulty remembering or because you/they experience periods of confusion?" 1 = Yes 2 = No (reference category)
Total family income	1 = Less than $35 000 2 = $35 000–$74 999 3 = $75 000–$99 999 (reference category) 4 = $100 000 or more 5 = Don't know or refused
Metropolitan statistical area (MSA) status	A metropolitan statistical area (MSA) is defined as a county or group of contiguous counties that contain at least one urbanized area of 50 000 population or more. Adults were defined as living in the central or principal city of an MSA (MSA, central city), in an MSA but not in the central city (MSA, non-central city), or not in a MSA. "Not in a MSA" indicates that the adult lives in a nonmetropolitan area, defined as an area that does not include a large urbanized area; these areas are generally thought of as more rural. 1 = MSA, central city 2 = MSA, non-central city 3 = Not in MSA (reference category)
Region of residence	1 = Northeast 2 = South 3 = West 4 = Midwest (reference category)
Sample adult interview conducted primarily by telephone	1 = Yes 2 = No (reference category)
Householder(s) made refusal-like statements	This variable is based on contact history data collected with the case. A case would be coded "yes" on this variable if the interviewer recorded that one or more householders made one or more of the following statements: "not interested," "hang-up/slams door on interviewer," or "hostile or threatens interviewer." 1 = Yes 2 = No (reference category)

Table 7.1 (Continued)

Variable	Description and coding
Householder(s) mentioned privacy concerns or made statements about the survey content	This variable is based on contact history data collected with the case. A case would be coded "yes" on this variable if the interviewer recorded that one or more householders made one or more of the following statements: "survey is voluntary," "privacy concerns," "local/state/federal government concerns," or "asks questions about the survey." 1 = Yes 2 = No (reference category)
Householder(s) mentioned time constraints	This variable is based on contact history data collected with the case. A case would be coded "yes" on this variable if the interviewer recorded that one or more householders made one or more of the following statements: "too busy," "interview takes too much time," or "scheduling difficulties." 1 = Yes 2 = No (reference category)
Interviewer performance measures	
Item nonresponse rate for preceding calendar quarter	This measure is based on 226 questions asked in the Family, Sample Child, and Sample Adult Core interviews, and was defined as: total number of don't know and refused responses for the preceding calendar quarter divided by total number of questions (of the 226) asked in the preceding calendar quarter. The counts were taken from sufficient partial or fully completed interviews. The resulting distribution of item nonresponse rates across interviewers was highly skewed. Therefore, the measure was recoded into four discrete categories using quartiles as the cut points: 1 = 0.00%–<0.31% (reference category) 2 = ≥0.31%–<0.57% 3 = ≥0.57%–<0.94% 4 = ≥0.94%
Cooperation rate for preceding calendar quarter	The interviewer cooperation rate was defined as: total number of interviews completed by the interviewer in the preceding calendar quarter divided by total number of in-scope cases for the interviewer in the preceding calendar quarter. The resulting distribution of interviewer cooperation rates was highly skewed. Therefore, the measure was recoded into four discrete categories using quartiles as the cut points: 1 = <73.91% 2 = ≥73.91%–<83.87% 3 = ≥83.87%–<91.30% 4 = ≥91.30% (reference category)

Table 7.1 (Continued)

Variable	Description and coding
Pace of sample adult interview for preceding calendar quarter	The interviewer pace of sample adult interview was defined as: total seconds spent on sample adult questions (sufficient partial and fully complete sample adult interviews) in the preceding calendar quarter divided by the total number of sample adult questions asked in the preceding calendar quarter. To guard against inflation of interview pace, we took the longest visit to a question when multiple visits occurred, and question times outside the 99th percentile of the time distribution for that question were recoded to the mean for that question. Since the resulting interview pace measure was highly skewed, it was recoded into one of the following four categories: 1 = <6.84 seconds per item 2 = ≥6.84–<8.14 seconds per item 3 = ≥8.14–<9.88 seconds per item 4 = ≥9.88 seconds per item (reference category)

dependent variables.[3] We assume that an interviewer's recent past performance is a predictor of current performance. More detailed information on the creation and coding of these measures can be found in Table 7.1, while quarterly descriptive statistics are presented in Table 7.3.

7.2.3 Data Structure

The data have a complex nested structure. Response time and item nonresponse are measured at the question-visit level. Question visits (level 1) are uniquely nested within questions (level 2), respondents (level 2), and interviewers (level 3). Respondents are also uniquely nested within interviewers. Questions, however, are not uniquely nested within respondents or interviewers, resulting in a cross-classified data structure. Since questions and respondents are treated as level 2 units in this analysis, response times and item nonresponse (level 1) are said to be cross-classified with questions and respondents. Figure 7.1 depicts the cross-classified data structure. The multilevel models described here account for the cross-classification.

For 2014, there were 36 697 adults included in the Sample Adult data file. Including all of them in the analysis yielded over 7 million visits to the 270 questions. The large data file and complex models produced estimation errors

3 If an interviewer worked during the fourth calendar quarter of 2014, the performance measures would be based on work completed during the third calendar quarter. For data collected during the first calendar quarter of 2014, interviewer performance measures were based on work completed during the fourth calendar quarter of 2013.

Table 7.2 Descriptive statistics for question and respondent/case characteristics: NHIS sample adult interview, 2014 (unweighted).

Question characteristics	Number of questions	Percent
Factors affecting respondent task complexity		
Unfamiliar technical terms		
Yes	162	60.0
No	108	40.0
Sensitivity score		
10	125	46.3
11–12	57	21.1
13–17	49	18.2
18–37	39	14.4
Factors affecting respondent processing efficiency		
Definition(s) and/or clarifying text		
Yes	51	18.9
No	219	81.1
Factors affecting interviewer task complexity		
Help screen		
Yes	58	21.5
No	212	78.5
Optional text		
Yes	98	36.3
No	172	63.7
Necessary question features		
Type of question		
Factual/demographic	224	83.0
Attitudinal/subjective	46	17.0
Response option format		
Yes/No	149	55.2
Integer	40	14.8
Pick one from a list	70	25.9
Text	6	2.2
Enter all that apply	5	1.9

Table 7.2 (Continued)

Question characteristics	Number of questions	Percent
Question length		
<70 characters	89	33.0
70–<103 characters	72	26.7
103–<143 characters	57	21.1
≥143 characters	52	19.3
Respondent/case characteristics	**Number of respondents/cases**	**Percent**
Age		
18–24	913	9.8
25–44	3078	33.0
45–64	3114	33.3
65+	2231	23.9
Sex		
Male	4169	44.7
Female	5166	55.3
Race/ethnicity		
Hispanic	998	10.7
Non-Hispanic white	6230	66.7
Non-Hispanic black	1289	13.8
Non-Hispanic other	818	8.8
Education		
Less than high school	1108	11.9
High school diploma/GED	2393	25.7
Some college/AA degree	3054	32.9
Bachelor's degree or higher	2741	29.5
Nativity		
US born	8121	87.0
Foreign born	1213	13.0
Reported health status		
Poor/fair	1327	14.2
Good	2466	26.4
Very good/excellent	5536	59.3

Table 7.2 (Continued)

Respondent/case characteristics	Number of respondents/cases	Percent
Cognitive difficulties		
Yes	411	4.4
No	8919	95.6
Total family income		
Less than $35 000	3381	36.2
$35 000–$74 999	2650	28.4
$75 000–$99 999	1001	10.7
$100 000 or more	1609	17.2
Unknown (refused/don't know)	695	7.4
MSA status		
MSA, central city	3006	32.2
MSA, non-central city	3967	42.5
Non-MSA	2363	25.3
Region of residence		
Northeast	1417	15.2
Midwest	2136	22.9
South	3053	32.7
West	2730	29.2
Sample adult interview conducted primarily by telephone		
Yes	1822	19.5
No	7514	80.5
Householder(s) made refusal-like statements		
Yes	1186	12.7
No	8145	87.3
Householder(s) mentioned privacy concerns or made statements about the survey content		
Yes	1459	15.6
No	7872	84.4
Householder(s) mentioned time constraints		
Yes	2790	29.9
No	6541	70.1

Table 7.3 Interviewer performance measures, lagged one calendar quarter ($n = 186$ interviewers for all of 2014): NHIS sample adult interview, 2014 (unweighted).

Interviewer performance measure	Quarter 1		Quarter 2		Quarter 3		Quarter 4	
	Number of interviewers	Percent	Number of interviewers	Percent	Number of interviewers	Percent	Number of interviewers	Percent
Item nonresponse rate (don't know and refused)								
0.00%–<0.31%	44	26.8	45	25.1	41	22.4	40	23.1
≥0.31%–<0.57%	43	26.2	47	26.3	49	26.8	38	22.0
≥0.57%–<0.94%	39	23.8	42	23.5	41	22.4	49	28.3
≥0.94%	38	23.2	45	25.1	52	28.4	46	26.6
Cooperation rate								
0.00%–<73.91%	41	25.0	53	29.6	45	24.6	42	24.3
≥73.91%–<83.87%	41	25.0	38	21.2	43	23.5	49	28.3
≥83.87%–<91.30%	32	19.5	32	17.9	44	24.0	36	20.8
≥91.30%	50	30.5	56	31.3	51	27.9	46	26.6
Sample adult interview pace								
<6.84 seconds per item	42	25.6	34	19.0	44	24.0	41	23.7
≥6.84–<8.14 seconds per item	40	24.4	39	21.8	36	19.7	34	19.7
≥8.14–<9.88 seconds per item	42	25.6	53	29.6	51	27.9	44	25.4
≥9.88 seconds per item	40	24.4	53	29.6	52	28.4	54	31.2

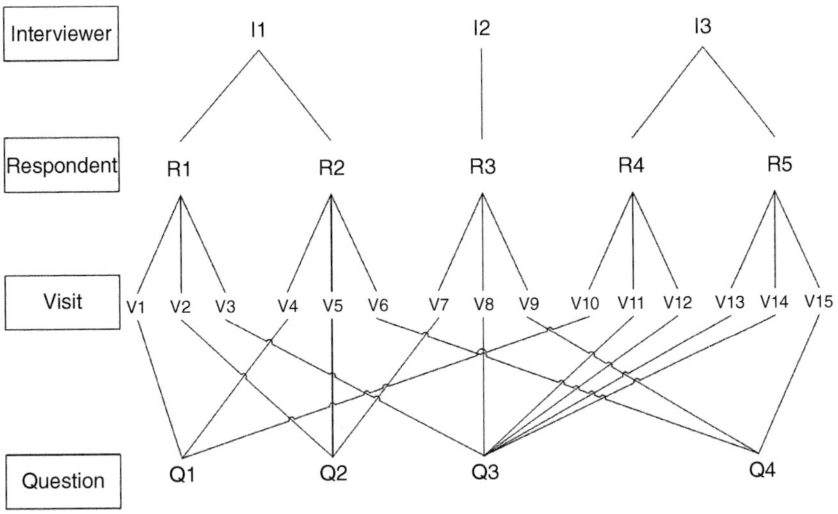

Figure 7.1 Network graph depicting level two cross-classification of visits (response time, item nonresponse) by questions and respondents, with respondents nested within interviewers.

in initial analyses. To address these problems, we took three 33% samples of the 562 interviewers who completed 20 or more sample adult interviews in 2014. Three sample replicates were randomly drawn using PROC SELECT in SAS (V9.4). We then randomly sampled one of the three replicates to serve as an initial sample and a second replicate to serve as a validation sample. All respondents and associated question visits were retained from each selected interviewer. This served two purposes: (i) it greatly reduced the number of question visits for the analysis (to improve analytic performance), and (ii) it produced a more stable data structure by ensuring that at least 20 sample adults were nested within each interviewer. In addition, the number of respondents, and therefore question visits, were further reduced by dropping all interviews conducted in Spanish. (The aforementioned Question Understanding AID, QUAID coding [described in Table 7.1 for the measure of unfamiliar technical terms] was based solely on the English version of question text.)

Table 7.4 shows the breakdown of the units of analysis used in the multilevel models of response time and item nonresponse. A total of 1 751 355 visits to 270 unique questions were included in the analysis. The question visits were made by 186 interviewers across 9336 respondents. The mean number of respondents per interviewer was 50.2. In addition, each of the 9336

Table 7.4 Number of observations in the analysis.

Number of visits to questions	1 751 355
Number of questions	270
Number of respondents	9 336
Number of interviewers	186
Mean visits per question	6 486.5
Mean questions per respondent	187.6
Mean respondents per interviewer	50.2

respondents received, on average, 187.6 questions, and the average number of visits per question was 6486.5.[4]

7.2.4 Statistical Analyses

For response time, we estimate a three-level cross-classified random effects model with response times cross-classified by respondents and questions, and response times nested within respondents nested within interviewers. Following notation provided by Beretvas (2010) and presented in Olson and Smyth (2015), the base model predicts the natural logarithm of response time ($Y_{i(j1,\,j2)k}$) as a function of an overall mean (γ_{0000}), a random effect due to the question (u_{00j10k}), a random effect due to the respondent (u_{0j2k}), a random effect due to the interviewer (v_{000k}), and a residual term ($e_{i(j1,\,j2)k}$), where u_{0j10k}, u_{00j2k}, and v_{000k} are normally distributed with mean zero and variance σ_{uj10}, σ_{uj2}, and σ_{uk}, respectively, and $e_{i(j1,\,j2)k}$ is normally distributed with mean zero and variance $\sigma_e{}^2$:

$$Y_{i(j1,j2)k} = \gamma_{0000} + u_{0j10k} + u_{00j2k} + v_{000k} + e_{i(j1,j2)k}$$

To estimate the proportion of variance in the log of response time attributable to questions versus respondents versus interviewers, the base model is used to compute intra-class correlation coefficients (ICCs).[5] The ICC for questions is calculated as

$$\rho_{question} = \frac{\sigma_{uj10}}{\sigma_{uj10} + \sigma_{uj2} + \sigma_{uk} + \sigma_e^2}$$

4 Since not all sample adults were eligible for all questions, the average number of visits per question is less than the number of respondents included in the analysis.

5 The intra-class correlation coefficient or ICC is a measure of the correlation between units clustered within the same level (e.g. respondents, questions, interviewers). In the multilevel model setting, it represents the proportion of the total unexplained variation in the dependent variable that is attributable to differences between the levels.

The equation is modified with the appropriate variance term in the numerator for ICCs for respondents and interviewers (Raudenbush and Bryk 2002).

The base model is then expanded to include all covariates at the question, respondent, and interviewer levels:

$$Y_{i(j1,j2)k} = \gamma_{0000} + \sum_{m-1}^{p} \beta_m \text{Question\_char}_{j10} + \sum_{s=1}^{q} \beta_s \text{Respondent\_char}_{j2} + \sum_{t=1}^{r} \beta_t \text{Interviewer\_char}_k + v_{000k} + u_{0j10k} + u_{00j2k} + e_{i(j1,j2)k}$$

For item nonresponse, we estimate a two-level, cross-classified random effects model (logistic regression) with item nonresponse cross-classified by questions and respondents. Preliminary analysis revealed that little to no variance in item nonresponse was attributable to the interviewer; hence, random effects for interviewer were excluded from the model. Using the previous notation, the base model predicts item nonresponse $\left(\ln\left(\frac{p_{i(j1,j2)}}{1-p_{i(ji,j2)}}\right)\right)$ as a function of an overall mean (γ_{000}), a random effect due to the question (u_{0j10}), a random effect due to the respondent (u_{00j2}), where u_{0j10} and u_{00j2} are normally distributed with mean zero and variance $\sigma^2_{u_{j10}}$ and $\sigma^2_{u_{j2}}$, respectively. The equation for the base model, therefore, is:

$$\ln\left(\frac{p_{i(j1,j2)}}{1 - p_{i(j2,j2)}}\right) = \gamma_{000} + u_{0j10} + u_{00j2}$$

As with response time, we calculate the ICCs at each level to estimate the proportion of variance in item nonresponse attributable to questions versus respondents. Because item nonresponse is a dichotomous dependent variable, we set the level 1 (visit) variance to 3.29, which is the variance of the underlying standard logistic distribution (Snijders and Bosker 1999). The following formula is used to approximate the value of the ICC for questions:

$$\rho_{question} = \frac{\sigma^2_{u_{j10}}}{\sigma^2_{u_{j10}} + \sigma^2_{u_{j2}} + 3.29}$$

To produce the ICC for respondents, the equation is modified by replacing the question variance component with the respondent variance component in the numerator.

The base model is then expanded to include all covariates at the question and respondent levels:

$$\ln\left(\frac{p_{i(j1,j2)}}{1-p_{i(j1,j2)}}\right) = \gamma_{000} + \sum_{m-1}^{p} \beta_m \text{Question\_char}_{j10} + \sum_{s=1}^{q} \beta_s \text{Respondent\_char}_{j2} + u_{0j10} + u_{00j2} + e_{i(j1,j2)}$$

The response time and item nonresponse models were estimated on both the initial and validation samples. While tabled results are presented for the initial sample and discussed in Section 7.3, we also note differences in findings across the two samples. All analyses are unweighted. The descriptive statistics were produced using SAS PROC FREQ (v9.4). The three-level, cross-classified regression of the log of response time was estimated with SAS PROC MIXED (v9.4) using restricted maximum likelihood estimation. The two-level, cross-classified logistic regression of item nonresponse was estimated with SAS PROC GLIMMIX (v9.4) using pseudo-likelihood estimation.

7.3 Results

In this section we describe the results from the multilevel models predicting log-transformed response times and item nonresponse. In general, the results were consistent between the initial sample and the validation sample. We note the key differences in the direction of a coefficient or large changes in levels of significance.

Table 7.5 shows the results of the base and full models predicting log-transformed response times. Focusing on the base model, there are significant variance terms for the question, respondent, and interviewer, indicating significant variability in response time at all three levels (see Random Effects at the bottom of the table). Furthermore, the computation of ICCs reveals that there is considerably more variation in response time at the question level compared to the interviewer and respondent levels. The ICC for questions is .329, indicating that 32.9% of the variance in response time is due to questions compared to 14.7% due to respondents and 9.1% due to interviewers. Finally, moving from the base model to the full model of response times, and using Akaike's information criteria (AIC) as the criterion, the addition of the question, respondent, and interviewer measures improved model fit (lower AIC indicates better model fit).

Focusing on the fixed effects in the full model, factors affecting respondent task complexity, including unfamiliar technical terms and item sensitivity, were not significantly related to response times. Question characteristics affecting respondent processing efficiency, such as the presence of definitions

Table 7.5 Cross-classified multilevel OLS regressions of log-transformed response times with fixed effects for question and respondent characteristics, and interviewer performance measures: NHIS sample adult interview, 2014 (unweighted).

Parameter	Null model			Full model		
	Coeff.	Sig.	SE	Coeff.	Sig.	SE
Fixed effects						
Intercept	1.71	***	0.04	1.68	***	0.10
Question characteristics: factors affecting respondent task complexity						
Unfamiliar technical terms (ref = no)				−0.01		0.04
Sensitivity score (ref = 11–12)						
10				0.03		0.05
13–17				0.08		0.06
18–37				0.12		0.07
Question characteristics: factors affecting respondent processing efficiency						
Definition/clarifying text				0.42	***	0.05
Question characteristics: factors affecting interviewer task complexity						
Help screen				0.11	*	0.05
Optional text				−0.49	***	0.04
Question characteristic: necessary question features						
Type of question (ref = attitudinal/subjective)						
Factual/demographic				−0.17	*	0.07
Response option format (ref = yes/no)						
Integer				0.53	***	0.06
Pick one from a list				0.35	***	0.06
Text				0.95	***	0.14
Enter all that apply				0.78	***	0.15
Question length (ref = ≥143 characters)						
<70 characters				−0.03	***	0.01
≥70–<103 characters				−0.03	***	0.01
≥103–<143 characters				0.03	***	0.01
Item nonresponse				0.10	***	0.01
Respondent/case characteristic						
Age (ref = 45–64)						
18–24				−0.08	***	0.01
25–44				−0.06	***	0.01
65+				0.05	***	0.01

Table 7.5 (Continued)

Parameter	Null model			Full model		
	Coeff.	Sig.	SE	Coeff.	Sig.	SE
Male respondent				−0.02	***	0.004
Race/ethnicity (ref = non-Hispanic white)						
Hispanic				−0.01		0.01
Non-Hispanic black				0.000 2		0.01
Non-Hispanic other				0.01		0.01
Education (ref = bachelor's degree or higher)						
Less than high school				−0.03	***	0.01
High school diploma/GED				−0.01		0.01
Some college/AA degree				0.01		0.01
Foreign born				0.01		0.01
Reported health status (ref = very good/excellent)						
Poor/fair				0.07	***	0.01
Good				0.04	***	0.01
Cognitive difficulties				0.03	**	0.01
Total family income (ref = $75 000–$99 999)						
Less than $35 000				0.03	***	0.01
$35 000–$74 999				0.02	*	0.01
$100 000 or more				−0.01		0.01
Unknown (refused/don't know)				−0.04	***	0.01
MSA status (ref = non-MSA)						
MSA, central city				0.02		0.01
MSA, non-central city				0.01		0.01
Region of residence (ref = Midwest)						
Northeast				0.02		0.05
South				0.03		0.04
West				0.09		0.05
Sample adult interview conducted primarily by telephone				0.12	***	0.01
Householder(s) made refusal-like statements				−0.01	*	0.01

Table 7.5 (Continued)

Parameter	Null model			Full model		
	Coeff.	Sig.	SE	Coeff.	Sig.	SE
Householder(s) mentioned privacy concerns or made statements about survey content				−0.03	***	0.01
Householder(s) mentioned time constraints				−0.02	***	0.01
Interviewer performance measures						
Item nonresponse rate (ref = 0.00%–< 0. 31%)						
≥0.31%–<0.57%				−0.03	***	0.01
≥0.57%–<0.94%				−0.03	***	0.01
≥0.94%				−0.04	***	0.01
Cooperation rate (ref = ≥91.30%)						
0.0%–<73.91%				−0.003		0.01
≥ 73.91%–<83.87%				0.005		0.01
≥ 83.87%–<91.30%				0.000		0.01
Sample adult interview pace (ref = ≥9.88 seconds per item)						
<6.84 seconds per item				−0.06	***	0.01
≥6.84–<8.14 seconds per item				−0.03	**	0.01
≥8.14–<9.88 seconds per item				−0.01		0.01
Random effects						
$\sigma^2_{int: question}$	0.25	***	0.02	0.10	***	0.01
$\sigma^2_{int: \text{respondent (interviewer)}}$	0.04	***	0.001	0.04	***	0.001
$\sigma^2_{int: interviewer}$	0.07	***	0.01	0.06	***	0.01
σ^2 (residual variance)	0.40	***	0.0004	0.40	***	0.000 4
Model fit						
−2 RE/ML log-likelihood	3 377 647		3 376 000			
AIC	3 377 655		3 376 008			
N	1 736 008		1 736 008			

Note: Coeff. = coefficient, sig. = significant, SE = standard error, AIC = Akaike's information criteria.

*$p < 0.05$; **$p < 0.01$; ***$p < 0.001$.

or clarifying text, significantly increase response times. Two features that affect interviewer task complexity are related to response times. Help screens significantly increase response times, and optional text significantly decreases response times. There are also a number of necessary question features that affect response times. Factual/demographic questions significantly decrease response times compared to attitudinal or subjective questions. Questions with a yes/no response format take significantly less time to answer compared to questions using all other response formats. The findings in Table 7.5 suggest a somewhat complex relationship between question length and response times such that questions between 103 and 142 characters have longer response times than questions with 143 characters or more; however, this result was not replicated in the validation sample. Therefore, we believe this is a spurious result.

Some respondent characteristics are also related to response times. Response times increase with age, men have significantly shorter response times compared to women, and respondents with less than a high school education have shorter response times compared to respondents with a bachelor's degree or higher. Response times increase as health declines. Similarly, respondents with cognitive difficulty have longer response times than those without cognitive difficulties. Respondents with less than $35 000 in total family income have longer response times compared to those who have a household income between $75 000 and $99 999. However, respondents from families where the income information was refused or a "don't know" response was provided have shorter response times. Paradata from the interview were also used to predict response times. Interviews conducted primarily by telephone have longer response times than interviews administered primarily in person. Finally, respondents from households where householders made refusal-like statements, mentioned a privacy concern, or mentioned time constraints tend to have shorter response times.

Interviewer performance measures (lagged one calendar quarter) are also significantly related to response times. Response times are significantly lower for interviewers in the top three quartiles with respect to their overall item nonresponse rate compared to interviewers whose item nonresponse rate is in the lowest quartile. However, this result was not replicated in the validation sample where all of the coefficients for interviewer item nonresponse rate are nonsignificant ($p > 0.2$ for all). Not surprisingly, interviewers who had a faster average pace per item in the previous calendar quarter tend to elicit shorter response times.

Overall, the inclusion of the question, respondent, and interviewer covariates had the largest impact on the variance at the question level. Roughly speaking, the addition of the covariates reduced the variance due to questions by 60%. There was a 13% reduction in variance at the interviewer level and just under a 9% reduction in variance at the respondent level.

Table 7.6 shows the results of the base and full logistic regression models predicting item nonresponse. As with response time, significant variance terms are observed for questions and respondents; hence, item nonresponse varies significantly at both levels (see Random Effects at the bottom of the table). In addition, the ICCs show that the largest component of variance is at the question level. The ICC for questions is .484, indicating that 48.4% of the variance in item nonresponse is due to questions compared to just 1.5% for respondents. As noted earlier, preliminary analyses revealed little to no variance in item nonresponse at the interviewer level, so random effects for interviewers were excluded from the models.

Focusing on the full model, one factor affecting respondent task complexity – question sensitivity – was significantly related to item nonresponse, with significantly higher item nonresponse observed for questions with both high and low levels of sensitivity compared to questions with medium sensitivity scores. The lone question characteristic affecting respondent processing efficiency – definitions or clarifying text – is also significantly related to item nonresponse. Item nonresponse is significantly more likely for questions with definitions or clarifying text. The coefficient for definitions or clarifying text, however, was not significant in the validation sample ($p = 0.10$). Item nonresponse is significantly less likely in questions with optional text. We anticipated that optional text may increase interviewer task complexity and lead to higher item nonresponse so we are uncertain about the interpretation of this finding. Some necessary question features are also related to item nonresponse. For example, item nonresponse is more likely for questions with integers, pick one from a list, and text as responses compared to questions with yes/no answer choices. Item nonresponse was more likely for questions between 70 and 103 characters compared to questions with 143 characters or more. The results in the validation sample, however, indicate that item nonresponse is significantly less likely with questions with fewer than 70 characters compared to questions with 143 characters or more. Again, the finding for question length in the initial sample may be spurious.

Respondent characteristics are also significantly related to item nonresponse. Item nonresponse was more likely for younger respondents aged 18–24 compared to respondents aged 45–64. In the validation sample, however, there was no significant difference between these two age groups. In addition, adults aged 65 and over were significantly more likely to produce item nonresponse than adults aged 45–64 in the validation sample. Item nonresponse is more likely for respondents who are either non-Hispanic black or non-Hispanic other race compared to non-Hispanic white. Item nonresponse is more likely for respondents with cognitive difficulties compared to those without cognitive difficulties and for respondents in poor or fair health compared to those in very good or excellent health. Socioeconomic status is also related to item nonresponse. Item nonresponse is less likely for respondents with a high school diploma or

Table 7.6 Cross-classified multilevel logistic regression of item nonresponse with fixed effects for question and respondent characteristics: NHIS sample adult interview, 2014 (unweighted).

Parameter	Null model			Full model		
	Coeff.	Sig.	SE	Coeff.	Sig.	SE
Fixed effects						
Intercept	−5.87	***	0.12	−8.13	***	0.50
Question characteristics: factors affecting respondent task complexity						
Unfamiliar technical terms (ref = no)				0.08		0.23
Sensitivity score (ref = 11–12)						
10				0.76	**	0.29
13–17				0.49		0.35
18–37				1.25	**	0.38
Question characteristics: factors affecting respondent processing efficiency						
Definitions/clarifying text				0.65	*	0.28
Question characteristics: factors affecting interviewer task complexity						
Help screen				−0.33		0.29
Optional text				−0.47	*	0.23
Question characteristic: necessary question features						
Type of question (ref = attitudinal/subjective)						
Factual/demographic				0.30		0.40
Response option format (ref = yes/no)						
Integer				1.75	***	0.33
Pick one from a list				0.89	**	0.34
Text				2.52	**	0.78
Enter all that apply				1.19		0.82
Question length (ref = ≥143 characters)						
<70 characters				−0.03		0.08
≥70–<103 characters				0.44	***	0.10
≥103–<143 characters				0.23		0.12
Natural log of response time				0.04	***	0.01
Respondent characteristics						
Age (ref = 45–64)						
18–24				0.10	**	0.03
25–44				−0.06	**	0.02
65+				0.04		0.02

Table 7.6 (Continued)

Parameter	Null model			Full model		
	Coeff.	Sig.	SE	Coeff.	Sig.	SE
Male respondent				−0.01		0.02
Race/ethnicity (ref = non-Hispanic white)						
Hispanic				0.01		0.03
Non-Hispanic black				0.18	***	0.02
Non-Hispanic other				0.10	**	0.03
Education (ref = bachelor's degree or higher)						
Less than high school				0.03		0.03
High school diploma/GED				−0.07	**	0.02
Some college/AA degree				−0.13	***	0.02
Foreign born				0.27	***	0.03
Reported health status (ref = very good/excellent)						
Poor/fair				0.07	**	0.03
Good				−0.01		0.02
Cognitive difficulties				0.40	***	0.04
Total family income (ref = $75 000–$99 999)						
Less than $35 000				0.08	**	0.03
$35 000–$74 999				0.06		0.03
$100 000 or more				−0.01		0.03
Unknown (refused/don't know)				0.93	***	0.03
MSA status (ref = non-MSA)						
MSA, central city				0.10	***	0.02
MSA, non-central city				0.09	***	0.02
Region of residence (ref = Midwest)						
Northeast				0.53	***	0.03
South				0.04		0.02
West				0.25	***	0.02
Sample adult interview conducted primarily by telephone				0.24	***	0.02
Householder(s) made refusal-like statements				0.27	***	0.02
Householder(s) mentioned privacy concerns or made statements about survey content				0.60	***	0.02

Table 7.6 (Continued)

Parameter	Null model			Full model		
	Coeff.	Sig.	SE	Coeff.	Sig.	SE
Householder(s) mentioned time constraints				0.10	***	0.02
Random effects						
$\sigma^2_{int:\ question}$	3.18	***	0.29	2.77	***	0.27
$\sigma^2_{int:\ \mathrm{respondent}}$	0.10	***	0.02	0.09	***	0.02
Model fit						
−2 residual log pseudo-likelihood		15 040 634	15 038 893			
Pseudo-AIC		15 040 638	15 038 897			
N		1 736 008	1 736 008			

Note: Coeff. = coefficient, sig. = significant, SE = standard error, AIC = Akaike's information criteria.
$^*p < 0.05$; $^{**}p < 0.01$; $^{***}p < 0.001$.

GED or some college/AA degree compared to respondents with a bachelor's degree or higher, and more likely for respondents who report a family income of less than $35 000 and those who do not report income compared to those who have family incomes between $75 000 and $99 999. Respondents who are foreign born also have a higher likelihood of item nonresponse. Respondents from MSAs have a higher likelihood of item nonresponse compared to those who are not from a MSA. Finally, item nonresponse is higher for respondents from the Northeast and West compared to the Midwest. Some paradata indicators were also significant predictors of item nonresponse. Interviews conducted primarily by telephone have a higher likelihood of item nonresponse. Interviews where the householder(s) made refusal-like statements, mentioned privacy concerns, or mentioned time constraints also have a higher likelihood of item nonresponse. Finally, longer response times increase the likelihood of item nonresponse.

Unlike the full model results for response time, the addition of question and respondent covariates had a roughly similar impact on variance at each level, with a 13.1% reduction in variance at the question level and a 16.1% reduction in variance at the respondent level.

7.4 Discussion

In this chapter we explored the joint effects of question characteristics, respondent characteristics, and interviewer performance on response times and item nonresponse. Our models found significant predictors of variability in both response times and item nonresponse that are largely consistent with previous studies using multilevel models to predict response times. Table 7.7 summarizes the main findings from the literature on response time and the current study. There were no other studies to directly compare our results to with respect to the impact of question characteristics on item nonresponse.

Our findings have implications for questionnaire design and evaluation. The finding that question characteristics account for the majority of variance in data quality is reassuring because this is one part of the survey process that survey designers can control. These findings can help inform survey designers to allocate resources in the effort to improve data quality (Olson and Smyth 2015; Couper and Kreuter 2013). We explained more than half of the variability in response times at the question level; however, many of the question characteristics that explain this variability are necessary features of the question that are difficult for a question designer to change (e.g. question type, response option format, etc.). We were less successful at explaining question-level variability for item nonresponse, but there were still a number of question characteristics that were significant predictors.

Respondent characteristics were also associated with response times and item nonresponse, although not as strongly as question characteristics. This indicates the potential importance of controlling for respondent-level characteristics when evaluating questions with existing data. For example, it is important to ensure that any findings with respect to a problematic question are not simply due to the composition of the sample. In addition, it is possible for future research to explore cross-level interactions between question and respondent characteristics. This will help to identify questions that may be more difficult for certain subgroups of respondents. One contribution from our research is that measures of respondent cooperation are significant predictors of both response times and item nonresponse. These variables should be included in future multilevel modeling of survey data quality measures. Unfortunately, we could only include a few interviewer-level predictors in the model so we were not able to explain much of the variation caused by interviewers.

There are at least two different uses of multilevel models with regard to question evaluation. One approach is to create generalizable knowledge about characteristics of questions that influence response times or item nonresponse. This seems to be the goal of many existing studies. Some of the existing findings, including ours, show that the results across evaluations using multilevel models are likely to vary depending on the specific study. For example, studies using different modes of data collection might get different results. In our study, we found differences in response times and item nonresponse when comparing interviews conducted mostly by telephone and those conducted face to face. Similarly, Couper and Kreuter (2013) find differences in response times between ACASI and CAPI. This will make it challenging to create generalizable findings about characteristics of questions

Table 7.7 Summary of literature using multilevel models to predict response times and comparison to current study.

Variable	Literature	Current study – response times	Current study – item nonresponse
Necessary question features			
Mode	• Response time is faster in ACASI compared to CAPI (Couper and Kreuter 2013)	• Response time is slower in interviews completed mostly by telephone compared to in-person	• Item nonresponse is more likely in interviews completed mostly by telephone compared to in-person
Position	• Response time gets faster as respondents get closer to the end of the questionnaire (Yan and Tourangeau 2008) • Response time gets slower as respondents get closer to the end of the questionnaire (Couper and Kreuter 2013)	• Not assessed	• Not assessed
Question length	• Response time increases with question length (Olson and Smyth 2015; Couper and Kreuter 2013; Yan and Tourangeau 2008)	• Response time is faster for questions with fewer than 103 characters compared to questions with 143 characters or more	• Inconclusive, but item nonresponse is consistently less likely for shorter questions in the initial and validation sample
Question type	• Response time is slower for behavior questions compared to attitude/opinion and demographic questions (Olson and Smyth 2015) • Response time is slower for factual and attitudinal questions compared to demographic questions (Yan and Tourangeau 2008)	• Response time is slower for attitudinal/subjective questions compared to factual/demographic questions	• Item nonresponse is not related to question type
Number of response options	• Response time increases with the number of response options (Olson and Smyth 2015; Yan and Tourangeau 2008)	• Not assessed	• Not assessed

Response option format	• Response time is slower for open-ended textual compared to open-ended numeric, closed ordinal, and yes/no questions (Olson and Smyth 2015) • Response time is slower for open-ended compared to fixed-choice, integer, and multiple response questions (Couper and Kreuter 2013)	• Response time is faster for yes/no questions compared to integer, pick one, open-ended textual, and enter all that apply questions	• Item nonresponse is more likely for integer, pick one, and open-ended textual compared to yes/no questions
Respondent task complexity			
Question complexity	• Response time increases with question reading level (Olson and Smyth 2015) • Response time increases with number of clauses and number of words per clause (Yan and Tourangeau 2008)	• Not assessed	• Not assessed
Sensitivity	• Response time is faster for sensitive questions compared to non-sensitive questions (Olson and Smyth 2015)	• Item sensitivity is not related to response times	• Item nonresponse is more likely for least sensitive and most sensitive questions
Interviewer task complexity			
Backing up in questionnaire	• Response time is slower for questions where the interviewer backs up (Olson and Smyth 2015)	• Not assessed	• Not assessed
Showcards	• Response time is slower for questions with showcards (Couper and Kreuter 2013)	• Not assessed	• Not assessed
Help screens	• Response time is slower for questions with help screens (Couper and Kreuter 2013)	• Response time is slower for questions with help screens	• Item nonresponse is not related to help screens
Question fills	• Response times are faster for women when questions have fills, but slower for men (Couper and Kreuter 2013)	• Not assessed	• Not assessed

Table 7.7 (Continued)

Variable	Literature	Current study – response times	Current study – item nonresponse
Interviewer instructions	• Response times are faster for questions with interviewer instructions (Couper and Kreuter 2013)	• Not assessed	• Not assessed
Optional text	• Not assessed	• Response time is faster for questions with optional text	• Item nonresponse is less likely for questions with optional text
Respondent processing efficiency			
Definitions	• Response times are slower for questions with definitions (Olson and Smyth 2015)	• Response times are slower for questions with definitions or clarifying text	• Item nonresponse is more likely for questions with definitions and clarifying text
Respondent characteristics			
Sex	• Sex is not related to response times (Olson and Smyth 2015)	• Response times are faster for men	• Sex is not related to item nonresponse
Age	• Response times increase with age (Olson and Smyth 2015; Couper and Kreuter 2013; Yan and Tourangeau 2008)	• Response times increase with age	• Item nonresponse is consistently more likely for respondents 65+ in the initial and validation sample
Employment	• Response times are faster for those who are employed (Olson and Smyth 2015)	• Not assessed	• Not assessed
Computer ownership	• Response times are faster for those with a computer (Olson and Smyth 2015) • Response times are faster for advanced web users (Yan and Tourangeau 2008)	• Not assessed	• Not assessed

Race	• Response times are slower for black and other race compared to white respondents (Couper and Kreuter 2013)	• Race is not related to response times	• Item nonresponse is more likely for non-Hispanic other race compared to non-Hispanic white
Marital status	• Response times are faster for married, cohabitating, and formerly married respondents compared to never-married respondents (Couper and Kreuter 2013)	• Not assessed	• Not assessed
Education	• Response times decrease as education increases (Couper and Kreuter 2013; Yan and Tourangeau 2008)	• Response times are faster for respondents with less than high school compared to those with a bachelor's degree or higher	• Inconsistent results between initial and validation samples
Language/nativity	• Response times are faster for English interviews compared to Spanish (Couper and Kreuter 2013)	• Foreign born is not related to response times	• Item nonresponse is more likely for foreign born
Health status	• Not assessed	• Response times are slower for respondents in fair, poor, or good health compared to very good or excellent health • Response times are slower for respondents with cognitive difficulties	• Item nonresponse is more likely for respondents in poor or fair health in initial sample but not in validation sample • Item nonresponse is more likely for respondents with cognitive difficulties
Income	• Not assessed	• Response times are slower for respondents with income less than \$35 K and faster for those who do not report income compared to those with income between \$35 K and \$75 K	• Item nonresponse is more likely for respondents with less than \$75 K and who do not report income compared to those with income between \$35 K and \$75 K

Table 7.7 (Continued)

Variable	Literature	Current study – response times	Current study – item nonresponse
MSA status	• Not assessed	• MSA status is not related to response time	• Item nonresponse is more likely for respondents in MSAs
Region	• Not assessed	• Region is not related to response time	• Item nonresponse is less likely for respondents in the Midwest compared to other regions
Interview cooperation	• Not assessed	• Response times are faster for respondents who initially refused the interview, mentioned privacy concerns, or mentioned time constraints	• Item nonresponse is more likely for respondents who initially refused the interview, mentioned privacy concerns, or mentioned time constraints
Interviewer characteristics			
Spanish interviewer	• Response times are slower for Spanish interviewers (Couper and Kreuter 2013)	• Not assessed	• Not assessed
Age of interviewer	• Response times increase with the age of the interviewer for male respondents (Couper and Kreuter 2013)	• Not assessed	• Not assessed
Interview order	• Response times are faster for cases later in the interviewer's case load (Olson and Smyth 2015)	• Not assessed	• Not assessed
Interviewer pace	• Not assessed	• Response times are shorter for interviewers with a faster overall pace	• Not assessed

that influence response times or item nonresponse. Furthermore, studies tend to use different question characteristics as predictors in the model. This will have to become more standardized across studies to better understand the consistency of findings.

Another approach is to use the results of multilevel models to aid in identifying questions that need further testing using other methods. Pickery and Loosveldt (2004) used similar multilevel models to predict item nonresponse. They then used the interviewer-level residuals from the model to identify interviewers whose mean "don't know" or "no opinion" rates deviated significantly from the general mean conditional on respondent-level variables included in the model. This enabled them to identify exceptional interviewers and interviewers who may need additional training. The same approach could be used to identify question characteristics associated with response times or item nonresponse rates that differ significantly from the general mean using a model like the one in this study. An analyst could then earmark questions with those characteristics for further evaluation. A method used by Kreuter (2002) to explore interviewer effects on responses to questions on fear of crime may also be applicable. For each question of interest, Kreuter first estimated the ICC for the entire set of interviewers. Next she estimated the ICC after dropping each individual interviewer, 18 in total, and all of his or her interviews from the analysis. She then plotted the 18 ICCs to identify interviewers who had a large effect on the ICC and, by extension, the design effect. If the ICC dropped significantly when an interviewer's cases were dropped, it meant that the interviewer had a substantial impact on the ICC and design effect for that question. One could use a similar process to determine which questions have a significant impact on the ICC or variability in response times and item nonresponse. For example, using the multilevel model presented here, we could estimate the question-level ICC for item nonresponse for the entire set of questions under analysis. Then, focusing on a subset of questions with characteristics related to high item nonresponse, each potentially problematic question would be dropped one at a time from the analysis and the ICCs plotted. Questions with a disproportionately large impact on the ICC or variability in item nonresponse would make good candidates for further evaluation using other question-evaluation techniques.

A challenge to identifying problematic questions with secondary data sources is the absence of true measures of reliability and validity, and previous research has relied on a single proxy measure of data quality (Yan and Tourangeau 2008; Couper and Kreuter 2013; Olson and Smyth 2015). We built upon this research by examining two measures (response times and item nonresponse), a broader range of question characteristics, and paradata measures. This may be important when a direct measure of quality is not available. For example, although many surveys have low levels of item nonresponse, there may be aspects of data quality that can be captured better through item nonresponse. We found no relationship between response times and question sensitivity, for example, but the most highly sensitive questions in our data did have higher levels of item nonresponse.

This research was subject to at least four limitations. First, the analyses were performed on a subset of NHIS interviewers and respondents. This was done to produce a stable data structure (i.e. nesting of sufficient numbers of respondents within interviewers) and to overcome estimation problems (i.e. insufficient processing memory) with the multilevel models when working with the full set of interviewers, respondents, and question visits. It is unclear if the observed results would be replicated on the full array of sample units. Second, we observed differences in findings when the models were estimated separately on the initial and validation samples. However, the differences were largely negligible when focusing on the impact of question characteristics, the primary focus of our work. Third, the level of agreement between raters with regard to the sensitivity rating items was moderate at best. To what extent this contributed to the unexpected curvilinear relationship between question sensitivity and item nonresponse is unknown. However, what is deemed sensitive by a given respondent (or interviewer) likely varies, and will be influenced by the context of the interview and the rapport established between interviewer and respondent. That the five raters did not always agree on which questions were sensitive is not surprising. And fourth, we lacked measures of interviewer characteristics such as age, race/ethnicity, education, and interviewing experience. While understanding the direct effects of interviewer characteristics on data quality is important, of particular interest is how these characteristics mediate the relationships between question characteristics and data quality. For example, more experienced interviewers may better convey the intent or alleviate the perceived sensitivity of some items, and/or elicit more substantive responses. Olson and Smyth (2015), for example, found that more experienced interviewers took longer to administer open-ended text questions and less time to administer yes/no questions than their less experienced counterparts.

In sum, this research shows the potential for using multilevel modeling to identify problematic questions that may need further evaluation. It is not possible to pretest every survey question prior to fielding surveys, and questions that performed well in pretests can develop problems later in production. Models such as the one used in the current study provide a way to utilize knowledge about question characteristics and data collected during production to identify questions that need further development. In the current survey-taking climate in which budgets are uncertain, if not shrinking, more efficient use of resources is a necessity. By utilizing by-product data (e.g. response times) of production data collection, survey organizations can more efficiently identify questions in need of more costly evaluations.

Biography

Aaron Maitland performed this work while employed at Westat. He is now employed by the National Center for Health Statistics and can be contacted at amaitland@cdc.gov.

Disclaimer

The views expressed in this chapter are those of the authors and do not necessarily represent the official views of the National Center for Health Statistics, the Centers for Disease Control and Prevention, or the U.S. Department of Health and Human Services.

References

Bassili, J.N. (1996). The how and the why of response latency measurement in telephone surveys. In: *Answering Questions: Methodology for Determining Cognitive and Communicative Processes in Survey Research* (eds. N. Schwarz and S. Sudman), 319–346. San Francisco, CA: Jossey-Bass.

Bassili, J.N. and Fletcher, J.F. (1991). Response time measurement in survey research: a method for CATI and a new look at nonattitudes. *Public Opinion Quarterly* 55: 331–346.

Bassili, J.N. and Scott, B.S. (1996). Response latency and question problems. *Public Opinion Quarterly* 60: 390–399.

Beatty, P. and Hermann, D. (2002). To answer or not to answer: decision processes related to survey item nonresponse. In: *Survey Nonresponse* (eds. R.M. Groves, D.A. Dillman, J.L. Eltinge and R.J.A. Little), 71–87. New York: Wiley.

Beretvas, S.N. (2010). Cross-classified and multiple-membership models. In: *Handbook of Advanced Multilevel Analysis* (eds. J.J. Hox and J.K. Roberts), 313–334. New York: Routledge.

Couper, M. and Kreuter, F. (2013). Using paradata to explore item level response times in surveys. *Journal of the Royal Statistical Society, Series A: Statistics in Society* 176 (Part 1): 271–286.

Draisma, S. and Dijkstra, W. (2004). Response latency and (para) linguistic expressions as indicators of response error. In: *Methods for Testing and Evaluating Survey Questionnaires* (eds. S. Presser, J.M. Rothgeb, M.P. Couper, et al.), 131–148. New York: Wiley.

Fazio, R.H., Sanbonmatsu, D.M., Powell, M.C., and Kardes, F.R. (1986). On the automatic activation of attitudes. *Journal of Personality and Social Psychology* 50: 229–238.

Graesser, A.C., Wiemer-Hastings, K., Kreuz, R. et al. (2000). QUAID: a questionnaire evaluation aid for survey methodologists. *Behavior Research Methods, Instruments, and Computers* 32: 254–262.

Heerwegh, D. (2003). Explaining response latencies and changing answers using client side paradata from a web survey. *Social Science Computer Review* 21: 360–373.

Kreuter, F. (2002). *Kriminalitatsfurcht: Messung und methodische Probleme.* Berlin: Leske and Budrich.

Krosnick, J.A. (1991). Response strategies for coping with cognitive demands of attitude measures in surveys. *Applied Cognitive Psychology* 5: 213–236.

Malhotra, N. (2008). Completion time and response order effects in web surveys. *Public Opinion Quarterly* 72: 914–934.

Mulligan, K., Grant, J.T., Mockabee, S.T., and Monson, J.Q. (2003). Response latency methodology for survey research: measurement and modeling strategies. *Political Analysis* 11: 289–301.

National Center for Health Statistics (2015). *2014 National Health Interview Survey (NHIS) Public Use Data Release: Survey Description.* Hyattsville, MD: National Center for Health Statistics.

Olson, K. and Smyth, J.D. (2015). The effect of CATI questions, respondents, and interviewers on response time. *Journal of Survey Statistics and Methodology* 3: 361–396.

Pickery, J. and Loosveldt, G. (2001). An exploration of question characteristics that mediate interviewer effects on item nonresponse. *Journal of Official Statistics* 17: 337–350.

Pickery, J. and Loosveldt, G. (2004). A simultaneous analysis of interviewer effects on various data quality indicators with identification of exceptional interviewers. *Journal of Official Statistics* 20: 77–89.

Raudenbush, S.W. and Bryk, A.S. (2002). *Hierarchical Linear Models: Applications and Data Analysis Methods,* 2e. Newbury Park, CA: Sage.

Shoemaker, P., Eicholz, M., and Skewes, E. (2002). Item nonresponse: distinguishing between don't know and refuse. *International Journal of Public Opinion Research* 14: 193–201.

Snijders, T. and Bosker, R. (1999). *Multilevel Analysis.* London: Sage.

Strack, F. and Martin, L.L. (1987). Thinking, judging and communicating: a process account of context effects in attitude surveys. In: *Social Information Processing and Survey Methodology* (eds. H.-J. Hippler, N. Schwarz and S. Sudman), 123–148. New York: Springer-Verlag.

Sudman, S. and Bradburn, N. (1974). *Response Effects in Surveys.* Chicago, IL: Adline Publishing Company.

Yan, T. and Tourangeau, R. (2008). Fast times and easy questions: the effects of age, experience and question complexity on web survey response times. *Applied Cognitive Psychology* 22: 51–68.

8

Response Burden: What Is It and What Predicts It?

Ting Yan[1], *Scott Fricker*[2], *and Shirley Tsai*[3]

[1] *Statistics and Evaluation Sciences Unit, Westat, Rockville, MD, USA*
[2] *Office of Survey Methods Research, Bureau of Labor Statistics, Washington, DC, USA*
[3] *Office of Technology Survey Processing, Bureau of Labor Statistics, Washington, DC, USA*

8.1 Introduction

Survey researchers have long worried about the burden their surveys place on respondents. As early as the 1920s, social scientists suggested that lengthy interviews would impose excessive burden (Chapin 1920; Sharp and Frankel 1983). Excessive demands violate the fragile social contract survey practitioners have with their respondents, and more recent research underscores the potential negative impacts respondent burden can have on survey participation and the quality of collected data. For example, response burden (variably defined) has been shown to be associated with a lower propensity to cooperate with survey requests (Groves et al. 1999; Rostald et al. 2011), higher attrition rates from panel surveys (e.g. Martin et al. 2001), higher break-off rates (e.g. Galesic 2006), and higher rates of missing data (Warriner 1991; Yan et al. 2014).

In his seminal paper on burden, Bradburn pointed out the challenges of studying burden: "The topic of respondent burden is not a neat, clearly defined topic about which there is an abundance of literature" (1977, p. 49). Nearly four decades later, burden still is "*not* a straightforward area to discuss, measure, and manage" (Jones 2012, p. 1). A review of the burden literature reveals that burden often is loosely defined and is measured in many different ways (see Table 8.1).

Only two studies have attempted to directly measure respondents' perceptions of burden. Galesic (2006) asked web-survey respondents to periodically report their level of experienced burden as they progressed through a survey questionnaire. The Consumer Expenditure Interview Survey (CE) similarly has incorporated direct measures of respondents' perceptions of burden in its production survey (Fricker et al. 2012; Yan et al. 2014). (This is the only U.S.

Advances in Questionnaire Design, Development, Evaluation and Testing, First Edition.
Edited by Paul C. Beatty, Debbie Collins, Lyn Kaye, Jose-Luis Padilla, Gordon B. Willis, and Amanda Wilmot.

Table 8.1 Conceptualization and measurement of burden in empirical research on burden.

Study	Purpose of study	Role of burden	Conceptualization of burden	Measurement of burden
Bergman and Brage (2008)	Effect of survey experience on future attitudes, intentions, and behaviors	Independent variable	Objective characteristic of survey	Time taken to complete earlier surveys
Filion (1981)	Impact of question wording and response burden	Independent variable	Objective characteristic of survey	Difficult survey items
Fricker et al. (2012)	Creating a summary burden score	Dependent variable	Subjective perception	Self-report of perceived burden
Geisen (2012)	Causes and effect of perceived burden	• Dependent variable • Independent variable	• Subjective perception as dependent variable • Objective characteristic as independent variable	• Subjective burden measured through perceived cognitive burden and perceived time burden • Objective burden measured through time taken to prepare for the survey and time taken to fill in the survey, # of items
Galesic (2006)	Effects of interest and burden on dropouts	Independent variable	Subjective perception	Self-report of perceived burden
Groves et al. (1999)	Effect of interview length, incentive, and refusal conversion on unit nonresponse	Independent variable	Objective characteristic of survey	Time taken to complete the interview
Hoogendoorn (2004)	Dependent interviewing	Dependent variable	• Objective characteristic of survey • Subjective perception	• Time spent on survey • Attitudes about survey (topic interesting; interview easy to do; questions clear; like layout; length of survey)

Hoogendoorn and Sikkel (1998)	Effect of burden on panel attrition	Independent variable	Objective characteristic of survey request	Number of survey requests within a period
Rostald et al. (2011)	Effect of questionnaire length on response rates	Dependent variable	"Effort required to answer a questionnaire"	Response rates
Sharp and Frankel (1983)	What affects burden	Dependent variable	Subjective perception	• Observation of respondent behavior (breakoff, unanswered item, signs of restlessness or discomfort) • Attitude about survey (time well spent; survey interesting; survey important; questions easy to answer; length of interview about right) • Behavioral intention (willingness to be re-interviewed; willingness to continue with survey for at least 15 minutes)
Singer et al. (1999)	Effects of incentive on response rates	• Control variable • Moderator variable	Objective character of survey	• Time taken to complete survey • Nature of survey request (diary, test, panel survey) • Presence of sensitive questions
Stocke and Langfeldt (2004)	Impact of survey experience on general attitudes toward survey	Independent variable	Subjective perception	• Perceived length of last interview • Feeling of exhaustion after last survey
Warriner (1991)	Accuracy of burdensome survey questions	Labeling the nature of survey items	Objective characteristic of survey items	Difficult survey items
Yan et al. (2014)	Effect of burden on data quality	Independent variable	Subjective perception	Self-report of perceived burden

government-sponsored, large-scale, household survey of which we are aware that collects such data.) Most of the studies, by contrast, measure burden indirectly in one of the three ways, as shown in Table 8.1.

The most common approach to measuring burden has been to examine objective survey features that are hypothesized to impose burden. Interview length and task difficulty are typical measures in these studies (Filion 1981; Warriner 1991; Hoogendoorn and Sikkel 1998; Groves et al. 1999; Singer et al. 1999; Hoogendoorn 2004; Rolstad et al. 2011). When interview length is used as a proxy measure of burden, it has been operationalized as number of pages for paper questionnaires, number of survey items, or time required to complete a survey. Other studies have approached burden by measuring respondents' attitudes and beliefs about surveys, such as how interesting or important they find the survey to be (e.g. Sharp and Frankel 1983; Hoogendoorn 2004; Stocke and Langfeldt 2004; Geisen 2012). The assumption is that respondents who are less interested in a survey or find the survey unimportant or not useful are more likely to find the survey burdensome. However, like the objective measures of burden, measures of respondents' attitudes about the survey itself likely are mediators of the perceptions of burden rather than direct measures of it. Respondents' attitudes and beliefs toward the survey (e.g. whether they perceive the survey as interesting or important) result in differential perceptions of burden for the same survey across respondents. Some authors simply infer burden from negative survey outcomes (e.g. refusing to be re-interviewed; break-offs;) and how respondents felt after the survey (Sharp and Frankel 1983; Stocke and Langfeldt 2004).

These very different measurements of response burden reflect both the lack of and the need for a well-developed conceptual framework on burden. As a result, empirical research on burden produces equivocal findings and over-simplified, if not misleading, conclusions and implications for the survey field. For instance, researchers and practitioners too often employ interview length as a proxy measure of burden (e.g. Groves et al. 1999; Rolstad et al. 2011) and believe that longer surveys impose greater burden and yield lower response rates. However, only 6 out of the 25 empirical studies included in a meta-analysis showed that longer surveys (as measured in number of pages) yield lower response rates (Rolstad et al. 2011). Results from the rest of the 19 surveys don't bear evidence for the assumed negative impact of survey length on response rates. Therefore, we do not know whether or not survey length actually does lead to burden and lower response rates. And it is not clear what the survey field should do with regard to survey length.

The lack of a clear conceptualization and consistent operationalization of burden prevents the survey field from understanding the factors that most contribute to respondent burden and operationally what steps should be taken do to reduce burden. Clearly, survey length itself and other objective characteristics of surveys are not the only defining features of burden. As a matter of fact, Bradburn explicitly stated that burden "is not to be an objective characteristic of the task, but is the product of an interaction between the nature of the task and the way in which it is perceived by the respondent" (1977, p. 49). To Bradburn, burden is a subjective phenomenon reflecting the influence of interview length, effort required of respondents, the frequency of interviewing, and the amount of stress on respondents, although Bradburn did not provide empirical evidence on the "interaction" of those variables or a measurement of burden.

Haraldsen (2004) outlines a model where the subjective perception of burden is explicitly shown as an intermediate variable between causes of response burden and data quality. Causes of response burden are further divided into survey properties and respondent characteristics. Haraldsen perceives burden as "what happens at the interface between the survey instrument and the respondent's ability to respond" (2004, p. 398). Applying this framework to internet business surveys, Haraldsen presented results from qualitative tests (e.g. tests combining cognitive interviewing and utility testing, usability tests, and pilot tests) to shed light on survey properties (in particular web survey tools) and their impact on data quality. The middle part – perceived burden assumed to happen at the interface between the survey and the respondent – was neither elaborated nor examined empirically.

This chapter extends Bradburn's and Haraldsen's work by defining burden as a subjective perception and feelings of burden perceived by respondents. We posit a path model that explicitly incorporates the direct and indirect effects of survey features, respondent characteristics, and respondents' perceptions of the survey on burden, in hopes of shedding light on which factors (or combination of factors) are most likely to result in response burden. We deliberately select respondent characteristics, survey and task characteristics, and respondent perceptions that are expected to be related to response burden and take advantage of structural equation modeling (SEM) to quantify the interaction of these variables. SEM is a powerful statistical technique that can model complex functional relationships between variables. A key advantage of SEM over regression models or path analysis is that the variables in SEM can be observed or not observed (that is, latent). SEM is appropriate for this analysis because some variables used in the SEM are latent. In addition, we make use of multiple group analysis (MGA) within SEM to test whether our models are equivalent across modes of data collection, as the mode of data collection could have differential impact on respondents' feelings of burden. We test both metric invariance and structural invariance across modes.

8.2 Methods

8.2.1 Data

Data used for this study come from the CE. The CE is a longitudinal survey sponsored by the Bureau of Labor Statistics (BLS). It collects comprehensive information on a wide range of consumers' expenditures and incomes, as well as the characteristics of those consumers. It adopts a rotation panel design, and sampled households are interviewed five times before retiring from the panel. Response rates are 70% for 2012 and 67% for 2013 (Department of Labor 2016), the two years for which these analyses were conducted.

At the end of their regular interview at Wave 5, panel members were asked additional questions about their perceptions of and attitudes about the CE in particular and surveys in general. One of the items directly asked how burdensome the survey was to respondents using a four-point item (i.e. the response categories were "not at all burdensome," "a little burdensome," "somewhat burdensome," and "very burdensome").

Analyses are limited to panel members who completed their fifth (and last) interview between October 2012 and March 2013, leading to a total of 6099 cases for the analyses after deleting cases with missing data on observed variables from the SEM estimation and analysis.

8.2.2 Analytic Method

The data were analyzed using SEM. We selected this method because it allowed us to test a model of burden that includes latent factors related to survey features, respondent perceptions, and other respondent characteristics, and to examine the causal relations (direct and indirect) between these factors and burden. We used the SAS procedure PROC CALIS to estimate the model. The weighted least square (WLS) estimation method is used because the WLS method does not make distributional assumptions and is recommended when data are not multivariate normal.

8.2.3 Burden Model

The SEM framework consists of two interrelated models: (i) the measurement models, which describe the assignment of the observed items (or indicators) to each unobserved latent factor; and (ii) the structural model, which describes the relationship among the set of latent factors. Both models are explicitly defined by the analyst and depicted in a path diagram.

For our structural model (see Figure 8.1), we examined one factor related to motivation, two factors related to task difficulty and survey effort, one intermediate factor related to respondent perceptions of the survey, and the key dependent variable – respondents' report of burden.

The measurement models that explain the operationalization of each of the latent factors used in the structural model are elaborated in the following sections. Table 8.2 displays the mean and standard deviation of the indicators for the latent factors.

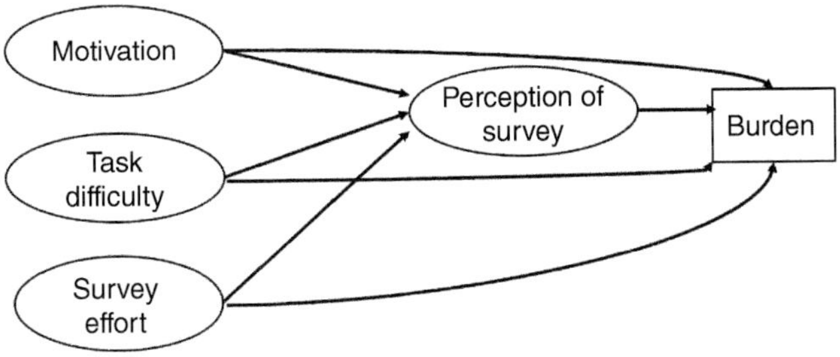

Figure 8.1 Structural model of response burden.

Table 8.2 Descriptives of indicator variables.

Indicator variable	Data source	Mean	Standard deviation
Burden	Self-report to survey item	2.19	0.98
Indicators of "motivation" factor			
High concern	Call records data	0.52	0.90
Number of refusals	Call records data	0.07	0.32
Sensitivity of CE	Self-report to survey item	2.41	1.05
Low trust	Self-report to survey item	1.96	0.97
Indicators of "task difficulty" factor			
Number of children in household	Self-report to survey items	0.33	0.51
Number of adult household members	Self-report to survey items	1.01	0.28
Less than 65	Self-report to survey item	0.78	0.42
Indicators of "survey effort" factor			
Duration of survey interviews	Call records data	2.53	1.11
Number of interviews completed	Call records data	3.93	1.35
Using information book	Paradata	1.01	1.46
Using records	Paradata	0.70	1.28
Indicators of "perception of survey" factor			
Survey not interesting	Self-report to survey item	2.50	0.99
Survey difficult	Self-report to survey item	1.94	0.80
Too many rounds of interviews	Self-report to survey item	0.55	0.50
Interview too long	Self-report to survey item	2.34	0.48

Table 8.3 Distribution of responses to burden item.

	Sample count	Percentage (%)
Very burdensome	645	10.6
Somewhat burdensome	1684	27.6
A little burdensome	1925	31.6
Not at all burdensome	1845	30.3
Total	6099	100

8.2.3.1 Measure of Burden

Burden was measured through a survey item asking respondents directly how burdensome they felt the survey was. Higher values indicate higher level of burden. Table 8.3 shows the percentages (and counts) of respondents who selected each answer category of the burden item. In the SEM, burden was treated as an observed variable with no measurement error.[1]

8.2.3.2 Measure of Motivation

Motivation in our model was measured as a latent construct with four indicators that are conceptually related to motivation. The first indicator, *high concern*, was a summary variable drawing from doorstep concerns data collected through the Contact History Instrument (CHI) before the actual interview started.[2] Following the methodology described in Yan and Tsai (2012), the indicator was calculated as a weighted sum of principal components extracted through principal component analysis on the CHI data and represents the level of concerns expressed by sampled respondents at the doorstep (see also Yan 2017). Higher values on this indicator denote higher level of concerns and are found to be associated with higher level of propensity

1 We acknowledge that this is a strong assumption, and almost certainly violated. We used an alternative approach suggested in Kline (1998, pp. 264–266) and Garson (2015) in which we re-specified our model by treating the observed burden variable as an observed indicator of a latent burden factor, fixing the loading of the observed burden indicator on the latent burden variable to be 1, and fixing the measurement error variance as the product of 1 minus the reliability of the burden indicator and the observed variance of the burden indicator. We used the Survey Quality Predictor (SQP) program (http://sqp.upf.edu) to obtain an estimate of the reliability of the burden item. Of course, the SQP estimate is at best only an approximation of the true reliability. We reran the SEM, and the model results are very similar whether we use the observed burden variable directly or as a single indicator of a latent factor. There are no changes in either the direction or the significance level of all factor loadings and effect sizes.

2 The CHI used for the CE is a stand-alone Blaise instrument. Interviewers are trained to enter a CHI record for each contact attempted with a sampled household. Interviewers are required to check (on a computer screen) one or more categories on a list of 21 verbal or nonverbal concerns that can be expressed by respondents during the survey introduction and interactions. For more information, refer to Yan (2017).

to refuse to the survey request and to provide missing data (Yan 2017). Thus, higher values of this variable indicate lower level of motivation.

The second indicator, *number of refusals expressed*, counts the number of times sampled respondents refused the survey request throughout their entire panel life (that is, across all five survey requests). It is obtained through call records data.

The next two indicators of motivation are self-reports to survey items that are collected at the end of the fifth interview. The third indicator, *sensitivity of CE*, represents how sensitive respondents considered the CE. The response options are "very sensitive," "somewhat sensitive," "a little sensitive," and "not at all sensitive." Higher values indicate higher level of sensitivity. The fourth indicator, *low trust*, denotes the extent to which respondents agreed with the statement that they trusted in the US Census Bureau to safeguard the information they provided. Higher values on this indicator are associated with lower level of trust.

8.2.3.3 Measure of Task Difficulty

Task difficulty was measured as a latent construct with three indicators. The first indicator, *number of children in household*, captures the number of children living in the same household with the sampled respondents. The second indicator, *number of adult household members*, represents the number of adults living in the household. We selected these two indicators because respondents living in large households (with more children or more adults) are likely to have more expenditures to report than those in small households, and because proxy reporting for other household members' expenditures may be more difficult (i.e. burdensome) than simply reporting for oneself. We did a log transformation on both count variables.

The third indicator, *less than 65 years old*, is a dummy variable coded from age where 1 means that the respondent is less than 65 and 0 means that he/she is 65 or older. Older respondents are considered to have reduced cognitive capacity (Salthouse 1991) and may find the same task more difficult and burdensome than their younger counterparts (Krosnick 1991).

8.2.3.4 Measure of Survey Effort

Survey effort was measured as a latent construct with four indicators. The first indicator, *duration of interviews*, divides respondents into four quartiles based on the sum of the duration of all interviews (in hours) completed by sampled respondents throughout their panel life. The second indicator, *number of interviews completed*, counts the number of interviews completed by sampled households. The third indicator, *using information book*, counts the number of times respondents used the information book always or almost always. The last indicator, *using records*, counts the number of times respondents resorted to records almost always when answering the expenditure questions.

8.2.3.5 Measure of Perception of Survey

Perception of survey was measured as a latent construct with four indicators. The four indicators are based on self-reports to single survey items that were administered at the end of the fifth interview. The first indicator, *survey not interesting*, looks at respondents' perception of the CE, and response options are "very interesting," "somewhat interesting," "a little interesting," and "not at all interesting." High values indicate that respondents considered the CE less interesting. The second indicator, *survey difficult*, represents the extent to which respondents considered the CE to be difficult. The response options for this question are "very easy," "somewhat easy," "somewhat difficulty," and "very difficult." Higher values denote higher level of difficulty. The third indicator, *too many rounds of interview*, reflects respondents' perception about the number of interview requests posed to them; they were asked to select either "too many rounds" or "a reasonable number." Higher values indicate that respondents considered that they have been asked to participate in too many rounds of interviews. The last indicator, *interview too long*, represents respondents' perception of the length of the CE. Response options are "too long," "too short," and "about right" with higher values indicating that they consider the survey too long.

8.3 Results

8.3.1 Model Fit Statistics

We examined several model fit statistics. The Chi-square test indicated a poor fit with the data (x^2 (90) = 1080, $p < 0.0001$). However, this measure tends to be an overly sensitive test of global fit with large sample sizes as we have in our study (Byrne 1998; Garson 2015; Kline 1998). The second index of overall fit, the standardized root mean square residual (SRMSR), was 0.086. According to O'Rourke and Hatcher (2013), SRMSR values less than 0.055 suggest a good fit and values less than 0.09 are suggestive of fair or adequate fit. Therefore, our SRMSR value indicates an adequate model fit.

We also looked at two parsimony indices. The root mean square error of approximation (RMSEA) was 0.04, indicating a good fit (O'Rourke and Hatcher 2013).[3] The adjusted goodness of fit index (AGFI) was larger than 0.90, reflecting a good fit (AGFI = 0.99). In addition, the Bentler comparative fit index (CFI), an incremental index, was 0.91 and was above the traditional 0.90 cut-off value, suggesting a good fit.

Looking across all indices, we considered our models to reflect a rather good fit to the data.

---

3 RMSEA values less than 0.055 indicates a good fit, and values less than 0.09 suggests a fair and adequate fit (O'Rourke and Hatcher 2013).

Table 8.4 Model estimates from the measurement models.

Measurement model		Standardized	Unstandardized		
Factor	**Indicator**	**Estimates**	**Estimates**	**S. E.**	***p*-value**
Low motivation	High concern	0.387	1.000		
	Sensitivity of CE	0.639	2.450	0.133	<0.0001
	Low trust	0.478	1.656	0.092	<0.0001
	Number of refusals expressed	0.163	0.019	0.019	<0.0001
Difficult task	Number of kids in household	0.636	1.000		
	Number of adults in household	0.356	0.303	0.303	<0.0001
	Respondent less than 65	0.474	0.598	0.598	<0.0001
Challenging survey effort	Duration of interviews	0.509	1.000		
	Number of interviews completed	0.360	0.867	0.023	<0.0001
	Using Information Book	0.687	1.768	0.177	<0.0001
	Using records	0.676	1.445	0.146	<0.0001
Negative perception of survey	Survey not interesting	0.626	1.000		
	Survey difficult	0.496	0.635	0.022	<0.0001
	Too many rounds of interviews	0.659	0.536	0.012	<0.0001
	Interview too long	0.617	0.487	0.012	<0.0001

8.3.2 Measurement Models

Estimates from our SEM measurement models are shown in Table 8.4. The unstandardized factor loadings for each item with its associated latent variable are statistically significant and the standardized loadings are generally sizeable (i.e. greater than 0.30). All of the loadings are in the expected direction. For example, *lower respondent motivation* was associated with respondents who consider the survey as sensitive, have low or no trust in the survey organization, express higher level of concerns at the doorstep, and have ever refused the survey request. Large households with more children, large households with more adult household members, and respondents younger than 65 all positively contribute to *task difficulty*. *Challenging survey effort* was positively associated with longer interviews, more interviews completed, using the information book, and using records during the interview. Negative *perceptions of the survey* were reflected in respondents complaining having too many rounds of interviews, interviews being too long, survey less interesting, and survey more difficult.

8.3.3 Structural Model

With our measurement models validated, we examined the hypothesized structure of our latent factors. We began by looking at the direct effects of each factor on the other model factors (see Table 8.5). All factors had significant direct effects on burden in the expected direction. Not surprisingly, *lower motivation, more difficult task, more challenging survey effort, and negative impressions of survey* all were associated with higher levels of perceived burden.

The remaining effects shown in Table 8.5 also are generally in the expected direction. For example, respondents with low motivation were more likely to have a negative impression of the survey than those who were more motivated. Similarly, more difficult task was associated with more negative perceptions of the survey, although the effect is small. The one puzzling finding is that more *challenging survey effort* (longer interviews, more surveys completed, using information book and records during the survey) was associated with less-negative feelings about the survey. One possible explanation for this finding is that individuals are more motivated with a task that is intricate, challenging, and enriching (e.g. Campbell 1988). Or, it could be that respondents expended more effort when they found the task to be engaging. Whatever the reason for the direction of the effects, the size of these effects is very small.

Table 8.5 Model estimates from the structural model of burden.

Structural model		Standardized	Unstandardized		
Factor	**Effect on**	**Estimates**	**Estimates**	**S. E.**	***p*-value**
Low motivation	Negative perception of survey	0.827	1.861	0.112	<0.0001
	burden	0.446	1.605	0.214	<0.0001
Difficult task	Negative perception of survey	0.064	0.121	0.039	0.002
	burden	0.077	0.234	0.046	<0.0001
Challenging survey effort	Negative perception of survey	−0.063	−0.068	0.022	0.002
	burden	0.047	0.083	0.024	0.001
Negative perception of survey	Burden	0.353	0.565	0.083	<0.001

Table 8.6 Decomposition of effects of latent factors on burden.

	Total effects	Direct effects	Indirect effects
Low motivation	0.739***	0.446***	0.292***
Difficult task	0.099***	0.077***	0.023**
Challenging survey effort	0.025	0.047**	−0.022**
Negative perception of survey	0.353***	0.353***	0

Note: *p < 0.05; **p < 0.01; ***p < 0.001.

We next estimated the indirect and total effects of our model factors. Indirect effects are mediated by at least one intervening variable; total effects are equal to the sum of the direct and indirect effects. Table 8.6 summarizes the results of this decomposition of effects. As shown in Table 8.6, *low motivation, difficult task,* and *negative perception of survey* had significant overall positive effects on burden. Contrary to views commonly held in the survey field, the usual-suspect causes of burden such as *challenging survey effort* had no significant overall effects on burden. The direct effects of *challenging survey effort* are positive and statistically significant at the 0.05 level. However, the indirect effects of this factor through *negative perception of survey* are negative and statistically significant. As a result, the sum of these two effects essentially canceled out each other, yielding small and non-significant total effects.

8.3.4 Burden Model by Mode of Data Collection

As the mode of data collection is thought to impose differential burden on respondents, we are interested in examining whether the same burden model (Tables 8.3 and 8.4) holds for different modes of data collection. Specifically, we examined whether the burden model is invariant or equivalent across people who were attempted mostly in-person and those who were attempted mostly over the phone for interviews. Although the survey protocol called for interviewers to contact respondents in person for the first and last rounds, actual practices varied, perhaps at least partially determined by respondent preferences. For the purpose of this analysis, for each sampled respondent, we summed up the number of contact attempts made in-person and the number of contact attempts made over telephone across all contact attempts and across all waves of interviews. Then we looked at the ratio of the sum of in-person contact attempts and the sum of phone contact attempts. Based on this ratio, we divided respondents into two groups. A total of 3584 respondents were classified as the "mostly in-person" group because they were attempted in-person more often than over the telephone. 2515 respondents were grouped together as the "mostly by phone" group as they were attempted over the phone more often than in-person. Cases who were attempted equally often in-person and by phone were removed from the analysis.

We conducted MGA to test measurement invariance at different levels. The first level of invariance tests the structure and direction of the overall relationships among indicators and latent factors. As shown in Table 8.7, the configural invariance model fits the data relatively well, evidenced by two model fit statistics (RMSEA and CFI), suggesting that the overall relationships among indicators and factors have the same structure and direction across respondents attempted in different modes of data collection. Subsequent analysis evaluated metric invariance, that is, whether *all* factor loadings were equivalent across the two groups of respondents, and found that some factor loadings were actually not equivalent across modes.[4] Having established partial metric invariance, the structural invariance test shows that the overall structural relationships did not differ across modes of data collection.[5]

8.4 Conclusions and Discussion

Prior research on response burden has relied on inadequate conceptualization and measures of burden. It is therefore not surprising that it is difficult, based on the small empirical literature that exists, to make firm predictions about survey features or respondent characteristics that are most likely to give rise to burden. In this study, we developed and tested a model that assumes that burden is a subjective phenomenon, affected by respondents' psychological

---

4 The fit of the metric invariance model is statistically different than the fit of the configural invariance model, suggesting that some factor loadings are actually not equivalent across modes. We therefore examined the equality constraints on factor loadings one at a time and identified nine factor loadings not equivalent by mode of data collection (specifically, loadings for *low trust, number of kids in household, number of adults in household, respondent less than 65, duration of interviews, number of interviews completed, using records, survey too long*, and *too many rounds of interview*). We then tested for partial metric invariance by allowing these nine factors loadings to be different by mode. The partial metric invariance model has a better fit than the configural invariance model (smaller RMSEA and larger CFI). Furthermore, the differences in model Chi-squares (between the partial invariance model and the configural invariance model) are not statistically significant, indicating that the partial invariance model is preferred over the configural invariance model.

5 Given the acceptance of the partial metric invariance model, we further tested whether the relationships among the latent factors in the model are equal across respondents attempted in different modes of data collection (structural invariance). The model with structural invariance fits the data well, and the model fit between the structural invariance and the partial metric invariance model is not statistically significant, indicating that the structural relationships among latent factors hold across modes of data collection.

Table 8.7 Multiple group analysis.

Models for comparison	χ^2	DF	*p*-value	RMSEA	CFI	$\Delta\chi^2$	ΔDF	*p*-Value
Configural invariance	1239	180	<0.0001	0.04	0.90			
Metric invariance	1389	198	<0.0001	0.04	0.89	151	18	<0.0001
Partial metric invariance	1247	185	<0.0001	0.04	0.90	8	5	0.12
Structural invariance	1257	192	<0.0001	0.04	0.90	9	7	0.23

responses to various elements of the survey. The survey data we used to test this model included direct measures of reactions from respondents themselves, interviewer assessments, and objective features of the survey. We used structural equation modeling to assess how well these data fit latent factors expected to be important contributors to burden, and then evaluated the impact those factors had on burden.

The results of this study validate our underlying measurement models – our indicators all are significantly related to their associated latent variables in the expected direction. Results of our structural model show that respondents' motivation, respondents' characteristics related to task difficulty, and respondents' subjective perceptions of the survey task had a significant direct impact on burden as well as significant overall effects on burden. Our results indicate that respondents with low motivation, difficult reporting tasks, and negative perceptions are more likely to have higher levels of burden than those with high motivation, easy reporting tasks, and positive perceptions. Although the objective survey features themselves have a significant direct impact on burden, this direct effect is canceled out by the indirect effect of respondents' perception of the survey, producing small and non-significant overall effects on burden. This finding explains the seemingly contradictory empirical evidence that a longer questionnaire sometimes leads to a lower response rate and sometimes does not.

We further conducted MGA to test whether the same burden model holds for people attempted in different modes. We found that some factor loadings are not equivalent for respondents attempted mostly in person and those attempted mostly by phone. However, the structural relationships among the latent factors hold whether respondents were attempted mostly in person or by phone. This is encouraging as the survey field worries that the mode of data collection affects how respondents feel about the burden of the survey. Our results demonstrate that the modes of data collection did not affect the paths leading to the perception of burden in this context; the same set of factors have the same impact on burden regardless of whether respondents were attempted mostly by phone or in person. Of course, the same conclusion may not hold in other contexts or for other surveys; future research is needed to explore the relation between modes of data collection and response burden.

Our findings support the notion of burden as a subjective, multidimensional phenomenon, and underscore that the impact of any given survey feature on burden will vary across respondents on the basis of measurable subjective reactions and attitudes. These results provide practical guidance on identifying respondents who may be more likely to experience burden, and point to possible strategies to reduce perceived burden. We note a few implications of our study.

First, respondents' perceptions not only have a direct impact on burden, but also mediate the impact of other factors on burden. Therefore, it is critical for survey researchers and practitioners to switch attention to strategies and methods that help reduce negative perception of respondents on surveys. Surveys are regularly evaluated on comprehension, recall, judgment, and response or mapping through question testing and evaluation methods (such as cognitive interviewing, expert reviews, and so on). But surveys are rarely evaluated on respondents' perception and attitudes. We think it is time that question testing and evaluation effort includes the evaluation of respondents' perceptions about the survey.

Our results demonstrate that respondents' perception of the survey (a latent factor) can be reasonably measured by assessing their perceptions of survey length and their interest in the survey. We recommend that when evaluating survey questionnaires, respondents' perceptions of survey length and survey interest also should be evaluated in addition to objective survey length (measured either as the number of survey items or the time taken to complete a survey).

Second, the indicators used in the measurement models suggest specific ways for survey researchers and practitioners to modify the essential survey conditions to reduce burden. For instance, respondents who expressed greater concerns about the survey request were associated with lower motivation and were more likely to report high levels of burden. Respondents with greater concerns could be identified in an adaptive design setting. They can then be provided a shorter questionnaire or given more confidentiality assurance in order to prevent them from eventually experiencing excessive burden. We recommend that adaptive designs take into consideration this burden model.

Third, our analysis was made possible because of the CE's unique datasets, which contain not only information about respondent reactions to the survey but also one survey item directly asking respondents how burdensome they felt. As a validation of the single item measurement of burden, we investigated the relationship between answers to this single burden item and some traditional measures of reluctance – the number of call attempts required to complete the fifth interview, whether or not respondents were a converted refusal at the fifth interview, the level of concerns expressed by respondents, and the number of Don't Know and Refused answers provided. As shown in Table 8.8, respondents

Table 8.8 Relationship between burden and measures of reluctance.

	Average contact attempts	Average level of concerns	% Converted refusers	Average # of Don't Know Answers	Average # of Refused Answers
Very burdensome	20.53	1.14	30.40%	1.20	0.37
Somewhat burdensome	18.43	0.61	13.70%	0.75	0.06
A little burdensome	17.8	0.44	8.00%	0.89	0.04
Not at all burdensome	16.12	0.31	4.90%	0.66	0.02
Significance test	$F(3, 6091) = 25$, $p < 0.0001$	$F(3, 6091) = 16$, $p < 0.0001$	$X^2(3) = 349$, $p < 0.0001$	$F(3, 3093) = 7$, $p = 0.0002$	$F(3, 3093) = 39$, $p < 0.0001$

reporting a higher level of burden tended to require more contact attempts, were more likely to have refused the survey request at some point in the life of the panel, had higher level of concerns, and provided more missing data. These significant relationships indicate that this single measure of burden does reflect respondents' level of reluctance and difficulty pretty well.

The approach of directly measuring the perception of burden with a single survey item is also used in Galesic (2006), who found that later web pages, pages where respondents spent longer time, and pages with more open-ended questions increased respondents' perception of burden and that burdened respondents were more likely to break off than those with a lower level of burden. Our results together with those of Galesic (2006) demonstrate that a single measure of burden can capture the level of reluctance and difficulty experienced by respondents, and that the approach of directly measuring burden through a single survey item is promising. When possible, we strongly encourage other surveys to include at least one survey item directly measuring the level of perceived burden, as is done in the CE.

Fourth, if it is not possible to add even one survey item to measure burden, we strongly encourage survey practitioners and researchers to take advantage of information that taps into the subjective and attitudinal evaluations in order to identify respondents at a greater risk of being burdened. Two kinds of paradata that most surveys routinely collect – doorstep concerns data (Yan 2017; Yan and Tsai 2012, 2015) and interviewer observations on respondents' attitudes and behaviors such as cooperation and interest in survey (Yan and Tsai 2013; Yan and Keusch 2015) – capture information reflecting either motivation or perception of the survey. Furthermore, task-evoked pupillary responses captured during eye-tracking also are found to measure burden unobtrusively (Yan et al. 2016). We encourage survey practitioners and researchers to make use of these data up front to identify those prone to feeling burdened and to take steps to increase their motivation or reduce their negative evaluation of the survey. We also recommend that survey practitioners include the single-item measurement of burden or survey items measuring motivation or perception as part of nonresponse follow-up studies.

One major limitation of this study is that burden was measured at the end of the fifth CE interview and is measured via a single survey item. We limited our analyses to those who completed all five interviews. As a result, the respondents included in our analyses are probably more cooperative and less burdened than those who attrited at an earlier round. Our analyses miss those who are most burdened, and our results underestimate the prevalence and the magnitude of burden experienced by CE respondents. However, we doubt that paths leading to burden as posited in our burden model would be very different for those who are the most burdened. Still, we encourage other researchers and surveys to replicate and extend our burden model by including people who are prone to being the most burdened. Future studies of burden could include other potential variables in the burden model, as well. For instance, objective measures of difficulties (as captured via item nonresponse, reading level, syntactical complexity, or request for definition or clarification) and measures of interaction trouble between respondents and interviewers (as revealed through behavior coding) could be included and tested in the burden model.[6] Future studies also could experimentally manipulate motivation and/or difficulty to empirically test whether or not burden is actually reduced. Lastly, future studies should examine further into the puzzling link between survey effort and negative feelings about the survey, as well as the indirect effects of survey effort through perception.

Acknowledgments

This research was funded by the ASA/NSF/BLS Research Fellowship awarded to Ting Yan. We thank American Statistical Association, National Science Foundations, and BLS for their financial support and CE staff for their help and valuable feedback to the research. We also thank the editor and reviewers for their useful comments and suggestions.

References

Bergman, L.R. and Brage, R. (2008). Survey experiences and later survey attitudes, intentions and behaviour. *Journal of Official Statistics* 24 (1): 99–113.

6 We thanked the reviewers for their suggestion of additional variables to be included in the burden model.

Bradburn, N. (1977). Respondent Burden. In: US Department of Health, Education and Welfare, National Center for Health Services Research, *Health Survey Research Methods, Research Proceedings from the 2nd Biennial Conference,* Publication No (PHS) 79-3207: 49–53.

Byrne, B. (1998). *Structural Equation Modeling with LISREL, PRELIS, and SIMPLIS: Basic Concepts, Applications, and Programming*. Mahwah, NJ: Lawrence Erlbaum Associates.

Campbell, D.J. (1988). Task complexity: a review and analysis. *Academy of Management Review* 13: 40–52.

Chapin, G. (1920). *Field Work and Social Research*. New York: The Century Co. Commission on Federal Paperwork.

Filion, F.L. (1981). Importance of question wording and response burden in hunter surveys. *The Journal of Wildlife Management* 45 (4): 873–882.

Fricker, S., Kreisler, C., and Tan, L. (2012). An exploration of the application of PLS path modeling approach to creating a summary index of respondent burden. Joint Statistical Meeting, San Diego, CA.

Galesic, M. (2006). Dropouts on the web: effects of interest and burden experienced during an online survey. *Journal of Official Statistics* 22: 313–328.

Garson, G.D. (2015). *Structural Equation Modeling*. Asheboro, NC: Statistical Associates Publishers.

Geisen, D. (2012). Exploring causes and effects of perceived response burden. International Conference on Establishment Surveys. http://www.amstat.org/meetings/ices/2012/papers/302171.pdf

Groves, R., Singer, E., and Corning, A. (1999). A laboratory approach to measuring the effects on survey participation of interview length, incentives, differential incentives, and refusal conversion. *Journal of Official Statistics* 15: 251–268.

Haraldsen, G. (2004). Identifying and reducing response burden in internet business surveys. *Journal of Official Statistics* 20: 393–410.

Hoogendoorn, A.W. (2004). A questionnaire design for dependent interviewing that addresses the problem of cognitive satisficing. *Journal of Official Statistics* 20: 219–232.

Hoogendoorn, A.W. and Sikkel, D. (1998). Response burden and panel attrition. *Journal of Official Statistics* 14: 189–205.

Jones, J. (2012). Response burden: introductory overview literature. International Conference on Establishment Surveys. http://www.amstat.org/meetings/ices/2012/papers/302289.pdf

Kline, R.B. (1998). *Structural Equation Modeling*. New York: Guilford Press.

Krosnick, J.A. (1991). Response strategies for coping with the cognitive demands of attitude measures in surveys. *Applied Cognitive Psychology* 5: 213–236.

Martin, E., Abreu, D., and Winters, F. (2001). Money and motive: effects of incentives on panel attrition in the survey of income and program participation. *Journal of Official Statistics* 17: 27–284.

O'Rourke, N. and Hatcher, L. (2013). *A Step-by-Step Approach to Using SAS for Factor Analysis and Structural Equation Modeling*. SAS Institutes.

Rostald, S., Adler, J., and Ryden, A. (2011). Response burden and questionnaire length: is shorter better? A review and meta-analysis. *Value in Health* 14: 1101–1108.

Salthouse, T.A. (1991). *Theoretical Perspectives on Cognitive Aging*. Hillsdale, NJ: Lawrence Erlbaum Associates.

Sharp, L.M. and Frankel, J. (1983). Respondent burden: a test of some common assumptions. *Public Opinion Quarterly* 47: 36–53.

Singer, E., Van Hoewyk, J., Gebler, N. et al. (1999). The effect of incentives on response rates in interviewer-mediated surveys. *Journal of Official Statistics* 15: 217–230.

Stocke, V. and Langfeldt, B. (2004). Effects of survey experience on respondents' attitudes towards surveys. *Bulletin de Methodologie Sociologique* 81: 5–32.

U.S. Department of Labor, Bureau of Labor Statistics. (2016). 2015 response rates: interview survey and diary survey, consumer expenditure public use microdata.

Warriner, G.K. (1991). Accuracy of self-reports to the burdensome question: survey response and nonresponse error trade-off. *Quality & Quantity* 25: 253–269.

Yan, T. (2017). Using doorstep concerns data to evaluate and correct for nonresponse error in a longitudinal survey. In: *Total Survey Error in Practice* (eds. P. Biemer, E. de Leeuw, S. Eckman, et al.), 415–434. Wiley and Sons.

Yan, T., Fricker, S., and Tsai, S. (2014). The impact of response burden on data quality in a longitudinal survey. International Total Survey Error Workshop, Washington, DC.

Yan, T. and Keusch, F. (2015). The effects of the direction of rating scales on survey responses in a telephone survey. *Public Opinion Quarterly* 79: 145–165.

Yan, T., Maitland, A., and Williams, D. (2016). Use of eye-tracking to measure response burden. Annual Conference of the American Association for Public Opinion Research.

Yan, T. and Tsai, S. (2012). Using doorstep concerns data to characterize and assess the level of reluctance of survey respondents. In: *JSM Proceedings, Survey Research Methods Section*, 5383–5393. Alexandria, VA: American Statistical Association.

Yan, T., and Tsai, S. (2013). Using doorstep concerns data to study the relationship between reluctance and measurement error. Annual Conference of American Association for Public Opinion Research.

Yan, T. and Tsai, S. (2015). Use of doorstep concerns to examine trade-offs between error and cost. Annual Conference of the American Association for Public Opinion Research.

9

The Salience of Survey Burden and Its Effect on Response Behavior to Skip Questions: Experimental Results from Telephone and Web Surveys

Frauke Kreuter[1], Stephanie Eckman[2], and Roger Tourangeau[3]

[1] *Joint Program in Survey Methodology, University of Maryland, College Park, MD, US; School of Social Sciences, University of Mannheim, Mannheim, Germany; Institute for Employment Research, Nuremberg, Germany*
[2] *RTI International, Washington, DC, US*
[3] *Westat, Rockville, MD, US*

9.1 Introduction

Survey questionnaires often include skip patterns, which allow respondents to skip over entire sections of a questionnaire or over a set of follow-up questions that do not apply to them. Such skip patterns allow respondents to proceed through the interview faster and reduce respondent burden. Past research demonstrated that respondents take advantage of these patterns and fail to endorse an item that applies to them, in order to avoid the follow-up questions. However skip patterns and filter questions become increasingly important as survey designers try to create short modular questionnaires in response to survey-taking on mobile devices and increased difficulties in recruiting respondents. We conducted a series of experiments to help with such design decisions.

Skip patterns that start with a filter question and are followed by a series of subsequent questions, if endorsed, can induce avoidance behavior, where respondents try to shorten the interview and avoid additional burden (Tourangeau et al. 2015, pp. 24–41). We refer to this format of filter and follow-up questions as interleafed. Once respondents learn that endorsing a filter question leads to a series of follow-up questions, they then have an incentive to misreport on the filter questions. To reduce the risk of motivated misreporting, some surveys administer the filter questions in a block, prior to asking any follow-up questions (grouped format). See Figure 9.1 for a visual

Advances in Questionnaire Design, Development, Evaluation and Testing, First Edition.
Edited by Paul C. Beatty, Debbie Collins, Lyn Kaye, Jose-Luis Padilla, Gordon B. Willis, and Amanda Wilmot.

Grouped version	Interleafed version
Purchased coat?	*Purchased coat?*
[....]	Description of item purchased
Purchased skirt?	For whom was it purchased?
Earlier you said you purchased a [...] ...	Month of purchase?
Description of item purchased	Cost of purchase?
For whom was it purchased?	[...]
Month of purchase?	*Purchased skirt?*
Cost of purchase?	

Figure 9.1 Questions used in the United States Consumer Expenditure Survey displayed in grouped format on the left and interleafed on the right, as fielded in the survey. Sketch of design used by Kreuter et al. (2010).

display of the two formats. Both formats are used in ongoing surveys (for example the National Survey on Drug Use and Health, U.S. Department of Drug Use and Health 2016).

Several studies have compared the two approaches – grouped and interleafed – in randomized experiments. Duan et al. (2007) found that with a set of mental health items, the grouped approach led to more affirmative answers to grouped filter questions than did the interleafed approach, confirming an earlier finding from Kessler et al. (1998). However, it is possible that respondents were not trying to shorten the interview so much as trying to avoid potentially painful follow-up questions on a sensitive subject. For this reason, Kreuter et al. (2010) conducted an experiment that compared the two formats (grouped versus interleafed) on a series of factual topics, including credit cards, consumer purchases, and leisure activities, none of them as sensitive as the mental health items. In addition to varying the organization of the questions (grouped filters versus interleafed), the order of the different topical sections was also varied to see whether the difference between the grouped and interleafed filter items increased as respondents worked their way through the questionnaire.

Overall, grouping the filter questions boosted the proportion of "yes" answers to those questions. The difference between grouped and interleafed format was affected by an item's position within each topical section. That is, within each section, the increase in "yes" responses to the filter items under the grouped arrangement tended to grow larger for each successive filter item. For example, looking at the Clothing panel in Figure 9.2, we see that skirts were the seventh filter item asked about in that section, and respondents were more than twice as likely to say they had purchased a skirt in the grouped than the interleafed format.

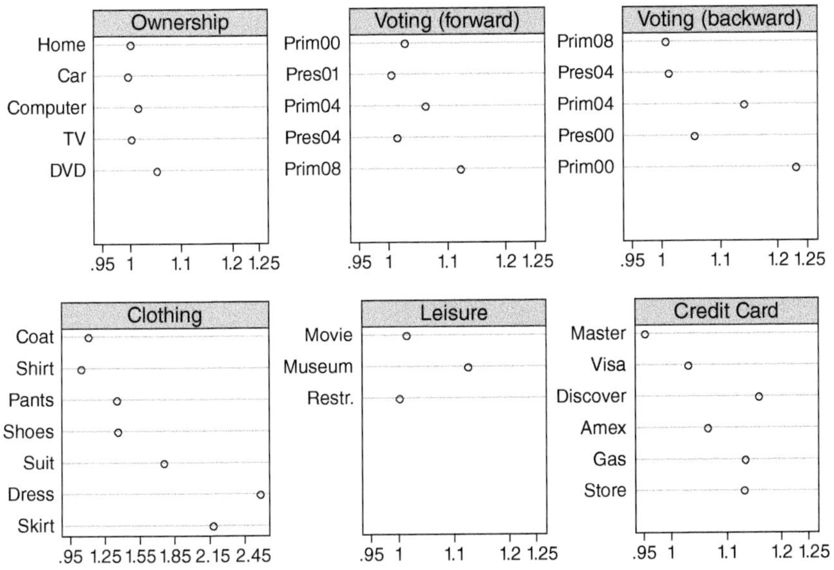

Figure 9.2 Kreuter et al. (2010) showed the ratio of endorsements in the grouped version to the interleafed presentation of filter and follow-up questions for six different topical sections. Values of 1 reflect equal endorsements across the two groups. Higher ratios show more endorsements in the grouped presentation.

What the experiments by Kreuter et al. (2010), Duan et al. (2007), and Kessler et al. (1998) lacked was clear evidence as to whether or not higher endorsement rates indicate better or worse response quality. Eckman et al. (2014) compared respondents' answers to administrative records and showed that grouping filter items produces more accurate responses to filter questions, with a significantly lower false negative rate. This finding supported the notion of motivated misreporting in the interleafed format.

Despite the evidence that it leads to motivated misreporting, many surveys continue to use the interleafed format for filter and follow-up items. The interleafed format is more conversational: respondents are asked about each topic only once and do not jump around between topics as in the grouped format. In particular, if the list of items is very long, asking all filter questions first, as in the grouped format, could create a very awkward survey experience. The more conversational organization of the interleafed format allows for easier cognitive processing: follow-up questions are asked right after the respondent retrieves an answer to the filter question and has thought about the topic (Tourangeau et al. 2000). Furthermore, thinking about surveys on mobile devices or other settings where respondents might break off or get interrupted, it is even more important to ask the follow-up questions right away to avoid the risk of not collecting that important information at all.

The current state of the research is thus that the grouped format collects more accurate responses, but the interleafed format has other advantages. The issue then arises: How can a series of filter and follow-up questions be presented to reduce the risk of motivated underreporting without having to ask all filter questions up front? What design features can be employed to mitigate the undesirable effects?

The answer might depend on how obviously the questionnaire design features signal increased burden as a function of certain prior answer choices. Couper et al. (2013) demonstrated the salience of question contingency; they found that placing filter and follow-up items into a single grid reduced filter endorsements relative to placing the follow-up items on multiple screens in a web survey. Kreuter et al. (2010) found that changing the topic reduced the difference between the two formats, leading them to suspect that a resetting effect takes place in respondents' minds (as shown earlier in Figure 9.2).

These results led us to design several experiments in which we varied the salience of the repetitive nature of filter and follow-up questions. Our goal was to empirically test the possibility of reducing the format effect, and subsequently improving the quality of the data collected via the interleafed format, through manipulation of questionnaire design features. Should we succeed, questionnaire designers could benefit from the conversational flow offered by the interleafed format without risking the errors induced by motivated underreporting.

9.2 Study Designs and Methods

To manipulate the salience of the consequences of endorsing filter questions, we carried out four surveys in which we experimentally varied different aspects of the filter and follow-up questions. Two of the studies were telephone surveys conducted in Germany on a random sample of adults registered in the social security system, and the second telephone survey was a follow-up to the first. The remaining two were web surveys, one conducted in the United States with two opt-in panels, and the other in the Netherlands with a probability-based web panel. Table 9.1 summarizes the most important features of each study. Throughout the text, we refer to the studies by the names give in the column headers.

All four studies randomized respondents to either grouped or interleafed versions, allowing us to replicate the earlier finding of a difference between these two formats. All surveys used filter and follow-up questions similar, if not identical, to those of the US Consumer Expenditure Survey, which asks about a series of purchases and specifics about each of these purchases. Each survey contained an additional experimental manipulation within the interleafed format of the salience of the burden imposed by the filter questions. These

Table 9.1 Summary of four surveys with design elements used here.

	US-Web	G-Phone1	G-Phone2	NL-Web
n Respondents	2484	1800	1325	5749
Mode	Web	Telephone	Telephone	Web
Data collection	Dec 2011	Aug – Oct 2011	Nov – Dec 2012	April 2012
Filter topics	Clothing	Clothing	Clothing	Household Activities
# of filters	6	6	6	13
Filter formats	Interleafed grouped	Interleafed grouped	Interleafed grouped	Interleafed grouped
Order of filters	Fully randomized	Forward vs. backward	Forward vs. backward	Forward vs. backward
Country	US	Germany	Germany	The Netherlands
Response rate	N/A[a]	19.4%	63.3%[b]	76.7%[c]

a) Response rate calculation not appropriate due to use of opt-in panel.
b) Response rate in wave 2, net of nonresponse in wave 1.
c) Response rate in this month of the panel, net of nonresponse in recruitment phases.

manipulations took the form of changes to the visual layout, the repetitiveness, and the number of questions asked in the follow-up format. We explain each of these in more detail in the next section.[1]

US-Web. This web survey was carried out by Market Strategies International (MSI) with 2427 respondents: 1204 from MSI's own online panel, Authentic Response, and 1223 from Survey Sampling's Dynamix sample. Because this survey used non-representative panels, no response rate calculation is possible. In the Authentic Response sample 14 132 cases were contacted, of which 1619 cases started the survey and 1204 completed, leading to a participation rate of 8.5%. More details on this study can be found in Tourangeau et al. (2011) and Couper et al. (2013).

G-Phone1. This telephone survey was conducted by the LINK Institute on behalf of the Institute for Employment Research (IAB) in Germany. A sample of named adults was selected from German administrative databases in three strata: those with who recently changed employers, those who were short-term unemployed, and those who were receiving welfare payments. The completed sample size was 1800 cases, with nearly equal sample sizes across the strata. (Note: The entire survey included 2400 completed cases, but not all are used

1 Each survey also contained other experiments that are fully crossed with the manipulations we focus on in this paper, or are related to other sections of the surveys and thus are not discussed here. We provide citations to papers discussing those manipulations.

in the analyses in this paper.) The questionnaire contained a section of six filter questions about clothing purchases in the last year. The six items were presented randomly in forward or backward order. One-third of the respondents received the filter questions in the grouped format, and two-thirds in the interleafed format, allowing for the additional manipulation described shortly. The format manipulation operated between respondents but within interviewers. Other aspects of this study and the questions themselves are provided in Eckman et al. (2014).

G-Phone2. The second telephone survey used respondents from G-Phone1 who agreed to be contacted again. A total of 1325 respondents, from the 2094 who consented to be recontacted, completed the second-wave survey, which was conducted between November and December in 2012. This study repeated the clothing filter and follow-up items.

NL-Web. The second web study was embedded in the Longitudinal Internet Studies for the Social Sciences (LISS) panel, a monthly survey carried out by the CentERData Institute in the Netherlands (Scherpenzeel 2011). The sample was recruited via telephone and face-to-face to participate in an online panel, and computers and internet access, as well as training, were provided to respondents who needed them. Thus, this survey suffers less from undercoverage due to lack of internet access or skills than many online panels do (Eckman 2016).

Most of the analyses presented from these studies were at the respondent level and controlled for the clustering of answers by respondents to correctly estimate the standard errors. Standard errors are estimated using logistic and linear regression models implemented in Stata. The analyses are unweighted because we do not wish to make inference to any population, but rather to compare experimental conditions. The dependent variable in most analyses is the endorsement rate, that is, the fraction of answers to the filter questions that were YES.

Before describing the different experiments carried out in each of these studies, we first confirm that the finding of more reports in the grouped format than in the interleafed format was replicated across countries and modes. Table 9.2 presents endorsements rates for the interleafed condition and the grouped condition in each of the four surveys. Although at different levels, all four studies show significantly higher endorsement rates for the grouped display of filter questions compared to the interleafed. In G-Phone1, an almost 10 percentage point difference was found between interleafed and grouped, similar to the difference found in the US-Web survey on the similar set of items. The format effect also appears in the G-Phone2 and NL-Web studies, although it is of smaller magnitude. These results reassure us the filter questions are operating in the expected way in the four studies.

Table 9.2 Percent of filter questions triggered, by filter format.

	US-Web	G-Phone1	G-Phone2	NL-Web
Interleafed	37.6 (1.17)	48.4 (1.08)	61.0 (1.74)	36.6 (0.38)
Grouped	49.0 (1.36)	59.8 (1.20)	65.6 (1.00)	42.9 (0.42)
Adjusted Wald test	40.80	7.03	5.18	11.24
p-Value	<0.000 1	<0.001	0.02	<0.000 1
n Filters	7 200	10 785[a)]	6 570[b)]	49 023[b)]
n Respondents	1 215	1 799	1 095	3 771

Standard errors, in parentheses, reflect clustering of items within respondents.
a) All cases receiving the same (non-varying) follow-up items.
b) All cases receiving two follow-up items.

9.3 Manipulating the Interleafed Format

Across the four studies, we carried out three different manipulations in the presentation of the interleafed format to empirically test if such design features can reduce format effects. The manipulations address (i) the visual layout of the web questionnaire, (ii) the repetitiveness of the follow-up questions, and (ii) the number of follow-up items.

The first manipulation was inspired by the findings from Couper et al. (2013) for grid questions. In that study, changing the boldness of the font as a result of prior answer choices within the grid questions affected respondents' answering behavior. This visual manipulation is directly related to visual salience of the consequences when answering a filter question. We manipulated the display of filter questions in a web survey in a comparable way. In addition, we directly manipulated the burden by presenting items on one versus multiple screens, again a manipulation that Couper et al. (2013) also experimented with. Multiple screens add burden through additional clicks.

While the manipulation of salience and burden can be directly operationalized in a web survey, for telephone surveys the options are somewhat more limited. We focused on two that seem easy to realize within the phone setting: altering the *repetitiveness of the follow-up questions* and the *number of follow-up questions.* We focused on the repetitiveness manipulation after taking a closer look into the construction of the follow-up items in (Kreuter et al. 2010). As revealed in Figure 9.2, the size of the format effect varies across topics. Aside from baseline endorsement rates for the varying events, a clear difference in the questions across sections in that study was that the clothing, voting, and credit card sections had exactly the same follow-up items within each section, whereas the ownership items and leisure items had varying follow-up questions. The larger format effect in the former sections could have been a function of perceived repetitiveness, adding to the burden of the task. In G-Phone1, we experimentally varied whether all follow-up questions in a given section were identical or whether there was some variation across topics.

The third set of experiments manipulated directly the burden of endorsing a filter question, by varying the number of follow-up items in the interleafed format. We implemented two versions of this manipulation, one on the phone and one in a web survey. The implementations and results of each of these manipulations are discussed next.

9.3.1 Visual Layout

US-Web. The questionnaire in the US-Web survey contained one section of six filter questions, each with four follow-up questions. The topic of the filters was clothing purchases. The order of the filter questions was fully randomized.

Respondents were randomly assigned to one of four conditions:

1 One quarter of the respondents received questions in a grouped format. These respondents first answered all six filter questions and then any triggered follow-up questions. Each question was on its own page.
2 The remaining three quarters all received different versions of follow-up displays within the interleafed format. These cases were further divided into three groups:
   a *Separate-screen.* For one-third of these respondents, each filter was immediately followed by its follow-up questions, if the filter was answered with a YES response. Each item in this condition appeared on a separate screen.
   b *Same-screen.* The second third saw both the filter question and its four follow-ups on the same screen. If a respondent answered NO to the filter question, he or she did not need to answer the follow-up questions, but that was not obvious on the screen.
   c *Same-screen/grayed out.* The final third within the interleafed condition saw a filter question and its follow-ups on one screen, but the follow-up question text turned from black to gray if the respondent clicked NO to the filter, indicating that the follow-ups need not be answered (see Figure 9.3).

In these last two conditions, the web instrument allowed respondents to answer the follow-up items even when they did not apply. In Condition Same-screen, where the follow-up items did not turn gray, 5.4% of respondents answered one or more follow-ups after saying NO to the filter. In Condition Same-screen/grayed out, where the follow-up items did turn gray, four respondents (<1%) answered one or more follow-ups after saying NO to the filter questions. (Although the web instrument displayed the questions in gray text, respondents could still answer them.) These respondents seem not to have understood how the filter and follow-up questions worked, and thus

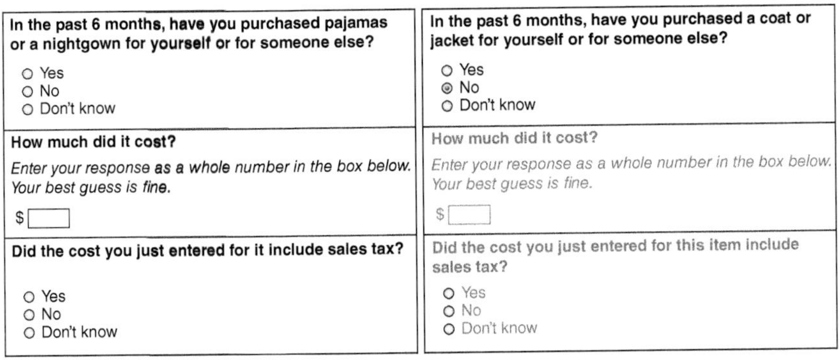
In the past 6 months, have you purchased pajamas or a nightgown for yourself or for someone else?
Yes
No
Don't know
How much did it cost?
Enter your response as a whole number in the box below. Your best guess is fine.
$
Did the cost you just entered for it include sales tax?
Yes
No
Don't know
In the past 6 months, have you purchased a coat or jacket for yourself or for someone else?
Yes
No
Don't know
How much did it cost?
Enter your response as a whole number in the box below. Your best guess is fine.
$
Did the cost you just entered for this item include sales tax?
Yes
No
Don't know

Figure 9.3 Display of filter and follow-up items in condition (Same-screen/grayed out).

Table 9.3 Sample sizes by experimental condition, US-WEB.

Format	Condition	All cases	Excluding problem cases
Grouped		624 respondents 3729 filters	624 respondents 3729 filters
Interleafed	Separate-screen	606 respondents 3626 filters	606 respondents 3626 filters
Interleafed	Same-screen	616 respondents 3676 filters	583 respondents 2999 filters
Interleafed	Same-screen/grayed out	664 respondents 3980 filters	660 respondents 3829 filters

we dropped those cases from our analyses. Table 9.3 gives the sample sizes for each of the experimental conditions both before and after dropping these cases.

The overall endorsement rate in the grouped version was 49.0% compared to 37.6% for those respondents who received the Separate-screen version of the interleafed format. This result together with its standard error is shown in Table 9.2. Looking into more detailed results in Table 9.4, when the items were displayed on one page and nothing in the display changed after endorsing the filter question (Condition Same-screen), endorsements rates were 50.6%, similar to the grouped format. When the follow-up items turn from black to gray text when the filter question was not endorsed (Same-screen/grayed out), respondents endorsed significantly fewer filter questions: 42.0% (adjusted Wald test $F(1, 2444) = 21.04$, $p < 0.0001$). These findings replicate the results from Couper et al. (2013), showing that respondents perceive and are affected by the way the questions are presented visually.

Table 9.4 Effects of visual layout – web.

Format	Percent of filters triggered (standard error)
Grouped	49.0 (1.36)
Interleafed, separate screen	37.6 (1.17)
Interleafed, same page	50.6 (1.45)
Interleafed, same page – grayed out	42.0 (1.19)
n Filters	13 905
n Respondents	2 445

Adjusted Wald test of difference between interleafed formats: $F(1, 2444) = 21.04$, $p < 0.0001$.

9.3.2 Repetitiveness

In the G-Phone1 study, we varied whether all the follow-up questions were identical or whether they varied. Each filter question had four follow-ups, which were triggered by a YES response to the relevant filter question. For all respondents in the grouped format (results shown earlier), as well as half of the respondents in the interleafed format, the follow-up questions were identical (monotonous format). That is, answering that one recently purchased a coat triggered the same four follow-up questions as answering that one recently purchased a shirt. For the other half of cases in the interleafed format, the follow-up questions varied with each of the filter questions (see Figure 9.4).

Table 9.5 shows the result of this experiment. The difference between interleafed and grouped is reduced by about four percentage points when the

Interleafed monotonous	**Interleafed varied**
Purchased coat?	*Purchased coat?*
Description of item purchased	Was it a winter coat or a summer coat?
For whom was it purchased?	Was this an impulse buy or did you plan it?
Month of purchase	Cost of item purchased
Cost of item purchased	Frequency of use
Purchased shirt?	*Purchased shirt?*
Description of item purchased	Material shirt is made out of
For whom was it purchased?	Dress shirt or sporty
Month of purchase	Is this a non-iron shirt or do you have to iron it?
Cost of item purchased	Cost of item purchased
AND SO ON FOR REMAINING ITEMS	*AND SO ON FOR REMAINING ITEMS*

Figure 9.4 Display of monotonous and varied follow-up questions to the clothing purchase items. Across all follow-up sections only one – the cost of the item purchased – was kept the same. The highlighting of this item is only for display purposes and was not in the actual interview.

Table 9.5 Effects of varied follow-up questions: reduced repetitiveness.

Format	Percent of filters triggered (standard error)
Grouped	59.8 (1.20)
Interleafed, monotonous	48.4 (1.08)
Interleafed, varied	52.3 (1.09)
n Filters	7201
n Respondents	1200

follow-up items were varied. Among respondents asked varying follow-up questions, the endorsement rate increased significantly to 52.3% from 48.4%. While not quite to the level of the grouped rate of 59.8% endorsements, this result does demonstrate respondents' responsiveness to a reduced repetitive nature of the filter and follow-up sequence.

Difference between two interleafed format was significant at the 5% level: $t = 2.52$, $p < 0.012$, accounting for clustering of items within respondents.

9.3.3 Number of Follow-Up Items

Both the G-Phone2 and the NL-Web studies contained experiments that varied the number of follow-up questions triggered by a YES response to the filter items. In the second round of the German phone study (G-Phone2), respondents were again asked the battery of filter questions and the non-varying follow-ups related to clothing purchases from the first wave (G-Phone1). Respondents were randomly assigned to either a grouped format, in which all six filter questions were asked first, or an interleafed format. All respondents in the grouped format received two follow-up questions when they answered YES to a filter question.[2] Respondents in the interleafed format received either two or four follow-up questions. Allocations to the two formats and to the two different follow-up sets was not equal, due to the need to keep the average administration time short in this study. The sample sizes in each of these experimental conditions are given in Table 9.6. As in G-Phone1, half of the cases received the filter questions in the forward order and half in the backward order, to allow for separation of question topic and learning effects.

2 In the grouped format, follow-up items are not asked until the end of the section. Thus the number of follow-up items cannot affect the responses to the filter questions. For this reason, neither of our experiments tested the upper-right cell in the experimental design shown in Tables 9.6 and 9.7.

Table 9.6 Sample sizes by experimental condition, G-Phone2.

	Number of follow-ups	
Format	**2**	**4**
Grouped	873 respondents 5238 filters	
Interleafed	222 respondents 1332 filters	230 respondents 1389 filters

Table 9.7 Sample Sizes by experimental condition, NL-web.

	Number of follow-ups	
Format	**2**	**6**
Grouped	1 928 respondents 25 064 filters	
Interleafed	1 843 respondents 23 959 filters	1 978 respondents 25 714 filters

We added a section of 13 filter questions about household purchases to the LISS panel study fielded in March 2012 (NL-Web). Respondents were randomly assigned to a grouped or interleafed format, and, as in G-Phone2, the number of follow-up questions also varied between respondents. All respondents in the grouped format received two follow-up questions, and respondents in the interleafed format received either two or six follow-up questions. Every question was displayed on a separate page in this web survey. Approximately one-third of all respondents were in each of these experimental conditions (see Table 9.7). As in G-Phone1 and G-Phone2, half received the filter questions in the forward order and half in the backward order.

In both the G-Phone2 and NL-Web studies, respondents who received more follow-up items were significantly less likely to answer YES and trigger follow-up questions than those who received fewer (Table 9.8).

9.4 Discussion and Conclusion

In this paper we replicated past findings, clearly demonstrating higher endorsement rates for filter questions when asked in a grouped, rather than an interleafed, format. Despite these results, the interleafed format offers

Table 9.8 Percent of filters triggered, by the number of follow-up questions.

Format	G-Phone2 Percent of filters triggered (standard error)	NL-Web Percent of filters triggered (standard error)
Interleafed, few	61.0 (1.74)	36.6 (3.82)
Interleafed, many	55.7 (1.88)	35.4 (3.66)
t-Test	4.29	2.17
p-Value	0.039	0.030
n Filters	2712	49 653
n Respondents	452	3820

In both surveys, the few condition used two follow-up items.
In G-Phone2, the many condition contained four follow-ups
In NL-Web, the many condition contained six follow-ups.

several substantial benefits over the grouped format, such as the improved conversational flow of a survey, reduced cognitive burden, and fewer missing data in the follow-up questions (Kreuter et al. 2010; Eckman and Kreuter 2015). For these reasons, we set out to test several design features that might allow practitioners to employ interleafed question formats without compromising endorsement rates. The experiments varied three question characteristics: The visual layout of filter and follow-up questions, the repetitiveness of the follow-up questions, and the number of follow-up questions. Each manipulation set out to make the filter-follow-up structure, and its related burden, more or less salient. Our results show that using slightly varied follow-up questions and reducing the repetitiveness of the tasks increased endorsements of filter questions in the interleafed mode and thus successfully reduced the effect of motivated underreporting. Similarly, not graying out the filter questions when respondents click "no" to the filter question was associated with higher endorsement rates. We cannot recommend the visual manipulation tested in Section 9.3.1, and encourage additional experimentation to find alternative visualizations that can increase the endorsement rate in the interleafed format further.

An obvious limitation of this study is the limited set of items, focusing on clothing purchases only. While the topic is important for the Consumer Expenditure survey, it is conceivable that effects are different for different topics. Eckman et al. (2014) demonstrated the sensitivity of the grouped versus interleafed effect for various topics, noting that reports about life circumstances show weaker format effects than reports about these purchasing patterns. Thus, replications of these experiments with other topics are highly encouraged.

Varying the follow-up questions is in itself quite intriguing and worked well in the experimental manipulation shown here. For actual data collection purposes, such manipulation is not always possible. Again, looking at the Consumer Expenditure survey, the follow-up items are all needed, and thus variation could only be achieved by adding items, or the purposeful application of a matrix design, where for some such items are imputed based on answers to other questions.

Nevertheless, our experimental results show that respondents are sensitive to design characteristics, and that design features of the survey items can elevate or reduce the motivated underreporting found in earlier studies, in particular when the salience of the filter-follow-up structure and its perceived burden is reduced. We thus conclude in suggesting the following design principles:

- Use as few follow-up questions as possible.
- Do not include follow-up questions just for the sake of symmetry or easier programming.
- Avoid (web) questionnaire designs that make the structure of the filter and follow-up question visually obvious.
- Vary the follow-up questions within a section when possible.

Biography

Frauke Kreuter is director of and professor in the Joint Program in Survey Methodology at the University of Maryland, USA; professor of statistics and methodology at the University of Mannheim, Germany; and head of the Statistical Methods Research Department at the Institute for Employment Research, Nuremberg Germany. Stephanie Eckman is a fellow at RTI International in Washington DC. Roger Tourangeau is a senior advisor at Westat in Rockville, Maryland, USA.

Acknowledgments

The work reported here is supported by the National Science Foundation (NSF) (SES 0850999 to F.K. and 0850445 to R.T.). The authors would like to thank the NSF Methodology, Measurement, and Statistics Program and Dr. Cheryl Eavey for their support. Any opinions, findings, conclusions, or recommendations expressed in this article are those of the authors and do not necessarily reflect the views of the National Science Foundation. Data collection was also supported by the Institute for Employment Research (IAB) and the Dutch LISS Panel.

References

Couper, M.P., Tourangeau, R., Conrad, F., and Zhang, C. (2013). The design of grids in web surveys. *Social Science Computing Review* 31 (3): 322–345.

Duan, N., Alegria, M., Canino, G. et al. (2007). Survey conditioning in self-reported mental health service use: randomized comparison of alternative instrument formats. *Health Service Research* 42: 890–907.

Eckman, S. (2016). Does the inclusion of non-internet households in a web panel reduce coverage bias? *Social Science Computing Review* 34 (1): 41–58.

Eckman, S. and Kreuter, F. (2015). Misreporting to looping questions in surveys: Recall, motivation and burden. IAB Discussion Paper, 29.

Eckman, S., Kreuter, F., Kirchner, A. et al. (2014). Assessing the mechanisms of misreporting to filter questions in surveys. *Public Opinion Quarterly* 78: 721–733.

Kessler, R.C., Wittchen, H.-U., Abelson, J.A. et al. (1998). Methodological studies of the Composite International Diagnostic Interview(CIDI) in the US National Comorbidity Survey (NCS). *International Journal of Methods in Psychiatric Research* 7: 33–55.

Kreuter, F., McCulloch, S., Presser, S., and Tourangeau, R. (2010). The effects of asking filter questions in interleafed versus grouped format. *Sociological Methods and Research.* 40 (1): 88–104.

Scherpenzeel, A. (2011). Data collection in a probability based internet panel: How the LISS panel was built and how it can be used. *Bulletin of Sociological Methodology* 109 (1) https://doi.org/10.1177/0759106310387713.

Tourangeau, R., Rips, L.J., and Rasinski, K. (2000). *The Psychology of Survey Response.* Cambridge: Cambridge University.

Tourangeau, R., M. Couper, F. Conrad, and R. Baker (2011). Web design experiment 9: 2011 (United States). Technical report, Interuniversity Consortium for Political and Social Research, Ann Arbor, MI (distributor). doi:10.3886/E55420V1.

Tourangeau, R., Kreuter, F., and Eckman, S. (2015). Motivated misreporting: Shaping answers to reduce survey burden. In: *Survey Measurements: Techniques, Data Quality and Sources of Error* (ed. U. Engel). Frankfurt: Campus Verlag.

United States Department of Health and Human Services. (2016). National survey on drug use and health, 2014. Substance Abuse and Mental Health Services Administration. Center for Behavioral Health Statistics and Quality. ICPSR36361-v1. Ann Arbor, MI: Inter-university Consortium for Political and Social Research [distributor]. https://doi.org/10.3886/ICPSR36361.v1.

10

A Comparison of Fully Labeled and Top-Labeled Grid Question Formats

Jolene D. Smyth and Kristen Olson

Department of Sociology, University of Nebraska-Lincoln, Lincoln, NE, USA

10.1 Introduction

The grid question format is common in mail and web surveys. In this format, a single question stem introduces a set of items, which are listed in rows of a table underneath the question stem. The table's columns contain the response options, usually only listed at the top, with answer spaces arrayed below and aligned with the items (Dillman et al. 2014). This format is efficient for respondents; they do not have to read the full question stem and full set of response options for every item in the grid. Likewise, it is space efficient for the survey researcher, which reduces printing and shipping costs in mail surveys and scrolling in web surveys.

However, grids also complicate the response task by introducing fairly complex groupings of information. To answer grid items, respondents have to connect disparate pieces of information in space by locating the position on the page or screen where the proper row (the item prompt) intersects with the proper column (the response option). The difficulty of this task increases when the respondent has to traverse the largest distances to connect items to response option labels (down and right in the grid) (Couper 2008; Kaczmirek 2011). This spatial connection task has to be conducted while remembering the shared question stem, perhaps after reading and answering multiple items. As a result, grid items are prone to high rates of item nonresponse, straightlining, and breakoffs (Couper et al. 2013; Tourangeau et al. 2004).

One way to possibly ease the burdens of grids in mail surveys is to repeat the response option labels in each row next to their corresponding answer spaces (Dillman 1978). Including response option labels near the answer spaces eliminates the need for vertical processing, allowing respondents to focus only on processing horizontally. However, fully labeling the answer spaces yields a

Advances in Questionnaire Design, Development, Evaluation and Testing, First Edition.
Edited by Paul C. Beatty, Debbie Collins, Lyn Kaye, Jose-Luis Padilla, Gordon B. Willis, and Amanda Wilmot.

more busy, dense display overall, which one can speculate might intimidate or overwhelm some respondents, leading them to skip the grid entirely.

In this chapter we report the results of a series of experimental comparisons of fully labeled versus top-labeled grid formats from a national probability mail survey, a convenience sample of students in a paper-and-pencil survey, and a convenience sample in a web-based eye-tracking laboratory study. For each experiment we compare mean responses, inter-item correlations, item nonresponse rates, and straightlining. In addition, for the eye-tracking experiment we also examine whether the different grid designs impacted how respondents visually processed the grid items. For two of the experiments, we conduct subgroup analyses to assess whether the effects of the grids differed for high and low cognitive ability respondents. Our experiments are conducted using both attitude and behavior questions covering a wide variety of question topics and using a variety of types of response scales.

10.1.1 Grid Items vs. Individual Items

The tension between the benefits and difficulties of grids has spurred interest in the quality of grids as a survey measurement tool. The bulk of research on grids has focused on the effect of asking about multiple items when they are presented individually, each with its own question stem and response options, to when they are presented as a set of items in a grid with a shared question stem and response options. Most of these studies have been conducted in web surveys (for an exception, see Iglesias et al. 2001) and many confound the separation of items with paging such that in individual item conditions, each item is on a separate page of the web survey rather than displaying them as individual items on the same page (Callegaro et al. 2009; Couper et al. 2001; Peytchev 2007; Stern et al. 2015; Thorndike et al. 2009; Toepoel et al. 2005; Tourangeau et al. 2004. For exceptions, see Bell et al. 2001; Richards et al. 2016; and Yan 2005).

Generally, mean scores across items displayed individually versus in a grid are not significantly different from each other (Bell et al. 2001; Iglesias et al. 2001; Peytchev 2007; Toepoel et al. 2005; Yan 2005). However, a consistent trend is for items displayed in the grid format to have higher inter-item correlations (Callegaro et al. 2009; Couper et al. 2001; Peytchev 2007; Toepoel et al. 2005; Tourangeau et al. 2004; Yan 2005; for an exception, see Iglesias et al. 2001), but the differences only reached statistical significance in studies by Peytchev (2007) and Tourangeau et al. (2004). Thus, the magnitude of the difference in the correlational structure is usually not large. Although increased correlations may be thought to reflect improved data quality, Peytchev (2007) showed that the increased inter-item correlation in the grid format is likely due to correlated measurement error among grid items, probably caused by increased straightlining in the grid format, not to improved data quality. Factor loadings consistently do not differ across items in a grid versus individual items

(Couper et al. 2001; Iglesias et al. 2001; Thorndike et al. 2009; Toepoel et al. 2005). Thus, compared to individual-item formats, the grid format appears to increase inter-item correlations due to shared method variance, but has little effect on other measurement outcomes like means and factor loadings.

Other data quality indicators show somewhat larger differences across the grid and individual item treatments. A consistent trend is for the grid format to increase item nonresponse, both to the entire grid (Richards et al. 2016) and to individual items within the grid (Iglesias et al. 2001; Peytchev 2007; Richards et al. 2016; Toepoel et al. 2005; for an exception, see Callegaro et al. 2009) and more so in grids with more items (Toepoel et al. 2005). Additionally, Couper et al. (2001) found that the grid format *decreased* the rate of "don't know" and "not applicable" responses in a web survey where, importantly, a response was required for every item. The grid format increases nondifferentiation and/or straightlining (i.e. a satisficing response behavior in which respondents provide the same or nearly the same response for all items [Krosnick 1991]) (Richards et al. 2016; Stern et al. 2015; Tourangeau et al. 2004; but see Couper et al. 2001 for an exception). Finally, items tend to be answered more quickly when they are displayed in the grid format than in an individual-item format (Bell et al. 2001; Callegaro et al. 2009; Couper et al. 2001; Peytchev 2007; Stern et al. 2015; Thorndike et al. 2009; Toepoel et al. 2005; Tourangeau et al. 2004).

Empirically, there is no consistent difference in respondent perceptions of grids versus single items, with some evidence that individual items are preferred over grids. Although two studies suggest that respondents view the questionnaire as more difficult with grids, evaluate the layout of the questionnaire more poorly with grids, or prefer individual items over grid formats (Thorndike et al. 2009; Toepoel et al. 2005), and even more so the more items were included in each grid (Toepoel et al. 2005), two other studies found no such differences in similar respondent perceptions (Callegaro et al. 2009; Yan 2005).

Taken together, the existing literature suggests that the grid format has little impact on substantive results, but poses some difficulty for respondents as they answer grid items more quickly and are more likely to skip items within the grid, straightline, or give nondifferentiated answers. Moreover, respondents may find the grid format more difficult and prefer the individual-item format, but evidence on this is mixed.

10.1.2 Dynamic Grid Features in Web Surveys

Several studies have attempted to find ways to make grids easier for respondents and improve data quality. For example, Kaczmirek (2008) experimented with two dynamic grid design features in a web survey – a postselection feature where each item in the grid was grayed out when answered so respondents could more easily differentiate answered and unanswered items; and a preselection feature in which the row and column over which the mouse hovered were shaded, creating a cross-hair to help respondents ensure they were clicking the correct answer space. Both of these methods were compared to a control treatment utilizing a white background and no dynamic shading or interactivity. The dynamic designs did not change response distributions or response time, but did affect item nonresponse. Seventeen percent of respondents skipped at least one item in the control version. The cross-hair shading increased that rate to 19.4% (perhaps because it distracted respondents), but graying out answered items decreased the rate to 11.8%. In a later study, Kaczmirek (2011) found similar results; preselection shading of table cells increased item nonresponse, but postselection graying of item rows decreased item nonresponse (and had no effect on nondifferentiation).

In another study, Couper et al. (2013) experimented with dynamic web design features in a matrix design where each row contained a type of fruit and two columns contained questions asking how often they eat each type of fruit and how much they usually eat. They tested graying out the "how much" question for fruits respondents reported never eating and graying out the entire row for fruits once both questions were fully answered. These dynamic features reduced item nonresponse and response time compared to a static version, but did not affect straightlining, which was rare in all of their treatments.

10.1.3 Easing Grid Question Burden in Mail Surveys

While such dynamic design features show promise, they cannot be used in mail surveys. However, the difficulty of responding to grids could be reduced in mail surveys by minimizing the need for respondents to work both horizontally and vertically to connect the relevant pieces of information. One way this might be done is by repeating the response option labels in every row of the grid as shown in the top panel of Figure 10.1 (Dillman 1978).

Several concepts from the vision sciences are relevant for understanding why this design may help respondents process grid items. First, according to the Gestalt psychologists' principle of continuity, items that appear to continue smoothly will be more easily perceived as belonging together (Ware 2004). In the fully labeled design, it should be much easier for respondents to group the items with the desired response option labels because the labels appear in the same horizontal line as the items. Respondents do not have to make the 90° upward turn required to process the response options in the top-labeled grid. Moreover, during attentive visual processing, we only attend to a narrow slice of the entire visual field, called the *useful field of vision* (Ware 2004). This includes the foveal view, which is made up two degrees of visual angle (i.e. 8–10 characters) and in which we can see very sharply, and an additional approximately 13° of visual angle in which we can detect visual elements, but our vision is much less sharp (Ware 2004). The useful field of view gets smaller when visual information is dense (Ware 2004), as in the case of grid designs. Visual elements

Fully Labeled Grid

21. In the past 12 months, how often did you experience each of the following?

	Never	Rarely	Sometimes	Often	Always
You had exciting new ideas or thoughts occurring to you one after the other.	ONever	ORarely	OSome	OOften	OAlways
You felt so confident, nothing could stop you.	ONever	ORarely	OSome	OOften	OAlways
You got much less sleep than usual but didn't really miss it.	ONever	ORarely	OSome	OOften	OAlways
You were so easily distracted that you had trouble staying on track.	ONever	ORarely	OSome	OOften	OAlways
You tended to show poor judgment (e.g., spending spree, sexual indiscretions, or impulsively quitting a job).	ONever	ORarely	OSome	OOften	OAlways
You thought you were being plotted against.	ONever	ORarely	OSome	OOften	OAlways
You were sure that everyone was against you.	ONever	ORarely	OSome	OOften	OAlways
You thought negative comments were being circulated about you.	ONever	ORarely	OSome	OOften	OAlways
You felt people were trying to make you upset.	ONever	ORarely	OSome	OOften	OAlways

Top Labeled Grid

21. In the past 12 months, how often did you experience each of the following?

	Never	Rarely	Sometimes	Often	Always
You had exciting new ideas or thoughts occurring to you one after the other.	O	O	O	O	O
You felt so confident, nothing could stop you.	O	O	O	O	O
You got much less sleep than usual but didn't really miss it.	O	O	O	O	O
You were so easily distracted that you had trouble staying on track.	O	O	O	O	O
You tended to show poor judgment (e.g., spending spree, sexual indiscretions, or impulsively quitting a job).	O	O	O	O	O
You thought you were being plotted against.	O	O	O	O	O
You were sure that everyone was against you.	O	O	O	O	O
You thought negative comments were being circulated about you.	O	O	O	O	O
You felt people were trying to make you upset.	O	O	O	O	O

Figure 10.1 Examples of top labeled and fully labeled grids.

that appear outside the foveal view are more likely to be overlooked. Including the response option labels in every row of the grid should eliminate wide areas without visual elements, keeping the string of visual elements all within two degrees of visual angle of each other. This should make it easier for respondents to visually track across the row, moving from one visual element to the next without mistakenly jumping to a different row. In sum, in a fully labeled grid design, all of the information respondents need is contained in a single row of the grid in a continuous stream.

Only two studies of which we are aware have previously examined fully labeled grids. In the first, Toepoel et al. (2005) examined the effects of presenting one item per screen versus grids consisting of 4, 10, or 40 items per screen. They fully crossed the 4, 10, and 40 items per-screen treatments with top versus fully labeled designs. The labeling had no effect in their study. However, their fully labeled design did not group the items, answer spaces, and response option labels together on one row as shown in Figure 10.1. Rather, they maintained a slightly more traditional grid design with the response option labels on the top row and the item and answer spaces one row below it for each item. Thus, each of their items appeared as a one-item top-labeled grid, and the need for vertical processing was not fully eliminated. Also, their respondents were web panel members who are likely very practiced at answering many types of survey questions, including grids.

In the second study, Smyth et al. (2014) compared item nonresponse and straightlining rates across top and fully labeled grid formats in a general population mail survey of Nebraska residents. With this sample and design, they found lower rates of item nonresponse in the fully labeled version, but no difference in straightlining rates. The current study attempts to replicate and extend this research.

10.1.4 Understanding How Respondents Process Items in Grids

Existing studies have compared response distributions, inter-item correlations, and a variety of data quality outcomes to understand how the grid format affects respondents' answers. These outcomes are indirect measures of underlying respondent processing. Eye-tracking methods provide a more direct measure of how respondents process survey questions by observing what they look at, for how long, and how their eyes move between visual elements (Galesic et al. 2008; Graesser et al. 2006; Redline and Lankford 2001). We take advantage of this capability to examine how respondents process grids. Insights from the eye-tracking study will allow us to better understand how grid format affects respondents and their answers.

10.1.5 Hypotheses

We report the result of 12 experimental comparisons of top versus fully labeled grid designs in paper-and-pencil and web surveys. First we test for differences in the substantive answers respondents provide and in data quality indicators for these answers. Then, for one of these experiments, which was conducted in an eye-tracking laboratory, we test for differences in how respondents visually processed the two types of grids.

We do not expect the repetition of the response options to affect how respondents understand the items or formulate their answers for them. That is, repeating the response option labels should not affect comprehension, retrieval, or judgment. It could affect mapping, but we do not expect a consistent effect across respondents (i.e. no biasing effect). Thus, *we hypothesize (H1) that there will be no difference across the top and fully labeled grids in means for individual items.*

Correlations between items in grids can be affected through nondifferentiation, sometimes called straightlining. If, as the research shows, respondents are more likely to give nondifferentiated responses when items appear in grids, the correlations between those items will increase (i.e. correlated measurement error) (Peytchev 2007). It follows that any design feature that reduces nondifferentiation within grids should also reduce correlations among items. We expect fully labeling the grid will ease response burden in grids, reducing motivation to shortcut by straightlining or giving nondifferentiated responses. Thus, *we hypothesize that the fully labeled grid will have (H2) lower rates of straightlining and (H3) lower correlations between items than the top-labeled grid.*

Because we expect the full labeling to reduce the difficulty of connecting information within the grid, *we hypothesize (H4) that the fully labeled grid treatments will produce lower rates of item nonresponse than the top-labeled grid treatments.* In particular, having the labels in each row should make it easier for respondents to answer without mistakenly getting off a row in either direction and inadvertently leaving items blank. However, while the fully labeled grid may be easier to complete (i.e. actual burden), it may initially be perceived as more burdensome because of its information-dense appearance. Thus, *we hypothesize (H5) that the fully labeled grid will produce higher rates of respondents skipping over the entire grid (*i.e. *not answering any items within the grid).*

In this study, we have a unique opportunity to evaluate how respondents are actually processing information in the grid through use of eye tracking. We anticipate that the full labeling should reduce or even eliminate the need for vertical processing. As a result, *we hypothesize (H6) that compared to respondents in the top-labeled version, respondents in the fully labeled version will spend less time looking (*i.e. *fixation duration) at the response option labels at the top of the grid columns.* In addition, *we hypothesize (H7) that those in the fully labeled version will spend more time than those in the top-labeled version fixating on areas internal to the grid.* With respect to entries (i.e. how many times respondents look at a specific area), *we hypothesize (H8) that respondents in the fully labeled version will look at the top row of labels fewer times than those in the top-labeled version.* We expect the manipulations here to impact the processing of the grid headings and answer spaces, but not the processing of the list of items in the left-most column in the grid. Thus, *we hypothesize there will be no difference across treatments in the amount of time (H9) or number of gaze entries (H10) into the item prompts.*

For reasons described earlier, responding to grid items is particularly difficult. We anticipate that it is even more difficult for those with low cognitive ability as their already limited cognitive resources are stretched further by the complicated demands of the grid format (Knäuper 1999; Krosnick 1991). As a result, we expect the full labeling to have a larger effect on these respondents. That is, *we hypothesize that the full labeling will reduce item nonresponse rates (H11), straightlining (H12), and nondifferentiation (H13) further for low-cognitive-ability respondents than for high-cognitive-ability respondents and that the fully labeled grid will increase rates of skipping the entire grid more for low- than high-cognitive-ability respondents (H14).*

10.2 Data and Methods

The comparisons between top and fully labeled grids in this chapter come from three different experiments in which we were able to test these ideas on both attitude and behavior items on a variety of topics and with a variety of types of response option scales. One experiment was conducted in the National Health, Wellbeing, and Perspectives Survey (NHWPS). NHWPS was a 12-page booklet questionnaire mail survey administered in summer 2015 with a random sample of 6000 addresses drawn from the USPS Postal Delivery Sequence File by Survey Sampling International. Households were randomly assigned to one of two experimental versions of the questionnaire ($n = 3000$ each) and asked to have the adult who would have the next birthday complete the survey. The American Association of Public Opinion Research (AAPOR) Response Rate 1 for NHWPS was 16.7% ($n = 1002$) (AAPOR 2016) and did not differ across the two questionnaire versions (Version 1: 17.4%, Version 2: 16.0%, $\chi^2 = 2.15$, $p = 0.143$). Respondent characteristics are shown in Table 10.1; they did not differ across the two experimental versions. Among the 77 questions in the NHWPS, seven were grid questions that were presented as top-labeled in one version and fully labeled in the other.[1] These included 6 behavior questions and 1 attitude question with a range of from 5 to 17 items per grid and response scales containing 5 points. The general topics of the item prompts and response option constructs for each of these grids are summarized in Table 10.2. Table 10.2 also shows whether the response options were presented fully (e.g. "Strongly Agree") or in abbreviated form in the fully labeled version (e.g. "SA") for each grid question.

The second experiment, the "Getting Along" survey, was a paper-and-pencil experiment carried out with a convenience sample of university students at a large Midwest university in Spring 2011. The survey contained 23 questions about student satisfaction with the university and diversity on campus. Two versions of the survey were developed with identical questions, but with one feature of each question experimentally varied across the versions. We focus here on three grid questions; one version had all three grids formatted as top-labeled, and the other fully labeled the grid questions. Two of the grids contained attitude questions; one asked about behaviors (see Table 10.2). Each of the grids contained five or six items. Prior to entering the classes, the two versions of the surveys were systematically arranged to alternate versions (fully-top-fully-top) in the set of surveys to distribute to a class so that quasi-random assignment could be achieved within classes. A member of the research team briefly introduced and handed out the survey to each class.

1 The NHWPS experimental design included 3 questionnaire cover treatments, 3 incentive treatments, and 2 questionnaire version treatments for a total of 18 fully crossed treatments. Here we focus only on the two questionnaire version treatments.

Table 10.1 Descriptive statistics for NHWPS, Getting Along, and Eye-Tracking studies overall and by version.

	Overall (%)	Top labeled (%)	Fully labeled (%)	Significance test	
				t or χ^2	p
NHWPS (n = 1002)					
Sex					
Male	39.0	41.8	36.0	1.76	0.078
Female	61.0	58.2	64.0		
Education					
Some college or less	47.4	49.7	44.9	1.46	0.144
Beyond BA degree	52.6	50.3	55.1		
Age					
18–64	62.4	63.3	61.3	0.61	0.540
65+	37.6	36.7	38.7		
Mean	57.1	57.3	57.0	−0.26	0.791
Getting Along (n = 512)					
Sex					
Male	43.1	46.5	39.6	−1.56	0.121
Female	56.9	53.5	60.4		
Class					
Freshman	43.3	42.9	43.7	3.89	0.273
Sophomore	27.1	28.7	25.4		
Junior	16.6	13.8	19.4		
Senior	13.0	14.6	11.5		
Age					
Mean	20.8	20.7	20.8	−0.59	0.556
Eye Tracking (n = 138)					
Sex					
Male	53.4	57.6	49.3	−0.96	0.340
Female	46.6	42.4	50.8		
Education					
Some college or less	74.1	80.9	67.6	3.24	0.198
Assoc. or BA	16.6	11.8	21.1		
Post graduate	9.4	7.4	11.3		
Age					
Mean	28.1	27.2	29.1	−1.05	0.294
Literacy (n = 94)					
Low	50.0	54.2	45.7	−0.82	0.415
High	50.0	45.8	54.4		

Table 10.2 Summary information about grid questions in three experiments.

Question number and concept(s) measured by item prompts	Number of item prompts	Type of response options	Type of question	Type of labeling[a)]
NHWPS				
9. Social support	11	Never/Always	Behavior	Full
12. Self-efficacy	16	Agree/Disagree	Attitude	Abb.
20. Depression and positive mental health	17	Never/Always	Behavior	Full
21. Mania and psychosis	9	Never/Always	Behavior	Full
22. Prosocial behaviors	9	Never/5 or More Times	Behavior	Full
23. Financial insecurity; time management	14	Never/Always	Behavior	Full
37. Alcohol consumption	5	Never/5 or More Times	Behavior	Full
Getting Along				
5. Time in activities	6	0 Hours/21+ Hours	Behavior	Full
10. Diversity commitment	5	Agree/Disagree	Attitude	Abb.
13. Diversity atmosphere	6	Satisfied/Dissatisfied	Attitude	Abb.
Eye Tracking				
15. Satisfaction with leisure spaces	6	Satisfied/Dissatisfied	Attitude	Abb.
29. Time in leisure spaces	7	Never/Very Often	Behavior	Full

a) "Full" indicates that the response options were fully written out (e.g. "Strongly Agree") in the answer area of the fully labeled version. "Abb." Indicates that the response options were abbreviated in the answer area of the fully labeled version (e.g. "SA"). In both versions, the response options were fully written out in the column headings of the grids.

Overall, 512 students completed the survey. Because this is a convenience sample and we do not have a count of how many students were in attendance on the days the survey was administered, we cannot calculate a response rate. Respondent characteristics did not differ across experimental versions for this experiment (Table 10.1).

The final experiment was a laboratory-based study with a convenience sample, using a web survey titled "Tourism and Recreation in Nebraska," containing 50 questions displayed across 44 web pages. For brevity, we refer to this study here as the "Eye-Tracking study." Two rounds of data collection occurred. The first round took place in Spring 2013 and included $n = 47$ university student participants who each received a \$5 incentive for participation. The second round took place from December 2013 to April 2014 and included 120 general population participants who each received $22 for participation. This resulted in a mix of 167 university students and general population members. In both rounds, participants were recruited through flyers, Craigslist advertisements, and word of mouth. Eligibility criteria for this study included being born in the United States, speaking English as a first language, and not wearing bifocals (a requirement for using the eye-tracking equipment). Participants were randomly assigned to receive one of two versions of the web questionnaire when they came to the laboratory in which features of individual questions, including grid labeling on two questions, were manipulated. After answering a brief in-person survey containing questions about technology use, literacy practices, and how they learned about the study, respondents completed the web survey while having their eye movements tracked. In the second round of this study (December 2013 to April 2014), respondents also completed the Wide Range Achievement Test 4 (WRAT4 – Wilkinson and Robertson 2006), a literacy assessment, as part of the in-person survey. Because of technical difficulties, eye-tracking data is not available for 28 cases. These are excluded from the analyses, resulting in an analytic sample size of 139. As with the other experiments, respondent characteristics did not differ across versions (Table 10.1). Table 10.2 provides details about the topics, response options, and question types for the grid questions in this experiment.

To record eye movements, we used Applied Science Laboratory's (ASL) D6 high-speed eye tracker, tracking eye movements at 120 Hz using a camera placed unobtrusively underneath the computer monitor. For this study, we defined a fixation as a gaze held for at least 60 milliseconds. This fixation length is shorter than that used other studies in the survey methodology field (e.g. Galesic et al. 2008; Galesic and Yan 2011), but is common practice in the vision sciences because people perceive information that influences their processing at this faster rate (Brunel and Ninio 1997; Sperling 1960). The eye tracker collects 120 measurements per second (e.g. 120 Hz), making the data fairly unwieldy (i.e. large and nonrectangular). Because of this, we use interest areas to define important areas of the web survey screen and then aggregate eye-tracking data within the interest areas, described in detail shortly. The eye-tracking data is aggregated into summary measures for each of these areas, yielding information such as total duration of fixations in each interest area or the number of times a respondent's gaze entered each area.

A number of small changes were made to the questionnaire for the second round of the Eye-Tracking study to improve the eye-tracking measurements. The relevant change on the grid questions is that padding was added around items and response options to create more clear distinction between individual items. As a result, interest areas had to be redrawn. We account for the different-sized interest areas (in square pixels of the interest area) between the two eye-tracking rounds in the analyses.

10.2.1 Measures and Analytic Plan

First we test for differences in mean responses to items in the grids using t-tests. We also use t-tests to test for differences in straightlining rates and nondifferentiation across the treatments. Our measure of strict straightlining is a dichotomous variable coded 1 if the respondent selected the same response option for every item they answered in the grid and coded 0 if they did not select the same response option for every item. Our measure of nondifferentiation is calculated as the standard deviation of each respondent's responses to all items within a grid. The mean of the respondent standard deviations are compared across the two treatments, where lower standard deviations are indicative of more nondifferentiation. Respondents who skipped over the entire grid are excluded from the straightlining and nondifferentiation analyses.

To examine differences in correlations across the two treatments, we start by testing for overall differences in the Pearson product-moment correlation matrix across treatments for each individual grid using a Jennrich chi-square test for equality of two correlation matrices (Jennrich 1970). For each grid, we then calculate the difference in each of the correlations between the top and fully labeled versions, testing for significant differences using Fisher's Z transformation (Cohen et al. 2003, p. 49). We do not evaluate factor structure here because not all of the grids contain established scales or measure an underlying latent construct.

We examine nonresponse in two ways. First, we generate a variable coded 1 for respondents who skipped the entire grid and 0 for those who answered at least one item within the grid. We test for differences across grid treatments in the proportion of respondents who skipped the entire grid using both large sample chi-square tests and Fisher's exact p-values given the low prevalence of this outcome. Second, we generate a variable that is a count of the number of items within each grid that each respondent left blank. We examine the mean number of items left blank in each grid among all respondents, testing for differences across experimental treatments using t-tests. We then repeat this same comparison of the mean number of items left blank, but exclude those who skipped the entire grid.

For the eye-tracking analyses, we start by defining the following interest areas, which are shown in Figure 10.2:

- The entire set of response option headings, labeled "Full Heading Area" in Figure 10.2
- Each individual response option heading, labeled "Individual Heading Areas" in Figure 10.2
- Headings and full answer area, labeled "Full Response Area with Headings" in Figure 10.2
- Each column of answer spaces within the grid, excluding the headings, labeled "Individual Response Columns" in Figure 10.2
- The column of item prompts, labeled "Item Prompts" in Figure 10.2

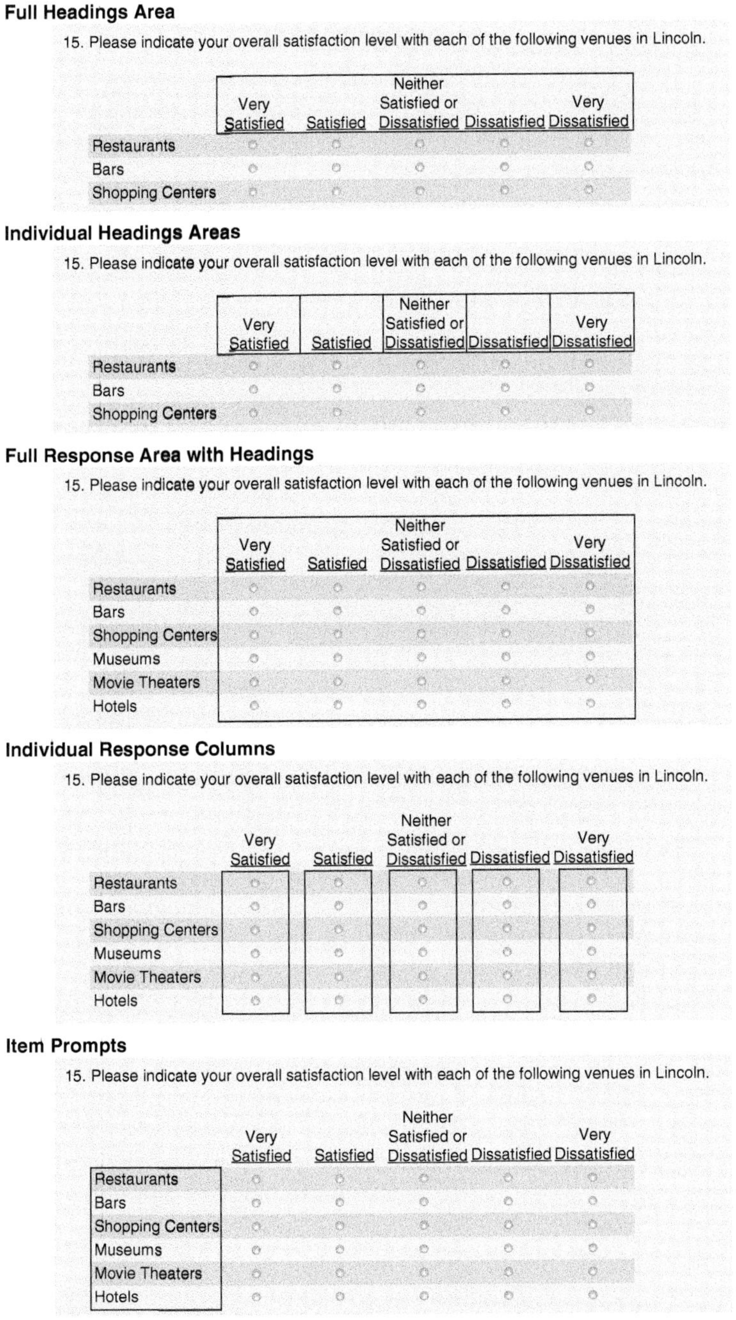

Figure 10.2 Illustration of interest areas from the Eye-Tracking study.

We define these interest areas separately for the top and fully labeled treatments. Because the spacing of the elements that made up the grids was different across the two treatments (e.g. the full labeling within the grid necessitated wider interest areas for the answer space columns) and because of the small spacing changes made between the two rounds of eye-tracking data collection, the size of the interest area varies slightly across rounds and across experimental treatments. This variation in area is accounted for in the analyses as described shortly.

After defining the interest areas, we then exported the total duration of all fixations each respondent made within each interest area and the number of times each respondent's gaze entered each interest area. We then log transform the duration variables, with zeros trimmed to the lowest observed value, to adjust for the typical skew of time-related data (Olson and Parkhurst 2013; Yan and Olson 2013). The duration analysis and the counts of gaze entries across experimental treatments use these areas as defined with one exception. For our duration analysis, to narrow down to just the full answer area without the headings, we subtract the "Full Headings Area" fixation duration from the "Full Answer Area with Headings" fixation duration. This subtraction is not possible for the entries outcome.

For each of the five resulting types of areas (full headings, individual headings, full answer area, individual answer columns, and item prompts), we test for differences across the experimental grid treatments by regressing (OLS) the log-transformed fixation duration variable on treatment (fully labeled = 1, top labeled = 0) and an area variable (square pixels in each interest area) that accounts for the differences in the size of interest areas across treatments and rounds of data collection. In the results tables, we report the raw mean durations for interpretability, but the significance tests are from the regression models that control for area. We use the same process to test for differences in the number of gaze entries across treatments, but use a negative binomial model for the significance tests rather than a linear regression model because our dependent variable is a count variable.

To test our hypotheses about the relationship between cognitive ability and our data quality outcomes, we conduct subgroup analyses using proxies for cognitive ability.[2] In the NHWPS, we use age and education as proxies for cognitive ability, a practice that is consistent with previous literature (Knäuper 1999; Knäuper et al. 2007; Krosnick 1991; Krosnick and Alwin 1987). We test the main effects of age and education and interaction effects for both of these variables with the grid format on item nonresponse (full grid and number of

2 We do not conduct subgroup analyses with the Getting Along data because, by virtue of being a convenience sample of university students, there was very little heterogeneity in age or education in this study.

items) and straightlining. Age and education are dichotomized (Age: 0 = under age 65, 1 = 65 or older; Education: 0 = BA or higher, 1 = some college or less).[3]

For the Eye-Tracking study, we do not have enough variation in age and education in this small sample to test our hypotheses (e.g. we only had one respondent age 65 or older). Thus, for this study we use literacy as our proxy for cognitive ability (Manly et al. 2004). Respondents' WRAT4 word reading and sentence comprehension scores were summed to calculate a composite score, which was then assigned a WRAT4 percentile rank (i.e. standardized to the US population) (see Wilkinson and Robertson 2006). The percentile ranks were then dichotomized with a median split into low literacy (coded 1) and high literacy (coded 0).

10.3 Findings

10.3.1 Substantive Outcomes

We start by assessing whether the top versus fully labeled grids produced different substantive responses (H1). The NHWPS included 81 individual items (across seven grids), the Getting Along survey included 17 individual items (across three grids), and the Eye-Tracking study included 13 individual items (across two grids). The average absolute value of the difference in means between the top and fully labeled versions was 0.05 for the NHWPS, 0.06 for Getting Along, and 0.1 for the Eye-Tracking study. These are all very small differences. Across all three surveys (i.e. 111 individual items), only three of these mean differences were statistically significant at the $p < 0.05$ level, and an additional four were moderately statistically significant. This is well within what we would expect by chance alone. Moreover, there is no clear trend in the direction of the differences; for 41% of items the fully labeled grid had a higher mean, for 51% of items it had a lower mean, and there was absolutely no difference for the remaining 8% of items (full results available from authors on request). Thus, our hypothesis (H1) that means would not differ across the top and fully labeled grids is supported.

We hypothesized that the fully labeled grid would have lower correlations between items than the top-labeled grid (H3). When we compare the overall correlational structures of the two formats, we see significant differences ($p < 0.05$) in 8 of the 12 grids (see Table 10.3), indicating that the grid format did change how items within the grids were related to one another overall.

3 Missing data for age (18.2% missing) and education (6.3% missing) were multiply imputed 10 times using sequential regression methods in Stata 13.1 (ice procedure).We attempted to impute using all of the grid items, but the imputation did not converge so a more limited imputation was done that excluded the grid items. As a result, the association between the data quality outcomes and the subgroup indicators may be slightly attenuated.

Table 10.3 Results of correlation matrix structure comparisons.

	Overall matrix structure Jennrich χ^2	# of items in grid	Number of correlations tested	Number of tested correlations that were sig. diff.	# of sig. diff. correlations with... Correlations in expected direction: Strongest in top labeled	Strongest in fully labeled	Correlations in unexpected direction
NHWPS							
Q9	91.20**	11	55	1	0	1	0
Q12	186.80***	16	120	12	4	6	2[a)]
Q20	163.58*	17	136	5	2	3	0
Q21	82.55***	9	36	5	3	2	0
Q22	36.54	9	36	2	1	1	0
Q23	126.47**	14	91	12	3	9	0
Q37	24.46**	5	10	2	0	2	0
Getting Along							
Q5	17.57	6	15	1	1	0	0
Q10	12.65	5	10	0	0	0	0
Q13	39.95***	6	15	3	3	0	0
Eye Tracking							
Q15	12.99	6	15	1	1	0	0
Q29	25.73*	7	21	2	2	0	0
TOTAL		**111**	**560**	**46**	**20**	**24**	**2**

Note: + $p < 0.100$, * $p \leq 0.050$, ** $p \leq 0.010$, *** $p \leq 0.001$

a) These two comparisons had the strongest correlation in the top-labeled version.

We then tested for differences in correlations across the grid formats between each possible pair of items in each grid. With 111 items in all of the grids, this yielded 560 tests for differences in correlations. Of these, 46 (about 8%) of the differences were statistically significant at a $p < 0.05$ level. This is about what we would expect by chance alone. When we looked more closely at the correlations that were significantly different at a $p < 0.05$ level across the grid treatments, we found that 44 of them were in the direction we would expect in both grid treatments based on the content of the items (e.g. we expect a positive correlation between items asking how often a respondent has people in their life with whom they have fun and with whom they enjoy doing things, and we expect a negative correlation between the statements, "I felt calm" and "I had trouble falling or staying asleep"). Of these, in 20 comparisons the correlation was strongest in the top-labeled format, and in 24 comparisons the correlation was strongest in the fully labeled format. Thus, while we know that the grid format produce significantly different correlational structures, our hypothesis that the fully labeled grid would reduce correlations between items (H3) is not supported.

10.3.2 Data Quality Indicators

Next we examine straightlining (providing identical responses to all questions in the grid) and nondifferentiation (standard deviation in responses) within the grids (H2). For our strict straightlining measure, there was very little difference across the top and fully labeled grids. The difference in the percent who straightlined only reached statistical significance for one grid in the Eye-Tracking study (Q29), where 6% of respondents straightlined in the top-labeled version and none straightlined in the fully labeled version ($p < 0.04$). Among the remaining 11 grids across all 3 studies, there were no statistically significant differences, nor was there a clear trend in direction of effect. These findings are consistent with those reported by Smyth et al. (2014) and suggest that the fully labeled grid does not reduce straightlining, perhaps because straightlining was generally rare. The results of the nondifferentiation analyses corroborate these findings. Differences across the treatments in the mean standard deviation were statistically significant for only 3 of the 11 grids ($p < 0.058$ for all three). For two of these (Q22 and Q23 in the NHWPS), there was more nondifferentiation in the fully labeled grid; and for one (Q29 in the Eye-Tracking study) there was more nondifferentiation in the top-labeled version. Thus there is no clear difference between these two formats in straightlining or nondifferentiation (H2).

Next we turn our attention to item nonresponse. We start by examining nonresponse to entire grids and find few differences across the two grid formats. In 7 of the 12 grids, the fully labeled format was skipped at higher rates than the top-labeled format as hypothesized (H5), but only one of these differences was large enough to be statistically significant (Q12 in NHWPS, $p = 0.041$). In the five remaining grids the differences were in the opposite direction, although also not statistically significant. Thus, fully labeling the grids does not appear to have a consistent negative impact on the rate of people skipping the grid entirely (H5).

We next assess item nonresponse to the individual items within a grid. We hypothesized that the fully labeled grid would have less item nonresponse (H4). We start by examining the mean number of items left blank among all respondents. We then exclude those who skipped the entire grid from the analysis, focusing only on those who answered at least one item in the grid. Among all respondents, the fully labeled version resulted in a higher mean number left blank in 7 of the 12 grids, but a lower mean number of items left blank in 5 grids. Moreover, only two of these differences were statistically significant. In Q12 in the NHWPS, the fully labeled version had a higher mean number of items left blank (0.53 vs. 0.23, $t = -2.24$, $p = 0.030$). In Q5 of the Getting Along survey the fully labeled version had a lower mean number of items left blank (0.04 vs. 0.06, $t = 2.43$, $p = 0.02$).

When those skipping the entire grid are excluded from the analyses, the results are similar in that the fully labeled grid format produced a higher mean number of items left blank in four of the grids and a lower mean number of items left blank in eight, but only one difference was large enough to be statistically significant (Q5 in the Getting Along survey, top-labeled = 0.06 versus fully labeled = 0.04, $t = 2.43$, $p = 0.020$). In fact, across all items, the average number of items left unanswered ranged from only 0.004 to 0.202. Thus, the fully labeled grid format does not appreciably reduce item nonresponse compared to the top-labeled format (H4).

10.3.3 Eye-Tracking Analyses

Next we turn our attention to how respondents visually process grid questions and the issue of whether processing patterns differ across the grid formats by examining the duration spent fixating on and the number of entries into key interest areas in the grid. Results for duration in question 15 can be seen in Table 10.4. On average, respondents spent 2.35 seconds fixating on the response option headings at the top of the grid, but looking at the two treatments separately reveals that those who answered in the fully labeled version spent 30% less time fixating on the headings (1.95 seconds) than those who answered in the top-labeled version (2.78 seconds, $t = 2.96$, $p = 0.004$), a finding that supports H6.[4] The pattern was in the same direction for question 29 (see Table 10.5), but the difference did not reach statistical significance. For both questions, this difference holds for each individual response option; that is, individual response option headings were fixated on less in the fully labeled than the top-labeled treatment with the differences reaching statistical significance for four of the five headings in question 15 and three of the five in question 29.

The analysis of fixation duration on individual headings also reveals that respondents spent more time fixating on the middle response option heading than any of the other response option headings. If all five response options in these grids were processed equally, we would expect respondents to spend about 20% of their fixation duration on each response option heading, but on Q15, respondents spent 38% of their total fixation duration fixating on the middle response option heading. The percent of time spent on each of the other response options ranged from 5–27%. This apparent anchoring happened in

4 Response option labels were provided at the top of the grids in both treatments because for some questions the full label could not be used in the answer area of the fully labeled version due to space limits. As a result, the full label was provided at the top and an abbreviated label was provided within the grid. Respondents likely used the top labels to help understand the abbreviated labels in this version or simply because they appeared within the reading navigational path as respondents moved from the introductory stem to the specific items.

Table 10.4 Mean number of seconds spent looking at response option headings and response option categories, Q15.

	Overall		Top labeled		Fully labeled		Diff.	|t|	p-Value
	Mean	sd	Mean	sd	Mean	sd			
Full headings area	2.35	2.35	2.78	2.41	1.95	2.23	0.83	2.96	**0.004**
Individual headings areas									
"Very satisfied" heading	0.50	0.57	0.58	0.64	0.42	0.49	0.16	1.44	0.15
"Satisfied" heading	0.40	0.55	0.44	0.48	0.36	0.60	0.08	2.21	**0.03**
"Neither satisfied nor dissatisfied" heading	0.69	0.08	0.88	1.00	0.52	0.74	0.36	2.93	**0.004**
"Dissatisfied" heading	0.14	0.32	0.19	0.31	0.10	0.32	0.09	3.42	**0.001**
"Very dissatisfied" heading	0.09	0.19	0.12	0.23	0.06	0.14	0.06	1.71	**0.09**
Full response area	6.68	4.24	6.78	3.94	6.59	4.52	0.19	0.98	0.33
Individual response columns									
"Very satisfied" column	1.26	1.49	1.05	1.37	1.45	1.58	−0.40	1.57	0.12
"Satisfied" column	1.75	1.61	1.95	1.85	1.56	1.34	0.39	1.06	0.29
"Neither satisfied nor dissatisfied" column	1.00	1.07	1.03	1.14	0.98	1.00	0.04	0.62	0.54
"Dissatisfied" column	0.35	0.63	0.28	0.60	0.42	0.66	−0.14	1.80	**0.07**
"Very dissatisfied" column	0.18	0.32	0.16	0.32	0.20	0.32	−0.04	1.30	0.20
Item prompt area	2.53	2.03	2.81	2.35	2.28	1.65	0.53	0.94	0.35

Note: Q15 question wording: Please indicate your overall satisfaction level with each of the following venues in Lincoln. Overall $n = 132$, top-labeled $n = 63$, fully labeled $n = 69$. Raw means and standard deviations are shown, but the statistical tests are estimated using log-transformed data with zeros trimmed to lowest observed value in a model controlling for the area (square pixels) in each interest area.

both grid treatments, with those in the top-labeled version spending 40% of their fixation duration on the middle response option and those in the fully labeled version spending slightly less at 36% of their total time. The same pattern occurs in Q29 where overall respondents spent about 29% of their total fixation duration fixating on the middle response option, but the values are 32% for the top-labeled version and 23% for the fully labeled version.

Table 10.5 Mean number of seconds spent looking at response option headings and response option categories, Q29.

	Overall		Top labeled		Fully labeled		Diff.	|t|	p-Value
	Mean	sd	Mean	sd	Mean	sd			
Full heading area	0.96	1.26	1.20	1.51	0.75	0.93	0.45	1.09	0.28
Individual headings areas									
"Very satisfied" heading	0.22	0.34	0.27	0.42	0.18	0.24	0.09	0.63	0.53
"Satisfied" heading	0.19	0.36	0.25	0.45	0.13	0.23	0.13	1.99	**0.05**
"Neither satisfied nor dissatisfied" heading	0.24	0.44	0.35	0.55	0.14	0.28	0.21	3.72	**0.00**
"Dissatisfied" heading	0.11	0.20	0.15	0.22	0.08	0.18	0.06	2.07	**0.04**
"Very dissatisfied" heading	0.08	0.21	0.09	0.21	0.07	0.22	0.02	0.90	0.37
Full response area	7.12	4.43	7.25	4.58	7.00	4.31	0.25	0.63	0.53
Individual response columns									
"Very satisfied" column	0.74	1.00	0.44	0.49	1.00	1.25	−0.56	0.87	0.39
"Satisfied" column	1.37	1.66	1.20	1.90	1.52	1.40	−0.31	0.54	0.59
"Neither satisfied nor dissatisfied" column	1.47	1.35	1.51	1.34	1.43	1.36	0.08	0.69	0.49
"Dissatisfied" column	1.08	1.19	0.99	1.23	1.17	1.16	−0.18	0.01	1.00
"Very dissatisfied" column	0.81	1.06	0.78	1.00	0.84	1.11	−0.06	0.44	0.66
Item prompt area	4.68	3.62	5.23	3.53	4.19	3.66	1.04	1.24	0.89

Note: Q29 question wording: How often do you use each of the following recreational facilities in Lincoln? Overall $n = 132$, top-labeled $n = 63$, fully labeled $n = 69$. Raw means and standard deviations are shown, but the statistical tests are estimated using log-transformed data with zeros trimmed to lowest observed value in a model controlling for the area (square pixels) in each interest area.

These findings suggest that the fully labeled version changes anchoring on the middle response option heading, perhaps because it encourages more direct left-to-right processing as respondents proceed from the items into the response options (i.e. processing the scale points in order).[5]

5 In other eye-tracking work, we have observed about a quarter of respondents process scales by starting in the middle of a horizontally displayed scale rather than at the first point in the scale. Those who do this are much more likely to then select the midpoint as their response.

Next we look at response options and answer spaces within the grid, excluding the column headings. The difference in fixation duration in the response area between the top and fully labeled treatments was very small and failed to reach statistical significance in both questions ($p > 0.05$). Moreover, the direction of the difference was opposite of what we hypothesized. Thus there is no support for our hypothesis that respondents would spend more time fixating in the response area in the fully labeled version (H7). Examination of individual columns within the response area reveals no significant differences, and no clear pattern of direction of effects. Thus, respondents do not differ in fixation duration in the response area of the grids across the two formats. There was also no significant difference in fixation duration on the item prompts themselves (i.e. the leftmost column) for either question in this experiment (supporting H9).

In addition to hypothesizing that respondents to the fully labeled grid treatment would spend less time overall looking at the column headings, we also hypothesized that they would look up to the heading area fewer times than those in the top-labeled treatment (H8). Our results generally support this hypothesis. Table 10.6 shows that on average, respondents' gaze entered the grid heading area 11 times for question 15, but that the mean number of entries differed significantly by grid type. In the top-labeled treatment, respondents' gaze entered the heading area an average of 12.8 times compared to 9.5 times in the fully labeled treatment ($t = 1.87$, $p = 0.06$). Moreover, each individual heading interest area was entered more times in the top than the fully labeled treatment, with two of the five differences reaching statistical significance. For question 29 (Table 10.7), the difference in the mean number of entries into the entire heading area did not reach statistical significance, although it was in the hypothesized direction. However, three of the five individual heading interest areas had a statistically lower mean number of gaze entries in the fully labeled treatment than the top-labeled treatment and a fourth was moderately statistically significant.

Further analysis revealed no significant difference for either question in the mean number of gaze entries into any of the interest areas capturing the individual response option columns. Nor were there any significant differences across the two treatments in the mean number of times respondents' gaze entered the interest area for the item prompts themselves (supporting H10).

10.3.4 Subgroup Analyses

For the NHWPS we also examine whether levels of each of our data quality indicators were affected by age or education overall as well as whether each of these proxies for cognitive ability moderated the effects of the experimental treatment.

Table 10.6 Number of entries into response option headings and response option categories, Q15.

	Overall		Top labeled		Fully labeled		Diff.	|z|	p-Value
	Mean	sd	Mean	sd	Mean	sd			
Full heading area	11.09	10.32	12.83	11.02	9.51	9.44	3.32	1.87	**0.06**
Individual headings areas									
"Very satisfied" heading	3.79	3.70	4.32	3.91	3.30	3.47	1.01	1.52	0.13
"Satisfied" heading	3.24	3.26	3.75	3.14	2.78	3.32	0.96	1.48	0.14
"Neither satisfied nor dissatisfied" heading	4.10	4.65	4.95	5.03	3.32	4.16	1.63	2.09	**0.04**
"Dissatisfied" heading	1.36	2.29	1.70	2.48	1.06	2.07	0.64	1.91	**0.06**
"Very dissatisfied" heading	0.92	1.57	1.11	1.57	0.74	1.56	0.37	1.26	0.21
Individual response columns									
"Very satisfied" column	10.83	6.74	11.43	7.12	10.29	6.38	1.14	1.06	0.29
"Satisfied" column	10.08	8.16	10.48	7.05	9.72	9.09	0.75	0.44	0.66
"Neither satisfied nor dissatisfied" column	6.19	7.46	6.44	7.64	5.96	7.33	0.49	0.55	0.58
"Dissatisfied" column	2.75	4.33	2.44	4.08	3.03	4.56	−0.58	0.87	0.38
"Very dissatisfied" column	1.51	3.27	1.54	4.11	1.48	2.30	0.06	0.85	0.40
Item prompt area	13.73	8.34	13.71	8.97	13.75	7.78	−0.04	0.01	0.99

Note: Q15 question wording: Please indicate your overall satisfaction level with each of the following venues in Lincoln. Overall $n = 132$, top-labeled $n = 63$, fully labeled $n = 69$. Statistical tests estimated using a negative binomial model to account for the count data comparing fully labeled to top labeled (reference category) and controlling for the number of square pixels in the interest area.

As Table 10.8 shows, education was not associated with the likelihood of skipping the entire grid or the mean number of items left blank when those who skipped the grid are included in the analyses. When those who skipped the grid entirely are excluded from the analyses, education is significantly associated with the mean number of items missing for three of the seven items (Q20, Q22, and Q23), such that individuals with some college or less have twice the rate of missing values in these grids than those with a BA or more. In addition, for two of the seven grids (Q22 and Q37), those with low education were more

Table 10.7 Number of entries into response option headings and response option categories, Q29.

	Overall		Top labeled		Fully labeled		Diff.	|z|	p-Value
	Mean	sd	Mean	sd	Mean	sd			
Full heading area	6.64	5.99	7.30	6.12	6.03	5.85	1.27	0.23	0.66
Individual headings areas									
"Very satisfied" heading	2.12	2.09	2.17	2.08	2.07	2.11	0.10	1.51	0.13
"Satisfied" heading	1.83	2.05	2.17	2.30	1.51	1.75	0.67	1.92	**0.06**
"Neither satisfied nor dissatisfied" heading	2.24	2.68	3.05	3.04	1.51	2.06	1.54	3.46	**0.00**
"Dissatisfied" heading	1.20	1.81	1.43	1.82	1.00	1.79	0.43	1.65	**0.10**
"Very dissatisfied" heading	0.79	1.69	1.14	2.11	0.46	1.11	0.68	2.20	**0.03**
Individual response columns									
"Very satisfied" column	9.52	5.95	9.16	5.74	9.86	6.16	−0.70	0.62	0.54
"Satisfied" column	10.54	8.06	10.81	8.83	10.29	7.35	0.52	1.52	0.13
"Neither satisfied nor dissatisfied" column	10.15	9.49	10.49	10.06	9.84	9.01	0.65	1.43	0.15
"Dissatisfied" column	7.20	7.02	6.87	6.49	7.51	7.51	−0.63	0.07	0.95
"Very dissatisfied" column	3.95	4.77	4.32	5.74	3.61	3.70	0.71	0.67	0.50
Item prompt area	18.14	12.06	17.92	9.93	18.33	13.79	−0.41	0.16	0.87

Note: Q29 question wording: How often do you use each of the following recreational facilities in Lincoln? Overall $n = 132$, top-labeled $n = 63$, fully labeled $n = 69$. Statistical tests estimated using a negative binomial model to account for the count data comparing fully labeled to top labeled (reference category) and controlling for the number of square pixels of the interest area.

likely to straightline; and for five of the seven (Q12, Q20, Q22, Q23, and Q37), those with lower education had lower mean standard deviations across items within single grids (i.e. more nondifferentiation). Contrary to our hypothesis, there were no significant interactions between education and grid format for any of these outcomes (H11, H12, H13, H14 – results available upon request).

The direct effects of age were also largely as hypothesized for the item nonresponse outcomes, but varied for straightlining (see Table 10.8). Those age 65 and older were twice as likely to skip the entire grid on four of the seven grid

Table 10.8 Regression results predicting data quality outcomes with education and age for NHWPS grid questions.

	Question number						
	Q9	Q12	Q20	Q21	Q22	Q23	Q37
Skipping entire grid (odds ratios)							
Education (<BA)	1.51	0.57	0.64	0.78	0.72	0.68	0.90
Age (65+)	1.75	1.38	2.19*	2.42**	2.14*	2.19*	1.06
Mean number left blank (all Rs) (incidence rate ratios)							
Education (<BA)	1.49	1.07	0.94	0.86	1.04	0.86	0.81
Age (65+)	1.98	1.45	2.09*	2.03+	2.16*	2.02+	1.42
Mean number of items left blank (excluding skipped entire grid) (incidence rate ratios)							
Education (<BA)	1.58	1.71	2.28*	1.14	2.99**	2.37*	0.91
Age (65+)	2.18+	1.41	2.29**	0.87	2.58**	1.77	4.61
Straightlining (odds ratios)							
Education (<BA)	0.93	4.10	—	1.17	12.66***	—	1.43**
Age (65+)	0.58**	1.54	—	0.69	1.06	—	2.07***
Nondifferentiation (coefficients)							
Education (<BA)	0.03	−0.09***	−0.06**	0.02	−0.22***	−0.16***	−0.15***
Age (65+)	0.04+	−0.10***	0.01	−0.02	0.08*	0.14***	−0.15***

Note: Only estimates for education and age are shown here. All models also controlled for experimental version. + $p \leq 0.100$; * $p \leq 0.050$; ** $p \leq 0.010$; *** $p \leq 0.001$.

items (Q20, Q21, Q22, and Q23) and left items left blank at a rate twice that of younger respondents on two of the seven grids when those who skipped entire grids were excluded from the analysis (Q20 and Q22). For two items, Q9 and Q37, respondents age 65 or older differed from their younger counterparts in their probability of straightlining, although the direction differed over the two grids. Older respondents had higher mean standard deviations across items (i.e. less nondifferentiation) for two grids (Q22 and Q23) and lower average standard deviations across items within grids for two of the grids (i.e. more nondifferentiation – Q12 and Q37). Contrary to our hypothesis, there were no significant interactions between age and the experimental treatments for any of these outcomes (H11, H12, H13, H14 – results available upon request).

Finally, for the Eye-Tracking study, we were able to examine the association of the data quality outcomes with respondent literacy as well as whether literacy interacted significantly with grid format. Literacy did not have any statistically significant main or interaction effects for any of the outcomes for either grid in this study (H11, H12, H13, H14).

10.4 Discussion and Conclusions

The grid format has a reputation for being difficult for respondents, and while findings are mixed, there is evidence that this format reduces response quality (i.e. increases item nonresponse and straightlining and increases correlated measurement error, thus impacting inter-item correlations; Couper et al. 2013; Peytchev 2007; Tourangeau et al. 2004). Several studies have demonstrated how dynamic feedback features can be used in web surveys to reduce these negative effects (Couper et al. 2013; Kaczmirek 2008, 2011), but no such features are available to assist respondents answering grid questions in paper-and-pencil surveys. We compared a traditional top-labeled grid design to a fully labeled grid design that was intended to reduce respondent burden by eliminating vertical processing and allowing respondents to process on a single continual horizontal row.

Overall, we found very few differences between the top and fully labeled grid designs on either responses or data quality indicators. There was no meaningful difference in mean responses to individual items or in straightlining, nondifferentiation, skipping the entire grid, or skipping items within the grid. These results held regardless of cognitive ability of respondents. That is, the grid treatments had virtually the same effect on the younger and older respondents, less- and more-educated respondents, and low- and high-literacy respondents. We did find, however, that the two grid formats produced different correlation matrix structures, but there is no clear evidence in our analyses as to why the correlations differed or which, if either, is better. We also could not look at whether data quality was improved or reduced when these same items were asked as individual items rather than in a grid. Future research should replicate these experiments and extend them by examining factor loadings for underlying traits in grid items and predictive validity of the grid items to try to ascertain the veracity and importance of the differences we found. Future research should also compare these treatments to questions asked as individual items rather than in a grid, and should continue to explore other design features that might reduce the difficulty of grids.

While the responses and data quality were very similar across the two grid formats, the type of labeling did seem to impact the way respondents processed the items. The eye-tracking analyses revealed that the fully labeled format required less vertical processing, as measured by the amount of time respondents spent looking at the column headings and the number of times their eyes moved to the headings. This reduced time spent looking at the headings in the fully labeled version was not made up by time looking at the answer area as there was no significant difference in the amount of time respondents spent looking at the answer area across the two treatments. Thus, respondents appeared to have visually navigated the headings and answer space more quickly in the fully labeled version without impacting their responses or data quality in any appreciable way. In addition, this format appeared to reduce the amount of anchoring on the middle response option (although this did not seem to affect endorsement of the middle option; analyses not shown). Moreover, the two formats did not differ in how much time or how many times respondents look at the item prompts.

The fact that the fully labeled grid format did not impact responses or diminish data quality suggests that it may be fruitful to explore whether this format can be used in web surveys with mobile devices. A major problem with the display of grid questions on small mobile devices is that respondents can typically see the item prompts if they hold their device vertically or the response option headings if they hold it horizontally, but can rarely see both at once. This results in increasing need to scroll vertically and horizontally to try to connect both pieces of the question. Most software that optimizes web surveys for mobile devices deals with this challenge by removing the grid format altogether and displaying the items one-by-one on mobile devices; however, this risks giving respondents on computers and mobile devices considerably different stimuli and may introduce device effects. Fully labeled grids, provided the labels are fairly short, may provide an alternative whereby the grid format can be maintained across both devices, but scrolling can be minimized and limited to only one direction (i.e. horizontal).

In sum, we come away from these experiments cautiously optimistic. The fully labeled grid design did not have the positive impacts we expected it to have on data quality indicators, but it may have reduced respondent burden by reducing reliance on the column headings. At the same time, more work is needed to understand the implications of the differences in correlation matrix structure that we found across the two treatments in terms of predictive validity, factor structure (where appropriate), and other relational measures.

Acknowledgments

NHWPS data collection was supported by funding provided by the Office of Research and Economic Development and the Department of Sociology at the University of Nebraska-Lincoln. Data processing and analysis was supported in part by funds provided to the University of Nebraska-Lincoln under a Cooperative Agreement with the USDA-National Agricultural Statistics service supported by the National Science Foundation National Center for Science and Engineering Statistics [58-AEU-5-0023 Jolene Smyth & Kristen Olson PIs]. The Eye-Tracking study was supported by funds provided by the Department of Sociology and the College of Arts and Sciences at the University of Nebraska-Lincoln. The authors would like to acknowledge and thank Josey Elliot for her efforts in in developing and fielding the Getting Along survey and Amanda Ganshert for assistance with data analysis.

References

AAPOR. (2016). Standard definitions: Final dispositions of case codes and outcome rates for surveys. https://www.aapor.org/AAPOR_Main/media/publications/Standard-Definitions20169theditionfinal.pdf.

Bell, D.S., Mangione, C.M., and Kahn, C.E. (2001). Randomized testing of alternative survey formats using anonymous volunteers on the World Wide Web. *Journal of the American Medical Informatics Association* 8 (6): 616–620.

Brunel, N. and Ninio, J. (1997). Time to detect the difference between two images presented side by side. *Cognitive Brain Research* 5: 273–282.

Callegaro, M., Shand-Lubbers, J., and Dennis, J.M. (2009). Presentation of a single item versus a grid: Effects on the vitality and mental health scales of the SF-36v2 health survey. American Association for Public Opinion Research Conference, Hollywood Florida.

Cohen, J., Cohen, P., West, S.G., and Aiken, L.S. (2003). *Applied Multiple Regression/Correlation Analysis for the Behavioral Sciences*, 3e. New York: Routledge.

Couper, M.P. (2008). *Designing Effective Web Surveys*. New York: Cambridge University Press.

Couper, M.P., Traugott, M.W., and Lamias, M.J. (2001). Web survey design and administration. *Public Opinion Quarterly* 65 (2): 230–253.

Couper, M.P., Tourangeau, R., Conrad, F.G., and Zhang, C. (2013). The design of grids in web surveys. *Social Science Computer Review* 31 (3): 322–345.

Dillman, D.A. (1978). *Mail and Telephone Surveys: The Total Design Method*. New York: Wiley.

Dillman, D.A., Smyth, J.D., and Christian, L.M. (2014). *Internet, Phone, Mail, and Mixed-Mode Surveys: The Tailored Design Method*. Hoboken, NJ: Wiley.

Galesic, M. and Yan, T. (2011). Use of eye tracking for studying survey response processes. In: *Social and Behavioral Research and the Internet: Advances in Applied Methods and Research Strategies* (eds. M. Das, P. Ester and L. Kaczmirek), 349–370. New York, NY: Taylor and Francis Group.

Galesic, M., Tourangeau, R., Couper, M.P., and Conrad, F.G. (2008). Eye-tracking data new insights on response order effects and other cognitive shortcuts in survey responding. *Public Opinion Quarterly* 72 (5): 892–913.

Graesser, A.C., Cai, Z., Louwerse, M.M., and Daniel, F. (2006). Question understanding aid (QUAID): a web facility that tests question comprehensibility. *Public Opinion Quarterly* 70 (1): 3–22.

Iglesias, C.P., Birks, Y.F., and Torgerson, D.J. (2001). Improving the measurement of quality of life in older people: the York SF-12. *QJM* 94 (12): 695–698.

Jennrich, R.I. (1970). An asymptotic $\chi 2$ test for the equality of two correlation matrices. *Journal of the American Statistical Association* 65 (330): 904–912.

Kaczmirek, L. (2008). Human-survey interaction: Usability and nonresponse in online surveys. Doctoral dissertation. University of Mannheim, Germany. https://ub-madoc.bib.uni-mannheim.de/2150/1/kaczmirek2008.pdf.

Kaczmirek, L. (2011). Attention and usability in internet surveys: effects of visual feedback in grid questions. In: *Social and Behavioral Research and the Internet: Advances in Applied Methods and Research Strategies* (eds. M. Das, P. Ester and L. Kaczmirek), 191–214. New York: Routledge.

Knäuper, B. (1999). The impact of age and education on response order effects in attitude measurement. *Public Opinion Quarterly* 63: 347–370.

Knäuper, B., Schwarz, N., Park, D., and Fritsch, A. (2007). The perils of interpreting age differences in attitude reports: question order effects decrease with age. *Journal of Official Statistics* 23 (4): 515–528.

Krosnick, J.A. (1991). Response strategies for coping with the cognitive demands of attitude measures in surveys. *Applied Cognitive Psychology* 5: 213–236.

Krosnick, J.A. and Alwin, D.F. (1987). An evaluation of a cognitive theory of response-order effects in survey measurement. *Public Opinion Quarterly* 51: 201–219.

Manly, J.J., Byrd, D., Touradji, P. et al. (2004). Literacy and cognitive change among ethnically diverse elders. *International Journal of Psychology* 39: 47–60.

Olson, K. and Parkhurst, B. (2013). Collecting paradata for measurement error evaluations. In: *Improving Surveys with Paradata: Analytic Uses of Process Information* (ed. F. Kreuter), 43–72. Hoboken, NJ: Wiley.

Peytchev, A.A. (2007). Participation decisions and measurement error in web surveys. Doctoral dissertation. University of Michigan.

Redline, C.D. and Lankford, C.P. (2001). Eye-movement analysis: A new tool for evaluating the design of visually administered instruments (paper and web). American Association for Public Opinion Research Conference, Montreal, Quebec, Canada.

Richards, A., Powell, R., Murphy, J. et al. (2016). Gridlocked: the impact of adapting survey grids for smartphones. *Survey Practice* 9 (3).

Smyth, J.D., Olson, K., and Kasabian, A. (2014). The effect of answering in a preferred mode versus a non-preferred survey mode on measurement. *Survey Research Methods* 8 (3): 137–152.

Sperling, G. (1960). The information available in brief visual presentations. *Psychological Monographs* 74 (11), Whole No. 498.

Stern, M., Sterrett, D., Bilgen, I. et al. (2015). *The Effects of Grids on Web Surveys Completed with Mobile Devices. Paper Presented at the American Association for Public Opinion Research Conference*. Florida: Hollywood.

Thorndike, F.P., Carlbring, P., Smyth, F.L. et al. (2009). Web-based measurement: effect of completing single or multiple items per webpage. *Computers in Human Behavior* 25 (2): 393–401.

Toepoel, V., Das, J.W.M., and van Soest, A.H.O. (2005). Design of web questionnaires: A test for number of items per screen. CentER Discussion Paper; vol. 2005-114. Tilburg University. https://pure.uvt.nl/ws/portalfiles/portal/776381/114.pdf.

Tourangeau, R., Couper, M.P., and Conrad, F. (2004). Spacing, position, and order interpretive heuristics for visual features of survey questions. *Public Opinion Quarterly* 68 (3): 368–393.

Ware, C. (2004). *Information Visualization: Perception for Design*. San Francisco, CA: Morgan Kaufmann.

Wilkinson, G.S. and Robertson, G.J. (2006). *WRAT4 Wide Range Achievement Test Professional Manual*. Lutz, FL: Psychological Assessment Resources Inc.

Yan, T. (2005). Gricean effects in self-administered surveys. Doctoral dissertation. University of Maryland, College Park, Maryland.

Yan, T. and Olson, K. (2013). Analyzing paradata to investigate measurement error. In: *Improving Surveys with Paradata: Analytic Uses of Process Information* (ed. F. Kreuter), 73–96. Hoboken, NJ: Wiley.

11

The Effects of Task Difficulty and Conversational Cueing on Answer Formatting Problems in Surveys

Yfke Ongena[1] and Sanne Unger[2]

[1] *Department of Communication studies, Centre for Language and Cognition, University of Groningen, Groningen, The Netherlands*
[2] *College of Arts and Sciences, Lynn University, Boca Raton, FL, USA*

11.1 Introduction

Interactions in surveys are expected to follow an established paradigmatic pattern with the interviewer asking the question exactly as worded, followed by a response formatted exactly as specified by the question (Maynard and Schaeffer 2002). Interaction analysis has shown how in telephone interviews, respondents often do not fulfill the expectation to provide adequately formatted responses (Ongena 2010). In this chapter we distinguish three types of formatting problems: (i) *mismatch answers* (i.e. uncodable answers that deviate from prescribed categories); (ii) *reports* (i.e. potentially relevant information from which an answer may be derived (Schaeffer and Maynard 2002)); and (iii) *qualified answers* (i.e. answers that, although they are adequately formatted, are accompanied by a qualification that may disqualify the adequacy of the answer). Mismatch answers, reports, and qualified answers are the most common formatting problems we have observed (Unger et al. 2016). Other formatting problems are less common – for instance, *serial extras* in which respondents use words like "also," "too," and "the same" to refer to earlier answers in a series of questions (see Unger et al. 2016), or *emphasizing answers*, such as answers that are repeated literally ("yes, yes," see Garbarski et al. (2016)) – and are not as problematic because they do not require interviewer intervention.

In this chapter, we focus specifically on the three most common types of formatting problems (mismatch answers, reports, and qualified answers). Examples of each of these are given in Table 11.1.

The main issue with the first type of formatting problem, a mismatch answer, is that the interviewer cannot record the respondent's answer in the specified

Advances in Questionnaire Design, Development, Evaluation and Testing, First Edition.
Edited by Paul C. Beatty, Debbie Collins, Lyn Kaye, Jose-Luis Padilla, Gordon B. Willis, and Amanda Wilmot.

Table 11.1 Three common formatting problems in interviewer-administered surveys: mismatch answers, reports, and qualified answers.

Example of survey question and response alternatives	Mismatch: Response is not formatted as specified by the question	Report: Respondent provides potentially relevant information from which an answer may be derived	Qualified answer: A qualification is added to an adequately formatted answer
The government should educate people on the consequences of second-hand smoking *Strongly agree – Agree – Neither agree nor disagree – Disagree – Strongly disagree*	"Yes," "No," "The middle one"	"I think they know that for themselves," "I would very much like that"	"I guess I agree," "I would probably strongly agree"
Do you consider your health to be *excellent, good, fair, or poor*?	"Fairly well," "Not too bad"	"I haven't been sick in 10 years"	"I think it is excellent," "I guess good"
How many days during a normal week do you drink alcoholic beverages? (*Response implied: a number between 0 and 7*)	"Most days"	"My doctor does not allow me to drink," "I drink a glass of wine every Sunday"	"I think it is five," "I am not sure, but let's say four," "about three"

format, which therefore requires probing. Standardization rules generally require interviewers to repeat all alternatives or to repeat the question (Fowler and Mangione 1990). However, conversational principles encourage interviewers to show that they are active listeners and not to ignore the respondent's attempt to answer (Clark 1985). Therefore, interviewers may be tempted to suggest an "appropriate" alternative to the respondent, to "repair" the mismatch answer. Studies suggest that mismatch answers are common. For example, Van der Zouwen and Smit's study (2004) showed that 65% of the administrations of a single question resulted in mismatch answers, and mismatch answers were observed for 27% of survey questions in Dijkstra and Ongena's study (2006). So it is not surprising that mismatch answers are the most common cause of suggestive interviewer probing (Smit et al. 1997). This breach of standardization lowers the reliability of survey outcomes (Fowler and Mangione 1990) and can increase interview time, interviewer and respondent burden, and interview costs.

With the second type of formatting problem, a report, respondents provide potentially relevant information, but it is not by itself an adequate response (Schaeffer and Maynard 2002). For example, the report shown in Table 11.1

("I haven't been sick in 10 years") provides pertinent information, but not enough to clearly indicate which answer is most appropriate. As Garbarski et al. (2016) note, a report "displays a problem for possible repair by the interviewer" (p. 16). The additional information may prompt the interviewer and the respondent to further discuss the respondent's situation in an attempt to arrive at the optimal answer – for example: does absence of sickness mean excellent health? From a conversational interviewing perspective (Beatty 2004; Schober and Conrad 2002), such reports may illuminate response problems more directly than mismatch answers, because they reveal difficulties in understanding question meaning.

The third type of formatting problem, a qualified answer, differs from mismatch answers and reports in the sense that the respondent has provided an adequately formatted answer, but a qualification is added that may undermine the validity of the answer. Such a response could also need interviewer probing to ascertain that the respondent has fully considered the answer. Similar to reports, qualified answers may reveal the reasons why respondents are having difficulty mapping their answer to a response.

Survey researchers may not notice these interactional problems because they cannot be observed within the final dataset (Beatty 2004) and interviewers generally do not report common interactional problems to their supervisors (Houtkoop-Steenstra 2000). To identify interactional problems, it is helpful to systematically observe the interviewer-respondent interactions by means of behavior coding, which can be a reliable (but relatively expensive) method of question evaluation (Fowler 2011). Behavior coding produces quantitative measures of interactional problems per question or per interviewer. Behavior coding can thus be used to assess data quality, and it can also be used to verify causes of interactional problems in experimental research on question wording (or as a manipulation check when interviewer behavior is experimentally manipulated, see Dijkstra 1987; Ongena and Haan 2016). By looking at interviewer and respondent behaviors, researchers obtain additional information to help them evaluate the effects of various questions (Beatty et al. n.d.). However, with conventional behavior coding, only a selection of behaviors are coded, and the sequence of the interaction is not usually taken into account (see also Ongena and Dijkstra 2006). *Interaction coding* is a method that elaborates on conventional behavior coding by examining a more detailed set of interview behaviors, including the order in which they occur.

In this chapter, we describe an experiment that uses interaction coding to test the effects of different types of question wording on the occurrence of formatting problems, specifically mismatch answers, reports, and qualified answers. First, we outline factors that may contribute to formatting problems. Next, we describe our experiment and survey data, followed by results, implications for questionnaire design, and recommendations for future studies.

11.2 Factors Contributing to Respondents' Formatting Problems

We propose that two main factors contribute to the occurrence of formatting problems in survey interviews. First is the difficulty of the response task. Tourangeau et al. (2000) describe four stages of survey response: comprehension, retrieval, judgment, and formatting. A survey question can cause difficulties in all of these four stages. For example, a question that does not clearly state the response alternatives increases task difficulty. When there is a mismatch between the question stem – for example, when the question is worded to imply a yes-no answer, but the response options consist of a five-point scale (see Smyth and Olson 2018) – respondents may provide a mismatch answer. The response task is especially difficult in phone surveys where response alternatives are presented orally, and respondents must hold the information in memory.

The second difficulty is the degree to which standard conversational conventions are used in the interaction. Since the interview in its basic form is a conversation – as reflected by the classic description of the interview as a "conversation with a purpose" (Bingham and Moore 1924; cited by Cannell and Kahn 1968 and Schaeffer 1991) – interviewers and respondents often use conversational conventions in interview interaction (Schaeffer 1991). See, for example, Dijkstra and Ongena's study (2006).

One convention that applies to ordinary conversations is that precise answers are usually considered unnecessary. Drew (2003) explains that the degree of exactness required in institutional interaction, such as between doctors and patients or in the courtroom, is not necessary in everyday interaction. Complying with Grice's (1975) quantity maxim, speakers provide only as much information as relevant (Drew 2003), which can lead to imprecise statements about how one feels (fine) or what time it is (10:30). These expectations are different in institutional conversations like survey interviews, but "to apply such standards of exactness in conversation might be inappropriate, disruptive, or even pathological" (Drew 2003, p. 923). As respondents rely on conversational conventions during survey interviews, they may apply the quantity maxim when a more precise answer is needed. In other words, mismatch answers can be seen as a type of conversational behavior in interviews, showing that the respondent is applying conversational conventions (Ongena and Dijkstra 2007).

Similarly, conversational conventions encourage respondents to give elaborated answers (i.e. provide reports or qualified answers). Houtkoop-Steenstra (2000) provides the example of yes-no questions, which in ordinary conversation tend to receive expanded responses. She writes: "If someone is asked, 'Do you have any children?' he or she may say, 'Yes, three,' or 'Yes, two boys.' Respondents, however, are not supposed to say more than 'yes' or 'no' in response to a yes-no question" (Houtkoop-Steenstra 2000, p. 65).

We argue that *conversational cueing*, by means of survey questions that allow for, or emphasize, such conversational conventions will improve the survey experience for participants because it allows respondents to use words that they use in everyday conversations rather than survey-specific language. To summarize, both task difficulty and conversational cueing affect the number of formatting problems when respondents answer survey questions. Combining these two factors creates four different situations.

The first, and least optimal, situation is created with high task difficulty and low conversational cueing. This can lead to a situation where respondents provide simple answers that are not adequately formatted (i.e. many mismatch answers) without additional information (i.e. reports or qualified answers). Such mismatch answers can lead to interviewers choosing an answer option without probing – possibly introducing error – or increase response burden by necessitating a probe sequence.

High task difficulty combined with high conversational cueing is also problematic. When respondents are not guided how to answer, task difficulty increases. When respondents also perceive high conversational cueing, they are more inclined to expand on their answer. In combination, this can lead to "chit chat satisficing," meaning that respondents provide expanded answers without putting in the cognitive effort required to map their answers to an appropriate response option (see Krosnick 1991). Therefore, mismatch answers, reports, and qualified answers are more likely.

The opposite situation is created with low task difficulty and low conversational cueing. Low task difficulty is expected to reduce the number of mismatch answers and, in combination with low conversational cueing, reduce respondents' tendency to do more than just answer survey questions, thus reducing the number of reports and qualified answers they provide. Because both task difficulty and conversational cueing are low, simple clipped interactions are more likely, with minimal deviations from the paradigmatic sequence. While this may reduce the likelihood of formatting problems, engagement may be low and respondents may make less effort to answer accurately (Dijkstra 1987).

Lastly, low task difficulty can be combined with high conversational cueing. Due to low task difficulty, respondents are able to provide adequately formatted answers (fewer mismatch answers), but due to high conversational cueing they may be more likely to elaborate on their answers. These elaborations can show respondent engagement (Garbarski et al. 2016), which, according to Dijkstra (1987), improves answer accuracy, although it may pose a challenge for maintaining standardization (Garbarski et al. 2016), as respondents may depart from the standard paradigmatic sequence more often.

In sum, to reduce all three types of formatting problems, we propose that task difficulty should be as low as possible. However, because conversational cueing can increase reports and qualified answers, conversational cueing should be high only when task difficulty is low, in order to avoid an increase in the number of formatting problems. In the next section, we will discuss how these factors can be influenced by question design.

11.2.1 Conversational Cueing and Formatting Problems: Colloquial or Formal Response Alternatives

One way to reduce formatting problems is to enhance conversational cueing by aligning the response alternatives to conversational conventions through use of commonly used words. Changing the formal language that is prevalent in surveys into language that people normally use enhances conversational cueing since it changes the interview into a more natural exchange where the respondent is not restricted through formal question wording. At the same time, language that is more familiar and easier to comprehend eases the response task, and thus reduces the number of mismatch answers. Ongena and Dijkstra (2010), in an experiment which systematically manipulated response alternatives, confirmed the assumption that respondents are more likely to give adequate answers when prescribed categories consist of informal words commonly used in conversations than when the categories consist of more formal words. This was especially the case for attitude questions asking about agreement. When presented with colloquial response alternatives ("yes"/"no") for these attitude questions, respondents gave far fewer mismatch answers than when they were presented with formal response alternatives ("agree/disagree"). In this chapter, we aim to replicate the findings from Ongena and Dijkstra's (2010) study, which was conducted in Dutch, by varying colloquial and formal response alternatives within an English language survey. We hypothesize that colloquial response alternatives lead to fewer formatting problems than formal response alternatives because they increase conversational cueing and reduce task difficulty.

11.2.2 Task Difficulty and Formatting Problems: Seemingly Open-Ended Questions or Delayed Processing Questions

Another way to reduce formatting problems is to reduce task difficulty by making the formatting task explicit. The structure of a survey question plays a major role in communicating the request to give a precise answer. However, scripted interviewers' utterances comprise more than the questions alone (Houtkoop-Steenstra 2002). In fact, interviewers often set up a question or series of questions, by first explaining the topic or format. After delivering the actual question component, they may then provide specifications and answer categories. Table 11.2 summarizes examples of various components mentioned by Houtkoop-Steenstra.

In some cases, more than one utterance may be communicated by the interviewer in a single turn. Such a construction is likely to conflict with the system

Table 11.2 Components of interviewer scripts.

Component	Abrev.	Indication of	Example
Action projection	APC	Type of action	"I will now ask some questions"
Question target	QTC	Topic of question	"...these will be about X"
Question delivery	QDC	Question	"How often do you do X"
Question specification	QSC	Definition	"...by X we mean..."

of turn-taking that respondents rely on for smooth distribution of turns within ordinary conversations (Clark 1985). Sacks et al. (1974) describe the rules of turn-taking that are universal for conversations. They argue that reliance on the organization of turn-taking minimizes simultaneous speech and silences. A key feature of this organization of turn-taking is the recognition of completion of turns, indicating that the recipient has the opportunity, or even the responsibility, to talk. Recognition of the completion signals a "transition relevance place" for changing speakers (Goodwin 1981).

Problems in the interaction may result when the transition relevance place is not signaled clearly. As a result, the requirement to select from a list of prescribed answers may be unclear, increasing task difficulty. For example, in many survey questions with a set of response alternatives, these alternatives are given after the question mark (Beatty et al. 2006). These *seemingly open-ended questions* (SOEQs) do not instruct respondents to wait, but appear to be complete before any of the response options have been heard, i.e. after the question delivery component (QDC, Houtkoop-Steenstra 2000, see Table 11.2). Therefore, respondents are unaware of the response options that follow (e.g. "How often do you do X? Never, sometimes, often, or always"). Respondents may begin the process of formulating their answer as soon as they have heard the QDC (Beatty et al. 2006), recognizing this as a completed turn-constructional unit and a transition relevance place where a change in turn-taking can occur. In other words, transition of the turn from the interviewer to the respondent becomes relevant as soon as the question turn-constructional unit is complete. Respondents may therefore provide an answer before the interviewer has finished reading the alternatives, interrupting the interviewer before being fully informed about the response alternatives. Consequently, they are more likely to provide inadequately formatted answers. Thus, the awkward position of response alternatives may increase task difficulty and increase the occurrence of mismatch answers.

A solution to this problem would be to construct questions in such a way that response alternatives are read *before* the QDC. However, such a construction is also awkward. For example, suppose the question in Table 11.2 ("How often do you do X?") included prescribed alternatives like "never, sometimes, often, or always," which preceded the QDC. This imposes a construction with dangling categories (Fowler and Cannel 1996) such as "Do you never, sometimes, often, or always do X?" It is difficult for respondents to memorize these four alternatives when the topic of the question has not been given yet. Fowler (1995) shows another instance of alternatives that precede the QDC:

Please tell me whether you consider each of the following to be a big problem, a small problem, or no problem at all.
- Pain in your bones or joints
- Difficulty breathing
- Any other health problems (Fowler 1995, p. 88)

Fowler (1995) explains that respondents will be confused by this question wording and therefore unable to answer. A clue to improve questions like this is illustrated in an earlier example that Fowler gives:

> Which of these categories best describes how likely you think you are to move in the next year: very likely, fairly likely, or not likely?
> Source: Fowler 1995, p. 88.

Although the alternatives are back at the end of the question, the question clearly projects the prespecified categories (Fowler 1995), reducing the chance that respondents interrupt the interviewer before all alternatives have been read. This linguistic structure ("which of the following…") is very common in surveys and is referred to as a *delayed processing question* (DPQ), since "the wording explicitly instructs respondents to wait until they have heard all response options before forming a judgment" (Holbrook et al. 2007, p. 327).

We argue that DPQs have lower task difficulty than SOEQs. Due to the delayed processing, the respondent is less likely to interrupt the interviewer and more likely to hear all of the response options, therefore reducing the likelihood of mismatch answers, compared to questions where the concept of prespecified categories is not as clearly communicated, as in SOEQs.

In addition, since the structure of a DPQ signals to respondents that the survey interview is not an ordinary conversation, their conversational cueing is decreased. DPQs should thus yield fewer reports and qualified answers than SOEQs since SOEQs have a structure more common in ordinary conversations.

11.2.3 Task Difficulty and Formatting Problems: Colloquial or Formal Wording for Open-Ended Quantity Questions

Formatting problems can also occur when task difficulty and conversational cueing are both high. For example, open-ended questions that require a numerical response without a list of prescribed answer alternatives (e.g. "How many times did you do X?") do not explicitly state that an exact number is required. As Ongena (2010) argues, this colloquial freedom may provoke mismatch answers that are too general (such as ranges or approximations). However, explicit reference to the required format may be included, such as in the question "What is the number of times you do X?" This is a formal question, in the sense that it is not the type of formulation used in everyday conversation (Ongena 2010). In contrast, "How many times do you do X?" is a more colloquial question without such an explicit reference. Ongena (2010) found some support for the hypothesis that questions with a more clearly communicated format (formal questions) reduce task difficulty and conversational cueing and generate fewer formatting problems than those with a more vaguely communicated response format (colloquial questions). Questions concerning the number of hours respondents watch television and the number of non-alcoholic beverages consumed on an average weekday indeed prompted numbers when the formal wording was used, whereas the colloquial wording often prompted a vague quantifier like "not a lot of hours" or "most days."

11.3 Hypotheses

To summarize, we have suggested three strategies to reduce formatting problems: using colloquial rather than formal response alternatives, using DPQs rather than SOEQs, and formally referring to the required format in open-ended quantity questions. Each of these strategies affects task difficulty and conversational cueing in different ways, as is summarized in Table 11.3.

In the first manipulation of our experiment, we compare colloquial and formal response alternatives. Colloquial response alternatives are adapted

Table 11.3 Predicted impacts of conversational cueing and task difficulty on formatting problems.

	High task difficulty	**Low task difficulty**
High conversational cueing	*SOEQ* *Colloquial quantity* "Chit-chat satisficing" Mismatch ↑ Report ↑ Qualified answer ↑	*Colloquial alternatives* "Conversational optimizing" Mismatch ↓ Report ↑ Qualified answer ↑
Low conversational cueing	*Formal alternatives* "Extreme satisficing" Mismatch ↑ Report ↓ Qualified answer ↓	*DPQ* *Formal quantity* "Clipped optimizing" Mismatch ↓ Report ↓ Qualified answer ↓

to align with the respondent's tendency to use colloquial words and hence provide higher conversational cueing and lower task difficulty. As a result, colloquial alternatives are expected to yield fewer mismatch answers, but more reports and qualified answers, than formal response alternatives, which provide lower conversational cueing and higher task difficulty.

In the second manipulation, we compare DPQs with SOEQs. A DPQ clearly communicates the concept that respondents should answer according to prespecified categories. This lowers task difficulty, and due to its unconventional structure also reminds respondents of the survey context to lower conversational cueing. However, a SOEQ seems to invite respondents to give descriptions in their own words, which increases conversational cueing and heightens task difficulty, and therefore is predicted to evoke more mismatch answers, reports, and qualified answers than a DPQ.

Lastly, we compare colloquial and formal wording for open-ended quantity questions. Similar to DPQs, an open-ended question that formally refers to implied response options (with the word "number") decreases task difficulty by clarifying the task, but due to its formal wording also lowers conversational cueing. An open-ended question that colloquially refers to the implied response options (with the phrase "how many") increases task difficulty by not specifying the format in which the answer should be presented, and increases conversational cueing due to the informal wording. The colloquial quantity question is therefore predicted to evoke more mismatch answers, reports, and qualified answers than an open-ended question that formally refers to implied response options, which reduces task difficulty as well as conversational cueing.

11.4 Method and Data

To test our hypotheses, an experiment was embedded in the Nebraska Annual Social Indicators Survey 2006 (NASIS), an omnibus telephone survey of quality of life among a representative sample of citizens of Nebraska (BOSR 2007). In this survey, 13 questions were included on the topic of Alzheimer's disease. Six of these 13 items were experimentally manipulated: 3 on question wording and 3 on the response alternatives. For each item, one of two versions was asked. Table 11.4 shows a complete overview of the manipulated questions. All experimental questions were asked after the interview had been running for approximately 15 min and were preceded by a request to audio record this section of the interview.

To make sure manipulations of the questions were not causing unwarranted differences in the ease of administration, all manipulated questions were also evaluated for reading ease, as measured by the Gunning fog index. The Gunning fog index is a rough measure of how many years of schooling it would take someone to understand the content (Gunning 1969). While

Table 11.4 Overview of manipulated questions.

Q1 Alzheimer definition			
DPQ (version 1)	**SOEQ (version 2 and 3)**	**Mismatch answers**	**Reports**
Which of the following categories would best describe Alzheimer's? 1 Mental illness 2 Neurological disorder 3 Natural effect of aging 4 Viral infection	What would be the best way to describe Alzheimer's? 1 Mental illness 2 Neurological disorder 3 Natural effect of aging 4 Viral infection	– Repeating only part of the response option ("Neurological", "aging") – Giving a number ("The first one") – Combining options ("Aging and neurological")	– Well you have a mental page – They just put me on pills for that – It's an old age disease – I know it just comes with aging

Q2 Number diagnosed			
Formal quantity question (version 1)	**Colloquial quantity question (version 2 and 3)**	**Mismatch answers**	**Reports**
Out of every hundred people age sixty-five or older, please estimate the number of people who currently have Alzheimer's.	Out of every hundred people age sixty-five or older, please estimate how many people currently have Alzheimer's.	– Range answer (two or more integers) – Open range ("more than five," "less than twenty") – Different reference than 100% ("one in twenty") – Vague quantifiers ("a lot," "many," "not so many") – A dozen	– It is increasing – It is getting more common – I know only one so it must be low – I have read about it – I work in a nursing home

Table 11.4 (Continued)

Q3 Worry self develop

Formal alternatives (version 2)	Colloquial alternatives (version 1 and 3)	Mismatch answers	Reports
I worry that I might personally develop Alzheimer's. 1 Agree 2 Neutral 3 Disagree	I worry that I might personally develop Alzheimer's. 1 Yes 2 Maybe 3 No	– "yes," "maybe," "no" – Repeating statement ("I do worry about Alzheimer's") – "sure," "right," "somewhat"	– Nobody in my family had it – Everybody does that – They thought my sister had it – Everybody thinks about it in the back of their minds

Q4 Worry family develop

I worry that a family member might develop Alzheimer's. 1 Agree 2 Neutral 3 Disagree	I worry that a family member might develop Alzheimer's. 1 Yes 2 Maybe 3 No	– "agree," "neutral," "disagree" – Repeating statement ("I do worry about Alzheimer's") – "sure," "right", "somewhat"	

Q5 Concerns everyone

Alzheimer's is a disease that concerns everyone. 1 Agree 2 Neutral 3 Disagree	Alzheimer's is a disease that concerns everyone. 1 Yes 2 Maybe 3 No		

Q6 Known number

Formal quantity question (version 1)	Colloquial quantity question (version 2 and 3)		
What is the number of people you know personally, if any, who have Alzheimer's disease? 0–100	How many people, if any, do you know personally who have Alzheimer's disease? 0–100		

readability formulas like the fog index are not valid indicators of question comprehensibility (Lenzner 2014), in our case they were helpful in predicting whether interviewers were equally able to read the two versions of questions exactly as worded.

To verify whether the questions were equally comprehensible they were also tested using the Question Understanding Aid (QUAID, http://quaid.cohmetrix.com). This tool can help identify problems that are not always detected by expert survey methodologists (Graesser et al. 2006). For most questions, a few problems were identified, such as one or two unfamiliar terms or vague or imprecise noun-phrases. The number and type of problems identified were the same for all experimental versions of the questions, except for the manipulated "quantity" questions. In the colloquial versions, the word "many" was identified as a vague or imprecise relative term. This is consistent with the intended manipulation of the question, as this vagueness reflects the colloquial nature of the tested question, which aimed to increase conversational cueing.

11.4.1 Colloquial vs. Formal Alternatives

To test the impact of colloquial and formal alternatives on the number of formatting problems, three questions (Q3 "Worry self develop," Q4 "Worry family develop," and Q5 "Concerns everyone") were manipulated. These questions measured the respondent's concern about Alzheimer's disease on a three-point Likert-type response scale. For colloquial alternatives, words were used that are common in ordinary conversation ("yes," "maybe," "no"), whereas for formal alternatives, less common words were used ("agree," "neutral," "disagree"). In this set of options, the colloquial middle option "maybe" was not positioned at exactly the middle (as the formal option "neutral" was), lacking the precision of formal response alternatives, but aligning more with colloquial expectations. The endpoints of the scale had more or less the same meaning as their formal equivalents[1].

11.4.2 Delayed Processing Questions (DPQs) vs. Seemingly Open-Ended Questions (SOEQs)

To assess the impact of the DPQs and SOEQs, one general knowledge question (Q1 "Alzheimer definition") was manipulated. The question asked respondents to describe Alzheimer's disease with one of four categories presented: mental illness, neurological disorder, natural effect of aging, or viral infection. For this question, we manipulated the way that the prespecified categories were communicated. The DPQ version includes the phrase "which of the following

1 For the questions and response alternatives, no problems were identified by the QUAID tool, except for the ambiguous noun phrase "family member" in question 4.

categories would best describe," whereas the SOEQ version uses the more open-ended equivalent "what would be the best way to describe."[2]

11.4.3 Colloquial vs. Formal Quantity Questions

To test the impact of colloquial versus formal wording for the quantity questions, two questions asking for a number between 0 and 100 (questions 2 and 6) were experimentally manipulated. In the formal version, the response format was communicated as "what is the number of," whereas in the colloquial question the response format was communicated as "how many."

Q2 ("Number diagnosed") asked respondents to estimate the number of people out of 100 people, aged 65 and over, who currently have Alzheimer's disease.[3] Q6 ("Known number") concerned personal facts: the number of people the respondent has known with Alzheimer's disease.[4]

11.4.4 Procedure

The data were collected from November 2006 until March 2007, as part of the NASIS. Through random digit dialing (RDD), telephone numbers were selected across the state of Nebraska. Respondents were randomly selected from all adults aged 19 or older living in the household. Interviews were completed with 1821 respondents (BOSR 2007). According to AAPOR's (2004) response rate definition RR1, the response rate was 33%, and the cooperation rate (COOP1) 42%. In the sample, the proportion of respondents with the lowest education level (no high school diploma) was lower (4.4%) than the US Census statistics (2006), which means that higher-educated people were overrepresented in this sample. Out of the full sample, 470 randomly selected respondents were asked for permission to audio record the interview, which 95% (n = 455) approved. Neither the interviewer nor the respondent knew which questionnaire version

2 The DPQ, due to the lengthy question stem (which is an indispensable requirement to delay the respondent's processing), also has a lower fog score (16.93, i.e. almost college graduate level), which confirms that it is difficult to read. The SOEQ scores much better on readability (fog score of 8.0 [middle school/junior high school]). The QUAID tool identified the word "best" (occurring in both versions) as a vague relative term and "Alzheimer," "neurological," "aging," and "viral" as unfamiliar technical terms. No other problems were identified with either version.

3 The fog scores of both versions indicated "difficult" (formal question fog = 13.9, colloquial question fog = 13.9). The QUAID tool identified the word "currently" as both an unfamiliar technical term and a vague or imprecise relative term for both questions. In addition, for the colloquial question, the word "many" was identified as a vague or imprecise relative term. This is in agreement with the intended manipulation of the question.

4 The fog scores of both versions indicated "fairly difficult" (formal question fog = 11.3, colloquial question fog = 11.4). The QUAID tool identified the word "many" as a vague or imprecise relative term for the colloquial version of the question. This finding confirms the intended operationalization of the colloquial question as imprecise terms are common in conversations.

Table 11.5 Number of respondents per question type.

	Questionnaire version		
	1	2	3
Q1 (Alzheimer definition)			
SOEQ (total $n = 138$, 46%)		105	33
DPQ (total $n = 161$, 54%)	161		
Q2, Q6 (Number diagnosed, known number)			
Colloquial question "How many" (total $n = 138$, 46%)		105	33
Formal question: "What number" (total $n = 161$, 54%)	161		
Q 3, 4, 5 (Statements: worry self develop, worry family develop, concerns everyone)			
Colloquial alternatives (total $n = 194$, 65%)	161		33
Formal alternatives (total $n = 105$, 35%)		105	

the computer-assisted telephone interview (CATI) system would generate, and no association was found between the approval of recording and the version administered ($\chi 2(2) = 3.9$, $p = 0.14$). The analyses reported in this chapter only include recorded and transcribed interviews ($n = 299$). Recordings were available for 16 different interviewers, who on average recorded 18 interviews each, ranging from 3 to 58 interviews per interviewer.

Three different questionnaires were compiled by the CATI program. Version 1 started with the DPQ, and was followed by a formal quantity question and then the questions with colloquial alternatives. In version 2, respondents received the exact opposite versions of questions. Version 3 was similar to version 2, except that it used the colloquial response alternatives as in version 1 (see Table 11.4). The CATI program ensured a random distribution of the question versions so that there was an equal distribution of respondents receiving each question manipulation (see Tables 11.4 and 11.5). The randomized assignment to conditions meant that respondents in each group were equivalent in their level of education and age. However, by chance, respondents did differ with respect to sex: more female respondents participated in the first questionnaire version (71%) than in the second (46%) and third versions (39%, $\chi^2 = 21.87$, df = 2, N = 296 $p < 0.001$).

11.4.5 Interaction Coding

Two coders coded the transcripts of the audio-recorded interviews in the Sequence Viewer (2017) program (Dijkstra 2017) using Dijkstra's multivariate coding scheme (Dijkstra 1999). All utterances of interviewers and respondents were systematically coded at the utterance level, and the sequence of the occurrence was preserved. Five different coding variables, each with a different set of values, were used in the coding scheme. Since the scheme is too complex to summarize in this chapter, we refer to Ongena (2010) for a more extensive description of the scheme. A reliability check revealed a weighted kappa[5] of 0.92. This kappa value is comparable to values found in earlier studies using the same coding scheme (Ongena and Dijkstra 2010) and can be considered almost perfect (Landis and Koch 1977).

For the current study, coding of mismatch answers, reports, and qualified answers was the main focus of our analysis. The final two columns of Table 11.4 provide examples of the kinds of mismatch answers and reports that could occur for each type of question. For an answer to be coded as qualified, it only needed to include phrases like "I guess," "I think," "it might be."

To test our hypotheses, eligible QA sequences were selected using the same procedure used by Ongena and Dijkstra (2010) as follows:

- Exclusion of QA sequences in which the question was (initially) not read exactly as worded, ignoring minor alterations such as additions of "and" and "uhm"
- Exclusion of QA sequences with additional interactions before, during, or after the interviewer's question reading
- Inclusion of mismatch answers, reports, and qualified answers when they were immediately (without interviewer intervention) followed by adequate answers

As a result of this procedure, 8% of the sequences were excluded. The exclusion varied significantly between interviewers ($F\ (15) = 6.46$, $p < 0.01$), with an average percentage of excluded sequences of 3–20%. However, exclusion did not vary significantly across experimental versions of the questions.

As was also reported by Haan et al. (2013), with Q1 ("Alzheimer definition") we found a distinctive pattern in one interviewer's reading of response alternatives. In 67% of her cases (8.6% of all cases), the interviewer systematically added numbers to the four response alternatives (i.e. reading them as: "One, mental illness, two, neurological disorder, three, natural effect of aging, four, viral infection"). This interviewer recorded 37 interviews, with an average percentage of mismatch answers across all questions except for Q1. The interviewer used this changed reading more often for the DPQ version of Q1 (n = 12 out of 14 interactions, 86%) than for the SOEQ version (n = 13 out

5 Since in the multivariate coding scheme a coding decision comprises five decisions per utterance (one for each code variable, yielding a five-variable code string per utterance), reliability can better be measured according to a weighted kappa, where disagreement is weighted according to the number of code variables within a code string coded differently. The unweighted kappa was 0.82.

of 23 interactions, 56.5%). Since this comprised a major change in question reading, all cases in which the interviewer changed question wording were omitted from analyses.

11.5 Results

We expected three question formats would decrease task difficulty: colloquial rather than formal response alternatives, DPQs rather than SOEQs, and the more formal format in open-ended numerical questions. In contrast, we expected conversational cueing to increase by using colloquial rather than formal response alternatives, using SOEQs rather than DPQs, and making no explicit reference to the implied format in open-ended numerical questions. In the next sections, we present results by first examining the number of formatting problems observed for each version (i.e. mismatch answers, reports, and qualified answers) and, where relevant, differences in the response distributions for each question.

11.5.1 Effects of Colloquial and Formal Response Alternatives on Formatting Problems

Mismatch answers occurred in 7% of the QA sequences of the three attitude questions. The questions with formal alternatives yielded many more mismatch answers (16%) than the questions with colloquial alternatives (3%, χ^2 (1) = 48.4, $p < 0.01$), replicating Ongena and Dijkstra's (2010) findings.

Reports occurred 5% of the time for the three attitude questions and were more frequent (although not significantly) with formal than with colloquial alternatives.

Qualified answers occurred in 1% of the sequences, and these occurred more frequently with formal alternatives than with colloquial alternatives, but this difference was not significant.

The response distributions for the three attitude questions indicated that respondents did not treat the middle options of both versions as equal. The colloquial option "maybe" was chosen relatively more often than the formal option "neutral," although this difference was only significant for the first two questions in the series of statements (Table 11.6).

11.5.2 Effects of SOEQ and DPQ Design on Formatting Problems

Mismatch answers occurred 28% of the time for Q1 "Alzheimer definition," but no significant difference was found in the occurrence of mismatch answers for the two versions of this question (Table 11.7).

Table 11.6 Comparison of formatting errors for questions with colloquial and formal response alternatives.

	Q3 Worry self develop		Q4 Worry family develop		Q5 Concerns everyone	
	Colloquial	**Formal**	**Colloquial**	**Formal**	**Colloquial**	**Formal**
Mismatch answer	2.6%	7.6%	2.6%	18.1%	3.6%	21.9%
	χ^2 (1) 1.38, p = 0.23		χ^2 (1) = 1.86, p = 0.17		χ^2 (1) = 0.40, p = 0.53	
Report	4.1%	8.6%	3.6%	7.6%	3.1%	8.6%
	χ^2 (1) = 2.5 p = 0.11		χ^2 (1) = 2.3, p = 0.13		χ^2 (1) = 4.30, p < 0.05	
Qualified answer	1.0%	2.9%	0.0%	0.9%	1.0%	1.9%
	χ^2 (1) 1.38, p = 0.23		χ^2 (1) = 1.86, p = 0.17		χ^2 (1) = 0.40, p = 0.53	
Sample	(n = 194)	(n = 105)	(n = 194)	(n = 105)	(n = 194)	(n = 105)
Responses						
Agree/Yes (%)	32.8	43.3	53.7	61.5	85.4	88.5
Neutral/Maybe (%)	26.0	5.8	18.8	2.9	5.2	2.9
Disagree/No (%)	41.2	49.0	27.1	31.7	9.4	8.7
Don't know (%)	0.0	1.9	0.5	3.9%	0.0	0.0
	χ^2 (3) = 21.33, p < 0.01		χ^2 (3) = 18.56, p < 0.01		χ^2 (2) = 0.94, p = 0.62	

Table 11.7 Comparison of formatting errors for SOEQ and DPQ design.

	Q1 Alzheimer definition	SOEQ	DPQ
Interaction behaviors	Mismatch answer	35.7%	29.2
	Report	9.6%	1.4%
	Qualified answer	0.8%	3.4%
		($n = 115$)	($n = 147$)

However, the number of reports (7% overall) differed across the two question versions. Reports occurred more frequently for the SOEQ version than for the DPQ version (χ^2 (1) 9.5, $p < 0.01$). With the SOEQ, respondents provided additional information, showing their thinking behind an answer (for example by saying, "you have a mental page," and referring to the Alzheimer's Association website to justify their choice of "mental disease" to describe Alzheimer's).

Qualified answers occurred in 2% of the sequences of Q1. Qualified answers occurred more often with the DPQ version of this question than with the SOEQ version of the question, but this was not significant.

In the analysis, we ignored instances where the question was not read exactly as worded (as noted in the final paragraph of the methods section).

11.5.3 Effects of Colloquial and Formal Quantity Questions on Formatting Problems

No significant differences were found in the occurrence of all three formatting problems for either version of the quantity questions (see Table 11.8). Mismatch answers occurred in only 2% of the Q2 "Number diagnosed" sequences and in

Table 11.8 Comparison of formatting errors for colloquial and formal quantity questions.

	Q2 Number diagnosed		Q6 Known number	
	Colloquial ("How many")	Formal ("What number")	Colloquial ("How many")	Formal ("What number")
Mismatch	2.2%	3.7%	11.6%	12.4%
	χ^2 (1) 1.98, p = 0.16		χ^2 (1) 0.005, p = 0.83	
Report	13.0%	8.1%	13.8%	10.6%
	χ^2 (1) 1.9, p = 0.16		χ^2 (1) 0.72, p = 0.40	
Qualified answer	25.4%	18.0%	13.8%	9.3%
	χ^2 (1) 2.4, p = 0.12		χ^2 (1) 1.46, p = 0.23	
	($n = 138$)	($n = 161$)	($n = 138$)	($n = 161$)

12% of the Q6 "Known number" sequences. In contrast, reports and qualified answers were relatively frequent for Q2 and Q6: 10% of Q2 sequences and 12% of Q6 sequences elicited reports, whereas 22% of Q2 sequences and 12% of Q6 sequences elicited qualified answers. In both cases, the colloquial quantity question yielded more reports and qualified answers than the formal version.

11.6 Discussion and Conclusion

In this study, we proposed that task difficulty (specifically the transparency of instructions on how to answer) and conversational cueing (the extent to which conventions of ordinary conversations are encouraged by the question format) would play a role in the survey interview interaction and the occurrence of formatting problems. In questionnaire design, we have the choice to influence each of these factors with two strategies that can impact on results in contradictory ways. When we focus on conversational cueing, we adjust the survey design to the respondent's conversational tendencies by choosing colloquial, i.e. familiar, wording for response alternatives. When we focus on task difficulty, we guide respondents to use the "language of surveys" by making it clear that the question is not open-ended so that they provide an answer in the required format.

The experiment discussed in this chapter comprised manipulation of six items on the topic of Alzheimer's disease, embedded in the NASIS 2006. We expected that high task difficulty and low conversational cueing (agree/disagree questions with formal alternatives) would result in many mismatch answers and few reports to help the interviewer. High task difficulty combined with high conversational cueing (SOEQ and colloquial quantity question) was also expected to result in many mismatch answers, but with more reports and qualified answers. In contrast, low task difficulty and low conversational cueing (DPQ and formal quantity question) was expected to reduce the number of mismatch answers and qualified answers, creating simple clipped interactions, with minimal deviations from the paradigmatic sequence. Finally, low task difficulty and high conversational cueing (agree/disagree questions with colloquial alternatives) was expected to result in few mismatch answers and an increase in reports and qualified answers. The results from this study confirmed some, but not all, of our expectations and these are discussed next.

11.6.1 Reducing Formatting Problems Through Conversational Cueing: Colloquial vs. Formal Response Alternatives

Similar to results of the earlier experiment conducted in Dutch (Ongena and Dijkstra 2010), the experimental study discussed in this chapter revealed support for the effectiveness of question formats that combine high conversational cueing with low task difficulty in reducing mismatch answers. By aligning our response alternatives with language that is common in ordinary conversations, we enhanced respondents' perceptions of being able to use language they are more accustomed to. Such colloquial response alternatives indeed yielded a lower number of mismatch answers than questions with formal response alternatives that forced respondents to use survey-specific language. Since in either type of question there are no additional instructions and warnings on the alternatives to be used, respondents may preempt the response task and "think up" their own answers (which they then need to map to the required response categories). Task difficulty is therefore increased in the case of formal alternatives, but not in the case of colloquial alternatives (where a respondent's spontaneous answer is more likely to align with the prescribed response categories).

In addition, we also expected that colloquial response alternatives would yield more reports and qualifications due to increased conversational cueing, because respondents would be stimulated to do more than just provide answers. Contrary to our expectation, we found that formal response choices yielded more reports and qualified answers than colloquial ones. This unexpected result could mean that formal alternatives increased task difficulty by heightening respondents' uncertainty about the task, making them more inclined to report and qualify their answers. Alternatively, high conversational cueing may cause respondents to treat the task more casually, and may be perceived as exempting respondents from the obligation to fully think through their answers. This reduction in cognitive effort may mean that colloquial response alternatives reduce task difficulty but enhance satisficing. To explore whether colloquial response alternatives indeed increase satisficing among respondents, further research is needed to assess the validity of answers, such as comparing answers to records in a record-check study.

Although our hypothesis on the effects of colloquial versus formal response alternatives was tested in three questions, for all three questions we used the same response alternatives. Response distributions showed that respondents did not treat the middle options ("neutral" in the formal response alternatives, and "maybe" in the colloquial options) as equal. The finding that qualified answers occurred more often (although not significantly) in the formal alternatives version (especially when "agree" was the final answer) may also be related to this difference, indicating that a "neutral" response may violate conversational expectations.

11.6.2 Reducing Formatting Problems through the Reduction of Task Difficulty: DPQs vs. SOEQs

We expected that questions that instruct respondents to wait until all response categories are read (DPQs) would evoke fewer formatting problems than questions that do not communicate the concept of prespecified response categories (SOEQs). We found mixed results: while we found no significant differences between the two question versions for mismatch answers, reports occurred more frequently with the SOEQ than with the DPQ, and qualified answers occurred more often with the DPQ than with the SOEQ.

With the more frequent reports in SOEQs, respondents provided more information than strictly required by the question, providing some indication of the thinking behind their answers. This is common in ordinary conversation (Houtkoop-Steenstra 2000). Thus, the more frequent occurrence of reports suggests that SOEQs, in addition to having a higher task difficulty, trigger a more colloquial style of responding than DPQs. However, our results are based on a single factual question, and it is worth further investigating whether the colloquial style of responding will be generated by SOEQ design for other types of questions.

For the DPQ and SOEQ, qualified answers did not occur frequently, but in contrast with our expectations they did occur more often with the DPQ than with the SOEQ, although not significantly. Qualified answers may be seen as respondents' attempts to clarify when response alternatives do not precisely reflect their position, and in the case of DPQs respondents are immediately aware of this discrepancy. DPQs specifically instruct respondents to listen to the list of response alternatives, and as a result respondents seriously consider the alternatives and are more likely to express their hesitation when response options do not adequately capture their situation. In contrast, with SOEQs, respondents may experience more conversational cueing and may be less oriented to the task of providing precise responses, and also less concerned with qualifying discrepancies. However, it is unclear whether the colloquial style of responding influences response validity.

When high conversational cueing is combined with *low* task difficulty (as is the case with colloquial response alternatives), we expected more formatting problems and *decreased* response effort because respondents would be more inclined to treat the task casually.

Such effects are more difficult to assess for high conversational cueing combined with *high* task difficulty (i.e. the colloquial response style generated through SOEQs). On the one hand, for questions with high task difficulty, the colloquial response style can add to interview length, increasing interviewer and respondent burden and interview costs. The colloquial style can distract respondents from providing adequately formatted answers, and this requires work from the interviewer. On the other hand, respondents may enjoy a more conversational approach that allows them to tell their story. This may add to rapport and increase respondents' engagement in the survey, which can enhance their effort and accuracy when answering the questions (Bell et al.2016; Garbarski et al. 2016).

An interesting observation from our study (although excluded from the results) was the interactions of one interviewer who added numbers when reading the response alternatives (also reported by Haan et al. 2013). Adding numbers may have communicated the distinctions between categories more clearly, thereby reducing task difficulty to significantly reduce the number of mismatch answers. More research is necessary to verify whether adding numbers to categorical response alternatives indeed decreases a respondent's task difficulty by improving their comprehension of the task.

11.6.3 Reducing Formatting Problems Through Reduction of Task Difficulty: Colloquial vs. Formal Quantity Questions

For quantity questions, we expected that a question that formally refers to implied response alternatives (with the word "number") would decrease task difficulty, compared to a question that colloquially refers to the implied response alternatives (with the words "how many"). For the two questions that were manipulated to verify this, there was a low incidence of mismatch answers and no difference in the occurrence of formatting problems was found. Ongena (2010) found a much larger incidence of mismatch answers for quantity questions that involved behavioral frequencies. Thus, the low incidence of mismatch answers in the current study might be due to the type of question, which involved estimates. Providing an estimate by its very nature implies that some inaccuracy is acceptable. However, questions that ask for a specific number imply that a greater degree of precision is expected. Since the two questions are very similar and the first of them (Q2 "Number diagnosed") specifically asks for an estimate, it is possible that respondents also assumed that an estimate was acceptable for the colloquial version of Q6 ("Known number"). It may be worthwhile to further investigate whether questions that ask for estimates generate fewer formatting problems than questions that ask for precisely recalled numbers.

11.7 Further Expansion of the Current Study

While our theoretical framework allowed for four different scenarios, in our study only three strategies were studied, each comparing only two versions of a question. It would be useful in future studies to compare four versions of the same questions in a 2 × 2 design to allow for analyses of interactions between conversational cueing and task difficulty. Furthermore, we did not vary the type of questions we tested for any of our three comparisons. We used a knowledge question with nominal response categories (SOEQ versus DPQ), three knowledge questions with numerical answers (colloquial versus formal quantity questions), and an attitude question with ordinal categories (colloquial versus formal alternatives). These questions differed in the number of formatting problems they yielded. Thus, the study should be expanded with different types of questions for each manipulation to see if these results can be replicated.

In this study, all experimental manipulations were conducted on questions in the middle of the questionnaire. By the time respondents reached the experimental part of the questionnaire, respondents may have had enough practice to become accustomed to answering with prespecified response alternatives. Interviewers may also have reminded them of this task several times and, as a consequence, the number of formatting problems may have decreased as the interview progressed. However, since interviews were only recorded for the middle part of the interview, it could not be verified whether formatting problems occurred more often at the beginning of the interviews. Consequently, we were unable to determine if practice effects reduced the number of mismatch answers we observed in our study. Systematically varying the location of the tested questions in future studies may help determine if mismatch answers increase or decrease depending on the questions' position within the questionnaire.

The location of questions may be especially relevant in long questionnaires. Comparing surveys that vary in length may also help to determine whether a lengthy questionnaire gives respondents more opportunities to practice and learn "the rules of the game," which may lead to fewer mismatch answers later on. However, a very long questionnaire can also generate fatigue and may demotivate respondents, which could lead to more formatting errors.

It would also be valuable to attempt to replicate our results by testing more similarly manipulated questions. This would require a more complicated experimental design, but it would be informative to determine whether repeated exposure to DPQs reduces formatting problems as respondents become more familiar with the format.

11.8 Conclusions

In summary, our study suggests that low task difficulty with high conversational cueing (colloquial alternatives) reduces mismatch answers.

To avoid mismatch answers, and thus format their answers correctly, respondents must use the words offered, and the more accustomed respondents are to using those words, the easier they will find it to use them. Using colloquial response alternatives may therefore reduce the number of mismatch answers. Especially when the research intends to measure the first answer that comes to the respondent's mind (as might be the case with attitude questions), we recommend using colloquial response alternatives, since they require least effort from respondents and least interviewer intervention.

Additionally, our expectation that high conversational cueing would increase the tendency of respondents to do more than just answer survey questions was only confirmed for SOEQs (high task difficulty with high conversational cueing), and only for reports in particular. However, colloquial alternatives (low task difficulty with high conversational cueing) elicited fewer reports than formal alternatives (high task difficulty with low conversational cueing), while qualified answers were rare throughout. These results suggest that, when task difficulty is low, increased conversational cueing may not pose a significant threat to standardization.

However, based on this study's results for DPQs and SOEQs, we cannot recommend one format consistently over the other, as both formats had formatting problems. Similarly, for the quantity questions tested in this study, we did not find convincing evidence that specifically instructing respondents to provide a number made a difference to the respondents. However, we do recommend using such instructions for behavioral frequency questions that require an exact number based on calculations, since they make the task more explicit.

Finally, as this endeavor and others (Beatty et al. n.d.; Ongena and Dijkstra 2010) have shown, experiments in which various options in questionnaire design are evaluated by means of behavior (or interaction) coding can go beyond the more common use of this method. Behavior coding normally identifies problems with question wording without providing solutions for improving question wording (Presser et al. 2004). The experimental study discussed in this chapter demonstrates the value of behavior coding to explore question effects and shows how it can be used to evaluate several different question formats to identify the most effective option.

References

AAPOR (2004). *Standard Definitions: Final Dispositions of Case Codes and Outcome Rates for Surveys*, 3e. Lenexa, KS: The American Association for Public Opinion Research.

Beatty, P. (2004). The dynamics of cognitive interviewing. In: *Methods for Testing and Evaluating Survey Questionnaires* (eds. S. Presser, J. Rothgeb, M.P. Couper, et al.), 45–66. New York: Wiley.

Beatty, P., Cosenza, C., and Fowler, F.J. (2006). Experiments on the structure and specificity of complex survey questions. In: *Proceedings of the Survey Methods Research Section of the American Statistical Association*, 4034–4040. Alexandria, VA: American Statistical Association.

Beatty, P., Cosenza, C., and Fowler, F.J. (n.d.). Experiments on the design and evaluation of complex survey questions. In: *Experimental Methods in Survey Research: Techniques that Combine Random Sampling with Random Assignment* (eds. P.J. Lavrakas, A. Holbrook, C. Kennedy, et al.). New York: Wiley.

Bell, K., Fahmy, E., and Gordon, D. (2016). Quantitative conversations: the importance of developing rapport in standardised interviewing. *Quality & Quantity* 50: 193–212.

Bingham, W. and Moore, B. (1924). *How to Interview*. New York: Harper and Row.
BOSR (2007). *Nebraska Annual Indicators Survey 2006 Methodology Report*. Lincoln, Nebraska: University of Nebraska-Lincoln.
Cannell, C.F. and Kahn, R.L. (1968). Interviewing. In: *The Handbook of Social Psychology*, vol. 2: *Research method* (eds. G. Lindzey and E. Aronson), 526–595. Reading, MA: Addison-Wesley.
Clark, H.H. (1985). Language use and language users. In: *The Handbook of Social Psychology*, 3e, vol. 2 (eds. G. Lindzey and E. Aronson), 179–232. New York: Random House.
Dijkstra, W. (1987). Interviewing style and respondent behavior: an experimental study of the survey-interview. *Sociological Methods and Research* 16 (2): 309–334.
Dijkstra, W. (1999). A new method for studying verbal interactions in survey interviews. *Journal of Official Statistics* 15: 67–85.
Dijkstra, W. (2017). Sequence Viewer. Computer software. www.sequenceviewer.nl.
Dijkstra, W. and Ongena, Y.P. (2006). Question-answer sequences in survey-interviews. *Quality & Quantity* 40: 983–1011.
Drew, P. (2003). Precision and exaggeration in interaction. *American Sociological Review* 68 (6): 917–938.
Fowler, F.J. (1995). *Improving Survey Questions. Design and Evaluation*. Thousand Oaks, CA: Sage.
Fowler, F.J. (2011). Coding the behavior of interviewers and respondents to evaluate survey questions. In: *Question Evaluation Methods: Contributing to the Science of Data Quality* (eds. J. Madans, K. Miller and A.G. Maitland), 7–22. New York: Wiley.
Fowler, F.J. and Cannell, C.F. (1996). Using behavioral coding to identify cognitive problems with survey questions. In: *Answering Questions. Methodology for Determining Cognitive and Communicative Processes in Survey Research* (eds. N. Schwarz and S. Sudman), 15–36. San Francisco: Jossey-Bass.
Fowler, F.J. and Mangione, T.W. (1990). *Standardized Survey Interviewing; Minimizing Interviewer-Related Error*. Newbury Park, CA: Sage.
Garbarski, D., Schaeffer, N.C., and Dykema, J. (2016). Interviewing practices, conversational practices, and rapport: responsiveness and engagement in the standardized survey interview. *Sociological Methodology.* 46 (1): 1–38. https://doi.org/10.1177/0081175016637890.
Goodwin, C. (1981). *Conversational Organization. Interaction between Speakers and Hearers*. New York: Academic Press.
Graesser, A.C., Cai, Z., Louwerse, M.M., and Daniel, F. (2006). Question understanding aid (QUAID): a web facility that tests question comprehensibility. *Public Opinion Quarterly* 70: 3–22.
Grice, H.P. (1975). Logic and conversation. In: *Syntax and Semantics*, vol. 3: *Speech Acts* (eds. P. Cole and J.L. Morgan), 41–46. New York: Academic Press.

Gunning, R. (1969). The fog index after twenty years. *Journal of Business Communication* 6: 3–13.
Haan, M., Ongena, Y., and Huiskes, M. (2013). Interviewers' question rewording: not always a bad thing. In: *Interviewers' Deviations in Surveys: Impact, Reasons, Detection and Prevention* (eds. P. Winker, N. Menold and R. Porst), 173–194. P.I.E.-Peter Lang (Schriften zur Empirischen Wirtschaftsforschung; vol. 22.
Holbrook, A.L., Krosnick, J.A., Moore, D., and Tourangeau, R. (2007). Response order effects in dichotomous categorical questions presented orally: the impact of question and respondent attributes. *Public Opinion Quarterly* 71: 325–348.
Houtkoop-Steenstra, H. (2000). *Interaction in the Standardized Survey Interview: The Living Questionnaire*. Cambridge: Cambridge University Press.
Houtkoop-Steenstra, H. (2002). Questioning turn format and turn-taking problems in standardized interviews. In: *Standardization and Tacit Knowledge: Interaction and Practice in the Survey Interview* (eds. D.W. Maynard, H. Houtkoop-Steenstra, N.C. Schaeffer and J. Van der Zouwen), 243–259. New York: Wiley.
Krosnick, J.A. (1991). Response strategies for coping with the cognitive demands of attitude measures in surveys. *Applied Cognitive Psychology* 5: 213–236.
Landis, J.R. and Koch, G.G. (1977). The measurement of observer agreement for categorical data. *Biometrics* 45: 255–268.
Lenzner, T. (2014). Are readability formulas valid tools for assessing survey question difficulty? *Sociological Methods & Research* 43 (4): 677–698.
Maynard, D.W. and Schaeffer, N.C. (2002). Standardization and its discontents. In: *Standardization and Tacit Knowledge: Interaction and Practice in the Survey Interview* (eds. D.W. Maynard, H. Houtkoop-Steenstra, N.C. Schaeffer and J. Van der Zouwen), 3–45. New York: Wiley.
Ongena, Y. and Dijkstra, W. (2010). Preventing mismatch answers in standardized survey interviews. *Quality & Quantity* 44 (4): 641–659. https://doi.org/10.1007/s11135-009-9227-x.
Ongena, Y. and Haan, M. (2016). Using interviewer-respondent interaction coding as a manipulation check on interviewer behavior in persuading CATI respondents. *Survey Practice* 9 (2) Retrieved from http://www.surveypractice.org/index.php/SurveyPractice/article/view/352.
Ongena, Y.P. (2010). *Interviewer and Respondent Interaction in Survey Interviews: Empirical Evidence from Behavior Coding Studies and Question Wording Experiments*. Saarbrücken: LAP Lambert Academic Publishing.
Ongena, Y.P. and Dijkstra, W. (2006). Methods of behavior coding of survey interviews. *Journal of Official Statistics* 22 (3): 419–451.
Ongena, Y.P. and Dijkstra, W. (2007). A model of cognitive processes and conversational principles in survey interview interaction. *Applied cognitive psychology* 21: 145–163.
Presser, S., Couper, M.P., Lessler, J.T. et al. (2004). Methods for testing and evaluating survey questions. *Public Opinion Quarterly* 68: 109–130.

Sacks, H., Schegloff, E., and Jefferson, G. (1974). A simplest systematics for the organization of turn-taking in conversation. *Language* 50: 696–735.

Schaeffer, N.C. (1991). Conversation with a purpose—or conversation? Interaction in the standardized interview. In: *Measurement Errors in Surveys* (eds. P.P. Biemer, R.M. Groves, L.E. Lyberg, et al.), 367–391. New York: Wiley.

Schaeffer, N.C. and Maynard, D.W. (2002). Occasions for intervention: interactional resources for comprehension in standardized survey interviews. In: *Standardization and Tacit Knowledge: Interaction and Practice in the Survey Interview* (eds. D.W. Maynard, H. Houtkoop-Steenstra, N.C. Schaeffer and J. Van der Zouwen), 261–280. New York: Wiley.

Schober, M.F. and Conrad, F.G. (2002). A collaborative view of standardized survey interviews. In: *Standardization and Tacit Knowledge: Interaction and Practice in the Survey Interview* (eds. D.W. Maynard, H. Houtkoop-Steenstra, N.C. Schaeffer and J. van der Zouwen), 67–94. New York: Wiley.

Sequence Viewer [Computer software]. (2017). Retrieved from sequenceviewer.nl

Smit, J.H., Dijkstra, W., and Van der Zouwen, J. (1997). Suggestive interviewer behaviour in surveys: an experimental study. *Journal of Official Statistics* 13: 19–28.

Smyth, J.D. and Olson K. (2018). The Effects of Mismatches between Survey Question Stems and Response Options on Data Quality and Responses. *Journal of Survey Statistics and Methodology*. 7: 34–65.

Tourangeau, R., Rips, L.C., and Rasinski, K. (2000). *The Psychology of Survey Response*. Cambridge: Cambridge University Press.

Unger, S., Ongena, Y.P. & Koole, T. (2016). Expanded and Nonconforming Answers in Standardized Survey Interviews. *Paper presented at the International conference on questionnaire design, development, evaluation and testing (QDET2)*. Miami, Florida.

US Census. 2006. Women and men population in the United States: 2006. https://www.census.gov/population/www/socdemo/men_women_2006.html.

Van der Zouwen, J. and Smit, J.H. (2004). Evaluating survey questions by analyzing patterns of behavior codes and transcripts of question-answer sequences: a diagnostic approach. In: *Methods for Testing and Evaluating Survey Questionnaires* (eds. S. Presser, J. Rothgeb, M.P. Couper, et al.), 109–130. New York: Wiley.

Part III

Improving Questionnaires on the Web and Mobile Devices

12

A Compendium of Web and Mobile Survey Pretesting Methods

Emily Geisen[1] *and Joe Murphy*[2]

[1] *RTI International, Research Triangle Park, NC, USA*
[2] *RTI International, Chicago, IL, USA*

12.1 Introduction

The early twenty-first century has seen dramatic changes in the survey climate and landscape. Advances in technology along with changes in coverage for different survey modes have led to an environment where many surveys are now being conducted in whole or in part via self-administered web questionnaires. Even within web surveys, the ways respondents are accessing the questionnaire are changing. A 2015 report by the Pew Research Center found that mobile device ownership in the United States is now almost on par with ownership of traditional computers: 73% of US adults own a desktop or laptop computer and 68% own smartphones. Canada, Japan, South Korea, Australia, and many countries in Western Europe have similarly high smartphone ownership (Poushter 2016). In line with this trend, respondents are increasingly choosing to complete web surveys on mobile devices. Recent estimates show that the proportion of respondents completing a survey on a mobile device can be 30% or more for some surveys (Lugtig et al. 2016; Saunders 2015). Mobile apps are also being used by survey respondents who are panel members and by interviewers to administer household screening surveys.

As surveys become increasingly self-administered, web-based, and mobile, pretesting becomes increasingly important in assessing the potential for measurement error. Measurement error occurs when the response captured in the survey is different from the respondent's true value (Groves 1989). Measurement error can be caused by a variety of sources including the respondent, an interviewer, the survey mode, and the survey instrument itself.

Traditional pretesting methodologies that do not assess the technology used by the survey itself may miss important quality concerns these newer modes introduce (mode effects). And while traditional pretesting methods

Advances in Questionnaire Design, Development, Evaluation and Testing, First Edition.
Edited by Paul C. Beatty, Debbie Collins, Lyn Kaye, Jose-Luis Padilla, Gordon B. Willis, and Amanda Wilmot.

have provided great insight into the errors to be addressed in surveys prior to fielding, they can be costly, time consuming, and limited in the range of perspectives gained. In this chapter, we introduce emerging survey pretesting methodologies and compare these with traditional methods to consider whether these methods can be combined to improve today's web and mobile survey needs. We begin by reviewing traditional pretesting methods such as expert review, cognitive interviewing, and pilot testing. Next, we introduce emerging pretesting methods including usability testing, eye tracking, and online pretesting. We conclude with a discussion of the optimal combination of traditional and newer methods for pretesting modern web surveys.

12.2 Review of Traditional Pretesting Methods

Pretesting is used to identify potential sources of error in surveys so they can be addressed before fielding. In this section, we briefly describe the three most common survey pretesting methodologies that are used on all types of surveys: expert review, cognitive testing, and pilot testing. For each method we discuss the benefits and limitations of using these methods in today's web survey landscape.

12.2.1 Expert Review

For many surveys, questionnaire development includes expert review by survey methodologists or subject matter specialists (Esposito and Rothgeb 1997; Olson 2010; Presser and Blair 1994; Willis et al. 1999). Expert review assists in writing and designing survey questions that are free of major flaws (e.g. double-barreled, response options that are not mutually exclusive) and measure the underlying concepts in an unbiased manner. There is a range in the formality and rigor with which expert review is conducted. For some surveys (e.g. a repeat administration of an annual cross-sectional study), a proofing or "run-through" of the draft instrument by survey researchers may be sufficient to identify any items that may result in measurement error. For new items or potentially problematic content, a standardized protocol for review may be necessary. One such approach is the Question Appraisal System (QAS) – a structured, standardized instrument review methodology designed to assist experts in evaluating how respondents will likely understand and respond to survey questions (Willis and Lessler 1999). A subsequent enhancement, the QAS-04, included updates for cross-cultural instrument development. (Dean et al. 2007). By following the steps of the QAS, the reviewer examines each item in the questionnaire according to specific question characteristics and determines whether the features of the question will likely cause problems.

The reviewer then develops recommendations for correcting each potential problem.

A primary limitation of the QAS is that there is currently no particular focus on visual design features that may introduce errors. This is especially important in the case of self-administered web surveys, where no interviewer is present to clarify the meaning or instructions. However, even interviewer-administered surveys can be affected by the screen design (Edwards et al. 2008). Indeed, the "visual language" of surveys is often considered as important as the wording of the survey questions when it comes to potential sources of measurement error (Christian 2003; Christian et al. 2005, 2009; Tourangeau et al. 2004, 2007). Although there exist a number of best practices, guidelines, and recommendations on visual design for web surveys (Dillman et al. 2009; Couper 2008), there is little information about how these guidelines hold up in a changing landscape that relies on smaller survey screens and continuous technological enhancements. Therefore, researchers are often faced with designing aspects of questionnaires for which no best practice guidelines exist.

12.2.2 Cognitive Interviews

Cognitive interviewing is a qualitative evaluation of how respondents understand and answer survey questions to determine whether changes are needed to improve data quality. Typically, a researcher trained in cognitive testing methods administers the survey questions in-person to participants (volunteers who are representative of the survey's target population). Participants are often instructed to think aloud as they are answering the survey questions so that the cognitive interviewer can understand how participants interpreted the question and came up with their answer (Willis 2005). Once a respondent answers the question, interviewers will administer follow-up questions (i.e. verbal probes) such as "What does this question mean in your own words?" or "How did you come up with your answer?" to further understand participants' cognitive processes as they answer survey questions. Verbal probes for a given question can be administered immediately after the participant answers a survey question (concurrent verbal probing) or they can be administered after all survey questions or a block of survey questions have been answered (retrospective verbal probing).

Cognitive interviewing is one of the predominant methods for pretesting survey questionnaires. Although few studies have empirically demonstrated that cognitive testing a survey can improve data quality (Beatty et al. 2006; Willis and Schechter 1997), there is a general consensus that cognitive interviewing is valuable. There are numerous examples that demonstrate that the practice of cognitive interviewing can identify a number of potential problems in survey questions (Willis 2005, 2015; Miller 2014; Collins 2015). To the extent that researchers can resolve and address these problems before fielding a survey, there is potential for improvements to data quality.

A general best practice is to administer the cognitive interviews, or at least a portion of the cognitive interviews, using the same mode that will be used for fielding the survey in order to replicate the respondent experience (Willis 2005). This can prove challenging for web or application-based surveys because they can be completed on a variety of devices from desktops to laptops to tablets to mobile devices. In addition, due to the time and cost associated with programming surveys, many organizations choose to cognitively test surveys based on paper specifications prior to programming. However, the visual aspects of a survey questionnaire can affect participants' comprehension of survey questions. For example Tourangeau et al. (2004) found that when respondents answered survey questions, they often used the visual midpoint of the scale instead of the conceptual midpoint of the scale. Any potential errors related to the interaction between the visual design of the survey and the wording of the questions may not be identified via orally administered cognitive testing or if the visual design used during testing does not match the design used when fielding the survey (e.g. tested on a laptop, completed on a mobile device).

Even when cognitive testing is conducted using the same mode as the final survey, the testing focus is primarily on cognitive issues with comprehension, retrieval, judgment, and response, and frequently on additional elements that may affect measurement error such as question logic and cultural appropriateness. In self-administered surveys, it is also critical to test the way that respondents use and interact with surveys and any errors associated with that interaction (Jenkins and Dillman 1995). Although respondents interact with paper surveys (e.g. marking boxes, writing out responses, following skip logic), this is increasingly important with web and mobile surveys that are dynamic and incorporate interactive features. For example, a respondent may understand the survey question and response options, but have difficulty entering an answer in the correct format on the computer and resolving any error messages received.

Although cognitive interviewing does not specifically target many of the quality concerns related to surveying respondents via the web and mobile surveys, changes in how we use technology has affected the way that we can conduct cognitive interviews in three key ways addressing some of the current limitations. It has (i) introduced additional participant recruit methods, (ii) provided the ability to conduct cognitive interviews remotely, and (iii) led the way to remote, unmoderated cognitive testing methods conducted online.

Traditionally, cognitive interview participants have been recruited via in-person intercept methods, newspaper advertisements, flyers, snowball sampling, participant databases, and list samples, and by working directly with organizations such as community or advocacy groups (Head et al. 2015).

With the advent of the web, pretesting recruitment methods began leveraging its speed and low cost. Use of online advertising tools such as Craigslist, Facebook, Twitter, Instagram, Google Ads, and Amazon Mechanical Turk introduced new methods for participant recruitment. Studies comparing these methods found that online recruitment methods can provide a pool of cognitive interview participants quickly and easily and that these participants often differ demographically from participants recruited via traditional methods (Murphy et al. 2015; Sage 2014; Antoun et al. 2015). Antoun et al. (2015) categorized the online recruitment methods into two groups: those that "pull in" individuals looking for paid opportunities (Craigslist and Mechanical Turk) and those that "push out" an advertisement to a broader online community such as social media. The "pull-in" method allowed for quicker and less-expensive recruitment while the "push-out" method resulted in a more demographically diverse recruitment pool. Identifying demographically diverse samples becomes increasingly important when the goal is to test a survey with a large sample that includes a variety of subgroups. This has been shown to identify additional errors compared with smaller samples (Blair and Conrad 2011).

In addition, technological advances have allowed researchers to conduct remote or "virtual" cognitive interviews using videoconferencing software (GoToMeeting, WebEx, Skype) (Cook et al. 2016). Virtual methods also allow for a more geographically dispersed sample since participants do not need to be in the same physical location as the cognitive interviewer. In addition, remote cognitive interviewing allows participants to complete the study in a more natural setting (compared with traditional lab-based cognitive testing) and frequently relies on participants using their own devices.

A primary limitation of traditional cognitive interviewing is that it relies on small convenience samples. Therefore, the results may not be representative of the full population. However, recent technological developments have also provided the opportunity to conduct online cognitive interviewing in which no moderator is present. This allows for a greater number of participants to be included at relatively low costs. This approach, which is often referred to as *web probing*, is discussed in more detail under 12.3 Emerging Pretesting Methods.

12.2.3 Pilot Testing

A traditional form of survey pretesting, the pilot test, has been in use since the mid-1930s (Presser et al. 2004). The pilot test is a small-scale version of a full-scale study and serves as a dry run for the survey. It is used to evaluate administrative procedures as well as item nonresponse rates, skip patterns, and response frequencies. Pilot testing was the predominant form of pretesting until the 1980s, when questionnaire design became heavily influenced by cognitive psychology theory, creating a field of study known as the Cognitive Aspects of Survey Methodology (CASM) (Tourangeau 1984).

Pilot tests can be used to collect preliminary survey data to evaluate whether questions are working well. Questions with high rates of item nonresponse may indicate that respondents did not understand the question, did not know the answer, or were not willing to answer the questions. Questions with very little variability from each other may indicate that the questions are measuring the same concept and one can be removed. Some pilot studies include additional evaluation tools such as respondent debriefings and behavior coding. In respondent debriefing, researchers embed follow-up questions in a standardized interview setting, typically at the end of the interview, to evaluate how well respondents understand the survey questions (Esposito et al. 1991). Behavior coding was initially developed as a method for evaluating interviewer performance such as speed, articulation, and reading verbatim (Cannell et al. 1968, 1975). It has also been used to assess difficulties with question comprehension such as respondent requests for clarification (Cannell and Robinson 1971; Morton-Williams and Sykes 1984).

A benefit of the pilot test in this changing technology landscape is that it can provide information on the proportion of the population that is completing a survey on a mobile device and often what type of mobile device. A pilot test may reveal that participants are having a difficult time accessing the web survey or using their login information correctly. A pilot test can also provide important paradata for web surveys such as the amount of time spent on each question or screen, the number of error or warning messages activated, and the number of times the back or previous button was pushed. This information can be used to determine which questions are not performing well. A limitation, though, is that it is harder to determine why a question is not performing well, which is needed to resolve the issue. Is it a usability concern or privacy issue, or do participants not understand the survey questions?

12.3 Emerging Pretesting Methods

In addition to the traditional pretesting methodologies identified in the previous section, there are now a number of emerging pretesting methods available to survey researchers. Specifically, we focus on three key emerging pretesting methodologies: (i) usability testing, (ii) eye tracking, and (iii) online pretesting (remote, unmoderated pretesting conducted online). For each methodology we provide a description of the approach, explain its advantages over traditional pretesting methodologies for web and mobile surveys, and discuss the key limitations of these methods.

Although some of these methods such as usability testing and eye tracking are not new, we consider them to be emerging due to advancements in technology that make these methods more critical for evaluating potential data quality concerns in modern surveys and changing how we implement these pretesting methods to evaluate surveys.

12.3.1 Usability Testing

Usability testing consists of watching a user such as a potential survey respondent or interviewer as they complete a task or goal (e.g. logging on to a survey, navigating between questions, answering a question, completing a roster of household members). The researcher observes the user to evaluate how well they are able to achieve their goals in terms of accuracy, efficiency, and satisfaction. Researchers are also typically interested in how easy the survey was to use and how easy the survey was to learn (for interviewer-administered surveys or panel survey applications). It is typically an iterative process conducted with small samples of participants throughout the survey development cycle, which allows researchers to use the results of usability testing to guide the design and development of the survey instrument.

As with cognitive interviewing, participants are often instructed to think aloud. Usability test participants are instructed to explain what they are thinking while completing tasks. Ideally participants explain not just what they are doing (e.g. "I'm clicking the Next button") but what they are thinking as they are doing it (e.g. "I want to see if I can continue without providing a response, so I am going to click the Next button"). Verbal probing can also be used, often retrospectively, to gather additional information about why a participant completed a task in a certain way. However, usability testing tends to rely less on verbal probing than cognitive interviewing, particularly concurrent verbal probing, as this can interfere with test performance and collection of usability metrics (e.g. eye tracking, task accuracy, and time to complete task) (Geisen and Romano Bergstrom 2017). Researchers' observations of clicks and mouse movements are then combined with participant comments and other metrics such as task success rates and task completion times to identify potential usability issues that would affect data quality.

Usability testing was first used in survey research in the 1990s as data collection methods became increasingly dependent on computer-assisted interviewing (CAI) methodologies. Initially, usability testing was primarily used to evaluate interviewer-administered computerized surveys and to a lesser extent self-administered computerized surveys (Hansen et al. 1997; Marquis et al. 1998; Couper 2000). Despite the surge in self-administered web surveys since 2000, there was not a corresponding rise in the use of usability testing as a pretesting strategy. However, as the landscape for survey data collection methodology changes again with the new technological features being incorporated into web surveys (e.g. videos, GPS, maps, dynamic functions) and increased reliance on mobile devices, usability testing is becoming increasingly helpful in identifying data quality concerns when conducting web and mobile surveys.

Usability testing allows researchers to observe and understand how respondents use and interact with surveys and how this may affect survey quality. While pilot testing focuses on logistics and procedures and cognitive testing focuses on how respondents *comprehend* surveys, usability testing focuses on how respondents *use* surveys. Geisen and Romano Bergstrom (2017) propose a conceptual model for understanding the usability of web and mobile surveys. This process includes three key aspects of usability that can be evaluated during usability testing as described in Table 12.1: (i) interpreting the design, (ii) completing actions and navigating, and (iii) processing feedback.

Geisen and Romano Bergstrom (2017) explain that how respondents interpret the survey is important to evaluate during user testing because a respondent may understand question wording, but not understand how to correctly input their response in the survey. Respondents have a mental model of how the survey should work based on the visual information presented to them, e.g. radio buttons allow you to select one response, check boxes allow you to select multiple responses. Usability testing allows researchers to evaluate how well the survey matches the respondent's mental model in practice. For example, the survey may use underlined text for emphasis, but respondents think that it is a hyperlink or pop-up.

Usability testing can also be used to evaluate the actions that participants make when completing the survey. As web surveys become more complex, respondents must interact more with the survey to provide their response. For example, they may need to access a definition, use the navigation menu, ensure a total adds up correctly, or use a menu bar to save their data and resume the survey later (see Figure 12.1a–d). Researchers must determine what the

Table 12.1 Usability model for surveys.

1. **Interpreting the design:**
   a. What meaning do respondents assign to visual design and layout?
   b. How do respondents believe the survey works?
2. **Completing actions and navigating:**
   a. How well does the survey support respondents' ability to complete tasks and goals?
   b. How well do respondents follow navigational cues and instructions?
3. **Processing feedback:**
   a. How do respondents interpret and react to the feedback provided by the survey in response to their actions?
   b. Does the survey help respondents identify, interpret, and resolve errors?

Source: Reproduced with permission from Geisen and Romano Bergstrom (2017).

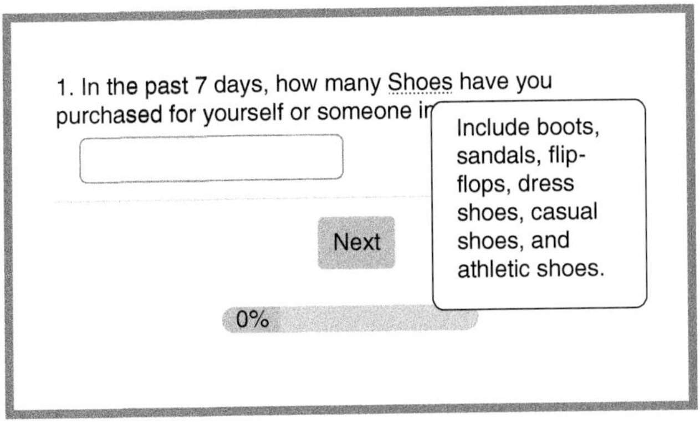

(a)

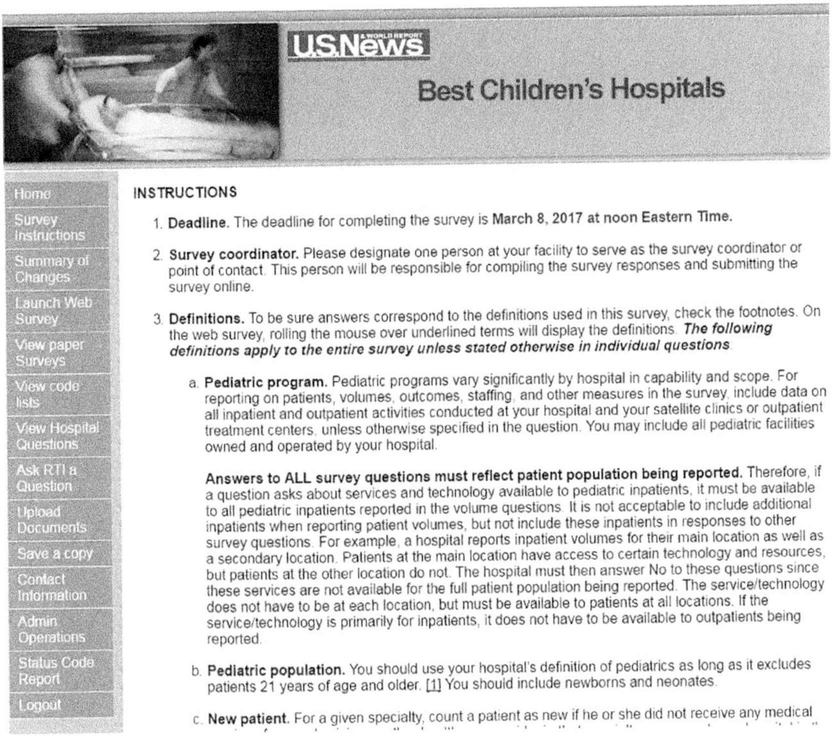
U.S.News & WORLD REPORT
Best Children's Hospitals
Home
Survey Instructions
Summary of Changes
Launch Web Survey
View paper Surveys
View code lists
View Hospital Questions
Ask RTI a Question
Upload Documents
Save a copy
Contact Information
Admin Operations
Status Code Report
Logout
INSTRUCTIONS
1. Deadline. The deadline for completing the survey is March 8, 2017 at noon Eastern Time.
2. Survey coordinator. Please designate one person at your facility to serve as the survey coordinator or point of contact. This person will be responsible for compiling the survey responses and submitting the survey online.
3. Definitions. To be sure answers correspond to the definitions used in this survey, check the footnotes. On the web survey, rolling the mouse over underlined terms will display the definitions. The following definitions apply to the entire survey unless stated otherwise in individual questions.
a. Pediatric program. Pediatric programs vary significantly by hospital in capability and scope. For reporting on patients, volumes, outcomes, staffing, and other measures in the survey, include data on all inpatient and outpatient activities conducted at your hospital and your satellite clinics or outpatient treatment centers, unless otherwise specified in the question. You may include all pediatric facilities owned and operated by your hospital.
Answers to ALL survey questions must reflect patient population being reported. Therefore, if a question asks about services and technology available to pediatric inpatients, it must be available to all pediatric inpatients reported in the volume questions. It is not acceptable to include additional inpatients when reporting patient volumes, but not include these inpatients in responses to other survey questions. For example, a hospital reports inpatient volumes for their main location as well as a secondary location. Patients at the main location have access to certain technology and resources, but patients at the other location do not. The hospital must then answer No to these questions since these services are not available for the full patient population being reported. The service/technology does not have to be at each location, but must be available to patients at all locations. If the service/technology is primarily for inpatients, it does not have to be available to outpatients being reported.
b. Pediatric population. You should use your hospital's definition of pediatrics as long as it excludes patients 21 years of age and older. [1] You should include newborns and neonates.
c. New patient. For a given specialty, count a patient as new if he or she did not receive any medical

(b)

Figure 12.1 (a) Example of a hover-over definition that requires the respondent to move the mouse over the word "shoes" to access the definition. (b) Navigation menu used on the Pediatric Hospital Survey conducted as part of the Best Children's Hospitals rankings. *Note*: The Pediatric Hospital Survey is conducted as part of the Best Children's Hospitals rankings, which are published by *US News & World Report* and have been conducted by RTI International since 2004. For more information, visit http://www.rti.org/besthospitals. (c) Example of a survey that requires a respondent to enter values that sum to 100%. (d) Example of a menu bar with the ability to save and log out.

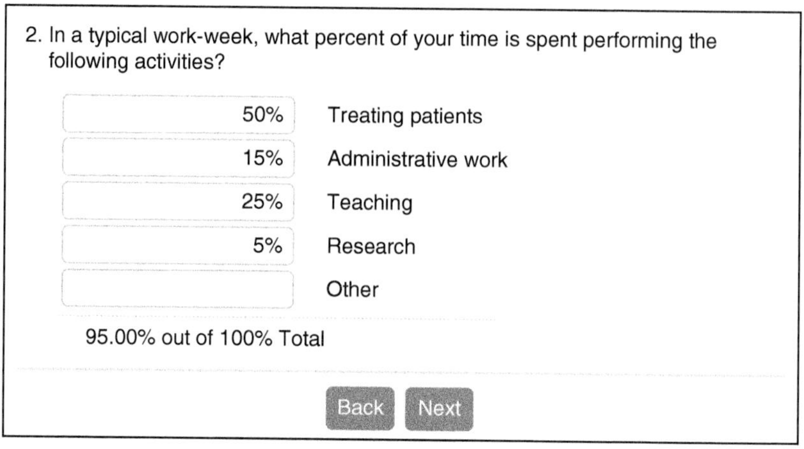

(c)

(d)

Figure 12.1 (*Continued*)

respondents' goals are, and whether the design of the survey supports them in completing these goals or whether it gets in their way making it harder to answer accurately and easily.

A unique aspect of web or mobile surveys compared with self-administered paper surveys is that the computer or device provides feedback to the respondent (Geisen and Romano Bergstrom 2017). The respondent uses this feedback to determine how to proceed in a survey. If a respondent selects a response option, but the radio button is marked or the option is not highlighted, the respondent will assume it is not selected. When the feedback provided by the survey is not clear, it interferes with respondents' ability to complete the survey. Providing useful feedback is increasingly important when respondents make errors. Surveys, like other platforms, should be designed so that they are error-tolerant (Quesenbery 2001). This means that the survey tries to prevent errors, but also helps respondents recover from any errors that are encountered. The survey must help the respondents understand when an error occurred, why an error occurred, and how to resolve the error.

Advances in technology have also made it easy and inexpensive to evaluate the usability of web and mobile surveys. The most useful tools are screen-recording software, screen-sharing software, and a web camera. Most web-conferencing software combines these capabilities, allowing a researcher to use one system to record a participant's computer screen, their face (if using a web camera), and audio while live streaming to allow observers to view the sessions in real time.

Geisen and Romano Bergstrom (2017) recommend two methods for conducting and recording usability sessions on mobile devices: (i) mobile sleds and (ii) remote, mobile screen sharing.

12.3.1.1 Mobile Sled

A mobile sled is a small, portable device with an attached digital camera. This allows usability test participants to hold their device as they normally would. The attached camera is used to record the screen and a participant's hands. Figure 12.2 shows a mobile sled being used to record and share a participant using a mobile device. Some sleds, such as the one shown in Figure 12.2, are adjustable to allow for recording of larger smartphones and tablets. By connecting the camera to a computer via a USB cord, the session can be recorded and shared with observers. A second camera can be used to capture the participant's face, if desired. The primary benefit of this method is that it does not require a participant to download and install any software on their mobile device. In addition, this setup is extremely versatile because it is portable, will work with almost any mobile device, and can be set up quickly. Another benefit is that this setup does not rely on internet access. Therefore application-based surveys can be tested in settings without internet access.

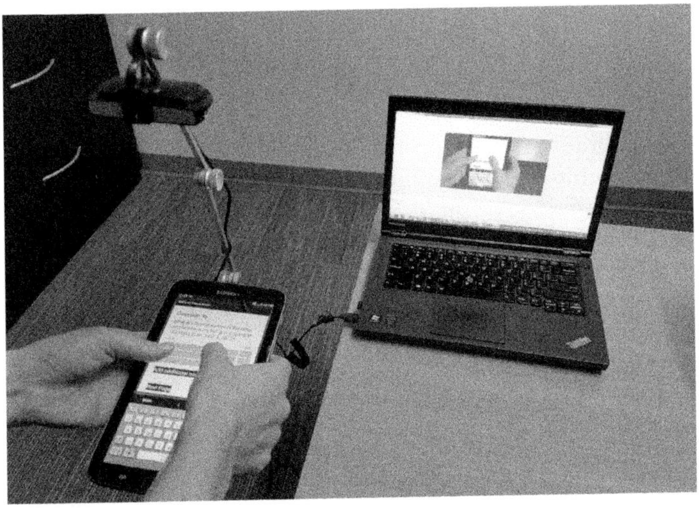

Figure 12.2 Example of a mobile sled, which allows researchers to view the participant's mobile screen on a computer for ease of recording or sharing.

12.3.1.2 Remote, Mobile Screen Sharing

Advances in web conferencing technology have now enabled the use of remote, mobile screening sharing allowing a mobile device screen to be viewed on a computer. Test participants would need to download an app that works for their phone (different apps may be needed for Android versus Apple OS versus other operating systems), and then share their screen with the moderator. An advantage of mobile screen *sharing* is that moderators and observers can easily view the device's screen when it is displayed on the laptop or desktop screen as it is being shared. Screen-recording software on the computer (or via the web) can then be used to record the session. The primary benefit of this approach is that the participant and moderator do not need to be in the same location, allowing the possibility for remote testing. For in-person testing, though, there are several limitations of remote, mobile screen sharing compared with the use of a mobile sled. Although the moderator or observer can see the participant's screen, they cannot see the participant's hands. Being able to see a participant's hands is helpful for identifying certain usability concerns. For example, a participant may be tapping a part of the screen that is not "tappable" or a participant may select an incorrect response due to the size of his fingers. Second, there can be a lag between the mobile device and the video on the computer. As a result, the image can sometimes appear pixelated depending on internet speed. Another limitation is that users may not want to download software to their device and may have privacy concerns with allowing a moderator to view their screen.

A study by Romano Bergstrom (2016) empirically showed that usability testing can improve a survey through a split-sample experiment comparing the results of a survey that was revised based on usability testing with the original survey. Although the survey design had been optimized for mobile devices, usability testing on mobile revealed a significant number of problems that participants had completing the survey on mobile devices. These included difficulty with drop-down menus, item nonresponse, and navigation. Both the mobile and non-mobile version of the survey were revised. Qualitative comments and observations were used to evaluate differences in the distributions of the survey estimates for the revised and unrevised survey to determine that the revised survey led to more accurate survey responses.

Although usability testing can provide a number of insights about the design of a survey and how that design may impact quality, a limitation – as with cognitive interviewing – is that results are primarily qualitative and may not be generalizable due to the small sample sizes. However, when repeated usability studies show that a certain issue is problematic on a number of different types of surveys, we can draw conclusions for usability testing about survey best practices. In addition, usability testing may not be necessary or feasible for all surveys, particularly those that are short and simple without any web-centric features (e.g. navigation menus, hover-over definitions). Geisen and Cook

(2011) pose that there is a usability testing "continuum" where some surveys require more testing and other surveys require less testing.

12.3.2 Eye Tracking

Eye-tracking equipment uses near-infrared technology and cameras that have been built into computer monitors or stand-alone devices to track where an individual looks within a visual field. Eye tracking records *fixations*, which are when the eye pauses on a specific area within the visual field, and *saccades*, which are rapid eye movements between fixations Previously, eye-tracking equipment required users to have a fixed head position and used uncomfortable equipment such as bite bars or chin rests. It also required a lengthy calibration process. Today's eye-tracking technology has been incorporated into computer equipment or as wearable glasses, and no longer needs to be connected to the user in an obtrusive manner. Although eye-tracking equipment can still be complex, technological advances have made eye tracking easier to set up, calibrate, and analyze; plus, reductions in costs have made it more accessible (Romano Bergstrom and Schall 2014).

When respondents complete surveys, we are often interested in knowing what they read or looked at on the screen. Did they read the instructions? Did they see the navigation menu or the Help button or the Calculate Total button? Yet, asking cognitive or usability test participants if they read or noticed this information is not always reliable (Nichols 2016). A primary advantage of using eye-tracking data to evaluate a survey is that it indicates what a participant actually looked at or paid attention to as opposed to what they say they look at or paid attention to. In the following, we indicate what information eye tracking provides and how that can be used to evaluate survey questionnaires in a number of ways (Geisen and Romano Bergstrom 2017; Olmsted-Hawala et al. 2014):

- *What people look at when completing a survey.* We can also learn what respondents do not look at – such as the survey instructions.
- *How many times they look at various things.* Do respondents need to read the question multiple times? Do they repeatedly look back and forth between different elements on the screen?
- *The order in which they look at things.* What first catches participants' attention? Do they complete the survey in the order we expect?
- *How long they look at things.* Respondents will spend more time looking at aspects of the screen that interest them or aspects that confuse them.

Eye tracking has revealed that the way that respondents read web screens with survey questions on them differs from how they read screens with other text on them. Everdell (2014) finds that users typically read web content in an F-shaped pattern. Users scan the first few lines of text or content at the top of the page, then they move down the left side of the page and scan horizontally to the right but not as far as they did at the top of the page. They continue to scan down the left side of the page. Jarrett and Romano Bergstrom (2014) and Romano and Chen (2011) found that when users encountered surveys or forms, however, their attention was drawn immediately to the input slots or response fields. Eye-tracking studies have also provided the following insights regarding how respondents complete web surveys: Respondents look for navigation buttons (Previous and Next) directly below response options, and prefer the Next button on the right. (Romano and Chen 2011). Text box labels should be near the field, preferably above the label, and presented in a single column (Penzo 2006). Text should not be placed inside the response boxes, and error messages should be placed next to the errors and where users will see them (Jarrett 2010). Even on computers, participants do not scroll vertically to see all response options with long lists (Romano and Chen 2011). Many of these findings offer additional evidence and support for existing questionnaire design guidelines such as placing survey instructions exactly where they are needed to increase likelihood of being seen and used (Dillman et al. 2009). It is important to note that eye tracking is usually conducted as part of usability testing, rather than as a stand-alone study, so that the data can be combined with participant commentary and researcher observations (Geisen and Romano Bergstrom 2017).

A primary limitation of eye tracking for evaluating surveys is that it does not work when the participant's attention and gaze are not together (Jarrett and Gaffney 2008). For example, when respondents have to think of answers on the spot (a password with complex rules or their opinion on something they have not considered before), their gaze is likely toward the screen, but their attention is elsewhere, which can make eye-tracking data unreliable. Similarly, there are other situations when a participant's gaze is not on the screen at all: for example, to answer a question about how much money was spent on clothes last month, the participant may need to look at a receipt or ask another household member for information. When the participant's gaze is not on the screen, eye-tracking data will be missing.

12.3.3 Online Pretesting

Just as the internet has opened new possibilities for reaching and conducting surveys with a variety of populations, the new era of web development (i.e. the Web 2.0 age) has made possible a type of pretesting on a scale well beyond the numbers typically used for in-person pretesting: remote, unmoderated pretesting conducted online – more commonly referred to as online pretesting. Online pretesting does not require an interviewer, and all items are self-administered. Consequently, there is no travel requirement for researchers or participants and no direct interaction between the two. This allows for large quantities of tests to be conducted quickly and inexpensively compared to moderated pretesting methods.

Online pretesting involves a geographically dispersed participant pool, often recruited through an online platform such as online panels (Holzberg et al. 2016; Lenzner and Neuert 2017; Meitinger and Behr 2016; Scanlon, Chapter 17 in this book). Another platform used for survey pretesting due to the low recruitment cost is Amazon Mechanical Turk, which has over 750 000 workers ("turkers"), about three-quarters of whom are located in the United States. Workers typically complete tasks in their own homes, and for a very small amount of money in a very short amount of time. Social researchers have recently used these types of online platforms to conduct surveys (Christenson and Glick 2013; Chunara et al. 2012; Behrend et al. 2011) and, as we discuss here, pretest survey questions.

The most common type of online pretesting is web probing (or online probing), which is somewhat akin to traditional cognitive interviewing or usability testing, yet on a bigger, faster, and cheaper scale. With web probing, probes are embedded in a self-administered web survey to delve into the participants' process of comprehension and retrieval when answering survey questions, gather reactions to images or other stimuli, and assess perceptions of confidentiality and privacy. For example, after answering the target question, "In the past 7 days, how many shoes have you purchased?" participants will then be probed on what they were thinking of when the word "shoe" was included in the survey question, Participants typically type in their responses to the online probes. A comprehensive guide for conducting web probing to evaluate questionnaires is provided by Behr et al. (2017).

Some tools used in online pretesting can also record participants' screens, keystrokes, and mouse movements, as well as their verbal responses to online probes or think-aloud commentary provided as they complete a survey. Researchers can also embed scenarios and tasks for the participants to complete them (e.g. "You had a banana, toast and coffee for breakfast. Please enter this into the food diary.") to evaluate how well participants can complete certain aspects of the study. Online tools with these additional features also allow for online usability testing (remote, unmoderated usability testing), which allows researchers to evaluate not only how participants understand the survey questions, but how they navigate and use the survey instruments as well. For example, Edgar et al. (2016) used a platform called TryMyUI to pretest a web survey. The software recorded the audio of participants' reactions to the questions and video of the screen itself.

The idea of embedding probes into a questionnaire outside the setting of a cognitive laboratory is not new. Schuman (1966) randomly selected survey questions to probe on after the interview to better understand responses. Converse and Presser (1986) discuss embedded probing in field interviews to identify problematic items. More recently web probing has been used to pretest substantive questions in web surveys (Behr et al. 2013b, Meitinger and Behr 2016; Paškvan and Plate 2017; Allum et al. 2017), evaluate survey question complexity (Slavec 2017), test alternate versions of questions that did not perform well in pilot tests (Murphy et al. 2015), and evaluate the comparability of items in cross-cultural surveys (Behr et al. 2012; Behr 2016; Meitinger 2017).

Many studies have shown similar results for traditional, in-person pretesting compared with web probing. Mockovak and Kaplan (2016) found that in-person and web probing gave similar results, but they found that the web probing missed one major problem that was identified in in-person testing and did not uncover any problems that had not also been identified by in-person testing. However, in this study, they conducted an equal number of in-person interviews and online interviews. This suggests that for web probing to be advantageous, it likely needs many more participants than typically used for in-person testing. Meitinger and Behr (2016) compared the results of 20 in-person cognitive interviews with over 500 web probing participants. They found that while in-person cognitive interviews were generally higher in quality, the ability to complete hundreds rather than tens of interviews online uncovered several otherwise missed error types and themes. A further discussion of the optimal sample size for web probing is provided in Meitinger et al. (2017).

Lenzner and Neuert (2017) also found that web probing participants were less likely to answer probes meaningfully compared to traditional cognitive interview counterparts. They compared 508 online participants with 20 face-to-face participants and found that suboptimal responding (e.g. item nonresponse, uninterpretable responses, or cut-and-paste definitions from the web) occurred for 14% of web probing participants. Despite this, they found that the two methods identified similar types of problems in the questionnaire, which led to identical revisions to the survey questions.

Edgar et al. (2016) also compared web probing to traditional lab-based cognitive interviews. They found comparable results for more straightforward, simple items like asking a respondent to provide examples of sportswear or define the flu season. More complex tasks, such as explaining the thought process behind remembering all clothes purchased in the last few months, were much better suited for traditional cognitive testing with an interviewer present. The differences in cost, timeliness, and geographic reach between traditional and online methods, however, were striking. Even without considering the interviewer labor necessary for cognitive interview preparation and conduct, the cost for participant incentives was five times cheaper for the online participants recruited through Mechanical Turk. In terms of timeliness, Mechanical Turk was able to recruit over 1000 participants in a matter of days. Typically, in cognitive interview settings, it can take days to recruit and confirm participants in the single digits. Further, web probing allowed for the recruitment of a much more geographically diverse population than traditional methods given the same time and cost restraints (see Figure 12.3).

Another key aspect of web probing is that the pretesting mode is self-administered, which matches the survey-administration mode for web surveys. In-person surveys are more prone to social desirability affects than self-administered surveys (Bowling 2005), which can affect data quality on sensitive items. Social desirability can also affect the results of survey pretesting. Holzberg et al. (2016) compared traditional in-person cognitive interviewing with web probing using both probability samples and two non-probability samples (Census opt-in panel and Mechanical Turk). While they found no major differences on comprehension of survey items between the in-person and online cognitive interview participants, the online participants (from both probability and non-probability samples) were more negative in their responses. Web probing may have an advantage over in-person methods on obtaining candid feedback from participants.

Despite the advantages, there are some potential drawbacks with web probing. Since demographic information is self-reported, the accuracy of the sample composition and responses is dependent on the honesty of participants. Participants who misrepresent themselves to meet the recruitment criteria for a given project can provide misleading results. Further, the fact that some online panel members are involved in so many research studies raises the concern of "professional respondents" (Marder 2015). One solution is to exclude panel members from your study if they have participated in similar studies in the past 12 months.

Another limitation of web probing studies is that they are less interactive than traditional cognitive interviewing. There is no interviewer to probe further on inadequate responses and web probing does not capture as many spontaneous comments by participants (Meitinger and Behr 2016; Lenzner and Neuert 2017). Web probing using only typed responses will not capture think-aloud verbalizations, which can be a rich source of information in traditional cognitive interviews. Lenzner and Neuert (2017) found that only 5% of their online participants noted that the term "civil disobedience" was unfamiliar to them compared with 30% of their face-to-face participants. In the face-to-face interviews, participants spontaneously commented on aspects of the question they did not understand before the interviewer even administered probes.

Online responses tend to have lower quality responses compared to traditional pretesting responses – they are often shorter and a larger portion is irrelevant (see Chapter 18 of this book). Yet, this is rarely a problem because online pretesting typically produces a higher volume of responses more quickly and easily than can be produced with traditional pretesting. Although many responses can be obtained quickly, the resulting data is qualitative. Coding and analyzing large sets of qualitative data can be time-consuming and labor-intensive. However, advances in computer algorithms that rely on natural

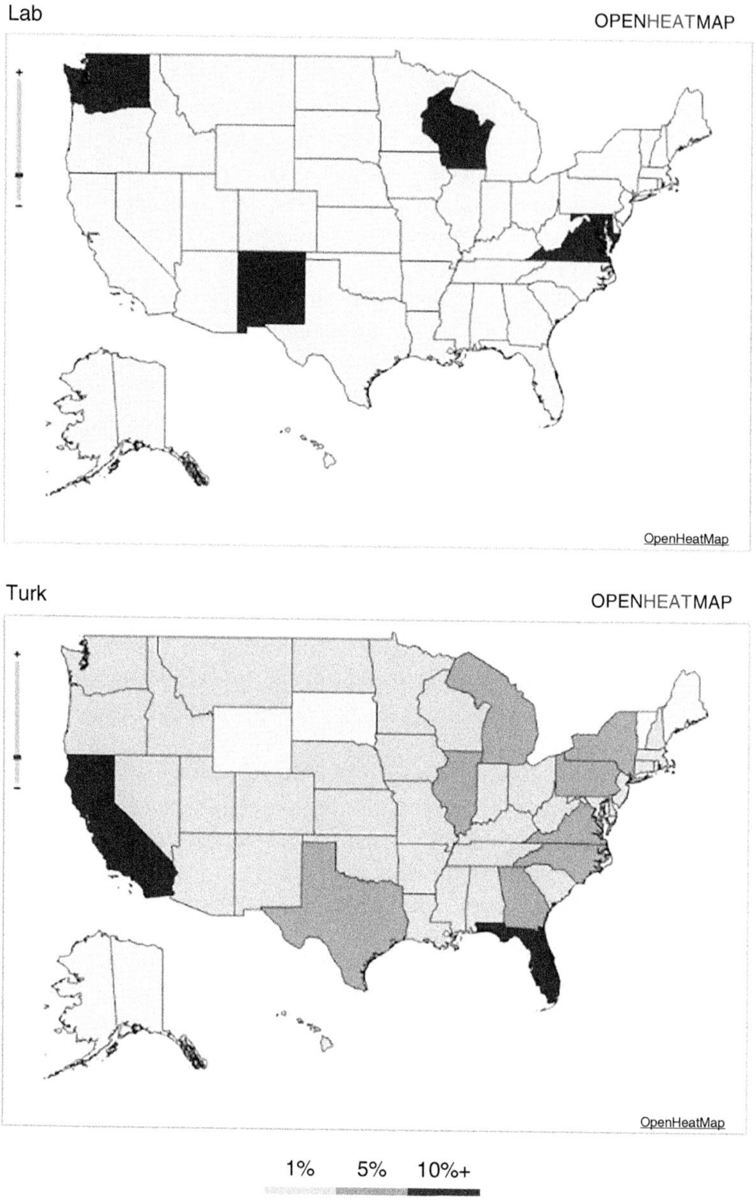

Figure 12.3 Geographic diversity of pretesting cases from traditional pretesting and web probing using Mechanical Turk. Source: Reproduced with permission from Murphy et al. 2014.

language processing (NLP) or machine learning techniques may, in the future, offer assistance in reducing qualitative data to more meaningful information (Card and Smith 2015). Researchers are also exploring the use of close-ended web probing methods (Scanlon, Chapter 17 in this book; Allum et al. 2017).

It is clear that methods and uses for online pretesting are still emerging. For example, Behr et al. (2013a) evaluated the optimal size for answer boxes for web probe responses. Kaczmirek et al.(2017) have used web probing methods to detect and reduce item-missing rates in open-ended survey questions. Meitinger et al. (2016) discuss how to administer multiple web probes in an order that will enhance quality. Mechanical Turk has also been used as an online pretesting method to evaluate order effects and test-retest reliability in a health survey (Fowler et al. 2016).

12.3.4 Summary and Conclusions

Emerging pretesting methods can be used to evaluate data quality aspects related to technological advancements that may be missed using traditional pretesting methods. In this chapter, we discussed usability testing, eye tracking, and online pretesting, and demonstrated how they could be used to address specific quality concerns in web and mobile surveys. Table 12.2 summarizes the primary advantages and limitations of these methods for evaluating data quality concerns in web and mobile surveys that may be missed using traditional pretesting methods only.

These emerging methods are not intended to replace traditional pretesting methodologies. Instead, we believe that multimethod pretesting approaches may work better because each method offers additional insights that might not be seen using another method. Figure 12.3 shows a diagram of how the different pretesting methods could potentially interact to identify potential issues with survey quality. The image on the left provides an example where each pretesting method identifies an equal share of problems with the survey. However, depending on the specific survey, one or two pretesting methods may offer more of an advantage compared to other methods in identifying the majority of issues as depicted in Figure 12.4 (right). The specific choice in pretesting methodologies will depend on the availability of participants, the stage of testing, the goals of pretesting, schedule and budget.

More research is needed to determine the optimal mix of methods for any given pretesting need. We expect that the emerging methods highlighted here will be especially important to consider when pretesting web and mobile surveys since they can best replicate the environment and setting in which the survey is being conducted.

Table 12.2 Advantages and limitations of emerging pretesting methods.

Approach	Advantages	Limitations
Usability testing	• Iterative process that guides design • Behavioral data (how participants use and interact with survey) • Performance measures (task accuracy, completion items) • Can be applied to mobile surveys	• Small sample sizes • Findings are mostly qualitative
Eye tracking	• Implicit data not affected by self-report • Can be applied to mobile surveys	• Requires specialized equipment to conduct • Does not work when participants' gaze and attention are not together
Online pretesting	• Geographic diversity • Larger samples • Obtain quantitative data • Quick to conduct and implement • Less expensive	• Shorter responses to probes • More missing responses • Potential for bias depending on how participants recruited • Less interactive • Tasks must be short

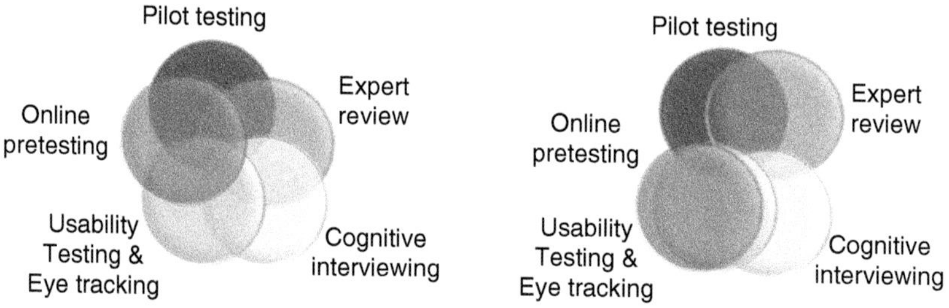

Figure 12.4 (Left) Each pretesting method identifies an equal share of potential problems. (Right) Some pretesting methods will identify the same types of problems.

References

Allum, N., Shapley, M., Curtis, J., and Pilley, S. (2017). Use of closed probes in a probability panel to validate cognitive pretesting. European Survey Research Association Conference, Lisbon, Portugal.

Antoun, A., Zhang, G., Conrad, F., and Schober, M. (2015). Comparisons of online recruitment strategies for convenience samples: Craiglist, Google Adwords, Facebook, and Amazon Mechanical Turk. *Field Methods* 28 (3) https://doi.org/10.1177/1525822X15603149.

Beatty, P., Fowler, F.J., and Cosenza, C. (2006). Do questionnaire design recommendations lead to measurable improvements? Some experiments with alternate versions of complex survey questions. *Proceedings of the Q2006 European Conference on Quality in Survey Statistics.*

Behr, D. (2016). Cross-cultural web probing and how it can enhance equivalence in cross-cultural studies. International Conference on Questionnaire Design, Development, Evaluation, and Testing (QDET2), Miami.

Behr, D., Kaczmirek, L., Bandilla, W., and Braun, M. (2012). Asking probing questions in web surveys: which factors have an impact on the quality of responses? *Social Science Computer Review* 32: 487–498.

Behr, D., Bandilla, W., Kaczmirek, L., and Braun, M. (2013a). Cognitive probes in web surveys: how the text box size can affect response quality. 66th Annual Conference of the World Association of Public Opinion Research (WAPOR), Boston.

Behr, D., Braun, M., Kaczmirek, L., and Bandilla, W. (2013b). Testing the validity of gender ideology items by implementing probing questions in web surveys. *Field Methods* 25 (2): 124–141. doi: 10.1177/1525822X12462525.

Behr, D., Meitinger, K., Braun, M., and Kaczmirek, L. (2017). Web probing – implementing probing techniques from cognitive interviewing in web surveys with the goal to assess the validity of survey questions. Mannheim, GESIS–Leibniz-Institute for the Social Sciences (GESIS–Survey Guidelines).

Behrend, T.S., Sharek, D.J., Meade, A.W., and Weibe, E.N. (2011). The viability of crowdsourcing for survey research. *Behavior Research Methods* 43: 800–813.

Blair, J. and Conrad, F. (2011). Sample size for cognitive interview pretesting. *Public Opinion Quarterly* 75: 636–658.

Bowling, A. (2005). Mode of questionnaire administration can have serious effects on data quality. *Journal of Public Health* 27 (3): 281–291.

Cannell, C.F. and Robinson, S. (1971). Analysis of individual questions. In: *Working Papers on Survey Research in Poverty Areas* (eds. J. Lansing, S. Withey and A. Wolf). Ann Arbor, MI: Institute for Social Research.

Cannell, C.F., Fowler, F.J., and Marquis, K. (1968). The influence of interviewer and respondent psychological and behavioral variables on the reporting in household interviews. In: *Vital Health and Statistics*, vol. 2 (26), 1–65. Washington, DC: U.S. Government Printing Office.

Cannell, C.F., Lawson, S.A., and Hausser, D.L. (1975). *A Technique for Evaluating Interviewer Performance.* Ann Arbor, MI: Survey Research Center, University of Michigan.

Card, D. and Smith, N.A. (2015). Automated coding of open-ended survey responses. https://www.ml.cmu.edu/research/dap-papers/DAP_Card.pdf.

Christenson, D.P. and Glick, D.M. (2013). Crowdsourcing panel studies and real-time experiments in MTurk. *The Political Methodologist* 20: 27–33.

Christian, L.M. (2003). The influence of visual layout on scalar questions in web surveys. Master's thesis. Washington State University. https://research.libraries.wsu.edu/xmlui/bitstream/handle/2376/134/l_christian_120803.pdf?sequence=1&isAllowed=y.

Christian, L.M., Dillman, D.A., and Smyth, J.D. (2005). Instructing Web and Telephone Respondents to Report Date Answers in a Format Desired by the Surveyor. Technical Report #05-067. Social & Economic Sciences Research Center Pullman, Washington State University. Retrieved on http://survey.sesrc.wsu.edu/dillman/papers.htm

Christian, L.M., Parsons, N.L., and Dillman, D.A. (2009). Designing scalar questions for web surveys. *Sociological Methods and Research* 37: 393–425.

Chunara, R., Chhaya, V., Bane, S. et al. (2012). Online reporting for malaria surveillance using micro-monetary incentives, in Urban India 2010–2011. *Malaria Journal* 11: 43.

Collins, D. (2015). *Cognitive Interviewing in Practice.* Los Angeles, CA: Sage.

Converse, J.M. and Presser, S. (1986). *Survey Questions: Handcrafting the Standardized Survey Questionnaire.* Newbury Park, CA: SAGE.

Cook, S.L., Sha, M., Murphy, J., and Lau, C.Q. (2016). *Technology options for engaging respondents in self-administered questionnaires and remote interviewing.* RTI Press Publication No. OP-0026-1603. Research Triangle Park, NC: RTI Press. doi: https://doi.org/10.3768/rtipress.2016.op.0026.1603.

Couper, M. (2000). Usability evaluation of computer-assisted survey instruments. *Social Science Computer Review* 18 (4): 384–396.

Couper, M.P. (2008). *Designing Effective Web Surveys.* New York: Cambridge University Press.

Dean, E., Caspar, R., McAvinchey, G. et al. (2007). Developing a low-cost technique for parallel cross-cultural instrument development: the question appraisal system (QAS-04). *International Journal of Social Research Methodology* 10 (3): 227–241.

Dillman, D., Smyth, J., and Christian, L. (2009). *Internet, Mail, and Mixed-Mode Surveys: The Tailored Design Method.* New York: Wiley.

Edgar, J., Murphy, J.J., and Keating, M.D. (2016). Comparing traditional and crowdsourcing methods for pretesting survey questions. *SAGE Open*: 1–13. July–September 2016.

Edwards, B., Schneider, S., and Brick, P. (2008). Visual elements of questionnaire design: experiments with a CATI establishment survey. In: *Advances in Telephone Survey Methodology* (eds. Lepkowski, M., Clyde Tucker J., Michael Brick Edith D., de Leeuw Lilli Japec, Paul J. Lavrakas, Michael W. Link, Roberta L. Sangster). Hoboken, NJ: Wiley. https://onlinelibrary.wiley.com/doi/book/10.1002/9780470173404

Esposito, J.L. and Rothgeb, J. (1997). Evaluating survey data: making the transition from pretesting to quality assessment. In: *Survey Measurement and Process Quality* (eds. L. Lyberg, P. Biemer, M. Collins, et al.), 541–571. New York: Wiley.

Esposito, J.L., Campanelli, P.C., Rothgeb, J.M., and Polivka, A.E. (1991). Determining which questions are best: methodologies for evaluating survey questions. In: *The American Statistical Association's Proceedings of the Section of Survey Methods Research*. Alexandria, VA: American Statistical Association, 46–55.

Everdell, I. (2014). Web content. In: *Eye Tracking in User Experience Design* (eds. R. Bergstrom and A. Schall). Waltham, MA: Morgan Kaufmann.

Fowler, S., Willis, G., Moser, R., et al. (2016). Use of Amazon MTurk online marketplace for questionnaire testing and experimental analysis of survey features. Proceedings of the Federal Committee on Statistical Methodology.

Geisen, E. and Cook, S. (2011). Model for incorporating usability testing as a survey pretesting method. European Survey Research Association conference, Lausanne, Switzerland.

Geisen, E. and Romano Bergstrom, J. (2017). *Usability Testing for Survey Research*. Hoboken, NJ: Morgan Kaufmann.

Groves, R. (1989). *Survey Errors and Survey Costs*. New York: Wiley.

Hansen, S.E., Fuchs, M., and Couper, M.P. (1997). CAI instrument and system usability testing. Annual conference of the American Association for Public Opinion Research, Norfolk, VA.

Head, B., Dean, L., Flanigan, T. et al. (2015). Advertising for cognitive interviews: a comparison of Facebook, Craigslist, and snowball recruiting. *Social Science Computer Review doi* https://doi.org/10.1177/0894439315578240.

Holzberg, J., Morales, G., Fobia, A., and Hunter Childs, J. (2016). A comparison of cognitive testing methods and sources: In-person versus online nonprobability and probability methods. International Conference on Questionnaire, Design, Evaluation and Testing, Miami, FL.

Jarrett, C. (2010). Avoid being embarrassed by your error messages. UXmatters. https://www.uxmatters.com/mt/archives/2010/08/avoid-being-embarrassed-by-your-error-messages.php.

Jarrett, C. and Gaffney, G. (2008). *Forms that Work: Designing Web Forms for Usability*. Amsterdam: Elsevier.

Jarrett, C. and Romano, B. (2014). Forms and surveys. In: *Eye Tracking in User Experience Design* (eds. R. Bergstrom and A. Schall). Waltham, MA: Morgan Kaufmann, 111–137.

Jenkins, C. and Dillman, D. (1995). Towards a theory of self-administered questionnaire design. In: *Survey Measurement and Process Quality* (eds. Lyberg L., Biemer P., Collins M., et al.). New York: Wiley-Interscience, 165–196.

Kaczmirek, L., Meitinger, K., and Behr, D. (2017). Higher data quality in web probing with EvalAnswer: a tool for identifying and reducing nonresponse in open-ended questions. GESIS Papers 2017/01. Köln: GESIS – Leibniz-Institut für Sozialwissenschaften.

Lenzner, T. and Neuert, C. (2017). Pretesting survey questions via web probing – does it produce similar results to face-to-face cognitive interviewing? *Survey Practice* 10 (4): 1–11.

Lugtig, P., Toepoel, V., and Amin, A. (2016). Mobile-only web survey respondents. *Survey Practice* 9 (4): 1–8.

Marder, J. (2015). The Internet's hidden science factory. http://www.pbs.org/newshour/updates/inside-amazons-hidden-science-factory.

Marquis, K., Nichols, E., and Tedesco, H. (1998). Human-computer interface usability in a survey organization: Getting started at the census bureau. *Proceedings of the Survey Research Methods Section*, American Statistical Association.

Meitinger, K. (2017). Necessary but insufficient: why measurement invariance tests need online probing as a complementary tool. *Public Opinion Quarterly* 81 (2): 447–472.

Meitinger, K. and Behr, D. (2016). Comparing cognitive interviewing and online probing: do they find similar results? *Field Methods* https://doi.org/10.1177/1525822X15625866.

Meitinger, K., Braun, M., and Behr, D. (2016). Sequence matters: The impact of the order of probes on response quality, motivation of respondents, and answer content. Second International Conference on Survey Methods in Multinational, Multiregional and Multicultural Contexts (3MC 2016), Chicago.

Meitinger, K., Kaczmirek, L., and Braun, M. (2017). Finding the optimal sample size for online probing: How many respondents are enough? European Survey Research Association conference, Lisbon, Portugal.

Miller, K. (2014). *Cognitive Interviewing Methodology*. Hoboken, NJ: Wiley.

Mockovak, B. and Kaplan, R. (2016). Comparing face-to-face cognitive interviewing to unmoderated online cognitive interviewing with concurrent and retrospective probing International Conference on Questionnaire, Design, Evaluation and Testing, Miami, FL.

Morton-Williams, J. and Sykes, W. (1984). The use of interaction coding and follow-up interviews to investigate comprehension of survey questions. *Journal of the Market Research Society* 26: 109–127.

Murphy, J.J., Keating, M.D., and Edgar, J. (2014). Crowdsourcing in the cognitive interviewing process. In: *Proceedings of the 2013 Federal Committee on Statistical Methodology Research Conference*, vol. H-1, 1–11.

Murphy, J., Mayclin, D., Richards, A., and Roe, D. (2015). A multi-method approach to survey pretesting. Federal Committee on Statistical Methodology Research Conference.

Nichols, B. (2016). Cognitive probing methods in usability testing – pros and cons. American Association for Public Opinion Research, Austin, TX.

Olmsted-Hawala, E., Holland, T., and Quach, V. (2014). Usability testing. In: *Eye Tracking in User Experience Design* (eds. R. Bergstrom and A. Schall). Waltham, MA: Morgan Kaufmann, 49–80.

Olson, K. (2010). An examination of questionnaire evaluation by expert reviewers. *Field Methods* 22 (4): 295–318.

Paškvan, M. and Plate, M. (2017). Online probing of the LFS questionnaire. European Survey Research Association Conference, Lisbon, Portugal.

Penzo, M. (2006). Label placement in forms. UXmatters. http://www.uxmatters.com/mt/archives/2006/07/label-placement-in-forms.php.

Poushter, J. (2016). Smartphone ownership and internet usage continues to climb in emerging economies. Pew Research Center. http://www.pewglobal.org/2016/02/22/smartphone-ownership-and-internet-usage-continues-to-climb-in-emerging-economies.

Presser, S. and Blair, J. (1994). Survey pretesting: do different methods produce different results? *Sociological Methodology* 24: 73–104.

Presser, S., Couper, M., Lessler, J. et al. (2004). Methods for testing and evaluating survey questions. *Public Opinion Quarterly* 66 (1): 109–130.

Quesenbery, W. (2001). What does usability mean: Looking beyond 'ease of use.' Proceedings of the 48th Annual Conference, Society for Technical Communication. http://www.wqusability.com/articles/more-than-ease-of-use.html.

Romano Bergstrom, J.C. (2016). Empirical evidence for the value of usability testing surveys. International Conference on Questionnaire Design, Development, Evaluation, and Testing (QDET2), Miami, FL.

Romano Bergstrom, J.C. and Schall, A. (2014). *Eye Tracking in User Experience Design*. Waltham, MA: Morgan Kaufmann.

Romano, J.C. and Chen, J.M. (2011). A usability and eye-tracking evaluation of four versions of the online National Survey for College Graduates (NSCG): Iteration 2. Statistical Research Division (Study Series SSM2011-01). U.S. Census Bureau. https://www.census.gov/srd/papers/pdf/ssm2011-01.pdf.

Sage, A. (2014). The Facebook platform and the future of social research. In: *Social Media, Sociality, and Survey Research* (eds. C. Hill, E. Dean and J. Murphy). Hoboken, NJ: Wiley, 87–106.

Saunders, T. (2015). Improving the survey experience for mobile respondents. http://www.marketingresearch.org/article/improving-survey-experience-mobile-respondents.

Schuman, H. (1966). The random probe: a technique for evaluating the validity of closed questions. *American Sociological Review* 31 (2): 218–222.

Slavec, A. (2017). Evaluating survey questions with different levels of comprehension difficulty using crowdsourced online probes. European Survey Research Association Conference, Lisbon, Portugal.

Tourangeau, R. (1984). Cognitive science and survey methods. In: *Cognitive Aspects of Survey Design: Building a Bridge Between Disciplines* (eds. T. Jabine,

M. Straf, J. Tanur and R. Tourangeau), 73–1000. Washington, DC: National Academy Press.

Tourangeau, R., Couper, M., and Conrad, F. (2004). Spacing, position, and order interpretive heuristics for visual features of survey questions. *Public Opinion Quarterly* 68 (3): 368–393.

Tourangeau, R., Couper, M., and Conrad, F. (2007). Color, labels, and interpretive heuristics for response scales. *Public Opinion Quarterly* 71: 91–112.

Willis, G.B. (2005). *Cognitive Interviewing. A Tool for Improving Questionnaire Design*. Thousand Oaks, CA: Sage Publications, Inc.

Willis, G.B. (2015). *Analysis of the Cognitive Interview in Questionnaire Design*. Oxford: Oxford University Press.

Willis, G.B. and Lessler, J.T. (1999). *Question appraisal system BRFSS-QAS: A guide for systematically evaluating survey question wording*. Report Prepared for CDC/NCCDPHP/Division of Adult and Community Health Behavioral Surveillance Branch). Rockville, MD: Research Triangle Institute.

Willis, G. and Schechter, S. (1997). Evaluation of cognitive interviewing techniques: do the results generalize to the field? *Bulletin de Methodologie Sociologique* 55: 40–66.

Willis, G.B., Schechter, S., and Whitaker, K. (1999). A comparison of cognitive interviewing, expert review, and behavior coding: What do they tell us? *Proceedings of Survey Research Methods Section of the American Statistical Association*. Washington, DC: American Statistical Association.

13

Usability Testing Online Questionnaires: Experiences at the U.S. Census Bureau

Elizabeth Nichols[1], *Erica Olmsted-Hawala*[1], *Temika Holland*[2], *and Amy Anderson Riemer*[2]

[1] *Center for Behavioral Science Methods, US Census Bureau, Washington, DC, USA*
[2] *Data Collection Methodology & Research Branch, Economic Statistical Methods Division, US Census Bureau, Washington, DC, USA*

13.1 Introduction

As more questionnaires and forms move online, pretesting methodology has evolved to address the effectiveness of the survey question and response options along with the associated design elements (e.g. navigation buttons). While the use of web surveys can result in lower costs and quicker data collection compared to traditional methods (Couper 2000), there are challenges. Web surveys are visual. Respondents often extract meaning and seek guidance not only from the question wording, but from how questions and response options are presented (Dillman et al. 2014; Christian et al. 2009; Couper 2008; Peytchev et al. 2006). If an instrument is poorly worded or designed, or has technical issues, inaccurate reporting can occur (Couper 2000). Therefore, a survey instrument must communicate effectively to the respondent to obtain quality response data.

Many organizations employ *usability testing* to pretest their automated surveys to make sure they are easy to understand and complete. According to the International Organization for Standardization (ISO), usability testing evaluates how a user interacts with an interface focusing on measures of efficiency, effectiveness, and user satisfaction while accomplishing tasks (ISO Standard 9241-11: 1998). Such tasks could be buying an airline ticket online, making sure air traffic control panels are optimal, and the like. When usability testing is applied to online surveys, the task is the user's completion of the survey itself. Government organizations, market researchers, and forms designers have engaged in usability testing to ensure the effectiveness of their online survey instruments, forms, and websites (Couper 2008; Dillman et al. 2005; Fox

Advances in Questionnaire Design, Development, Evaluation and Testing, First Edition.
Edited by Paul C. Beatty, Debbie Collins, Lyn Kaye, Jose-Luis Padilla, Gordon B. Willis, and Amanda Wilmot.

2001; Jarrett and Gaffney 2008; Health and Human Services 2006; Penzo 2006; Potaka 2008; Wroblewski 2008). Courses and books on how to incorporate usability into survey design have also emerged in recent years, highlighting the benefits of testing procedures (Geisen and Romano Bergstrom 2017; Geisen and Jarrett 2011). In practice, usability testing typically involves one-on-one in-person sessions with a test participant and a user-experience researcher. Survey organizations, like the US Census Bureau, often take advantage of these one-on-one sessions to expand the usability research goals to include other kinds of assessments, such as cognitively testing questions or obtaining feedback from respondents on their likely motivation to respond or their beliefs about the confidentiality of their responses (Romano Bergstrom et al. 2013).

As mentioned earlier, when testing online or automated surveys for usability, the participant's task is to complete the survey while navigating a graphical user interface. The three ISO measures of usability testing (i.e. efficiency, effectiveness, user satisfaction) may be ideal for measuring the usability of websites (Hornbæk 2006); they are not always directly applicable to testing surveys. For example, are people really "satisfied" when they have completed a mandatory government survey? And if they are "satisfied," what does that really mean? This chapter shares our experience of usability testing self-administered surveys and censuses at the US Census Bureau. We focus on both the techniques that offered insights into measuring and improving usability, and those that did not. In this chapter, we begin with a brief history of how usability testing started at the Census Bureau followed by a summary of current best practices, before discussing them in more detail. While the practices described in this chapter are not mandated by the Census Bureau as official usability testing methodology per se, they represent the current state of practice in use at the Census Bureau for both business and household surveys. Currently the Census Bureau has two separate usability staff groups, one focused on household surveys and one focused on business and government surveys, hereafter collectively referred to as the Census usability staff.

13.2 History of Usability Testing Self-Administered Surveys at the US Census Bureau

Usability testing of self-administered surveys at the Census Bureau began in the late 1990s with the advent of the first online business surveys (Sweet and Russell 1996). Testing of household surveys and censuses began in earnest in the mid-2000s as internet penetration rates rose in residences and made online household surveys feasible (for a sample of earlier usability reports, see Norman and Murphy 2004; Sweet et al. 1997; Bates and Sweet 1997; Marquis et al. 1998). By 2003, the US Census Bureau pretesting policy was updated to include usability testing – and emphasized the importance of testing all electronic survey applications before public release (US Bureau of the Census 2003). Now, over 20 years after the first usability test of an online survey, nearly 60 business surveys and censuses, and public sector surveys and censuses, offer online reporting options. For the first time, the decennial census will also offer an online reporting option in 2020 (US Census Bureau 2016) with usability testing prior to fielding.

When Census staff first began conducting usability testing, they consulted user experience (UX) pioneers such as Jakob Nielsen, Ginny Redish, and other UX designers (Dumas and Redish 1999; Nielsen 1993a; Redish 2010, 2007). Interestingly, the methods of these UX pioneers included think-aloud protocols, which were very similar to cognitive testing methodology (US Bureau of the Census 1998; DeMaio et al. 1993). During the last 20 years, usability-testing methodology at the Census Bureau has evolved as new technologies have become available to both researchers and respondents. For example, the growth of mobile device use has required researchers to develop new tools and methods to capture data on respondents' use of these devices that are distinct from the tools and methods used to capture data on respondents' use of desktops/laptops. Watching the mobile device screen as people complete their survey is critical to identifying solutions to the challenges these devices pose to survey designers, such as longer completion times, higher break-off rates, and missing data rates (Nichols et al. 2015; de Bruijne and Wijnant 2013; Callegaro 2010; Horwitz 2015; McGeeney 2015). As these testing experiences accumulate and technology evolves, an array of different techniques beyond think-aloud protocols are incorporated to measure usability and are shared in this chapter.

13.3 Current Usability Practices at the Census Bureau

The practice of usability testing at the Census Bureau differs slightly between our household and business surveys. However, for both survey types, it is important that the survey owner/sponsor and the usability team first agree on objectives and methods before testing starts. While the sponsor typically sets the general objectives, the usability team recommends the methodology.

For household surveys, in-person usability sessions with a participant and at least one researcher are recommended as the most effective way to evaluate a product and obtain rich information about the user experience (Olmsted and Gill 2005). Sessions are conducted at the Census Bureau's usability laboratory or at public locations such as libraries. Participants either bring in their device to use in the study or are provided a government-owned device. Bringing their own device allows for a diversity of browsers and operating systems and can highlight places where the user interface needs additional work.

Participants are representative of the typical respondents of the survey. They are *not* Census Bureau employees or people familiar with the survey or survey methodology. The number of participants and rounds of testing depend on the schedule, budget, and the questionnaire development stage.

A protocol guides the usability session. This typically includes having the participant complete the survey with their actual (versus hypothetical) information as they would at home, while thinking aloud. At the beginning of each session, researchers provide the participant with materials that mimic what the respondent would receive in real life – whether by mail, email, or advertising – and observe what they do: how they get to the login page and begin their survey. Audio, screen, and eye-tracking data are recorded. Little to no concurrent probing is used during the survey-completion portion of the session. Post-survey data collected include responses to a satisfaction questionnaire, vignettes, and comments made during retrospective debriefing. Sessions last between 1 and 1.5 hours. When possible, and when funding allows, having more than one researcher present during each usability session is beneficial because the type and amount of usability problems identified increases with multiple researchers (Molich et al. 2007).

For surveys offered in multiple languages, whether the survey translations are developed simultaneously or sequentially, it is most efficient to develop one protocol, in our case in English, and then translate that protocol into other languages, rather than developing separate protocols for each language (see Wang et al. 2017; Meyers et al. 2017). When separate protocols are developed, there are bound to be gaps, making comparison of usability-testing results across languages difficult. In our experience, non-English usability sessions also take longer than English sessions, so it is important for researchers to agree a priori on what to eliminate if sessions are exceeding the promised time limit.

Shortly after test completion, a PowerPoint presentation summarizing the main usability findings and recommendations is prepared and delivered to the sponsor. Using the presentation as a guide, researchers, developers, and sponsors meet to discuss the results. If the schedule permits, changes are made to the instrument and another round of testing is conducted to confirm that the changes improve the usability of the survey. A final written report is prepared after all testing is complete. The report contains a detailed description of the methods and problems, including screen shots highlighting the usability issues, recommendations for improvement, and data associated with eye-tracking results, vignettes, and satisfaction scores.

Usability testing of business surveys is much like household surveys with a few exceptions. Like household surveys, the sessions are usually in-person, but they often involve a participant who has been an actual survey respondent. These sessions involve travel to the business location because business surveys usually rely on the respondent accessing company records from their work PC to obtain the data needed to complete the survey. They complete the testing during their workday; so bringing them into a laboratory would be a deterrent to participating. Observing the interaction between respondents, their records, and the data collection instrument has provided valuable insight into the effort needed to respond to business surveys. When record look-up is not possible or would take too much time or effort during the usability test, business respondents are encouraged to make reasonable estimates and provide information about the record look-up process. Subject matter experts often accompany researchers when visiting businesses in order to assist with answering specific questions respondents have about the survey.

Because travel to employment locations is resource-intensive, remote testing was introduced using an online meeting platform (e.g. WebEx) for usability testing of non-confidential public sector surveys, starting with the State Government Research & Development Survey (Stettler 2013). This method was successful in that it gathered information using audio and video screen capture from dispersed respondents across the United States who normally would not be able to participate in the usability study. It also met a shortened timeframe required by the sponsor, and reduced costs as researchers did not have to travel to test participants. However, drawbacks included that the system setup was not completely intuitive for participants or researchers; the ability to collect certain observational data, such as facial expressions and body language that can indicate problems or confusion, was not available in the WebEx system at the time, nor was it possible to observe respondents accessing records.

Probing techniques are also slightly different when evaluating business surveys. Concurrent probing is often used as the participant moves throughout the screens. Because of the complexity of the survey and the fact that it requires data input from sources like spreadsheets, it is sometimes beneficial to probe concurrently while particular information is visible rather than spending time backtracking within the instrument after the completion of the survey. Occasionally, the researcher will ask the participant to perform a specific task on the screen in order to evaluate different features of the online survey, such as accessing help information. Another typical concurrent probe is to gather information about the edit messages. All business surveys contain edit checks to ensure that data entered are valid and meet certain criteria. It is important to evaluate the edit message wording as well as the layout of the messaging onscreen to ensure that participants can locate the invalid data and understand how to fix it. If participants do not trigger an edit message as they enter data, researchers will direct them to enter values that would trigger a message in order to have participants react to and discuss what they see on the screen.

Currently, usability testing with business surveys typically involves only one round of testing because of the expense of traveling to the business locations. In addition, the testing is usually conducted on business-owned PCs. Eye tracking is not usually used in testing primarily because the software would need to be installed on the businesses' PC. However, in one case, eye tracking using census-owned devices was used to evaluate revised language messaging on legally required informed consent statements for a monthly business survey (Olmsted-Hawala et al. 2016).

In the rest of this chapter we discuss in more detail the features of usability-testing practice at the Census Bureau. In particular, we discuss the participants to recruit for usability testing, how often to conduct such testing, techniques for assessing the three measures of usability – accuracy, efficiency, and satisfaction – participant debriefing strategies, communicating results with the development team, and assessing whether changes made based on usability findings improve the usability of the final questionnaire.

13.4 Participants: "Real Users, Not User Stories"

Crucial to usability testing is having a real (or potential) respondent of the survey complete the questionnaire during the usability evaluation. This includes answering the survey questions as they pertain to their real lives, while the researcher(s) observe. This is the fundamental basis for any usability test of a survey. Usability testing is not user acceptance testing (UAT), which is a type of testing used to make sure the system meets specified system requirements without any major errors (Hambling and van Goethem 2013). However, the "user story" terminology associated with UAT can be confused with "usability testing." In UAT testing, people close to the development of the instrument or application make sure all the paths work as specified using *user stories*. These stories do not necessarily cover all situations that a real respondent encounters. The testers generally already know how the paths should go because they helped develop them. In contrast, usability testing lets a typical respondent interact with the survey. The value in bringing in participants with no prior relationship to the product is that their actions are not predictable. Because they are unfamiliar with the instrument their interactions are more representative of the typical user. This is shown by the following example.

For the 2016 Census Test online survey, which is a household survey conducted in preparation for the 2020 Census, a UAT user story specified that users would receive a thank-you page after submitting their survey. From there, users could navigate away from the survey. UAT was successful for this system specification user story as demonstrated by the thank-you page in Figure 13.1. However, in usability testing, one participant submitted her data and received the thank-you page, but instead of understanding that she was done with the survey and should either navigate to a different page or close the browser, selected Save and Logout on the thank-you page (the arrow in the upper-right corner of Figure 13.1 points to Save and Logout).

Selecting that option brought her back to the login screen (Figure 13.2) where she then attempted to log back in to the survey, but the system would not allow her back in because she had already submitted her data. The experience

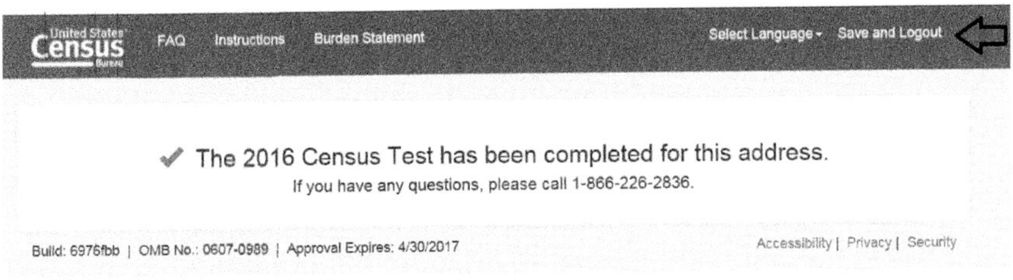

Figure 13.1 PC version of the web page the participant received after submitting the survey. Notice the Save and Logout link in the upper-right corner implying the survey is not done. Source: 2016 Census Test usability testing.

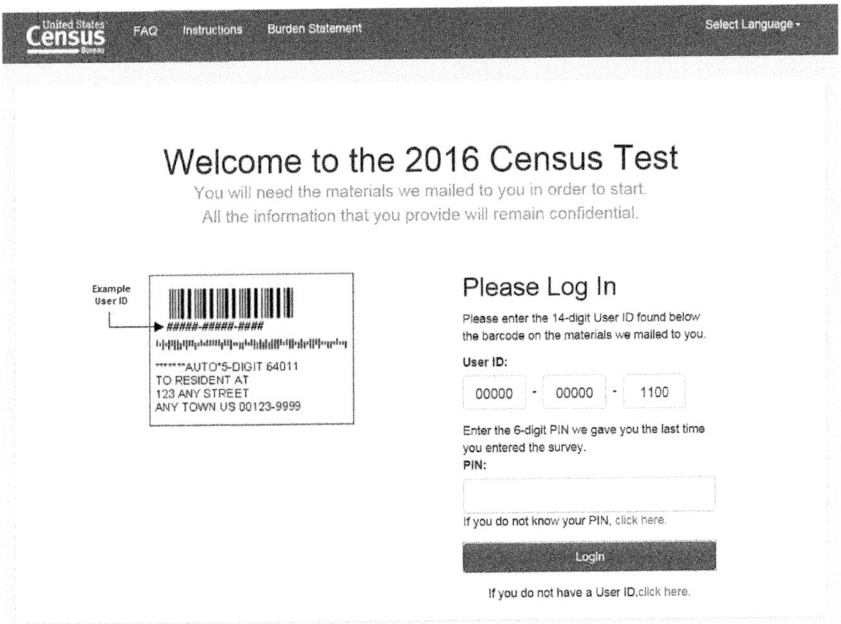

Figure 13.2 PC version of the login web page for the 2016 Census Test. Source: 2016 Census Test usability testing screen as it was redesigned after usability recommendations were implemented.

was confusing for this participant, and the recommendation from the Census usability staff was to remove the Save and Logout option on the thank-you page. An in-house user-story tester would not have made the unexpected action of clicking on Save and Logout because he or she would have already known that the survey was complete. Only real users uncovered the confusion that arose from having the words Save and Logout on the thank-you page.

For household and business surveys, Census staff typically conduct usability testing with 10–20 participants in each round. While the number is well over the five participants suggested by usability researchers such as Virzi (1992) and Nielsen and Landauer (1993), it is similar to the number used in cognitive testing in the United States and is more in line with what Faulkner (2003) suggests when she investigated the "five-user assumption" for usability testing.

The number of participants also depends on the number of user groups. A *user group* or *quota* is a set of respondents with particular characteristics (Collins and Gray 2014). For example, if an education survey has different paths depending upon the education level of the respondent, then recruiting people with different levels of education is important to ensure all questions are tested. If a new question or a new response category is being developed and tested for the first time, one user group would include people who should choose the new response category, and another user group would include people who should not. If the survey will be accessed on mobile devices and PCs, then user groups could be defined by device use. Similarly, business survey respondents are recruited based on their industry classification, size, or other business characteristics. Regardless of the user groups, recruiting some older individuals is also important as there are age-related differences in task completion of automated instruments and other web-based tasks (Olmsted-Hawala and Holland 2015; Tullis 2007). To keep participant numbers low, one participant can fulfill more than one user group.

The development life-cycle stage and the schedule also affects the number of participants. If the instrument is not fully functional, Census usability staff begin testing with what is available first, such as the PC version or a particular navigational path. As more paths become active, they are incorporated into the testing. A reduced or accelerated schedule also affects the number of participants to include. Finally, testing can stop earlier than planned when researchers feel they have reached saturation with the usability problems identified.

Recruiting for usability testing begins two weeks ahead of the first session. For household surveys, local ads are placed online in venues such as Craigslist or listservs or by using paper fliers placed in community centers. The ads specify the characteristics of interest for a given survey. An effective screening question is to ask people who call why they think they qualify for the study (Chisnell 2011). "No shows" are less of a problem with reminder calls a day in advance and clear emailed instructions on where and who to meet for the session. Contact information and general demographics for these participants are added to an internal database, which our recruiter can access for other cognitive or usability tests. Limiting the number of usability sessions that the same person can participate in during a calendar year reduces the "experienced test user" problem that sometimes plagues qualitative research. Because business surveys are often completed year after year by the same individuals, calls are made to the last known respondent of the survey being tested. Occasionally, a general business number may be used to recruit staff in particular offices such as Payroll or Financial Accounting who are typically responsible for completing the survey request. Scheduling testing to end a few days earlier than expected allows the recruiter time to schedule any backup participants as needed.

13.5 Building Usability Testing into the Development Life Cycle

Incorporating usability testing throughout the software development life cycle of a product allows for usability feedback and recommendations to be incorporated into the design (Göransson et al. 2003). Multiple rounds of testing (i.e. iterative rounds) are added to the schedule based on discussions between the usability staff and the development team. Issues identified in the first round of testing can be addressed and the changes tested in subsequent rounds (Nielsen 1993b). Iterative testing can continue until the questionnaire is deployed (Medlock et al. 2005). While very little empirical research to date has proven the benefits of iterative testing (Romano Bergstrom et al. 2011), there has been overwhelming anecdotal evidence and case studies (Sy 2007; Karat 1989), including our own experience at the Census Bureau, as shown by the following example, that suggest this approach is beneficial.

In the first round of testing the mobile-optimized version of the 2015 Census Test, a household survey created in preparation for the 2020 Census, we observed participants selecting incorrect choices to the question on relationship among household members because the radio buttons were too close together (see Figure 13.3). We also observed several participants touching the underline in that question (see "James C Doe is Jane Doe's ____" in Figure 13.3).

Participants appeared to think that the underline was a text box where they could type or that it would bring up a list of choices, as they commented that they would fill in their answer on that line. We recommended that the developers add more space between the response options and change the underline to an ellipsis. Changes were made, as shown in Figure 13.4, and in the next round of testing, we did not observe anyone accidently tapping on wrong radio buttons or on the ellipsis.

In this example, iterative usability testing successfully contributed to a better design of the mobile screens prior to production, and we were able to confirm that there were no other problems with the comprehension of the response options offered for test participants. Had we not corrected the errors between iterations, we would not have been able to find other possible problems as our attention and the participant's attention would have stayed focused on the original usability issue.

Iterative testing can be built into the schedule no matter the software development model, e.g. waterfall or agile. In the *waterfall* model, the survey is

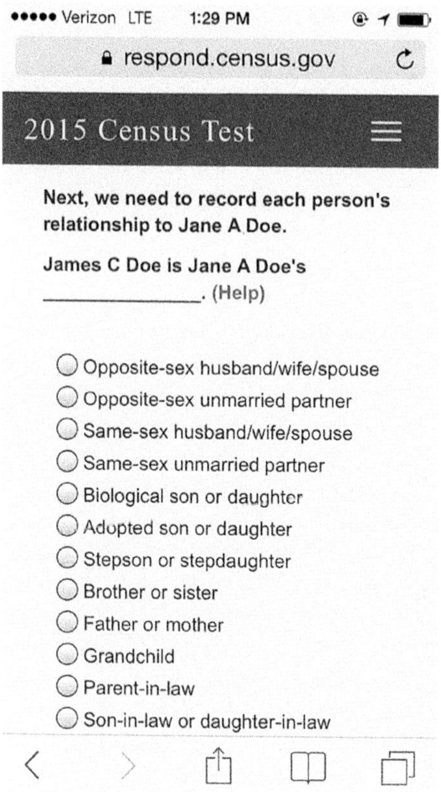

Figure 13.3 Mobile version of the 2015 Census Test relationship question during Round 1 of usability testing. Notice the underline implying the respondent needs to fill in the blank and the lack of space between the radio buttons. Source: 2015 Census Test usability testing, Round 1.

nearly complete before any testing is conducted, meaning that usability testing is typically conducted toward the end of the life cycle (for information on how usability testing fits into the waterfall method, see Costabile (2001)). In the *agile* model, usability testing can occur throughout the schedule because the programmers develop the interface in segments, called *sprints*. Usability testing can occur after each sprint, or after an accumulation of sprints. Although some researchers have found agile development to allow incorporation of user-centered design strategies to be more effective (see for example, Sy 2007), others have found that the compressed timing in an agile environment can leave little room for usability testing (Deaven et al. 2016).

With either method, in theory there is much a usability team can do prior to any code being written, such as showing participants PowerPoint mockups, or screens developed with other tools that allow different screen layouts and graphical features to be compared (Virzi et al. 1996). For example, in a recent follow-up household survey developed for a small screen (e.g. mobile phone), a designer on the team was able to create screen mockups that were fairly rudimentary without any working code, but which allowed the team

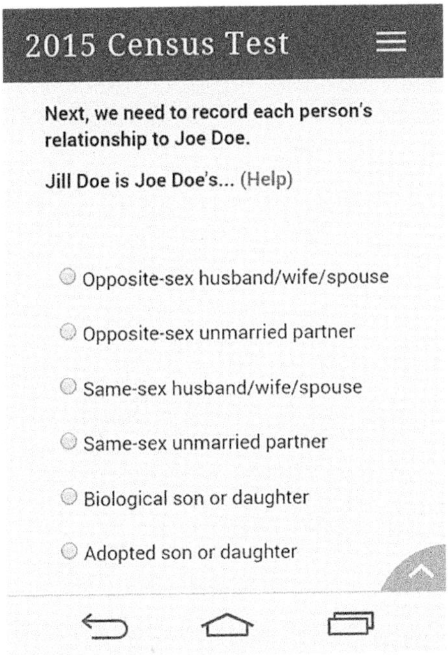

Figure 13.4 Mobile version of the 2015 Census Test relationship question during Round 2 of usability testing. Notice the underline changed to an ellipsis, and more space was added between the radio buttons. Source: 2015 Census Test usability testing, Round 2.

to see which of three different graphical designs users preferred and better understood.

With both software development models, Census usability staff have experienced the situation in which there is only enough time for a single round of testing, and sometimes there is not even enough time to make changes to the product before release. In this situation, iterative testing can occur across survey releases, as shown in the next example.

In the 2015 National Content Test (an updated version of the 2015 Census Test), we observed several participants selecting the Click Here link to log in to the survey after they had typed in their User ID, as shown by the arrow in Figure 13.5. This was a suboptimal path. They should have selected the Login button, which was under the User ID label. We speculated that as the participant typed the User ID and got to the third data-entry box, the Click Here link was almost directly under that box, within the participant's visual field of view. Additionally, the Login button itself was small and set to the left side of the screen, which could have contributed to it being less visible. The usability finding was communicated to the developer, but there was no time to make any changes to that survey. In the next iteration of the test (called the 2016 Census Test), the developers changed the design of the Login button to solve the usability problem by making it more prominent. The Login button spanned the entire User ID field, as shown in Figure 13.6. During usability testing all participants correctly selected Login, and thus the usability problem was rectified.

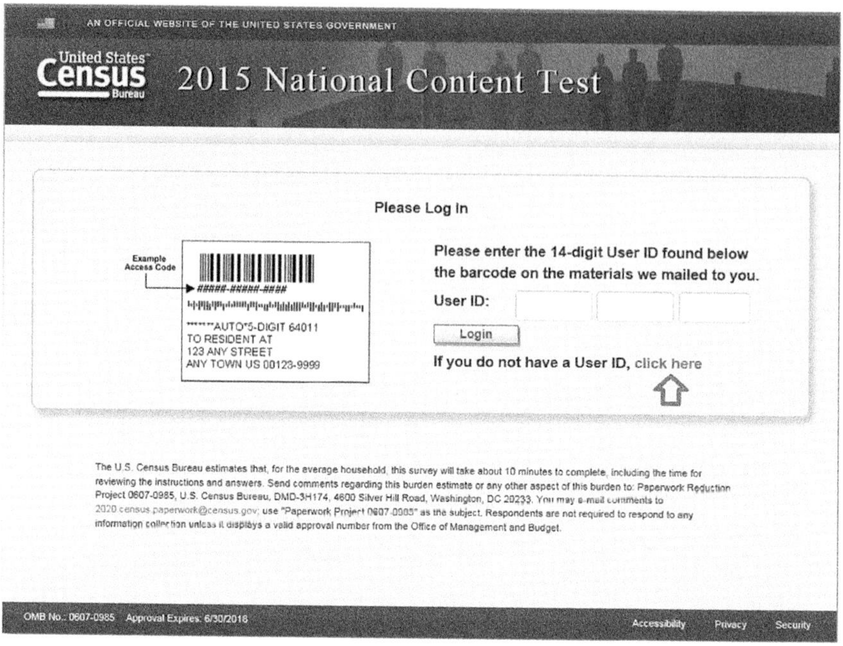

Figure 13.5 PC version of the login page for the 2015 National Content Test. Notice the small Login button below label for the User ID field and the Click Here link below the last input field for the User ID. Source: 2015 National Content Test usability testing.

Due to the complexities of business surveys, those surveys may not be fully functional or available for testing. Consequently, business surveys rely on iterative testing across survey releases and on low-fidelity or paper prototypes (Douglass and Hylton 2010). For example, in preparation for the 2017 Economic Census, Census usability staff involved respondents in identifying user needs in order to develop requirements that were then translated into low-fidelity prototypes. These prototypes were used during several phases of usability testing to help refine the visual layout of the interface and identify necessary functionality for the online survey (Beck et al. 2016). The prototypes used to test early versions of the 2017 Economic Census forms were initially drafted on paper, but later prototypes were programmed using HTML and resembled a functioning online survey. Respondents were able to provide more substantive feedback when using the HTML compared to the paper prototypes. We found that some respondents felt uncomfortable "pretending" with paper screen shots. This is consistent with the literature that suggests some components of the user-interaction design require more functionality than paper mockups provide (Snyder 2003).

Whether the software development is waterfall or agile, or whether the survey is for households or businesses, having usability team members present

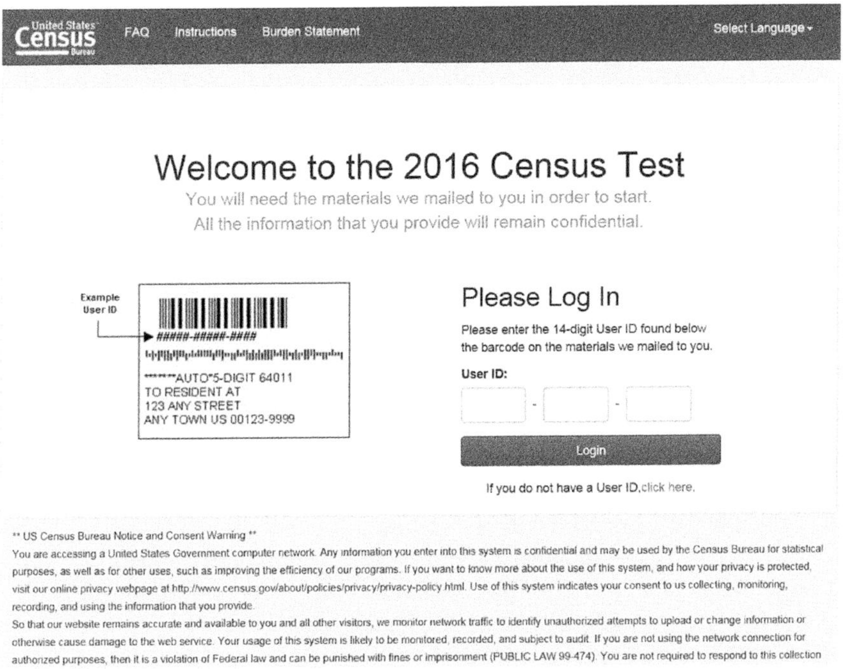

Figure 13.6 PC version of the login web page for the 2016 Census Test with the Login button that spans the length of the User ID field. Source: 2016 Census Test usability testing.

when design discussions are held and having usability testing built into the development schedule are crucial to a successful user-centered design.

13.6 Measuring Accuracy

Accuracy as a measure of effectiveness is one part of usability testing. To measure accuracy, one approach is to evaluate whether the participant completes the task successfully without error. When testing household surveys, the task Census usability staff instruct participants to accomplish is "Answer the survey questions as they apply to you in real life." However, most participants finish the survey during the session. Therefore, instead of using only survey completion as the measure, other effectiveness measures are used, such as spontaneous comments indicating that participants would exit the survey prematurely, whether they can start or resume the questionnaire, whether they submit their data, whether they successfully resolve any edit messages they receive, and, when possible, whether they answer the questions accurately. For the more labor-intensive business surveys, Census usability staff give participants discrete tasks to complete, and ask them to use past response data to complete a portion of the survey.

Sometimes, the participant cannot complete the task because of fatal flaws in the programming. These flaws, such as infinite loops, missing questions, inability to submit data, illogical skip sequences, and inability to access and complete surveys using particular devices or operating systems, could result in task failure or, in survey terms, *non-response*. Other non-fatal survey errors such as duplicate questions, incorrect question fills, misspellings, dead links, and other types of UAT errors are also found and can be corrected during usability testing.

The following three sections discuss the main techniques Census usability staff use to measure accuracy during usability testing of surveys.

13.6.1 Observation Methods

Observation is one of the primary techniques used to evaluate accuracy, which is why screen recording is valuable during testing. Instead of hovering over the participant as they complete the survey, it is more effective to transmit the screen recording to another device, which the researcher can watch in real time. Observable usability problems can be as simple as participants having difficulty finding the survey online. For example, we have observed participants making keying errors when typing in a long URL or conducting an online search of the survey name with their favorite search engine (e.g. "a Google search") and not finding the correct survey-landing page. Both of these would have been difficult to observe without a screen recording. Other observations, such as the format of the login credentials on the survey notification not matching the format on the interface, do not necessarily need a screen recording to capture the usability problem.

One of the more difficult practices during observation is to be patient with the participant's actions and behaviors. An instinct the researcher might have is to fill in any awkward silences or respond immediately if the participant starts to experience problems. But this is the exact opposite of good observational practice. Instead, it is important to let the survey session play out to its conclusion, from the participant's perspective, e.g. when the participant appears to believe they have submitted their completed survey and logged out. Often, waiting a few seconds at natural stopping points during the survey before interacting with the participant allows unexpected issues to be uncovered. In fact, this is how we discovered the usability issues described in Section 13.4 relating to Figures 13.1 and 13.2. In that session, instead of immediately concluding the session and taking control of the computer when the participant navigated to the final screen, we waited. That is when, instead of turning to us and saying that she was finished, she selected Save and Logout, which looped her back to the login screen and resulted in her confusion and our take-away of a usability issue.

13.6.2 Cognitive Probing Techniques

Although concurrent or verbal probing (as described by Willis 2015) is used frequently in cognitive testing protocols, Census usability staff have moved away from that type of protocol during usability tests of household surveys because it occasionally was a catalyst to change participant behavior. For example, in a usability test of an educational survey, a participant changed her answer to a question about working when we asked, "What does this question mean in your own words? What does *working* mean to you?" She had originally answered "No" that she did not work, but after the probe, changed her answer to "Yes" and said, "Yeah I was caregiving to someone, but without pay. It is a job but it has other benefits." At that point, the participant went down the "wrong" path and subsequent questions did not make sense to her situation. For this usability test, the verbal probing also changed participants' behavior so much that sometimes they would pause after answering each question to make sure we did not have an additional question to ask them before they proceeded to the next one. Other usability researchers have also found that verbal probing is disruptive, leads to longer sessions, and imposes a higher mental workload on participants than necessary, in addition to the potential for asking leading questions that can change participants' behavior (Boren and Ramey 2000; Ericsson and Simon 1993; Fox et al. 2011; Nørgaard and Hornbæk 2006). A similar technique of coaching (e.g. "Is that what you mean?" or "Did you notice this link here?") had been shown to lead to a higher task success rate and higher satisfaction, but no difference in task time (Olmsted-Hawala et al. 2010).

So, instead of using cognitive probes concurrently as participants complete the survey, Census usability staff rely on the think-aloud protocol to uncover the likely cause behind a user's struggle. Think aloud provides an approach to understanding the user's cognitive process while completing a given task. The goal is to get the user to verbally express their expectations and expressions of surprise when things do not go as expected and with minimal intervention by the researcher. The running commentary of the participant's thought-action process helps highlight struggles with both the design and the question wording. Other usability researchers have found the think-aloud protocol to be an effective tool (Dumas and Redish 1999; McDonald et al. 2012; Nielsen 1993a; Nørgaard and Hornbæk 2006; Rubin 1994; Van Den Haak et al. 2007). Jakob Nielsen, a pioneer in usability research, states, "Thinking aloud may be the single most valuable usability engineering method" (Nielsen 1993a, p. 195).

Like others, Census usability staff introduce the concept of thinking aloud by having the participant practice before the survey begins. An effective practice question is, "How many windows are in your home?" (Dillman et al. 2014; Ericsson and Simon 1993). If the participant struggles in thinking aloud while answering that question, it is good practice to provide an example of how to answer the question while thinking aloud and then have them answer the question again. Using the think-aloud protocol is not without challenges. For instance, some participants fall silent soon after being reminded to keep talking or they mumble and are difficult to understand. Here, a gentle instruction to "keep talking" helps if participants fall silent. Using backchannel continuers such as "mm-hm" or "uh-huh" keeps the participant talking at appropriate points in conversation and helps to stimulate commentary (Boren and Ramey 2000).

The value of the think-aloud protocol is apparent when one compares usability testing with think aloud to usability testing without think aloud (in silence). Across two iterative rounds of usability testing on the American Community Survey, a survey with about 50 questions for each person in the household, nine question-wording problems were identified when the usability evaluation was completed in silence. In a subsequent release of the survey for usability testing, with two iterative rounds of testing, this time using the think-aloud protocol, 35 question wording problems were found (Nichols 2016). Although cognitive problems with the questions were not the main purpose of the usability sessions, the think-aloud data uncovered unobservable thought processes and sources of measurement error.

Another valuable attribute of think aloud is that it does not affect eye-tracking data, except with older users. Romano Bergstrom and Olmsted-Hawala (2012) found that there were no differences in fixations between think aloud and silence for younger adults. However, older adults had fewer fixations than young adults when thinking aloud. This is a consideration to be aware of when planning the methodology, protocol, and user groups for a usability test.

13.6.3 Vignettes

Use of vignettes is another technique for measuring accuracy in usability testing. As Martin (2004) describes, vignettes are used to identify problems and test solutions. A "story" or a "situation" is presented to a person, and the person reacts. During usability testing, Census usability staff use vignettes either during the survey or at the end to get reactions to particular scenarios. Pretending to break off is a common vignette that can be used during usability testing of surveys. A *break-off* occurs when the respondent exits the survey prematurely before finishing. Typical Census Bureau household survey break-off rates range between 4% and 27% depending upon the device used and the length of the survey, with the percent increasing for longer surveys and smaller devices (Nichols et al. 2015; Horwitz 2015). For survey organizations, minimizing survey break-offs is important, but it is equally important to ensure that respondents can resume the survey with ease when this happens. During usability testing of household surveys, we often stop a participant halfway through the survey and tell him or her to pretend that he or she needs to "leave the survey" because there is an appointment to attend. Neutral task wording is important so that task success is not biased by the words used in the vignette. "Leaving the survey" does not include typical words used for links within the survey itself, such as "exit" or "logout." Once the participant is out of the survey, we then have them "get back in," again being careful not to use words or provide information that would guarantee success. Of course, if the participant fails to get back in after a few attempts, we assist them so that they can finish the survey, making a note about why they were unable to get back into the survey without the test administrator's assistance.

Vignettes can be used retrospectively, after survey completion, to test response options or different question wording using static screen shots. They can also be used to test other infrequent features within the survey instrument, such as accessing help screens or using links. Edit messages (i.e. messages indicating an error has occurred in the data collection) are a good example of an important design feature that does not always occur naturally during a session. Successful task completion occurs when the participant sees and reads the edit message and then enters accurate data. A common vignette we use in household surveys to test an edit message consists of a scenario where age and date of birth of a household member would not be known by the respondent. For example, we provide the following situation to a participant, "Pretend you recently had a new roommate move in named Jamie Doe. You do not know Jamie's age, but Jamie recently went out to celebrate her birthday this month. She also recently graduated from college. Please show me what you would do to answer this question for Jamie and tell me why you are answering this way." Then, the date of birth screen for Jamie is presented to the participant. We observe and take notes on what the participant says and does on the screen. In this way, we can assess and report on whether the participant makes up data, accesses help, and reads the edit message if they receive one, and what they do after the edit is triggered.

Usability testing in business surveys also makes use of vignettes to test edit message layout and wording, but does so concurrently by requesting that respondents enter erroneous data on specific screens. If respondents have not accessed help information, Census usability staff ask them to look up information related to particular questions or terms, again concurrently. Concurrent techniques are used in business surveys because these surveys tend to be complex and time consuming. Often respondents only complete a portion of the survey, and therefore it makes more sense to accomplish all goals related to a particular section before moving on.

13.7 Measuring Efficiency

The second measure of usability, efficiency, frequently operationalized as time-on-task, is a useful measure when comparing different survey designs using a split-ballot test with enough samples to conduct meaningful statistical tests. In a usability test of a survey, however, it is not clear that comparing the time participants take in Round 1 to Round 2 is a useful measure for household surveys because most surveys have different numbers of questions dependent upon the skip sequence or number of people in the household. The larger the household, the more time the survey will take. For economic surveys and censuses, there are extraneous tasks, such as accessing records (electronic or paper) during usability sessions, which vary by company. Consequently, timing an in-person, usability test session of both household and economic surveys is not a meaningful measure, and therefore Census usability staff do not capture that data. However, eye-tracking data, discussed in the next section, may provide an indicator of efficiency.

13.7.1 Eye Tracking

Eye tracking has been on the rise in usability testing of online surveys. A number of recent usability studies on electronic surveys show how eye tracking is being used to better understand the user experience of surveys; see for example Galesic et al. (2008) and Romano Bergstrom et al. (2016). Eye-tracking data can provide additional insights into how a participant interacts with the electronic survey interface. Eye tracking records where the eyes are relatively still (e.g. fixations), and the stillness can be used as an indicator of an area of attention. Eye tracking also records the duration of these fixations and can indicate how long a participant looks at a question or an answer category. A gaze plot records where the participant looked on a page over a period of time in a sequential pattern (Ehmke and Wilson 2007). Eye tracking can help researchers understand what part of a survey interface draws participants' attention, e.g. whether they see the navigation elements of Next and Previous buttons, or whether they read an edit message that pops up on the page. Eye tracking can also indicate if participants overlook instructions, if they are looking back and forth at various answer options, whether they notice a help icon, and so forth. In an example from a household survey (the American Community Survey), the gaze plot in Figure 13.7 shows the sequence of where one participant looked. The larger circles are where the eyes paused for a longer period of time. If this two-part question was read once from top to bottom, the numbers should be in a sequential pattern from top to bottom. However, the fact that higher numbers appear alongside lower numbers on the first part of the question means that the participant's eyes jumped back to the first part of the question after reading the second part of it, indicating there was some confusion.

Although beneficial, the use of eye tracking in usability testing of surveys is still a relatively new trend. Some researchers use eye-tracking fixation data as a type of efficiency measure for surveys. For example, when comparing two designs, Penzo (2006) concludes that the design with fewer fixations is

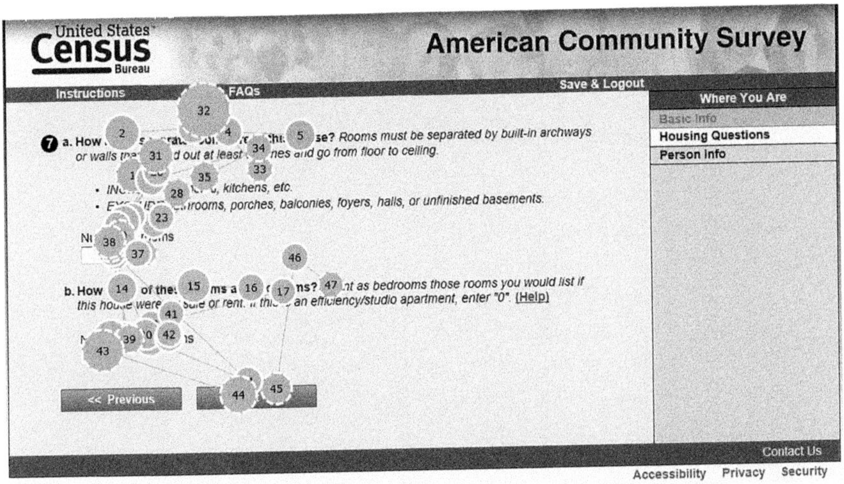

Figure 13.7 PC version of the "number of rooms" question with the scan path for one participant. The numbers in the circles show the order the eyes scanned the web page. Once the second part of the question was answered, the eyes moved back to the beginning of the question. Source: American Community Survey number of room question, usability testing 2013.

preferred. However, there is no consensus on how to interpret eye-tracking data. For example, long eye-fixation durations could reflect either confusion with or engagement in the material (Jacob and Karn 2003; Poole and Ball 2005). Analyzing eye-tracking data in combination with other user behaviors and metrics may aid in its interpretation and can lead to a more complete understanding of the issues with the interface (Olmsted-Hawala et al. 2016). It also may provide quantitative evidence of a problem verbalized through the think-aloud protocol, as the next example demonstrates. In a household survey usability evaluation, we heard several participants mention that they did not know what the label "Unit Designation" meant when looking at a screen that collected address information (see Figure 13.8 with that label circled).

Eye-tracking data allowed us to show the survey designers that participants were spending relatively more time looking at (or fixating on) the "Unit Designation" label compared to other parts of the screen (Olmsted-Hawala and Nichols 2014). Based on both the think-aloud data and the eye-tracking data, we concluded that the "Unit Designation" label was confusing. It was changed to "Apt/Unit," which in the United States means apartment or unit number.

Census usability staff collect eye-tracking data, but do not rely on it as the main source for identifying usability issues in household surveys. In fact, depending upon the schedule, eye-tracking data are not analyzed for every page. Instead, eye tracking is used for screens where the sponsor wants additional insight, where there are a high number of errors, where participants

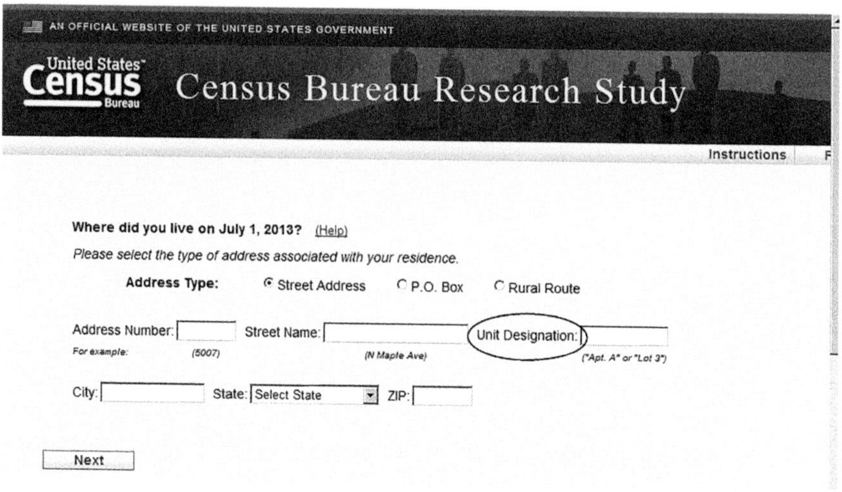

Figure 13.8 PC version of the address question with the confusing field label "Unit Designation" circled. Source: Census Bureau Research Study, usability testing 2013.

make changes on the screen prior to moving forward in the instrument, or for other predetermined problematic screens. All of these are clues that something sub-optimal may be happening on the page, and indicate that eye tracking might provide a better understanding of what is causing problems.

Census usability staff have been using eye tracking to test surveys running on desktop/laptop computers, and more recently on mobile devices such as smartphones and tablets. Usability sessions with eye tracking on desktops/laptops give consistently higher quality data (e.g. user calibration and sampling rate is better on a laptop/desktop than on mobile), and data analysis is more efficient than on mobile devices (Olmsted-Hawala 2016). This occurs even though the same tracking equipment is used. The weaker mobile data likely happens because the "map" of the user's fixation is necessarily compressed on a mobile device's smaller screen. It is also easier to look away from the small device, which decreases the sampling rate. In addition, depending on the software analysis tool, eye-tracking data, such as depicted in the large ball that covers about half of the screen in Figure 13.9, do not really lend insight to what the user interaction is with the instrument. Thus, the goals of the study should drive the decision on whether it is worthwhile to conduct eye tracking on mobile devices.

As mentioned previously, at this time Census usability staff do not collect eye-tracking data on business surveys because the software to run the eye tracking would have to be installed on the participant's PC, which is not practical due to time constraints and the complexity of adding and removing the licensed software for each use.

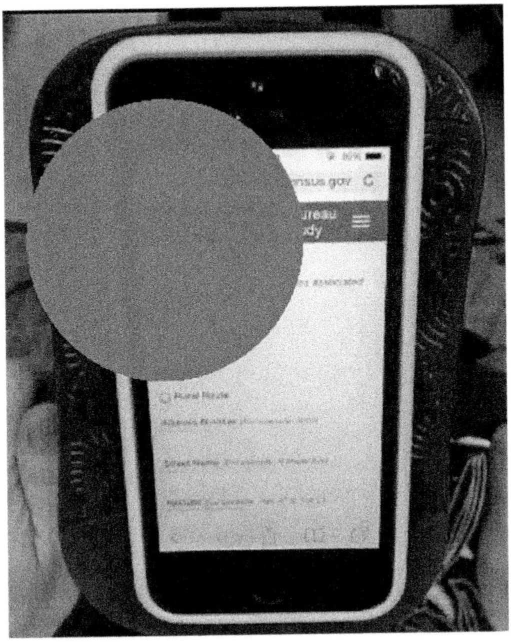

Figure 13.9 Mobile screenshot of one participants' eye-tracking fixation results imposed over the mobile screen. Notice how the results (in the form of the circle) cover much of the top of the screen. Source: Census Bureau Research Study, usability testing 2016.

13.8 Measuring Satisfaction

Measuring a user's satisfaction with a product is the third piece of usability (see Frøkjaer et al. 2000; ISO Standard 9241-11: 1998). There are some common industry satisfaction questionnaires, including the System Usability Scale (SUS) (Brooke 1996) and the Questionnaire for User Interaction Satisfaction (QUIS) (Chin et al. 1988). The SUS, widely used in the usability industry, is a short 10-item questionnaire with questions asking for the participant's agreement or disagreement with statements about the features of the system such as confidence, consistency, and ease of use. Each question has five response options with the endpoints labeled with the words "Strongly Agree" at one end and "Strongly Disagree" at the other end. However, Census usability staff found that the SUS worked better for evaluating users' experiences with a website rather than with an online survey. The QUIS, on the other hand, is a very long questionnaire with a variety of items that aim to measure system satisfaction with 12 specific interface factors. For each item, it uses a nine-point, end-labeled scale, with the labels varying based on the factor. For most usability studies involving surveys, Census usability staff modified the QUIS to use a 7-point scale with a subset of the questions. However, user satisfaction responses are not always indicative of the users' experience. For example, we have observed users having difficulty with a survey and heard spontaneous comments expressing this difficulty, but then watched as the users provide positive satisfaction scores. Another challenge with current satisfaction questions is to disentangle the participant's satisfaction with the survey questions from their satisfaction with the survey's graphical user-interface design. Some users may be rating their satisfaction with the survey questions, while others may be rating the interface design. Additional research on satisfaction questionnaires for usability evaluations of surveys is needed to improve their measurement.

While Census usability staff have not yet found a more fitting satisfaction questionnaire for surveys, the scores do allow tracking trends across repeated studies. Untangling the meaning of the trends, however, relies on qualitative data collected during the sessions. A good example of this is the iterative testing conducted in preparation for the 2020 Census. For the 2015 Census Test usability testing, two rounds of testing occurred with 15 participants in each round. The primary objective of the testing was to evaluate the optimized design for mobile devices, and thus 28 of the 30 participants used their own smartphones or tablets to answer the online questionnaire. Navigational elements in the survey did not change between rounds, yet participants reported lower satisfaction in the QUIS with the forward navigation in the second round, as shown by the darker colors in the Round 2 bar in Figure 13.10.

What differed between rounds was that the survey pages loaded more slowly during the second round of testing than they had in the first, and participants in the second round commented on the slow load times as they navigated through the instrument. The lower QUIS scores for the forward navigation question in the second round of testing therefore are most likely a reflection of the slower load times, which is a common source of user frustration with poor web design for mobile phones (Johansson 2013). The darker the color implies more difficulty with forward navigation. The darker bands in Round

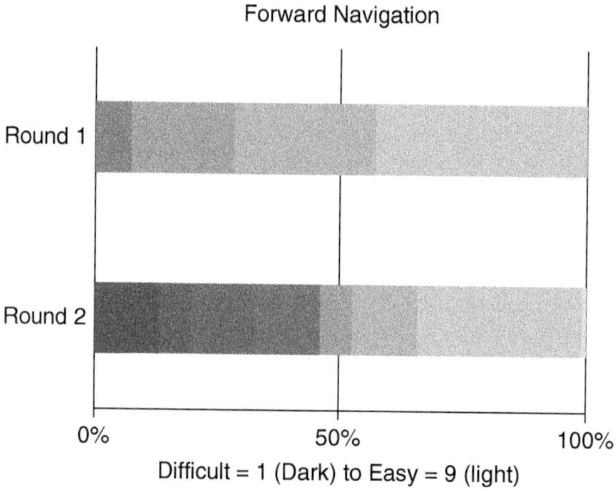

Figure 13.10 Example of how to display satisfaction ratings across rounds of testing. Source: 2015 Census Test usability testing.

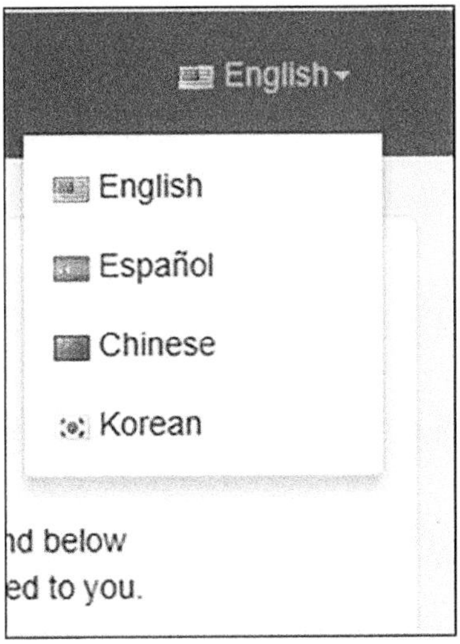

Figure 13.11 Example of the language toggle with flags of different countries next to the words in English. Source: 2016 Census Test usability testing at the beginning of the testing window.

2 compared to Round 1 imply forward navigation was more difficult in Round 2.

13.9 Retrospective Probing and Debriefing

Retrospective debriefing probes are administered in usability tests if the sponsor requests more information about a particular question or design, or if researchers want to follow up on something that occurred during the session. Census usability staff have found that a premade PowerPoint document containing screen shots of the instrument reorients the participant to the appropriate place in the questionnaire when administering the probes. At each screen, administering scripted questions or ad hoc probes helps to find out more about a particular topic. Although the retrospective probes provide some value, memory error plagues this type of probing. Even when the survey experience was within the past hour, some participants cannot recall or will incorrectly remember whether they received a particular question, how they answered it, or whether they had difficulty with a certain part of the survey. For example, in the debriefing associated with a household survey, the debriefing screen shot showed to the participant (see Figure 13.11) had a picture of four different languages with flags next to each of them. However, during the actual test sessions, the design had been changed and no longer had the flags; rather, it had the words for English, Spanish, Chinese, and Korean in their respective language (see Figure 13.12).

Although the debriefing probe was not meant to be a memory test of the participant (we did not have the opportunity to update the debriefing screen

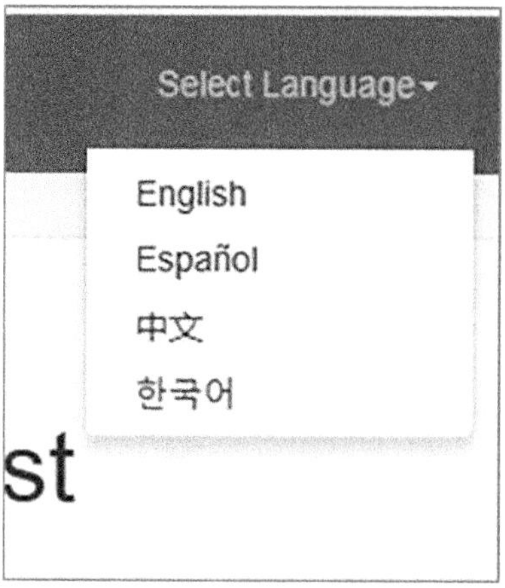

Figure 13.12 Example of language toggle in the correct languages and without the flags. Source: 2016 Census Test usability testing after an update during the middle of usability testing.

shot prior to testing), several participants who did not see the version with the flags, said that they did. Due to reliability issues such as these, we are cautious when using retrospective feedback for design decisions. For the same reason, caution must also be exercised when asking participants to remember why they did something or if they read or saw something on the screen.

Retrospective debriefing probes are more valuable when they do not rely on the participant's memory. For example, Census usability staff have found value in asking debriefing probes about features that might not generate spontaneous comments during the session or asking preference questions with A/B designs. In the same usability evaluation where the language option design changed, participants were randomly assigned to one of two questionnaire paths, which included two different versions of the race category for Black or African American. One of the paths used an abbreviation for American (circled in Figure 13.13) while the other did not (Figure 13.14).

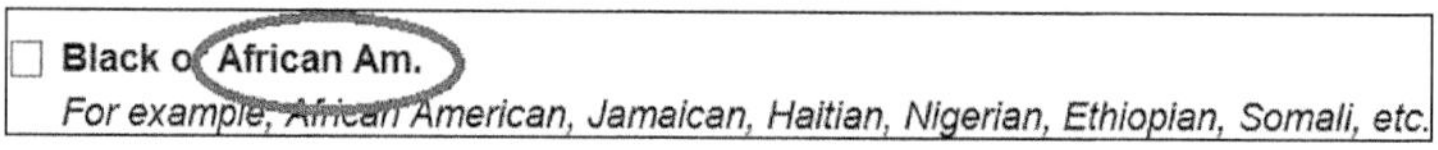

Figure 13.13 Example of abbreviation of American to Am. in the race category Black or African American. Source: 2016 Census Test usability testing.

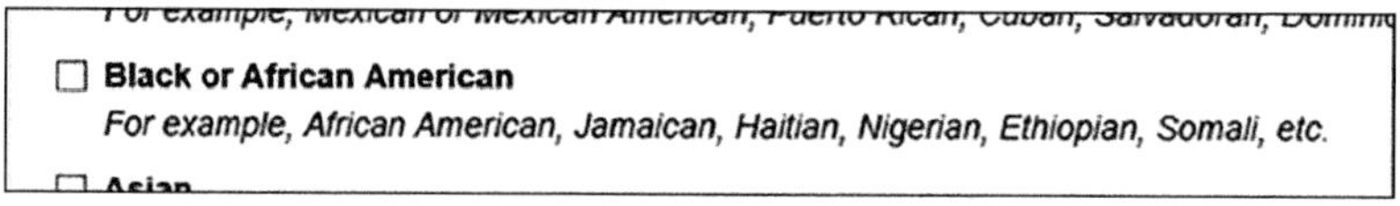

Figure 13.14 Example of the race category Black or African American spelled out. Source: 2016 Census Test usability testing.

Participants saw only one of the paths during the session, and there were no spontaneous comments about the abbreviation or lack of abbreviation during any of the sessions. However, when shown side-by-side in the debriefing and asked specifically to comment on the differences, participants were able to comment extensively on the abbreviation and overwhelmingly said they preferred the full spelling of the word American, as seen in Figure 13.14.

13.10 Communicating Findings with the Development Team

Our method of communicating testing results has evolved over time. Initially, Census usability staff would write a long and detailed report that could sometimes take a month or more to compose. It followed the Common Industry Format, created by the National Institute of Standards and Technology, which has been incorporated into an ISO standard (ISO/IEC 25062: 2006). This is a useful report format when presenting usability results at the end of the project or when the sponsor needs to create a baseline to measure any future design changes. However, this type of report is inadequate for a development team that needs to know "yesterday" what the main issues are and what they need to do to fix the problems prior to production. The issues raised and recommendations often will not be timely or succinct enough to be considered or implemented in the design of the instrument. Like others, we learned that time constraints often dominate in the production setting; and in order to affect change, usability test findings need to be available quickly and in a format that is suitable for the development team (see, for example, Hornbæk and Frøkjær 2005; Molich et al. 2007; Nørgaard and Hornbæk 2009).

A useful tactic in communicating findings is to invite the development team to observe the usability sessions firsthand so they can see how the participant interacts with the instrument. Having the development team for household or business surveys observe sessions provides a sense of being part of the usability process, and they can receive immediate feedback on items that concern them. Due to the time pressure of communicating results quickly to the development teams, Census usability staff have found it useful to spend 15 minutes at the conclusion of each session, after the participant has left, discussing the highlights of the session among ourselves and with observers if present. The synergy and brainstorming effect of the multi-researcher conversation often elicits the preliminary findings quickly and thoroughly.

An email message to the sponsors containing the relevant aspects of the session (device type, general participant characteristics without revealing any sensitive information) and the important findings serves several purposes. It provides the development team with immediate feedback on the session, it ensures that the results are not a surprise, and it is a useful reference when writing the final report. Analysis generally focuses on usability and functionality issues with the design, and content issues with the questions and response choices when measurement error is observed. The qualitative data gathered during each session, including observations and think-aloud verbalizations, are the main source of data for the usability findings. Researcher notes during the session are also very useful in the event of equipment malfunction, including audio issues, the microphone being too far away and inaudible commentary, or blurry video screen capture.

In addition to the email summaries, Census usability staff present initial usability results in the form of a PowerPoint presentation with screen shots and call-outs that help identify issues and recommended changes. After sending the PowerPoint presentation to the sponsor, Census usability staff meet with developers and sponsors to walk through the findings and determine what can and cannot be implemented within time and resource constraints or technological limitations. A final written report is provided at a later date. It includes more detailed analysis including any satisfaction metrics collected or eye-tracking data. To prepare the final document, researchers watch all videos to ensure that nothing was missed. Re-watching videos often leads to new insights.

13.11 Assessing Whether Usability Test Recommendations Worked

When time permits, a best practice is to conduct a summative usability test to verify that the recommended design changes solve the usability problems identified in earlier iterations of testing. However, from our experience in a production setting, there is rarely the extra time to conduct a confirmation test. Paradata from completed surveys in the field offer an alternative measure of the success of changes made as a result of findings from usability testing. *Paradata* are process data, including all keystrokes, time stamps, and events within the survey (see Kreuter 2013 for more information on paradata). Comparing paradata across survey iterations can confirm usability problems and design solutions. For example, analysis of paradata from the American Community Survey over several years showed a reduced triggering of edit messages as a result of redesign work on the survey based on findings from usability testing (Ashenfelter et al. 2012; Horwitz et al. 2013; Zelenak 2016).

Occasionally, patterns in paradata are difficult to interpret. In these instances, usability testing is often helpful in making sense of the paradata. For example, when looking at the navigational paradata, a sponsor noticed frequent backward navigation within a particular section of a household survey. During usability testing, Census usability staff were able to observe this behavior and to identify its cause: a confusing skip sequence with the section.

13.12 Conclusions

In this chapter, we have shared current practices in usability testing online surveys at the Census Bureau. Our practices have developed in response to technological change, sponsors' requirements, our wider reading of the UX literature and its application to social research, and our own experiences and reflections on our practice. The chapter highlights areas of agreement between our experience and empirical research evidence: on think-aloud methodology and on the optimal number of participants. In other areas, we have identified gaps in the empirical evidence and challenges to applying usability concepts to the testing of social survey data-collection tools. For example, developing a robust satisfaction score for survey design that actually matches observed participant behavior could enhance our ability to evaluate the effectiveness of subsequent survey iterations as well as allow comparisons across different surveys. Developing standard templates for using eye-tracking data and developing theory about what eye-tracking data really mean with regard to question and survey design could provide insights into creating an effectiveness measure. Systematic use of paradata to pre-identify areas in the questionnaire that need additional investigation during usability testing, through either observation or vignettes, could result in more focused, tasked-based usability testing of surveys.

While usability-testing practices may still be evolving, this does not mean they do not afford survey designers valuable insights about design features and potential sources of survey error. By observing what users do and listening to what they say, researchers can arrive at a deeper, more nuanced understanding of what the usability issues are and what changes need to be made to the survey to make the instrument work better for the respondent.

Disclaimer and Acknowledgments

The views expressed in this chapter are those of the authors and not necessarily those of the US Census Bureau. The authors thank Lin Wang, Joanne Pascale, Diane Willimack, and Paul Beatty from the Census Bureau for their reviews and helpful suggestions, and our editor, Debbie Collins.

References

Ashenfelter, K.T., Holland, T., Quach, V. et al. (2012). ACS internet 2011 project: Report for rounds 1 and 2 of ACS wireframe usability testing and round 1 of ACS internet experiment mailing materials cognitive testing. Survey Methodology #2012–01. http://www.census.gov/srd/papers/pdf/ssm2012-01.pdf.

Bates, N. and Sweet, E. (1997). Results from usability test of OPM math stat job application. U.S. Census Bureau Human-Computer Interaction Report Series #2.

Beck, J., Tuttle, A., and Salyers, E. (2016). Using a collaborative approach and multi-method research to re-engineer a Census of Businesses. Fifth International Conference on Establishment Surveys, Geneva, Switzerland. http://ww2.amstat.org/meetings/ices/2016/proceedings/ICESV_TOC.pdf.

Boren, M.T. and Ramey, J. (2000). Thinking aloud: reconciling theory and practice. *IEEE Transactions on Professional Communication* 43 (3): 261–278.

Brooke, J. (1996). SUS – a quick and dirty usability scale. In: *Usability Evaluation in Industry* (eds. P.W. Jordan, B. Thomas, I.L. McClelland and B. Weerdmeester), 20. London: Taylor & Francis.

de Bruijne, M. and Wijnant, A. (2013). Can mobile web surveys be taken on computers? A discussion on a multi-device survey design. *Survey Practice* 6 (4)) Retrieved from: http://www.surveypractice.org/index.php/SurveyPractice/rt/printerFriendly/238/html.

Callegaro, M. (2010). Do you know which device your respondent has used to take your online survey*? Survey Practice* 3 (6) Retrieved from: http://www.surveypractice.org/index.php/SurveyPractice/article/view/250/html.

Chin, J.P., Diehl, V.A., and Norman, K.L. (1988). Development of an instrument measuring user satisfaction of the human-computer interface. In: *Proceedings of CHI '88: Human Factors in Computing Systems*, 213–218. New York: ACM Press.

Chisnell, D. (2011). Bonus research: Do the recruiting yourself [online forum]. http://usabilityworks.com/?s=recruiting.

Christian, L.M., Parsons, N.L., and Dillman, D.A. (2009). Designing scalar questions for web surveys. *Sociological Methods & Research* 37 (3): 393–425.

Collins, D. and Gray, M. (2014). Sampling and recruitment. In: *Cognitive Interviewing Practice* (ed. D. Collins), 80–100. London: Sage.

Costabile, M.F. (2001). Usability in the software life cycle. In: *Handbook of Software Engineering and Knowledge Engineering* (ed. S. Chang), 179–192. NJ: World Scientific.

Couper, M.P. (2000). Web surveys: a review of issues and approaches. *The Public Opinion Quarterly* 64 (4): 464–494. Retrieved from: https://academic.oup.com/poq/article-abstract/64/4/464/1920309/Review-Web-Surveys-A-Review-of-Issues-and?redirectedFrom=fulltext.

Couper, M.P. (2008). *Designing Effective Web Surveys*. New York, NY, US: Cambridge University Press https://doi.org/10.1017/CBO9780511499371.

Deaven, K., Fox, J., Sun, N., et al. (2016). Good GovUX: Improving the user experience of federal government digital products through public-private collaboration. Panel Discussion at the annual meeting of the Human Factors and Ergonomics Society International, Washington, DC.

DeMaio, T., Mathiowetz, N., Rothgeb, J., et al. (1993). Protocol for pretesting demographic surveys at the Census Bureau. ASA '93, Alexandria, VA. https://www.census.gov/srd/www/abstract/sm93-04.html.

Dillman, D.A., Gertseva, A., and Mahon-Haft, T. (2005). Achieving usability in establishment surveys through the application of visual design principles. *Journal of Official Statistics* 21 (2): 183–214.

Dillman, D.A., Smyth, J.D., and Christian, L.M. (2014). *Internet, Mail, and Mixed-Mode Surveys: The Tailored Design Method*, 4e. Hoboken, NJ, US: Wiley.

Douglass, R. and Hylton, K. (2010). Get it RITE: Rapid iterative testing and evaluation (RITE). *User Experience* 9 (1) http://uxpamagazine.org/get_it_rite.

Dumas, J.S. and Redish, J.C. (1999). *A Practical Guide to Usability Testing* (Revised ed.). Portland, OR: Intellect Books.

Ehmke, C. and Wilson, S. (2007). Identifying web usability problems from eye-tracking data. In: *Proceedings from HCI '07: People and Computers XXI – HCI... but Not as We Know It*, vol. 1. Tracking Usability Issues (eds. L.J. Ball, A. Sasse, C. Sas, et al.). London, UK: BCS Retrieved from: http://www.bcs.org/upload/pdf/ewic_hc07_lppaper12.pdf.

Ericsson, K.A. and Simon, H.A. (1993). *Protocol Analysis: Using Verbal Reports as Data*. Cambridge, MA: The MIT Press.

Faulkner, L. (2003). Beyond the five-user assumption: benefits of increased sample sizes in usability testing. *Behavior Research Methods, Instruments, & Computers* 35 (3): 379–383. https://doi.org/10.3758/BF03195514.

Fox, J.E. (2001). Usability methods for designing a computer-assisted data collection instrument for the CPI. Federal Committee on Statistical Methodology (FCSM) Research Conference. https://nces.ed.gov/FCSM/pdf/2001FCSM_Fox.pdf.

Fox, M.C., Ericsson, K.A., and Best, R. (2011). Do procedures for verbal reporting of thinking have to be reactive? A meta-analysis and recommendations for best reporting methods. *Psychological Bulletin* 137 (2): 316–344.

Frøkjaer, E., Herzum, M., and Hornbaek, K. (2000). Measuring usability: Are effectiveness, efficiency and satisfaction correlated? SIGCHI Conference on Human Factors in Computing Systems.

Galesic, M., Tourangeau, R., Couper, M.P., and Conrad, F.G. (2008). Eye-tracking data: new insights on response order effects and other cognitive shortcuts in survey responding. *Public Opinion Quarterly*. 72 (5): 892–913.

Geisen, E. and Jarrett, C. (2011). Introduction to usability testing for survey research. SAPOR short course lecture slides. http://www.slideshare.net/cjforms/introduction-to-usability-testing-for-survey-research

Geisen, E. and Romano Bergstrom, J. (2017). *Usability Testing for Survey Research*. Cambridge, MA: Morgan Kaufmann.

Göransson, B., Gulliksen, J., and Boivie, I. (2003). The usability design process – integrating user-centered systems design in the software development process. *Software Process Improvement and Practice* 8: 111–131.

Hambling, B. and van Goethem, P. (2013). *User Acceptance Testing: A Step-by-Step Guide.* Swindon, U.K.: BCS Learning and Development Ltd.

Health and Human Services (2006). *Research-Based Web Design & Usability Guidelines.* Washington, DC: U.S. Government Printing Office. ISBN: 0-16-076270-7) Chapter 18: Usability Testing Retrieved from: https://webstandards.hhs.gov/guidelines/.

Hornbæk, K. (2006). Current practice in measuring usability: challenges to usability studies and research. *International Journal of Human-Computer Studies* 64 (2): 79–102. https://doi.org/10.1016/j.ijhcs.2005.06.002.

Hornbæk, K. and Frøkjær, E. (2005). Comparing usability problems and redesign proposal as input to practical systems development. In: *Proceedings of ACM Conference on Human Factors in Computing Systems (CHI 2005)*, 391–400. New York, NY: Association for Computing Machinery.

Horwitz, R. (2015). Usability of the American Community Survey Internet Instrument on Mobile Devices. U.S. Census Bureau #ACS15-RER-04. http://www.census.gov/content/dam/Census/library/working-papers/2015/acs/2015_Horwitz_01.pdf.

Horwitz, R., Tancreto, J.G., Zelenak, M.F., and Davis, M. (2013). Use of paradata to assess the quality and functionality of the American Community Survey internet instrument. U.S. Census Bureau. https://www.census.gov/library/working-papers/2013/acs/2013_Horwitz_01.html.

International Standards of Organization (ISO). (1998). ISO 9241-11:1998 Ergonomic requirements for office work with visual display terminals (VDTs) – Part 11: Guidance on usability. http://www.iso.org/iso/catalogue_detail.htm?csnumber=16883.

International Standards of Organization (ISO). (2006). ISO/IEC 25062:2006 Software engineering -- Software product Quality Requirements and Evaluation (SQuaRE) -- Common Industry Format (CIF) for usability test reports. http://www.iso.org/iso/catalogue_detail.htm?csnumber=43046.

Jacob, R.J.K. and Karn, K.S. (2003). Eye tracking in human-computer interaction and usability research: ready to deliver the promises. In: *The Mind's Eye: Cognitive and Applied Aspects of Eye Movements* (eds. J. Hyönä, R. Radach and H. Deubel), 573–605. Amsterdam: Elsevier.

Jarrett, C. and Gaffney, G. (2008). *Forms that Work: Designing Web Forms for Usability.* Amsterdam: Elsevier.

Johansson, J. (2013). How to make your websites faster on mobile devices. *Smashing Magazine.* https://www.smashingmagazine.com/2013/04/build-fast-loading-mobile-website.

Karat, C. (1989). Iterative usability testing of a security application. *Proceedings of the Human Factors and Ergonomics Society* 33 (5): 273–277.

Kreuter, F. (2013). Improving surveys with paradata. In: *Analytic Uses of Process Information* (ed. F. Kreuter). Hoboken, NJ: Wiley.

Marquis, K. Nichols, E., and Tedesco, H. (1998). Human-computer interface usability in a survey organization: getting started at the census bureau. U.S. Census Bureau. https://www.census.gov/srd/papers/pdf/sm98-04.pdf.
Martin, E. (2004). Vignettes and respondent debriefing for questionnaire design and evaluation. In: *Methods for Testing and Evaluating Survey Questionnaires* (eds. S. Presser, J.M. Rothger, M. Couper, et al.), 149–172. NJ: Wiley.
McDonald, S., Edwards, H.M., and Zhao, T. (2012). Exploring think-alouds in usability testing: an international survey. *IEEE Transactions on Professional Communication* 55 (1): 2–19.
McGeeney, K. (2015). Tips for creating web surveys for completion on a mobile device. Pew Research Center. http://www.pewresearch.org/2015/06/11/tips-for-creating-web-surveys-for-completion-on-a-mobile-device.
Medlock, M.C., Wixon, D., McGee, M., and Welsh, D. (2005). The rapid iterative test and evaluation method: better products in less time. In: *Cost Justifying Usability: An Update for the Internet Age* (eds. R.G. Bias and D.J. Mayhew), 489–517. San Francisco: Morgan Kaufman.
Meyers, M., Trejo, Y., and Lykke, L. (2017). The performance of vignettes in focus groups and cognitive interviews in a cross-cultural context. *Survey Practice* 10 (3) Retrieved from http://surveypractice.org/index.php/SurveyPractice/article/view/413.
Molich, R., Jeffries, R., and Dumas, J.S. (2007). Making usability recommendations useful and usable. *Journal of Usability Studies* 2 (4): 162–179.
Nichols, E. (2016). Cognitive probing methods in usability testing – Pros and cons. In: *American Association of Public Opinion Research (AAPOR) annual conference*. U.S. Census Bureau: Austin, TX.
Nichols, E., Olmsted-Hawala, E., Horwitz, R., and Bentley, M. (2015). *Optimizing the Decennial Census for Mobile, – A Case Study*. Washington, D.C.: Federal Committee on Statistical Methodology Retrieved from: http://fcsm.sites.usa.gov/files/2016/03/I2_Nichols_2015FCSM.pdf.
Nielsen, J. (1993a). *Usability Engineering*. Cambridge, MA: Academic Press.
Nielsen, J. (1993b). Iterative user interface design. *IEEE Computer* 26 (11): 32–41.
Nielsen, J., & Landauer, T. K. (1993, May). A mathematical model of the finding of usability problems. Proceedings of the INTERACT'93 and INTERCHI'93 Conference on Human Factors in Computing Systems (206–213). Retrieved from: https://pdfs.semanticscholar.org/2450/61d94c228abdc9bda2d9a10679e897526465.pdf
Nørgaard, M. and Hornbæk, K. (2006). What do usability evaluators do in practice? An explorative study of think-aloud testing. In: *Proceedings of the 6th Conference on Designing Interactive Systems*, 209–218. New York, NY, USA: ACM Retrieved from: https://www.researchgate.net/profile/Kasper_Hornbaek/publication/200553218_What_do_usability_evaluators_do_in_practice_an_explorative_study_of_think-aloud_testing/links/57157cdf08ae1a840264fd5d.pdf.

Nørgaard, M. and Hornbæk, K. (2009). Exploring the value of usability feedback formats. *International Journal of Human-Computer Interaction* 25 (1): 49–74.

Norman, K.L. and Murphy, E.D. (2004). Usability testing of an internet form for the 2004 overseas enumeration test: iterative testing using think-aloud and retrospective report methods. *Proceedings of the Human Factors and Ergonomics Society* 48 (13): 1493–1497.

Olmsted, E., & Gill, M. (2005). In-person usability study compared with self-administered web (remote—different time/place) study: does mode of study produce similar results? Poster presented at the International Usability Professionals' Association, Montreal, Canada. https://www.academia.edu/981317/In-person_usability_study_compared_with_self-administered_web_remote-different_time-place_study_Does_mode_of_study_produce_similar_results.

Olmsted-Hawala, E. (2016). Association of eye tracking with other usability metrics. In: *American Association of Public Opinion Research (AAPOR) annual conference.* U.S. Census Bureau: Austin, TX.

Olmsted-Hawala, E. and Holland, T. (2015). Age-related differences in a usability study measuring accuracy, efficiency, and user satisfaction in using smartphones for census enumeration: fiction or reality? In: *Human Aspects of IT for the Aged Population* (eds. J. Zhou and G. Salvendy), 475–483. Switzerland: Springer.

Olmsted-Hawala, E., Murphy, E.D., Hawala, S., and Ashenfelter, K.T. (2010). Think-aloud protocols: a comparison of three think-aloud protocols for use in testing data dissemination web sites for usability. In: *In Proceedings of the SIGCHI Conference on Human Factors in Computing Systems (CHI '10)*, 2381–2390. New York: ACM Press.

Olmsted-Hawala, E. and Nichols, E. (2014). Using eye tracking to evaluate email notifications of surveys and online surveys collecting address information. In: *American Association of Public Opinion Research (AAPOR) annual conference.* U.S. Census Bureau: Los Angeles, CA.

Olmsted-Hawala, E., Wang, L., Willimack, D.K. et al. (2016). A pilot investigation of the association between eye-tracking patterns and self-reported reading behavior. In: *Universal Access in Human-Computer Interaction. Methods, Techniques, and Best Practices,* Lecture Notes in Computer Science, vol. 9737 (eds. M. Antona and S. Stephanidis), 442–453. Switzerland: Springer.

Penzo, M. (2006). Label placement in forms. UX Matters. http://www.uxmatters.com/mt/archives/2006/07/label-placement-in-forms.php.

Peytchev, A., Couper, M.P., McCabe, S.E., and Crawford, S.D. (2006). Web survey design. *Public Opinion Quarterly* 70 (4): 596–607.

Poole, A. and Ball, L.J. (2005). Eye tracking in human-computer interaction and usability research: current status and future prospects. In: *Encyclopedia of Human Computer Interaction* (ed. C. Ghaoui), 211–219. PA: Idea Group.

Potaka, L. (2008). Comparability and usability: key issues in the design of internet forms for New Zealand's 2006 census of populations and dwellings. *Survey Research Methods* 2 (1): 1–10.

Redish, J. (2007). *Letting Go of the Words: Writing Web Content that Works.* San Francisco, CA: Morgan Kaufman.

Redish, J. (2010, November). *Letting Go of the Words–Writing Web Content that Works.* Invited talk presented at Census Usability Day. Washington, D.C.: U.S. Census Bureau.

Romano Bergstrom, J., Childs, J., Olmsted-Hawala, E., and Jurgenson, N. (2013). The efficiency of conducting concurrent cognitive interviewing and usability testing on an interviewer-administered survey. *Survey Practice* 6 (4) Retrieved from: http://www.surveypractice.org/index.php/SurveyPractice/article/view/79.

Romano Bergstrom, J. and Olmsted-Hawala, E. (2012). Effects of age and think-aloud protocol on eye-tracking data and usability measures. International conference EyeTrackUX, Las Vegas, NV. https://www.slideshare.net/ Jennifer-RomanoBergstrom/eye-trackux-romanobergstromolmstedhawalajune12.

Romano Bergstrom, J.C., Erdman, C., and Lakhe, S. (2016). Navigation buttons in web-based surveys: respondents' preferences revisited in the laboratory. *Survey Practice* 9 (1) Retrieved from: http://surveypractice.org/index.php/SurveyPractice/article/view/303/html_51.

Romano Bergstrom, J.C., Olmsted-Hawala, E.L., Chen, J.M., and Murphy, E. (2011). Conducting iterative usability testing on a web site: challenges and benefits. *Journal of Usability Studies* 7 (1): 9–30.

Rubin, J. (1994). *Handbook of Usability Testing: How to Plan, Design, and Conduct Effective Tests.* New York, NY: Wiley.

Snyder, C. (2003). *Paper Prototyping: The Fast and Easy Way to Design and Refine User Interfaces.* New York: Morgan Kaufmann.

Stettler, K. (2013). Using Web Ex to conduct usability testing of an on-line survey instrument. QUEST Workshop, Washington, DC.

Sweet, E., Marquis, K., Sedivi, B., and Nash, F. (1997). Results of expert review of two internet R&D questionnaires. Human-Computer Interaction Report Series, 1.

Sweet, E.M and Russell, C.E. (1996). A discussion of data collection via the internet. Proceedings of the Joint Statistical Meetings, Chicago, IL. https://www.census.gov/srd/papers/pdf/sm96-09.pdf.

Sy, D. (2007). Adapting usability investigations for agile user-centered design. *Journal of Usability Studies* 2 (3): 112–132.

Tullis, T.S. (2007). Older adults and the web: lessons learned from eye-tracking. In: *Universal Access in Human Computer Interaction: Coping with Diversity,* Lecture Notes in Computer Science (ed. C. Stephanidis), 1030–1039. New York: Springer.

U.S. Bureau of the Census. (1998). Census Bureau standard: pretesting questionnaires and related materials for surveys and censuses. Washington, DC: U.S. Department of Commerce.

U.S. Bureau of the Census. (2003). Pretesting policy and options: demographic surveys at the Census Bureau, Washington, DC. Washington, DC: U.S. Department of Commerce. https://www.census.gov/srd/pretest-standards .html.

U.S. Census Bureau. (2016). Are you in a survey? http://www.census.gov/ programs-surveys/are-you-in-a-survey/about-business-surveys.html.

Van Den Haak, M.J., de Jong, M.D.T., and Schellens, P.J. (2007). Evaluation of an informational web site: three variants of the think-aloud method compared. *Technical Communication* 54 (1): 58–71.

Virzi, R., Sokolov, J. L., and Karis, D. (1996). Usability problem identification using both low- and high-fidelity prototypes. Conference on Human Factors in Computing Systems (CHI 1996), Vancouver, Canada.

Virzi, R.A. (1992). Refining the test phase of usability evaluation: how many subjects is enough? *Human Factors* 34 (4): 457–468.

Wang, L., Sha, M., and Yuan, M. (2017). Cultural fitness in the usability of U.S. census Internet survey in Chinese language. . Special issue: cross-cultural and multilingual research. *Survey Practice* 10 (3) ISSN: 2168-0094.

Willis, G.B. (2015). *Analysis of the Cognitive Interview in Questionnaire Design*. New York: Oxford University Press.

Wroblewski, L. (2008). *Web Form Design: Filling in the Blanks*. Brooklyn, NY: Rosenfeld Media.

Zelenak, M.F. (2016). 2014 American Community Survey internet test. Decennial Statistical Studies Division 2016 American Community Survey Memorandum Series. #ACS16-MP-05. U.S. Census Bureau.

14

How Mobile Device Screen Size Affects Data Collected in Web Surveys

Daniele Toninelli[1] *and Melanie Revilla*[2]

[1] *Department of Management, Economics and Quantitative Methods, University of Bergamo, Bergamo, Italy*
[2] *RECSM, University Pompeu Fabra, Barcelona, Spain*

14.1 Introduction

In the last few years, methodologists working with web surveys have faced the new phenomenon of "unintended mobile respondents" (de Bruijne and Wijnant 2014; Peterson 2012), i.e. respondents who participate in web surveys through mobile devices, even if this was not intended by the researchers and the surveys were not optimized for small screens. This kind of survey participation has spread quickly and is not negligible anymore (Callegaro 2010; Stern et al. 2014). For example, in the Netherlands-based LISS probability panel, the percentage of participants using a smartphone or a tablet grew from 3% in March 2012 to 11% in September 2013; similarly, in the CentERpanel, mobile participation also increased from 3%, in February 2012, to 16%, in October 2013 (de Bruijne and Wijnant 2014). These levels are confirmed by data coming from the German GESIS panel: in 2014, 17.9% of online respondents participated in the panel's surveys using mobile devices (Struminskaya et al. 2015). Similar trends were observed in the Netquest panel surveys in Spain, Portugal, and some Latin American countries: panelists showed not only an increasing ownership of mobile devices (or of a combination of multiple devices, counterbalancing a decreasing trend in laptop PC ownership (Revilla et al. 2015)), but also an increasing preference for using such devices for participating in surveys, mostly when they are mobile-optimized (Revilla et al. 2016b). We can reasonably expect a further rapid increase of this phenomenon that reflects the observed growth of mobile web usage: worldwide, the mobile share was 8.99% in March 2012, 19.08% in September 2013, 31.82% in December 2014, and 50.02% in November 2017, whereas the desktop share decreased from 91.01% to 45.68% in the same period (StatCounter Global Stats 2017).

Advances in Questionnaire Design, Development, Evaluation and Testing, First Edition.
Edited by Paul C. Beatty, Debbie Collins, Lyn Kaye, Jose-Luis Padilla, Gordon B. Willis, and Amanda Wilmot.

Mobile devices, mostly smartphones, differ from PCs (desktops and laptops) at several levels (Chae and Kim 2004; Parush and Yuviler-Gavish 2004; Sweeney and Crestani 2006): they are provided with smaller screens (reduced visibility) and with virtual keyboards (rather than physical), their portability is higher (usable in a wider range of places), they could have reduced computing power and a more instable and/or a slower web connection, they are battery-dependent, they have limited memory capabilities, they are provided with interesting features (sensors, camera), etc. As a consequence, web survey participation using mobile devices has stimulated a lot of methodological interest. This interest generated work about the increasing importance of the phenomenon (Revilla et al. 2015, 2016b), about the quality and comparability of data collected through mobile devices (de Bruijne and Wijnant 2013a; Mavletova 2013; Mavletova and Couper 2013; Stapleton 2013; Wells et al. 2014; Toninelli and Revilla 2016a), about the adaptation of questionnaires for mobile participation (Boreham and Wijnant 2013), and about other more specific aspects linked to participation that can be affected by the use of mobile devices, such as completion times (Couper and Peterson 2015), response rates (Millar and Dillman 2011, 2012), and the screen-size effect (Fischer and Bernet 2014, 2016).

The main purpose of this chapter is to evaluate if and how the screen size of mobile devices can affect data quality on web surveys, measured by four indicators. We use data from a two-wave crossover experiment implemented in 2015 within the opt-in online panel Netquest, in Spain. Following a review of relevant literature, we introduce our hypotheses, then present details of our data collection methodology and analytic approach. We conclude by presenting our particular findings and outlining broader implications for surveys using mobile devices.

14.2 Literature Review

It is already well established that data collection using mobile devices is different than that using other modes, in several important ways (e.g. Couper and Peterson 2015; Fischer and Bernet 2014; Liebe et al. 2015). Most mobile devices have a virtual keyboard, which affects how respondents navigate throughout the questionnaire and response options. Internet connection speeds are often slower on mobile devices than on PCs. The reduced sizes of mobile devices increases the devices' portability, allowing respondents to participate in surveys from any place (Brick et al. 2007). The wider variety of locations and of situations can affect the level of effort of respondents and their cognitive processing, for example causing a higher social desirability bias (mostly for sensitive topics), when participating in public places (Mavletova and Couper 2013; Toninelli and Revilla 2016a). Multitasking can also vary by device (Toninelli and Revilla 2016b), affecting the quality and comparability of the data.

However, many researchers have considered the screen size as the key factor to explain potential differences across devices (e.g. Buskirk and Andrus 2012; Couper and Peterson 2015, 2016; de Bruijne and Wijnant 2013a; Liebe et al. 2015; Williams et al. 2015). Smaller screen sizes of mobile devices, particularly smartphones, result in reduced visibility. Thus, if researchers try to render paper questionnaires on an electronic device in a similar format, only a limited amount of information is available directly on the screen, in comparison to PCs. As a consequence, if the questionnaire is not optimized, the web pages may extend beyond the actual size of the screen (Peytchev and Hill 2008), requiring users to frequently scroll horizontally and vertically in order to read the full information. This can affect both readability (if smaller fonts are used) and comprehension. Thus, survey participation through a mobile device "still requires more effort from the respondent (due to [...] more difficult task handling)" (de Bruijne and Wijnant 2013a, p. 494), potentially increasing breakoffs and reducing response rates (see, e.g. Baker-Prewitt 2013; Bosnjak et al. 2013; Buskirk and Andrus 2014; Peytchev 2009). Moreover, the need for scrolling and zooming significantly increases completion time relative to other modes, whereas the effects of variables such as browser and connection speed are minimal (Mavletova 2013; Fischer and Bernet 2014; Toepoel and Lugtig 2014; Andreadis 2015; Couper and Peterson 2015, 2016). Liebe et al. (2015) also detected significantly different survey lengths between respondents using tablets and smartphones. A consequence of longer completion times is a higher respondent burden, which can lead to a more negative evaluation of the survey experience (although in one study, de Bruijne and Wijnant [2013a] detected no difference in terms of respondent satisfaction). The higher burden can also introduce "some undesirable differences in responses" (Peytchev and Hill 2008, p. 4298), affecting the quality and the comparability of collected data. In 2010, the same authors confirmed that respondents using mobile devices showed differences in responses and were less likely to put effort in writing text, if the questions extended beyond the visible area of the screen. More recently, Williams et al. (2015, p. 16) found "generally little evidence of differences by device type," whereas Liebe et al. (2015, p. 28) found that "surveys completed on smartphones are somewhat counter intuitively associated with higher survey quality" (in terms of longer interview length and lower acquiescence bias).

Previous studies generally tended to attribute differences in the data collected, their quality, and/or the survey participation process to the different screen sizes of the devices used (see e.g. Couper and Peterson [2016, p. 4]: "the size of mobile devices is a frequently mentioned source of the differences in completion times"). Nevertheless, only a few studies specifically measured the devices' exact screen size and used this information to evaluate whether size is what really determines these differences. Instead, most studies compared two or three wide categories of devices, with smaller or larger screens (Fischer and Bernet 2014; Mitchel 2014; Revilla et al. 2016a). One exception is the study by Couper and Peterson (2016): the authors found that completion time decreases by 0.2 minutes for each additional 100 pixels in screen size. Liebe et al. (2015) also directly studied the screen size (diagonally measured in centimeters), focusing on mobile respondents only. Within this group, they found a negative link between screen size and completion time and a positive correlation between screen size and acquiescence tendency. The authors also studied the error variance, a measure indirectly correlated with the consistency in respondents' choices (potentially due to respondents' distraction or survey interruptions). They found a U-shaped relationship between error variance and screen size: the highest error variances are observed for smaller and for larger screens; the error variance exponentially decreases for screens up to 17.45 cm and exponentially increases for screens bigger than this threshold. In addition, among mobile respondents, de Bruijne and Wijnant (2014) found clear differences in terms of device usage that also affect the likelihood of participating: this is higher if respondents have more-advanced input interfaces and use smartphones frequently for tasks such as reading emails.

In order to address some of the issues linked to the usage of small screens, the literature suggests optimizing the questionnaire for mobile participation (e.g. de Bruijne and Wijnant 2013b) by using responsive fonts and larger buttons, and automatically adapting the layout to the screen size in order to avoid horizontal scrolling. This enhances readability, favors consistent results for different question types (Fischer and Bernet 2014), and reduces straightlining and the presence of conflicting answers (Mitchel 2014).

14.3 Our Contribution and Hypotheses

14.3.1 Contribution

This chapter explores the impact of screen size on the mobile survey participation. Compared to previous works, it differs in several respects. First, we focus on mobile devices only. All mobile devices are by definition highly portable, and almost all of them include virtual keyboards. Nevertheless, they are quite diverse in terms of screen size: the biggest tablets can be three or four times the size of the smallest smartphones. We presume that visibility and readability are likely to vary between a 4 and 7 in. screen. Thus, our goal is to investigate if and how these variations affect different performance indicators. A comparison of PCs and mobile devices using the same data is available in other papers (for a summary, see Revilla et al. 2016a).

Second, in order to take into account the differences across mobile devices, we study the impact of the exact screen size (measured in inches) of each device used, whereas most of the previous works evaluated the screen-size effect comparing only broad classes of devices (such as smartphones, tablets, and PCs;

Couper and Peterson 2016; Fischer and Bernet 2014; first part of Liebe et al. 2015; Mitchel 2014; Revilla et al. 2016a; Williams et al. 2015).

Third, previous studies focused on the effect of the screen size on a single aspect of survey data quality, such as respondents' acquiescence tendency (Liebe et al. 2015) or completion time (Couper and Peterson 2016). In order to obtain a more complete view of the phenomenon, we study four indicators of quality: (i) completion *times*, (ii) the proportion of respondents that *failed* answering to an *Instructional Manipulation Check* (IMC), (iii) the *answers' consistency* over two survey waves, and (iv) evaluations of the *survey experience*, from the panelist's perspective. To our knowledge, the relationship between screen size and answer consistency between waves, or the failures in answering an IMC, has not been previously studied.

Fourth, we compare optimized and non-optimized questionnaire layouts, whereas most previous works considered only a non-mobile-optimized (Peytchev and Hill 2008) or only a mobile-optimized layout (Fisher and Bernet 2016; Liebe et al. 2015).

14.3.2 Hypotheses on the Impact of the Screen Size

In this research, we study the impact of the mobile device's screen size on four different aspects of use. Following, we indicate our expectations for each of them.

14.3.2.1 Completion Time

The use of completion time to evaluate data quality is quite controversial (Revilla and Ochoa 2015). For example, Liebe et al. (2015, p. 18) concluded that "it is not clear whether longer interviews are always 'better' and associated with higher quality such as greater response consistency (i.e. similar answers to similar questions)" or if they are caused, for example, by distracting factors that diminish respondents' focus on the survey task. Nevertheless, because completion time is linked to various aspects of the survey experience (e.g. response burden and motivation, or willingness to participate again in a survey), it has been an important variable in numerous studies (Andreadis 2015; Couper and Peterson 2015, 2016; Mavletova 2013; Toepoel and Lugtig 2014). Moreover, Revilla and Ochoa (2015) found a significant correlation between completion time and other quality indicators. Thus, we decided to consider completion times as a potential indicator of data quality. We expect that the smaller the screen size, the higher the need for scrolling and zooming, which increase completion times (Couper and Peterson 2016). Thus, our first hypothesis (**H1**) is: *The smaller the screen size, the longer the completion times.*

Scrolling and zooming can also make survey participation more difficult, which, in turn, can decrease respondent motivation and lead to more satisficing and less careful processing of the information; this could, in contrast, reduce completion times. Nevertheless, we expect the lengthening effect to prevail. The other factors that usually cause longer completion times (i.e. the slow Wi-Fi networks of mobile devices, multitasking, or the use of a virtual keyboard) should not be relevant in our study, since mobile devices should all be similar in that regard.

14.3.2.2 Failure in Answering an IMC

An IMC is a "question embedded within the experimental materials [that] asks participants [...] to provide a confirmation that they have read the instruction" (Oppenheimer et al. 2009, p. 867). This measure is very common in opt-in panels to detect (and, sometimes, exclude) respondents who are not putting sufficient effort in the survey task. We expect smaller screens to lower visibility and to increase the respondents' burden and fatigue, diminishing participants' effort. Thus our second hypothesis (**H2**) is: *The smaller the screen size, the higher the proportion of respondents who will fail to properly follow the IMC.*

14.3.2.3 Answer Consistency

Answer consistency is evaluated by comparing the answers provided by the same respondent to the same questions in two survey waves. A higher consistency indicates a higher quality of answers. However, memory effects or systematic patterns of answers (e.g. always selecting the middle response category) could also lead to high answer consistency. For similar reasons as before, we expect screen size to increase errors in selecting an answer category. Thus, our third hypothesis (**H3**) is: *The smaller the screen size, the lower the answer consistency.*

14.3.2.4 Survey Experience

The respondent's evaluation of the survey experience is a crucial factor, in particular for panels, since it can affect willingness to participate in future surveys (Loosveldt and Joye 2015). Because we expect reduced visibility to increase the required effort, leading to a higher burden, our fourth hypothesis (**H4**) is: *The smaller the screen size, the more negative and difficult the survey experience.*

14.3.3 Hypotheses on the Impact of Questionnaire Optimization

Besides the effect of screen size on the four indicators of quality just listed, we also consider the effect of optimizing the questionnaire layout for small screens on these same indicators. Questionnaire optimization improves readability. Thus, we expect optimization to reduce the effects of a small screen size on completion time (**H1-sub**), on the failures in answering to the IMC (**H2-sub**), on answer consistency (**H3-sub**), and on the survey experience evaluation (**H4-sub**).

14.4 Data Collection and Method

Our data are a subset of a dataset produced in a crossover experiment (see, e.g. Toninelli and Revilla 2016a) implemented using the Netquest opt-in online panel for Spain. Netquest (www.netquest.com) is an online fieldwork company accredited with the ISO 26362 quality standard. Panelists are rewarded for each survey completed depending on the estimated length of the questionnaire.

14.4.1 Experimental Design

In this experiment, the same questionnaire was administered twice to the same group of panelists. We set a one-week lag between the two waves of the survey, in order both to limit the possibility of changes in the surveyed opinions and to avoid memory effects. For each wave, panelists were randomly assigned to one of the following survey conditions: participation through PCs; through a smartphone, using a mobile-optimized version of the questionnaire; or through a smartphone, using a non-optimized version of the questionnaire. In Figure 14.1, one can see the layouts of the same question for the two survey conditions analyzed in this chapter: note that for the mobile non-optimized condition, it is

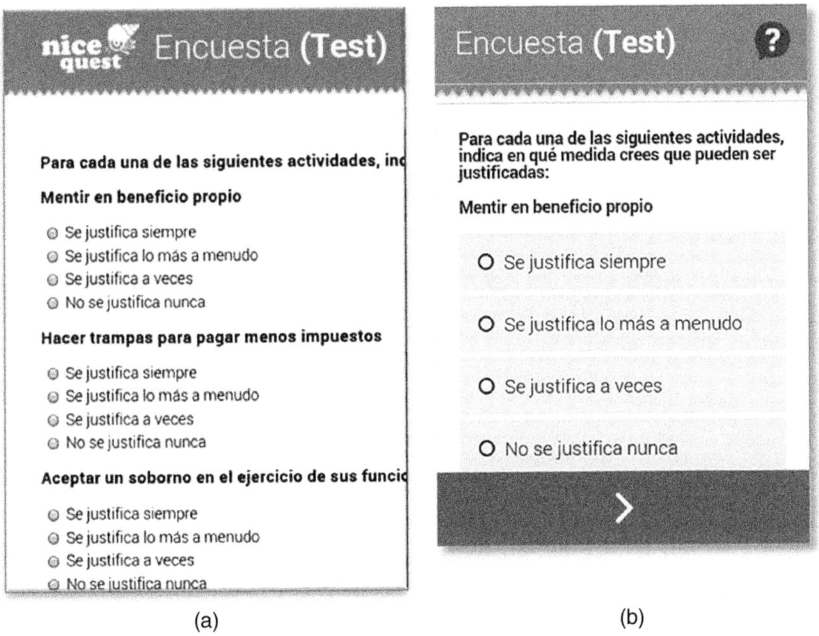

Figure 14.1 Layout of the same question by survey condition. (a) Mobile non-optimized; (b) mobile optimized.

necessary to scroll both horizontally and vertically, whereas for the mobile optimized questionnaire the layout is self-adapting to the screen size, unnecessary elements are avoided, and buttons are bigger (no horizontal scrolling is needed).

At the beginning of each survey, we set an automatic detection of the device used by respondents. If the device did not correspond to the one assigned (PCs or mobile devices), panelists were stopped by a warning until they connected through the requested device. If they did not come back using the right device type, they were screened out. The survey was originally designed to include only PC or smartphone respondents, but some respondents managed to participate using tablets.

14.4.2 Fieldwork

The first wave took place from February 23 to March 2, 2015, and the second wave from March 9 to March 18 of the same year. In order to obtain a group of participants with a distribution similar to the one of Netquest panelists, we used cross quotas on age and gender.

For wave 1, we contacted 3317 panelists; some of them were screened out for not being in-scope (e.g. for not using the requested device type or for not respecting other selection criteria, such as the age range). 1800 (54.3%) completed the first wave survey. These 1800 panelists were invited again, and 1608 completed the wave 2 survey (89.3%, i.e. 48.5% of contacted panelists). In this chapter we focus on respondents who participated through mobile devices (i.e. smartphones or tablets) in both waves, that is on 719 units (44.7% of the 1608 who completed the second wave).

14.4.3 Questionnaire

Part of the questionnaire was modeled after the one used by Mavletova and Couper (2013). The complete questionnaire can be found at the following links: http://goo.gl/5jF2vr (mobile optimized group); http://goo.gl/4c9d1C (mobile non-optimized group). Table 14.1 provides an overview of the sets of questions studied in this chapter.

14.4.4 Method

14.4.4.1 Measuring the Screen Size

Our analyses focus on the screen size of mobile devices used to participate in the survey. Screen height and width (in pixels) were automatically collected as paradata when respondents accessed the survey. However, the number of pixels is not directly linked to the size of a screen, as the latter depends on the density of pixels.[1] Density can vary a lot: in our study it ranged from

1 The pixel density is computed as follows: pixel density $= \frac{\sqrt{(\text{screen height})^2 \cdot (\text{screen width})^2}}{\text{screen size}}$, where the height and width are measured in pixels and the screen size in inches.

Table 14.1 Sets of questions studied in this chapter.

	Content	No. of items	Layout	Scale
Set1 – Deviant behavior: justified	"For each of the following activities, to what extent do you think they are justified?" – Lying in your own benefit, prostitution, etc.	15	4 pages/3 or 4 separate items on each	4-point; fully labeled (1 = "always" to 4 = "never")
Set2 – Deviant behavior: done	"Have you ever done the following?" – Steal something in a shop, cheat on your partner, etc.	15	2 pages/7 or 8 separate items on each	"yes"/"no"
Set3 – Immigration	"To what extent do you agree/disagree with the following statements?" – Spain should allow more people of the same race or ethnic group than the majority of Spanish people to come and live here, etc.	14	1 page/grid [*IMC* included]	5-point; fully labeled (1 = "completely agree" to 5 = "completely disagree")
IMC	*"To confirm that you are reading this text, please do not select any answer, but click here:"*	*1*	*Included in Set3 grid (11th position)*	*Button*
Set4 – Alcohol: consumption	"Please, indicate how frequently you drank alcoholic drinks in the last 30 days"	1	1 page/1 item	7-point; fully labeled (1 = "never" to 7 = "every day")
Set5 – Alcohol: done	"Please, answer the following questions" – Have you ever been drunk during several days?, etc.	9	2 pages/4 or 5 separate items on each	"yes"/"no"
Set6 – Alcohol: judgment	"Please, indicate to what extent you think that the following behaviors are bad or good:" – Drink alcohol being pregnant, etc.	14	1 page/grid	11-point/only the extremes labeled (0 = "completely bad" to 10 = "completely good").
Set7 – Background variables	Income/frequency of internet access by mobile devices	2	2 pages/1 item in each	6-point; fully labeled (recoded into €500–€5000)/6-point; fully labeled (recoded into 1 = "once a month" to 30 = "every day")
Set8 – Sensitivity questions	Perceived sensitivity of questions	1	1 page/1 item	4-point; fully labeled (1 = "very sensitive" to 4 = "not sensitive at all")
Set9 – Survey experience	Easiness/liked the survey	2	2 pages/1 item in each	4-point; fully labeled (1 = "very difficult"/"I did not like it at all" to 4 = "very easy"/"I liked it a lot")

a minimum of 117.5 to a maximum of 539.1 pixels/in. (average: 262.5). The collected paradata also provided us with information about the brand and the model of the mobile devices. This information, combined with the detected height/width, allowed us to find the exact device screen size from the internet, looking up different mobile-devices-specialized websites. This resulting variable "screen size" is measured, diagonally, in inches.

14.4.4.2 Measuring Data Quality

We consider four indicators of data quality: completion time, IMC failures, answer consistency, and survey experience. Each of these measures has limits (for instance, for completion time, see Section 14.3.2). However, taking all indicators together, we expect to get a reasonable picture of the impact of screen size on data quality.

The completion time considered in this chapter is the "client side" completion time, i.e. it excludes the downloading time of the web page on the device. Because some extremely long completion times occurred (suggesting that some respondents left the survey open while doing other tasks), we first dealt with the outliers. According to the completion-time distribution observed for all the pages, we set a general threshold: for each page, the top 1% completion times were substituted using the average completion time of the remaining 99% of cases. Then, the total completion time by respondent was computed, in seconds, summing up completion times observed for single questions.

The IMC was included in a grid of 14 items at the 11th position. The instruction was: "In order to confirm that you are reading this text, please do not select any answer, but click *here*." A higher proportion of respondents who fail in following this instruction indicates a lower data quality.

Answer consistency is measured computing an index based on all the sets of questions listed in Table 14.1. In a specific set of question b, we measured the *relative Answer Consistency* across waves (rAC), as follows:

$$rAC_{ikb} = 1 - \frac{|x_{ikb,w1} - x_{ikb,w2}|}{\max(x_{kb}) - \min(x_{kb})} \tag{14.1}$$

where: $x_{ikb,w\#}$ is the answer x provided by the ith respondent ($i = 1, 2, \ldots, n$) to the kth question ($k = 1, 2, \ldots, p$) of the bth set of items ($b = 1, 2, \ldots, m$), during a certain wave $w\#$ (*wave 1* or *wave 2*); $\max(x_{kb})$ and $\min(x_{kb})$ are, respectively, the maximum and the minimum levels of the evaluation scale used for the kth question of the bth set of items. In order to obtain an indicator of a respondent's consistency for a specific set of items b, we then compute a 0–1 index, called $\overline{rAC}_{ib}$ (*average relative Answer Consistency*). If the bth set includes p items, the average for this set and for the ith respondent is:

$$\overline{rAC}_{ib} = \frac{\sum_{k=1}^{p} rAC_{ikb}}{p} \tag{14.2}$$

We next computed a general index, based on all questions: ($\overline{rAC_i}$). This is the weighted average of the $\overline{rAC_{ib}}$, where the weights correspond to the number of items n_b included in the bth set:

$$\overline{rAC_i} = \frac{\sum_{b=1}^{m} (\overline{rAC_{ib}} \cdot n_b)}{\sum_{b=1}^{m} n_b} \tag{14.3}$$

This index provides a general idea of the consistency observed for a specific respondent ith across items.

The survey experience is measured by two questions, asked at the end of the survey (Set9 in Table 14.1): *Easy survey* ("Please indicate how easy it was to complete the survey") and *Like survey* ("How much did you like completing the survey?").

14.4.4.3 Analyses

The following analyses are divided, for each indicator, into two main phases: the first consists of an analysis by group according to the "screen size" distribution quartiles. *Q1* includes devices with a screen up to 4 in. (34.6% of respondents); *Q2* those with a screen larger than 4 and up to 4.5 in. (19.1%); *Q3* those with a screen larger than 4.5 and up to 5 in. (32.6%); *Q4* those with a screen larger than 5 in. (12.9%). In practice, this criterion places smartphones into three categories (i.e. devices with a screen size up to 5 in.) and unifies devices that are sometimes defined as phablets (5–7 in. screens, extremes excluded) and tablets (7+ in.[2]) into the fourth. In this analysis by group, we apply an ANOVA (or a chi-square test), checking, for each indicator, the null hypothesis of equality of means across the four quartile groups (or the independence between these groups and a categorical dependent variable). These tests are followed by a post-hoc test, in order to find the category (or the categories) that shows significantly different results. For answer consistency, this analysis by group is focused only on respondents showing no screen size change between the two waves.

The second step, a regression analysis, allows us to study the effect of size on the different indicators. In particular, we estimate: a multiple regression model for completion time (y = total completion time, in seconds) and for answer consistency (y = relative answer consistency, measured as shown in formula (14.3)); a logistic regression for the IMC (y = dummy; 1 = "failure in answering the IMC question"); and an ordinal logistic regression model for each of the two survey experience variables (y = how much the respondent found easy or liked the survey participation). Table 14.2 includes the list of independent variables used in the analyses.

2 For details about this classification: http://www.sellmymobile.com/blog/phablet-want-one.

Table 14.2 List of regressors (for completion time and failures in IMC regression models).

Variable	Description	Unit of measure	Collection
Screen size	Diagonally measured	Inches	Paradata + websites
Optimization	Questionnaire optimization	Yes/no	Paradata
Pixel density	Device screen density (quality of view)	Pixel/inches	Paradata + websites
How long acc. int.	How long the respondent has had mobile device internet access (familiarity with the device)	No. of years and months	Self-reported
Freq. acc. int.:	Frequency of mobile devices internet access (experience in surfing the web using mobile devices)	From "once a month" to "every day"	Self-reported
Fare-TimeUse:	Fare paid according to time/usage	Yes/no	Self-reported
Fare-Wifi	Use limited to Wi-Fi availability	Yes/no	Self-reported
Conn. speed satisf.	Satisfaction with connection speed during the survey	4-point; fully labeled (1 = "completely unsatisfied" to 4 = "completely satisfied")	Self-reported
Easy survey	Easy to participate	From "very difficult" to "very easy"	Self-reported
Like survey	How much the respondent liked participating	From "I did not like it at all" to "I liked it a lot"	Self-reported
Felt comfortable	How much the respondent felt comfortable during the survey despite the sensitive topics	4-point; fully labeled (1 = "definitely not comfortable" to 4 = "definitely comfortable")	Self-reported
Perceived sensit.	Perceived question sensitivity	See Table 4.1	Self-reported
Age	Age	Years	Self-reported
High educ.	Higher education level	Dummy (1 = "Graduation/Masters/PhD")	Self-reported

Our analysis of answer consistency necessarily includes data from both wave 1 and wave 2. The rest of our analyses use data from wave 1 only. This is because wave 2 is an uncommon survey situation, and administering the same survey a second time affects user performance in a number of important ways. For example, among the 719 respondents who participated via mobile device in both waves, the average completion time dropped from 16.38 to 13.36 minutes (and standard deviations decreased from 6.06 to 4.84). We suggest that participation in wave 1 is likely to be more representative of typical survey response.[3]

In the answer consistency model, for the "Optimization" variable we use three dummies showing the optimization survey conditions for a given respondent: "always optimized," "not optimized wave 1–optimized wave 2" and "optimized wave 1–not optimized wave 2" ("never optimized" is the reference category).[4]

Throughout the next sections, we use a significance level of 5% ($\alpha = 0.05$), unless otherwise specified.

14.5 Main Results

The data collected on mobile-device respondents within the two waves show a screen size varying from 2.44 to 10.50 in. (mean = 4.65 in.). Does this large variation in screen sizes across mobile devices affect completion time, failures to IMC, answer consistency, and survey experience?

14.5.1 Screen Size and Completion Time

Studying the link between completion time and screen size, we found a significant but low negative correlation (Pearson coefficient = −0.155; $p < 0.001$). Applying an ANOVA to the four quartile groups introduced in Section 14.4.4.3, we found significantly different completion time averages ($p < 0.001$). A Tukey post-hoc test[5] confirmed ($p < 0.001$) that the Q1 completion time average is significantly different from the others groups' averages: the average completion time for Q1 (18.0 minutes) is 22.1% longer than the Q4 one (14.8 minutes). Moreover, the smaller the screen, the stronger the link between completion time and screen size. Indeed, the correlation is negative and significant for Q1

3 Results regarding wave 2 are discussed and reported in a file available upon request from the corresponding author (daniele.toninelli@unibg.it).

4 For all regression models, no problem of multicollinearity is detected between the regressors. In order to check the robustness of estimated models, using the Enter procedure, we also developed the regression analyses using the PASW Forward and Backward procedures, obtaining very similar results, for what concerns the objectives of this chapter. For more information about the Enter, Forward, and Backward procedures, see: http://www.statisticssolutions.com/selection-process-for-multiple-regression.

5 See Tukey (1949).

Table 14.3 Multiple regression model (dependent variable: completion time).

Variables	Coeff.	Std. error	Std. coeff.	*p*-values
(Constant)	**1173.7**	**139.1**	—	**<0.001**
Screen size	**−37.9**	**11.9**	**−0.120**	**0.002**
Optimization	−36.0	27.5	−0.050	0.190
How long acc. int.	1.8	2.5	0.026	0.487
Freq. acc. int.	**−7.7**	**2.7**	**−0.112**	**0.004**
Fare-TimeUse	6.7	71.9	0.003	0.926
Fare-Wifi	103.3	57.5	0.068	0.073
Conn. speed satisf.	14.3	16.7	0.033	0.393
Easy survey	**−50.8**	**24.0**	**−0.094**	**0.035**
Like survey	25.0	24.4	0.046	0.307
Felt comfortable	**38.0**	**18.8**	**0.088**	**0.044**
Perceived sensit.	9.6	22.4	0.017	0.668
Pixel density	−0.3	0.1	−0.068	0.077
Age	**6.3**	**1.3**	**0.189**	**<0.001**
High educ.	−18.3	27.5	−0.025	0.506

Note: Bold = significant variables ($\alpha = 0.05$). $R^2 = 0.116$; adjusted $R^2 = 0.097$.

($-0.209, p < 0.01$) and for Q2 ($-0.198, p = 0.02$), whereas for Q3 and Q4 the correlation is not significant ($p = 0.817$ and $p = 0.703$, respectively). This confirms that very small screens more significantly affect completion times.

Regressing completion time on the independent variables listed in Table 14.2, we obtained the results shown in Table 14.3.[6]

Screen size has a significant effect on completion time: for each additional inch, we observe a reduction of about 38 seconds. These findings support *H1*: the smaller the screens, the longer completion times are.

Optimization has a negative but non-significant coefficient. Thus, optimizing the survey does not significantly reduce completion times. This suggests that horizontal scrolling and zooming are not as important as other factors (e.g. vertical scrolling) in affecting completion times. It is also possible that participants avoided zooming and/or scrolling, losing part of the available information. The *H1-sub* hypothesis is not supported by our data.

6 No preliminary transformation was needed: the dependent variable showed a normal distribution. The same analysis was developed using the standardized versions of the variables, but results are very similar to the ones shown in Table 14.3. Despite the quite low level of R^2 (non-adjusted: 0.116; adjusted: 0.097), for the model showed in Table 14.3, there are no problems with the assumptions about residuals (normally distributed, mean = 0, independent) or for the ANOVA test ($F = 6.019, p < 0.001$).

14.5.2 Screen Size and Failures in Answering the IMC

Overall, 13.6% of mobile respondents in wave 1 failed to answer the IMC. By quartile, the percentage of IMC failures was 14.64% for Q1 (smallest screens), 10.95% for Q2, 16.44% for Q3, and 8.99% for Q4 (biggest screens). Although the percentage for Q1 is 62.9% larger than for Q4, there does not seem to be a general trend linking screen size to IMC failures. Moreover, the ANOVA confirms that the averages by quartile group of the variable "Fail IMC" are not significantly different ($F = 1.381$; $p = 0.247$). However, there is a significant lower proportion of IMC failures in the group with optimized layouts (9.78%) than in the group with non-optimized layout (17.45%), a reduction of 44.4%. The Beasley and Schumacker (1995) post-hoc test confirms that *H2* is not supported ($ps \geq 0.156$), whereas *H2-sub* is supported by our data ($p = 0.012$).

Logistic regression results are shown in Table 14.4. The Cox & Snell R^2 is 0.046 (Cox and Snell 1989), and the Nagelkerke R^2 is 0.085 (Nagelkerke 1991). The average percentage of correct classification is quite high (86.8%); nevertheless the model has difficulties predicting IMC failures (i.e. the model is able to predict 99.8% of IMC correct answers, but it predicts just 1 out of 87 cases of IMC failures, that is 1.1%; thus, it tends to predict all cases as correct answers to the IMC).

Table 14.4 Logistic regression model (dependent variable: rate of IMC failures).

Variables	Coeff.	Std. error	*p*-values
Constant	−2.012	1.363	0.140
Screen size	*−0.365*	*0.193*	*0.059*
Optimization	**−0.530**	**0.248**	**0.033**
How long acc. int.	−0.008	0.030	0.777
Freq. acc. int.	*0.047*	*0.028*	*0.099*
Fare-TimeUse	*0.873*	*0.512*	*0.088*
Fare-Wifi	−0.005	0.515	0.992
Conn. speed satisf.	−0.098	0.137	0.476
Easy survey	*−0.387*	*0.202*	*0.056*
Like survey	0.044	0.217	0.838
Felt comfortable	−0.098	0.173	0.572
Perceived sensit.	*0.382*	*0.198*	*0.054*
Pixel density	0.002	0.002	0.147
Age	*0.022*	*0.011*	*0.045*
High educ.	−0.209	0.242	0.387

Note: Bold = significant variables for $\alpha = 0.05$; Italic = significant variables for $\alpha = 0.10$.

Screen size is significant at a level of $\alpha = 0.1$, so *H2* is supported, at this confidence level, by the estimated model. The IMC failure rate is also affected by optimization; this variable is significant ($p < 0.05$). Thus, *H2-sub* is supported by our data: optimizing the survey significantly reduces the proportion of failures to the IMC by more than half. The IMC failure rate is also affected (at the $p < 0.10$ level) by other variables: frequency of access to the internet, connection cost based on the time of use, perceived sensitivity of the questions and age (all these variables show a positive linkage with the failure rate), and by perceived easiness of the survey (with a negative relation with the dependent variable).

14.5.3 Screen Size and Answer Consistency

The third aspect studied is answer consistency across waves. Within the 719 respondents, we have 638 valid observations (i.e. 93.0% of the 686 respondents that participated in both waves using a mobile device with the change of size data not missing); among these, we observed 48 changes of device size (7.0%). In 14 cases (29.2%) the screen size was reduced (average: −2.13 in.), whereas in 34 cases (70.8%) it was increased (average: +3.31 in.). The overall average increase was of 0.12 in. (mode/median = 0).

The average answer consistency over all items and for all respondents is 0.901, meaning that panelists participating in the two waves answered consistently 90.1% of the time. The analysis by the four size groups focuses on respondents for whom the screen size did not change between the two waves (590 cases). Results show a very slightly decreasing trend, for the answer consistency percentage, over the four quartile groups: 90.4% for Q1, 90.1% for Q2 and Q3, and 89.6% for Q4. The differences are very small, and the ANOVA suggests not rejecting equality of means across groups ($F = 0.490$; $p = 0.689$).

Regarding optimization, the answer consistency percentage is 91.0% for respondents who were twice in the optimized group; 90.0% for respondents assigned twice to the non-optimized group; and 89.9% for respondents who switched between optimized and non-optimized and vice versa. The ANOVA test for the equality of the means of the four quartile groups confirms that there are no significant differences at the significance level we set ($p = 0.089$).

Pearson's correlation coefficient between the average answer consistency by respondent and screen size is negative (−0.025), as is the correlation with "screen size change" (−0.038), but not significant ($p = 0.530$ and $p = 0.350$, respectively). We also computed the correlation coefficients between the average answer consistency and the two size variables ("screen size" and "screen size change") for each set of items listed in Table 14.1. None of these correlations were significant (all $ps \geq 0.152$).

Finally, we conduct a multiple regression analysis with answer consistency as dependent variable (see Section 14.4.4.2). In estimating the regression model, we use all the variables listed in Table 14.2 from wave 1.[7] Nevertheless, since answer consistency takes into account a rate of change in the answers, for some variables we also include the difference between waves 1 and 2 within the model. This group of additional regressors (marked with "(*w2-w1*)" in

Table 14.5 Multiple regression model (dependent variable: answer consistency between wave 1 and wave 2).

Variables	Coeff.	Std. error	Std. coeff.	*p*-values
(Constant)	**0.896**	**0.021**	—	**<0.001**
Screen size	−0.001	0.002	−0.028	0.510
Screen size change (w2-w1)	*−0.004*	*0.002*	*−0.077*	*0.096*
Optimized both waves	*0.009*	*0.006*	*0.084*	*0.097*
Optimized w1-Not optimized w2	−0.001	0.006	−0.013	0.795
Not optimized w1-Optimized w2	−0.001	0.006	−0.008	0.868
Pixel density	0.000	0.000	−0.039	0.354
Pixel density (*w2-w1*)	0.000	0.000	−0.070	0.122
How long acc. int.	0.000	0.001	−0.025	0.552
Freq. acc. int.	0.000	0.000	−0.008	0.861
Fare-TimeUse	*−0.020*	*0.011*	*−0.074*	*0.070*
Fare-Wifi	−0.010	0.009	−0.048	0.257
Satisfaction with connection speed	0.003	0.003	0.048	0.350
Satisfaction with connection speed (w2-w1)	−0.003	0.003	−0.059	0.229
Easy survey	**0.010**	**0.005**	**0.137**	**0.037**
Easy survey (*w2-w1*)	**0.008**	**0.004**	**0.132**	**0.025**
Like survey	−0.005	0.004	−0.060	0.276
Like survey (*w2-w1*)	0.001	0.004	0.013	0.796
Felt comfortable	0.005	0.003	0.081	0.156
Felt comfortable (*w2-w1*)	**0.007**	**0.003**	**0.098**	**0.049**
Perceived sensitivity	**−0.012**	**0.004**	**−0.164**	**0.002**
Perceived sensitivity (*w2-w1*)	**−0.010**	**0.003**	**−0.133**	**0.006**
Age	0.000	0.000	0.054	0.213
High educ.	*0.007*	*0.004*	*0.076*	*0.069*

Note: (*w2-w1*) = change wave 2-wave 1; all other variables = values observed for wave 1; Bold = significant variables for $\alpha = 0.05$; Italic = significant variables for $\alpha = 0.10$.

7 We expect that these variables do not show any change between wave 1 and wave 2; or, if a change is observed, it is not relevant for our study (because it is not linked with the data quality/survey evaluation).

Table 14.5) includes: screen size, pixel density, satisfaction with connection speed, easy survey, like survey, felt comfortable, and perceived sensitivity.

The R^2 is quite low (non-adjusted: 0.083; adjusted: 0.046), but we found no problems with the residuals' assumptions (approximately normally distributed, with mean = 0, independent), or with the model's ANOVA test ($F = 2.226$, $p = 0.001$).

Screen size does not have a significant effect on answer consistency. Thus, *H3* is not supported by our data. In addition, "Screen size change" between the two waves only affects answer consistency at the $\alpha = 0.1$ level. Furthermore, when taking into account questionnaire optimization, consistency only increases (for $\alpha = 0.1$) when the questionnaire is optimized in both waves. Thus, *H3-sub* is only partially supported.

Other variables have significant effects on answer consistency: consistency in filling the questionnaire seems to be directly linked to the perceived easiness of the survey task and with whether the respondents "Felt comfortable" (despite the sensitive nature of the questions), whereas the "Perceived sensitivity" of the questions negatively affects answer consistency. In addition, "High education" and "Fare Time-Use" were significant at the $\alpha = 0.1$ level.

As a robustness check, we applied the same regression analysis to the subsets of questionnaire items introduced in Table 14.1. Results in general were similar (e.g. screen size was not a significant predictor of answer consistency), with a few exceptions. For Set1, higher screen size is associated with lower answer consistency ($p = 0.027$). "Screen size change" is also significantly ($p = 0.040$) associated with lower answer consistency. Although optimization of the survey generally is not a significant predictor of answer consistency, the variable "Optimized both waves" positively affects answer consistency for particular grids (i.e. Set3, with $p = 0.040$, and Set6, with $p = 0.016$, of Table 14.1); this means that when both waves were optimized, consistency increased; thus a shift in the questionnaire format seems to affect observed responses. As with the overall questionnaire, *H3-sub* is only partially supported for these specific questions. Moreover, going deeper into the analyses by sets of survey item, it is interesting to note that the specific optimization dummy "Not optimized w1-Optimized w2" significantly ($p = 0.001$) and negatively affects answer consistency for the Set9 (ease/difficulty of survey) variables. Thus, shifting from a non-optimized to an optimized condition can negatively affect answer consistency, in this case when respondents self-report their survey experience. Moreover Set9 and Set1 questions both use fully labeled four-point scales (without a "neutral" central point), whereas all other sets have scales with a higher number of points (and usually, excluding Set6, with a central point). It may be that larger screens, displaying questions with fewer response options, would be more prone to inconsistency. The reason for that is not completely clear and warrants additional research.

14.5.4 Screen Size and Survey Experience

The last aspect considered is the survey experience. We found significant and positive (but also very low) Pearson coefficients between screen size and the variables "Easy survey" (0.089, $p = 0.012$) and "Like survey" (0.085, $p = 0.014$). Applying ANOVA to the four screen-size quartile groups, we found that the bigger the screen, the easier respondents perceive participation to be (the null hypothesis of equality of means across groups is rejected: $p < 0.01$). The post-hoc Tukey test detects a significantly lower level of easiness for Q1, in comparison to Q3 ($p = 0.011$) and Q4 ($p = 0.017$). There are also differences by group, at the 10% significant level, about how much the survey is liked ($p = 0.099$).

Estimating two ordinal logistic regression models using, respectively, "Easy survey" and "Like survey" as dependent variables and the variables listed in Table 14.4 as regressors,[8] we obtain the results shown in Table 14.6.

Screen size does not have a significant effect on "Easy survey" or on "Like survey." These findings do not support our *H4*.

Table 14.6 Ordinal logistic regression models for survey experience variables.

	Y = *Easy survey*			*Y* = *Like survey*		
Variables	**Coeff.**	**Std. err.**	***p*-values**	**Coeff.**	**Std. err.**	***p*-values**
Screen size	0.052	0.074	0.479	0.132	0.072	0.068
Optimization*	**−0.645**	**0.166**	**0.000**	−0.081	0.164	0.622
Pixel density	**0.002**	**0.001**	**0.048**	−0.001	0.001	0.191
How long acc. int.	−0.009	0.016	0.566	0.034	0.019	0.077
Freq. acc. int.	**0.040**	**0.016**	**0.012**	−0.006	0.016	0.720
Conn. speed satisf.	**0.484**	**0.099**	**0.000**	0.193	0.099	0.052
Like/easy survey	**1.463**	**0.142**	**0.000**	**1.511**	**0.144**	**0.000**
Perceived sensit.	**0.342**	**0.134**	**0.010**	**0.561**	**0.127**	**0.000**
Age	0.008	0.008	0.314	−0.003	0.008	0.653
High educ.	0.282	0.166	0.090	0.007	0.164	0.967
Pseudo R^2						
Cox and Snell	0.283			0.250		
Nagelkerke	0.332			0.290		

Note: Bold = significant variable ($\alpha = 0.05$); * = Reference category: "0 = not optimized."

8 In the two ordinal logistic regression models, we excluded from the list of regressors the two dummies referring to the connection cost, or fare ("Fare-TimeUse" and "Fare-Wifi") and the degree to which respondents felt at ease during the survey ("Felt comfortable").

Optimizing the questionnaire has a positive and significant effect on the variable "Easy survey," whereas the effect is not significant for "Like survey." Thus, optimization makes survey participation significantly easier, but it does not increase how much participants like the survey. This means that *H4-sub* can be accepted for only one of the two survey experience variables. Nevertheless, the post-hoc Tukey test shows significant differences for both variables ($p < 0.001$ and $p = 0.038$) across survey conditions (i.e. optimized versus non-optimized layout).

14.6 Discussion

In this chapter, we analyzed data from an experiment implemented in 2015 in Spain using the Netquest online panel. We administered the same questionnaire to the same respondents in two consecutive waves of data collection, separated by a one-week break. In this work we used a subset of the complete dataset, focusing on 719 respondents who participated in the survey both times using a mobile device. Our main objective was to study if the screen size affects four different indicators: completion time, failure to answer an IMC question, answer consistency, and evaluation of the survey experience. Moreover, we aimed at evaluating if questionnaire optimization reduces the potential effects of using smaller screens, helping in obtaining more comparable data. Our results explore two indicators that have not been previously studied in relation to the screen size (to our knowledge): the proportion of failures to the IMC, and answer consistency. They also provide new insight into the effect of the screen size on evaluation of the survey experience.

Our results confirm an overall finding about completion times from prior studies: the smaller the screen size, the longer the completion time. This supports our *H1*. Moreover, this phenomenon is clearer for the smallest screens, whereas for devices with bigger screens the link becomes weaker. However, questionnaire optimization significantly helps in reducing completion times, even when smaller devices are used (*H1-sub* is supported). This also means that longer completion times are most likely due to the need to zoom and scroll in non-optimized versions of the questionnaire, as suggested by previous research (e.g. Couper and Peterson 2015, 2016).

Through both a quartile analysis and a logistic regression model, we found no significant link between screen size and failure to answer an IMC (*H2* is not overall supported: for $\alpha = 0.1$, we found support in the logistic regression analysis, but the ANOVA analysis did not support the hypothesis); but, again, optimization of the questionnaire significantly helps in lowering these failures (*H2-sub* is supported). We did not find any link between answer consistency across waves and screen size (*H3* not generally supported), and optimization was only significant for particular question formats (i.e. grids; thus,

H3-sub is partially supported). To evaluate the survey experience, we analyzed two questions about how easy respondents found the survey experience and how much they liked participating. Although ANOVA analyses support *H4*, regression analysis suggests no significant link between screen size and survey experience. Optimization affects only perceptions of ease, but not attitudes of liking the survey experience (*H4-sub* is partially supported).

One potential concern is that data from respondents using different screen sizes might not be comparable in terms of data quality. We did find some effects of smaller screen sizes on the response process, but mainly in terms of completion times. Our findings suggested that the reduced size of the screen does not considerably affect the data quality, at least using our indicators. We also confirmed that an optimized questionnaire significantly reduces this risk. Also, given that optimization increases perceptions of ease, it may also affect willingness to participate again in surveys. Nevertheless, given rapid evolution of the technology, we suggest periodically monitoring the effect of device size on data quality. Further research could use enhanced indicators or more detailed paradata (e.g. paradata directly collected about the real screen size of the device, changes in the responses selected, interruptions of the survey, and so on). This will lead to a more precise evaluation of data quality and of how certain participation characteristics can affect perception of the survey experience.

This study has some limitations. First, it used respondents of a non-probability based online panel. It would be useful to repeat the study using a population-based probability sample. Second, our data are from one country (Spain); replication in other areas would also be useful. Also, we found no significant links between the studied indicators and the respondents' frequency of internet access, but in subsequent studies we recommend including a more direct item aimed at measuring respondents' level of experience with mobile devices, which is likely to play an important role in moderating the effects of different types of screens and devices.

Acknowledgments

We are very grateful to Netquest for providing us with the necessary data, to the University of Bergamo (this research has been partially supported by 60% University funds), and to Carmelita Mercuri (for help collecting the screen-size data). This work has been realized with the support of the Grants for Visiting Professor and Scholar named ITALY – Italian Talented Young Researchers program (University of Bergamo, 2014/15). We are thankful also to the anonymous referees of this book who significantly ($p < 0.001$) contributed to enhancing the quality of this chapter. Finally, we thank very much the organizers of the International Conference on Questionnaire Design, Development, Evaluation, and Testing (QDET2) (for inviting us to present this work), and, in particular, Paul Beatty (who, with invaluable effort – and cutting a lot of useless "the"s – enhanced the general quality of the text and helped us in clarifying several of its parts with useful comments and remarks); "last," but of course not "least," we would like to thank the whole QDET2 Publications Committee for giving us the chance of being published in this book.

References

Andreadis, I. (2015). Comparison of response times between desktop and smartphone users. In: *Mobile Research Methods: Opportunities and Challenges of Mobile Research Methodologies* (eds. D. Toninelli, R. Pinter and P. de Pedraza), 63–79. London, UK: Ubiquity Press https://doi.org/10.5334/bar.

Baker-Prewitt, J. (2013). Mobile Research Risk: What Happens to Data Quality When Respondents Use a Mobile Device for a Survey Designed for a PC? CASRO Online Research Conference, San Francisco, US.

Beasley, T.M. and Schumacker, R.E. (1995). Multiple regression approach to analyzing contingency tables: post hoc and planned comparison procedures. *Journal of Experimental Education* 64 (1): 79–93. https://doi.org/10.1080/00220973.1995.9943797.

Boreham, R. and Wijnant, A. (2013). Developing a web-smartphone-telephone questionnaire. In: *Proceedings of the 15th International Blaise Users Conference, IBUC*, 145–160. Heerlen: Statistics Netherlands.

Bosnjak, M., Poggio, T., Becker, K.R., et al. (2013). Online survey participation via mobile devices. 68th Annual Conference of the American Association for Public Opinion Research (AAPOR), Boston, Massachusetts, US.

Brick, J.M., Brick, P.D., Dipko, S. et al. (2007). Cell phone survey feasibility in the U.S.: sampling and calling cell numbers versus landline numbers. *Public Opinion Quarterly* 71 (1): 23–39.

de Bruijne, M. and Wijnant, A. (2013a). Comparing survey results obtained via mobile devices and computers: an experiment with a mobile web survey on a heterogeneous group of mobile devices versus a computer-assisted web survey. *Social Science Computer Review* 31 (4): 482–504.

de Bruijne, M. and Wijnant, A. (2013b). Can mobile web surveys be taken on computers? A discussion on a multi-device survey design. *Survey Practice* 6 (4): 1–8. Retrieved from: http://surveypractice.org/index.php/SurveyPractice/article/download/238/pdf.

de Bruijne, M. and Wijnant, A. (2014). Mobile response in web panels. *Social Science Computer Review* 32 (6): 728–742.

Buskirk, T.D. and Andrus, C. (2012). Smart surveys for smartphone: exploring various approaches for conducting online mobile surveys via smartphones. *Survey Practice* 5 (1): 1–11.

Buskirk, T.D. and Andrus, C. (2014). Making mobile browser surveys smarter: results from a randomized experiment comparing online surveys completed via computer or smartphone. *Field Methods* 26 (4): 322–342. https://doi.org/10.1177/1525822X14526146.

Callegaro, M. (2010). Do you know which device your respondent has used to take your online survey? *Survey Practice* 3 (6): 1–12.

Chae, M. and Kim, J. (2004). Do size and structure matter to mobile users? An empirical study of the effects of screen size, information structure, and task complexity on user activities with standard web phones. *Behaviour & Information Technology* 23: 165–181.

Couper, M.P. and Peterson, G.J. (2015). Exploring why mobile web surveys take longer. General Online Research (GOR) conference. http://www.websm.org/uploadi/editor/doc/1436889306Couper-Exploring_Why_Mobile_Web_Surveys_Take_Longer-113.pdf.

Couper, M.P. and Peterson, G.J. (2016). Why do web surveys take longer on smartphones? *Social Science Computer Review* https://doi.org/10.1177/0894439316629932.

Cox, D.R. and Snell, E.J. (1989). *Analysis of Binary Data*, 2e. London, UK: Chapman & Hall.

Fischer, B. and Bernet, F. (2014). Device effects: How different screen sizes affect answer quality in online questionnaires. General Online Research (GOR) conference, Cologne, Germany. http://www.websm.org/uploadi/editor/1396100898Fischer_Bernett_Device_Effects.pdf.

Fisher, B. and Bernet, F. (2016). Device effects – How different screen sizes affect answers in online surveys. General Online Research (GOR) conference. http://www.websm.org/uploadi/editor/doc/1470831394Fischer_Bernet_2016_Device_Effects_-_How_different_screen_sizes_affect_answers.pdf.

Liebe, U., Glenk, K., Oehlmann, M., and Meyerhoff, J. (2015). Does the use of mobile devices (tablets and smartphones) affect survey quality and choice behaviour in web surveys? *Journal of Choice Modelling* 14: 17–31. https://doi.org/10.1016/j.jocm.2015.02.002.

Loosveldt, G. and Joye, D. (2015). Defining and assessing survey climate. In: *The SAGE Handbook of Survey Methodology* (eds. C. Wolf, D. Joye, T. Smith, et al.), 67–76. London, UK: SAGE.

Mavletova, A. (2013). Data quality in PC and mobile web surveys. *Social Science Computer Review* 31 (4): 725–743.

Mavletova, A. and Couper, M.P. (2013). Sensitive topics in PC web and mobile web surveys: is there a difference? *Survey Research Methods* 7 (3): 191–205.

Millar, M.M. and Dillman, D.A. (2011). Improving response to web and mixed-mode surveys. *Public Opinion Quarterly* 75 (2): 249–269.

Millar, M.M. and Dillman, D.A. (2012). Encouraging survey response via smartphones: effects on respondents' use of mobile devices and survey response rates. *Survey Practice* 5 (4) Retrieved from: http://www.surveypractice.org/index.php/SurveyPractice/article/view/19/html.

Mitchel, N. (2014). When it comes to mobile respondent experience and data quality, survey design matters. Quirk's Media. http://www.quirks.com/articles/2014/20140825-3.aspx.

Nagelkerke, N.J.D. (1991). A note on a general definition of the coefficient of determination. *Biometrika* 78 (3): 691–692.

Oppenheimer, D.M., Meyvis, T., and Davidenko, N. (2009). Instructional manipulation checks: detecting satisficing to increase statistical power. *Journal of Experimental Social Psychology* 45 (4): 867–872.

Parush, R.E. and Yuviler-Gavish, N. (2004). Web navigation structures in cellular phones: the depth/breadth trade-off issue. *International Journal of Human-Computer Studies* 60 (5–6): 753–770.

Peterson, G. (2012). Unintended mobile respondents. CASRO Technology Conference, New York, NY, US.

Peytchev, A. (2009). Survey breakoff. *Public Opinion Quarterly* 73 (1): 74–97.

Peytchev, A. and Hill, C.A. (2008). Experiments in mobile web survey design. 63rd Annual Conference of the American Association for Public Opinion Research (AAPOR).

Revilla, M. and Ochoa, C. (2015). What are the links in a web survey among response time, quality, and auto-evaluation of the efforts done? *Social Science Computer Review* 33 (1): 97–114. https://doi.org/10.1177/0894439314531214.

Revilla, M., Toninelli, D., Ochoa, C., and Loewe, G. (2015). Who has access to mobile devices in an opt-in commercial panel? An analysis of potential respondents for mobile surveys. In: *Mobile Research Methods: Opportunities and Challenges of Mobile Research Methodologies* (eds. D. Toninelli, R. Pinter and P. de Pedraza), 119–139. London, UK: Ubiquity Press https://doi.org/10.5334/bar.

Revilla, M., Toninelli, D., and Ochoa, C. (2016a). Personal computers vs. smartphones in answering web surveys: does the device make a difference? *Survey Practice* 9 (3): 1–6. Retrieved from: http://surveypractice.org/index.php/SurveyPractice/article/view/338.

Revilla, M., Toninelli, D., Ochoa, C., and Loewe, G. (2016b). Do online access panels need to adapt surveys for mobile devices? *Internet Research* 26 (5): 1209–1227. https://doi.org/10.1108/IntR-02-2015-0032.

Stapleton, C.E. (2013). The smartphone way to collect survey data. *Survey Practice* 6 (2): 1–7.

StatCounter Global Stats (2017), available at: http://gs.statcounter.com (accessed December 2017).

Stern, M.J., Bilgen, I., and Dillman, D.A. (2014). The state of survey methodology: challenges, dilemmas, and new frontiers in the era of the tailored design. *Field Methods* 26 (3): 284–301.

Struminskaya, B., Weyandt, K., and Bosnjak, M. (2015). The effects of questionnaire completion using mobile devices on data quality. Evidence from a probability-based general population panel. *Methods, Data, Analyses* 9 (2): 261–292.

Sweeney, S. and Crestani, F. (2006). Effective search results summary size and device screen size: is there a relationship? *Information Processing & Management* 42 (4): 1056–1074.

Toepoel, V. and Lugtig, P. (2014). What happens if you offer a mobile option to your web panel? Evidence from a probability-based panel of Internet users. *Social Science Computer Review* 32 (4): 1–17.

Toninelli, D. and Revilla, M. (2016a). Smartphones vs PCs: does the device affect the web survey experience and the measurement error for sensitive topics? – A replication of the Mavletova & Couper's 2013 experiment. *Survey Research Methods* 10 (2): 153–169. https://doi.org/10.18148/srm/2016.v10i2.6274.

Toninelli, D and Revilla, M. (2016b). Is the smartphone participation affecting the web survey experience? 48th Scientific Meeting of the Italian Statistical Society (SIS), Salerno, Italy. http://hdl.handle.net/10446/71456.

Tukey, J. (1949). Comparing individual means in the analysis of variance. *Biometrics* 5 (2): 99–114.

Wells, T., Bailey, J.T., and Link, M.W. (2014). Comparison of smartphone and online computer survey administration. *Social Science Computer Review* 32 (2): 238–255.

Williams, D., Maitland, A., Mercer, A., and Tourangeau, R. (2015). The impact of screen size on data quality AAPOR conference, Hollywood, Florida, US. http://dc-aapor.org/2015%20conference%20slides/Williams.pdf.

15

Optimizing Grid Questions for Smartphones: A Comparison of Optimized and Non-Optimized Designs and Effects on Data Quality on Different Devices

Trine Dale and Heidi Walsoe

Department of Research and Development, Kantar TNS, Oslo, Norway

15.1 Introduction

Mobile design and use has evolved immensely in the last decade (e.g. Link et al. 2014), and we are in the middle of a paradigm shift in survey research, invoked by what Conrad et al. (2017) have termed "the mobile revolution." Indeed Miller has argued that "... the future of surveys will depend on their adaptation to changing communication technology" (Miller 2017, p. 208). The mobile revolution poses opportunities and challenges to survey designers.

For example, despite an increasing number of respondents using mobile devices to participate in surveys, web surveys are often not designed with smartphone users in mind, and this can lead to what some researchers have termed "unintended mobile respondents" (Peterson 2012), i.e. respondents who respond via mobile devices even if the survey is optimized for PCs. When questionnaires are not designed with smartphones in mind, some questions, such as matrix grids, do not fit properly on the screen. In such cases, data quality may suffer (e.g. Richards et al. 2016; Callegaro et al. 2015). Adopting a responsive and intelligent questionnaire design approach, where the visual presentation of content automatically adjusts to the device's screen size and browser, is seen as a way in which such problems can be addressed.

In this chapter we investigate how different designs, optimized and non-optimized for smartphones, affect results in grid questions on different devices with different screen sizes. In a quasi-experiment implemented in 2015, we compare response patterns in different grid designs. Our main findings suggest that design choices with regard to visual presentation and functionality in grids may impact data quality, both for smartphone respondents and respondents using other devices. Before presenting our findings and discussing their implications, we consider arguments for adapting questionnaires and grid questions specifically for mobile devices, summarizing what

Advances in Questionnaire Design, Development, Evaluation and Testing, First Edition.
Edited by Paul C. Beatty, Debbie Collins, Lyn Kaye, Jose-Luis Padilla, Gordon B. Willis, and Amanda Wilmot.

is known about the performance of grid questions on different devices and setting out the rationale for our research and the design of the study.

15.2 The Need for Change in Questionnaire Design Practices

The proportion of respondents using mobile devices to complete surveys is relatively low, but rising (e.g. Richards et al. 2016; Toepoel and Lugtig 2016; Struminskaya et al. 2015). In the Netherlands, about 30% of respondents occasionally use mobile devices to answer surveys (Lugtig et al. 2016). In Nordic Kantar surveys, the average share of respondents using mobile devices varied between 24% (Finland) and 46% (Sweden) on average in 2017.[1]

Traditional web survey practices, including questionnaire design, need to be reviewed to take account of the increasing use of mobile devices, particularly smartphones, and how they may affect response behavior. This requires a mobile-friendly approach to web survey design. Indeed some researchers predict that in the near future, respondents will expect or demand to be able to use the device of their choosing when participating in surveys (Conrad et al. 2017). In the Nordics, this future is already upon us.

There is growing evidence that web surveys do not work well on smartphones if questionnaires are not optimized for mobile devices (e.g. Couper 2013, 2016; Revilla et al. 2016; Dillman et al. 2014). If questions are not optimized, they do not fit properly on the screen, making excessive scrolling – vertical and horizontal – necessary in order to read and understand the questions, and to record answers. Response rates are also reported to be lower, and completion times longer, on smartphones compared to PCs when questions are not optimized (Andreadis 2015; de Bruijne and Wijnant 2013b; Mavletova and Cooper 2013), and higher break-off rates have been observed (Couper 2016; Bosnjak et al. 2013). These factors can lead to a very poor respondent experience, requiring more cognitive effort from respondents to respond to non-optimized questionnaires on a mobile device (de Bruijne and Wijnant 2013a). This higher response burden may lead to more break-offs and to respondents taking cognitive shortcuts when responding to questions (Buskirk and Andrus 2014; Bosnjak et al. 2013; Peytchev 2009; Dale and Haraldsen 2007).

Ignoring mobile device users is no longer an option, and some researchers argue that designing for smartphones also has benefits for non-mobile device users. An example is the use of an item-by-item design to address many of the recognized challenges grids pose for respondents (e.g. Andreadis 2015; Tourangeau et al. 2013).

---

1 Kantar key performance indicator (KPI) data can be made available by authors. The low rate in Finland is due to a high share of Windows-based phones.

15.2.1 Designing Grids for Smartphones

Grids or matrices are typically used to present a series of questions or items that use the same response scale, where the questions are listed in the rows of a grid and the response options are listed in the columns. The use of grid questions in self-completion questionnaires is widespread, yet their use is not without problems. Research on traditional matrix grid designs used in paper and web self-completion questionnaires completed on a PC suggests that the grid structure (of rows and columns) can be cognitively difficult, and that this difficulty can be added to if items are long and complex (Dillman et al. 2009; Couper et al. 2013; Couper 2008). There is evidence to suggest traditional grid designs can suffer from higher break-off rates, i.e. the respondent leaves the interview without completing it (Couper 2016; Tourangeau et al. 2013; Peytchev 2009); satisficing in the form of straightlining or non-differentiation of answers given (Couper 2016; Schonlau and Toepoel 2015; Zang and Conrad 2013), and item-missing data (Couper 2016; Tourangeau et al. 2013). These types of challenges are often cited as evidence that grids should be avoided (e.g. Dillman et al. 2009, 2014; Tourangeau et al. 2013).

Recently, researchers have started to look at how grid questions perform on different devices, particularly smartphones (e.g. Toninelli and Revilla, Chapter 14 of this book; Couper 2016; Richards et al. 2016). Matrix grid designs are potentially problematic for smartphone users because not all necessary information (e.g. response options/scales, column headings/labels, items to be rated) may be visible to respondents at the same time due to the small screen size, necessitating both vertical and horizontal scrolling to view the entire grid (Tourangeau et al. 2013). With scales, which are often associated with grids, Mavletova (2013) and Peytchev and Hill (2010) found some evidence that respondents struggled when recording answers. Radio buttons and check boxes are often small and grouped so closely together that it is difficult to record answers, and errors can occur. These problems can make the response task more difficult for respondents using smartphones (Antoun et al. 2017; Tourangeau et al. 2013).

There is a growing body of research looking at how different grid designs – traditional matrix and optimized – affect survey results, using a range of quality indicators. However, evidence from these studies is mixed. For example, some studies show little or no response effects of mobile optimization. Mavletova (2013) found no differences between mobile and PC users with regard to non-substantive responses and primacy effects. de Bruijne and Wijnant (2014) found no evidence that mobile respondents avoided an open "other" category to avoid typing. Toepoel and Lugtig (2016) found that most satisficing behavior did not change when respondents switched device from PC to a mobile phone. Antoun et al. (2017) found that smartphone respondents were at least as likely to disclose sensitive information as PC respondents, and that smartphone users typed longer answers in open-ended questions. However, they also found that respondents had some problems using their fingers to accurately move a small-sized slider handle and date-picker wheel, and suggest that smartphone respondents should be accommodated in web surveys. Antoun and colleagues recommend that the survey instrument should be optimized for small screens, but that features like sliders and date-picker wheels should be avoided (Antoun et al. 2017).

Other studies indicate that an optimized item-by-item approach performs better on certain quality indicators than a non-optimized design. For instance, Richards et al. (2016) found less item nonresponse and straightlining in optimized grids. Couper (2016) found somewhat lower break-off rates, and less straightlining. He also found that the item-by-item approach resulted in longer response times, which can be seen as a negative effect. However, it can be argued that longer response times help respondents give more accurate answers and thus contribute to improved data quality in grids (Couper et al. 2013). Mavletova et al. (2017) found evidence of higher measurement error and lower response quality on several measures in a matrix design when compared to an item-by-item design, and Liu and Cernet (2017) found differences in favor of optimizing in straightlining and longer scales. Making sense of these mixed findings is hindered by the fact that descriptions (including screenshots) of the grid designs being measured and compared are often not detailed enough to determine the extent or nature of the optimization applied. Also, while some researchers recommend an item-by-item instead of a matrix grid design (Tourangeau et al. 2013), there are many ways to break a grid down into an item-by-item design, and both visual design and functionality applied may vary greatly. Research comparing different item-by-item designs is needed.

15.3 Contribution and Research Questions

The main objective of this chapter is to explore some of the issues raised by the mixed findings in published research described in Section 15.2. Specifically, we report on a quasi-experiment that compared different optimized and non-optimized grid designs using three indicators of quality: (i) the extent of straightlining, (ii) scale use, and (iii) item nonresponse. We chose these three indicators for the following reasons:

Straightlining – Several studies demonstrate that straightlining, or non-differentiation, often occurs in grids, with respondents giving identical ratings (same column or response option) to a series of questions/statements. Lack of response differentiation is seen as an indicator of poor response quality (Schonlau and Toepoel 2015; Zang and Conrad 2013).

Clustering – According to Dillman et al. (2009), there are many considerations to be made when using scales in questions. For closed-ended ordinal scales, like the ones used in this quasi-experiment, it is important that the visual presentation is linear (vertical or horizontal), that the spacing between options is even, and that all scale points carry equal weight (spacing, labels, end labels) and are visible in a self-completion questionnaire. This requirement may be a challenge for grid questions on smartphones, particularly with longer scales (seven points or more), because of limited screen space. The clustering of responses around end points or the scale midpoint is often an indicator of poor data quality, as it is likely that respondents did not take the time to read all statements or keep track of the meaning of different scale points, or they may not have bothered to do the necessary scrolling to see the whole scale.

Item nonresponse – Sometimes referred to as *item missing data,* this refers to whole questions or parts of questions not being answered: for instance, one or more rows in a grid. Item nonresponse can be caused by design effects (e.g. difficult, unclear question wording; poor visual display; scrolling; etc.), because all question items don't show on the screen and respondents may not realize or make the effort to read them all, or by respondent characteristics (e.g. does not wish to answer, takes shortcuts). The literature states that item nonresponse is one of several quality issues in grids (Dillman et al. 2009; Couper 2008), and studies indicate that item nonresponse may be influenced by grid design and device, even if the results are not conclusive (see for instance Tourangeau et al. 2013 for a review of grid designs).

Device use, response rates, and break-off rates are also included when reporting results, to provide context and allow comparison with other studies. Device use is important when studying the impact of different designs on data quality, specifically between PCs and smartphones, due to differences in screen size, functionality, and answer modes (touchscreen versus mouse). Since grid matrices display poorly on smartphones, much effort is going into finding alternative designs – for instance, different item-by-item designs, and using interactive features to facilitate response (e.g. Puleston 2014; Tourangeau et al. 2013).

Most studies have hitherto focused on how a non-optimized grid design performs on different devices with regard to data quality measurements, or on how an optimized design for smartphones affects data quality compared with a non-optimized (matrix grid) for PCs. Several studies have also focused on whether an item-by-item design for smartphones and a traditional matrix design for PCs produce different results and differences in data quality (Couper 2016; Fisher and Bernet 2016; Richards et al. 2016). However, little research has tested whether applying optimized design principles to grids improves data quality irrespective of device used, despite some researchers advocating optimization by using an item-by-item approach regardless of device (e.g. Dillman et al. 2014). Furthermore, there is limited research that compares different optimized grid designs to assess whether certain designs produce better quality data than others. If optimization produces better-quality data regardless of device use, this would help researchers tackle the well-known issues around grids in web and mobile web surveys. It would also reduce or eliminate the need for different question layouts for different devices, with its associated increase of costs and risks to data comparability. Our research contributes to knowledge in these areas.

15.3.1 Research Questions

Our hypothesis is that grids optimized for mobile will lead to better survey data quality than non-optimized grids on all devices, particularly on smartphones.

Our research questions are:

1) How does redesigning a 10-point scale matrix grid to an item-by-item loop with a slider affect response quality on different devices?
2) How does redesigning a seven-point scale matrix grid to an item-by-item dynamic grid with large buttons affect response quality on different devices?
3) How does redesigning an item-by-item loop with check boxes to a dynamic grid with large buttons affect response quality on different devices?

We explore questions 1 and 2 more closely by testing the effects of the aforementioned grid designs on straightlining, use of end points in scales, and item nonresponse. For research question 3, we use number of answers per statement and item-nonresponse as quality indicators.

15.4 Data Collection and Methodology

The data for our grid study is from a survey on summer holiday experiences implemented in September 2015 in GallupPanelet, Kantar TNS Norway's probability-based online panel.

15.4.1 Quasi-Experimental Design

As stated earlier, the focus of our study is the impact of questionnaire (grid) design – optimized versus non-optimized – on data quality. We chose a high-interest topic, questions that everyone should be able to respond to, and response tasks with relatively small cognitive challenges that should not be too burdensome for the respondent, in an effort to limit external influences on the data. A low-interest topic or very demanding response tasks could lead to more break-offs or poor response quality regardless of design. We used a quasi-experimental design in which the same questions were asked in two different formats (optimized and non-optimized) in two independent subsamples.

We embedded three quasi-experiments in the survey, in which we manipulated the questionnaire (grid) design. Two versions of the questionnaire were

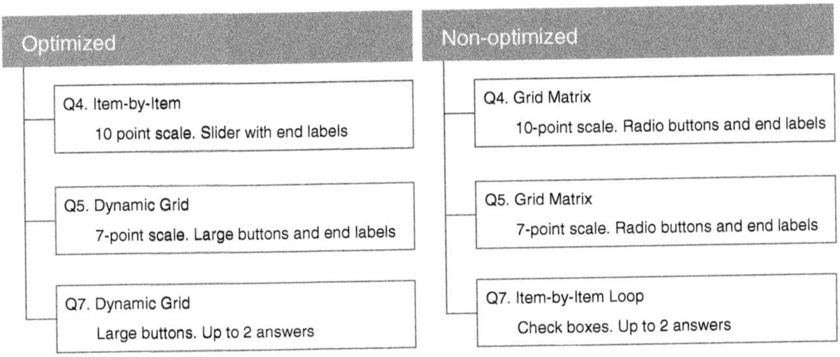

Figure 15.1 Design of the quasi-experiment.

produced: one in which the grids were optimized for mobile; and one where more traditional designs were used (see Figure 15.1).

The wording of the questions tested was identical across designs. The survey[2] included 18 questions, but in this analysis we focus on the 3 grid questions (Q4, Q5, and Q7). We excluded a fourth grid (Q9), as the design was identical to Q7, and preliminary analysis demonstrated similar results. Screenshots of the grids included in the analysis are displayed and described briefly in Section 15.4.4.

Respondents did not know they were taking part in an experiment, and were free to use the device of their own choosing. They were randomly assigned to one of the two versions of the questionnaire (optimized/non-optimized) regardless of which device they used. We chose this approach as respondents in GallupPanelet are accustomed to using whichever device they prefer, since all surveys are optimized, and the design at the time was adaptive (programmed to "fit" certain devices and screen sizes). The questionnaire included automated device detection, which identifies which device is used by which respondent, and this information was used in the analysis.

15.4.2 Fieldwork and Sample Design

As previously mentioned, the sample source was Kantar TNS's probability-based online panel, GallupPanelet.[3] The panel is the only access panel in Norway accredited with the ISO 26362:2009 quality standard, and the majority of its members are recruited from CATI surveys. The data was collected during a seven-day field period. Invites were distributed via email, and one reminder was sent four days after launch. Median interview length was seven minutes. Respondents were rewarded 1 credit point (equal to 1 NOK) per minute for a

2 See the appendix for an overview of the questions and English wording.
3 https://www.gallup-panelet.no. The panel book in English can be provided by the authors.

completed interview (about $0.12 USD), which is in accordance with the panel's incentives policy.

The gross sample totaled 1200 respondents, 600 for each questionnaire condition (optimized and non-optimized). The overall response rate was 54%,[4] which produced a net sample of 651 respondents: 328 on the optimized, and 323 on the non-optimized version of the questionnaire. An overview of sample distribution is presented in Table 15.1.

The sample was representative for Norwegians aged 18 years and older with internet access. Prior to sampling, the panel was stratified by geography, gender, and age, producing 32 strata. Within each stratum, simple random samples were selected in such a manner that the sample sizes within the strata were proportional to the population size (see Table 15.1). As previously mentioned, the sample was randomly split into two subsamples. Since the randomization was done by the questionnaire script after the interview had started, it was not possible to analyze survey nonresponse separately for the two subsamples. Response was voluntary, both to the questionnaire and to individual questions, and as such some item nonresponse was expected.

Respondents and non-respondents were compared by gender, age, education level, and geography. Age was the only variable that was significantly different between the two groups, with respondents ($M = 43$, $SD = 16.2$) being on average six years older than non-respondents ($M = 49$, $SD = 16.1$) – ($t(1198) = 6.43$, $p < 0.001$).

15.4.3 Data Analysis

Given that the measurement scales of our dependent variables are nominal and ordinal, the optimized and non-optimized samples were compared by using Chi-square and Wilcoxon rank-sum tests. When analyzing Q4 and Q5, we used the same approach: for all 12 holiday types we compared the respondents' use of the most-positive response category in the two questionnaire conditions by using a first-order Rao-Scott modified Chi-square. To analyze straightlining, we used a traditional Chi-square test. For Q7 we created two new dichotomous variables: (i) those who responded to all eight statements, and (ii) those who responded to all eight statements and reported for two holiday types on all statements. These two variables were analyzed by using a Chi-square test. For all three questions, we controlled for confounding effects from devices by applying stratified analysis. We used the Mantel-Haenszel statistic to test the association between the analysis variable and design after controlling for device used. This included significance testing for each analysis variable and design after stratification for device used.

Respondents and non-respondents were compared by demographic factors with Chi-square or Student t-test. We used a CHAID (Chi-square automatic

4 $RR = \frac{Net}{Gross} \cdot 100\%$

Table 15.1 Sample distribution by demographics and device type.

	Gross	Nonresponse	Break-off	Net	Optimized	Non-optimized
	1200	513	36	651	328	323
Sample	100	43%	3%	54%	27%	27%
Gender and age						
Male 18–29	10.3	14.8	11.7	6.7	7.3	7.1
Male 30–44	14.3	18.2	10.0	11.3	12.8	9.6
Male 45–59	13.1	10.0	16.7	15.5	14.3	16.1
Male 60+	12.6	8.6	18.3	15.6	18.3	12.7
Female 18–29	10.0	10.8	10.0	9.3	8.5	10.5
Female 30–44	13.1	14.3	10.0	12.2	11.9	12.4
Female 45–59	12.3	11.9	8.3	12.8	12.2	13.6
Female 60+	14.3	11.4	15.0	16.6	14.6	18.0
Region						
Central	25.1	25.9	29.1	24.5	24.4	24.5
East	27.1	27.3	30.8	26.8	27.1	26.6
South and West	33.5	30.0	27.3	36.2	29.3	31.9
Mid and North	14.3	16.8	12.8	12.5	19.2	17.0
Education level						
Primary and secondary	29.8	33.2	25.0	27.2	29.9	24.8
Vocational school	15.7	15.4	15.0	15.9	17.4	14.2
University level	54.5	51.4	60.0	56.9	52.7	61
Personal income (000) NOK						
0–299	22.5	24.5	21.7	20.7	22	19.5
300–599	50.8	48.8	53.3	50.8	49.1	52.4
600+	15.9	14.3	21.7	16.4	15.2	17.6
Missing	10.8	12.3	3.3	12.1	13.7	10.5
Device						100
PC			51.1	64.9	64.7	64.7
Tablet			22.3	14.9	13.1	16.7
Smartphone			26.6	20.2	21.6	18.6
Other					0.6	

interaction detection) decision tree to study relationships between device used and demographic factors. CHAID is based on adjusted significance testing and is useful when looking for patterns in datasets, as the relationships can be easily visualized (Kass 1980).

15.4.4 Grid Design Features

For the optimized grids, we used two different designs: item-by-item with a slider (Q4) and dynamic grids (Q5 and Q7). A dynamic grid is an item-by-item based carousel feature, where the items being rated are placed in "cards" at the top of the screen, and the response alternatives are placed in large answer boxes below. The first dynamic grid (Q5) only allowed one answer per statement, and moved automatically to the next statement when an answer was recorded. The second dynamic grid (Q7) allowed up to two answers per item and did not have an automated forward function. For the non-optimized grids in Q4 and Q5, we applied a traditional matrix design, while Q7 was designed as an item-by-item loop. The items (holiday types) being rated in these grids were filtered from Q1 and Q2, which asked about holidays taken and considered (see Appendix 15.A).

15.4.4.1 Grid Matrix Design vs. Item-by-Item Design with Slider Bar – Q4

Respondents were asked to rate holiday types they had used and/or considered relevant on a 10-point scale (very poor to very good). In the non-optimized matrix grid, the horizontal scale was presented as radio buttons with labels on the end points and numbers on the between points on all devices. In the optimized grid, the scale was presented as a slider with labels on the end points and numbers on the between points on all devices. The cursor was placed within the slider at the most negative point. Cursor placement can impact responses (Couper 2008), in this case leaning toward the negative end of the scale as the scale goes from negative to positive. The left-side anchoring of the cursor was standard at that time. Screenshots of the optimized and non-optimized grid designs are displayed in Figure 15.2.[5]

15.4.4.2 Grid Matrix Design vs. Dynamic Grid Design – Q5

Respondents were asked to rate the quality of different holiday types by evaluating statements on a seven-point scale (disagree completely to agree completely) for each holiday type used or considered relevant. In the non-optimized matrix grid, the horizontal scale was presented as radio buttons with the end points labeled and the between points numbered on all devices. In the optimized dynamic grid, the response alternatives appear as large buttons – vertical on mobile, horizontal on PC, with labels on the end points and numbers on the between points on all devices. Screenshots of the optimized and non-optimized grid designs are displayed in Figure 15.3.

5 For English wording, see the Appendix 15.A.

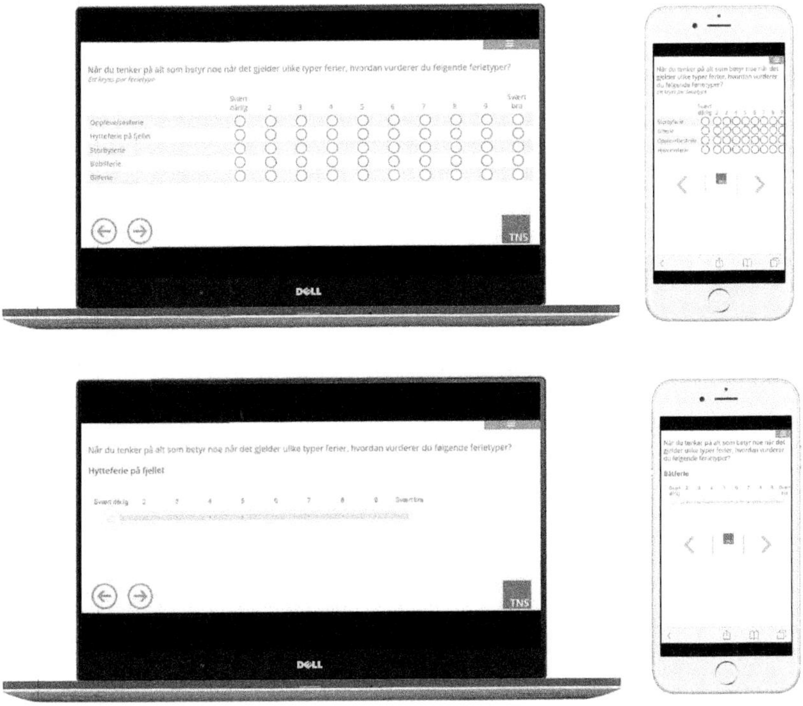

Figure 15.2 Q4 Grid matrix design (non-optimized) and item-by-item design with a slider bar (optimized) on PC and IPhone 6.

The visual appearance of the matrix grids on smartphones was not very appealing. They were also not very practical or respondent friendly. It was practically impossible to register an answer without zooming, as the radio buttons were too close together, horizontally as well as vertically. When zooming, the items to be rated and/or the column labels were lost beyond the visible edges of the screen. In addition, the distance between response alternatives was not uniform, with the end points being slightly further away from the other response options to accommodate the end-point labels. This may accentuate the end points from the other response options, making them more prominent, which in turn could influence respondents' answers and lead them to favoring the end points. In both the optimized grids (Q4 and Q5) the intention was to make the response task easier than in the matrix design, with items being presented one at a time (item-by-item), the response options being visible at all times, and the response buttons being larger, more distinguishable, and easier to use. In addition, in the optimized designs the spacing between response options was uniform. The visual presentation of the interactive designs worked well on all devices.

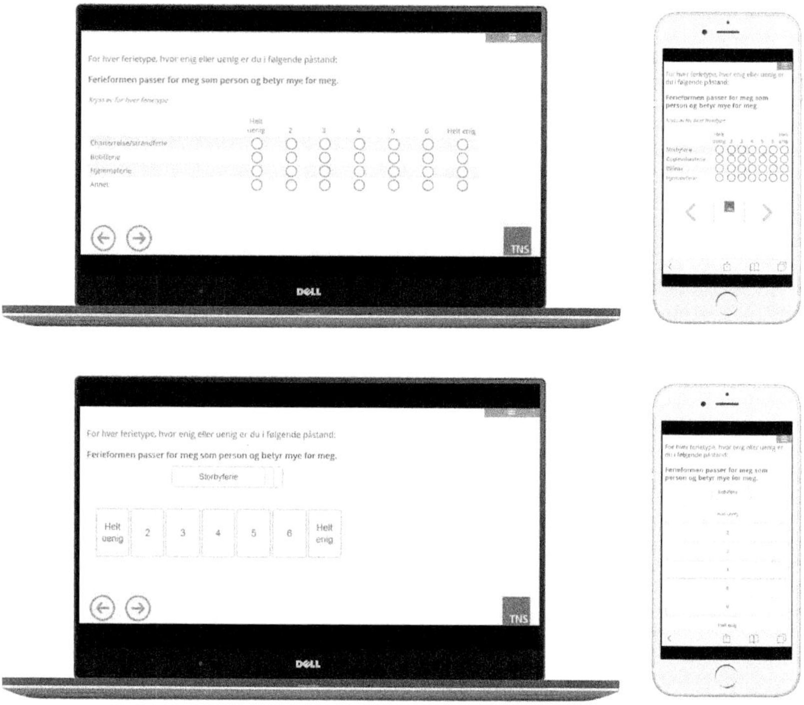

Figure 15.3 Q5 Grid matrix design (non-optimized) and dynamic grid (optimized) on PC and IPhone 6.

15.4.4.3 Item-by-Item Loop Design vs. Dynamic Grid Design – Q7

In this test we chose an item-by-item approach for both the optimized and non-optimized grids. The non-optimized grid was designed as an item-by-item loop, where the statements were presented one at a time, and answers were recorded in check boxes on all devices. The optimized version was a dynamic grid, with large answer buttons and the response option written inside the button on all devices. As in the previous dynamic grid (Q5), the response options were vertical on smartphones and horizontal on PCs. The question consisted of eight statements in both questionnaire conditions, and respondents were asked to choose which holiday types fit the statements best – up to two responses per statement were allowed, see Figure 15.4.

15.5 Main Results

We explored the impact of the grid design on data quality, looking at straightlining and scale use for Q4 and Q5, and item nonresponse and number of answers

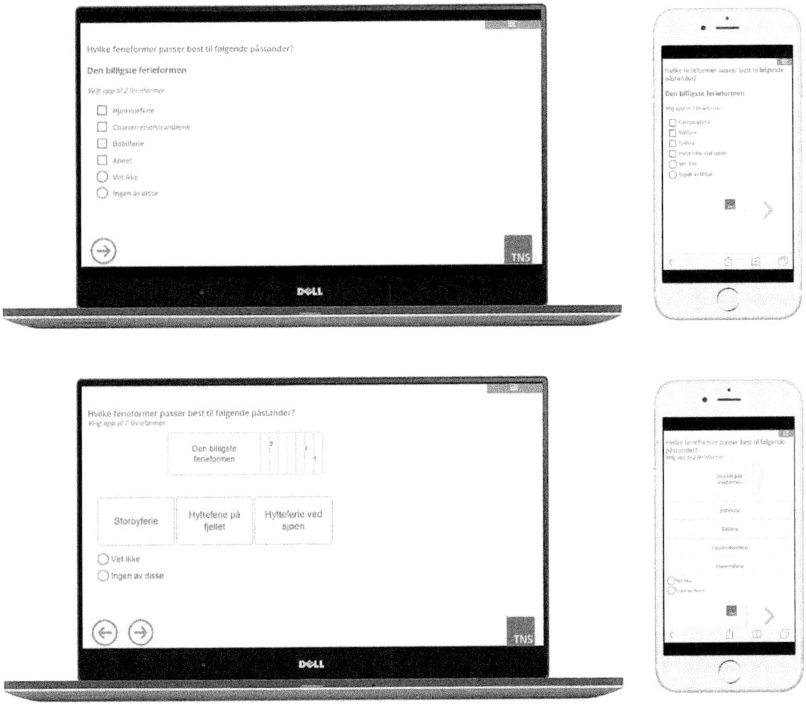

Figure 15.4 Q7 Item-by-item loop design (non-optimized) and dynamic grid (optimized) on PC and IPhone 6.

in Q7. In Q4 and Q5 we defined as straightliners those respondents who gave four or more identical answers to the items included in the questions (internal variation = 0), since it was expected that the answers would be on the positive end of the scale for a large portion of the respondents. The topic was of high interest to most people and normally associated with positive experiences.

15.5.1 Analysis of Grid Matrix vs. Item-by-Item With a 10-Point Slider Scale – Q4

With the 10-point scale grid, we found significantly more straightlining in the non-optimized matrix design than in the item-by-item slider design (($\chi^2(12\,N = 649) = 26.1, p < 0.001$) (Table 15.2). Placement of the cursor in the slider at the most negative point within the scale in the optimized design did not seem to have any effect, since most of the answers were on the positive end of the scale both in the optimized and non-optimized designs.

Despite the need for horizontal scrolling by smartphone users in the matrix grid to see the most positive scale point, we found a significantly higher

Table 15.2 Q4 (10-point scale) Proportion of cases exhibiting straightlining and using the most-positive response category, by design and device.

	Optimized				Non-optimized				
	All units (%)	PC (%)	Tablet (%)	Smartphone (%)	All units (%)	PC (%)	Tablet (%)	Smartphone (%)	p-value
SL	2	2	2	0	9	9	7	10	*<0.001*
MPC	15	15	13	17	25	23	31	23	*0.02*

SL, straightlining; MPC, most positive scale category.

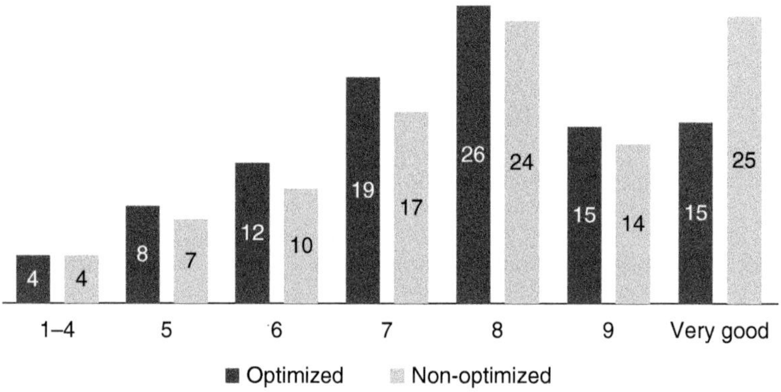

Figure 15.5 Comparison of response distribution (%) by scale point for grid designs with a 10-point answer scale.

frequency of use of the most-positive scale point in the non-optimized design $((\chi^2(12N = 649) = 23.4, p = 0.02)$ on all devices (Table 15.2). The results indicate that for this grid question, its design affected both straightlining and scale use, resulting in satisficing in the non-optimized grid. Device had no significant impact on the results, and there was no confounding effect when controlling for device.

Distribution of responses in the optimized and non-optimized design is shown in Figure 15.5.[6] Respondents to the non-optimized matrix grid were 1.7 times more likely to choose the most-positive scale category for all holiday types than respondents to the optimized matrix grid. Again, there were no significant differences between devices.

6 The first four points on the scale (very poor + points 2, 3, and 4) were collated, as very few respondents used them.

Table 15.3 Q5 (7-Point scale) Proportion of cases exhibiting straightlining and using the most-positive response category, by design and device.

	Optimized				Non-optimized				
	All units (%)	PC (%)	Tablet (%)	Smart-phone (%)	All units (%)	PC (%)	Tablet (%)	Smart-phone (%)	p-value
SL	6	6	9	4	10	11	11	7	*0.08*
MPC	24	23	35	25	29	28	32	30	*0.98*

SL, straightlining; MPC, most positive scale category.

15.5.2 Analysis of Grid Matrix vs. Item-by-Item Dynamic Grid with a 7-Point Scale – Q5

With the 7-point scale grid there was also a tendency to more straightlining in the non-optimized matrix design than in the optimized grid, though this difference was not significant. There was also no significant difference in the use of the most-positive scale point between the designs, contrary to what we found in Q4. The distribution of the results on straightlining and use of the most-positive scale point is displayed in Table 15.3.

Distribution of responses in the optimized and non-optimized design is shown in Figure 15.6.[7] Respondents to the non-optimized grid tend to use the two most positive scale points, while the respondents to the optimized grid are slightly more negative, but the results were not significant.

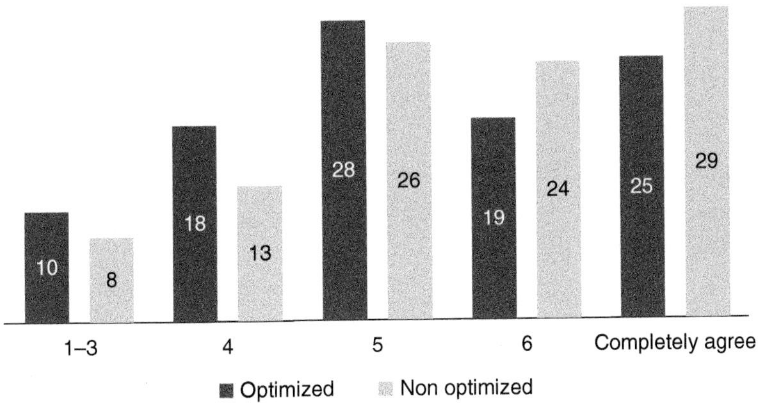

Figure 15.6 Comparison of response distribution (%) by scale point for grid designs with a 7-point answer scale.

7 The first three points in the scale (completely disagree + points 2 and 3) were grouped together for the analysis, as very few respondents used them.

Table 15.4 Q7 Respondents who responded to all eight statements and reported two holidays on all eight statements by grid design and device.

	Optimized				Non-optimized				
	All units	PC	Tablet	Smart-phone	All units	PC	Tablet	Smart-phone	p-value
Respondents	325	211	43	71	322	208	54	60	
All eight statements	256	162	30	64	257	168	43	46	
	79%	77%	70%	90%	80%	81%	80%	77%	*0.74*
All eight statements and two holidays	24	12	2	10	38	24	8	6	
	7%	6%	5%	14%	12%	12%	15%	10%	*0.05*

15.5.3 Analysis of Item-by-Item Loop Design vs. Dynamic Grid Design – Q7

Respondents were asked to mark which holiday types best fit a description in eight statements. Two answers were allowed per statement. As can be seen in Table 15.4, the proportion of respondents reporting two holidays on all statements was significantly higher in the non-optimized loop design ($\chi^2(1647) = 3.70$, $p = 0.054$). PC and tablets users reported two holiday types on all eight statements in the non-optimized loop twice as often as occurred in the optimized dynamic grid. However, for smartphone users, the picture was different: they were 1.4 times more likely to report two holiday types on all eight statements in the optimized dynamic grid.

With regard to item nonresponse, there were no significant differences in the number of statements responded to by design or device (Figure 15.7). However, a significantly higher proportion of smartphone users responded to all eight statements in the optimized design (χ^2 1,N = 325) = 7.02, $p = 0.008$).

15.5.4 Effects of Design on Break-off Rate, Completion Rate, Item Nonresponse, and Device Use

The break-off rate was low regardless of design and device used (6% in the optimized, 4% in the non-optimized design), and there were no significant differences or patterns to the break-off, except that it all occurred before any questions had been answered. Hence, design and device use did not seem to impact the completion rate. Median interview length was also very similar across designs and devices. An overview of the distribution of sample, completes, break-offs, and interview length, as well as number of responses for each grid, by design and device, is presented in Table 15.5. The small discrepancy between completes on the whole survey and the different grids

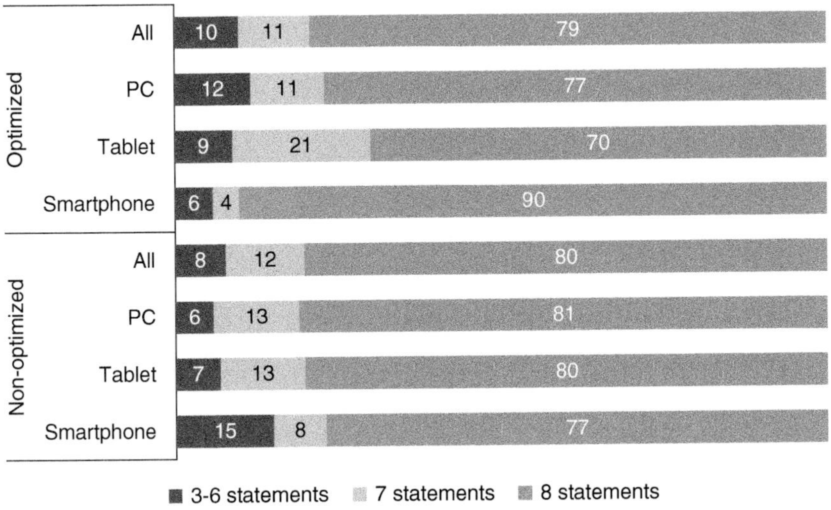

Figure 15.7 Distribution (%) of number of statements responded to by design and device.

Table 15.5 Overview of distribution of response, break-off, interview length, and grid questions by design and device.

	Optimized[a)]			Non-optimized		
	PC	Tablet	Smartphone	PC	Tablet	Smartphone
No. of respondents starting survey	220	49	79	217	58	62
No. of breakoffs	8	6	8	8	4	2
No. of respondents completing survey	212	43	71	209	54	60
Q4: Rating of holiday activities used or preferred	208	43	71	208	54	60
Q5: Rating of quality of different holiday types	209	43	71	208	54	60
Q7: Holiday types that fit statements. 8 statements, 1–2 responses per statement	211	43	71	208	54	60
Median interview length survey	7	7	8	6	8	7

a) Two respondents are excluded due to lack of information on device.

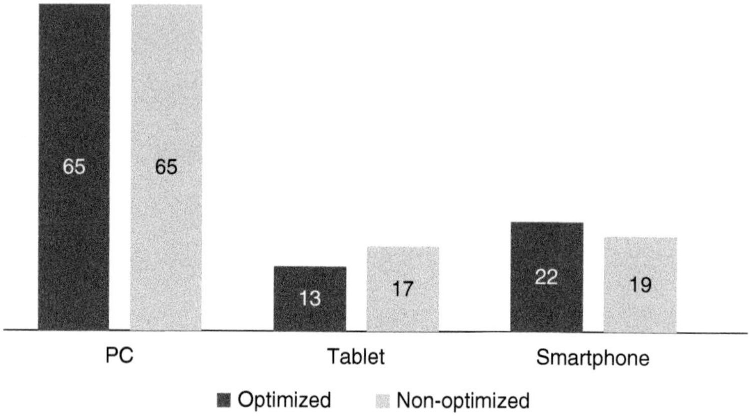

Figure 15.8 Comparison of response rates (%) by questionnaire condition (optimized and non-optimized) and device.

is explained by item nonresponse. Respondents were able to skip questions, but very few used this opportunity: four PC respondents skipped Q4, three Q5, and one Q7 in the optimized design. In the non-optimized design, one PC respondent skipped each question. There was no item nonresponse by tablet and smartphone respondents in any of the designs.

We found no significant differences in device use between questionnaire conditions, which implies that the routing to the questionnaire designs worked as intended. Just over a third of respondents used a mobile device in the optimized and non-optimized designs (35% and 36%, respectively). The distribution of response rate by questionnaire condition and device used can be seen in Figure 15.8. No respondents switched devices while responding.

As mentioned in Section 15.4.3, device use was discovered by CHAID. Age, and to a lesser extent gender, were the most important variables determining device use, and are the same factors that determined response rate. CHAID resulted in three age groups: 18–44, 45–69, and 70+. Smartphone users were more likely to be in the youngest age group (38%) and female (45%). The average age of smartphone users was 37.3 ($SD = 12.6$), while PC and tablet users were older, with an average age of 52 ($SD = 16$); see Figure 15.9.

15.6 Discussion

Our research was designed to compare optimized and non-optimized grid designs on multiple devices, testing whether an optimized design produced data of a higher quality on any or all devices, but particularly on smartphones. We explored this through studying straightlining, scale use, number of

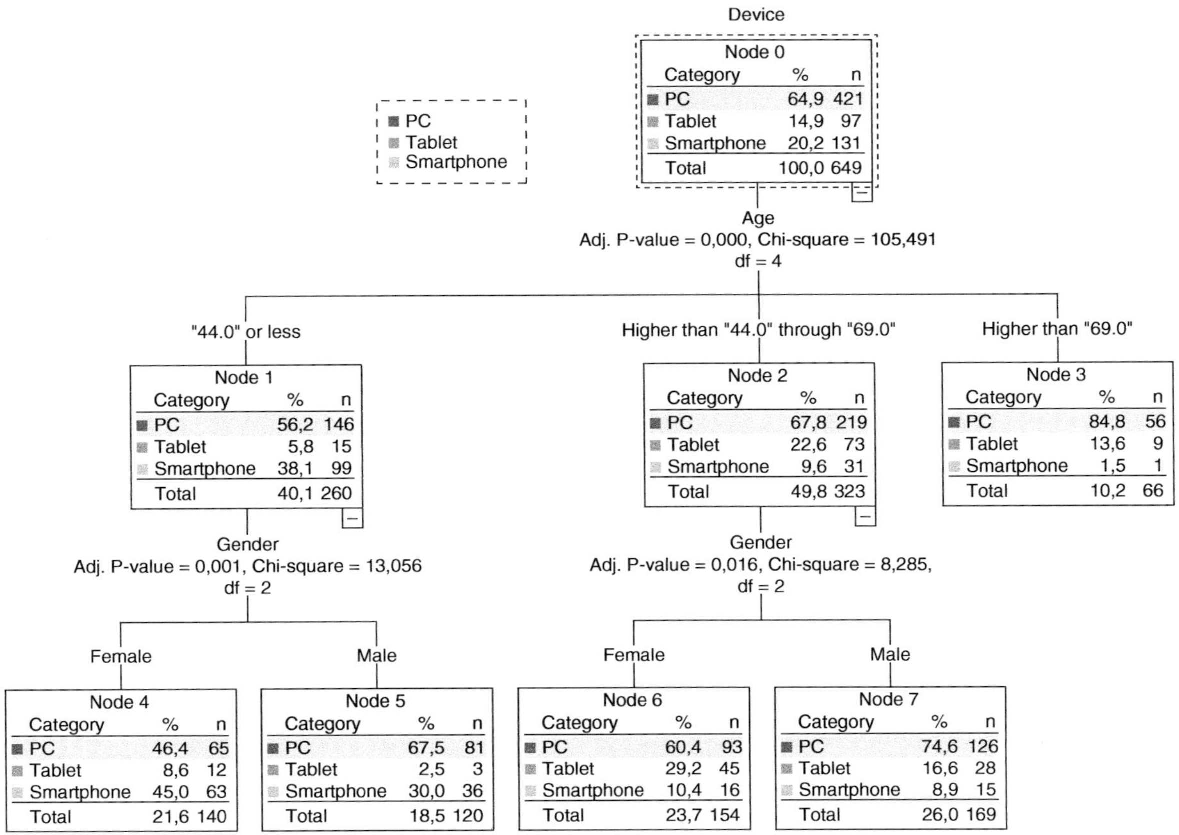

Figure 15.9 CHAID tree with device as a dependent variable.

answers, and item nonresponse. We also explored the impact of design and device use with regard to completion and break-off. The grid designs compared were grid matrix versus loop with a slider; grid matrix versus dynamic grid, and item-by-item loop versus dynamic grid. All the optimized questions were presented using an item-by-item design. The wording and content of the questions were identical in both questionnaire conditions.

The results were mixed. We found some significant differences in response behavior between the optimized and non-optimized designs for some questions formats and not others. In the matrix versus item-by-item grid design with a 10-point scale (Q4), there was significantly less straightlining and clustering (non-differentiation) in the optimized item-by-item design on all devices. A significantly higher proportion of smartphone users also responded to all statements in the optimized design. When comparing a matrix with an item-by-item dynamic grid with a 7-point scale (Q5), we found no significant differences between the designs, although there was a tendency toward more straightlining in the non-optimized design. These results may indicate that for longer scales, at least, an optimized item-by-item approach may improve data quality on all devices, but particularly for smartphone users. However, there are other possible explanations for these findings. One explanation might be that the response pattern reflected actual attitudes rather than a lack of effort (i.e. satisficing). As we pointed out earlier, the summer holiday is a positive experience for most people, and it is therefore likely that many respondents deliberately used the most-positive response alternative. Another explanation might be that the design and/or content made it easier to differentiate answers in one of the grids compared to the other. In the 10-point scale matrix grid, the most-positive response alternative was not visible without scrolling for smartphone users, but they used this option just as often as PC respondents who did not have to scroll. Understanding response behaviors to this question could help explain this finding. A third explanation might be that 10-point scales are cognitively more challenging for respondents. There is some evidence of this from other studies: Mavletova et al. (2017) found indications of differences in measurement equivalence in sets with seven or more scale points, and Liu and Cernet (2017) found indications of lower data quality in longer scales. Further research is needed to explore the results from our study and explanations proposed here.

In the comparison of a non-optimized and optimized item-by-item design (Q7), PC and tablet users reported a higher number of holidays in the non-optimized loop design, whereas smartphone users were more likely to report a higher number of holidays in the optimized dynamic grid. The results indicate that the optimized design improves response quality for smartphone users, even when an item-by-item approach is used in the non-optimized design, whereas the non-optimized design improved data quality for PC and tablet users. These results should be treated with caution, however. Due to the small number of respondents giving two answers on all statements; further testing is recommended.

15.6.1 Limitations of the Study

There are some limitations to this study due to challenges with doing experiments in a real survey and the study's design:

Experimental design. We did not allocate respondents to a certain device. This may undermine the results some, but in an access panel where respondents are used to being able to use the device of their choice it is tricky allocating them to certain devices. Several studies have shown that respondents do not necessarily follow instructions to use a certain device, and that forcing them to do so leads to more break-off and other errors. Consequently, many researchers (e.g. Conrad et al. 2017; Link et al. 2014) do not recommend this approach. Also, in an access panel with 45 000 members, it would not be possible to provide devices for respondents. Smartphone penetration in Norway is among the highest in the world, and all (or close to all) panel members probably have access to a smartphone and could use one if they wanted to. Since we did not control device allocation, it is difficult to determine to what extent observed differences are due to grid design or characteristics of the respondents and their choice of device. However, we believe the results give valuable insights to the response behavior in grid questions in real life.

Study design. Due to a misunderstanding, only the grids had different designs in the two questionnaire conditions. This is likely to have influenced the low break-off and item nonresponse rates, particularly for smartphone users, as respondents are less likely to leave a survey once they have started. It may also have influenced response times. We believe the differences between designs would have been more prominent if the design conditions had been applied throughout the questionnaire. In retrospect, it would also have been better to draw two separate samples and maybe even script two different questionnaires instead of splitting the sample and design within the questionnaire script. Using one sample and one script saves time and costs, but it also makes it more difficult to keep track of response rates and separate different effects. Moreover, the wording in some of the questions could have benefited from a redesign, particularly with regard to smartphone optimization. However, we chose to keep the wording as close to the original template[8] as possible despite the challenges because we wanted to see how the template worked with this type of questionnaire.

8 The questionnaire was based on a template with a fixed set of questions used for measuring brand strength.

15.6.2 Suggestions for Future Work

Although it has limitations, our study supports earlier findings that optimization improves data quality for smartphone users on several indicators. This is in line with the latest research on grids (e.g. Toninelli and Revilla, Chapter 14 of this book; Mavletova et al. 2017; Liu and Cernet 2017). Our study also indicates that optimization may improve data quality on PCs/tablets when matrix grids are replaced with an item-by-item design. One design for all devices may help reduce mode effects, design effects, and costs. It is our belief that an optimized design will also improve usability and the respondents' survey experience. However, it is still unclear which type of optimization should be used – paging or scrolling; dynamic or non-dynamic; sliders or no sliders; optimization for all devices or a split design. It is also not clear how features such as the number of items in a grid, different scale lengths, and different survey lengths impact data quality. We suspect that some of the discrepancies in the results of existing research on grids cited in this chapter are related to differences in the question designs – visual and content – that were used. More research comparing different grid designs, interactive tools, question types (e.g. sensitive/not sensitive, cognitively complex/not complex), scale lengths, and survey lengths on different devices is needed.

One of the major concerns with replacing matrices with an item-by-item design is increased response times, but in shorter surveys like ours, with fairly low complexity questions used in the grids, the increase is marginal. Increased response time might even be an indication that respondents actually engage in deeper cognitive processing of the questions/response alternatives. In longer surveys and/or in surveys with more complex grid questions, the differences in response times might be bigger. More research is needed to understand the relationship between response latency and data quality in different grid designs.

We need to understand better what happens when respondents participate in real surveys. Our study sheds some light on this, but we can also learn from our own experience in running a panel, where we capture data on respondents' behavior, performance, and feedback on different surveys. For example, we also know a lot about panelists' use of and performance on different devices, and are able to monitor quality measures regularly. Since we started optimizing our surveys in 2014, we have seen a drop in break-off rates on all devices, but particularly on smartphones. We have also received fewer complaints from respondents, and we have seen increased response rates in younger age groups. With optimization, our surveys have become shorter; and when we occasionally do longer surveys or non-optimized surveys, we see lower response rates increased break-off rates, and receive more complaints and more negative feedback from respondents. We see the same trend when, in some very rare cases, respondents have only been allowed to respond on PC/Mac, or where for reasons beyond our control, we have not been able to optimize. Many respondents trying to access a survey first on their smartphone will not log back in on the assigned device and complete the survey. A combination of qualitative and quantitative studies, paired with paradata from quantitative surveys, would likely add to our understanding of respondent behavior, and how different designs work in practice on different devices. We encourage other researchers with "behind the scene" access to panels to collect and publish such data.

Acknowledgments

We would like to thank our editors, Debbie Collins and Jose-Luis Padilla, for their excellent help and guidance throughout the process of writing this chapter! Thank you also to Kantar TNS for financing and supporting our project, and to anonymous readers and helpers at Kantar TNS.

References

Andreadis, I. (2015). Web surveys optimized for smartphones: are there differences between computer and smartphone users? *Methods, Data, Analyses* 9 (2): 213–228. https://doi.org/10.12758/mda.2015.012.

Antoun, C., Couper, M.P., and Conrad, F.C. (2017). Effects of mobile versus PC web on survey response quality. *Public Opinion Quarterly* 81 (Special Issue): 280–306.

Bosnjak, M., Poggio, T., Becker, K.R., et al. (2013). Online survey participation via mobile devices. 68th Annual Conference of the American Association for Public Opinion Research, Boston, MA. http://www.websm.org/uploadi/editor/1378470537Bosnjak_et_al_2013_Online_Survey_Participation.pdf.

de Bruijne, M. and Wijnant, A. (2013a). Comparing survey results obtained via mobile devices and computers: an experiment with a mobile web survey on a heterogeneous group of mobile devices versus a computer-assisted web survey. *Social Science Computer Review* 31 (4): 482–504.

de Bruijne, M. and Wijnant, A. (2013b). Can mobile web surveys be taken on computers? A discussion on a multi-device survey design. *Survey Practice* 6 (4): 1–8.

de Bruijne, M. and Wijnant, A. (2014). Improving response rates and questionnaire design for mobile web surveys. *Public Opinion Quarterly* 78 (4): 951–962.

Buskirk, T.D. and Andrus, C. (2014). Making mobile browser surveys smarter: results from a randomized experiment comparing online surveys completed via computer or smartphone. *Field Methods* 26 (4): 322–342.

Callegaro, M., Lozar, K., and Vehovar, V. (2015). *Web Survey Methodology*. Sage Publications.

Conrad, F., Shober, M., Antoun, C. et al. (2017). Respondent mode choice in a smartphone survey. *Public Opinion Quarterly* 81 (Special Issue): 307–337.

Couper, M.P. (2008). *Designing Effective Web Surveys*. New York: Cambridge University Press.

Couper, M.P. (2013). Is the sky falling? New technology, changing media, and the future of surveys. *Survey Research Methods* 7 (3): 145–156.

Couper, M.P. (2016). Grids versus item-by-item designs on smartphones. General Online Research (GOR) Conference, Cologne.

Couper, M.P., Tourangeau, R., Conrad, F., and Zang, C. (2013). The design of grids in web surveys. *Social Science Computer Review* 31 (3): 322–345. Retrieved from https://www.ncbi.nlm.nih.gov/pmc/articles/PMC4172361.

Dale, T. and Haraldsen, G. (eds.) (2007). *Handbook for Monitoring and Evaluating Business Survey Response Burden*. European Commission/Eurostat Retrieved from http://ec.europa.eu/eurostat/documents/64157/4374310/12-handbook-for-monitoring-and-evaluating-business-survey-resonse-burden.pdf.

Dillman, D.A., Smyth, J.D., and Christian, L.M. (2009). *Internet, Mail, and Mixed-Mode Surveys. The Tailored Design Method*, 3e. Hoboken, NJ: Wiley.

Dillman, D.A., Smyth, J.D., and Christian, L.M. (2014). *Internet, Phone, Mail, and Multi-Mode Surveys: The Tailored Design Method*, 4e. Hoboken, NJ: Wiley.

Fisher, B. and Bernet, F. (2016). Device effects: How different screen sizes affect answers in web surveys. General Online Research (GOR) Conference, Cologne.

Kass, G.V. (1980). An exploratory technique for investigating large quantities of categorical data. *Applied Statistics* 29 (2): 119–127. Retrieved from: ftp://public.dhe.ibm.com/software/analytics/spss/documentation/statistics/23.0/en/client/Manuals/IBM_SPSS_Decision_Trees.pdf.

Link, M., Murphy, J., Shober, M. et al. (2014). Mobile technologies for conducting, augmenting and potentially replacing surveys. Executive summary of the AAPOR task force on emerging technologies. *Public Opinion Quarterly* 78 (4): 779–787.

Liu, M. and Cernet, A. (2017). Item-by-item versus matrix questions: a web survey experiment. *Social Science Computer Review*. First Published November 30, 2016. https://doi.org/10.1177/0894439316674459.

Lugtig, P., Toepoel, V., and Amin, A. (2016). Mobile-only web survey respondents. *Survey Practice* 9 (4): 1–8.

Mavletova, A. (2013). Data quality in PC and mobile web surveys. *Social Science Computer Review* 31 (6): 725–743. https://doi.org/10.1177/0894439313485201.

Mavletova, A. and Cooper, M.P. (2013). Grouping of items in mobile web questionnaires. *Field Methods* 28 (2): 170–193.

Mavletova, A., Couper, M.P., and Lebedev, D. (2017). Grid and item-by-item formats in PC and mobile web sruveys. *Social Science Computer Review*. First Published Online 25. October 2017. doi: https://doi.org/10.1177/0894439317735307.

Miller, P. (2017). Is there a future for surveys? *Public Opinion Quarterly* 81: 205–212.

Peterson, G. (2012). What we can learn from unintentional mobile respondents. CASRO Technology Conference. http://www.insightsassociation.org/sites/default/files/misc_files/casro-2013_journalfinal_pdf.pdf.

Peytchev, A. (2009). Survey breakoff. *Public Opinion Quarterly* 73 (1): 74–97.

Peytchev, A. and Hill, C. (2010). Experiments in mobile web survey design. similarities to other modes and unique considerations. *Social Science Computer Review* 28 (3): 319–335.

Puleston, J. (2014). Designing bonsai surveys: The small but perfectly formed survey experience to meet the needs of the modern day mobile-based survey consumer. Esomar Asia Pacific.

Revilla, M., Ochoa, C., and Toninelli, D. (2016). PCs vs. smartphones in answering web surveys: does the device make a difference? *Survey Practice* 9 (4) https://doi.org/10.29115/SP-2016-0021.

Richards, A., Powell, R., Murphy, J. et al. (2016). Gridlocked: the impact of adapting survey grids for smartphones. *Survey Practice* 9 (2) https://doi.org/10.29115/SP-2016-0016.

Schonlau, M. and Toepoel, V. (2015). Straightlining in web survey panels over time. *Survey Research Methods* 9 (2): 125–137.

Struminskaya, B., Weyandt, K., and Bosnjak, M. (2015). The effects of questionnaire completion using mobile devices on data quality. Evidence from a probability-based general population panel. *Methods, Data, Analysis* 9 (2): 261–292.

Toepoel, V. and Lugtig, P. (2016). The use of PCs, smartphones and tablets in a probability based panel survey. Effects on survey measurement error. *Social Science Computer Review* 34 (1): 78–94.

Tourangeau, R., Conrad, F.G., and Couper, M.P. (2013). *The Science of Web Surveys*. Oxford University Press.

Zang, C. and Conrad, F. (2013). Speeding in web surveys: the tendency to answer very fast and its association with straightlining. *Survey Research Methods* 8 (2): 127–135.

15.A Questionnaire Items

Order: Question Wording

1. Which type of holiday did you have this summer?
   charter/beach holiday; hiking in mountains; camping; city holiday; campervan holiday; boating holiday; adventure holiday; cabin holiday in mountains; cabin holiday at the ocean; road trip; at home holiday; no summer holiday; other; don't know
2. Which holiday types are relevant to you?
   charter/beach holiday; hiking in mountains; camping; city holiday; campervan holiday; boating holiday; adventure holiday; cabin holiday in mountains; cabin holiday at the ocean; road trip; at home holiday; no summer holiday; other; don't know; none of these
3. How many days did you spend on each of these holiday types this summer?
   charter/beach holiday; hiking in mountains; camping; city holiday; campervan holiday; boating holiday; adventure holiday; cabin holiday in mountains; cabin holiday at the ocean; road trip; at home holiday; other
4. When you think about everything meaningful to you with regards to different holiday types, how do you rate the following holiday types?
   1 = Very poorly, 10 = Very good
   charter/beach holiday; hiking in mountains; camping; city holiday; campervan holiday; boating holiday; adventure holiday; cabin holiday in mountains; cabin holiday at the ocean; road trip; at home holiday; other
5. For each holiday type, how much do you agree or disagree with the following statement: This holiday type suits me as a person and means a lot to me
   1 = Completely disagree, 7 = Completely agree
   charter/beach holiday; hiking in mountains; camping; city holiday; campervan holiday; boating holiday; adventure holiday; cabin holiday in mountains; cabin holiday at the ocean; road trip; at home holiday; other
6. When you think about different types of holiday, how important to you think the choice of holiday type is?
   extremely important; very important; somewhat important; not particularly important; not at all important
7. Which holiday type fit the following statements the best? (Choose up to two holiday types):
   A. The cheapest holiday type
   B. Easiest to organize
   C. The holiday type my travel companions prefer
   D. What we can afford
   E. Tradition/what we always do
   F. Shortest travel time

G. Suits my/our life situation
H. Most practical given our available time
charter/beach holiday; hiking in mountains; camping; city holiday; camper-van holiday; boating holiday; adventure holiday; cabin holiday in mountains; cabin holiday at the ocean; road trip; at home holiday; other; don't know; none of these

8. Which of the following conditions influence on your choice of holiday?
stable weather conditions; can meet up with family; can meet up with friends; many activities for the kids; economy; tradition/habit; activities for the whole family; wants to experience other cultures; sense of adventure; family obligations, access to cabin; experience nature; interesting activities; short distance; new experiences; relaxation
9. Which holiday types do you associate with the factors you said are important to you?
stable weather conditions; can meet family; can meet friends; many activities for the kids; economy; tradition/habit; activities for the whole family; wants to experience other cultures; sense of adventure; family obligations, access to cabin; experience nature; interesting activities; short distance; new experiences; relaxation
charter/beach holiday; hiking in mountains; camping; city holiday; camper-van holiday; boating holiday; adventure holiday; cabin holiday in mountains; cabin holiday at the ocean; road trip; at home holiday; other; don't know; none of these
10. Who did you spend your holiday with this summer?
spouse/companion; girl-/boyfriend; my children; my partners children; children's- /partners children's family; friends; alone; other; don't know; prefer not to answer
11. Where did you spend your holiday this year?
Norway; the Nordics; Southern Europe; other Europe; North- or South-America; Asia; Africa; Oceania; don't recall; prefer not to answer
12. How much of the time did you have access to the Internet while on holiday?
all the time; parts of the time; no access; don't know; prefer not to answer
13. Which social media are you using?
Facebook; Twitter; Snapchat; Instagram; YouTube; Kik; Google+; Nettby; Kiwi; Other; None of these; Prefer not to answer
14. How often did you use social media while on holiday?
multiple times per day; appx once per day; 5–6 times per week; 2–4 times per week; once per week or less often; never; don't know; prefer not to answer
15. Which of these devices did you bring with you on holiday?
smart phone; other mobile; tablet/ipad; PC/Mac; e-reader (ex. Kindle); other; none; don't know/prefer not to answer

16. Have you checked your work email while on holiday?
    yes, every day; yes, now and then; yes, but seldom; no; don't know/prefer not to answer
17. Why did you not have a summer holiday this year?
    just started a new job/changed jobs; used up all my vacation earlier this year; prefer to go on vacation during other seasons; can't afford it; other; don't know; prefer not to answer
18. Are you planning on going on vacation later this year?
    yes; maybe; no

16

Learning from Mouse Movements: Improving Questionnaires and Respondents' User Experience Through Passive Data Collection

Rachel Horwitz[1], *Sarah Brockhaus*[2,3], *Felix Henninger*[3,4], *Pascal J. Kieslich*[3,5], *Malte Schierholz*[6], *Florian Keusch*[3,7], *and Frauke Kreuter*[3,6,7,8]

[1] *US Census Bureau, Washington, DC, USA*
[2] *Institute for Statistics, Ludwig Maximilian University of Munich, Munich, Germany*
[3] *Mannheim Centre for European Social Research (MZES), University of Mannheim, Mannheim, Germany*
[4] *Department of Psychology, University of Koblenz-Landau, Landau, Germany*
[5] *Department of Psychology, School of Social Sciences, University of Mannheim, Mannheim, Germany*
[6] *Institute for Employment Research, Nuremberg, Germany*
[7] *Department of Sociology, School of Social Sciences, University of Mannheim, Mannheim, Germany*
[8] *Joint Program in Survey Methodology, University of Maryland, College Park, MD, US*

16.1 Introduction

Over the past few decades, web surveys have become a standard, and often preferred, mode of survey administration. Their popularity is due to lower costs compared to other data collection modes, and because the technological capabilities associated with this mode present numerous advantages over other modes. At a basic level, web surveys often include automated skips, text fills, and edit messages to help respondents complete a survey. While these features help all respondents by guiding them efficiently through the survey, some features can be used to target specific respondents who may need assistance on particular questions. One such feature that has been used in this way is the ability to track mouse movements. An analysis of mouse movements is commonly used in web design and e-learning as a method to help identify user uncertainty (Cox and Silva 2006; Zushi et al. 2012) and user interest (Mueller and Lockerd 2001; Rodden et al. 2008).

In survey research, early work in mouse tracking focused on the total distance a respondent moved the mouse on a particular screen, measured in pixels. Stieger and Reips (2010) found lower data quality associated with questions on which a respondent's mouse movements exceeded by two standard

Advances in Questionnaire Design, Development, Evaluation and Testing, First Edition.
Edited by Paul C. Beatty, Debbie Collins, Lyn Kaye, Jose-Luis Padilla, Gordon B. Willis, and Amanda Wilmot.

deviations the average length respondents move their mouse. Building on Stieger and Reips' (2010) work, Horwitz et al. (2017) used a laboratory study to relate mouse movements to difficulty answering survey questions. Rather than looking at the total distance the mouse traveled, they applied specific patterns of mouse movements identified in the web design literature to a survey setting, including hovering over particular areas of interest (AOIs) and moving the mouse back and forth on the vertical and horizontal axes. In a laboratory experiment using factual questions from the American Community Survey, human coders identified several patterns of movement that were predictive of respondent difficulty.

The current study builds on Horwitz et al. (2017) to develop automated (as opposed to hand-coded) procedures for detecting and quantifying mouse movements associated with uncertainty or difficulty, and applies them in an online survey as opposed to a laboratory. Automation provides many benefits compared to hand coding, including reduced bias and the ability to analyze larger datasets quickly, potentially even in real time. The study also builds on recent methodological advances in psychological research that use mouse-tracking measures for assessing the tentative commitments to, and conflict between, response alternatives over time (Freeman et al. 2011; Koop and Johnson 2011; Kieslich and Henninger 2017).

In this study, we monitored and logged participants' activity as they completed an online questionnaire. The survey was designed to reflect a typical questionnaire in social science research, and included factual questions, opinion questions, and problem-solving questions with a variety of response formats, such as radio buttons and sliders. This variety of question types and formats allowed us to assess the degree to which selected mouse-movement indicators were predictive of difficulty in each format, and to determine whether specific movements were of greater relevance for specific question types.

This chapter reviews the background of mouse tracking, provides an overview of existing studies, and reports the results from our new study. Our results constitute initial steps toward a fully automated analysis of web survey paradata based on mouse movements (Kreuter 2013), which could be used to detect respondent's difficulties in online questionnaires and adapt surveys to resolve such difficulties in real time, ultimately leading to more accurate questionnaire data.

16.2 Background

In this section, we review the existing literature on how mouse movements have been used in psychology, e-learning, web design, and survey research.

16.2.1 Results from Psychology and Other Disciplines

Completing tasks online can be difficult for users if an interface is not intuitive; the task is unclear; they are hesitant to answer personal questions; or they are uncertain what they are looking for or what the correct choice is. Mouse movements have been studied in a variety of disciplines, including psychology, e-learning, and web design, to help understand what users are thinking as they complete tasks and to identify interfaces or tasks that might be difficult. Researchers have found that people not only move the mouse once they have made a decision, but are already moving the mouse during the decision-making process, making mouse movements "real-time motor traces of the mind" (Freeman et al. 2011, p. 2).

For example, response competition can be reflected in the way the user moves the mouse. Freeman and Ambady (2009) presented participants with pictures of sex-typical (e.g. masculine male) and sex-atypical (e.g. feminine male) faces and asked them to identify which of two adjectives was the stereotypically appropriate one for each face (one adjective was stereotypically masculine, one feminine). For sex-atypical faces, the trajectory of the mouse was more curved toward the opposite-gender stereotype, indicating a cognitive conflict between the two possible responses. In addition, Freeman (2014) presented participants with pictures of morphed faces ranging from very male to very female. They were asked to click on a "male" or "female" response button (displayed in the top left and right corners of the screen) to categorize the face presented at the bottom center of the screen. They found that when the pictures were more ambiguous, participants changed the direction of the cursor along the x-axis more frequently. In other words, they changed more often from moving the mouse toward one sex category to the other. Similarly, Duran et al. (2010) found that when participants were instructed to answer questions falsely (versus truthfully) by moving a cursor through a Wii Remote[1] controller to the response options in the top left and right corners of a screen, cursor movements displayed a greater number of changes in horizontal movement direction. These studies suggest that response competition, or cognitive conflict, can manifest in people's hand movements, and that these can be traced, for example, by tracking mouse movements. A review of mouse-tracking applications in psychological research is provided by Freeman et al. (2011).

In e-learning, researchers have also found that mouse movements can be used to detect uncertainty. Longer, arched trajectories were found to be associated with greater user uncertainty (Zushi et al. 2012). Similarly, in usability testing for web design, users who were less certain of how to complete tasks on a web page moved the mouse more slowly in an arched trajectory,

1 A Wii Remote is a controller with motion-sensing capability that allows users to navigate items on a screen through wrist and arm movements.

while researchers found more direct, fast movements associated with certainty (Arroyo et al. 2006; Huang et al. 2011; Schneider et al. 2015). In a psychological study, Dale et al. (2008) asked participants to learn randomized pairs of unfamiliar symbols. Participants were presented with a symbol at the bottom center of the screen and a correct match in either the upper left or right corner and a random incorrect symbol in the other corner. They had to indicate their response by moving the cursor using a Wii Remote and received feedback about the correctness of their response afterward. Originally, answers to the pairs were guesses, but as participants saw them multiple times, they were able to learn which symbols were matches. Among other things, Dale et al. found that participants changed the direction of the cursor fewer times along the x-axis when they answered correctly compared to incorrectly, suggesting more uncertainty.

Psychologists have also used mouse tracking to study the cognitive processes in social dilemma situations. In social dilemmas, people are faced with the choice between two courses of action. They can decide to either defect and maximize their individual payoff, or cooperate to maximize the payoff overall (Dawes 1980). Importantly, defection can be tempting for the individual as it yields individually superior payoffs. However, if each individual defects, all are worse off than if everyone had cooperated. Kieslich and Hilbig (2014) used the curvature of participants' mouse movements in social dilemmas as an indicator of cognitive conflict. They found that in defection decisions, mouse trajectories deviated on average more toward the non-chosen option than in cooperation decisions, and that this was especially the case for participants whose personality traits reflected a higher tendency to actively cooperate with others across situations.

Given the insight mouse movements can provide into understanding the thought process people are going through when working on a computer, web designers have tracked mouse movements to see which parts of websites people are interested in (Mueller and Lockerd 2001; Rodden et al. 2008; Huang et al. 2011). For example, Mueller and Lockerd (2001) were able to predict which web search option was the user's second choice by looking at the options users hovered over with the cursor.

Survey respondents can also experience cognitive conflict and may feel some indecision or uncertainty, such as between two competing response options, so it is plausible that their mouse trajectories and patterns of movement would reflect this difficulty.

16.2.2 Mouse Movements in Survey Research

Respondent difficulty taking surveys is particularly problematic because it often leads to incorrect responses and reduced accuracy which increases measurement error (Schober and Conrad 1997; Conrad and Schober 2000;

Schober et al. 2004; Ehlen et al. 2007). Therefore, the design of a survey and its questions often undergo multiple rounds of testing and great scrutiny before questions are finalized, with the aim of making them easy to understand and answer. Despite these efforts, respondents can still struggle with surveys. Online data collection can relieve some of this difficulty. For example, rather than respondents navigating skip instructions on paper, which can lead to errors, web surveys can automate these skips so that respondents are only presented with the intended questions. Web surveys often include text fills and edit messages to help personalize the experience and assist respondents in completing the instrument. During data collection, web surveys also provide the opportunity to adapt what respondents see to help them. For example, Conrad et al. (2007) offered real-time help to respondents whose response time on a specific question exceeded a predetermined threshold. Similarly, Conrad et al. (2017) detected respondents who answered questions so quickly that they could not have given much thought to their answers. When respondents answered faster than a minimum response time threshold, a pop-up appeared, asking them to answer carefully and take their time. These features can improve data quality while making the survey easier to understand and improving respondents' experience.

While these features can help respondents and make the survey process easier, they do not guarantee that respondents will understand the questions. A lack of understanding can stem from not understanding particular terms in the question or response options, being unsure of how one's experience maps onto the response options available, the structure of the question, or an unfamiliarity with the question format, among other sources of confusion (Tourangeau et al. 2000), which can all lead to response error. These types of difficulty are not mode-specific and have been widely studied across all modes (Heerwegh 2003; Schober and Bloom 2004; Ehlen et al. 2007). What is mode-specific, however, is how these types of difficulties are detected. In interviewer-administered surveys, for example, interviewers listen for hedges, pauses, fillers, changes in intonation, and long response latencies to determine when a respondent is having trouble answering a question (Smith and Clark 1993; Brennan and Williams 1995; Heerwegh 2003; Schober and Bloom 2004; Ehlen et al. 2007; Yan and Tourangeau 2008). In web surveys, where these verbal cues are not available, researchers have identified response times (Heerwegh 2003; Ehlen et al. 2007; Conrad et al. 2007; Lind et al. 2001) and mouse movements (Stieger and Reips 2010; Horwitz et al. 2017) as helpful indicators to identify when respondents are experiencing difficulty and may answer a question incorrectly.

Researchers often analyze response times because research shows that response times can often predict respondent difficulty (Heerwegh 2003; Ehlen et al. 2007; Yan and Tourangeau 2008). This measure is more commonly used than mouse movements because response times are easier to collect (as many online software packages automatically provide them). While mouse movements are more difficult to collect and examine, they allow for a more robust analysis, particularly when combined with response times, because they guarantee that the respondent is engaged with the survey. For example, Horwitz et al. (2017) found that while response latencies were predictive of respondent difficulty, they were even more predictive when combined with specific mouse-movement indicators. Horwitz et al. expected this increased precision would result in fewer false positives, or instances in which respondent difficulty was incorrectly assumed.

The first study that examined mouse movements in survey research found lower data quality associated with questions on which a respondent's mouse movements exceeded, by two standard deviations, the average length of respondent mouse movements (Stieger and Reips 2010). Building on that work, Horwitz et al. (2017) used a lab study to relate mouse movements to difficulty answering survey questions. Rather than looking at the total distance traveled, they used the web design literature to identify specific patterns of mouse movements that users frequently engaged in and applied them to a survey setting. Three movements were found to be significantly predictive of difficulty (*Hover* – holding the mouse over the question text for two or more seconds, *Marker* – holding the mouse over a radio button or response option text for more than two seconds, and *Regressive* – moving the mouse back and forth between two AOIs, specifically, between question text, white space, Next button, and response options). While this study was successful in identifying a set of movements that were predictive of difficulty, the analysis was very time consuming because the classification of movements was not automated but rather performed manually. Additionally, this study only used factual questions with a standard radio button response format. Investigating different question types, such as opinion questions, and a variety of response formats, including check boxes, may uncover different relationships between mouse movements and cognitive processing.

16.2.3 Research Gaps

From the body of research on mouse tracking summarized here, it appears that respondents' conflict between response options can be measured through the curvature of the response trajectory and the number of regressive movements they make. However, much of the psychological mouse-tracking literature uses a specific methodological setup that may not be directly applicable to the survey context. Rather than presenting choices as questionnaire items, the screen layout presents response options as buttons in the (typically upper left and right) screen corners, and participants are required to move the cursor upward to their preferred option from a central starting point in the lower center of the screen. This differs considerably from the typical setup in a survey design, where several choice alternatives are often presented as a vertically arranged list of radio buttons.

In addition, holding the cursor over an item on a page can indicate uncertainty, or possibly interest. Interest and uncertainty can sometimes be related concepts. When there is choice, there is also the potential for indecision. For example, given a list of response options, holding the cursor over a certain item may reflect an interest in and preference for choosing that particular option. However, even after finding an option the respondent is interested in, she may continue to scan the list for a better fit than the option she originally held the cursor over. If the respondent finds another option she is interested in, but neither is a perfect match to her personal situation, her initial interest in one option may become uncertainty as she decides which is the best of multiple options. This could result in the respondent hovering over one of the options she is interested in, but unsure about.

Finally, the current state of research does not tell us whether people engage in different movements when they face different types of cognitive difficulty, such as not understanding specific words or the concept behind a question that is unfamiliar. For example, when people are unsure of how to categorize something, the research shows that they move the mouse toward the second-choice option before selecting the correct option. But what mouse movements do respondents display when they do not understand the question? Do they move the mouse between the various response categories, or do they hover over the question text that is confusing them?

Research on how to use mouse movements to improve internet surveys is still in its infancy. Results have either come from laboratory experiments or relied on simple indicators like the total distance traveled. The study presented in this chapter operationalized mouse-movement collection on a large scale outside of a laboratory setting and aimed to confirm that the mouse movements identified in the literature work in a similar way for surveys – and to determine whether survey respondents engage in different types of mouse movements when presented with different types of difficulty. Our goal was to demonstrate whether we can detect, in real time, when a respondent was having trouble answering a question and attempt to isolate what was causing their confusion (e.g. question, response options, format, terminology, etc.). By successfully researching these issues, mouse-movement indicators could be used to deliver targeted interventions to respondents by providing them with the help they need while they are answering questions, thereby increasing data quality and reducing burden.

16.3 Data

The data for this study come from a survey conducted by the Institute for Employment Research (IAB) in Nuremberg, Germany. The sample consisted of employees, unemployed persons, job seekers, welfare recipients, and participants in active labor market programs. A previous wave of this survey was conducted in winter 2014/2015. From that wave, 1627 respondents agreed to future contacts and were contacted again for the current study. The current data were collected from September through October 2016.

The survey contained questions on a number of topics, including employment and demographics. The employment section covered questions such as employment status, employer, changes in employment, and income. The demographic section collected basic data including gender, age, and nationality.

The only response mode available for respondents was internet, and on the survey's login page, we asked respondents not to use a mobile device[2] – such as a smartphone or a tablet computer – as in these cases no continuous cursor movements would be observed. Any respondent that did use a mobile device was therefore excluded from the analysis. The web survey was designed in SoSci Survey software (Leiner 2014), and the response data were collected in a database. The mouse movements were collected via a JavaScript library (Henninger and Kieslich 2016) that was integrated into the web survey, and which logged cursor coordinates and timestamps as well as a representation of the page layout so that mouse tracks could be associated with survey content.[3] Sample cases that did not have JavaScript enabled in their browser could not be included in the analysis.

16.4 Methodology

16.4.1 Mailout Procedures

Invitation letters with a €5 incentive were sent on September 5, 2016 to 1527[4] sample cases. Sample cases that did not respond to the initial invitation received a reminder email, if an email address was on file (370 individuals), and two postal reminders. The email reminder was sent on September 15, 2016 and the postal reminders were sent on September 22 and October 4, 2016. The email reminder focused on a request for help and the second reminder encouraged participation so that results would be representative. The final reminder informed the sample members that it was their last chance to participate. The survey instrument was open for eight weeks. In total, 1201

2 Text in the invitation letter and on the welcome screen stated, "Because this study was not designed for smartphones and tablet computers, we ask you to use your laptop or desktop computer."

3 There was additional information gathered, including mouse clicks and typing input, that is not evaluated as part of this research.

4 An additional 100 individuals who agreed to future contacts were emailed in a preliminary test to ensure the mouse-tracking software was working properly and to assess the need for incentives.

sample cases responded, including 30 persons who did not complete the questionnaire, resulting in a response rate of 76.7% (AAPOR RR1). An additional 49 respondents from a preliminary mailout were included in the analysis as well.[5] Although 1250 people responded in total, only 1213 completed the questionnaire. Of those, 886 reported using a mouse (the majority of those participants who did not use a mouse reported using a trackpad or touch screen), and mouse-tracking data were collected continuously for 853 of these. Only these respondents are included in the following analysis.

This final sample consisted of people with a mean age of 51.5 years (SD = 10.9 years, 60% range from 35–63 years); 51% of participants were female, 49% male. At the time of the interview, 89% were employed. Additionally, 12% had completed comprehensive school ("Hauptschule"), 34% had completed an intermediate tract ("Mittlere Reife"), and 51% had a university entrance qualification.[6] We did not find any substantial differences in demographic characteristics (sex, age, nationality, education) between participants who did and did not use a mouse.

16.4.2 Hypotheses and Experimental Manipulations

The survey consisted of a maximum of 36 questions; respondents may have received fewer questions depending on automated skip patterns that varied based on participants' responses. For several of the questions, there were two versions intended to manipulate the difficulty respondents would experience when formulating an answer, which are the focus of this chapter. Table 16.1 describes the two versions for each type of manipulation and the predicted difference in mouse movements across the two versions. The measures are defined in detail here.

The order of the response options was manipulated for two questions: highest educational attainment and employees' level of responsibility at work. The education question had 11 response options while the employee question had 4. In the ordered version of each question, response options were ordered progressively from low to high levels of education and low to high levels of responsibility at work. In the unordered versions, response categories were randomized so that they appeared in no logical order. We thus expected more vertical regressions in the unordered list as respondents looked for the appropriate response as well as a greater total distance traveled to account for this searching.

The next manipulation asked respondents to name the reasons they quit their previous job. In one version, participants were required to select Yes or No for each response option, while in the other they could check every response

5 The same mailout procedure consisting of one email and two postal reminders was tested beforehand, beginning July 20, 2016 when an invitation not containing an incentive was sent to the 100 sample cases that received the preliminary test.
6 The remaining respondents had foreign, no, or other qualifications.

Table 16.1 Description of manipulations and the mouse-movement prediction.

Version 1	Version 2	Hypothesis
Ordered response options	Unordered response options	H1a: Respondents' mouse movements will cover a greater distance when the response options are unordered compared to ordered H1b: A higher number of vertical regressions will be observed when the response options are unordered compared to ordered
Check all that apply	Yes/No (y/n)	H2a: Respondents' mouse movements will cover a greater distance when the response options are in a yes/no format than a check all that apply H2b: A higher number of horizontal regressions will be observed when the response options are in a yes/no format compared to a check all that apply
Straightforward response options	Complex response options	H3: Respondents will hover more when there are complex terms in the response options compared to straightforward terms

"H" represents a hypothesis, "1" represents the first type of manipulation, and "a" represents the first hypothesis within the type of manipulation.

option that applied. The yes/no version required additional effort to answer and required more movement of the mouse, so we expected to see more horizontal regressions and a greater total distance traveled.

A common issue with survey questions is that respondents do not always understand them. Therefore, we manipulated the complexity of the response options for a question that asked about the respondent's type of employer (if the respondent was employed). The straightforward version used simple, concise language, whereas the complex version added repetitive, bureaucratic, technical information that is commonly included in questions to be explanatory, but can lead to confusion. As can be seen in Figure 16.1, the text of the response options in the complex version is considerably longer. We expected respondents to have difficulty understanding the complex version of the question, resulting in increased hovering while they read the text and tried to process the information to select the best-fitting option.

For each of these manipulations, all sample cases were randomly assigned one version of each question. The randomization was done within the SoSci Survey software using an internal randomization function, such that for each question approximately half of the respondents received each version.

Sind Sie derzeit...
Falls Sie momentan in mehreren Beschäftigungsverhältnissen stehen, denken Sie bitte bei dieser und den folgenden Fragen an Ihre hauptsächliche Beschäftigung. Das ist diejenige Erwerbstätigkeit, in der Sie die meisten Stunden arbeiten.

- angestellt bei einem privaten, profitorientierten Unternehmen
- angestellt bei einem privaten, nicht profitorientierten Unternehmen
- angestellt im öffentlichen Dienst des Bundes
- angestellt im öffentlichen Dienst eines Bundeslandes
- angestellt im öffentlichen Dienst einer Kommune
- selbständig in einem freien Beruf
- selbständig in Handel, Gewerbe, Industrie, Dienstleistung, Landwirtschaft
- Freier Mitarbeiter/Freie Mitarbeiterin
- Mithelfende/r Familienangehörige/r

(a)

Sind Sie derzeit...
Falls Sie momentan in mehreren Beschäftigungsverhältnissen stehen, denken Sie bitte bei dieser und den folgenden Fragen an Ihre hauptsächliche Beschäftigung. Das ist diejenige Erwerbstätigkeit, in der Sie die meisten Stunden arbeiten.

- angestellt bei einem privaten, profitorientierten Unternehmen oder einer Privatperson, das/die regelmäßig Löhne oder Gehälter auszahlt
- angestellt bei einem privaten, nicht profitorientierten Unternehmen, welches gemeinnützig ist oder von Steuervergünstigungen profitiert
- angestellt im öffentlichen Dienst des Bundes (Arbeitgeber ist die Bundesrepublik)
- angestellt im öffentlichen Dienst eines Bundeslandes (Arbeitgeber ist ein Bundesland)
- angestellt im öffentlichen Dienst einer Kommune (Arbeitgeber ist zum Beispiel eine Gemeinde, eine Stadt, ein Landkreis, ...)
- selbständig in einem freien Beruf (z.B. Arzt/Ärztin, Rechtsanwalt/Rechtsanwältin oder Architekt/in)
- selbständig, jedoch nicht in einem freien Beruf im Bereich Handel, Gewerbe, Industrie, Dienstleistung, Landwirtschaft
- Freier Mitarbeiter/Freie Mitarbeiterin, der aufgrund eines Dienst- oder Werkvertrags für ein Unternehmen Aufträge persönlich ausführt
- Mithelfende/r Familienangehörige/r ohne Bezahlung

(b)

Figure 16.1 Screenshot of the survey question about the type of employer. We randomly varied the complexity of the question, using either a straightforward version (a) or a complex one (b).

Together, these manipulations aimed to determine whether respondents exhibit the same mouse movements when they are confused or experiencing difficulty, or whether specific types of difficulty yield different types of mouse movements.

In addition to testing the hypotheses included in Table 16.1, we also conducted a post hoc analysis on two related questions that involved evaluations. The first question asked respondents how they felt about Germany's economic situation (general evaluation) and the second asked about their personal economic situation (self evaluation). The rating scale ranged from 1 – very bad, to 5 – very good, displayed vertically. While we did not have specific hypotheses for this set of questions, we were curious to see how mouse movements differed between the questions. For example, did respondents familiarize themselves with the question format on the first question so they engaged in fewer movements on the second question? Or alternatively, were general evaluations more difficult to make than personal evaluations, potentially resulting in more movements for the general question? Given there were no manipulations for this question and all respondents received the questions in the same order, we cannot isolate the cause of differences between the two questions, but we can postulate different reasons that can be further tested in future research.

16.4.3 Mouse-Movement Measures

The JavaScript code embedded in the survey that collected mouse movements resulted in a string of coordinates, reflecting where on the screen a respondent's cursor was positioned at different points in time. As the layout of the survey differs depending on the browser and the window size, the interpretation of absolute coordinates is different for each user. For this analysis, however, we translated the coordinates to relative coordinates so that a single coordinate pair identified the same location in the browser window of the survey for every respondent. The raw data were imported into R (R Core Team 2016) and preprocessed and analyzed using the mousetrap library (Kieslich et al. 2019). A number of trial-level indices were computed (see the following).

Hovers are a common indicator that measures interest (Mueller and Lockerd 2001). This could include respondents hovering over one response option they think is the best fit while scanning through the rest of the options or holding the mouse over a difficult word or concept. To assess hovers, we measured the overall time the mouse remained in place (excluding the initial phase without movement). If the mouse stayed in one location without any movement anywhere on the screen for more than two seconds, this was counted as a hover. We also computed the total time spent hovering.

Other measures we used were horizontal and vertical regressions. Depending on the layout of the response options, horizontal or vertical, this movement can indicate indecision between competing response options (Horwitz et al. 2017).

To measure regressions, we counted the number of times there was any change of direction along the horizontal or vertical axis, respectively.

Finally, indirect trajectories are another indicator that can suggest uncertainty or conflict (Arroyo et al. 2006; Duran et al. 2010). To this end, we computed the total (Euclidian) distance traveled by the mouse (in pixels) on each page.

16.4.4 Analysis Procedures

Before analyzing the data for each question, we removed cases that did not provide an answer and cases for which mouse-movement data was unavailable. Additionally, we excluded cases that exceeded a time limit of seven minutes for answering the individual question (most respondents took less than 30 seconds to answer a question).

The hypotheses described in Table 16.1 focus on several mouse-movement indicators that were included in the following analyses: vertical and horizontal regressions, total distance traveled, and hovers. To make the analyses robust against outliers, we used rank-based statistics and tests.

Although we postulated specific hypotheses for each manipulation (see Table 16.1), we compared all mouse-movement indicators for each question type to be able to link specific movements to specific types of difficulty. This method allowed for a more explorative analysis beyond our hypotheses.

16.5 Results

In this section, we present the results for all of the mouse-movement indicators for each question type as well as each of the hypotheses outlined in Table 16.1.

16.5.1 Ordered vs. Unordered Response Options (H1a and H1b)

When presented with the unordered list of response options, respondents moved the mouse a greater distance and engaged in more horizontal and vertical regressions compared to the standard, ordered list for both questions containing the order manipulation. Table 16.2 provides the mean and median values for each indicator per condition and question, along with the corresponding statistical test and p-value.

While we hypothesized that there would be more vertical regressions and a greater distance traveled when the response options were unordered, we did not expect that the list order would have an impact on hovers or horizontal regressions. While there was no difference in the number of hovers or the time spent hovering for the education question, there was significantly more and longer hovering for the responsibility at work question in the unordered condition.

Table 16.2 Comparison of the mouse-movement indicators for ordered and unordered lists.

	Mouse movement		Ordered	Unordered	Wilcoxon rank-sum test	P-value
Educational attainment	Total distance traveled (pixels)	Mean	2057.1	2387.9	74 524	0.002
		Median	1608.4	1804.8		
	Horizontal regressions	Mean	11.3	14.0	74 060	0.002
		Median	8	10		
	Vertical regressions	Mean	11.8	13.9	75 297	0.005
		Median	8.5	11.0		
	Number of hovers	Mean	1.2	1.3	81 896	0.363
		Median	1	1		
	Hover time (s)	Mean	8.7	7.1	79 863	0.133
		Median	2.9	3.8		
Responsibility at work	Total distance traveled (pixels)	Mean	2157.5	2298.5	46 280	0.039
		Median	1228.3	1373.5		
	Horizontal regressions	Mean	10.0	11.5	45 432	0.015
		Median	6	7		
	Vertical regressions	Mean	11.9	13.7	45 740	0.021
		Median	7	9		
	Number of hovers	Mean	1.8	2.2	44 866	0.006
		Median	1	2		
	Hover time (s)	Mean	10.8	15.2	43 678	0.001
		Median	7.5	11.2		

For the Educational Attainment question, 412 respondents answered the sorted and 412 the unsorted version. For the responsibility at work question, 307 respondents answered the sorted version and 333 answered the unsorted version.

16.5.2 Check All That Apply vs. Yes/No Response Formats (H2a and H2b)

We initially found no difference between horizontal regressions or distance traveled between these response formats. However, upon further investigation, we discovered that 70% of respondents in the yes/no format condition actually treated the yes/no format as a check all that apply question. That is, they only used the Yes column where applicable instead of systematically answering Yes or No for each statement. Figure 16.2 shows example response behaviors for respondents under the check all that apply format, the yes/no format that was

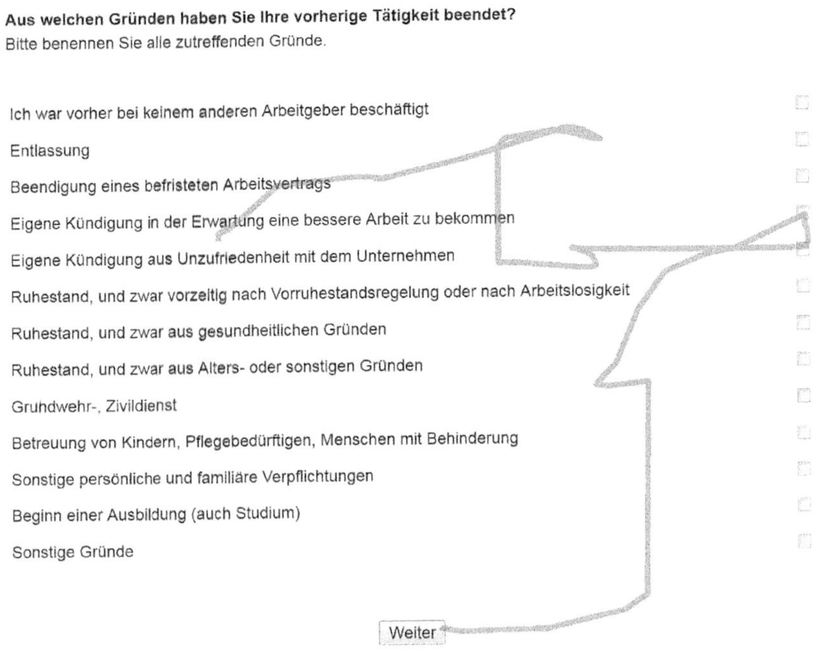
Aus welchen Gründen haben Sie Ihre vorherige Tätigkeit beendet?
Bitte benennen Sie alle zutreffenden Gründe.
Ich war vorher bei keinem anderen Arbeitgeber beschäftigt
Entlassung
Beendigung eines befristeten Arbeitsvertrags
Eigene Kündigung in der Erwartung eine bessere Arbeit zu bekommen
Eigene Kündigung aus Unzufriedenheit mit dem Unternehmen
Ruhestand, und zwar vorzeitig nach Vorruhestandsregelung oder nach Arbeitslosigkeit
Ruhestand, und zwar aus gesundheitlichen Gründen
Ruhestand, und zwar aus Alters- oder sonstigen Gründen
Grundwehr-, Zivildienst
Betreuung von Kindern, Pflegebedürftigen, Menschen mit Behinderung
Sonstige persönliche und familiäre Verpflichtungen
Beginn einer Ausbildung (auch Studium)
Sonstige Gründe
Weiter

(a)

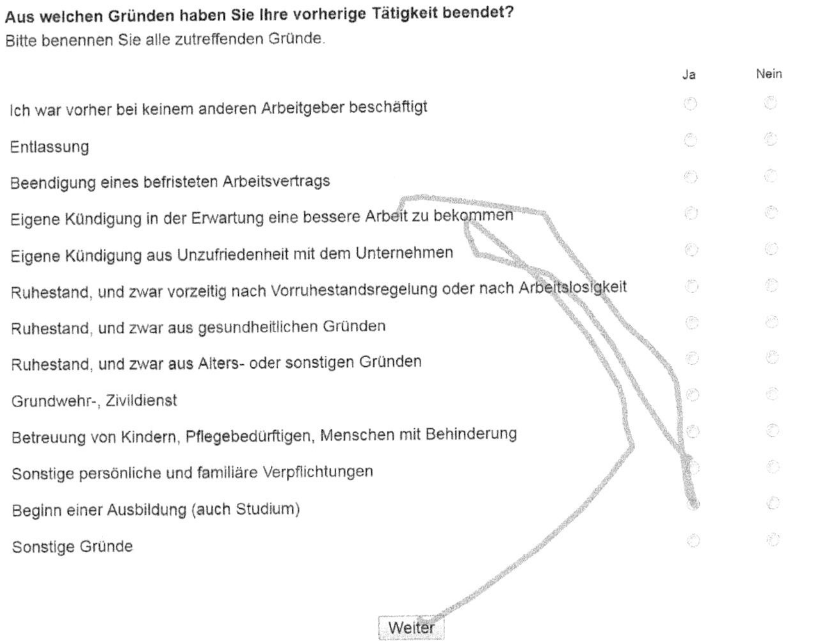
Aus welchen Gründen haben Sie Ihre vorherige Tätigkeit beendet?
Bitte benennen Sie alle zutreffenden Gründe.
Ja
Nein
Ich war vorher bei keinem anderen Arbeitgeber beschäftigt
Entlassung
Beendigung eines befristeten Arbeitsvertrags
Eigene Kündigung in der Erwartung eine bessere Arbeit zu bekommen
Eigene Kündigung aus Unzufriedenheit mit dem Unternehmen
Ruhestand, und zwar vorzeitig nach Vorruhestandsregelung oder nach Arbeitslosigkeit
Ruhestand, und zwar aus gesundheitlichen Gründen
Ruhestand, und zwar aus Alters- oder sonstigen Gründen
Grundwehr-, Zivildienst
Betreuung von Kindern, Pflegebedürftigen, Menschen mit Behinderung
Sonstige persönliche und familiäre Verpflichtungen
Beginn einer Ausbildung (auch Studium)
Sonstige Gründe
Weiter

(b)

Figure 16.2 Exemplary mouse trajectories for the check all that apply condition (a), the yes/no condition when participants checked only applicable items (b), and the intended use of the yes/no condition (c).

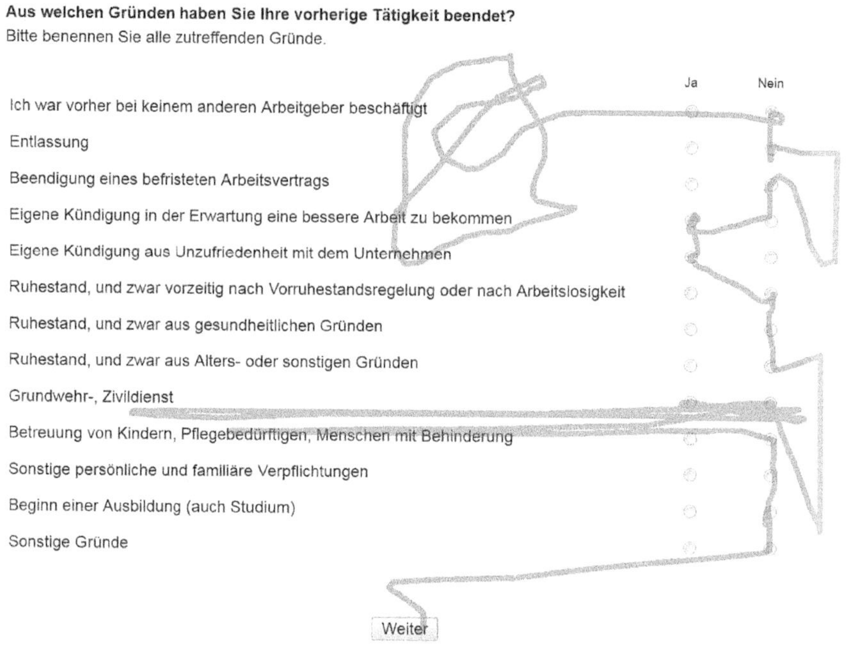
Aus welchen Gründen haben Sie Ihre vorherige Tätigkeit beendet?
Bitte benennen Sie alle zutreffenden Gründe.
Ja
Nein
Ich war vorher bei keinem anderen Arbeitgeber beschäftigt
Entlassung
Beendigung eines befristeten Arbeitsvertrags
Eigene Kündigung in der Erwartung eine bessere Arbeit zu bekommen
Eigene Kündigung aus Unzufriedenheit mit dem Unternehmen
Ruhestand, und zwar vorzeitig nach Vorruhestandsregelung oder nach Arbeitslosigkeit
Ruhestand, und zwar aus gesundheitlichen Gründen
Ruhestand, und zwar aus Alters- oder sonstigen Gründen
Grundwehr-, Zivildienst
Betreuung von Kindern, Pflegebedürftigen, Menschen mit Behinderung
Sonstige persönliche und familiäre Verpflichtungen
Beginn einer Ausbildung (auch Studium)
Sonstige Gründe
Weiter

(c)

Figure 16.2 (*Continued*)

treated as a check all that apply, and the yes/no format that was answered as expected.

We therefore separated the analysis into these three groups instead of the initial two groups. Once we differentiated by how respondents were answering the yes/no format, we found significant differences for both distance traveled and the number of horizontal regressions across the three treatment groups as hypothesized, with higher values in the yes/no condition when it was treated as such (Table 16.3). In addition to the hypothesized effects, respondents in the yes/no condition (when it was treated as such) also moved back and forth more frequently on the vertical axis, hovered more, and spent more time hovering.

16.5.3 Straightforward vs. Complex Response Options (H3)

Table 16.4 shows indices for the mouse-movement behavior in the straightforward and complex versions of the type of employee question. As hypothesized, when respondents were presented with response options that were wordy and unclear (i.e. the complex response option format), they hovered more frequently and for longer than when response options were clear and concise (i.e. the straightforward response option format). In addition to the hypothesized

Table 16.3 Comparison of the mouse-movement indicators for yes/no and check all formats ($N = 537$).

Mouse movement		Check all (n = 276)	Yes/No as check All (n = 180)	Yes/No (n = 81)	Kruskal-Wallis rank-sum test	P-value
Total distance traveled (pixels)	Mean	3833.1	4266.0	6073.5	27.91	<0.001
	Median	2899.7	2778.0	4241.4		
Horizontal regressions	Mean	18.2	22.5	45.2	104.03	<0.001
	Median	14	14.5	37		
Vertical regressions	Mean	20.3	24.1	46.2	85.12	<0.001
	Median	15	16	38		
Number of hovers	Mean	2.5	2.7	4.8	32.48	<0.001
	Median	2	2	4		
Hover time (s)	Mean	14.1	16.2	22.2	11.84	0.003
	Median	10.5	11.5	18.6		

Table 16.4 Comparison of the mouse-movement indicators for straightforward and complex response options ($N = 716$).

Mouse movement		Straightforward (n = 342)	Complex (n = 374)	Wilcoxon rank-sum test	P-value
Total distance traveled (pixels)	Mean	2741.2	3351.4	71 560	0.006
	Median	2065.4	2363.0		
Horizontal regressions	Mean	13.5	15.8	69 123	0.061
	Median	10	11		
Vertical regressions	Mean	14.5	18.9	73 912	<0.001
	Median	11	14		
Number of hovers	Mean	1.6	2.3	77 306	<0.001
	Median	1	2		
Hover time (s)	Mean	10.3	16.2	75 430	<0.001
	Median	5.8	8.6		

effects, respondents also moved the mouse up and down more often and traveled a greater distance with complex response options.

16.5.4 Self Evaluations vs. General Evaluations (H4)

The general evaluation resulted in significantly higher values for all mouse-movement indicators compared to the self-evaluation version. Note

Table 16.5 Comparison of the mouse-movement indicators for a general and self-evaluation question ($N = 778$).

Mouse movement		General evaluation	Self evaluation	Wilcoxon signed-rank test	P-value
Horizontal regressions	Mean	7.7	4.6	35 615	<0.001
	Median	6	3		
Vertical regressions	Mean	7.8	5.2	47 320	<0.001
	Median	6	4		
Number of hovers	Mean	0.6	0.3	13 237	<0.001
	Median	0	0		
Hover time (s)	Mean	3.3	1.7	16 854	<0.001
	Median	0	0		

778 respondents answered both the general-evaluation and self-evaluation questions.

that for these analyses a Wilcoxon signed-rank test was used as the comparison was within participants. Table 16.5 provides the results of each hypothesis and outcomes for the other indicators.[7] We saw more vertical regressions and hovering in the general evaluation than the self evaluation, suggesting that either respondents were more certain or they could more easily estimate their feelings toward their own economic situation than Germany's, or that they used the general evaluation to learn how to answer rating scale-type questions so they were more efficient when answering about themselves.

16.6 Discussion

Determining when users experience difficulty with a task and finding ways to measure and respond to it has been studied for decades in a variety of disciplines. It is practically relevant to survey research because signs of confusion or indecision may be a precursor of erroneous or invalid responses. Recently, new technologies have made available new indicators for respondent difficulty. In this study, we collected mouse-tracking data in an online survey to identify measures that can provide meaningful indications of data quality in a survey context. We examined several different potential indicators of difficulty to see if respondents behaved differently when questions were manipulated to be more or less challenging, and took a first step toward linking specific movements to specific types of difficulty.

7 We did not compare the total distance traveled for this question because the starting point of the mouse for the general version was lower on the screen than the starting position for the personal question. Thus, respondents necessarily had a longer trajectory for the general version.

Overall, we found several consistent indicators of difficulty based on mouse-tracking measures. Specifically, we found that unordered response options were associated with greater distance traveled by the cursor and more vertical and horizontal regressions than ordered response options. The yes/no response format (when people attended to all options) also resulted in greater distance traveled, more horizontal and vertical regressions, more hovers, and more time spent hovering than the check all that apply format. Similarly, when the response options were complex compared to straightforward, higher values were observed for all mouse-movement indicator variables, except for horizontal regressions. Our hypotheses were all supported, and beyond that, the effects were much more general than we expected. While we expected to see specific effects, we found that our manipulations had a significant influence on more of the mouse-movement indicators than we had originally anticipated.

The differences in mouse movements observed between the straightforward and complex versions of response options and between the ordered and unordered response options support the findings from prior research that there is a relationship between users' mouse movements and their experience of difficulty (Stieger and Reips 2010; Horwitz et al. 2017). A secondary aim of our research was to associate specific types of difficulty with specific movements. Given the general measures we used in this study, we were not able to differentiate between the various types of difficulty. One way that future research may help define these relationships is by focusing on AOIs. For example, regressions in this analysis equate to any change in direction along the x- or y-axis. If we were to redefine regressions to be more specific, such as making each response option an AOI, an alternative definition of a vertical regression would be moving back and forth between two AOIs. This could be a better indicator of indecision between two response options. Additionally, most of our manipulations only applied to a single question. Future studies should test the same hypotheses on multiple questions to build a general profile of the movements that correspond to different types of difficulty.

While mouse movements can inform when a respondent is experiencing difficulty, they can also help us learn how respondents are interacting with an instrument. When analyzing the yes/no style question, item nonresponse is high in our data. By looking at the mouse movements for this type of question, we saw that many respondents did not attend to the No column and, consequently, only selected items from the Yes column. The ability to understand what information respondents are attending to and how they are answering questions can help designers to write streamlined questions that are easy for respondents to navigate and answer.

In addition to the types of difficulty respondents experience, we also explored how respondents answered a general- and self-evaluation question using rating scales. Respondents engaged in all of the movements more often when evaluating Germany's economic situation compared to their own. We offer two possible explanations for this difference. The general evaluation was the second question in the survey and introduced a new type of question, after which it was succeeded by the self-evaluation question. It is possible that respondents engaged in more movements in the general evaluation because they were learning how to navigate a new response format. By the time they saw the self-evaluation question, they were already familiar with the format so they were able to answer with fewer movements. Another possible explanation is that respondents may not have already formed a judgment or did not have a strong opinion on Germany's economic situation but were very familiar with their own. In this case, answering the general evaluation would be more difficult, so the increase in movements could reflect that difficulty. Given the question order, it is also possible that the increased movements in the general evaluation is a result of both of these phenomena combined. As we have already seen that mouse movements are associated with difficulty, it would be interesting to see in a future study if these movements can reflect learning as well.

The research presented in this chapter demonstrates that a number of mouse movements are associated with difficulty. We found that our manipulations had a significant influence on more mouse-movement indicators than originally anticipated, which might be explained by correlations between the indicators. For example, the time and effort respondents spend on a page to answer a question might influence their mouse movements. Our analysis was thus rather global in nature, in the sense that our manipulations and indicators were related to complete questions and not specific to some part of a question. To continue to uncover the relationship between mouse movements and difficulty, however, it is important to explore where respondents are hovering and which AOIs they are regressing between. Focusing on more specific AOIs could help identify the specific parts of questions people are struggling with. For example, hovers in the question stem could suggest the question is not clear whereas hovers in the response options could indicate uncertainty or indecision between several options.

Once these relationships have been defined, mouse movements can be used as a diagnostic tool to identify people that are having difficulty answering a question. Before a survey is fielded, they can be used in pretesting to isolate and identify problem questions. Typically, pretesting only occurs with a handful of respondents due to time and cost constraints. An analysis of mouse movements allows for an inexpensive way to obtain information from a large number of people, which could give researchers greater confidence that the results they find are likely to apply broadly across their population of interest. This can lead to questionnaire improvements and greater confidence in the quality of data.

Mouse movements could also be analyzed in real time, while respondents are taking a survey. At this point, we have not determined specific thresholds for the number of regressions or the time spent hovering, that – if surpassed – indicate a critical amount of difficulty, but studying movements on many types of questions may aid in determining this. Once these thresholds are identified, survey developers might provide interventions, such as dynamic help text or the option to online chat with an interviewer, if respondents reach the threshold set for any particular movement or combination of movements.

After a survey closes out, mouse movements could also be used to help explain unexpected results. For example, if reliability is low between two items where it was expected to be high, mouse movements could show whether respondents had difficulty with one of the questions that may have led to measurement error.

As this study shows, mouse movements can be a valuable tool to evaluate questionnaires. Continuing this work to test more specific hypotheses relating types of respondent difficulty to specific mouse movements will help build a body of knowledge on how respondents interact with web questionnaires so that we can make questions easier for respondents to answer and increase the quality of the data they provide.

References

Arroyo, E., Selker, T., and Wei, W. (2006). Usability tool for analysis of web designs using mouse tracks. In: *CHI'06 Extended Abstracts on Human Factors in Computing Systems* (eds. R. Grinter, T. Rodden, P. Aoki, et al.), 484–489. Montreal: ACM.

Brennan, S.E. and Williams, M. (1995). The feeling of another's knowing: prosody and filled pauses as cues to listeners about the metacognitive states of speakers. *Journal of Memory and Language* 34: 383–398.

Conrad, F.G. and Schober, M.F. (2000). Clarifying question meaning in a household telephone survey. *Public Opinion Quarterly* 64 (1): 1–28.

Conrad, F.G., Schober, M.F., and Coiner, T. (2007). Bringing features of human dialogue to web surveys. *Applied Cognitive Psychology* 21 (2): 165–187.

Conrad, F.G., Tourangeau, R., Couper, M.P., and Zhang, C. (2017). Reducing speeding in web surveys by providing immediate feedback. *Survey Research Methods* 11 (1): 45–61.

Cox, A.L. and Silva, M.M. (2006). The role of mouse movements in interactive search. In: *Proceedings of the Cognitive Science Society* (ed. R. Sun), 1156–1161. Vancouver: Curran Associates, Inc.

Dale, R., Roche, J., Snyder, K., and McCall, R. (2008). Exploring action dynamics as an index of paired-associate learning. *PLoS One* 3 (3): e1728.

Dawes, R.M. (1980). Social dilemmas. *Annual Review of Psychology* 31 (1): 169–193.

Duran, N.D., Dale, R., and McNamara, D.S. (2010). The action dynamics of overcoming the truth. *Psychonomic Bulletin and Review* 17 (4): 486–491.

Ehlen, P., Schober, M.F., and Conrad, F.G. (2007). Modeling speech disfluency to predict conceptual misalignment in speech survey interfaces. *Discourse Processes* 44 (3): 245–265.

Freeman, J.B. (2014). Abrupt category shifts during real-time person perception. *Psychonomic Bulletin and Review* 21: 85–92.

Freeman, J.B. and Ambady, N. (2009). Motions of the hand expose the partial and parallel activation of stereotypes. *Psychological Science* 20 (10): 1183–1188.

Freeman, J.B., Dale, R., and Farmer, T.A. (2011). Hand in motion reveals mind in motion. *Frontiers in Psychology* 2: 59.

Heerwegh, D. (2003). Explaining response latencies and changing answers using client-side paradata from a web survey. *Social Science Computer Review* 21 (3): 360–373.

Henninger, F. and Kieslich, P.J. (2016). Beyond the lab: collecting mouse-tracking data in online studies. In: *Abstracts of the 58th Conference of Experimental Psychologists* (eds. J. Funke, J. Rummel and A. Voß), 125. Lengerich: Pabst.

Horwitz, R., Kreuter, F., and Conrad, F.G. (2017). Using mouse movements to predict web survey response difficulty. *Social Science Computer Review* 35 (3): 388–405.

Huang, J., White, R.W., and Dumais, S. (2011). No clicks, no problem: using cursor movements to understand and improve search. In: *Proceedings of the SIGCHI Conference on Human Factors in Computing Systems* (eds. D. Tan, G. Fitzpatrick, C. Gutwin, et al.), 1225–1234. ACM.

Kieslich, P.J. and Henninger, F. (2017). Mousetrap: an integrated, open-source mouse-tracking package. *Behavior Research Methods* 49 (5): 1652–1667.

Kieslich, P.J. and Hilbig, B.E. (2014). Cognitive conflict in social dilemmas: an analysis of response dynamics. *Judgment and Decision Making* 9 (6): 510–522.

Kieslich, P.J., Henninger, F., Wulff, D.U. et al. (2019). Mouse-tracking: A practical guide to implementation and analysis. In: *A Handbook of Process Tracing Methods.* (eds. M. Schulte-Mecklenbeck, A. Kühberger and J. G. Johnson), 111–130. New York, NY: Routledge.

Koop, G.J. and Johnson, J.G. (2011). Response dynamics: a new window on the decision process. *Judgment and Decision Making* 6 (8): 750–758.

Kreuter, F. (ed.) (2013). *Improving Surveys with Paradata: Analytic Uses of Process Information*. New York: Wiley.

Leiner, D.J. (2014). SoSci Survey (version 2.5.00-i). Computer software. www.soscisurvey.com.

Lind, L.H., Schober, M.F., and Conrad, F.G. (2001). Clarifying question meaning in a web-based survey. In: *Proceedings of the American Statistical Association, Section on Survey Research Methods*. Alexandria, VA: ASA.

Mueller, F. and Lockerd, A. (2001). Cheese: tracking mouse movement activity on websites, a tool for user modeling. In: *CHI'01 Extended Abstracts on Human Factors in Computing Systems* (ed. M. Tremaine), 279–280. New York, NY: ACM.

R Core Team (2016). *R: A Language and Environment for Statistical Computing*. Vienna, Austria: R Foundation for Statistical Computing https://www.R-project.org.

Rodden, K., Fu, X., Aula, A., and Spiro, I. (2008). Eye-mouse coordination patterns on web search results pages. In: *CHI'08 Extended Abstracts on Human Factors in Computing Systems* (eds. M. Czerwinski, A. Lund and D. Tan), 2997–3002. Florence: ACM.

Schneider, I.K., van Harreveld, F., Rotteveel, M. et al. (2015). The path of ambivalence: tracing the pull of opposing evaluations using mouse trajectories. *Frontiers in Psychology* 6: 996.

Schober, M.F. and Bloom, J.E. (2004). Discourse cues that respondents have misunderstood survey questions. *Discourse Processes* 38 (3): 287–308.

Schober, M.F. and Conrad, F.G. (1997). Does conversational interviewing reduce survey measurement error? *The Public Opinion Quarterly* 61 (4): 576–602.

Schober, M.F., Conrad, F.G., and Fricker, S.S. (2004). Misunderstanding standardized language. *Applied Cognitive Psychology* 18: 169–188.

Smith, V.L. and Clark, H.H. (1993). On the course of answering questions. *Journal of Memory and Language* 33: 25–38.

Stieger, S. and Reips, U.D. (2010). What are participants doing while filling in an online questionnaire: a paradata collection tool and an empirical study. *Computers in Human Behavior* 26 (6): 1488–1495.

Tourangeau, R., Rips, L.J., and Rasinski, K. (2000). *The Psychology of Survey Response*. Cambridge, UK: Cambridge University Press.

Yan, T. and Tourangeau, R. (2008). Fast times and easy questions: the effects of age, experience and question complexity on web survey response times. *Applied Cognitive Psychology* 22 (1): 51–68.

Zushi, M., Miyazaki, Y., and Norizuki, K. (2012). Web application for recording learners' mouse trajectories and retrieving their study logs for data analysis. *Knowledge Management and E-learning* 4 (1): 37–50.

17

Using Targeted Embedded Probes to Quantify Cognitive Interviewing Findings

Paul Scanlon

Collaborating Center for Questionnaire Design and Evaluation Research, National Center for Health Statistics, Centers for Disease Control and Prevention, United States Department of Health and Human Services, Hyattsville, MD, USA

17.1 Introduction

The evaluation of survey questions and questionnaires is typically either a qualitative or quantitative endeavor: relying on methods such as cognitive interviewing or metrics such as item nonresponse, but doing so separately. This methodological separation is largely due to cost. While cognitive interviewing projects are relatively cheap to conduct as compared to staging full survey field tests, they rely on purposive samples; attempting to conduct cognitive interviews across a statistical sample of a population would be infeasible. Similarly, inserting cognitive probes or question experiments into survey field tests – such as in what Converse and Presser (1986) call "Embedded Probing," or the use of random, close-ended probe questions as described by Schuman (1966) – is rare, given the amount of planning and sample needed in full-fledged household and random digit dial (RDD) surveys. The advent of online surveying, however – particularly the development and maturation of standing, commercially available web panels of potential survey respondents – is building a mixed-methodology bridge between these two styles of question evaluation. Two "lanes," or methodological approaches, are under construction: using either open-ended or closed-ended cognitive probes on web surveys to understand how respondents answer survey items. The first method is an attempt to deal with the limitations of the small, purposive samples used in cognitive interviewing by collecting similar data but in self-report web surveys, for which much larger samples can be used. The goal of this approach is to supplement, or even replace, face-to-face cognitive interviews by using open-ended probing questions that mimic the types of probes traditionally used in cognitive interviews. The second method, which

Advances in Questionnaire Design, Development, Evaluation and Testing, First Edition.
Edited by Paul C. Beatty, Debbie Collins, Lyn Kaye, Jose-Luis Padilla, Gordon B. Willis, and Amanda Wilmot.

is the focus of this chapter, is not an attempt to replace face-to-face cognitive interviews, but rather is designed to take cognitive interviewing findings and expand them to a survey population primarily using closed-ended probing questions embedded in a web survey alongside the items under evaluation.

After briefly discussing the development of cognitive probing using web surveys, this chapter will provide examples of how these "targeted, embedded probes" may contribute to question evaluation. These examples are not full question evaluations in and of themselves, but rather serve as an overview of how this probing methodology can be used to supplement existing evaluation methods. This will be followed by some methodological considerations and outstanding questions about how these probes can be best formatted and implemented in web surveys.

17.1.1 The Development of Web Probing Methodology

As cognitive interviewing methodology developed – particularly in labs and centers established in governmental statistical agencies – variations began to emerge, not only in protocol, but also in the scope and purpose of the interviews themselves. Willis (2005) touched on this in a practical way, directing researchers to consider whether conducting interviews using "think aloud" prompts, verbal probes, or some combination of the two best fit their research purposes. Alongside these variations in the conduct of cognitive interviews, the focus of the interviews themselves began varying as well. While some practitioners focused on eliciting cognitive problems (and then retesting the "fixed" questions), others applied a more sociological approach and focused on using the interviews to tease apart what specific patterns of interpretation respondents use to answer to each question, as well as the social constructs each item captured – whether the interpretations are determined to be "in-scope" in that they measure the constructs the question designers hope to capture, or not. This latter variation of the method gives researchers and methodologists much more flexibility in how and when they can use cognitive interviewing findings. These findings can be used to not only revise survey items, but also to provide context to the future quantitative analysis of the survey data itself and to allow for the robust design of follow-up cognitive probes (see Miller et al. 2014).

Embedding probe questions in surveys emerged out of conducting respondent debriefings during survey field tests. Relatively simple, structured probe questions, such as "Could you tell me more about that?" following a survey item of interest proved to be a relatively efficient way to identify problematic questions within the survey environment. Schuman (1966) embedded probe directives in a survey of factory workers in Pakistan, with interviewers instructed to select 10 random questions per respondent to ask semi-structured follow-up questions immediately after the selected question was administered. These probe responses were then analyzed both qualitatively and quantitatively (by assigning a code indicating whether the probe response accurately "predicted" the respondents' original survey responses) in an effort to determine question validity. Converse and Presser (1986) suggest using open-ended, embedded probes to validate close-ended survey questions – again randomly distributed across the entire survey instrument. Both Schuman and Converse and Presser advise researchers to randomly assign probes in order to "cover" the entire survey instrument under evaluation. On the other hand, Cannell et al. (1989) used what they called "special probes" that targeted the evaluation on a specific set of questions. By using a split sample design on a field test, they were able to evaluate over 20 questions for respondent difficulties. These probes were administered in two ways: either directly following the survey item under evaluation (i.e. "embedded probes") or at the end of the completed questionnaire (i.e. retrospective probes). While they found that these special probes were good at uncovering comprehension difficulties, they suggested that would not be useful for other areas of question evaluation.

Nonetheless, the idea of using embedded probes began to take off in the last two decades with the advent of internet mode surveys (Willis 2014). Typically, these probes have been administered as open-ended questions with text fields, which attempt to obtain the same sort of information that face-to-face cognitive interviews can provide – a method that Meitinger and Behr (2016) refer to as "Online Probing." For example, Behr et al. (2013) asked open-ended category-selection probes about two items on an internet mode survey conducted by members of two opt-in web panels in Germany about gender roles in the household. They found that the probes were successful in uncovering the same forms of potential response errors that a previous multimode cognitive interviewing project (Braun 2008) had discovered. Most works examining online probing have noted that the method is limited by the fact that a web survey cannot tailor the follow-up probes to an individual respondent in the same way that an interviewer can in a face-to-face interaction (see for instance Behr et al. 2014, p. 525); they argue that the larger sample sizes that web surveys provide in comparison to cognitive interviewing partially makes up for this. While Behr et al. (2012) recommend that embedded probes could supplement traditional face-to-face cognitive interview studies, recent work (i.e. Edgar 2013; Murphy et al. 2014) has suggested that such open-ended, embedded cognitive probes could supplant face-to-face interviews in some cases. Fowler et al. (2015) note that in the health field in particular, it can be costly and difficult to gather the necessary sample to conduct face-to-face cognitive interviews for specialized, national health surveys, and thus online probing may provide a feasible alternative.

While the use of open-ended probes in web-format surveys has shown some promise in discovering problematic survey items and concepts, a more targeted approach is necessary if the goal is to quantify the spread of in- or out-of-scope interpretative patterns across a sample. In order to accomplish this, the qualitative findings of in-depth cognitive interviews must be used to develop targeted, *close-ended* embedded probes that can be administered directly following a survey item under evaluation (Baena and Padilla 2014; Miller and Maitland 2010). Close-ended probes, in contrast to open-ended ones, are used for two reasons. First, while open-ended probes may more closely replicate the traditional cognitive interviewing method, using them extensively in a field test not only places additional burden on respondents (thus increasing the likelihood of item nonresponse and break-offs), but also requires a large amount of effort on the researchers' parts to clean and code the data. Closed-ended probes, on the other hand, function just like any other item in the survey – so neither respondents nor analysts are over-burdened. Beyond the issues of respondent and researcher effort, however, close-ended probes allow question evaluators to target their examination of a question on specific terms or aspects of the question-response process via the framing that the close-ended answer categories provide.

One of the clearest examples of using targeted, embedded probes in a question-evaluation project was during the development of the Washington Group on Disability Statistics' Functioning Module (which is included on the National Health Interview Survey [NHIS]; see Miller 2016 for an overview of the entire design and evaluation process). In an effort to study proposed anxiety questions for a multinational survey of disability, Miller et al. (2011) used cognitive interviews to uncover the various patterns of interpretation of a series of questions including one measuring intensity: "Thinking about the last time you felt anxious, how would you describe the level of anxiety? (A lot, A little, Somewhere in between a little and a lot)." Most of the patterns that emerged from this evaluation appeared to be "in-scope" – relating in some way to how the phenomena of extreme anxiety or nervousness could limit a respondent's ability to function in everyday situations. However, testing in the United States and Canada also uncovered a problematic, "out-of-scope" interpretation – anxiety and nervousness were understood by some respondents to be a positive force that helped them focus on tasks at hand and get things done. During the field test phase of this evaluation project, a close-ended probe asking about these patterns was included alongside the actual Washington Group questions in an effort to determine whether this pattern existed outside of America and Canada, and how common it was (Miller and Maitland 2010). By administering a targeted probe question alongside the other anxiety survey questions, differences in the extent of each of these patterns of interpretations became clear across both countries and respondents' answers to the survey questions themselves (Loeb 2016). For instance, they found that 82% of respondents in Mongolia understood anxiety to be a positive feeling, and that across all countries the likelihood of respondents employing the positive feeling pattern of interpretation decreased as they reported a higher intensity of anxiety. Alerted by this country-specific finding, the Washington Group design team revisited and improved the Mongolian translation.

17.2 The NCHS Research and Development Survey

In order to continue refining this technique, and to better understand how findings from cognitive interviews can be expanded to a wider population, in late 2015 the National Center for Health Statistics (NCHS) launched a survey dedicated to methodological research that explored the use of these targeted, embedded probes. The Research and Development Survey (RANDS) was a self-administered web-mode survey that used the Gallup Panel as its frame. Behr et al. (2014) noted that one of the key limitations of their study was the use of an opt-in web panel, where there is no control over who signs up for the panel, and thus final results cannot be even *theoretically* related back to an original frame (and therefore extrapolated to a population). On the other hand, as a recruited panel, the Gallup Panel is constructed primarily by recruiting respondents to a dual-frame cell-phone and landline RDD survey (supplemented by occasional recruitment efforts via an address-based sample (ABS) survey that uses a commercially available frame of all US postal addresses); by combining both RDD and ABS methods in their recruitment, Gallup attempts to create a panel that is representative of the US adult population (Gallup 2016). A panel sample such as this is clearly limited as compared to a full ABS design (such as used by the NHIS), and results should not be understood as truly representative of the county (or, since RANDS was only administered over the internet, of all US adult internet users). However, given its basis in statistical sampling, we expect the Gallup Panel to provide more reliable information about the range of potential American survey respondents than an opt-in panel would.

17.2.1 The Questionnaire

The RANDS questionnaire consisted of a subset of 88 NHIS questions (primarily from the NHIS' sample adult questionnaire) and a set of 21 targeted, embedded probe questions. The targeted, embedded probes were designed using the findings of three iterative rounds of cognitive interviewing that evaluated the 88 NHIS questions (Scanlon 2017). The focus of these cognitive interviews was to understand how respondents understood and answered the survey questions. The various patterns of interpretations respondents used while answering each item were uncovered, and during the systematic analysis of these patterns, response schemata were outlined. These schemata then served as the basis for the design of the targeted, embedded probe questions – with each pattern of interpretation typically becoming one of the probe questions' answer categories. For instance, the evaluation of the NHIS question on general perceptions of health ("Would you say your health in general is excellent, very good, good, fair, or poor?") found that respondents were using seven separate patterns of interpretations when thinking about their general health (Scanlon 2017, p. 6):

1. Diet and nutrition habits, including the type and amounts of food consumed
2. The amount and frequency of their physical activity
3. Whether or not, and how much, a respondent smoked, drank alcohol, or used other drugs
4. The presence or absence of chronic and acute health conditions and diseases
5. How often a respondent visited a healthcare provider
6. The amount and frequency of pain or fatigue
7. Conversations about the respondent's health with their healthcare providers

Because the face-to-face cognitive interviews already provided information on the composition of the "general health" domain, the focus of the targeted, embedded probe questions was not on trying to replicate this information, but rather to determine the frequency of each of these patterns across the survey population. As such, these questions were designed to ask respondents about which of the patterns already discovered during the face-to-face interviews they used when answering the survey item under evaluation. For instance, using the previous seven patterns of interpretation, the targeted, embedded probe for the NHIS general health question was:

When you answered the previous question about your health, what did you think of? (*Select all that apply*)

1. My diet and nutrition
2. My exercise habits
3. My smoking or drinking habits
4. My health problems or conditions
5. The amount of times I seek health care
6. The amount of pain or fatigue that I have
7. My conversations with my doctor

After being cognitively tested themselves, each probe question was then embedded in the RANDS questionnaire following the specific question (or, in a few cases, set of questions) it asked about.

17.2.2 Survey Respondents

8232 members of the Gallup Panel were invited to participate in the survey, and 2480 complete responses were obtained, for an American Association of Public Opinion Research (AAPOR) Response Rate 1 of 30.1% (an additional 148 respondents started, but did not complete, the questionnaire). Table 17.1 shows

Table 17.1 Percent distribution of sample and responder demographic characteristics.

Characteristic	Non-responders (n = 5603)	Incomplete cases (n = 148)	Complete cases (n = 2480)	Total sample (n = 8231)
Female	42.07	42.57	45.52	43.12
Male	57.93	57.43	54.48	56.88
Black	24.22	17.57	20.28	22.91
White	53.83	85.14	54.72	57.06
Asian	2.55	1.35	2.26	2.44
Hispanic	16.88	16.22	11.85	15.36
Some other race	2.25	1.35	1.09	1.88
Less than 36 years old	47.17	58.78	29.31	42.00
36–49 years old	22.76	16.89	21.17	22.17
50–64 years old	24.45	18.24	33.67	27.12
65 years old or greater	5.62	6.08	15.85	8.71

Note. Data from rounds 1 and 2 of NCHS RANDS.

the distribution of the non-responders, incomplete cases (partial responses), and responders by gender, race, and age category.

Unless otherwise noted, the weighted data are presented here, using the post-stratification weights developed by Gallup that took into account the overall panel composition and the survey response. Based on suggestions from Gallup staff, independence was assumed for all statistical tests described. The weights used in these analyses are normalized such that the weighted sample size is equal to the unweighted sample size, and therefore test statistics and standard errors were calculated using the sample size as n. The examples are presented as a broad overview of the potential of these probes – more complex analyses (including examining respondents' response behaviors to the probes themselves and exploring how responses to the probes may influence or relate to responses to other survey questions) are possible, but not presented in this chapter.

17.3 Findings

Because they are structured as any close-ended survey items, targeted, embedded probes can be directly analyzed alongside other survey variables. While there are a multitude of ways that their data can be used in question evaluation, this paper addresses three potential uses of this type of probes: (i) determining the extent and distribution of a pattern of interpretation across a survey population, (ii) examining whether or not different subgroups of a survey population interpret the same survey item in different ways, and (iii) directly comparing the way in which multiple questions perform. Data from the RANDS survey will be used to illustrate each of these approaches.

17.3.1 The Distribution of Patterns of Interpretation

One of the most basic applications of targeted, embedded probes is to use them to determine the distribution and extent of patterns of interpretation across a survey population. As noted earlier, one of the primary limitations of cognitive interviewing is that findings cannot be extrapolated from the purposive sample used to conduct the interviews to the wider population. Therefore, while the analysis of cognitive interviews can uncover all the various patterns of interpretation that a respondent might use to answer a survey question (assuming that enough cognitive interviews have been conducted to reach the point of theoretical saturation), it cannot say which pattern is more common outside of the purposive sample itself. However, by directly asking a statistical sample of respondents about which pattern or patterns they used, determining the extent of the patterns is possible.

The RANDS questionnaire included not only the Washington Group anxiety questions mentioned earlier, but also the probe question that Miller and Maitland (2010) used to examine the cross-national patterns of interpretation of these items. This full set of questions is shown in Table 17.2.

The probe at the end of this series of questions asks the respondents about whether they used any of the four patterns of interpretation that emerged during cognitive testing. Respondents were able to select more than one pattern of interpretation when answering this probe question. Just looking at the weighted distributions of these four patterns – that anxiety is an intense feeling (13.79%), that anxiety interferes with the respondent's life (49.27%), that anxiety refers to a diagnosed disorder (21.95%), and that anxiety is a positive emotion that helps the respondent get things done (48.97%) – it is clear that the out-of-scope "positive" interpretation was used by a large proportion of respondents when answering the anxiety questions. Of the respondents who chose the positive pattern, most (76.76%) *only* chose that one, while the remaining 23.42% indicated *both* the positive and one or more of the other three patterns (most commonly the "intense feeling" pattern).

On the face of this, and from a question-evaluation perspective, this distribution is troubling because the clearly out-of-scope pattern of interpretation is either the first- or second-most commonly used across the sample. The difference between the proportion of respondents using the "positive" and "interfere" patterns is not statistically significant at an alpha level of 0.05 (used for all statistical tests throughout this chapter), with $\chi^2(1,N = 1663) = 0.02$, $p = 0.89$. However, because the responses to probe questions can be cross-tabulated with other survey questions, it is possible to examine how each pattern relates to

Table 17.2 Washington Group anxiety questions on the RANDS questionnaire.

Variable	Question text	Answer categories
ANX_1	How often do you feel worried, nervous, or anxious?	Daily Weekly Monthly A few times a year Never
ANX_2	Do you take medication for these feelings?	Yes No
ANX_3	Thinking about the last time you felt worried, nervous, or anxious, how would you describe the level of these feelings? Would you say you felt a little this way, a lot this way, or somewhere in between?	A little A lot Somewhere in between a little and a lot
PROBE21	Which of the following statements, if any, describes your feelings? (*Select all that apply*)	Sometimes the feelings can be so intense that my chest hurts and I have trouble breathing These are positive feelings that help me to accomplish goals and be productive The feelings sometimes interfere with my life, and I wish that I did not have them I have been told by a medical professional that I have anxiety

the answers to the survey questions under evaluation. First, by crossing the anxiety frequency question by the probe question (collapsed in Figure 17.1 to a dichotomous variable comparing the respondents who *only* used the positive, out-of-scope interpretation, with those who used at least one of the in-scope interpretations) a straightforward pattern emerges.

It is clear that as the frequency of a respondent's anxiety increases, the likelihood of them employing the out-of-scope pattern decreases. Beyond simple cross-tabulations however, it is possible to explore how a respondent's answer is related to patterns they choose to use when considering the question. In the Washington Group scheme, anxiety is not measured by frequency alone, but by scaling both frequency and intensity together (see Loeb 2016; Miller 2016). By running a series of logistic regressions, it is possible to see which interpretations contribute to the various levels of anxiety. For these regressions, the outcome variables were dichotomous flags indicating whether or not the respondent answered in each of the 12 combination of the frequency and

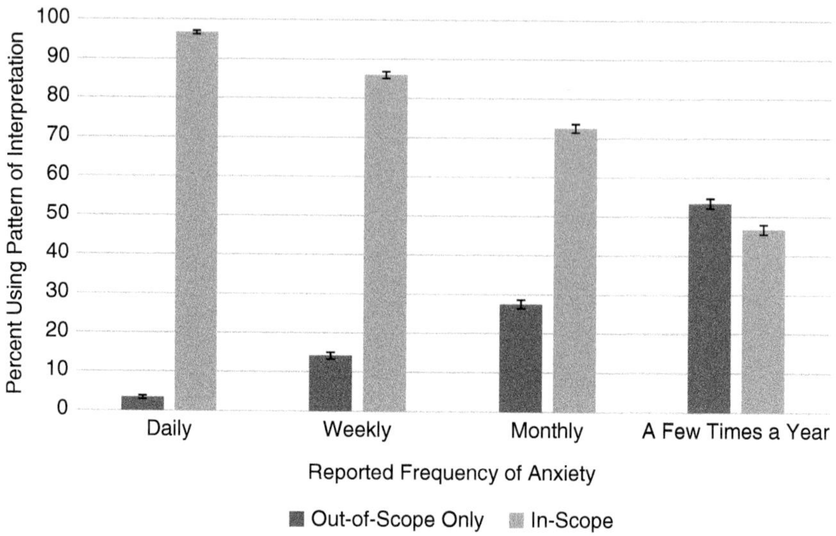

Figure 17.1 Weighted percent of respondents' patterns of interpretation by frequency of anxiety. This figure illustrates the percent of RANDS round 2 respondents who used either in-scope or out-of-scope interpretations when answering the Washington Group for Disability Statistics' anxiety questions, by their anxiety frequency.

intensity question's answer categories, and the independent variables were the four patterns of interpretation and the respondents' levels of educational attainment. Each of the 12 regressions shown in Table 17.3 included all four patterns of interpretation together, since respondents were able to select more than one

Table 17.3 Results of logistic regressions.

	Frequency			
Intensity	**A few times a year**	**Monthly**	**Weekly**	**Daily**
A little	−Intense*** +Positive*** −Interfere*** −Diagnosed***	+Positive** +Interfere**	+Diagnosed*** +Interfere*	+Interfere** +Diagnosed**
Somewhere in between	+Positive***	+Interfere***	+Interfere*** +Diagnosed***	+Interfere*** +Diagnosed***
A lot	+Interfere*	+Interfere***	+Interfere*** +Intense* −Positive*	+Intense*** +Interfere*** +Diagnosed***

Notes. Data from round 2 of NCHS RANDS. Signs indicate whether a pattern of interpretation positively (+) or negatively (−) correlated to a respondent indicating if they were in each cell of the matrix. *$p < 0.05$. **$p < 0.01$. ***$p < 0.001$.

answer to the probe question. The results of these regressions are shown in Table 17.3. Unweighted data were used for this analysis.

As can be seen in Table 17.3, using the positive pattern of interpretation indicates that a respondent has a low level of anxiety, while the other patterns of interpretation tend to indicate that a respondent has a higher level of anxiety. So, while it is clear that the out-of-scope pattern is not only present in the population, but also frequently used, it appears that respondents who employ this pattern are not in the analytic population of interest: people who experience high levels of anxiety.

17.3.2 Differences Across Subgroups

Knowing the distribution and extent of various patterns of interpretation of a survey question is certainly helpful in question evaluation, but in many cases, it is more informative to go a step further and explore whether any differences exist across subgroups. In short, the question moves from "Do respondents think about this question in different ways?" to "Are different groups effectively responding to different questions?" Again, because targeted, embedded probes function as any other survey question, we can cross the probe responses themselves by various demographic and other characteristics in order to determine if subgroup differences in interpretation exist.

A good illustration of the examination of subgroup differences comes from the self-reported health question ("Would you say your health in general is excellent, very good, good, fair, or poor?") and its structured probe question explained earlier. A simple cross of the question by gender reveals that men and women do not answer the question itself differently, as seen in Table 17.4.

However, the same cannot be said about how men and women *respond and think* about the question. As seen in Table 17.5, men and women base their answer on different phenomena when formulating their answer to the general health question.

Table 17.4 General health status by gender.

Reported status	Males	Females	χ^2	df	p
Excellent	0.14	0.12	8.76	5	0.12
Very good	0.36	0.37			
Good	0.37	0.34			
Fair	0.09	0.12			
Poor	0.03	0.04			
Missing	0.01	0.01			

Note. Data from round 2 of NCHS RANDS. Weighted proportions are presented.

Table 17.5 General health patterns of interpretation by gender.

Pattern of interpretation	Males	Females	χ^2	df	p
Diet*	0.57	0.53	3.40	1	0.03
Exercise*	0.54	0.49	5.41	1	0.01
Vices	0.24	0.22	2.36	1	0.06
Health problems***	0.62	0.68	8.05	1	0.00
Sought care*	0.24	0.27	3.90	1	0.02
Pain or fatigue	0.37	0.37	0.03	1	0.43
Conversation with doctor	0.17	0.15	1.28	1	0.13

Note. Data from round 2 of NCHS RANDS. Weighted proportions are presented. *$p < 0.05$. **$p < 0.01$. ***$p < 0.001$.

Men are significantly more likely to answer thinking about their diet and exercise habits than are women, while women are more likely to base their responses on health problems and whether or not they had sought care from a health provider. It should be noted that while seven patterns of interpretation of the general health question were uncovered during cognitive testing and subsequently included in the web probe, this does not indicate a problem with the validity of the question itself: all seven patterns are in-scope and relate directly to the overall phenomenon of general health.

Differences are also present across how respondents of different levels of educational attainment both answer ($\chi^2(10, N = 2461) = 96.90, p < 0.001$) and think about ($\chi^2(12, N = 2457) = 37.533, p < 0.001$) the general health question. Figure 17.2 illustrates the latter of these and shows the variation in which patterns of interpretation respondents of different levels of educational attainment used while answering this question.

In addition to conducting subgroup analysis by demographic characteristics such as gender and educational attainment, it is also possible to explore whether respondents who *answer the question certain ways* differ from one another – for instance, do respondents who report being in excellent, very good, or good health differ in their interpretations from those who report being in fair or poor health? Table 17.6 shows that respondents reporting different levels of health do indeed tend to base their answers on different phenomenon.

The real power of conducting subgroup analysis using targeted, embedded probes surrounds out-of-scope or questionable patterns of interpretation. Out-of-scope patterns emerge frequently in cognitive interviewing studies, but because of the sample limitations, the question always remains as to whether the problematic interpretation is significant across either a population or subgroup.

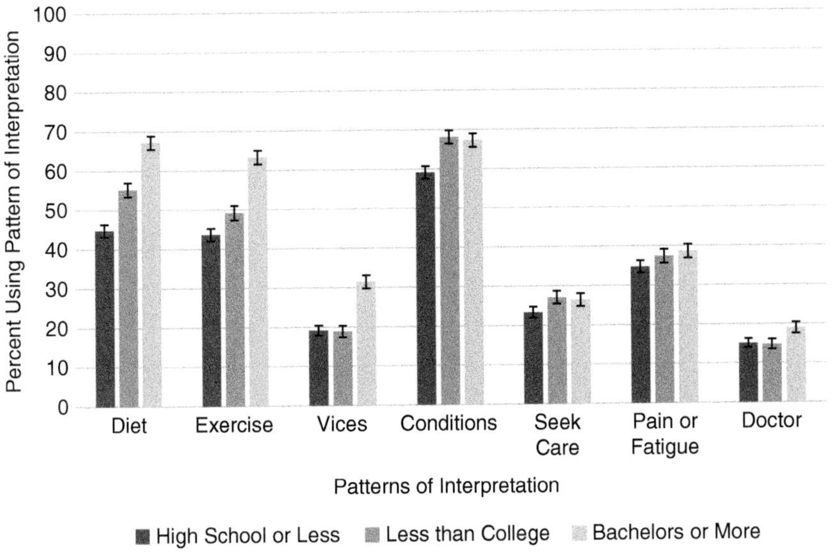

Figure 17.2 Weighted percent of respondents' general health patterns of interpretation by respondent. This figure illustrates the percent of RANDS round 2 respondents who used each of seven patterns of interpretations when answering the self-reported health question, by their educational attainment.

Table 17.6 General health patterns of interpretation by survey response.

Pattern of interpretation	Excellent to good health	Fair or poor health	χ^2	df	p
Diet	0.57	0.53	3.04	1	0.08
Exercise	0.54	0.49	1.02	1	0.31
Vices*	0.24	0.22	6.37	1	0.01
Health problems***	0.62	0.68	42.73	1	0.00
Sought care*	0.24	0.27	4.38	1	0.04
Pain or fatigue***	0.37	0.37	85.49	1	0.00
Conversation with doctor***	0.17	0.15	11.66	1	0.00

Note. Data from round 2 of NCHS RANDS. Weighted proportions are presented. *$p < 0.05$. **$p < 0.01$. ***$p < 0.001$.

The RANDS questionnaire includes the Kessler Physiological Distress Scale (K6 Scale, Figure 17.3), which is a set of questions about how frequently a respondent experienced any of six negative feelings or emotions in the past 30 days, and which is designed to measure the latent variable of nonspecific psychological distress (Kessler et al. 2003).

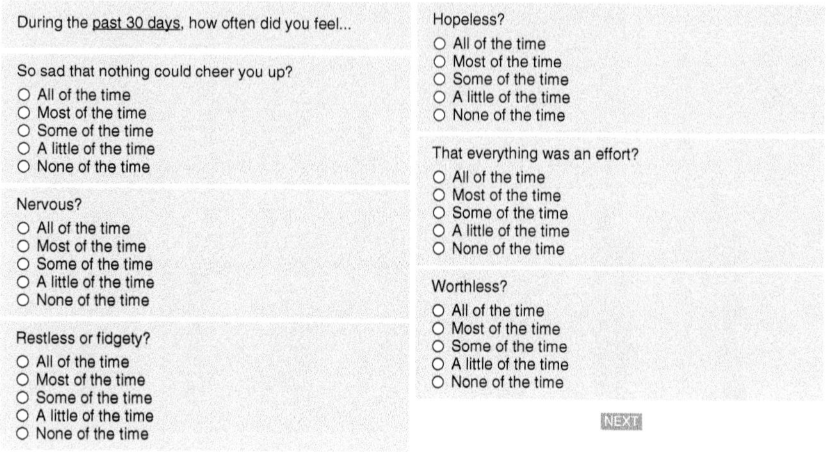
During the past 30 days, how often did you feel...

So sad that nothing could cheer you up?
- O All of the time
- O Most of the time
- O Some of the time
- O A little of the time
- O None of the time

Nervous?
- O All of the time
- O Most of the time
- O Some of the time
- O A little of the time
- O None of the time

Restless or fidgety?
- O All of the time
- O Most of the time
- O Some of the time
- O A little of the time
- O None of the time

Hopeless?
- O All of the time
- O Most of the time
- O Some of the time
- O A little of the time
- O None of the time

That everything was an effort?
- O All of the time
- O Most of the time
- O Some of the time
- O A little of the time
- O None of the time

Worthless?
- O All of the time
- O Most of the time
- O Some of the time
- O A little of the time
- O None of the time

NEXT

Figure 17.3 Kessler 6 Psychological Distress Scale, as presented on the RANDS web survey instrument.

During cognitive interviewing, there was a significant split in how respondents interpreted the K6's "effort" question (Scanlon 2017, p. 95). Whereas some respondents understood this question to be asking about a negative feeling – that they felt burdened, and that everything they did required extra effort – many others thought the question was asking about a positive attribute. In short, these respondents understood the question to be asking how frequently they thought they *put in an effort*, or gave something "their all." Thus, for respondents who used the first of these patterns, a more frequent response indicated a problem (which was the intent of the question in this *psychological distress* scale). However, for the respondents who used the second interpretation, a more frequent response indicated a positive character trait. In the cognitive interviewing sample, each of these patterns emerged in about half of the respondents. Using a targeted, embedded probe on RANDS allowed us to ascertain whether this nearly 50/50 distribution simply an artifact of the purposive sample, or did it hold across the population?

In order to determine this, a forced-choice cognitive probe was embedded following the effort question, which read:

Would you consider everything being an effort to be a good thing or a bad thing?

1. Good thing
2. Bad thing
3. Neither good nor bad

Just looking at the extent of these three interpretations across the survey population (Figure 17.4), it turns out that the half-and-half split seen in the

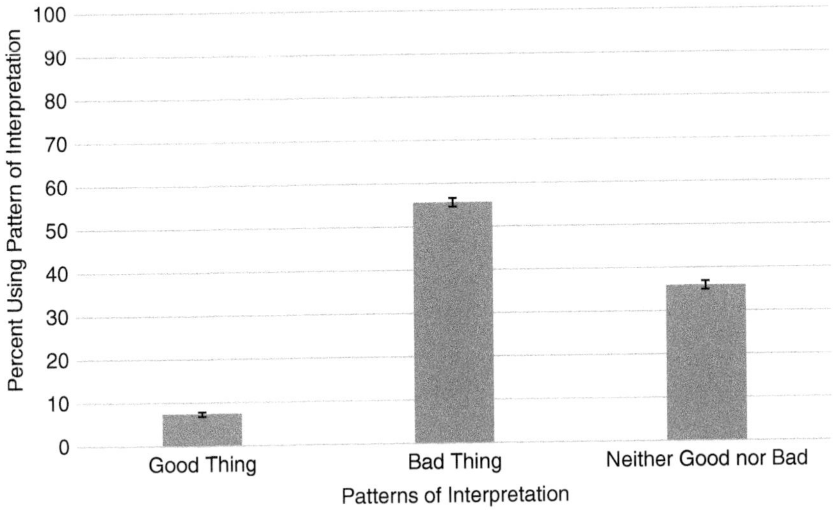

Figure 17.4 Weighted percent of respondents' patterns of interpretation for the K6 Effort question. This figure illustrates the percent of RANDS round 2 respondents who used each of the three patterns of interpretation when answering the K6 Effort question.

cognitive sample was indeed a result of the non-statistical sample. In the representative sample, only about 7% of respondents indicated that they were thinking about the effort question in the positive sense of "putting in an effort."

So, while from a construct validity standpoint, it is reassuring that the extent of the problematic pattern is not as large as the cognitive interviews suggested, the next question the targeted, embedded probe allows us to answer is whether or not this pattern is evenly distributed across subgroups of the population. Figure 17.5 shows the distribution of these patterns of interpretation across three categories of educational attainment (high school degree or less, less than a four-year college degree, and four-year college degree or more). Respondents who have less than a high school diploma are more likely than both respondents with less than a college degree ($\chi^2(1,N = 182) = 28.71, p < 0.001$) and respondents with at least a four-year college degree ($\chi^2(1,N = 182) = 24.09, p < 0.001$) to use this out-of-scope, positive interpretation.

So, while the problematic pattern is not very common across the survey population as a whole, the fact that nearly a third of low-education respondents use it indicates that the question's construct validity does not stand across all population subgroups.

17.3.3 Head-to-Head Comparisons of Questions

One specific use of targeted, embedded probes – particularly in the context of "field testing" new survey questions – is the ability to directly compare

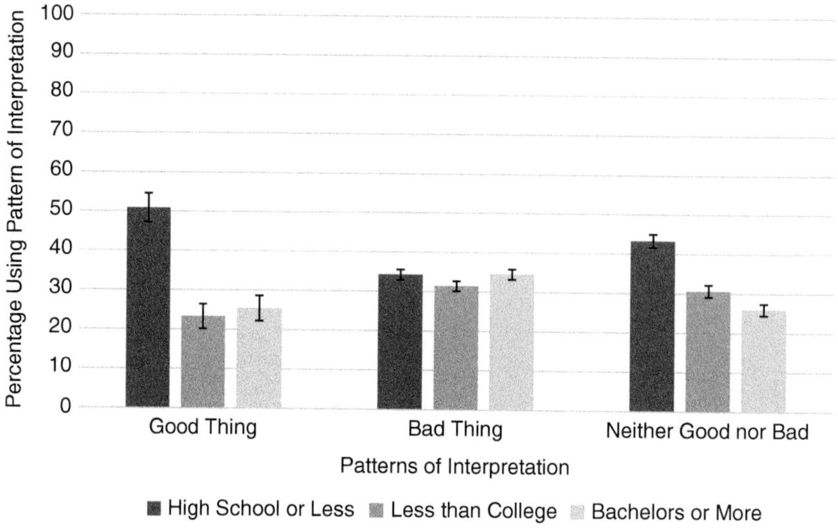

Figure 17.5 Weighted percent of respondents' patterns of interpretation the for K6 Effort question by educational attainment.

the constructs that two similar questions capture. Conventional methods of pretesting survey items by field testing them use quantitative metrics such as item nonresponse and missing (i.e. refusals and don't know responses), response latency, behavior codes, and interviewer debriefing (Presser et al. 2004) to compare and contrast survey items. Of these, the first two rely on an assumed correlation between item comprehension and survey response (or nonresponse), while the latter two largely rely on the observational skills of field interviewers. However, the assumption behind the use of targeted, embedded probes is the same as with all other survey items – simply that they actually capture the intended construct – and this assumption can be examined through cognitive interviewing. Therefore, while traditional quantitative evaluation methods rely on either proxies or a third party, embedding cognitive probes alongside survey items under evaluation allows respondents to directly report how they are interpreting each question. This then allows the constructs that each question (or version of a question) captures to be directly compared. The requirement for this type of comparison is of course that the same targeted, embedded probes be used alongside the questions one wants to compare. By doing so, it is possible to tell if the patterns of interpretations that respondents use when answering the questions have comparable distributions.

For example, physical activity is measured on the NHIS by asking separately about "vigorous" (VIGNO in Table 17.7) and "moderate" physical activity (MODNO in Table 17.7). One potential way to shorten the survey – and reduce respondent burden – would be to combine these questions and ask only a

Table 17.7 Approaches to asking about physical activity tested on RANDS.

Variable	Question
	Current NHIS approach
VIGNO	How often do you do vigorous leisure-time physical activities for at least 10 min that cause heavy sweating or large increases in breathing or heart rate?
MODNO	How often do you light or moderate leisure-time physical activities for at least 10 min that cause only light sweating or a slight to moderate increase in breathing or heart rate?
	Proposed alternative approach
NEWPHYSACT	In the past week, on how many days have you done a total of 30 min or more of physical activity, which was enough to raise your breathing rate? This may include sports, exercise, and brisk walking or cycling for recreation or to get to and from places, but should not include housework or physical activity that may be part of your job.

single item about physical activity (NEWPHYSACT in Table 17.7). RANDS provided an opportunity to compare the phenomena these two approaches collect head-to-head.

Cognitive interviewing revealed that respondents consider a wide, but seemingly overlapping, range of activities in both approaches. A targeted probe was designed using the combined schemata across all three physical activity questions, and was embedded in the RANDS questionnaire following all three of these questions. The probe contained 10 answer categories corresponding to the patterns of interpretation that emerged during the cognitive testing:

Which of the following types of physical activities, if any, did you include when you answered the previous question? (*Select all that apply*)

1. Running
2. Jogging
3. Walking or hiking for exercise
4. Walking to or from and activity
5. Walking at work
6. Working out with exercise equipment
7. Cycling, swimming, or other aerobic exercises
8. Yoga or stretching
9. Playing sports
10. Housework or yard work

The first question in this evaluation was to determine whether or not respondents understood that the existing approach to be asking about two separate phenomena – that is, do they interpret moderate and vigorous physical activity

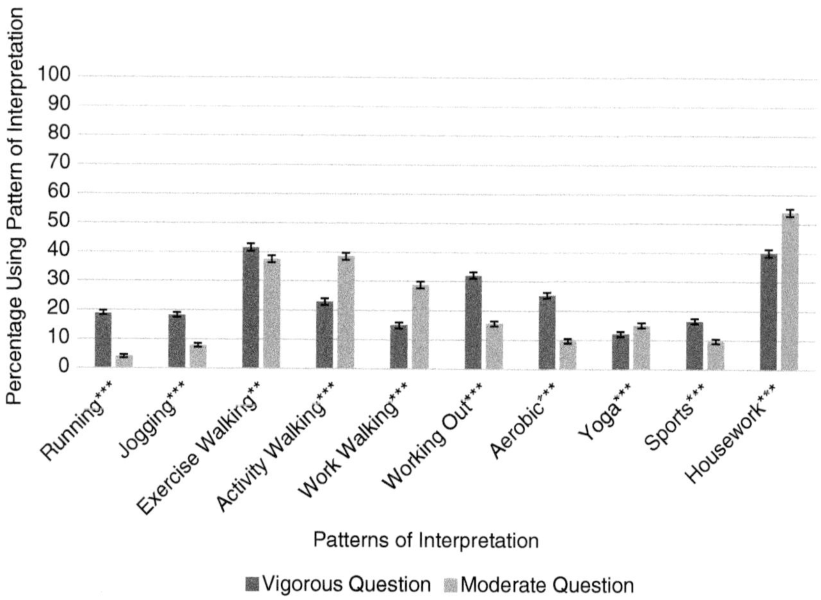

Figure 17.6 Weighted percent of respondents' patterns of interpretation for the vigorous and moderate physical activity questions.

in the same or in different ways? Figure 17.6 shows the distribution of how the respondents used the 10 patterns of interpretation across both the vigorous and moderate physical activity questions. A series of chi-square tests were used to test whether or not respondents interpreted the questions differently (all with a $df = 1$).

Cognitive interviewing indicated that respondents counted both jogging and some forms of walking as both vigorous and moderate exercise, but when the findings are taken out of the cognitive sample and expanded to a larger sample, this "double counting" does not exist. Respondents were more likely to consider running, jogging, working out with exercise equipment, aerobic exercise, and playing sports as vigorous activities; on the other hand, they were more likely to think about walking for work or to activities, yoga, and housework as moderate activities. The fact that these two questions largely captured different physical activities, and that the activities that they did pick up appear to correspond with the correct levels of physical effort, indicate that the current NHIS question design works as intended.

In order to feel comfortable changing approaches, the new, "combined" physical activity question should therefore capture the same constructs that the current dual-phenomena approach does. Figure 17.7 shows the distribution of how the respondents interpreted the new question versus how they interpreted *both* the vigorous and moderate activity questions (as measured

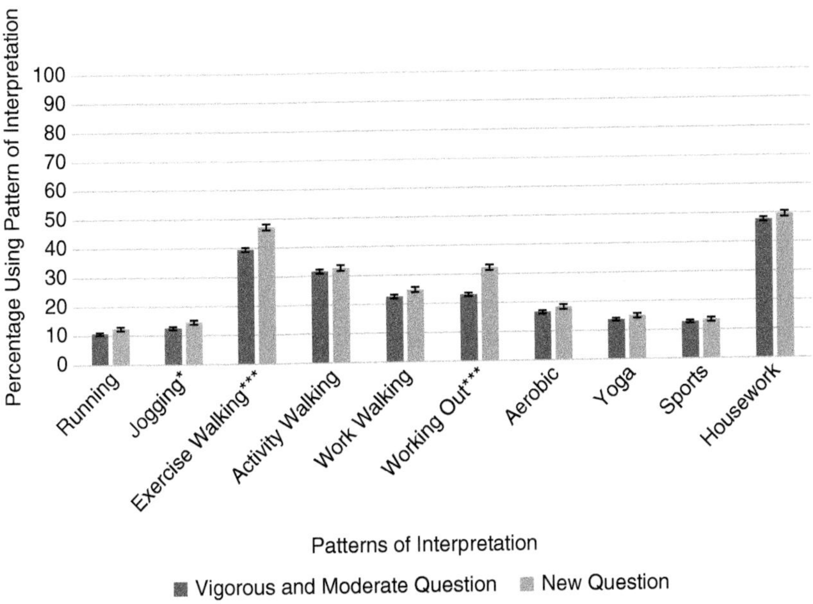

Figure 17.7 Weighted percent of respondents' patterns of interpretation for the existing NHIS approach to measuring physical activity and the proposed combined question.

by a flag indicating whether they choose a given pattern for either the vigorous or the moderate question). Again, a series of chi-square tests were conducted to examine whether respondents considered these two approaches differently.

The new question only captures slightly more jogging (($\chi^2(1, 5198) = 4.06$, $p = 0.02$), exercise walking ($\chi^2(1, 5198) = 27.96$, $p < 0.001$), and working out with exercise equipment ($\chi^2(1, 5198) = 51.23$, $p < 0.001$) than the old approach. Analyzing the embedded probe data for all three of these questions side-by-side shows that the proposed new approach of asking a single physical activity question largely captures the same distribution of activities as does the current approach of asking about vigorous and moderate activity separately. While performing an analysis such as this is limited in that it does not explore the respondents' answers to the two approaches (i.e. considering whether or not they quantitatively produce similar estimates of physical activity duration and frequency), it does indicate that the new approach may be a viable way forward in that it captures the same phenomena as does the current approach.

17.4 Discussion

The use of cognitive web probes, though a relatively recent methodological development, shows promise in the evaluation of survey questionnaires and questions. Methods such as cognitive interviewing are good at uncovering how respondents interpret and respond to a question because they allow the researcher to obtain detailed information from the respondents through a semi-structured conversation. That they are not used to generalize findings across a population is not simply a matter of the sample used, but in a larger part due to the fact that the strength of the method stems from the flexibility the researcher has in conducting the interview – probing and following up on parts of the respondents' narratives as he or she sees fit. Web probes, such as the ones described in this chapter, are specifically intended to answer a question that more traditional qualitative methods of question evaluation do not – that of the extent of patterns of interpretation and validity issues in a survey sample. This study specifically examined close-ended probe questions, which were designed using the results of cognitive interviewing and are therefore not primary sources of qualitative information themselves. Because the use of these probes has been limited until recently, deeper examination of their strengths and weaknesses as question-evaluation tools is necessary moving forward.

The three applications of using targeted, embedded probes in web surveys described here illustrate the power this mixed-method approach can bring to question and questionnaire evaluation studies. This method does not try to *replace* cognitive interviewing, but rather expands the in-depth, qualitative findings of cognitive interviewing studies to a larger sample by creating and embedding probe questions that ask specifically about the patterns of interpretation a respondent use when answering a survey item. While such close-ended cognitive probes have been used in a limited number of cases in the past, the cost of embedding questions into a field test – and taking up valuable questionnaire "real estate" – is typically prohibitive. Using web panels – particularly commercially available, probability-based, recruited ones – as vehicles not only for question evaluation in general, but for employing targeted probes specifically, is much less cost-prohibitive and could be used regularly with great effect. This assumes that the web panel can truly be weighted to the population in a statistically meaningful way. Couper (2000) reviews this and other assumptions about web surveys and panels that researchers must take into account.

One of the primary strengths of close-ended probes that is illustrated in the examples is their structure. Though designed specifically to elicit cognitive information about how respondents interpret and answer survey items, these questions are structured and function as any other survey question. They can therefore be cross-tabulated and analyzed like any other item on a survey. This allows researchers to not only examine the extent of a given pattern of interpretation across the sample (or a population, depending on the sample used), but also explore how these patterns vary by subgroups. The ability to evaluate how certain classes of respondents interpret questions, and how they differ from others, is particularly important when considering the construct validity of items in large surveys where not all groups will be represented in a cognitive interviewing sample. For example, the analysis of the K6 Effort question presented earlier showed that while the out-of-scope interpretation of the question was not as prevalent across the larger survey sample as the cognitive interviews indicated it may be, it was used disproportionately by respondents with lower levels of education – indicating that the construct validity of that particular item may not extend across the entire population.

This ability to focus in directly on the interpretation of the question (and thus the item's validity) is another strength of close-ended cognitive probes. Cognitive interviewing respondents will always provide some amount of superfluous information that does not directly answer the interviewer, who must then use more targeted probes to focus the respondents' comments. This is even more true in open-ended web survey responses where follow-up probes cannot be used to target a respondent's comments to a specific matter (such as how they interpret a question). However, because the targeted, embedded probes discussed in this chapter are (i) based on already-analyzed qualitative data and (ii) framed explicitly by their answer categories, they focus the respondent directly on the concept under question. While there is no way to completely limit the interpretative variation of a question, concise answer categories are an effective way to do so. The question, of course, then becomes whether or not web probes – close-ended or otherwise – do a good job of actually capturing the cognitive process for which they are intended. To ensure that they do, these probe questions need to be evaluated in the same manner as any other survey item. The probes presented earlier were all tested and revised using iterative cognitive interviews (Scanlon 2017).

However, while the power of using these probes is clear, the methodology should be refined through more methodological work. A number of outstanding questions about how best to design and deploy these close-ended probe questions remain, including:

1. Do they burden respondents more than other survey items, and if so, what are the best practices for determining the upper limit of probe questions in one questionnaire?
2. Are certain types of targeted, embedded probes more effective – that is, do probes that attempt to uncover the constructs respondents use versus ones that attempt to uncover other parts of the question-response process, such as the recall stage, all produce high-quality data?
3. Is there an optimum number of answer categories, or should there be a limit on the number of answer categories in each probe question? Relatedly, is there any evidence that randomly rotating answer categories is either necessary or beneficial?

The single round of RANDS used in the analysis presented in this chapter does not allow for a complete examination of these questions. For instance, while there was an observable drop in the number of people who responded to the last probe question as compared to the first one, more work is necessary to see whether this decline corresponds to the overall loss of response across the 109-item questionnaire.

While this methodology is still being refined (Scanlon 2019), the data presented here show that this technique is an excellent way of combatting the sampling limitations of traditional face-to-face cognitive interviews and expanding their findings to a wider population. In doing so, the analysis of these close-ended, targeted, embedded probes allow for a more robust, complete question-evaluation process than either qualitative or quantitative evaluation methods do alone.

References

Baena, I. and Padilla, J.L. (2014). Cognitive interviewing in mixed research. In: *Cognitive Interviewing Methodology* (eds. K. Miller, S. Willson, V. Chepp and J.L. Padilla), 133–152. Hoboken, NJ: Wiley.

Behr, D., Kaczmirek, L., Bandilla, W., and Braun, M. (2012). Asking probing questions in web surveys: which factors have an impact on the quality of responses? *Social Science Computer Review* 30 (4): 487–498.

Behr, D., Braun, M., Kaczmirek, L., and Banilla, W. (2013). Testing the validity of gender ideology items by implementing probing questions in web surveys. *Field Methods* 25 (2): 124–141.

Behr, D., Bandilla, W., Kaczmirek, L., and Braun, M. (2014). Cognitive probes in web surveys: on the effect of different text box size and probing exposure on response quality. *Social Science Computer Review* 32 (4): 524–533.

Braun, M. (2008). Using egalitarian items to measure men's and women's family roles. *Sex Roles* 59: 644–656.

Cannell, C., Fowler, F., Kalton, G., et al. (1989). New quantitative techniques for pretesting surveys. 47th International Statistical Institute, Paris, France.

Converse, J. and Presser, S. (1986). *Survey Questions: Handcrafting the Standardized Questionnaire*. Newbury Park, CA: Sage.

Couper, M. (2000). Web surveys: a review of issues and approaches. *Public Opinion Quarterly* 64: 464–494.

Edgar, J. (2013). Self-administered cognitive interviewing. 68th AAPOR Conference, Boston, MA.

Fowler, S., Willis, G., Moser, R., et al. (2015). Use of Amazon MTurk Online Marketplace for questionnaire testing and experimental analysis of survey features. Federal Committee on Statistical Methodology Research Conference, Washington, DC.

Gallup (2016). *Gallup Panel Whitepaper Brief*. Washington, DC: The Gallup Organization Available by contacting GallupPanel@Gallup.com.
Kessler, R.C., Barker, P.R., Colpe, L.J. et al. (2003). Screening for serious mental illness in the general population. *Archives of General Psychiatry* 60 (2): 184–189.
Loeb, M. (2016). Development of disability measures for surveys: the Washington group extended set on functioning. In: *International Measurement of Disability: Purpose, Method and Application. The Work on the Washington Group* (ed. B.M. Altman), 97–122. Cham, Switzerland: Springer International Publishing.
Meitinger, K. and Behr, D. (2016). Comparing cognitive interviewing and online probing: do they find similar results? *Field Methods* 28 (4): 363–380.
Miller, K. (2016). Summary of Washington group question evaluation studies. In: *International Measurement of Disability: Purpose, Method and Application. The Work on the Washington Group* (69–84) (ed. B.M. Altman). Cham, Switzerland: Springer International Publishing.
Miller, K. and Maitland, A. (2010). A mixed-method approach for measurement construction for cross-national studies. Joint Statistical Meetings, Vancouver, Canada.
Miller, K., Mont, D., Maitland, A. et al. (2011). Results of a cross-national structured cognitive interviewing protocol to test measures of disability. *Quality and Quantity* 45 (4): 801–815.
Miller, K., Willson, S., Chepp, V., and José Luis Padilla. (2014) Cognitive Interviewing Methodology. John Wiley and Sons: Hoboken, NJ.
Murphy, J., Edgar, J., and Keating, M. (2014). Crowdsourcing in the cognitive interviewing process. 69th AAPOR Conference, Anaheim, CA.
Presser, S., Couper, M., Lessler, J. et al. (2004). Methods for testing and evaluating survey questions. *Public Opinion Quarterly* 68 (1): 109–130.
Scanlon, P. (2019). The effects of embedding closed-ended cognitive probes in a web survey on survey response. *Field Methods* 31(4): 328–343.
Scanlon, P. (2017). *Evaluation of the 2015–2016 Research and Development Survey*. Hyattsville, MD: National Center for Health Statistics.
Schuman, H. (1966). The random probe: a technique for evaluating the validity of closed questions. *American Sociological Review* 31 (2): 218–222.
Willis, G. (2005). *Cognitive Interviewing: A Tool for Improving Questionnaire Design*. Thousand Oaks, CA: Sage.
Willis, G. (2014). *Analysis of the Cognitive Interview in Questionnaire Design*. New York, NY: Oxford University Press.

18

The Practice of Cognitive Interviewing Through Web Probing

Stephanie Fowler[1] *and Gordon B. Willis*[2]

[1] *Office of the Director, All of Us Research Program, National Institutes of Health, Bethesda, MD, USA*
[2] *Behavioral Research Program, Division of Cancer Control and Population Sciences, National Cancer Institute, Bethesda, MD, USA*

18.1 Introduction

As survey research and its related technologies evolve, the set of methods used in support of questionnaire development and evaluation must keep pace. For example, cognitive interviewing has been an accepted evaluation method for several decades, but as surveys target increasingly diverse populations, cognitive interviewing studies have tended to employ larger and more varied samples to ensure cross-cultural comparability (Willis 2015). Further, technological developments, and especially the establishment of internet-based systems and social-media applications that facilitate access to a wide range of prospective respondents, have opened new methodological strategies for data collection.

These challenges and opportunities have led survey researchers to consider alternative approaches to questionnaire pretesting, particularly the adaptation of *web probing* as a variant of cognitive interviewing. Web probing is a modernization of the decades-old procedure generally labeled *embedded probing* (Converse and Presser 1986; note that a related procedure named Random Probing was described by Schuman 1966). Embedded probes may be used to collect information concerning the survey response process within a field (production) environment, in fully standardized form, to either the full sample or a subsample of respondents. Embedded probes can be read by a field interviewer for an interviewer-administered survey, or included within a self-administered survey instrument. For example, Murphy et al. (2014) evaluated the item "Since the first of May, have you or any other member of your household purchased any swimsuits or warm-up or ski suits?" by following up with the open-ended probe: "What types of items did you think of when you read this question?"

Because it can be logistically challenging to embed probes in field settings, embedded probing has, to our knowledge, been used infrequently in

Advances in Questionnaire Design, Development, Evaluation and Testing, First Edition.
Edited by Paul C. Beatty, Debbie Collins, Lyn Kaye, Jose-Luis Padilla, Gordon B. Willis, and Amanda Wilmot.

survey pretesting, as it has proved more practical to conduct small cycles of cognitive interviews early in the questionnaire development process. However, embedded probing has recently experienced a resurgence (e.g. Behr et al. 2014; Baena and Padilla 2014). Further, internet data collection, respondent panels, and online labor marketplaces such as Amazon.com's Mechanical Turk (MTurk), have made web probing more operationally and logistically feasible. Platforms such as MTurk, along with web-based question administration systems like Qualtrics, SurveyGizmo, and SurveyMonkey, can be engaged quickly and often very inexpensively – which enhances the feasibility of their incorporation into field pretesting, prior to full-scale implementation. As such, several authors (e.g. Edgar et al. 2016; Fowler et al. 2015) have investigated the utility of engaging embedded probing, in the form of web probing, into pretests of internet surveys.

18.2 Methodological Issues in the Use of Web Probing for Pretesting

There are several challenges to relying on web probing as a pretesting or evaluation method. To our knowledge, only one study (described by Edgar et al. 2016; Murphy et al. 2014) has evaluated web probes for purposes of pretesting draft survey questions. That study, however, was mainly intended as an evaluation of alternative web-based crowdsourcing platforms, rather than of web probes specifically. As such, the usefulness of web probes for evaluating and pretesting survey questions – especially compared to more standard forms of pretesting – has not been well established.

In particular, optimal application of web probes has not been investigated. One important question is whether probes should be placed concurrently or retrospectively (Willis 2005) – that is, should researchers probe immediately following the target item, or wait until the end of the standard questionnaire to ask about item comprehension and interpretability? For self-administered surveys, embedding a probe immediately following a target item (normally termed *concurrent probes*), as opposed to waiting until the end *(retrospective probes)*, could disrupt the natural flow of the survey. Further, concurrent probing might influence responding to later items. However, responses to retrospective probes may be biased by cognitive processing of intervening items, and memories of responding may be less precise as time passes (Willis 2005).

A second methodological issue regarding web probing concerns the utility of different variants of cognitive probes. As probe questions used in more traditional forms of cognitive testing, web probes can be general or specific, and either *expansive* (Beatty et al. 1997), in that they request that the respondent to explain the circumstances surrounding their response in more detail; or *directive*, which ask the respondent to explain some aspect of their response process (e.g. "Were you thinking of X, or Y, or something else, when you answered that question?"). However, we know of no systematic investigation of these varying probe types, and certainly not in the context of web probing.

The current chapter is an empirical investigation of the following methodological issues concerning web probing – specifically:

1) The effects of probe placement. We discuss results of an experiment that assigned respondents to either concurrent probing (i.e. probes placed immediately following the targeted question) or fully retrospective probing, and explored differences in responses both qualitatively (type of themes that emerged, and relevance) and quantitatively (word count, item nonresponse, time to completion, differences in how they responded to survey items).
2) The utility of several variants of web probes, overall and relative to standard cognitive interviews, as assessed through thematic coding, including a comparison with responses from standard cognitive testing.
3) Finally, we synthesize the results of these studies to produce a set of integrated conclusions concerning the practice of web probing. Specifically, we will (i) summarize the strengths, limitations, and best uses of this procedure; (ii) recommend specific techniques, as a means toward development of best practices; and (iii) suggest avenues for further research on web probing techniques.

18.3 Testing the Effect of Probe Placement

18.3.1 Testing Platform

We selected Amazon's Mechanical Turk as the survey administration platform because it has demonstrated the capacity to produce high-quality data with use of point-in-time (cross-sectional) designs (e.g. Berinsky et al. 2012; Buhrmester et al. 2011). Mechanical Turk is an online labor market composed of a community of internet users who log in to the platform to complete tasks (*crowdsourcing*). Administered by Amazon.com, the onlinec labor market allows requestors to post tasks requiring human intelligence (human intelligence tasks [HITs]), and in turn, respondents (*workers*, in MTurk), are compensated through small amounts of payment per task. Participation in MTurk is therefore through self-selection rather than probability samples. Although there are approximately 500 000 MTurk workers registered worldwide, we limited our study to only workers in the United States.

18.3.2 Participant Characteristics

A total of 646 respondents completed the study, with 478 in the concurrent condition and 168 in the retrospective condition. Due to administrative clearance processes, the concurrent condition was fielded first, followed by the fielding of the retrospective condition three weeks later. Furthermore, due to resource and time constraints, fewer participants were tested in the

Table 18.1 Demographic variables for the MTurk sample compared to the population of US adults as recorded by the U.S. Census Bureau (2010).

Variable ($N = 646$)		MTurk	US
Sex (%)			
	Female	48.2	50.8
	Male	51.8	49.2
Race (%)			
	White	84.5	78.3
	Black	6.2	13.1
	All other races	7.6	8.6
	Don't know	1.7	
Ethnicity (%)			
	Hispanic	6.4	16.9
Education (%)			
	Less than bachelor's degree	46.6	69.8
	Bachelor's degree or higher	54.6	31.2
Age		Mean = 33.4, SD = 11.2	

Note: There were no significant differences between concurrent and retrospective conditions on any of the included demographic variables.

concurrent ($n = 478$) than in the retrospective ($n = 168$) condition. As depicted in Table 18.1, respondents were mostly white (84.5%) and were overall younger (mean = 33.4, SD = 11.2), slightly more male (52%), and more educated (54.6%) than the US adult population (United States Census Bureau 2010). Demographic variables were roughly eventually distributed across the two experimental conditions involving probe placement ($p > 0.10$).

18.3.3 Instruments and Procedure

Survey items asked about walking, taken from the National Center for Health Statistics' Health Interview Survey Cancer Control Supplement (2016), and adapted for web-based, self- administration. The National Health Interview Survey (NHIS) walkability module included three major sections: demographics, walking behavior, and perceptions of neighborhood walkability. First, respondents self-reported their age, sex, race, ethnicity, age, and education level. They then completed a 6-item section of yes/no questions about how much they walk for transportation and leisure. An example item reads "In the past 7 days, did you walk to some place that took you at least 10 minutes?" Lastly, respondents completed a 9-item section of yes/no questions asking

Table 18.2 Web probes administered to examine cognitive processing of questions on walkability.

Probe 1 (Expansive). When answering the question, "Where you live, are there roads, sidewalks, paths or trails where you can walk?" Please say more about what you were thinking when answering this question.
Probe 2 (Expansive). We asked some questions about whether there are places in your neighborhood that you can walk to. Please say more about how you decided whether or not you can walk there.
Probe 3 (Directive). When answering the questions about places in your neighborhood that you can walk to, were you thinking of places that YOU might walk to, or places that OTHERS might walk to, or both?
Probe 4 (Directive). When answering the questions about places in your neighborhood that you can walk to, were you thinking about what you actually DO, or about what you COULD do if you wanted to?

them to report their perceptions of neighborhood walkability. An example question reads "Where you live, are there roads, sidewalks, trails, or paths where you can walk?" Four web probes followed the walkability items: two probes were *expansive* in nature, and two were *directive* (see Table 18.2).

The placement of these four web probes served as the independent variable. Respondents in the *concurrent* probing condition were presented with the web probes immediately following the target question(s) the probes were intended to evaluate (i.e. concurrent probing). Respondents in the *retrospective* condition were presented the web probes following completion of the survey (i.e. debriefing). Across both conditions, participants responded to all four web probes, in the same order.

Prior to administering the survey, clearances from the National Institutes of Health's Office of Human Subjects Research Protections (OHSRP) and the Office of Management and Budget (OMB) were obtained, and all study procedures were developed in accordance with American Psychological Association's ethical guidelines. Eligibility criteria included being a US resident aged 18 years or older. Respondents were compensated $0.30 for their completion of the study, which took approximately 10 minutes to complete. Complete details concerning MTurk administration of the survey are described by Fowler et al. (2015).

18.4 Analyses of Responses to Web Probes

Table 18.3 contains examples of the types of responses MTurk respondents provided to each of the four web probes from both concurrent and retrospective web probe administration conditions. The content of probe responses appeared to be similar across concurrent and retrospective conditions.

Table 18.3 Example verbatim responses to probes across both conditions.

Probe 1 (Expansive) responses	"I was thinking about a place where I used to live in New York State, where there were no sidewalks and people looked at you funny if you weren't driving a car."
	"Most areas have sidewalks, even bike lanes, so pretty walkable. The specific route I take is for going out on weekends, and avoiding the highway, there's plenty of sidewalk. The back of my apartment has trails cutting through wooded areas leading to the exit."
Probe 2 (Expansive) responses	"There are two convenience stores. Unsavory people linger around outside the store. To walk to these convenience stores I would have to walk on the unimproved side of the narrow street. The closest convenience store is across the intersection of US98. There is a blinking caution lite at the corner. Crashes occur at this intersection weekly."
	"If it was a reasonable (10 to 20 minutes) to walk there. And if it is a place like a store that I could carry my purchases back with me."
Probe 3 (Directive) responses	"I am thinking of places where everyone can walk to."
	"I was thinking of places I might walk to."
Probe 4 (Directive) responses	"I mostly think about what I could do, most of the time I walk at the gym or I go to a park to walk, I don't really walk around the neighborhood."
	"I was considering the route I actually walk on, not where I could walk."

18.4.1 Prevalence and Length of Responses to Probes

One concern about web probes is that respondents may tend to find them burdensome or uninteresting (see Aoki and Elasmar 2000), leading to either nonresponse, or brief and unhelpful responses. MTurk respondents were reimbursed a very small amount to complete the survey (30 cents), which may serve as inadequate motivation. However, of the 646 respondents in this study, only 7 respondents failed to respond to any of the 4 probes.

Linguistic Inquiry and Word Count (LIWC: Pennebaker et al. 2001) was used to examine word count for each free-text response, across the four web probes. The mean number of words provided to each probe generally ranged from 10 and 20 words (see Table 18.4). Responses to *Probes 1 and 2* (expansive), which asked for open-ended explanations, were on average almost twice as long as those to *Probes 3 and 4* (directive), which offered respondents several categories (e.g. were they thinking about walkability from their own perceptive, from the perspective of others, or both); although they were free to respond in any manner they chose within the open-text box provided. Variation in response length

Table 18.4 Word count as a function of placement condition (concurrent versus retrospective).

	Overall ($N = 646$)	Concurrent condition ($n = 478$)	Retrospective condition ($n = 168$)
Probe 1			
Mean	20.8	17.9	21.8
SD	13.2	11.9	13.5
Probe 2			
Mean	22.5	21.3	23.0
SD	13.2	15.2	14.2
Probe 3			
Mean	10.5	9.9	10.7
SD	10.5	8.5	11.2
Probe 4			
Mean	11.5	11.9	11.3
SD	11.2	11.6	10.9

a) Significant between-subjects effects at $p < 0.05$.

could be attributable to both the nature of the probe (expansive/directive) and the degree of open-endedness in the probe.

18.4.2 Effects of Probe Placement on Number of Words Produced

For the primary analysis, we ascertained the effect of probe placement on word count: Table 18.4 includes the means and standard deviations for word count for all four probes as a function of placement condition. We observed a slight trend in which the retrospective condition was more productive. However, we also conducted additional linear regressions (one for each probe) to determine the extent to which probe placement affected the volume of responses to probes in the context of other potentially influential variables. Probe placement (concurrent versus retrospective) served as the independent variable and was entered on the first step of the regression equation. Also included on the first step were the control variables which included age (continuous), race (White, non-White), sex (female, male), and education (bachelor's degree or higher, less than a bachelor's degree):

Probe 1. A significant probe-placement effect emerged, $t = -3.42$, $p < 0.01$, $R^2 = 0.075$, with respondents in the retrospective condition producing a greater number of words in their free-text responses than did respondents in the concurrent condition. Of the demographic control variables, sex, age, and race were also significantly correlated with word count, all *ps* < 0.01. Specifically, age ($t = 2.5$), identifying as female versus male ($t = 3.53$), and identifying as White versus non-White ($t = 2.86$) were all positively associated with greater word count.

Probe 2. There were no significant differences in word count between placement conditions, $t = -1.40$, $p = 0.15$, $R^2 = 0.061$. However, for the demographic control variables, significant associations emerged between several demographic factors (age, sex, race) and word count, *ps* < 0.05. Identical to the pattern that emerged for word count for *Probe 1*, age ($t = 2.52$), identifying as female versus male ($t = 3.79$), and identifying as White versus non-White ($t = 2.33$) were positively associated with greater word count.

Probe 3 and Probe 4. There were no significant probe-placement effects at the 0.5 level, nor were any of the demographic control variables significant predictors of word count.

Based on only a single probe for which placement affected word count (*Probe 1*), it appears that concurrent versus retrospective placement did not emerge as an especially potent influence on amount of information provided by respondents.

18.4.3 Effects of Probe Placement on Responses to Questionnaire Items

It has been hypothesized that concurrent probing may produce reactivity effects, causing respondents to modify their cognitive processing of later substantive items – for example, if they believe that they should answer more carefully in anticipation of further probing – or probes make them think about the items more intensively (Willis 2005). In order to determine if probe placement affected responses to the nine questions about perceptions of walkability, we first computed a Walkability Index by summing across the nine items concerning perceptions of neighborhood walkability (items that were negatively worded were reverse scored such that higher scores reflected a greater perception of walkability). Walkability Index scores ranged from 0 to 9 (mean $= 6.91$, $SD = 2.04$).

To test whether probe placement affected responses to scores on the Walkability Index, we conducted a linear regression with probe placement entered on the first step as the independent variable. In addition, the same demographic control variables used in the previous analyses were also entered on the first step as covariates in the analysis. Walkability Index scores served as the dependent variable. We found a significant effect of probe placement such that respondents in the concurrent condition (mean $= 7.17$, $SD = 1.88$) reported slightly enhanced perceptions of neighborhood walkability compared to the retrospective condition (mean = 6.81, SD = 2.07), as captured by the Walkability Index, $t = 2.2$, $p = 0.05$, $R^2 = 0.042$.

18.5 Qualitative Analysis of Responses to Probes

18.5.1 Web-Probe Analysis

Arguably, quantitative metrics such as word count are incomplete measures of the utility of web probes. Therefore, we also conducted qualitative analyses of responses to the probes in order to understand their content. Due to the size of the overall sample ($n = 646$), we selected a subset of respondents, $n = 121$ (retrospective condition, $n = 61$; immediate condition, $n = 60$) from the original sample for which we balanced sex, age, race, ethnicity, and education level, and that was representative of the larger MTurk sample. The qualitative analysis sample size was determined based on the methods developed by Fugard and Potts (2015) for qualitative coding of research results. Two independent raters (the authors) coded the responses; coding disagreements were adjudicated through follow-up discussions. Table 18.5 summarizes the results of the coding analysis for each probe.

18.6 Qualitative Coding of Responses

Probe 1 pertained to the single item asking participants if there were roads, sidewalks, trails, or paths that they could walk on in their neighborhoods. We coded which of these components each respondent mentioned in his or her responses to the web probes (Kappas ranged from 0.93 to 1.0, all *ps* < 0.01). Of the 121 respondents in the coding analysis, 13.2% reported 0 of these components of walking infrastructure, 34.7% reported 1 of them, 35.5% reported 2 in any combination, 12.4% reported 3, and 4.2% mentioned all 4. Next, a linear regression with the same set of predictors and controls used to examine word count was conducted to determine if probe placement affected the number of walking components mentioned in the free-text responses. There were no significant effects of probe placement, nor did any of the demographic control variables predict the number of walking infrastructure components mentioned in *Probe 1*, all *ps* > 0.05.

Probe 2 asked respondents how they decided whether there were places in their neighborhood they could walk to. Following the procedures of Edgar et al. (2016), the same two independent raters as previously coded responses to this item based on whether respondents had provided information relevant to the basis for their answer to the survey question – specifically, whether the answer provided to the probe described information that conveyed the process

Table 18.5 Results of qualitative coding of web probes.

Probe	Wording (*N* = 121)		
1	When answering the question, "Where you live, are there roads, sidewalks, paths or trails where you can walk?" – Please say more about what you were thinking when answering this question?	Percentage of respondents mentioning how many of the four categories: 0 of 4: 13.2% 1 of 4: 34.7% 2 of 4: 35.5% 3 of 4: 12.4% 4 of 4: 4.2%	Inter-rater reliability (Kappa) for: Roads: 0.93 Sidewalks: 1.00 Trails: 0.96 Paths: 1.00
2	We asked some questions about whether there are places in your neighborhood that you can walk to. Please say more about how you decided whether or not you can walk there	Percentage of respondents providing responses that were: Highly relevant response: 46.3% Moderately relevant response: 52.9% Irrelevant Response: 0.8%	Inter-rater reliability (Kappa) for judgment of relevant versus not relevant: 0.97
3	When answering the questions about places in your neighborhood that you can walk to, were you thinking of places that YOU might walk to, or places that OTHERS might walk to, or both?	Percentage of respondents providing responses that were about: Themselves: 40.3% Others: 1.7% Both: 33.6% Miscellaneous: 24.4%	Inter-rater reliability (Kappa) for judgment for the four categories: 0.89
4	When answering the questions about places in your neighborhood that you can walk to, were you thinking about what you actually DO, or about what you COULD do if you wanted to?	Percentage of respondents providing responses that were what they: Actually do: 47.9% Could do: 23.2% Both: 18.2% Miscellaneous: 10.7%	Inter-rater reliability (Kappa) for judgment for the four categories: 0.91

Note: Presented values are pooled over value of probe placement (concurrent versus retrospective), because placement was not found to affect code distribution.

concerning how they decided how or if they could walk to some place. All but one of the respondents (99.2%) provided a free-text response to the question. To determine *quality of response*, we coded the responses based whether they were minimally versus highly relevant (based on Edgar et al. 2016). Minimally relevant responses included language that addressed the question in some capacity but did not directly answer the question – "Not much to walk to, but there are few things." Highly relevant responses directly answered the question by providing a basis or rational for the response: "I consider a place within walking distance if I can get there in 10 minutes." Overall, 52.9% of responses were coded as minimally relevant and 46.3% were coded as highly relevant (Kappa = 0.97, $p < 0.01$). To determine if probe placement affected quality of responses, we conducted a logistic regression with probe placement and the demographic variables on the first step, and whether the response was minimally or very relevant (dichotomous) as the dependent variable. Quality of responses was not related to probe condition or any of the demographic correlates, all $ps > 0.05$.

Probe 3 also pertained to all nine items in the perceptions of walkability as a set, and asked if they were thinking about places (i) that they could walk to or (ii) that others could walk to, or (iii) if they were answering the questions from both perspectives. We coded free-text responses by assessing whether respondents mentioned themselves (40.3%), others (1.7%), or both (33.6%), Kappa = 0.886, $p < 0.01$. Note that 24.4% of respondents did not answer the probe in a manner that provided information concerning the key element of self-versus-other perspective, which suggests that some may resist the requirements of the directive probe approach that focuses on elements defined only by the researchers. Due to multiple response categories, we conducted a Chi-square test of independence to determine if coding distribution differed across probe-placement condition. There were no significant differences ($p = 0.67$).

Probe 4 also pertained to all nine items in the perceptions of walkability survey. For this analysis (involving the same two raters; Kappa = 0.92, $p < 0.01$), we found that 47.9% of respondents answered based on what they actually did, 23.1% answered based on what they potentially could do (23.1%), and 18.2% answered based on both of these. The remaining 10.7% of responses were coded as not relevant. Coding also established that respondents included both actual and hypothetical perspectives in their responses, although actual ones were more common. Based on a Chi-square test of independence, code distribution did not vary according to the probe-placement condition ($p < 0.65$), possibly because, as with the previous probe, it appeared following a series of nine items, and so in effect represented a hybrid between concurrent and retrospective approaches.

18.6.1 Thematic Coding

In order to make an overall determination concerning respondent interpretation of "walkability," the lead investigator reviewed the open-text responses to

Probe 2: "Please say more about how you decided whether or not you can walk there?" using bottom-up, grounded theory–based coding in which coding categories were developed from the data rather than preassigned. Coding was done until saturation was achieved (that is, the point at which no substantially new information was gained through additional coding), which required 76 interviews. Several dominant themes emerged in the free-text responses concerning the perceived nature of walkability: the most frequently mentioned reasons for deciding whether something is walkable were distance (in miles or blocks) and the time it would take to walk there (usually in minutes). Secondarily, there were several mentions of safety due to traffic and crime, the weather, topography, and feasibility/logistics of walking. Overall, we did not observe any differences in the themes that emerged between probe-placement conditions.

18.6.2 Comparison to Standard Cognitive Interviews

Of particular interest was the question of whether web probing produces responses similar to those obtained through more standard cognitive testing. In order to address this question, we compared our qualitative findings with those obtained through previous, standard cognitive testing of the same item by staff of the National Center for Health Statistics (NCHS), where 19 interviews had been conducted. In their written report summarizing the testing results, the NCHS researchers reported that for the perceptions of walkability scale, two general patterns of interpretation emerged: whether it was physically possible to walk, or whether one actually did walk to the places asked about by the survey question. Further, similarly to the current study, two sub-themes emerged for respondents who thought about whether or not they could physically walk to places: some thought about distance in terms of time, while others thought about physical distance (in terms of miles or blocks). In summary, despite the different approaches to item evaluation, dominant themes concerning question functioning were assessed as qualitatively similar between the current and NHIS studies (notwithstanding the fact that it was not possible to assess the absolute magnitudes of these themes, or of more specific or nuanced trends).

18.7 Current State of the Use of Web Probes

To reiterate, the current study evaluated the use of web probes as a means for obtaining clear and useful information concerning item function. To do so, we addressed several research questions:

1) *Do web probes produce a sufficient volume of information?* In terms of gross productivity, we found that the open-text responses to web probes were

generally full sentences. They are likely not as lengthy as those produced through standard cognitive interviews (e.g. as reported by Murphy et al. 2014), but far more than the simple yes-no responses of the types provided to standard survey questions. Overall, responses to web probes were more extensive than those to survey questions, suggesting that respondents do view them as different tasks, and likely as soliciting meaningful reflection, as intended. Of course, this conclusion is limited to the current web survey administration platform used (MTurk). One next step is to utilize other web-based and crowdsourcing platforms such as Facebook and web-based probability panels (especially when respondents receive no incentive payment), as these platforms may differ with respect to either respondent characteristics or their motivation to complete tasks carefully, either of which might affect quality of responses to web probes.

2) *Do different types of probes produce different amounts of information?* Expansive probes – designed to produce additional explanation – produced roughly twice the gross output of that produced by more focused, directive probes, which are designed to orient the respondent toward a specific feature of the item or its cognitive processing. This finding suggests that web probing may have utility as a means for methodological investigations of specific cognitive interviewing practices. As a caveat, we note that our results were complicated by the fact that the two directive probes followed the expansive ones, and to the extent that individuals provide less information to later probes because they believe they have already provided relevant information, one might expect responses to the later ones to be more brief. On the other hand, our four probes asked for fundamentally different types of information that were unlikely to be captured by the others. Still, further experimentation featuring web probing could examine this issue in depth, as well as methodological questions concerning the nature of probes on response tendencies. In particular, we advocate further study of probe characteristics: for example, the effectiveness of hypothetical probes; of those that ask for explanations in terms of "why" one decided to answer a question in a particular way as opposed to just "tell me more"; and of those that vary in terms of global versus specific level of inquiry (Willis 2015).

3) *Does web probing provide a means for dividing responses into discrete categories concerning "what the question captures,"* as advocated by Miller et al. (2014)? All four of our web probes, whether expansive or reorienting in nature, produced responses that were reliably codeable into categories that appeared to reflect alternative interpretations of the items. One might argue that for our directive probes, this is not surprising, as the specific nature of the probe (e.g. were you thinking of places that YOU might walk to, or places that OTHERS might walk to, or both?) leads respondents directly to one of

those interpretations. However, it was not clear at the outset that participants within an online labor market such as MTurk would answer these in a coherent or codeable manner – yet most did so, and it was frequently observed that they added explanatory detail to support their responses (for example, describing where they actually walk, or explaining who they considered "others" to be). We also note that responses to our expansive probes, such as "Please say more about how you decided whether or not you can walk there" also could be coded into discrete categories representing alterative interpretations of the same item (e.g. time, distance). Overall, especially due to the relatively large sample sizes obtainable, web probing provides an intriguing way to more fully address the notion of question function in the sense described by Miller et al. (2014), as it provides a *descriptive* measurement of question interpretation in addition to the *reparative* focus that has come to dominate cognitive pretesting (Willis 2015). In other words, the strength of web probing may not only be in "fixing" flawed questions, but in more fully chronicling how they function generally, perhaps across a broad and representative cross-section of the relevant population.

4) *Do web probes provide information similar to that obtained in standard cognitive interviews?* We have only partial data on this point, but for the items that were tested at both the NCHS cognitive lab using standard procedures, and within the current study, the degree of conceptual overlap was considerable. Despite the wide disparity of techniques, concerning sample selection and size, degree of probing flexibility, and overall amount of probing possible to conduct (much more restricted in web probing), the overall picture of "what walkability consists of" was similar across these investigations. Of course, a more comprehensive examination of this issue would require a study featuring elements that ours admittedly did not incorporate. In particular, we advocate an assessment of whether the results obtained from web probing are actionable, or can be found to contribute to data quality, in a manner that demonstrates their utility, especially relative to more standard cognitive testing.

5) *What effect does probe placement (immediate versus retrospective) have?* The large, split-sample approach possible with the use of web probing within a web-based platform such as MTurk makes it possible to begin to address this heretofore unresolved issue. Several findings are intriguing. First, it did not appear that the gross amount of information provided by respondents differed markedly as a function of where probes were placed. It could be hypothesized that probes that are embedded concurrently might appear burdensome as the respondent completes the interview, and end up neglected or minimally attended to, relative to a set placed at the end, where the respondent knows she/he is almost finished and perhaps more motivated to "finish strong." This, however, was not observed.

We did find some limited evidence for differences in responses as a function of probe placement, however. First, respondents in the immediate condition produced a higher overall mean for the key measure of walkability, relative to the retrospective case. Presumably this difference is attributable to reactivity effects of immediate probing, although the reasons for the effect are unclear. Even so, these effects were not large, and do not seem to constitute a substantial impediment to the use of concurrent probes. In addition, respondents in the concurrent probing condition were found to provide substantially more highly relevant responses to one of our four probes than did respondents in the retrospective condition (56% versus 36%, respectively). This result is consistent with the hypothesis that those probed soon after responding to a target item are more likely to retain reasons for their answers in memory, and are better able to provide a veridical description when probed (Willis 2005). Overall, however, we found few response differences as a function of cognitive probe placement, and cannot therefore strongly advocate the use of either approach in lieu of the other.

18.8 Limitations

As a caveat, we acknowledge that our manipulation of probe placement in particular was not an especially powerful one, given the relatively short interview (10 minutes), and the fact that several of our probes pertained not to a single target item, but rather to a concept (walkability) that had been the focus of a series of preceding items. Hence, the differences between concurrent and retrospective conditions may have been less than those typically found in cognitive interviewing studies. On the other hand, as mentioned earlier, we did find some evidence of more-relevant responses to concurrent than retrospective probing, even given this limitation. As a second limitation, the investigation was limited to a short survey on a single topic.

To better assess the generality of our results, we therefore conducted an additional experiment using a multi-topic survey with substantially more items. For that investigation, 616 respondents (again recruited via MTurk) completed the National Cancer Institute's Health Information National Trends Survey (HINTS), which required approximately 30 minutes to complete and covered a range of topics including questions on energy balance, family history of cancer, and patterns of seeking health information. Using a split-sample survey design, the 57-item survey contained five web probes either embedded immediately following the target item or retrospectively following completion of all items. For this follow-up investigation, no significant placement effects occurred on quantity or quality of responses (Fowler et al. 2016). Overall, the results of that investigation reinforce the conclusion that probe placement has minimal effect on gross productivity or qualitative nature of the responses obtained.

18.9 Recommendations for the Application and Further Evaluation of Web Probes

Given the current findings, we suggest that researchers strongly consider the use of embedded forms of web probes as a means for question development, testing, and evaluation – particularly for online-based self-administered questionnaires. Further, it is relatively straightforward to conduct split-sample experiments, such as those pertaining to probe placement, or probe type, to develop our methods further. In comparison with other pretesting procedures, we do not necessarily consider web probing to be "better" or "worse" than other methods, or as a potential substitute for those. Rather, we view web probing as one more tool within our methods toolbox that can be used in conjunction with other tools. A particularly fruitful approach may be to combine standard cognitive testing and web probing within the same investigation. For example, it is possible to first conduct standard cognitive testing to develop hypotheses concerning item function, and then in turn to develop web probes; followed by a web-based testing component in which the web probes are embedded. This process allows for the development of web probes that are free of substantial problems themselves, an important concern given that such probes are scripted and therefore do not exhibit the flexibility normally incorporated within the context of standard cognitive testing. The practice of following cognitive interviewing with web probing also allows the investigators to identify particularly important issues or items to target for evaluation, given that one can generally not test nearly the same amount through web probing as in standard cognitive testing.

Concerning the nature of future evaluation efforts, there are several other considerations to bear in mind as researchers utilize online and crowdsourcing platforms for the conduct of web probing:

18.9.1 Device Type

First, and arguably one of the most important areas for future research on web probing, is examining if type of technological device relates to the quality and quantity of responses to web probes. Given the difference in screen size, keyboard type, and mobility of smartphones relative to laptops, desktops, or tablets, there may be nuances associated with fielding web probes in a manner that yields the highest quality of output for these types of technological devices. We attempted to provide data on this issue within the subsidiary HINTS study briefly summarized earlier. Specifically, we asked a demographically representative subsample ($N = 236$) of respondents which device they were using while completing the survey. Most (98%) indicated they were on either a laptop (45%) or PC/desktop computer (53%), and only 2% reported using a tablet/iPad or smartphone. Due to the low number of smartphone users, we were not able to examine device differences on web probe response quantity or quality. However, future research should address this question, as use of mobile devices becomes more widespread.

18.9.2 Sample Size

Another consideration is sample size. Traditionally, questionnaire evaluators must consider whether the sample is large enough to identify problems (Blair and Conrad 2011) or identify themes (Fugard and Potts 2015). Web probing now allows researchers to rapidly collect a large number of responses in a very short period of time (e.g. $n = 1000$ in 10 hours on MTurk). The target must be chosen carefully – it is possible to collect more cases than can be practically processed and analyzed, especially where the analysis is qualitative in nature and involves human-based coding. On the other hand, the large sample sizes provided through web-survey platforms offer extensive opportunities to conduct mixed-method evaluation. For example, the researcher can first conduct standard cognitive interviewing to determine the presence of a potential phenomenon or problem (Blair and Conrad 2011), and then web probing can be conducted to ascertain its quantitative extent (see Scanlon, Chapter 17 of this volume). Of course, this latter objective depends fundamentally on issues of appropriate coverage and population representation afforded by the web survey vehicle, a further issue that is vital, but beyond the scope of the current chapter.

18.9.3 Selection of Target Items to Probe

Another consideration for the further development and evaluation of web probing – as for any form of cognitive testing – is determining whether it is appropriate to probe all types of questionnaire items. There may be higher-level processes involved in the survey response process that people are not able to articulate (Nisbett and Wilson 1977), and web probes (or for that manner, even standard cognitive probes) would therefore not be expected to produce useful information. Given that it is impractical to probe on all items, it is important to make careful choices about where to target our attention.

18.9.4 Use of Web Probing to Investigate Questionnaire Structure

Finally, web probes might be used to provide feedback on the structure of the online survey itself, as opposed to features of individual survey questions. For example, Behr et al. (2012b) demonstrated that the box size provided for text responding to web probes can impact responses. Other visual features of the online survey, such as vertical versus horizontal response formats, may influence the way respondents perceive the question. Researchers can assess the utility of web probes for purposes of obtaining feedback on the usability and organization of the survey that may impact the validity or reliability of results.

18.10 Conclusion

We conducted the research in this chapter to evaluate the potential of web probing to provide useful insight into the meaning of survey responses. On the basis of this research, we conclude that this method has utility as a component of the questionnaire development, evaluation, and testing (QDET) toolbox. Like other questionnaire evaluation tools, it has attractive features, and some natural limits, such that it most likely augments, rather than replaces, traditional cognitive interviewing. We look forward to further development and application of web probing, as well as further evaluations of its effectiveness, as the world of survey methods continues to evolve.

Acknowledgments

We are grateful to Dr. Paul Scanlon, at the National Center for Health Statistics, for conducting the in-person cognitive interviews and for conducting a qualitative analysis on the responses provided by the respondents. The lead author would also like to thank the National Cancer Institute's Cancer Prevention Fellowship Program.

References

Aoki, K. and Elasmar, M. (2000). Opportunities and challenges of a web survey: A field experiment. 55th Annual Conference of American Association for Public Opinion Research. Portland, Oregon.

Baena, I.B. and Padilla, J.L. (2014). Cognitive interviewing in mixed research. In: *Cognitive Interviewing Methodology* (eds. K. Miller, S. Willson, V. Chepp and J.L. Padilla). Hoboken, NJ: Wiley.

Beatty, P., Willis, G.B., and Schechter, S. (1997). Evaluating the generalizability of cognitive interview findings. In: *Seminar on Statistical Methodology in the Public Service: Statistical Policy Working Paper 26*, 353–362. Washington, DC: Federal Committee on Statistical Methodology, Office of Management and Budget.

Behr, D., Kaczmirek, L., Bandilla, W., and Braun, M. (2012b). Asking probing questions in web surveys: which factors have an impact on the quality of responses? *Social Science Computer Review* 32: 487–498.

Behr, D., Braun, M., Kaczmirek, L., and Bandilla, W. (2014). Item comparability in cross-national surveys: results from asking probing questions in cross-national surveys about attitudes towards civil disobedience. *Quality and Quantity* 48 (1): 127–148.

Berinsky, A.J., Huber, G.A., and Lenz, G.S. (2012). Evaluating online labor markets for experimental research: Amazon.com's Mechanical Turk. *Political Analysis* 20: 351–368.

Blair, J. and Conrad, F.G. (2011). Sample size for cognitive interview pretesting. *Public Opinion Quarterly* 75 (4): 636–658.

Buhrmester, M., Kwang, T., and Gosling, S.D. (2011). Amazon's Mechanical Turk: a new source of inexpensive, yet high-quality, data? *Perspectives on Psychological Science* 6 (1): 3–5.

Converse, J.M. and Presser, S. (1986). *Survey Questions: Handcrafting the Standardized Survey Questionnaire*. Newbury Park, CA: Sage.

Edgar, J., Murphy, J., and Keating, M. (2016). Comparing traditional and crowdsourcing methods for pretesting survey questions. *SAGE Open* 6: 1–14.

Fowler, S.L., Willis, G.B., Moser, R.P., et al. (2015). Reliability testing of the Walking Environment Module. Annual meeting of the Federal Committee on Statistical Methodology, Washington, DC.

Fowler, S.L., Willis, G., Moser, R.P., et al. (2016). Web probing for Question Evaluation: The effects of probe placement. 71st American Association for Public Opinion Research Meeting, Austin, TX.

Fugard, A.B.J. and Potts, H.W.W. (2015). Supporting thinking on sample sizes for thematic analyses: a quantitative tool. *International Journal of Social Research Methodology* 18 (6): 669–684. https://doi.org/10.1080/13645579.2015.1005453.

Miller, K., Willson, S., Chepp, V., and Padilla, J.L. (2014). *Cognitive Interviewing Methodology*. New York: Wiley.

Murphy, J., Edgar, J., and Keating, M. (2014). Crowdsourcing in the cognitive interviewing process. Annual Meeting of the American Association for Public Opinion Research, Anaheim, CA.

Nisbett, R.E. and Wilson, T.D. (1977). Telling more than we can know: verbal reports on mental processes. *Psychological Review* 84 (3): 231–259.

Pennebaker, J.W., Francis, M.E., and Booth, R.J.K. (2001). *Linguistic Inquiry and Word Count (LIWC): A Computerized Text Analysis Program*, 2e. Mahwah, NJ: Erlbaum.

Schuman, H. (1966). The random probe: a technique for evaluating the validity of closed questions. *American Sociological Review* 21: 218–222.

U.S. Census Bureau. 2010. Quick facts. http://www.census.gov/quickfacts/table/PST045215/00.

Willis, G.B. (2005). *Cognitive Interviewing: A Tool for Developing Survey Questions*. Thousand Oak, CA: Sage.

Willis, G.B. (2015). *Analysis of the Cognitive Interview in Questionnaire Design*. New York: Oxford University Press.

Part IV

Cross-Cultural and Cross-National Questionnaire Design and Evaluation

19

Optimizing Questionnaire Design in Cross-National and Cross-Cultural Surveys

Tom W. Smith

NORC, University of Chicago, Chicago, IL, USA

19.1 Introduction

Questionnaire development in cross-national and cross-cultural surveys starts with the same challenges that exist for monolingual and monocultural surveys. The individual measures need to be reliable and valid and need to function well among the target population (e.g. understandable, answerable, consistently understood across respondents). Furthermore, when multi-item measures are to be used, the scales must measure well the concepts of interest (e.g. high Cronbach's alpha, strong and appropriate factor loadings). However, for cross-national/cultural surveys, the process becomes more complicated and more challenging. The individual measures and the scales need to achieve a high level of functional equivalence across surveys and to minimize comparison error.

Achieving this is not easy. First, one needs use the total survey error (TSE) paradigm to rigorously design surveys and optimize comparability across surveys in general (Smith 2011, 2019). Second, by applying rigorous translation procedures such as the TRAPD model (Translation, Review, Adjudication, Pretesting, and Documentation) and by the analysis of results for differences due to measurement variation resulting from errant or at least suboptimal translations wording, comparability can be optimized. Third, one needs to consider whether there are structural or cultural differences (besides language) that undermine the comparability of items. Finally, when multi-item scales are involved, one needs to test for equivalence. Then if, as is often the case, equivalence is not established, one needs to identify the source of the non-equivalence. There are numerous possible reasons for this such as (i) poorly operating scales, even within countries; (ii) poor translations; (iii) structurally or culturally based differences in specific items; and (iv) true and meaningful differences in the configuration of values and attitudes related to a

Advances in Questionnaire Design, Development, Evaluation and Testing, First Edition.
Edited by Paul C. Beatty, Debbie Collins, Lyn Kaye, Jose-Luis Padilla, Gordon B. Willis, and Amanda Wilmot.

concept due to substantive variation in how societies assess the concept. The source for the non-equivalency must be ascertained, since the steps that need to be taken to deal with it depends on its source and nature.

Unless the difficult methodological challenges of achieving functional equivalence and minimizing comparison error are met, cross-national/cultural research is scientifically of little value and more likely to misinform than to enlighten. As Linnaeus noted, "method…is the very soul of science" (Pulteney 1805). Operational and methodological techniques have been developed to help deal with these challenges. These include (i) using the TSE paradigm to minimize comparison error in cross-national/cultural surveys; (ii) adopting best-practices for doing and evaluating translations; (iii) design-to-archiving metadata protocols to maximize the careful development and documentation of the surveys (e.g. the Questionnaire Design Documentation Tool and Translation Management Tool [TMT]); (iv) coordination across data collections to fully implement and enforce input harmonization from the conceptual level down to the nitty-gritty of coding, cleaning, and other details of surveys; (v) items development and testing aids such as Survey Quality Predictor (SQP) and Question Understanding Aid (QUAID); and (vi) scale-evaluation procedures such as item response theory (IRT), confirmatory factor analysis (CFA), Bayesian approaches, and other statistical techniques.

This chapter examines (i) the use of the TSE paradigm to assess and minimize comparison error in cross-national/cultural surveys; (ii) guidelines and other resources for designing and conducting research; (iii) best practices for doing and evaluating translations, including the consideration of linguistic, structural, and cultural dimensions; (iv) developing comparative scales; (v) the use of focus groups and pretesting; (vi) tools for developing and managing cross-national/cultural surveys; (vii) resources for developing and testing cross-national measures; and (viii) pre- and post-harmonization.

19.2 The Total Survey Error Paradigm and Comparison Error

TSE is the sum of all the myriad ways in which survey measurement can go wrong (Smith 2011). TSE (i) distinguishes two types of error: (a) variance or variable error, which is random and has no expected impact on mean values, and (b) bias or systematic error, which is directional and alters mean estimates; and (ii) classifies error into branching categories in which major categories are subsequently subdivided until presumably all notable survey-error components are separately delineated. Figure 19.1 illustrates a standard model of TSE. It has two error flows (variable with a solid lines and systematic with dashed lines) from each error source. It has 35 components (the right-most boxes in each path).

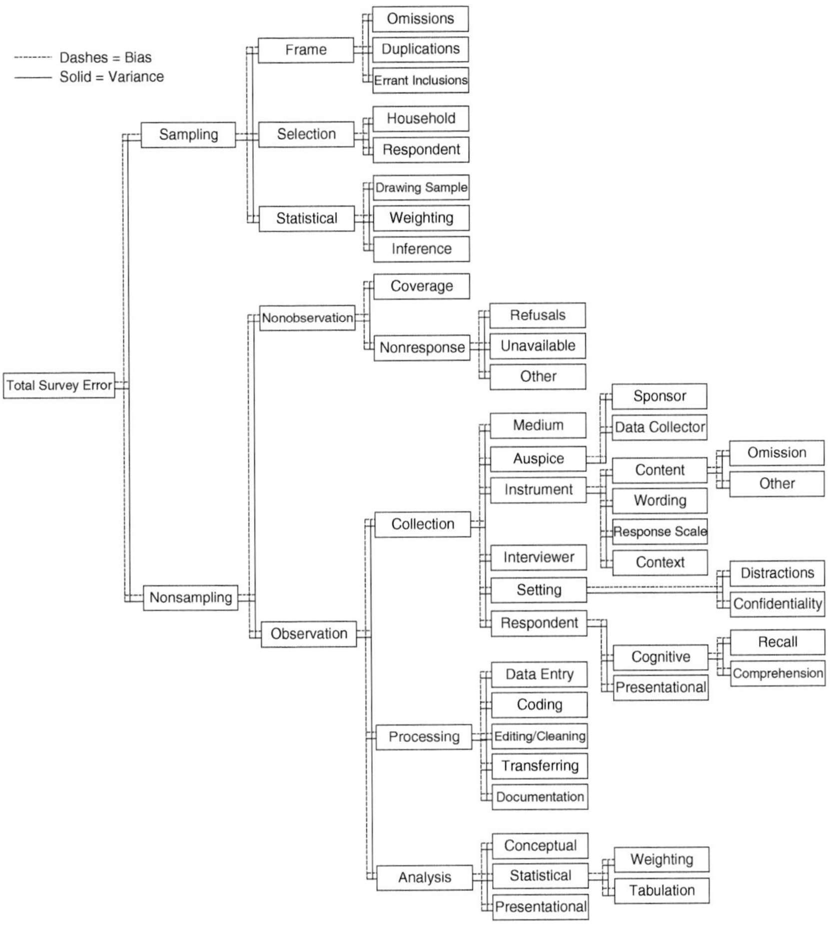

Figure 19.1 Total survey error.

Traditionally, TSE was used to describe the error structure of a single survey. But much of survey research involves the use of two or more surveys (e.g. cross-sectional time series, multi-wave panels, and cross-national comparisons). Fortunately, TSE can be easily adapted to apply to and improve such multi-survey research (Smith 2011, 2019). In effect, for each country in a cross-national/cultural survey, there would be a parallel set of boxes duplicating those shown in Figure 19.1. There would be a stack of Figure 19.1's with a new set of figures for each country in the comparative research. The interaction of errors across cross-national/cultural surveys then leads to comparison error (Smith 2011). Comparison error occurs both for each component and in the aggregate across all components. For example, errors due to mistranslations are comparison errors that are interactions between the question-wording components of each study. The TSE paradigm indicates that one should consider all of the many components of comparison error across surveys including both the individual comparison errors from each component and the cumulative comparison error across all components.

Comparison error is especially likely in studies involving a large number of countries and societies that are very different from one another (e.g. varying greatly on languages, structures, cultures). More countries mean a larger number of components (e.g. research teams, field staffs, translations) that must be planned and coordinated. The larger number also means that the goal of achieving functional equivalence across all countries is harder since more bilateral comparisons must be optimized and steps to make two countries more similar will often draw one or both of the countries away from still other societies. In a bilateral study, there is one comparison per box. For 10 countries, there would be 45 bilateral comparisons per box. Multiply by the 35 boxes, and there are 1575 bilateral comparisons. If interactions across error components (i.e. boxes) are considered, tens of thousands of comparisons are generated.

The five boxes under Instrument represent the main errors sources related to QDET (questionnaire design, development, evaluation, and testing). There are content errors consisting of (i) essential questions that are omitted, (ii) models that are misspecified, (iii) wording errors in which questions have one or more problems (e.g. double barreled, low understandability, factual mistakes), (iv) response scale errors such as confusing, inappropriate, or incomplete response options, and (v) context errors in which prior questions create an order effect that distorts answers to subsequent questions.

In addition, these instrument errors can interact with many other error components. For example, poor interviewer training will usually aggravate wording errors, many questions will function differently across modes, and respondents with lower cognitive ability will have more difficulty with complex, demanding questions than with simpler items. Any of these interaction effects can vary across countries, thus contributing to comparison error. Because of such interaction, the whole TSE structure needs to be kept in mind even when the focus is on the instrument component and item and scale development (Smith 2011, 2019).

Applying the TSE paradigm and systematically examining all sources of survey error in comparative surveys greatly enhances their traditional emphasis on achieving functional equivalence. The adoption of the TSE paradigm ensures that a rigorous and comprehensive consideration of the total error structure is considered and that comparison error can be better understood and reduced. Also, using the functional equivalence approach augments the utilization of TSE by focusing on the goal of achieving similarity across surveys and their measurements (i.e. reducing comparison error), concentrating on the components that most often contribute to comparison error (e.g. wording differences due to the difficulty of developing translations that are equivalent linguistically, structurally, and culturally), and drawing the rich, detailed experiential lessons from past comparative research (Smith 2019).

19.3 Cross-Cultural Survey Guidelines and Resources

To implement the TSE paradigm and assess and minimize comparison error, there are a number of valuable guidelines and resources available. The best overall resource for conducting cross-national survey research is the Cross-Cultural Survey Guidelines (CCSG – http://ccsg.isr.umich.edu). CCSG developed out of the International Workshop on Comparative Survey Design and Implementation (CSDI – https://www.csdiworkshop.org/index.php). The website version has 18 comprehensive chapters plus a glossary and references:

1. Study Design and Organizational Structure
2. Study Management
3. Tenders, Bids, and Contracts
4. Sample Design
5. **Questionnaire Design**
6. **Instrument Technical Design**
7. **Translation**
8. **Adaptation**
9. **Pretesting**
10. Interviewer Recruitment, Selection, and Training
11. Data Collection
12. Paradata and Other Auxiliary Data
13. **Data Harmonization**
14. Data Processing and Statistical Adjustment
15. Data Dissemination
16. Statistical Analysis
17. Survey Quality
18. Ethical Considerations

These cover all major components of conducting cross-national survey research. CCSG is frequently updated, and a comprehensive revision was released in 2016. The sections in **bold** in this list (Questionnaire Design, Instrument Technical Design, Translation, Adaptation, Pretesting, and Data Harmonization) most directly cover the main QDET elements for cross-national/cultural surveys.

There are also several excellent sites for cross-national projects that go well beyond providing documentation about each specific project and include methodological and operational information applicable to cross-national/cultural survey research in general. These include: (i) the

European Social Survey (ESS) (http://www.europeansocialsurvey.org) – for general information on the methodology of the ESS, see Jowell et al. (2007) and Fitzgerald (2015b); (ii) the International Social Survey Programme (ISSP) (http://www.issp.org and http://www.gesis.org/issp/home); (iii) the World Bank Living Standards Measurement Study (LSMS) (http://surveys.worldbank.org/lsms); and (iv) the Study of Health, Aging, and Retirement in Europe (SHARE) (http://www.share-project.org). A number of the particular resources from these projects are referenced in the following sections.

For a discussion of the different types of cross-national surveys and links to other major cross-national projects, see Smith and Fu (2016).

19.4 Translation

Translation is inherently qualitative, but survey research is largely quantitative. Rigorous methods including experiments and other quantitative techniques have to be used both to optimize translations in a particular application and to improve the art and science of translation in general. First, to minimize comparison error in wordings, one needs to utilize a rigorous and comprehensive procedure for item development, translation, and adaption (ESS 2016c,d; Harkness 2007; Kessler et al. 2013; Smith 2002, 2004, 2007a,b, 2009; World Mental Health Initiative 2015). The best approach developed to date for achieving this is the TRAPD model (Harkness et al. 2004; Willis 2015) developed for the ESS.

Second, the steps in the TRAPD process have to be carefully documented and coordinated so translation is optimized and consistent across the many languages and countries that are commonly part of cross-national studies. Improved tools for doing this are described in Section 19.7. Pretesting is an important, but underutilized, part of good translation. Translations both need to achieve functional equivalence and to work well in each target language and country, and pretesting is especially important to achieve the latter. Improved comparative pretesting procedures are described in Section 19.6.

Third, one needs to recognize that achieving item equivalence is not limited to language alone, but must also take into consideration differences in structure and culture. Translation is not a matter of changing words in a master source questionnaire into words in each of the target languages, but of fully considering not only the linguistic meaning and connotations of words and phrases, but also linguistic matters such as level of formality and tone, word difficulty, syntax complexity, etc. Beyond the linguistic dimension, translations must consider differences in social structure and culture, how these affect the functional equivalence of translations, and what can be done to minimize comparison error from these sources. This point is discussed in depth in Smith (2019).

Fourth, quantitative methods must be used to augment the best qualitative procedures for implementing translations and evaluating the outcomes. A series of quantitative steps can and should be taken at both the developmental stage and as part of post-hoc assessment to test and improve translation and therefore minimize wording-comparison error. Smith (2007a,b, 2010) describes a series of techniques such as (i) rating of proposed translations by evaluators, (ii) applying IRT models and other statistical comparison techniques at the developmental stage, (iii) pretesting of wordings both within and across samples of bilinguals, (iv) experimental randomization testing different wordings/translations, (v) the measurement of the intensity of various response options, and (vi) post-hoc evaluation of results across language subgroups.

As an example, in selecting ordinal response options, specific response options need to have the same strength and the same interval between response options across surveys. A useful procedure for achieving this is to have members of each target language group rate the strength or intensity of the offered response options. Within a country, this can establish that the chosen terms are ordered as intended and what the interval or distance is between response options. Across languages and countries, the technique can establish that response options have comparable strengths, and that they mark attitudes at the same point along an underlying continuum. For example, a five-point Likert scale using Strongly Agree, Agree, Neither Agree nor Disagree, Disagree, Strongly Disagree would want versions in other languages to have their categories represent the same points on the totally agree to totally disagree continuum as the English points did. Smith et al. (2005) conducted experiments in the United States, Germany, and Japan that show (i) how the strength of response options can be calibrated and the interval between them measured, and (ii) how the comparability of the response options across countries/languages can be assessed and adjusted as needed.

Also, in regard to post-hoc evaluations, the 2006 General Social Survey (GSS) was done in Spanish and English with respondents selecting their strongest language to do the survey in (Smith 2009, 2013). In addition, respondents indicated their language ability in the language not selected for the interview. Among Hispanics, this process identified four groups: English monolinguals; bilinguals doing it in English; bilinguals doing it in Spanish; and Spanish monolinguals. The analysis compared responses across these four groups with focus on cases in which there were no differences in distributions between the two English groups, no difference between the two Spanish groups, and differences across the bilingual groups using English and Spanish. Then controls were introduced for assimilation and socio-demographic variables to see if the apparent language differences had other explanations. When a few suspect items were identified by this procedure, a new team of translators assessed the flagged items. They either identified clear translation errors or at least proposed alternative translations seen as more equivalent.

Then the original and alternative translations were fielded on random samples on the 2008 GSS. The results indicated that the alternative translations did not show the suspicious shift between Hispanic bilinguals interviewed in Spanish or English, thus indicating that the new translations were more equivalent. While not as strong as experimental designs, this method has the advantages of being usable without special data collection and of using the larger and more representative samples associated with final studies rather than the less-generalizable and smaller samples typically utilized in pretests and other development work. It has the disadvantage of detecting translation problems only after final data collection.

Finally, translation and item development need to be tailored to the specifics of the comparative study design. There is a bias favoring the language and countries that use the master/source languages. The wording and content of questions are optimized for the source language, and for other languages/countries there is an inherent challenge to have the translated and adapted items perform as well as the master, source-language items. If there is a bilingual survey in two countries, this problem can be minimized by the joint development and testing of items so that there are in effect mutual, co-equal source languages (sometimes referred to as *decentering*). This iterative approach becomes impractical when the studies involve dozens of languages and countries. But the advantages of joint, iterative development can be approximated by extensive collaboration across all participants from the earliest conceptualization stage through item development, pretesting, data collection, and analysis. For example, the ISSP takes two years to design each annual survey with three annual plenary meetings of all participants successively selecting the topic for the survey, then the themes and subtopics, and finally the exact wordings and item-level guidelines (e.g. translation notes) for all items and with a drafting group of typically six nations doing the detailed design work and pretesting in between the plenary meetings.

19.5 Developing Comparative Scales

While words are the building blocks of questions, questions are the fundamental components of scales (i.e. multi-item measures). Scales are essential for measuring most constructs and are even more important in comparative research than in monolingual/cultural research. For both substantive and methodological reasons, items in a scale and the scale itself will often not work in a functionally equivalent manner across all countries/languages. With single-item measures it is virtually impossible to determine whether cross-national differences (or non-differences) are substantive, methodological, or a combination of both. Having well-constructed, multi-item measures both allows for the detection of non-equivalence across countries/languages and assists in figuring out why the non-comparability is occurring. During both the developmental/pretesting phase and the post-data collection stage of cross-national survey research, appropriate statistical techniques should be employed to test the comparability of scales and of the individual measures within scales. These can be used to help avoid comparison error both during the design stages involving Content and Wording of TSE and the later Analysis stage. Chief among these are the use of IRT models, structural equation modeling (SEM), and CFA (Davidov et al. 2015; Kline 2011; MacIntosh 1998a,b; Wright and Masters 1982). Their appropriateness depends first on the specific statistical and analytical assumptions and constraints that are incorporated into each technique. For example, IRT tests for an ordering of items in a scale, and parametric and non-parametric versions of SEM and factor analysis vary according to the assumed measurement level of the constituent variables. The details of each statistical technique need to be fully understood and carefully considered in any application. No detailed comparison of the exact statistical assumptions and nature of these various techniques is considered here.

If CFA or related statistical tests of equivalence indicate non-equivalence, this may be because (i) the statistical standards are too strict; (ii) the scales are weak in all countries; (iii) there are translation errors; (iv) measurement differences due to connotation, structure, cultural factors, and/or other errors are occurring; or (v) real, substantive differences exist. While it is often difficult to do, it is important to sort out the basis for the indicated non-equivalence, since the course of action taken depends on this assessment (Schwarz et al. 2010; Smith 2019).

Regarding the first possibility, Davidov et al. (2015) propose a Bayesian approach that substitutes approximate measurement equivalence for the exact matching required by standard CFA. (i) If the less stringent test is deemed appropriate and it indicates equivalence, then one merely proceeds to use the scale. See also the perspective on comparative analysis presented in Welzel and Inglehart (2016). (ii) Items that perform marginally or poorly based on existing national research are unlikely to be satisfactory when fielded in multiple countries. Such weak scales should not be utilized in cross-national surveys and when they in fact prove to be comparatively problematic, the scale should be dropped or notably redesigned. (iii) CFA and other analytic assessments can be useful to identify translation and other measurement error problems such as when an ESS question about being "wealthy" wrongly used "healthy" when translated into Italian (Rother 2005). Ideally, as discussed earlier, translation errors are minimized by using rigorous procedures such as TRAPD, but some outright errors and other less-than-optimal translations will of course still occur. Careful post-hoc review of translations should be done when an item is flagged as non-equivalent. If mistranslation is confirmed as the cause, a correct/improved translation needs to replace it, and the errant item needs to be deleted from analysis. (iv) If some other measurement error is identified as the problem, one needs to consider redesigning the scale or dropping problematic items. The latter option is often not possible since there may be too many aberrant items bilaterally or the aberrant items in one country may not match the outliers in other countries (MacIntosh 1998a,b). There is also the danger that dropping items to improve model matching can dismiss true variation as a measurement artifact and thus overly simplify our understanding of national differences. (v) If no comparison error due to mistranslation or some other discernable measurement error is detected, one needs to decide whether to accept the differences as real and substantive or artifactual due to some unidentified measurement artifact. Often the tendency is to reject the scale as non-equivalent and not use it in comparative analysis, but that is frequently a mistake. In an IRT analysis, MacIntosh (1998b) noted that "If the observed measures across groups are not related in the same way to the latent trait or attitude, group differences in means or patterns of correlations are **potentially** artifactual and **perhaps** substantively misleading (emphasis added)." The research bias has been to forget the qualifiers "potentially" and "perhaps" and just accept the model's decision of non-equivalence. Differences in IRT scores and factor loadings may indicate real, meaningful substantive differences and not measurement error (Smith 2019).

If a particular scale with appropriate factor loadings is determined to be a definitive measure of a construct, then a country statistically deviating from it is often deemed to have failed to "measure up" or, alternatively, the scale itself is deemed to not be comparatively valid. But if the measures making up the scale have been found to be individually equivalent and yet the scales across countries are not equivalent, this may instead indicate a meaningful cross-national variation in the substantive structure of the construct and that the observed differences are valid comparative findings and not dismissible as errant due to the statistical deviations in the models. When an item fails to function as the model specifies or an expected factor in general does not emerge, this indicates that closer inspection and evaluation of the deviation is needed and does not support the automatic assumption that measurement error and measurement non-equivalence has occurred. Such situations are one of the reasons why more items are often needed in cross-national surveys than in monocultural, monolingual surveys. In some cases the result may indicate that combined *emic* (referring to those that are culture specific or close to being societally unique) and *etic* (which describes aspects seen as universal that are "understood in a consistent manner across cultures and national boundaries [i.e. to the extent that they have interpretive equivalence]" (Johnson 1998)) measurement approaches are needed to adequately measure the topic cross-nationally (Smith 2019). For example, a study of obedience to authority in the workplace in the United States and Poland had five common items plus three country-specific items in Poland and four in the US (Slomczynski et al. 1981). This allowed both direct cross-national comparisons as well as more valid measurement of the construct within countries (and presumably better measurement of how that construct works in models).

19.6 Focus Groups and Pretesting in Cross-National/Cultural Surveys

Notable advances have been made in recent years in adapting questionnaire development and testing procedures originally designed for a single monolingual survey to work with multiple, multilingual surveys. For example, research shows that interactions within focus groups often differ across cultural and language groups (Colucci 2008; Halcomb et al. 2007; Huer and Saenz 2003; Lee and Lee 2009; Pan et al. 2016). Given the differences in verbal styles, deference practices, and other cultural-linguistic features, focus groups operate differently across these groups. If these variations are understood and adjusted for, valuable and functionally equivalent information can be collected from focus groups covering different cultures and languages (Pan et al. 2016).

Similarly, considerable progress has been made to optimize pretesting in general and cognitive pretesting in particular for cross-nation/culture studies (ESS 2016a,b,2016d; Fitzgerald 2015a; Fitzgerald et al. 2011, 2014; GESIS 2016; Kessler et al. 2013; Miller et al. 2011; Smith 2010). Key points of pretesting for cross-national/cultural surveys are:

1) Doing pretests in different countries/languages. Ideally these should be done in all countries/languages involved in a cross-national/cultural study (Kessler et al. 2013).
2) Careful coordination to insure that the pretests and their results are equivalent (e.g. comparable pretest populations, harmonization of interviewer recruitment and training, consistent evaluation procedures).
3) "Triangulation" of results combining both quantitative analysis with qualitative assessment by the various parties involved (e.g. translators, interviewers, substantive experts, national/group specialists). Using multiple assessments is crucial for cross-national/cultural surveys since they are more challenging than monolingual/cultural surveys and involve more components.
4) Using the pretests to improve both the translated instruments and the original source questionnaire. This last point is particularly important. The cross-national error source typology (CNEST) of the ESS has identified three major sources for comparison error: (i) poor source question design, (ii) translation problems, and (iii) cultural portability (i.e. that a concept or situation either does not exist in all countries or is "in such a different form as to make the use of the same source question impractical" (Fitzgerald et al. 2014)). While weaknesses in source language questions can and should be detected by initial monolingual testing, multilingual/cultural

pretests can disclosure problems in the source questions not previously revealed. Translation problems can involve simple mistranslation, but may also relate to problems in the source questions. In general, source questions that rely on language-specific metaphors or country-specific situations will be difficult to translate to achieve functional equivalence. Also, concepts may not travel well across cultures and languages. An ISSP question asking about "hard work" was mistranslated in one country into a phrase meaning doing work that was difficult. But this mistranslation also revealed that the concept "hard work" was not easily and uniformly understood in other cultures and languages and needed explanation to facilitate accurate translation. Cultural portability can often be anticipated from the start by considering if a particular concept or situation exists in a similar manner in the other countries to be studied. But multilingual/cultural pretesting can disclose additional importability issues that were not initially anticipated.

19.7 Tools for Developing and Managing Cross-National Surveys

The Data Service Infrastructure for the Social Sciences and Humanities (DASISH) in 2012–2014 brought together five European research infrastructure initiatives in the social sciences and humanities including the ESS, SHARE, and the Consortium of European Social Science Data Archives (CESSDA) (DASISH, http://dasish.eu/; Marker 2014). It sought to create a multi-language questionnaire development and dissemination system that could be utilized for cross-national surveys in general. It had as a central goal interoperability across tools and followed the protocols of the Data Documentation Initiative (DDI), an international standard for describing statistical and social science data (http://www.ddialliance.org). The interacting components that it developed include:

1) The TMT (Balster and Martens 2014; Dorer and Martens 2015) is designed to allow centralized management of the translation process for cross-national and multi-lingual surveys, to do the translation with a web application, to create a permanent, retrievable record of each step in the process, and to implement the TRAPD approach.
2) The Questionnaire Design and Documentation Tool (QDDT) (Bakkmoen et al. 2014; Orten et al. 2014; Prestage 2014) serves to facilitate and document questionnaire development covering the whole flow of questionnaires from beginning to end, including information on design, pretesting, translation, and question quality and the rationale behind the development of individual items, scales, and entire modules. This builds on a paper-based questionnaire design template developed by the ESS (Fitzgerald 2015b). For a related tool, see Joye et al. (2015).

3) The Question Variable Data Base (QVDB) (Orten et al. 2014; Prestage 2014) is to build and maintain an item-level database allowing Boolean searches on words and other coded fields.

In addition, these three tools were to be compatible with and used along with the separately developed SQP, which is discussed in the next section.

Synergies for Europe's Research Infrastructures in the Social Sciences (SERISS) 2015–2019 is continuing the work of DASISH. Besides the ESS, SHARE, and CESSDA, it involves the Generations and Gender Programme, the European Values Survey, and the WageIndicators Survey (SERISS, http://seriss.eu/). Among other things, it will build on DASISH tools to (i) develop general, online technological tools to "facilitate greater harmonization of data collection, analysis and curation" and (ii) maximize translation equivalence. The conceptual basis for these tools is presented in Fitzgerald et al. (2014). The first SERISS training course was held October 24–25, 2016 in Ljubljana, Slovenia. It featured a course on Designing Questionnaires for Cross-Cultural Survey by Ana Villar and Dorothee Behr.

While the management and coordination tools described here go well beyond the questionnaire development function, having this overall coordination and comparative structure is essential for the development of reliable, valid, and functionally equivalent items and scales because the instrument interacts with all other TSE components and cannot succeed as if it was an isolated, self-contained component. For example, developing optimally comparable scales will not ensure measurement equivalence if interviewer training or data-entry quality control are inadequate in some countries.

19.8 Resources for Developing and Testing Cross-National Measures

The minimization of TSE in general and comparison error in particular is facilitated by several programs designed to assess the quality of survey questions and/or aid in their design (Smith 2015). (i) Survey Quality Predictor 2.1 (http://sqp.upf.edu) analyzes question text and various metadata fields about the items and produces quality predictions related to reliability and validity (Saris and Galhoher 2007). It was developed in part using the ESS, and Willem Saris and Daniel Oberski were awarded the Mitofsky Innovators Award by the American Association for Public Opinion Research in 2014 for its creation. In collaboration with the ESS, the SQP2.1 is being extended and developed further (ESS 2016a). (ii) The QUAID is a program to assess question wordings (http://quaid.cohmetrix.com). It identifies problematic features of items that may confuse respondents and lower reliabilities. QUAID is designed to identify items with unfamiliar technical terms, vague or imprecise relative terms, vague or ambiguous noun phrases, complex syntax structures, and high demands on working memory. QUAID is based solely on English examples from American surveys (505 questions on 11 surveys developed by the US Census Bureau) and thus is not rooted in cross-national/cultural data. It was primarily developed by Andrew C. Graesser et al. (2006). (iii) A different approach is taken by Q-Bank of the National Center for Health Statistics (http://wwwn.cdc.gov/qbank/home.aspx). It allows one to search a database of existing questions and provides references to methodological research that examined the items. It also is not cross-nationally oriented. Q-Bank has a database of about 4600 question wordings and 150 methodological reports related to the items and the surveys in which the questions appeared. One can either search for relevant questions and be linked to methodological reports that relate to these or directly search the reports. The items are all from US government surveys, but some do relate to English/Spanish comparisons. Finally, another example is the World Bank Living Standards Measurement Study of the World Bank (http://surveys.worldbank.org/lsms), which includes an archive of cross-national questionnaires on this topic, a compilation of methodological experiments, and guides to survey design and analysis. While it includes questionnaires from about 109 surveys in 40 countries, individual items are not searchable.

In addition to these tools, which either help to assess questions or provide methodological information about items, there are various guides for searching for individual questions. These include (i) ZACAT – the GESIS Online Study Catalogue (http://www.gesis.org/en/services/research/daten-recherchieren/zacat-online-study-catalogue), (ii) the Variable and Question Data Bank of the UK Data Service (https://discover.ukdataservice.ac.uk/variables), (iii) Polling the Nations (http://www.orspub.com), (iv) the Social Science Variables Database of the Interuniversity Consortium for Political and Social Research (https://www.icpsr.umich.edu/icpsrweb/ICPSR/ssvd/index.jsp), (v) the CESSDA data catalog (https://datacatalogue.cessda.eu), and (vi) iPOLL of the Roper Center for Public Opinion Research – US data only (https://ropercenter.cornell.edu/CFIDE/cf/action/home/index.cfm).

19.9 Pre- and Post-Harmonization

To achieve functional equivalence and maximize comparability across surveys, pre- and/or post-harmonization are needed (also known as *input* and *output* harmonization). When designing a new cross-national study, the focus is on pre-harmonization (the standardization of items and study design before the surveys are conducted). Researchers design items, scales, and questionnaires to be comparable across countries and languages. When dealing with already-collected data, post-harmonization is used to create a cross-national file. Often both pre- and post-harmonization need to be employed. For example, the ISSP uses pre-harmonization in the design of the 60 substantive questions developed for each annual module, but uses post-harmonization to convert national demographics to ISSP cross-national standards. Even studies like the ESS that include pre-harmonization of demographics, need a post-harmonization phase to deal with nation-specific measures such as geographic variables and education (https://www.europeansocialsurvey.org/docs/round6/survey/ESS6_appendix_a1_e02_2.pdf).

There are also post-harmonization projects that merge and/or make more comparable studies not originally designed for comparative purposes, such as the following:

Cross-national Equivalent File: https://cnef.ehe.osu.edu
Democratic Values and Protest Behavior: Data Harmonization, Measurement Comparability, and Multi-Level Modeling in Cross-National Perspective http://dataharmonization.org
Eurostat: http://ec.europa.eu/eurostat
Harmonized European Time Use Study: https://www.h6.scb.se/tus/tus/
Integrated Public Use Microdata Series International (IPUMS-I): https://international.ipums.org/international
International Stratification and Mobility File: http://www.harryganzeboom.nl/ismf/index.htm
Luxembourg Income Study (LIS): http://www.lisdatacenter.org
United Nations: http://unstats.un.org

In addition, there are variable-level harmonization efforts rather than those designed to make major studies comparable (Hoffmeyer-Zlotnik and Wolf 2003). One example is the International Standard Classification of Occupations (ISCO), which was developed by the International Labour Organization in 1958 and updated most recently in 2008 (http://www.ilo.org/public/english/bureau/stat/isco). It can be used to directly code occupations cross-nationally, or nation-specific occupational codes can be converted to ISCO for international comparability. For example, there is a crosswalk between ISCO08 and the US-based 2010 Standard Occupational Classification (http://www.bls.gov/soc/soccrosswalks.htm). Another example is the International Standard Classification of Education (Briceno-Rosas 2016; Ortmanns 2016; Schneider 2008, 2009, 2016). Currently, a project at GESIS, CAMCES (Computer-Assisted Measurement and Coding of Educational Qualifications in Surveys, http://www.leibniz-education.de/projekt.html?id=72&lang=en), is finalizing a tool to measure educational qualifications in surveys by accessing a large, international database of educational degrees and assigning educational codes based on detailed national educational qualifications.

Finally, harmonizations may consist of only preparing aggregated data in the form of tables and reports (e.g. in many UN reports), while in other cases microlevel datasets are produced (e.g. LIS and IPUMS-I).

Notable advances in harmonization have appeared in recent years. Winters and Netscher (2016) at GESIS have developed a tool to better document data harmonization. Moreover, the Democratic Values and Protest Behavior: Data Harmonization, Measurement Comparability, and Multi-Level Modeling in Cross-National Perspective (Breustedt 2015; Democratic Values 2016; Dubrow 2015; Lillard 2015) has advanced harmonization as a general object for development rather than just being applied for a creation of a project-specific merged file. It has a journal: *Harmonization: Newsletter on Survey Data Harmonization in the Social Sciences* (http://consirt.osu.edu/newsletter).

19.10 Conclusion

Cross-national/cultural studies are a massive conceptual, methodological, and organizational challenge. The TSE paradigm provides a powerful, conceptual structure for planning and evaluating comparative research. But the tasks involved in designing, organizing, executing, and evaluating are so large and complex that well-developed methodologies, protocols, and tools are needed to conduct comparative research in a scientifically reliable and valid manner. Fortunately, increasingly this challenge is being met by the development of a series of resources and tools to ease, systematize, and improve cross-national/cultural research. Their rigorous use and close examination of the resulting data will both ease the design and collection of cross-national/cultural data and enhance their reliability and validity.

References

Bakkmoen, H.V., Orten, H., and Prestage, Y. (2014). DASISHIl: Keeping track of the questionnaire design process. European DDI User Conference, London.

Balster, E. and Martens, M. (2014). Translation management tool: Ensuring cross-cultural equivalence. Final DASISH Conference, Gothenburg.

Breustedt, W. (2015). The barometer surveys: insights into the quality of the harmonized political trust items. *Harmonization: Newsletter on Survey Data Harmonization in the Social Sciences* 1 (2): 7–10.

Briceno-Rosas, R. (2016). Asking about education in multinational, multiregional, and multicultural contexts: Results from cognitive pretesting in two countries. Second International Conference on Survey Methods in Multinational, Multiregional and Multicultural Contexts (3MC 2016), Chicago.

Colucci, E. (2008). On the use of focus groups in cross-cultural research. In: *Doing Cross-Cultural Research* (ed. P. Liamputtong), 233–252. Amsterdam: Springer.

Davidov, E., Cieciuch, J., Meuleman, B. et al. (2015). The comparability of measures of attitudes towards immigrants in the European social survey: exact versus approximate measure equivalence. *Public Opinion Quarterly* 79: 244–266.

Democratic Values. (2016). Democratic values and protest behavior: Data harmonization, measurement comparability, and multi-level modelling in cross-national perspective. Data Harmonization: About the Project. http://dataharmonization.org/about/.

Dorer, B. and Martens, M. (2015). Translation Management Tool (TMT). SCDO Workshop, London.

Dubrow, J.K. (2015). A brief history of survey data harmonization projects. *Harmonization: Newsletter on Survey Data Harmonization in the Social Sciences* 1 (1): 2–4.

European Social Survey (2016a), Data Quality Assessment. http://www.europeansocialsurvey.org/methodology/ess_methodology/data_quality.html.

European Social Survey. (2016b), Source Questionnaire Development, http://www.europeansocialsurvey.org/methodology/ess_methodology/source_questionnaire/source_questionnaire_development.html.

European Social Survey. (2016c), Translation. http://www.europeansocialsurvey.org/methodology/ess_methodology/translation/.

European Social Survey. (2016d), Translation Assessment. http://www.europeansocialsurvey.org/methodology/ess_methodology/translation/translation_assessment.html.

Fitzgerald, R. (2015a). Sailing in unchartered waters: Structuring and documenting ross-national questionnaire design. GESIS paper.

Fitzgerald, R. (2015b). Striving for quality, comparability, and transparency in cross-national social survey measurement: Illustrations from the European Social Survey (ESS). Doctoral dissertation. City University of London.

Fitzgerald, R., Widdop, S., Gray, M., and Collins, D. (2011). Identifying sources of error in cross-national questionnaires: application of an error source typology to cognitive interview data. *Journal of Official Statistics* 27: 569–599.

Fitzgerald, R., Winstone, L., and Prestage, Y. (2014). A versatile tool? Applying the cross-national error source typology (CNEST) to triangulated pre-test data. FORS working paper.

GESIS. (2016). GESIS Pretest Database. http://pretest.gesis.org/Pretest/en.

Graesser, A.C., Cai, Z., Louwerse, M., and Daniel, F. (2006). Question understanding aid (QUAID): a web facility that helps survey methodologists improve the comprehensibility of questions. *Public Opinion Quarterly* 70: 3–22.

Halcomb, E.J., Gholizadeh, L., DiGiacomo, M. et al. (2007). Literature review: considerations in undertaking focus group research with culturally and linguistically diverse groups. *Journal of Clinical Nursing* 16: 1000–1011.

Harkness, J. (2007). Improving the comparability of translations. In: *Measuring Attitudes Cross-Nationally: Lessons from the European Social Survey* (eds. R. Jowell, C. Roberts, R. Fitzgerald and G. Eva). London: Sage.

Harkness, J., Pennell, B.-E., and Schoua-Glusberg, A. (2004). Survey questionnaire translation and assessment. In: *Methods for Testing and Evaluating Survey Questionnaires* (eds. S. Presser, M.P. Couper, J.T. Lessler, et al.). Hoboken, NJ: Wiley.

Hoffmeyer-Zlotnik, J.H.P. and Wolf, C. (eds.) (2003). *Advances in Cross-National Comparison: A European Working Book for Demographic and Socio-Economic Variables*. New York: Kluwer Academic.

Huer, M.B. and Saenz, T.I. (2003). Challenges and strategies for conducting survey and focus group research with culturally diverse groups. *American Journal of Speech and Language Pathology* 12: 209–220.

Johnson, T.P. (1998). Approaches to equivalence in cross-cultural and cross-national survey research. *ZUMA-Nachrichten-Spezial* 3: 1–40.

Jowell, R., Kaase, M., Fitzgerald, R., and Eva, G. (2007). The European Social Survey as a measurement tool. In: *Measuring Attitudes Cross-Nationally: Lessons from the European Social Survey* (eds. R. Jowell, C. Roberts, R. Fitzgerald and G. Eva). London: Sage.

Joye, D., Luisier, V., Metral, G. et al. (2015). A small tool to generate structured survey questionnaires with a Libreoffice (or Excel) sheet in a mixed tool environment. Unpublished report, University of Lausanne.

Kessler, R.C., Harkness, J., Heeringa, S.G. et al. (2013). Methods of the world mental health surveys. In: *The Burdens of Mental Disorders: Global Perspectives from the WHO World Mental Health Survey* (eds. J. Alonso, S. Chatterji and Y. He). Cambridge: Cambridge University Press.

Kline, R. (2011). *Principles and Practice of Structural Equation Modeling*, 3e. New York: Guilford.

Lee, J.-J. and Lee, K.-P. (2009). Facilitating dynamics of focus groups interviewers in East Asia: evidence and tools by cross-cultural study. *International Journal of Design* 1: 17–28.

Lillard, D. (2015). The Cross-National Equivalent File: harmonized panel survey data in eight countries. *Harmonization: Newsletter on Survey Data Harmonization in the Social Sciences* 1 (1): 4–6.

MacIntosh, R. (1998a). A confirmatory factor analysis of the affect balance scale in 38 nations: a research note. *Social Psychology Quarterly* 61: 83–95.

MacIntosh, R. (1998b). Global attitudes measurement: an assessment of the world values survey postmaterialism scale. *American Sociological Review* 63: 452–464.

Marker, H.J. (2014). DASISH results overview at a glance. Final DASISH Conference, Gothenburg.

Miller, K., Fitzgerald, R., Padilla, J.-L. et al. (2011). Design and analysis of cognitive interviews for comparative multinational testing. *Field Methods* 23: 379–396.

Orten, H., Bakkmoen, H.V., Prestage, Y. et al. (2014). New tools from complex survey: The DASISH Questionnaire Design and Development Tool and the Question Variable Data Base. Final DASISH Conference, Gothenburg.

Ortmanns, V. (2016). The CAMCES Database and its cross-national education coding system. Second International Conference on Survey Methods in Multinational, Multiregional and Multicultural Contexts (3MC 2016), Chicago.

Pan, Y., Park, H., Sha, M. et al. (2016). Cross-cultural comparison of focus groups as a research method. RTI unpublished report.

Prestage, Y. (2014). Questionnaire Design and Development Tool (QDDT) and the Question Variable Data Base (QVDB). Final DASISH Conference, Gothenburg.

Pulteney, R. (1805). *A General View of the Writings of Linnaeus* (ed. W.G. Maton). Cambridge: Cambridge University Press, (reprinted in 2011).

Rother, N. (2005). Measuring attitudes towards immigration across countries with the ESS: potential problems of equivalence. In: *Methodological Aspects in Cross-National Research* (eds. J.H.P. Hoffmeyer-Zlotnik and J.A. Harkness), 109–125. Mannheim: ZUMA.

Saris, W.E. and Galhoher, I. (2007). Can questions travel successfully? In: *Measuring Attitudes Cross-Nationally: Lessons from the European Social Survey* (eds. R. Jowell, C. Roberts, R. Fitzgerald and G. Eva), 53–78. London: Sage.

Schneider, S. (2016). Computer Assisted Measurement and Coding of Educational Qualifications in Multicultural Surveys (CAMCES): A new set of survey tools. Second International Conference on Survey Methods in Multinational, Multiregional and Multicultural Contexts (3MC 2016), Chicago.

Schneider, S.L. (ed.) (2008). *The International Standard Classification of Education: An Evaluation of Content and Criterion Validity for 15 European Countries*. Mannheim: MZES.

Schneider, S.L. (2009). Confusing credentials: The cross-nationally comparable measurement of educational attainment. Doctoral dissertation. University of Oxford.

Schwarz, N., Oyserman, D., and Peytcheva, E. (2010). Cognition, communication, and culture: implications for the survey response process. In: *Survey Methods in Multinational, Multiregional, and Multicultural Contexts* (eds. J. Harkness, M. Braun, B. Edwards, et al.), 177–190. New York: Wiley.

Slomczynski, K.M., Miller, J., and Kohn, M.L. (1981). Stratification, work, and values: a polish-United States comparison. *American Sociological Review* 46: 720–744.

Smith, T.W. (2002). Developing comparable questions in cross-national surveys. In: *Cross-Cultural Survey Methods* (eds. J. Harkness, F. van de Vijver and P.P. Mohler), 69–92. London: WileyEurope.

Smith, T.W. (2004). Developing and evaluating cross-national survey instruments. In: *Methods for Testing and Evaluating Survey Questionnaires* (eds. S. Presser, M.P. Couper, J.T. Lessler, et al.), 431–452. New York: Wiley.

Smith, T.W. (2007a). *An Evaluation of Spanish Questions on the 2006 General Social Survey. GSS Methodological Report No. 109*. Chicago: NORC.

Smith, T.W. (2007b). Integrating translation into cross-national research. Midwest Association for Public Opinion Research, Chicago.

Smith, T.W. (2009). A translation experiment on the 2008 General Social Survey. International Workshop on Comparative Survey Design and Implementation, Ann Arbor.

Smith, T.W. (2010). Surveying across nations and cultures. In: *Handbook of Survey Research*, 2e (eds. P.V. Marsden and J.D. Wright), 733–764. Bingley: Emerald.

Smith, T.W. (2011). Refining the total-survey error perspective. *International Journal of Public Opinion Research* 23: 464–484.

Smith, T.W. (2013). An evaluation of Spanish questions on the 2006 and 2008 U.S. General Social Surveys. In: *Surveying Ethnic Minorities and Immigrant Populations: Methodological Challenges and Research Strategies* (eds. J. Font and M. Mendez), 219–240. Amsterdam: IMISCOE-Amsterdam University Press.

Smith, T.W. (2015). Resources for conducting cross-national survey research. *Public Opinion Quarterly* 79: 404–409.

Smith, T.W. (2019). Improving cross-national/cultural comparability using the total survey error paradigm. In: *Advances in Comparative Survey Methods: Multinational, Multiregional, and Multicultural Contexts* (eds. B.E. Pennell, T.P. Johnson, I.A.L. Stoop and B. Dorer), 13–44. New York: Wiley.

Smith, T.W. and Fu, Y.-C. (2016). The globalization of surveys. In: *The Sage Handbook of Survey Methodology* (eds. C. Wolf, D. Joye, T.W. Smith and Y.-C. Fu), 680–692. Thousand Oaks, CA: Sage.

Smith, T.W., Mohler, P.P., Harkness, J., and Onodera, N. (2005). Methods for assessing and calibrating response scales across countries and languages. *Comparative Sociology* 4: 365–415.

Welzel, C. and Inglehart, R.F. (2016). Misconceptions of measurement equivalence: time for a paradigm shift. *Comparative Political Studies* 49: 1068–1094.

Willis, G.B. (2015). The practice of cross-cultural cognitive interviewing. *Public Opinion Quarterly* 79: 359–395.

Winters, K. and Netscher, S. (2016). Proposed standards for variable harmonization documentation and referencing: a case study using QuickCharmStats 1.1. *PLoS One* 11 (2) https://doi.org/10.1371/journal.pone.0147795.

World Mental Health Survey Initiative. (2015). http://www.hcp.med.harvard.edu/wmh/index.php.

Wright, B.D. and Masters, G.N. (1982). *Rating Scale Analysis*. Chicago: MESA.

20

A Model for Cross-National Questionnaire Design and Pretesting

Rory Fitzgerald[1] *and Diana Zavala-Rojas*[2]

[1] *ESS ERIC HQ. City, University of London, London, UK*
[2] *Department of Political and Social Sciences, Universitat Pompeu Fabra, RECSM, Barcelona, Spain*

20.1 Introduction

Questionnaire design and pretesting are critically important parts of the survey life cycle, helping to create an instrument that captures valid and reliable answers while allowing the investigators to tailor the instrument to answer theoretical research questions. Without effective design it is unlikely the data collected will reflect the views, behavior, or circumstances of the target population. Designing such an instrument for a cross-national survey has challenges in addition to those of a single nation, having to cover multiple languages and cultures. In this chapter, we consider this additional challenge in more detail, looking at how a large, cross national survey – the European Social Survey (ESS) – develops its questionnaire in order to provide high-quality and comparable measures. The process has become more extensive in recent rounds, from having previously involved a 7-stage approach in 2001 to include 17 stages in 2016. The weaknesses of the initial approach will be discussed and the enhanced approach introduced and outlined. This enhanced design and pretesting model is then evaluated for its "fitness for purpose" using the Logical Framework Approach (LFA) (Coleman 1987). Critical examples will be used to highlight strengths and weaknesses. The authors conclude that the new approach is fit for purpose for the ESS and may also serve as a point of reference for other similar surveys.

20.2 Background

In this volume, Smith demonstrates that for a cross-national survey, the number of error sources from the Total Survey Error (TSE) framework are essentially

Advances in Questionnaire Design, Development, Evaluation and Testing, First Edition.
Edited by Paul C. Beatty, Debbie Collins, Lyn Kaye, Jose-Luis Padilla, Gordon B. Willis, and Amanda Wilmot.

multiplied by the number of countries included in the study (see Chapter 19 by Smith). The elements of TSE most relevant to questionnaire development, evaluation, and testing (QDET) are Instrument and Respondent. Smith (2011) highlights how in a cross-national study you need to ensure not only high-quality comparable measurement across all groups answering your questionnaire within a single country, e.g. men and women, but also comparable measurement between countries and also across countries, e.g. men in country X to men in country Y. This means that questions need to be understood in a similar way in all countries regardless of group membership, and respondents need to use the answer categories provided in a similar manner (Saris and Gallhofer 2014).

While in principle it is possible to simultaneously develop the questionnaire in all languages, the far more usual model for large scale surveys is a *sequential* approach that starts with a source questionnaire, which is subsequently translated into all target language versions (Harkness et al. 2010a). This means that the cross-national survey researcher has to develop a questionnaire that works effectively in the source language and that also serves as the basis for translation into all target languages and cultures (Fitzgerald 2015). This is a considerable challenge, and cross-national survey researchers need to make sure they are well-equipped with the skills and resources required.

The study of the design of survey questionnaires has a long tradition, dating from the start of public opinion polling in the early twentieth century (Bradburn and Sudman 2004). Literature on questionnaire design has focused on the elements that affect survey questions, e.g. the formulation of response scales or order effects (for a review, see Smyth 2016) as well as on measurement theory and practice (Billiet 2016; Saris and Gallhofer 2014; Hox 1997; Revilla et al. 2016). Literature on the deliberate design of comparable questions for cross-national surveys is less well-developed but has been evolving quickly as cross-national research has increased. Harkness et al. (2003: pp. 20–23) describe the possibilities for cross-national questionnaire design based on two decisions. The first is whether existing questions from national surveys are to be used or whether new questions will be developed. If new questions are being developed, the second question is whether the questions will be harmonized cross-culturally, i.e. adopting the Ask the Same Question (ASQ) model, or if questions will be adapted to fit cultural nuances, i.e. adopting the Ask Different Questions (ADQ) model. Finally, the decision on whether a sequential, parallel, or simultaneous approach will be followed needs to be determined (see Harkness and Schoua-Glusberg 1998). However, the different options presented in the literature – sequential, parallel, and simultaneous approaches under an ASQ or ADQ strategy – are descriptors of current best practice. To our knowledge there is no quantitative evidence of which options help to produce the most comparable measurement instruments. One of the first assessments of such differences will be examined via a series of experiments looking at the impact of translation adaptation of quality and comparability (see www.seriss.eu).

Another area that is underdeveloped is a blueprint of how to best organize the process of questionnaire design and pretesting overall. Implicit in the literature of comparative questionnaire design and pretesting is that the actions taken are part of an *iterative process*. For instance, Billiet (2016: p. 198) suggests that in order to achieve *measurement validity,* i.e. when relationships predicted by the theory are not rejected by an analysis of the collected data, and *conceptual validity,* i.e. the general quality of the correspondence between the theory and the measures, survey researchers should iteratively go back and forward to tune the specification of the concepts and the required measurement testing. However, there is no clear best practice in terms of how this should be undertaken in practical terms, notwithstanding that each study has individual requirements. A key challenge is that the measurement tests suggested by Billiet would require samples of hundreds at each step, becoming prohibitively expensive to conduct across many countries.

The literature implicitly suggests that questionnaire designers should clarify throughout the *process* which theoretical elements are covered by the questions and which are left out, as well as specifying the potential measurement challenges derived from the chosen formulation, e.g. acquiescence response style when agree/disagree scales are used. However, as a field, insufficient attention has been focused thus far on how to evaluate the questionnaire design and pretesting life cycle from a project evaluation perspective. Since the configuration of the questionnaire design and pretesting stages has an enormous impact on the quality of the overall output of a survey, evaluation of procedures used in combination are a rather critical but under researched area. This chapter tries to address that gap in a comparative context by using the Logical Framework Matrix (LFM).

After briefly introducing the ESS, we will then go on to describe two questionnaire design and pretesting life cycles that have been implemented in its history.

20.3 The European Social Survey

The ESS (www.europeansocialsurvey.org) is a cross-national survey of attitudes and behavior that aims to chart change and stability in European countries, using a rigorous scientific methodology among a representative sample of the general public in each country taking part (Jowell et al. 2007). The survey has taken place biennially since 2002 in over 30 countries (with at least 20 in each round). In 2013 the ESS became organized as an intergovernmental research consortium known as ESS European Research Infrastructure Consortium led by its director and Core Scientific Team (CST).

Collectively, they are responsible for the design and coordination of each round of the survey, working with national coordinators (NCs). The average duration of the interview is 60 minutes in British English. Data is deposited to a data archive and made freely available for noncommercial use (see www.europeansocialsurvey for more information).

20.4 ESS Questionnaire Design Approach

The ESS questionnaire design model was revised after Round 3 (R3) because it was not sufficiently robust. There was evidence of insufficient cross-cultural input. The ESS model for cross-national questionnaire design and pretesting now has 17 stages whereas in the first 3 rounds there were just 7. The development of the questionnaire consists of triangulation of pretesting results throughout several iterations of qualitative and quantitative pretesting. In this chapter we examine whether the ESS 17-step questionnaire design process minimizes survey error and helps to produce comparable, valid measures.

As noted, in the first three rounds of the ESS, a seven-stage questionnaire design process was implemented (Saris and Gallhofer 2007). Stage 1 involved specifying the concepts and questions and operationalizing them into question items and included consideration of comments from national teams. Stage 2 involved checking or predicting the quality of items. For existing items, reliability and validity were checked using existing data. Where new items were drafted, the Survey Quality Predictor (SQP) program (Revilla et al. 2016; Saris and Gallhofer 2014) was used to predict quality by mapping new items to old ones based on their formal characteristics to try to predict the quality of the new item. Predictions of SQP are based on a meta-analysis of the relationships between the measurement quality estimates of survey questions obtained through multitrait-multimethod (MTMM) experiments and the characteristics of such questions in those experiments, e.g. the measurement properties, layout of visual aid, and indicators of linguistic complexity such as number of words and nouns. MTMMs are within-subject designs that allow studying the construct validity of survey questions using several correlated concepts asked using different formulations (Campbell and Fiske 1959; Saris and Andrews 1991). Other measurement considerations such as anticipated item nonresponse and social desirability were also considered. Stage 3 saw an initial translation from the source language (English) into one other language for the purpose of a pilot. Stage 4 involved a two-nation pilot (400 cases per country), which also contained a number of split-run experiments on question wording alternatives. In Stage 5 the pilot was analyzed to assess both the quality of the questions and the distribution of the substantive answers. Problematical questions, whether they displayed weak reliability or validity, deviant distributions, or weak scales, were revised, but any new items developed after that time were not piloted. It was on the basis of these pilot analyses that the final source questionnaire was produced. Stage 6 saw the source questionnaire translated into multiple languages. During expert review meetings or as a result of pilot feedback where it was identified that items had specific meanings difficult for non-native speakers to understand, they were annotated to provide greater definition. The translation process followed an innovative committee approach to try to ensure that the questions in all languages were functionally equivalent (Harkness 2003; Harkness et al. 2010b). Stage 7 saw MTMM experiments included in the ESS main questionnaire, designed to measure the extent of random and systematic error in different countries and possibly allow corrections later. Although used for testing psychological scales, this seventh stage was rather innovative and remains rare in a repeat cross-sectional cross-national survey.

20.5 Critique of the Seven-Stage Approach

An assessment of the seven-stage approach (Fitzgerald 2015) suggested that improvements needed to be made. In part this was because early analysis of data collected on the ESS suggested mixed success in developing cross-national questionnaires that measure attitudes equivalently (Saris and Gallhofer 2007, p. 71). In addition, there was only a single quantitative test (in two countries), and this was late in the process. Furthermore, apart from comments from NCs in all countries, the pilot was limited to the UK and one other country, meaning that the design had insufficient cross-national empirical input. This often led to major changes being requested by various countries after the source questionnaire had been "finalized" and once national teams started detailed translation. These requests often resulted in last-minute changes to the source questionnaire, sometimes meaning they were not implemented in all countries (Fitzgerald 2015). Two further weaknesses were also identified. (i) Pretesting was almost entirely quantitative (the sole exception being interviewer debriefs from the pilot study), which, while effective at identifying problematic questions, rarely provided information on why items did not work. (ii) There was only a single translation during the design of the source questionnaire (for the pilot), meaning insufficient attention was placed on those challenges. Stage 1 also rarely included adequate documentation of the design aims for modules and probably resulted in frequent "specification errors" (Biemer 2010), meaning it is likely (although unknown) that some of the questions fielded did not measure the intended concepts.

20.6 A Model for Cross-National Questionnaire Design and Pretesting

In response to the deficiencies outlined here, a new model for ESS questionnaire design and pretesting was established (Fitzgerald 2015). The aim was to "decenter" the overall design approach by adding greater cross-national input and include qualitative input in early design stages to improve the cognitive dimension. Additionally the process was to become more iterative to enable changes to be tested prior to implementation. In addition, procedures were put in place to reduce the risk of specification error.

Figure 20.1 (Fitzgerald 2015) shows the ESS model for cross-national questionnaire design and pretesting used to develop new items in Round 8. The earlier seven-stage model is now replaced with a more intensive 17-stage model used for new ESS modules. The stages are as follows:

Stages 1–2 (proposals from designers and expert review of questions). The first stage is similar to Stage 1 in the previous model where designers have to operationalize concepts and items that are then subject to expert review. However, Stage 2 now complements this by requiring the completion of a questionnaire design template used throughout the design process (Fitzgerald 2015). This template (see Appendix 20.A) structures the development of the concepts/constructs, facilitates clear specification, and documents the various iterations of the questionnaire. As under the seven-stage model, expert review involves a multinational group of substantive questionnaire designers (the Questionnaire Design Team) and a methodological group of experts from the CST.

Stage 3 (use of the SQP software). This stage, which was also previously included, involves the testing of new items using the SQP (Saris and Gallhofer 2014) to check whether they have acceptable reliability and validity. For repeat modules, this stage tends to include an assessment based on data collected in earlier wave(s).

Stage 4 (revised proposals from designers). As in the past, question designers are required to explain any changes being implemented and to document this in the template. The reasons for rejecting either expert review suggestions or data findings from Stage 3 are noted.

Stage 5 (consultation with ESS NCs). The questionnaire design template, including the latest version of the question items, is sent to NCs for further expert review, as under the seven-stage model. In addition to the usual academic input, coordinators try to foresee the barriers to achieving equivalence.

Stage 6 (omnibus testing and cognitive interviewing). This new stage sees early quantitative and qualitative testing of potential new questions. First, commercial omnibus surveys are used in a few countries to test items that most obviously require checks for reliability and validity, distribution, item nonresponse, social desirability, and relationship with other variables. The UK is always included to test the questions in the source language, with the others selected to take account of linguistic or conceptual challenges already identified when designing the module (as well as price and omnibus availability). Items are translated according to the usual ESS translation committee approach (Harkness et al. 2004) to maximize the results from the

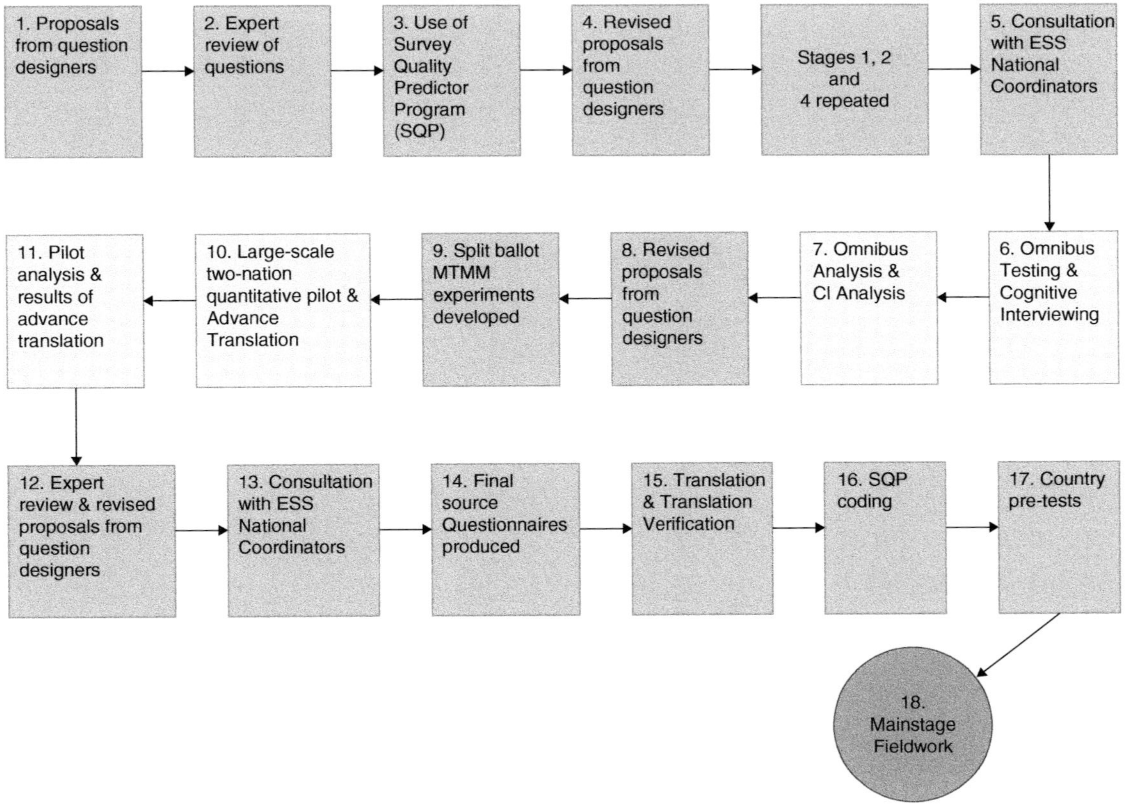

Figure 20.1 ESS 17-stage questionnaire design model. Source: (Fitzgerald 2015).

experiment. In addition, around 10 selected items are tested using cognitive interviewing in 3–4 countries using a methodology developed specifically for cross-national question testing (Miller et al. 2011). The items selected are those identified in the expert review phase as being most likely to pose challenges related to cognitive processing.

Stages 7–8 (omnibus analysis and cognitive interviewing analysis and revised proposals from designers). Detailed analysis of both the omnibus test and cognitive interview data is conducted. The analysis is interpreted using the Cross National Error Source Typology (CNEST) (Fitzgerald et al. 2011) developed specifically for the ESS to help identify errors and suggest remedial action. The sources include: poor source questionnaire design, translation problems, and cultural portability. The analysis and reporting are included in the template, which is then updated with revised proposals.

Stages 9–10 (split-ballot MTMM experiments developed; large-scale two-nation quantitative pilot and advance translation). Experiments are designed to help ascertain the quality of selected items included in the two-nation pilot (UK and one other country, around $n = 400$ per country) alongside the other items. A short respondent debrief is sometimes used, and interviewers are always debriefed. These additional sources of evidence help in assessing the questions and explaining the quantitative results. Alongside the pilot, a selection of two–three countries translate the questionnaire in *advance* (see Dorer 2015). Challenges encountered are noted and possible solutions listed.

Stages 11–12 (pilot analysis and results of advance translation; expert review and revised proposals from designers). The pilot analysis, interviewer debrief feedback, and advance translation findings are triangulated and positioned within the CNEST, which did not take place in earlier rounds. Experts discuss the findings and agree on a revised set of proposals.

Stage 13 (consultation with ESS NCs). ESS NCs provide further comments with a particular focus on translation.

Stage 14 (final source questionnaire produced). Following intense discussions, the source questionnaire is finalized. The questionnaire is "annotated" to help translators understand any British English peculiarities.

Stages 15–16 (translation and translation verification; SQP coding). A detailed committee approach to translation is undertaken (Harkness 2003). Queries arising during the translation phase are answered by translation experts, survey methodologists, and the substantive experts. Once national teams have completed their translations, they are forwarded to a specialist translation organization for verification (Dept et al. 2010). The verifiers were added as a step in more recent ESS rounds, and they point out areas that don't appear to be equivalent to the source. Alongside, national teams code their questionnaires using the SQP program, which helps to compare the formal characteristics of the target and source questionnaires.

Stage 17 (country pretests). Each country conducts a small pretest ($n = 30$) to test for basic understanding, routing, and questionnaire flow.

During Stages 15–17, issues may arise that require the source questionnaire to be amended. In those instances an "alert" is issued so all countries are made aware. The documentation from all stages is made publicly available along with a final version that accompanies the questionnaire implemented. Questionnaires in all target language versions are also made available along with showcards.

20.7 Evaluation of the Model for Cross-National Questionnaire Design and Pretesting Using the Logical Framework Matrix (LFM)

In this section we evaluate the extent to which the 17-stage model just described ensures the effective design of the source instrument. In order to evaluate the 17-stage process, we use the LFA (Cordingley 1995), a methodology developed by the United States Agency for International Development in the 1970s. This is widely used in the monitoring and evaluation of international development projects, which are similar to the ESS questionnaire design and pretesting model. Both have several interlinked implementation phases as well as an international dimension. Therefore, the LFA allows an analytical framework to assess the success of a project, not only developmental ones.

We applied the LFA methodology to the ESS questionnaire design and pretesting life cycle by populating an LFA matrix using the ESS procedures (Table 20.1). This included information about the components of the project in a clear, concise, logical, and systematic way by defining the "objectives," "measurable indicators," "means of verification," and "important assumptions."

The matrix is split into three major sections: Goals, Purpose, and Outputs.[1] Taking each row in the table in turn, we focus on the Means of Verification column in order to evaluate the 17-stage ESS cross-national questionnaire design and pretesting model using a critical example approach to show whether the model is efficient and complete.

20.7.1 Goals

The goals of a project, in this case designing the questionnaire in the ESS, are its overall aims. To what extent this process is successful can be measured in terms of whether the goals have been met.

1 The matrix can also include Activities, but these are excluded here as they are beyond the scope of this chapter.

Table 20.1 A model for cross-national questionnaire design and pre-testing (logical framework matrix).

	Objectives	Measurable indicators	Means of verification	Important assumptions
	Goals			
1	Minimizing *instrument* and *respondent* components of total survey error.	Individual questions with high measurement quality.	Assessment of the measurement quality of some questions with SQP and MTMM experiments.	Assessment methods are correct tools for assessing quality and equivalence.
2	Producing questions that result in equivalent measures.	Simple and complex concepts found to be measurement invariant across participating countries.	Assessment of the quality of some complex concepts.	Concepts being measured actually exist across countries.
	Purpose			
3	Producing an effective questionnaire allowing comparable measurement across countries.	Use of an array of pretesting methods to indicate question quality, validity, and cross-national equivalence (expert review, omnibus testing, cognitive interviewing, advanced translation, two-nation pilot).	Pretesting methods and findings used to improve questions. Methods developed in the process are accepted by the academic community, e.g. published in peer-reviewed journals. Some approaches used by the ESS for pretesting still need to be fully verified by the academic community e.g. advanced translation and SQP not yet peer reviewed.	Resources available for CST and external pretesting. NCs translate the questions correctly during pretesting. NCs to implement pretesting. National teams engage fully in the process, providing comments when requested.
4	Ibid.	Combination of pretesting methods to indicate question quality, validity and cross-national equivalence, e.g. omnibus survey, cognitive interviewing and SQP, and later the pilot with advanced translation and SQP. Comments from experts included at each stage.	Triangulation of findings actioned and results used to improve questions.	Triangulation performed rigorously.

5	An effective questionnaire allowing comparable measurement across countries.	Changes to draft questions made on an informed evidence based basis.	Template documents decisions for changing questions and provides reasons for making these linked to evidence from the pretests.	Questionnaire design teams and CST have adequate willingness to seek changes and improvements derived from the evidence.
	Outputs			
6	Theoretical, conceptual, and items specification finalized in template and updated when necessary.	Documentation of the entire questionnaire design process finalized and published for end users (each iteration and final template reflecting source questionnaire fielded).	Concepts are clearly defined and are developed following multiple iterations in each round. Concepts effectively measured by questions identified as indicators of them.	Concepts being measured actually exist across countries. Concepts can be measured using ASQ across countries.
7	Questions finalized for cognitive interviewing.	Cognitive interviewing topic guide finalized and issued to national teams.	Questionnaire design template documenting the process shows that questions were changed based on findings from cognitive interviewing.	Cognitive interviewing and related analysis conducted to high quality standards.
8	Questions finalized for omnibus survey.	Questionnaire finalized and sent to omnibus fieldwork agencies.	Questionnaire design template documenting the process shows that questions were changed based on findings from omnibus testing.	Omnibus and related analysis conducted to high quality standards.
9	Questions finalized for pilot study.	Questionnaire finalized and sent to pilot fieldwork agencies.	Questionnaire design template documenting the process shows that questions were changed based on pilot.	Pilot study and related analysis conducted to high quality standards.

Table 20.1 (Continued)

	Objectives	Measurable indicators	Means of verification	Important assumptions
10	Questions finalized for advanced translation.	Questionnaire finalized and sent to NCs.	Questionnaire design template documenting the process shows that questions were changed based on advanced translation. Source questionnaire designed to facilitate translation.	Translation conducted to a high standard and produces suggestions that are helpful for improving the source questionnaire and resulting target questionnaires in all languages.
11	Optimal source questionnaire released.	Optimal question formulation chosen based on evidence collected during the process, e.g. position of the items, response scales. QDT and CST decide which items are part of the module based on evidence from the design process.	Academic community accepts the module as state-of-the-art for that topic, e.g. peer-reviewed articles on health inequalities and trust in justice modules. Policy makers and government trust outputs from the ESS e.g. dissemination at the OECD.	

20.7.1.1 Row 1 – Minimizing Instrument and Respondent Components of Total Survey Error

The overall goal of questionnaire design and pretesting in the ESS is twofold: (i) to minimize *instrument* and *respondent* components of TSE while producing questions that result in equivalent measures across countries, and (ii) to ensure these measures meet the theoretical aims of the project.

In order to achieve these goals, it is a precondition that individual questions have high measurement quality (low measurement error) (Saris and Gallhofer 2014). One way ESS achieves this is to draw on experimental evidence to inform its questionnaire design, much of which is conducted as part of the ESS but which also works for other surveys. For example, Revilla et al. (2014) analyzed data from four ESS MTMM experiments implemented in R3 in 23 countries covering three topics. The authors showed that when "agree-disagree" scales are used, 5 answer categories yielded data with consistently higher quality than using 7 or 11 categories.

20.7.1.2 Row 2 – Producing Questions That Result in Equivalent Measures

High measurement quality is only one part of meeting the overall objective of questionnaire design and pretesting. On a repeat cross-sectional cross-national survey, one way of verifying that the questionnaire design and pretesting process is successful is if the questions fielded are equivalent cross-nationally. In each round, the ESS tests two complex concepts to assess cross-national equivalence.[2] Although the scalar invariance test – which allows for the comparison of means across groups – is a strict statistical test, it has in fact been established for most concepts tested so far. For instance, "Trust in political authorities" was scalar invariant in 4 rounds tested and 33 out of 33 language groups (Zavala-Rojas 2012), allowing mean scores to be compared. Longitudinal invariance has also been established in the ESS, e.g. "Interest in political issues presented in the media" was scalar invariant across 4 rounds tested in 29 out of 30 countries (Coromina et al. 2012). By testing their measurement models, questionnaire designers in the ESS get valuable information about which dimensions do not "travel well." Although this equivalence information is collected largely post hoc, it again helps to inform future design and replication decisions, providing a form of process verification.

Finally, the questionnaire design and pretesting process in the ESS allows for the verification of questions to see whether they measure the intended concepts. Billiet (2016) argues that measurement validity is a necessary but not sufficient requirement for conceptual validity, and also that conceptual validity cannot be assessed only by statistical analysis. A number of sources are put together in the question design template: the theoretical aims, the concepts

2 Scholars in a broader network have also published about measurement equivalence in the ESS (for example, see Davidov et al. 2008; Davidov and Meuleman 2012).

required to establish the theory, the items required to measure the concepts, pretesting evidence, and comments from expert review (Fitzgerald 2015). This facilitates inspection of the theoretical coverage of the questions in a bid to reduce specification error (Smith 2011). The questionnaire design and pretesting model in the ESS ensure that changes to draft questions are made on the basis of evidence from the pretesting and that questionnaire design decisions are documented. In this way, data users have information about the rationale behind the decisions for changing questions (e.g. taken from other surveys) linking reasoning to evidence.

One example shows how this specification and documentation tool helps to prevent a mismatch between the theoretical aims and the survey questions. Fitzgerald (2015) describes the design and documentation of the concept "Trust in police distributive fairness" being developed as part of the Trust in Police and Courts module (Jackson et al. 2011). The intended concept required respondents to evaluate whether police treat all members of society equally or not. At initial stages of the question design process, the items proposed were a list of three statements with a 5-point agree/disagree scale (asking whether people thought the police provided the same quality of services to all, consistently applied the law and made sure people got the outcomes they deserved) which had been used in a previous study on the same topic (Reisig et al. 2007).

After several iterations of expert review and qualitative and quantitative pretesting, evidence gathered suggested that the concept was not well reflected in the questions. The distributive angle, i.e. equal treatment of different groups, was not felt to be clear enough in the proposed items and the items were also found to have an overlap with other concepts. In addition the questions were felt to be vague or out of scope of the police mandate. An example of the latter is a statement that referred to "the police making sure people receive the outcomes they deserve under the law." However, this is largely a matter for prosecutors and the courts to achieve and not the police, so the item was not felt to be measuring the specified dimension. The final questions used focused specifically on how the police treated people when they were victims of crime, given the victims' income and ethnicity with the latter to focus attention on the distributive elements. "[…] When victims report crimes, do you think the police treat rich people worse, poor people worse, or are rich and poor treated equally?". "And when victims report crimes, do you think the police treat some people worse because of their race or ethnic group or is everyone treated equally?". The template enabled a check between the theory and proposed items and led to revised questions, in turn helping to prevent one source of measurement error.

20.7.2 Purpose

The second section of the LFA is the *purpose* of projects' goals. It breaks down the goals into their rationale reflecting the aims in more detail.

Row 3 and Row 4 – Producing an effective questionnaire allowing comparable measurement across countries

20.7.2.1 Row 3

The purpose of the questionnaire design and pretesting process is to produce an effective questionnaire allowing comparable measurement across countries. In order to achieve that, the ESS uses an array of pretesting methods including expert review, omnibus testing, cognitive interviewing, advance translation, two-nation pilot, SQP coding, and so on. Such methods have been developed to meet the needs of the ESS and have been accepted by the academic community. For instance, the ESS approach to advance translation (Dorer 2015), cognitive interviewing (Fitzgerald et al. 2011), and MTMM experiments (Saris et al. 2004) have all been published in refereed journals. Other areas have either appeared in specialist books, e.g. SQP software (Zavala-Rojas et al. 2018, Revilla et al. 2016; Saris and Gallhofer 2014), or have not been published yet but have been presented at conferences, e.g. translation verification used in the ESS context specifically (Dorer et al. 2013). The aim going forward is to ensure that every step has been published in a peer-reviewed publication.

20.7.2.2 Row 4: Having an Array of Pretesting Methods Has a Clear Advantage

It makes it possible to triangulate findings, strengthening the decisions made. For instance, the decision to drop an item intended to be asked in R6 was made after sufficient quantitative and qualitative evidence suggested there was a "source question problem" as classified through the CNEST (Fitzgerald et al. 2011). The question "How difficult or easy do you think it is for immigrants to get the right to vote in national elections in [country]? (0 = far too difficult; 10 = far too easy)" was asking about an issue that many respondents knew little about. The pilot study data showed that a substantial number of respondents gave a "don't know" or midpoint response in both the UK and Russia. Cognitive interviewing helped to explain those high figures: respondents in both countries interpreted this as a "knowledge" question, referring to their lack of knowledge about the legal situation for immigrants. Therefore, they could not judge the difficulty of getting the "right to vote" (Fitzgerald 2015).

Furthermore the SQP software is used to predict the measurement quality of survey questions: the strength of the relationship between the observed variable and the variable of interest (Saris and Gallhofer 2014, p. 188). SQP in the questionnaire design and pretesting process allows, in a cost effective way, assessment of how vulnerable questions are to measurement error, given specific changes in their properties. For instance, in ESS R7, a question measured the opinion of respondents on whether unemployment benefits should be cut "totally," "partially," or "not at all" given particular circumstances, such as rejecting a job offer when the position requires a much lower level of education.

The introduction, the response scales, and the description of the hypothetical scenarios required very long question stems. By comparing their predictions with those of "simpler" (but less precise) items, SQP provided evidence that the large length of the items caused by these specific features did not actually reduce measurement quality despite expectations to the contrary (European Social Survey 2016).

20.7.2.3 Row 5 – An Effective Questionnaire Allowing Comparable Measurement Across Countries

The blank ESS questionnaire design and pretesting template for Round 8 is reproduced in Appendix 20.A. It contains four sections: theoretical background; an outline of the concepts to be measured and the relationships between them; a section outlining the complex concepts to be measured – built up by several sub-concepts and requiring multiple items; and a section for simple concepts, which need a single item. Each complex concept, sub-concept, and simple concept is defined, and this definition may evolve as pretesting evidence is obtained. Finally, the template records the questions that are proposed to measure each concept and sub-concept, which often change during development.

The design of the template has three clear strengths: i) it facilitates the tracking and operationalization of the concepts into survey questions throughout the process in a structured way; ii) it requires the specification of predicted relationships between the variables encouraging testing at key stages, and iii) it records decisions made, reducing the chances of unnecessary circular discussions (see Fitzgerald 2015 for examples of how the template is used in the ESS).

20.7.3 Outputs

Finally, in the section *outputs,* the LFA assesses to what extent the deliverables of the project produce an effective English source questionnaire that minimizes measurement error and enhances data comparability across countries.

20.7.3.1 Row 6 – Theoretical, Conceptual, and Items Specification Finalized in Template and Updated When Necessary

It has been crucial for the success of the ESS that during the design and pretesting process, the source questionnaire is verified, in the sense that the concepts are effectively measured by the questions included for that purpose. By extension it is hoped that this can be replicated in the other country in the pilot via a translated instrument, thus providing an indication that this can potentially also be achieved in the other countries. The theoretical specification of the items is updated in the questionnaire design and pretesting template when necessary until the questions are finalized (Fitzgerald 2015). This facilitates a clear definition of concepts and items that is improved at each iteration as evidence is gathered. This documentation of the entire questionnaire design process is published for end users.

For instance, in ESS R8, a question asked about respondents' stance on the implementation of a "basic income scheme." Such a scheme provides an income for all citizens, regardless of their employment status or financial resources (ESS 2016). The first version of the question was "Some propose that government should provide all adult citizens, regardless of whether they work or not or whether they are poor or rich, with a basic income that is sufficient to cover their basic needs. Would you personally be against or in favor of this proposal?" It was suggested that this question might have cultural transportability problems because at the time of the ESS pilot, some countries were discussing it in the media (e.g. Germany), whereas in other countries the idea was rather new (e.g. the UK). Expert review from ESS NCs and the CST suggested that it would be necessary to provide a single definition to try to give an identical stimulus in all countries; otherwise people would be answering slightly different questions. Providing a detailed example allowed the "fixing" of the stimuli cross-nationally, minimizing the possibilities that a different scheme was being considered. The final question was "Some countries are currently talking about introducing a basic income scheme. [...] A basic income scheme includes all of the following:

- The government pays everyone a monthly income to cover essential living costs.
- It replaces many other social benefits.
- The purpose is to guarantee everyone a minimum standard of living.
- Everyone receives the same amount regardless of whether or not they are working.
- People also keep the money they earn from work or other sources.
- This scheme is paid for by taxes.

Overall, would you be against or in favor of having this scheme in [country]?"

20.7.3.2 Row 7 – Questions Finalized for Cognitive Interviewing

Introducing cognitive interviewing in the ESS questionnaire design and pretesting process has allowed in-depth checks on whether questions can be answered effectively in the source language country as well as whether the question can "travel" effectively. When such tests were not available in the past it was more likely that questions with significant meaning differences across countries were fielded in the final main survey. An example underlining the importance of including this stage came during preparations for R4, when the following question was tested:

"Using this card please tell me which of the three statements on this card, about how much working people pay in tax, you agree with most:

1) Higher earners should pay a greater proportion in tax than lower earners

2) Everyone should pay the same proportion of their earnings in tax
3) High and low earners should pay exactly the same amount in tax"

The question aimed to identify respondents' preference among three different tax-collection systems. This question was found to be problematic in all countries where it was tested cognitively (Bulgaria, Portugal, Spain, Switzerland, and the UK), with the source of error determined to be a cultural portability issue because the salience of the concept was directly related to different cultural contexts, as well as a source questionnaire problem as the question was simply challenging (Fitzgerald et al. 2011). Many respondents felt the question suggested that some level of tax-system knowledge was required. While this applied in all countries, it was regarded as a particular problem in Spain and Switzerland, where respondents felt less knowledgeable about tax. For example, respondents in Spain reported that option 2 reflected the tax system in their country (when in fact option 1 did), and respondents in Switzerland expressed low levels of confidence in answering. The Swiss research team pointed out that in Switzerland it is the sole responsibility of the head of the household to complete tax returns, therefore it is likely that other members of the household who are not involved in this will only have minimal knowledge of the tax system. This assumption was also reflected in the responses given by some Swiss participants. Cognitive interviewing proved to be particularly well suited to identifying the source of error (See Fitzgerald et al. 2011 for a discussion).

20.7.3.3 Row 8 – Omnibus (Questions Finalized for Omnibus Survey)

An *omnibus survey* is a method for large-scale collective data collection commonly used in market research. In omnibus surveys several (unrelated) topics are asked to respondents in the same interview with sections of the questionnaire purchased by different clients. Although, in general, no inferential conclusions can be made as these surveys typically use quota sampling, omnibuses are a cost-effective strategy to quantitatively test early question versions. In the ESS questionnaire design process, omnibus surveys help to make informed decisions on question formulation, and explore potential response effects that would harm the quality of the measures. For example, in R6, omnibus surveys in Great Britain, Hungary, and Portugal allowed the ESS to test if pair-wise or list-wise administration produced a difference in response patterns for some items to be included in the democracy module. Findings showed that list-wise administration provided greater differentiation in the answers than pair-wise, and this approach was adopted (see Winstone et al. 2016).

20.7.3.4 Row 9 – Questions Finalized for Pilot Study

A two-nation, large-scale pilot study is conducted in the ESS as the last pretesting method during questionnaire design ($n \geq 400$ cases per country). The pilot study is administered in the UK and in a second country chosen based on linguistic and cultural considerations, while also considering budget. Alongside new items, the questionnaire includes several other measures that are part of the core questionnaire and that are expected to be statistically related to the new items. The large sample size facilitates advanced statistical analysis but is limited to two countries.

The pilot study has served several objectives. For instance, in the ESS R7 "attitudes toward immigration" module, it helped to inform decisions about which items should be dropped given space constraints (ESS 2014). The large number of cases in the pilot study allows testing of whether the measurement models behind the questions are functioning as expected despite large institutional and cultural differences in the two countries participating in the pilot. For example, Jackson et al. (2011) established full measurement equivalence for the measures of "trust in police effectiveness" between the UK and Bulgaria in the R5 pilot. Finally, the pilot data has aided decisions regarding whether a formulation is likely to introduce large method effects that will potentially confound measurement artifacts with substantive findings. For instance, in ESS R7, a split-ballot experiment tested if item nonresponse patterns in a highly sensitive topic were affected by the response format. There were concerns that *Don't know/No answer* options would be more likely to be endorsed if items were asked using a direct dichotomous question: "Would you say that some cultures are much better than others or that all cultures are equal?" (answer options: "Some cultures are much better than others" or "All cultures are equal"). A second option was to ask, "To what extent do you agree or disagree that all cultures are equal?" (answer options: strongly disagree, disagree, neither disagree nor agree, agree, strongly agree), in which respondents could calibrate their answer. It turned out that the proportion of DK/NA was not significantly different between the two options, but that the latter yielded a large proportion of cases in the "neither disagree nor agree" category. As this neutral category represented a loss of information for substantive researchers, the dichotomous format was preferred (ESS 2014).

20.7.3.5 Row 10 – Questions Finalized for Advanced Translation

In each ESS round, the source questionnaire is translated into more than 20 languages. In the ESS, *advance translation* sees the pre-final version of the source questionnaire translated to detect comprehension problems that impair translation quality. As a result, the source questionnaire can be modified or annotations added to clarify terms (Harkness and Schoua-Glusberg 1998; Dorer 2015) since word-for-word translations mirroring the British version are not required. Dorer (2015) analyzed to what extent translators encountered comprehension problems by comparing a sample of translation versions conducted before and after advance translation. A first group of translators were given the pre-advance translation version: "To what extent do you feel that you have a sense of direction?" Comments highlighted that guidance was needed for *sense of direction*. When a second group translated the post-advance translation version, in which an annotation has been provided for *sense of direction*, they did not find translation problems. Amending the source questionnaire in these ways helps to reduce uncertainty for translators, increasing the likelihood of developing equivalent questions.

20.7.3.6 Row 11 – Optimal Source Questionnaire Released

At the end stage of the design and pretesting process, the optimal source English questionnaire is released. At this stage the QDT and CST have decided which items are part of the module and their optimal formulation. Both types of choices are based on careful consideration of the evidence.

Acceptance of the survey questions (and the data they produce) by the academic community through peer-review publication is an indicator that the questionnaire design and pretesting process has been successful. Peer-reviewed articles about the design of ESS module have been published, e.g. Eikemo et al. (2016) on "the social determinants of health inequalities" and Jackson et al. (2011) on the "trust in the police and criminal courts." As of August 2017 there were also over 100 000 registered users of the ESS and over 3000 publications. Furthermore, results have been presented to the Organisation for Economic Co-operation and Development (OECD), European and Italian Parliaments, UK Cabinet Office, and at other policy-relevant events, again indicating quality acceptance from peers and other users.

20.8 Conclusions

The ESS model for cross-national questionnaire design and pretesting provides a structured, multilayered, and culturally sensitive framework for questionnaire design on large-scale, cross-national projects. Its core elements of specification, documentation, multidisciplinary input, qualitative and quantitative testing, advance translation, and detailed data analysis provides a robust and rigorous environment for developing a source questionnaire that can then be translated. The committee approach to translation pioneered by Harkness (2003) is another key strength. The model builds on experimental work and knowledge accumulation too, which helps to inform questionnaire design in future.

Taking the Purpose from the LFM, we have demonstrated that the ESS is generally producing equivalent data that has low overall error stemming from the instrument and respondent components of TSE. This suggests that the model is meeting the objectives in terms of basic survey quality requirements and the additional cross-national dimension. The examination of the Purpose and Output rows together demonstrates the usefulness of specific components of the model and provides examples of their cumulative effect through triangulation.

Challenges remain, such as the long time period for development (currently 24 months), the rather limited pretesting of the final questionnaire in countries once the source questionnaire has been finalized (Stage 17), and the relatively large resources required. Cases where sources of evidence conflict on the same item or concept are also a challenge, as is the use of statistical analysis that assumes probability sampling but is used on non-probability samples.

It is difficult to compare this model directly with that implemented on other cross-national surveys as there is little published information available in a comparable format (Fitzgerald 2015). There are elements of the model that could be adopted by other similar surveys at relatively little cost, such as the use of the template to ensure conceptually structured development, the use of SQP for quality prediction, and advance translation as a means of decentering. What is clear is that it is necessary to think about the questionnaire design, pretesting, and translation procedures as interlinked processes, part of a whole as well as separate phases. Most important is to ensure that the process remains driven by the underlying measurement aims. In an age with an increasing variety of data sources available, what makes survey measurement stand out is the ability to design a bespoke measurement instrument that can answer theoretically relevant questions. However, this only works if the questions asked are high quality and measure the intended concepts. A rigorous approach to questionnaire design is therefore not an optional extra but essential.

Finally, the LFM approach has proved to provide a useful mechanism for assessing a cycle and combination of tasks that together have an overall purpose. There is a vast array of literature that assesses individual parts of the questionnaire design process and the results from specific pretests and experiments. What is missing are assessments of the complete process. It is unlikely there will ever be split tests to compare different overall approaches, due to cost and logistical constraints. The LFM therefore provides one way to assess the totality of the process, and using this for other parts of the survey life cycle could also be of interest in the future.

20.8.1 Future Developments

While the current model of cross-national questionnaire design is fit for purpose, there are a number of related developments that might augment or alter the model in future being conducted in a project with other European research infrastructures (www.seriss.eu). The ESS is experimenting with establishing a cross-national input harmonized web panel populated by respondents recruited at the end of the main face-to-face interview. This could provide a cheaper alternative to using commercial omnibus surveys for testing. Other developments include producing an interactive database to record the design rather than the paper-based templates, and an examination of the fitness for purpose of technological developments in translation and language sciences, e.g. machine translation.

References

Biemer, P.P. (2010). Total survey error: design, implementation, and evaluation. *Public Opinion Quarterly* 74 (5): 817–848. https://doi.org/10.1093/poq/nfq058.

Billiet, J. (2016). What does measurement mean in a survey context? In: *The SAGE Handbook of Survey Methodology* (eds. C. Wolf, D. Joye and T.W. Smith), 193–209. London: SAGE Publications Ltd https://doi.org/10.4135/9781473957893.n14.

Bradburn, N.M. and Sudman, S. (2004). The current status of questionnaire research. In: *Measurement Errors in Surveys* (eds. P.P. Biemer, R.M. Groves, L.E. Lyberg, et al.), 27–40. Hoboken, NJ: Wiley https://doi.org/10.1002/9781118150382.ch2.

Campbell, D.T. and Fiske, D.W. (1959). Convergent and discriminant validation by the multitrait-multimethod matrix. *Psychological Bulletin* 56 (2): 81.

Coleman, G. (1987). Logical framework approach to the monitoring and evaluation of agricultural and rural development projects. *Project Appraisal* 2 (4): 251–259.

Cordingley, D. (1995). Incorporating the logical framework into the management of technical co-operation projects. *Project Appraisal* 10 (2): 103–112.

Coromina, L., Saris, W. E., and Lilleoja, L. (2012). Measurement of concepts based on the media module of the ESS RECSM Working Paper Number 25. https://www.upf.edu/documents/3966940/3986764/RECSM_wp025.pdf.

Davidov, E. and Meuleman, B. (2012). Explaining attitudes towards immigration policies in European countries: the role of human values. *Journal of Ethnic and Migration Studies* https://doi.org/10.1080/1369183X.2012.667985.

Davidov, E., Schmidt, P., and Schwartz, S. H. (2008). Bringing Values Back In: The Adequacy of the European Social Survey to Measure Values in 20 Countries. *Public Opinion Quarterly* 72 (3), 420–445. https://doi.org/10.1093/poq/nfn035.

Dept, S., Ferrari, A., and Wäyrynen, L. (2010). Developments in translation verification procedures in three multilingual assessments: a plea for an integrated translation and adaptation monitoring tool. In: *Survey Methods in Multinational, Multiregional, and Multicultural Contexts* (eds. J.A. Harkness et al.), 157–173. Wiley https://doi.org/10.1002/9780470609927.ch9.

Dorer, B. (2015). Carrying out “advance translations” to detect comprehensibility problems in a source questionnaire of a cross-national survey. In: *In Translation and Comprehensibility* (eds. S.G. Maksymski and S. Hansen-Schirra), 77–112. Berlin: Frank & Timme.

Dorer, B., Widdop, S., and Fitzgerald, R. (2013). Translation verification in the ESS: a means for achieving equivalent translations in a cross-national survey? 5th Conference of the European Survey Research Association (ESRA), Ljubljana (Slovenia).

Eikemo, T.A., Bambra, C., Huijts, T., and Fitzgerald, R. (2016). The first pan-European sociological health inequalities survey of the general population:

the European social survey rotating module on the social determinants of health. *European Sociological Review* https://doi.org/10.1093/esr/jcw019.

European Social Survey (ESS). (2014). Report on R7: The measurement quality of pilot questions D4.11. London: City University. Available upon request: ess@city.ac.uk.

European Social Survey (ESS). (2016). SQP coding in the OMNIBUS survey and pilot study in preparation of the European Social Survey Round 8. London: City University. Available on request ess@city.ac.uk.

Fitzgerald, R. (2015). Sailing in unchartered waters: structuring and documenting cross-national questionnaire design. *GESIS Papers* no. 5. Mannheim. http://nbn-resolving.de/urn:nbn:de:0168-ssoar-462191.

Fitzgerald, R., Widdop, S., Gray, M., and Collins, D. (2011). Identifying sources of error in cross-national questionnaires: application of an error source typology to cognitive interview data. *Journal of Official Statistics* 27 (4): 569–599.

Harkness, J.A. (2003). Questionnaire translation. In: *Cross-cultural survey methods* (eds. J.A. Harkness, F.J.R. van de Vijver and P.P. Mohler), 35–56. Hoboken: Wiley.

Harkness, J.A. and Schoua-Glusberg, A. (1998). Questionnaires in translation. *ZUMA-Nachrichten Spezial* 3 (1): 87–127.

Harkness, J.A., Pennell, B.-E., and Schoua-Glusberg, A. (2004). Survey questionnaire translation and assessment. In: *Methods for Testing and Evaluating Survey Questionnaires*, 453–473. https://doi.org/10.1002/0471654728.ch22.

Harkness, J.A., Edwards, B., Hansen, S.E. et al. (2010a). Designing questionnaires for multipopulation research. In: *Survey Methods in Multinational, Multiregional, and Multicultural Contexts*, 31–57. Wiley https://doi.org/10.1002/9780470609927.ch3.

Harkness, J.A., Villar, A., and Edwards, B. (2010b). Translation, adaptation, and design. In: *Survey Methods in Multinational, Multiregional, and Multicultural Contexts*, 115–140. Wiley https://doi.org/10.1002/9780470609927.ch7.

Hox, J.J. (1997). From theoretical concept to survey question. In: *Survey Measurement and Process Quality*, 47–69. Hoboken, NJ: Wiley https://doi.org/10.1002/9781118490013.ch2.

Jackson, J., Bradford, B., Hough, M. et al. (2011). Developing European indicators of trust in justice. *European Journal of Criminology* 8 (4): 267–285. https://doi.org/10.1177/1477370811411458.

Jowell, R., Roberts, C., Fitzgerald, R., and Eva, G. (2007). *Measuring attitudes cross-nationally: Lessons from the European Social Survey*. London: Sage.

Miller, K., Fitzgerald, R., Padilla, J.-L. et al. (2011). Design and analysis of cognitive interviews for comparative multinational testing. *Field Methods* 23 (4): 379–396.

Reisig, M.D., Bratton, J., and Gertz, M.G. (2007). The construct validity and refinement of process-based policing measures. *Criminal Justice and Behavior* 34 (8): 1005–1028. https://doi.org/10.1177/0093854807301275.

Revilla, M.A., Saris, W.E., and Krosnick, J.A. (2014). Choosing the number of categories in agree–disagree scales. *Sociological Methods & Research* 49 (1): 73–97.

Revilla, M.A., Zavala-Rojas, D., and Saris, W.E. (2016). Creating a good question: how to use cumulative experience? In: *The SAGE Handbook of Survey Methodology* (eds. C. Wolf, D. Joye, T.W. Smith and Y. Fu), 236–254. London: SAGE Publications Ltd.

Saris, W.E. and Andrews, F.M. (1991). Evaluation of measurement instruments using a structural modeling approach. In: *Measurement Errors in Surveys* (eds. P.P. Biemer, R.M. Groves, N.A. Lyberg, et al.), 575–597. New York: Wiley.

Saris, W.E. and Gallhofer, I. (2007). Can questions travel successfully. In: *Measuring Attitudes Cross-Nationally: Lessons from the European Social Survey* (eds. R. Jowell, C. Roberts, R. Fitzgerald and E. Gillian), 1–31. London: Sage.

Saris, W.E. and Gallhofer, I. (2014). *Design, Evaluation, and Analysis of Questionnaires for Survey Research*, 2e. Wiley.

Saris, W.E., Satorra, A., and Coenders, G. (2004). A new approach to evaluating the quality of measurement instruments: the split ballot MTMM design. *Sociological Methodology* 34 (1): 311–347. https://doi.org/10.1111/j.0081-1750.2004.00155.x.

Smith, T.W. (2011). Refining the total survey error perspective. *International Journal of Public Opinion Research* 23 (4): 464–484.

Smyth, J.D. (2016). Designing questions and questionnaires. In: *The SAGE Handbook of Survey Methodology* (eds. C. Wolf, D. Joye and T.W. Smith), 218–235. London: SAGE Publications Ltd https://doi.org/10.4135/9781473957893.

Winstone, L., Widdop, S., and Fitzgerald, R. (2016). Constructing the questionnaire: the challenges of measuring attitudes toward democracy across Europe. In: *How Europeans View and Evaluate Democracy*. Oxford: Oxford University Press https://doi.org/10.1093/acprof:oso/9780198766902.003.0002.

Zavala-Rojas, D. (2012). Evaluation of the concepts "Trust in institutions" and "Trust in authorities. RECSM Working Paper 29. https://www.upf.edu/documents/3966940/3986764/RECSM_wp029.pdf.

Zavala-Rojas, D., Saris, W.E., and Gallhofer, I.N. (2018). Preventing differences in translated survey items using the Survey Quality Predictor. In: *Advances in Comparative Survey Methods: Multinational, Multiregional and Multicultural Contexts (3MC)* (eds. T.P. Johnson et al.), 357–384. https://doi.org/10.1002/9781118884997.ch17.

20.A ESS Question Module Design Template

Europen
Social
Survey

ESS Round 7

Question Module Design Template

Module Title:

Module Authors:

SECTION A: Theoretical background
Describe the theoretical background of the module, its aims and objectives

SECTION B. Briefly describe all the concepts to be measured in the module and their expected relationships, either verbally or diagrammatically. Give each concept a 5-8 digit working name. Identify each concept as simple (S) or complex (C). Specific details about the concepts and sub concepts should be specified in SECTION C.

SECTION C: Complex Concepts. For each complex concept listed in Section B, describe it in detail and specify the sub concepts as appropriate. Add more boxes to the template as required to describe all the complex concepts and relevant sub-concepts included in the module. **Once the conceptual structure is agreed with the CST add the question wording for the proposed item.**

COMPLEX CONCEPT NAME:

Describe the concept in detail, outlining the various sub concepts it comprises
Expected relationship with other complex and simple concepts

SUB CONCEPT NAME:

Describe the first sub concept in detail outlining any further sub concepts[1] or specifying that it can be measured directly

[1] If further sub concepts are foreseen these should be numbered i.e. 1ai, 1aii etc

Expected relationship with other sub concepts
Question item wording

SECTION D: Simple Concepts. For each simple concept listed in Section B, describe it in detail here. Add more boxes to the template as required to describe all of the simple concepts in the module. **Once the conceptual structure is agreed with the CST add the question wording for the proposed item.**

SIMPLE CONCEPT NAME:

Describe the concept in detail
Question item wording

SIMPLE CONCEPT NAME:

Describe the concept in detail
Question item wording

21

Cross-National Web Probing: An Overview of Its Methodology and Its Use in Cross-National Studies

Dorothée Behr[1], Katharina Meitinger[2], Michael Braun[1], and Lars Kaczmirek[3,4,5]

[1] *Survey Design and Methodology, GESIS – Leibniz Institute for the Social Sciences, Mannheim, Germany*
[2] *Department of Methodology and Statistics, Utrecht University, Utrecht, The Netherlands*
[3] *Library and Archive Services/AUSSDA, University of Vienna, Vienna, Austria*
[4] *ANU Centre for Social Research and Methods, College of Arts and Social Sciences, Australian National University, Australia*
[5] *Monitoring Society and Social Change, GESIS – Leibniz Institute for the Social Sciences, Mannheim, Germany*

21.1 Introduction

Cross-*national* surveys are increasingly being set up or used in analyses, and so are cross-*cultural* surveys in general, including those conducted in different ethnic groups within a single country (Smith 2010; Van de Vijver 2013). It is therefore more pressing than ever to have a sound methodology at one's disposal that allows a researcher to produce equivalent data that can be meaningfully compared across countries and cultures. After all, equivalence is the prerequisite for any sound conclusions. In the words of Johnson (1998, p. 30): "In addition to the traditional reliability and validity requirements for monocultural survey instruments, researchers conducting cross-cultural survey research have the added concern of equivalence. Indeed, cross-cultural research demands a commitment to the establishment of equivalence that is at least equal to the attention routinely reserved for the problems of reliability and validity." From a quantitative perspective and for multiple-item scales, equivalence is often discussed within a three-level framework that distinguishes between configural, metric, and scalar invariance (Meredith 1993; Steenkamp and Baumgartner 1998; Vandenberg and Lance 2000). The lowest level that is required for cross-cultural comparisons is *configural*

Advances in Questionnaire Design, Development, Evaluation and Testing, First Edition.
Edited by Paul C. Beatty, Debbie Collins, Lyn Kaye, Jose-Luis Padilla, Gordon B. Willis, and Amanda Wilmot.

invariance where the same constructs are measured in each cultural group. The next-higher level is *metric invariance*; here, the scales have the same unit of measurement even though their origins differ. The highest level is known as *scalar invariance*. With this type of invariance, both measurement unit and origin of scale are the same across cultural groups. Only when scalar invariance is confirmed can researchers study the latent means of the tested constructs. Different types of *bias*, that is, nuisance factors, can reduce equivalence or compromise it altogether. Construct bias, for instance, results from different culture-specific behaviors associated with a construct or only partial overlap of construct definitions across cultural groups. Method bias originates from the sample, from the administration or from the instrument, with bias from the instruments resulting from differential familiarity with survey material or from differential response styles (e.g. acquiescence or extreme response style) across cultural groups. Item bias can be caused by poor translations, ambiguous items or any cultural-specifics being brought to the response process by the respondents (Van de Vijver and Leung 2011). Several qualitative procedures help to prevent or reduce bias during the questionnaire development process – these procedures apply both to multiple-item scales and to single items, but for the latter they are even more crucial since invariance testing procedures such as confirmatory factor analysis cannot be applied to single items. In order to avoid construct bias, intercultural questionnaire development teams work together to operationalize constructs and draft items in such a way that, ideally, equivalence across countries can be achieved. As a next step in the development process, cross-national cognitive interviewing can be implemented. This relatively recent but already well-established procedure comprises the implementation of face-to-face cognitive interviews in several countries with the aim to identify generic, cultural, or linguistic problems early on in the development process (e.g. Miller et al. 2011; Lee 2014; Thrasher et al. 2011). It notably contributes to identifying construct bias or item bias. Even more recent is cross-national web probing, which is the implementation of probing techniques from cognitive interviewing in web surveys in several countries. Also with this procedure, construct or item bias can be identified. Cross-national web probing has been conceived by Braun et al. (2014) in order to collect qualitative data from a large number of respondents from different countries in a standardized and efficient way – data that can then be used for equivalence testing during questionnaire development but also after the fact to assess equivalence and support subsequent data analysis.

It needs to be acknowledged that at about the same time that Braun et al. (2014) developed cross-national web probing, Murphy et al. (2013) pursued web probing activities, too, even though without cross-national or cross-cultural use in mind. Furthermore, Mockovak and Kaplan (2015) explored potential online applications of cognitive interviewing. In their approach, they asked respondents to think aloud and respond verbally to scripted probes in a mono-cultural web survey. New cognitive pretesting methods or online extensions of cognitive interviewing have therefore been conceived by several research groups at about the same time (see also Edgar 2013).

This chapter focuses on cross-national web probing. We will present (i) the strengths and weaknesses of the method and additionally draw a brief comparison between web probing and cross-national cognitive interviewing, (ii) possibilities of access to respondents in different countries, (iii) the specifics of the web probing implementation, (iv) particularities of translating and coding cross-national probe answers, (v) a selection of substantive results, and finally (vi) an overview of different application scenarios throughout the survey life cycle.

In the following, the term *cross-national* web probing will be used due to the cross-national focus of the research projects that we are mainly summarizing in this chapter. The research projects (2010–2015) were funded by the German Research Foundation (DFG) and aimed at developing and optimizing web probing procedures for cross-national studies. However, the web probing method can be applied in a variety of contexts, including different ethnic groups within a single country or just one ethnic group or country.

21.2 Cross-National Web Probing – Its Goal, Strengths, and Weaknesses

Web probing in general, as we understand it, is "the implementation of probing techniques from cognitive interviewing in web surveys with the goal to assess the validity of survey questions" (Behr et al. 2017, p. 1). The probing techniques referred to in this definition are "additional, direct questions about the basis for responses" given to closed-ended items (Beatty and Willis 2007, p. 289). With the adoption of probing techniques, web probing shares a core feature with cognitive interviewing, at least when considering the probing paradigm of cognitive interviewing. Thus, it comes as no surprise that a set of studies have been set up to compare cognitive interviewing and web probing, albeit, to date, with an exclusive focus on monocultural, monolingual contexts. Meitinger and Behr (2016) found a large overlap between results in a pretesting study in Germany. While interactivity was found to be a great strength of cognitive interviewing, in particular spontaneous respondent comments, web probing helped to prevent local bias in themes by surveying a (geographically) more dispersed set of respondents. Murphy et al. (2013) compared, for a US study, cognitive interviewing with what they called "crowdsourcing in the cognitive interviewing process."[1] Comparing the traditional cognitive interview to web surveys that recruited respondents from three crowdsourcing platforms – TryMyUI, Amazon Mechanical Turk, and Facebook – they concluded that cognitive interviewing had its particular strength in enabling spontaneous probing, the fact of which would make it particularly useful for in-depth exploration of (new) items or constructs. The crowdsourcing platforms excelled in terms of speed, geographic dispersion, and partly also motivation of respondents. Thus, Murphy et al. (2013) came to similar conclusions as Meitinger and Behr (2016). While online recruitment and web probing itself certainly take less time, thorough development of a coding scheme and coding of potentially hundreds of answers are time-consuming.

When transferred to the cross-national context, web probing means that probing questions are implemented in cross-national web surveys with the goal to assess comparability of items. It is here where cross-national web probing resembles cross-national cognitive interviewing, but both methods have their own characteristics as well as strengths and weaknesses. Table 21.1 summarizes these, building on a general comparison of web probing versus cognitive interviewing (Behr et al. 2017; Meitinger and Behr 2016) and expanding it to include cross-national particularities.

On the positive side, cross-national web probing is characterized by potentially large sample sizes that allow assessment and comparison of answer patterns or errors across countries. The samples can often be recruited in such a way that respondents are geographically widely dispersed and cover a diverse set of sociodemographic groups. Web probing does not require cognitive interviewers, and hence no recruitment and training of interviewers is needed. Furthermore, the self-administered mode helps to ensure comparability through standardization. On the negative side, the recourse to the online mode means that certain population groups are excluded from this data-collection procedure right from the start. Without an interviewer present, a certain number of respondents may not be motivated enough to answer open-ended probes. The absence of an interviewer also means that, without further interactive follow-ups, spontaneously emerging issues in response behavior or probe answers cannot be dealt with. Moreover, only a limited number of items can be probed; otherwise researchers run the risk of increased nonresponse or even survey break-offs.

In the remainder of the chapter, we will focus on cross-national web probing and refrain from providing further comparisons between cognitive interviewing and web probing. For cross-national studies, an empirical comparison and further delineation between web probing and cognitive interviewing methods still needs to be done.

---

1 While the term *web probing* focuses on the technique of asking questions, crowdsourcing focuses on the means of recruiting respondents. These may be surveyed using web probing but also using other techniques such as recorded think-aloud.

Table 21.1 Strengths and weaknesses of cross-national cognitive interviewing and cross-national web probing.

Comparison	±	Cross-national web probing	Cross-national cognitive interviewing
Sample size	+	Large sample sizes and good assessment and comparison of answer patterns or errors possible	
	–		Typically small sample sizes, even though larger in the cross-national context than in the national context (Willis 2015)
Coverage of target groups	+		Special target groups, including illiterate, old, poor, ill, etc. persons can be reached
	–	Only online population can be reached	
Geographical and sociodemographic coverage	+	Broader coverage in a country as long as people have internet access	
	–		Typically limited to certain geographical areas and socio-demographic groups in a country
Cognitive interviewers	+	No interviewers are needed, and hence no recruitment or training is necessary	Interviewer can motivate respondents to provide an answer
	–	Interviewer cannot motivate respondents to provide an answer	Careful recruitment and training of interviewers necessary so that cognitive interviewing is done in a similar fashion across countries (Lee 2014; Willis 2015)
Probing	+	Standardized probes → comparability	Flexible, spontaneous probes possible, reacting toward unforeseen, even country-specific issues (Willis 2015)
	–	Standardized probes → potentially insufficient information	If flexible and spontaneous approach prevails → potential lack of comparability, not only within a country, but also across countries
Number of probes	+		Due to a motivating interviewer, more items can be probed
	–	Due to the lack of a motivating interviewer, fewer items can be probed	

Source: Adapted from Behr et al. (2017, p. 3).

21.3 Access to Respondents Across Countries: The Example of Online Access Panels and Probability-Based Panels

The great strength of web probing consists of easy and cost-efficient access to a large sample size. Online access panels are one source to turn to in the search for respondents. Online access panels provide a pool of respondents that have voluntarily signed up to take web surveys at more or less regular intervals. For a fee, researchers can invite these respondents to participate in their surveys. Online access panels are available in many countries, even though not everywhere and certainly with varying levels of quality and varying degrees of penetration in a society. With online access panels, representativity of the general population cannot be achieved since respondent selection typically follows nonrandom recruitment procedures and mostly does not include the offline population. Nevertheless, respondents for one's study can be selected according to quotas (education, age, region, etc.) and thus a balanced sample can be achieved or a specific group targeted. Especially in the cross-national context, quotas for the various country surveys help to ensure that the samples are set up in equivalent ways. Otherwise method bias through dissimilar quotas may impact and reduce the comparability of results obtained.

Given the nonrandom nature of online access panels and the resulting noncoverage of certain population groups (e.g. the elderly, migrants not speaking the language of the panel surveys), these panels are not a panacea. Furthermore, being in the panel does not automatically mean that every panelist is equally able or willing to answer open-ended probing questions. The education level becomes important in this regard, with more-educated respondents being in general more prone to answering open-ended questions and providing longer answers than lower-educated respondents (Oudejans and Christian 2010; Zuell et al. 2015). However, online panels are certainly a useful way for gaining access to respondents in various countries and for increasing sample size and geographical and sociodemographic scope in pretesting or follow-up studies.

Online access panels can especially be useful for random experiments and for general equivalence checks. We have so far used the panels as a kind of add-on study to representative population studies. However, before drawing inferences to these representative surveys, such as the International Social Survey Programme (ISSP), we compared quantitative results, mainly distributions of suspicious closed-ended items, from the ISSP with the online panel data. Only when similar patterns across countries emerged for the item(s) under investigation did we use the online panel data to clarify issues related to peculiarities we found in the ISSP data. Thus, researchers can build in some consistency checks before using panel data when attempting to retrospectively explain anomalies in regular survey data.[2]

In some European countries (the Netherlands, Germany, France), representative online panels have been set up that can be used by the research community (Blom et al. 2016); similarly for the United States (Callegaro et al. 2014) and South Korea (Cho et al. 2017). The cross-national web research endeavor is fostered by projects such as the Open Probability-Based Panel Alliance (http://openpanelalliance.org), uniting the mentioned panels in Germany, the Netherlands, South Korea, and the United States in one initiative. Any fielding of items in these probability-based panels requires a research proposal and a subsequent review process, which is why any quick access and turnaround is often impossible.

21.4 Implementation of Standardized Probes

Web surveys need to be carefully designed in order to maintain respondents' motivation and reduce detrimental response behavior such as nonresponse. In web surveys, wording, visual features, and overall design all contribute to the respondents' survey experience and are thus decisive in keeping up the motivation and ensuring response quality (Reja et al. 2003). Open-ended questions where respondents write their answers in their own words without any constraints on length (narrative answers) have seen a revival over the past two decades due to their relative ease of implementation in web surveys, and they have produced promising results, particularly when compared to open-ended questions in paper-and-pencil surveys (Oudejans and Christian 2010). Open-ended questions need to be particularly well designed since they are more cognitively demanding than closed-ended questions. They are associated with a higher response burden for respondents due to the lack of answer categories that could guide respondents in answering the question and due to the necessary typing activities. For these reasons, they are more prone to insufficient response in general and nonresponse in particular. Thus, it comes as no surprise that quite a number of studies have looked into design features of open-ended questions and their role for securing the quality of responses as well as other factors influencing response quality. The focus so far has been on different text box sizes, the use of motivational instructions and of follow-ups to open-ended questions, the use of clarification features or the impact of topic interest, and demographic characteristics on response quality (e.g. Denscombe 2008; Holland and Christian 2009; Metzler et al. 2015; Oudejans and Christian 2010; Smyth et al. 2009; Zuell et al. 2015). Since web probes are essentially open-ended questions, we combined findings from these

2 Behr et al. (2017) list further (nonrandom) sources for respondent recruitment.

studies and research strands to design and implement cognitive probes in web surveys. Our design and implementation decisions are presented next.

21.4.1 Probe Placement, Types, Presentation, and Text Box

21.4.1.1 Probe Placement

In our web probing studies, we implemented the embedded or concurrent approach (Willis 2005), that is, probes that are integrated into the usual questionnaire as a direct follow-up to closed-ended items. This approach is essentially a web-based implementation of Schuman's (1966) "random probes." Moreover, in order to disentangle the answering process for the closed-ended item from the answering process for the probe, we found it useful to present the probes on a separate screen. Thus, when answering the closed-ended item, the respondents were not affected by the probe to come, even though a learning process throughout the survey and anticipation of probes could not be ruled out (see Couper 2013, who shows response effects when systematic commenting is allowed). Fowler and Willis (Chapter 18 in this book) additionally tackled the retrospective method in which probes are asked after the web survey is completed.

21.4.1.2 Probe Types and Presentation

We used the following probe types in our studies to identify potential construct or item bias:

(1) *Category-selection probes* (Prüfer and Rexroth 2005) serve to gain insights into the reasons for a selected answer. An example of a category-selection probe is: "Please explain why you selected 'agree'." A category-selection probe is similar to what is called a *process-oriented probe* ("How did you arrive at that answer?") by Willis (2015), at least when considering attitude items. In our setup with a separate probing screen, we repeated the closed item and the chosen answer category on the probe screen to help recall the item and the answer. Thus, we managed to reduce respondent burden and increase response (Behr et al. 2012). In the case of a numbered scale, we provided respondents not only with the item referred to and the selected answer but also with the range of the answer scale so that the answer itself, being only a number, was put into the larger context.
(2) *Comprehension probes* serve to uncover the respondents' general understanding of a term. Examples of such probes are: "What ideas do you associate with the phrase 'civil disobedience'? Please give examples" and "What do you consider to be a 'serious crime'?" Comprehension probes are particularly useful in the case of "fuzzy concepts" (Ziegler et al. 2015, p. 1) that lack "clear cut demarcation lines" and thus can particularly affect cross-national research.

(3) *Specific probes* serve to gather additional information on a detail of an item. A specific probe may be worded as follows: “What particular civil rights did you have in mind when answering the question?” or “Which type of immigrants where you thinking of when you answered the question?”

Figure 21.1 provides screenshots of these three probe types. The use of web probing is certainly not restricted to these probe types or our chosen formulations. But there is one principle that applies to all probe formulations: Probes should be worded in such a way that respondents know what is expected of them (Züll 2016). A spontaneous rewording during the study if respondents do not provide answers as intended is not possible on the web. For cross-cultural cognitive interviewing, research finds that not all probe types work equally well across all cultural groups. For instance, a lack of focused answers and the avoidance of personal views were found among Chinese and Korean respondents in a study by Pan et al. (2010) (see Willis 2015, for an overview of probe types and challenges in the cross-cultural context). Such findings are important to consider and explore further when deciding on probes and when setting up web probing studies beyond the Western context and/or in countries where survey and opinion research is not widely known or used.

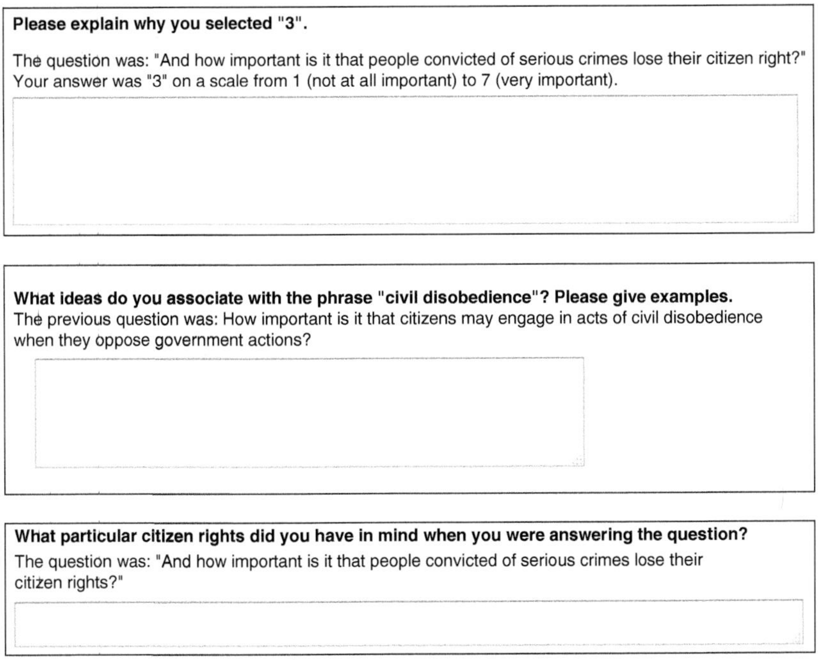
Please explain why you selected "3".
The question was: "And how important is it that people convicted of serious crimes lose their citizen right?" Your answer was "3" on a scale from 1 (not at all important) to 7 (very important).
What ideas do you associate with the phrase "civil disobedience"? Please give examples.
The previous question was: How important is it that citizens may engage in acts of civil disobedience when they oppose government actions?
What particular citizen rights did you have in mind when you were answering the question?
The question was: "And how important is it that people convicted of serious crimes lose their citizen rights?"

Figure 21.1 Screenshots of examples for category-selection, comprehension, and specific probes.

21.4.1.3 Text Box

An important parameter for open-ended questions is the size of the text box. Numerous studies showed the effect of text box size on answering behavior, with a larger text box producing more text than a smaller box (e.g. Couper et al. 2001; Smyth et al. 2009; Zuell et al. 2015). The same applies to probes as well so that the size of the text box should be adapted to the desired answer type, whether these are examples or more narrative-type answers without length restriction (Behr et al. 2014a). In Figure 21.1, for instance, the category-selection and comprehension probe (first and second) were assigned a larger text box while the specific probe was assigned a smaller text box to cue respondents on the desired length and type of answers.

21.4.2 Sequence of Probes

The aspect of the sequence of probes can be looked at from two different perspectives – sequence of probes for one specific item versus sequences of probes in the entire survey. Neither the cognitive interviewing literature nor research on open-ended questions provided us with answers to these issues so that we had to deal with this without prior input.

21.4.2.1 For One Specific Item

In one experiment, we investigated the best sequence if several probes need to be combined for one specific item. Meitinger et al. (2018) tested two combinations that differed in the sequences of probes: (i) category-selection probe, specific probe, and comprehension probe versus (ii) comprehension probe, specific probe, and category-selection probe. They found that a sequence that had the category-selection probe first increased response rate and motivation and decreased mismatching answers, that is, answers that did not fit the asked probe (e.g. replying with some sort of reasoning answer to a comprehension probe). Interestingly, however, not all effects were equally evident across the countries in our study, which were Germany, Great Britain, Mexico, Spain, and the United States. This at least warrants some caution regarding the uncritical transfer of findings with regard to questionnaire design established in one cultural and linguistic group to other cultural and linguistic groups. More cross-cultural research on probe sequence, nonresponse, and mismatching answer behavior is required.

21.4.2.2 For the Entire Survey

The sequence of probes throughout a survey should not be taken lightly either. Behr et al. (2014a) found that respondents habituated to a specific probe type (e.g. a category-selection probe) in relation to a specific text box size and overall layout when the probe came up repeatedly. The same visual outlook of a repeatedly occurring probe seemed to have suggested to the respondents the same – known – probe type. Rather than consciously reading the probe question of subsequent probes, respondents answered in terms of their expectations and thus ran the risk of missing any new probe type. In concrete terms: After having been exposed to four or five category-selection probes with identical layout and text box size, when the next probe (a comprehension probe) was reached, the same text box size and overall layout were taken as an indication for a category-selection probe. Thus, we received many responses that in fact matched a category-selection probe but were not a good match for our comprehension probe. We regarded these answers as "mismatching answers" (Meitinger et al. 2018) that were basically useless. In sum, efforts should be directed toward creating a survey where probe type, text box size, and overall layout encourage the respondents to actively read the probe question(s). This could require choosing different text box sizes or layouts for different probe types or conscious decisions on probe sequence. Mismatch conversion does not exist. Particularly with these mismatches, web probing currently reaches its limits. However, with regard to nonresponse we developed first solutions, as described below.

21.4.3 Nonresponse Follow-Ups

Given the high response burden of open-ended questions, probes are particularly prone to nonresponse. In our projects, automated solutions were developed to convert nonrespondents into respondents. The starting point were empirical corpora, first in German, later in English and Spanish, that contained original nonresponse answers to probes. The answers had been coded manually according to different nonresponse categories, as shown in Table 21.2. The various nonresponse answers were used to develop search patterns in the form of regular expressions, that is, generic text strings for automatically identifying patterns in respondents' answers.[3] For instance, the regular expression "^((be)?cause)? *i? *[a-z]* *do *n.?t *[a-z]* *k* *now*" finds several variants of English-language "don't know" answers. When these regular expressions are programmed into the survey software, automated follow-up probes to probe nonresponse can be triggered, possibly with fitting motivating sentences that encourage the respondent to answer the probe despite the first nonresponse. A first set of regular expressions were developed with iterative rounds of testing and validation for German, English, and Spanish. The set of regular expressions as well as the underlying script and technical details are publicly available (Kaczmirek et al. 2017).

3 E.g.: http://www.regular-expressions.info.

Table 21.2 Categories of nonresponse.

Category	Type of nonresponse
Category 1	Complete nonresponse: respondent leaves a text box blank
Category 2	No useful answer: response is not a word, e.g. "dfgjh"
Category 3	Don't know: e.g. "I have no idea," "DK," "I can't make up my mind"
Category 4	Refusal: e.g. "no comment," "see answer above"
Category 5[a)]	Other nonresponse: responses that are insufficient for substantive coding, e.g. "my personal experience," "it depends," "just do," "just what it is"
Category 6[a)]	One word only: respondent just writes a single word, e.g. "economy"
Category 7	Too-fast response: respondent takes less than two seconds to answer

a) Answers of categories 5 and 6 may for some research questions count as a substantive response.

21.5 Translation and Coding Answers to Cross-Cultural Probes

As we were, in our studies, mainly interested in interpretation patterns across countries, our analyses of web probing almost exclusively relied on a thematic approach. That is, rather than identifying whether certain errors occurred (DeMaio and Landreth 2004; Fitzgerald et al. 2011; Willis and Zahnd 2007), we coded themes that were mentioned in order to detect (non-)equivalent – or biased – patterns across countries: for instance, immigrant groups that respondents thought of in different countries or aspects that made respondents proud of their country. One of the key questions in the analyses was whether team members sufficiently understood all the languages of the study to both develop the coding scheme and code the responses. For certain languages in our first set of studies, this was not the case so that we commissioned professional translators for the task of translating open-ended probing answers into the project language (in our case, German). By specifically instructing the translators for the task (overall goal of research, leeway in translation, examples for required comments) and providing them with space for additional comments, we attempted to narrow the impact of translation on the coding (Behr 2015). We acknowledge, however, that we cannot fully exclude impact on the coding: Whether coding of translated responses leads to different conclusions than coding of original responses, and whether there is a best-practice approach is still an unsolved research question. Being able to understand the responses is needed not only for coding, but also for the development of the coding scheme – if this is question-specific rather than based on generic error types. Especially in an inductive approach where the responses suggest the categories of a coding scheme, understanding responses from *all* countries is needed in order to develop a balanced coding scheme that takes into account the respective country narratives. Otherwise categories and illustrative examples may favor some countries' themes and perspectives over others and thus introduce some form of bias.

Since coding scheme development and coding (and possibly prior translation) are time-consuming, (semi)-automatic approaches to coding are worthwhile to look at, even though they require a sufficiently large manually coded sample size (about 500 responses) to train a learning algorithm (Schonlau and Couper 2016). Automatic coding has the effect, though, that comments made by the translators, for instance on particularities of translated open-ended responses, cannot be taken into account in code assignments. For a "quick and dirty" problem spotting, one might also try out visualization or text/content analysis tools (see, e.g., wordle.net or tools listed under http://tapor.ca/tools) to gain a quick and rough overview of the data, for instance with regard to the frequency of words across countries. If the tools are used on an external server rather than on one's organization's server, data privacy and confidentiality should always be guaranteed, though.

21.6 Substantive Results

The research aim of our group, once the web design challenges were overcome, was to assess equivalence in cross-national surveys by using the collected web probing answers. We have so far tackled items in existing surveys that were identified as problematic during statistical equivalence testing (e.g. inconsistencies or lack of higher levels of invariance) or items that have repeatedly provoked calls in the research community for further research to elucidate their meaning in cross-national contexts.

21.6.1 Issues in Statistical Analyses

In this first line of research, that of shedding light on problematic and suspicious data, we conducted research on the "rights in a democracy scale" of the ISSP 2004, in particular on the item "How important is it that citizens may engage in acts of civil disobedience when they oppose government action?" In the ISSP data, this item showed both high item nonresponse and inconsistent results with regard to the other items in the six-item scale (response scale running from 1 to 7). The inconsistent results – for Canada, Denmark, Germany, Hungary, Spain, and the United States – were as follows: For the index (the response mean across variables) of the first five items in the battery (e.g. importance of all citizens having adequate standard of living or importance of protection of minority rights), the index was quite similar for all countries

(6.2–6.6). However, the last item, the civil disobedience item, markedly divided the countries into two groups, with the mean for Canada, Denmark, and the United States being particularly low (3.8–4.1) and the mean for the other countries ranging between 5.0 and 5.5. To understand what may have driven these results, we implemented the scale in our cross-national web survey in Canada, Denmark, Germany, Hungary, Spain, and the United States (*n* between 507 and 538) and had the scale followed by a probe split after the civil disobedience item. In each country, respondents were randomly assigned to receive one of two probe versions. Half of the respondents received a category-selection probe inquiring after the reasons for the selected answer and the other half received a comprehension probe asking for the ideas respondents associated with "civil disobedience." The striking pattern for the scale from the ISSP could be replicated in the web survey data, and the probe answers showed that lower support for civil disobedience in the United States and Canada was partly "real" (due to a higher level of trust in politicians) and partly a methodological artifact due to different associations with the concept of civil disobedience. Respondents from the United States and Canada in particular associated violent actions with civil disobedience, while this answer pattern was much less prevalent in the other countries. We concluded that item bias in the form of different meanings attached to the item's key term compromised equivalence (Behr et al. 2014b).

Meitinger (2017) took the ISSP 2013 as a starting point. She used multiple group confirmatory factor analysis (MGCFA) to test measurement invariance for five items measuring nationalism and constructive patriotism in five countries: Germany, Great Britain, Mexico, Spain, and the United States. Although the study could confirm metric measurement invariance, (partial) scalar invariance tests failed in MGCFA, which precluded a cross-national comparison of the latent means of the constructs. We then implemented the items in a web probing study and this enabled Meitinger to investigate how three of the five items were understood across countries, namely "And how proud are you of [country] with regard to the way democracy works?" "[...] to its social security system?" and "[...] to its fair and equal treatment of all groups in society?" Based on the probe answers, she found that the lack of scalar invariance could be explained by a major misunderstanding of the term "social security system"[4] [sistema de seguridad social] by 39% of Mexican respondents ("security on the streets" rather than "state benefits") and by differences in the perceived scope of the various terms for "social security" in the different languages, pushing respondents' understanding in one or the other direction.

---

4 A social security system is a set of measures that fulfill basic needs for citizens. It is put in place by the government and examples include (monetary) benefits for people who are unemployed, require health care, need welfare, are retired, or have children.

21.6.2 Testing Questionable Items

The second line of research, that of focusing on critical or questionable items or terms in general, was followed by Behr and Braun (2015) when looking into respondents' reasons for rating the functioning of democracy in their respective countries in a positive or negative way. In the European Social Survey, the item is worded as follows: "How satisfied are you with the way democracy works in your [country]?". The item, in this or a similar wording, is widely used in cross-national studies but is nevertheless highly controversial due to vagueness, context-dependency, and the fact that it measures a complex concept with just one item (Ariely and Davidov 2011; Canache et al. 2001; Linde and Ekman 2003). We implemented the item in a cross-national web survey in Canada, Denmark, Germany, Hungary, Spain, and the United States and followed up with a category-selection probe asking for the reasons that respondents had in mind when selecting their answers (on a scale from 1 [extremely dissatisfied] to 11 [extremely satisfied]). While a variety of reasoning patterns or dimensions could be found, most notably on the levels of government output, governance, and the political system as such, these dimensions played a role in *all* countries of our study. In particular, probe responses expressing dissatisfaction were strongly linked to output and governance, whereas the political system assembled probe responses that expressed both satisfaction and dissatisfaction. These differences also fitted to the respective country results regarding the closed-ended item. For instance, Denmark stood out as the most satisfied of the five countries, as measured by the closed-ended item. At the same time, satisfaction with the political system and, to a lesser degree, governance was mostly a probe theme put forward by Danish respondent. Given the results, we concluded that some form of comparability did indeed exist for the item even though the item itself could not be nailed down to a single dimension. If some countries in our study had exclusively relied on assessing government output and others on governance (regardless of satisfaction or dissatisfaction with these aspect), we would have been less positive in our conclusion.

In another study, Braun et al. (2013) looked into the meaning of "immigrants" in a cross-national context. Although the term "immigrant" can more or less easily be translated into other languages, this does not mean that the groups associated with "immigrants" are necessarily comparable across countries (see Heath et al. 2005, for a similar concern expressed toward the term "people from poorer countries"). In order to gain insights into respondents' understanding, we took items from the ISSP 2003 module on National Identity, here in particular the item scale on xenophobic attitudes: "immigrants increase crime rates, … are generally good for country's economy, … take jobs away from people who were born in [country], … improve society by bringing in new ideas and cultures." We rotated these items and asked for the first item in each experimental condition what immigrant groups the respondents had in mind when answering the closed-ended item. Braun et al. (2013) found for Canada,

Denmark, Germany, Hungary, Spain, and the United States that respondents thought of the most visible immigrant groups in their respective countries. For instance, in Germany, these were mostly the Turkish immigrants and, in Canada, these were immigrants of Asian origin. As such, immigrant reality was more or less captured in a comparable way, even though the nationalities or ethnic groups of migrants were different across countries.

These examples of substantive analyses show that the different social contexts determine in the end how a translation is understood. Meaning is nested in many contexts, the questionnaire context, the respondents themselves, and in particular in the socio-cultural context in each country: "Meanings and thought patterns do not spontaneously occur within the confines of a respondent's mind, but rather those meanings and patterns are inextricably linked to the social world […]" (Miller and Willis 2016, p. 212). This is why cognitive interviewing has become such a useful tool in cross-national questionnaire design and refinement of translations; this is also why web probing with its own strengths and weaknesses is likely to become a promising supplementary method in cross-national survey research.

21.7 Cross-National Web Probing and Its Application Throughout the Survey Life Cycle

In the research presented here, we have mainly implemented web probing as a follow-up study to a main survey to understand what may have caused non-equivalence in items. A use of web probing for pretesting purposes is equally possible. However, any use should be seen in light of and coordinated with well-established pretesting methods, in particular with cognitive interviewing with which web probing shares the probing questions. The integration and the interplay between web probing and cognitive interviewing is currently a matter of debate and testing in general survey methodology, as was described in Section 21.1 of this chapter (see also Behr et al. 2017). The current tendency, if both cognitive interviewing and web probing are possible, is to have cognitive interviewing first since it allows in-depth and interactive exploration of items. More targeted web probing may follow to assess the prevalence of themes or issues in a larger and more geographically dispersed population. For web probing in this scenario, the researcher should already have a thorough understanding of the item – and potential issues and hypotheses in mind – in order to determine and word the probe(s). This sequence, cognitive interviewing first, followed by web probing, can also be implemented with closed-ended probes. Scanlon (Chapter 17 in this book) follow an approach where the cognitive interviewing results help to word closed-ended probes for a subsequent web survey. These closed-ended probes have the advantage that demanding or burdensome open-ended questions, when seen from a respondent's perspective, and time-consuming coding scheme development and manual coding, when seen from a researcher's perspective, and are not needed. The combination of cognitive interviews and closed-ended probes in field tests (albeit not in web-based field tests) was also followed by Miller and Maitland (2010) (cited by Baena and Padilla 2014) to assess the range of meanings attached to anxiety and the respective prevalence among respondents in Kazakhstan, Cambodia, Sri Lanka, Maldives, Mongolia, and the Philippines.

Web probing as a pretesting method may also become interesting if cross-national cognitive interviewing is not viable (e.g. lack of cognitive interviewers in some countries, time constraints) or should be supplemented with web probing in additional countries to increase the spread of cultural and linguistic groups (Behr et al. 2017). It goes without saying, however, that (besides the practical constraints) the research questions, the desired probes one has in mind (including their complexity and likelihood to trigger follow-up probes) and the target group will eventually decide which method and in which combination to use.

If the main survey is a web survey, one could imagine implementing probes for selected items and selected respondents. Already in the mid-1960s, Schuman brought up the idea of "random probes" whereby randomly selected respondents receive probes for selected questions (e.g. 10 probes per respondents). Schuman argues that

> [t]hrough qualitative and quantitative review of random probe responses the survey researcher has an opportunity to increase his own sensitivity to what his questions mean to actual respondents [...] In research in other cultures—and under some conditions in one's own culture—it forms a useful supplement to standard attitude survey methods.
>
> (Schuman 1966, p. 222)

Response burden and potential effects on closed-ended items should be considered in such a design, though.

The use of web probing in cross-national follow-up studies was described earlier, notably in Section 21.5, where substantive analyses were presented. These follow-up studies can help to explain statistical inconsistencies and problems and gather additional qualitative data to aid analysis.

Regardless of the stage of use (pretesting, main study, follow-up study), the web probing data collected can serve to foster systematic mixed-methods approaches in cross-national research. Mixed-methods research refers to a combination of qualitative and quantitative methods and their integration to reach the research objective (Baena and Padilla 2014).

Respondent perceptions and quantitative approaches can be reconciled by this type of research and limitations associated with each research paradigm offset. Van de Vijver and Chasiotis (2010) have already made a plea for cross-cultural mixed-methods research in general; Baena and Padilla (2014) have echoed this plea, thereby focusing on cognitive interviewing in connection with quantitative methods. In this chapter, we are recommending web probing to be considered in mixed-method research, too.

21.8 Conclusions and Outlook

In this chapter, we have described the methodology and use of cross-national web probing by drawing on the major findings from two research projects conducted by the authors of this chapter. Where available, we supplemented our findings with approaches and applications by other researchers in order to guard against "project bias." However, cross-national web probing is a new endeavor; therefore, the literature both on theory and practice is still scarce. While the probing technique as such has been adopted from cognitive interviewing, design decisions have mainly been driven by advances in web survey design and here in particular current knowledge on the design of open-ended questions. Findings and innovations in these areas should be duly considered when embarking on web probing research activities.

We have implemented our studies mainly in Western countries (Mexico is an exception), and with a limited number of probe types. In terms of future research, it would be highly useful to test the web probing approach with a greater spread of countries and particularly in non-Western countries. The research questions should inquire whether web probing can elicit meaningful answers from a diverse set of cultural groups, and whether limitations for (certain) probe types in certain cultures exist (similarly to research conducted in cross-cultural cognitive interviewing, see Willis 2015). After all, what is ultimately needed is a method that, without introducing method bias itself, helps to confirm equivalence on the one side or uncover construct or item bias on the other side. Attention should also be paid to conducting comparative studies between cross-national cognitive interviewing and web probing to further delineate these types of methods and provide clearer guidance on when which of these methods can or should be used.

As Van de Vijver and Chasiotis (2010) point out, systematic mixed-methods studies in cross-national and cross-cultural research are still wanting. Due to its relative ease of implementation, web probing can contribute to increasing the number of these studies, whether at the pretesting, main study or follow-up stage. Cross-national web probing is in a unique position, besides cross-national cognitive interviewing, to take into account the socio-cultural contexts of respondents and its influence on understanding and answering (translated) survey questions.

Acknowledgments

This research was funded by the German Research Foundation (DFG) as part of the PPSM Priority Programme on Survey Methodology (SPP 1292) (project BR 908/3-1) and in a follow-up project (BR 908/5-1). Researchers on the projects: Michael Braun, Wolfgang Bandilla, Lars Kaczmirek (grant applicants), Dorothée Behr, and Katharina Meitinger.

References

Ariely, G. and Davidov, E. (2011). Can we rate public support for democracy in a comparable way? Cross-national equivalence of democratic attitudes in the World Value Survey. *Social Indicators Research* 104: 271–286. https://doi.org/10.1007/s11205-010-9693-5.

Baena, I.B. and Padilla, J.-L. (2014). Cognitive interviewing in mixed research. In: *Cognitive Interviewing Methodology* (eds. K. Miller, V. Chepp, S. Willson and J.-L. Padilla), 133–152. Hoboken, NJ: Wiley.

Beatty, P. and Willis, G. (2007). Research synthesis: the practice of cognitive interviewing. *Public Opinion Quarterly* 71: 287–311. https://doi.org/10.1093/poq/nfm006.

Behr, D. (2015). Translating answers to open-ended survey questions in cross-cultural research: a case study on the interplay between translation, coding, and analysis. *Field Methods* 27: 284–299. https://doi.org/10.1177/1525822X14553175.

Behr, D. and Braun, M. (2015). Satisfaction with the way democracy works: how respondents across countries understand the question. In: *Hopes and Anxieties. Six Waves of the European Social Survey* (eds. P.B. Sztabinski, H. Domanski and F. Sztabinski), 121–138. Frankfurt am Main: Lang.

Behr, D., Kaczmirek, L., Bandilla, W., and Braun, M. (2012). Asking probing questions in web surveys: which factors have an impact on the quality of responses? *Social Science Computer Review* 30: 487–498. https://doi.org/10.1177/0894439311435305.

Behr, D., Bandilla, W., Kaczmirek, L., and Braun, M. (2014a). Cognitive probes in web surveys: on the effect of different text box size and probing exposure on response quality. *Social Science Computer Review* 32: 524–533. https://doi.org/10.1177/0894439313485203.

Behr, D., Braun, M., Kaczmirek, L., and Bandilla, W. (2014b). Item comparability in cross-national surveys: results from asking probing questions in cross-national web surveys about attitudes towards civil disobedience. *Quality & Quantity* 48 (1): 127–148. https://doi.org/10.1007/s11135-012-9754-8.

Behr, D., Meitinger, K., Braun, M. et al. (2017). Web probing – implementing probing techniques from cognitive interviewing in web surveys with the goal to assess the validity of survey questions. Mannheim, GESIS – Leibniz-Institute for the Social Sciences (GESIS – Survey Guidelines). doi:https://doi.org/10.15465/gesis-sg_en_023.

Blom, A.G., Bosnjak, M., Cornilleau, A. et al. (2016). A comparison of four probability-based online and mixed-mode panels in Europe. *Social Science Computer Review* 34: 8–25. https://doi.org/10.1177/0894439315574825.

Braun, M., Behr, D., and Kaczmirek, L. (2013). Assessing cross-national equivalence of measures of xenophobia: evidence from probing in web surveys. *International Journal of Public Opinion Research* 25: 383–395. https://doi.org/10.1093/ijpor/eds034.

Braun, M., Behr, D., Kaczmirek, L., and Bandilla, W. (2014). Evaluating cross-national item equivalence with probing questions in web surveys. In: *Improving Survey Methods: Lessons from Recent Research* (eds. U. Engel, B. Jann, P. Lynn, et al.), 184–200. New York: Routledge.

Callegaro, M., Baker, R., Bethlehem, J. et al. (2014). Online panel research. History, concepts, applications, and a look at the future. In: *Online Panel Research: A Data Quality Perspective* (eds. M. Callegaro, R. Baker, J. Bethlehem, et al.), 1–22. Chichester: Wiley.

Canache, D., Mondak, J.J., and Seligson, M.A. (2001). Meaning and measurement in cross-national research on satisfaction with democracy. *Public Opinion Quarterly* 65: 506–528.

Cho, S.K., LoCascio, S., Lee, K.-O. et al. (2017). Testing the representativeness of a multimode survey in South Korea: results from KAMOS. *Asian Journal for Public Opinion Research.* 4 (2): 73–87. https://doi.org/10.15206/ajpor.2017.4.2.73.

Couper, M.P. (2013). Research note: reducing the threat of sensitive questions in online surveys? *Survey Methods: Insights from the Field* https://doi.org/10.13094/SMIF-2013-00008.

Couper, M., Traugott, M., and Lamias, M.J. (2001). Web survey design and administration. *Public Opinion Quarterly* 65: 230–253.

DeMaio, T.J. and Landreth, A. (2004). Do different cognitive interview techniques produce different results? In: *Methods for Testing and Evaluating Survey Questionnaires* (eds. S. Presser, J.M. Rothgeb, M.P. Couper, et al.), 89–108. Hoboken, NJ: Wiley.

Denscombe, M. (2008). The length of responses to open-ended questions: a comparison of online and paper questionnaires in terms of a mode effect. *Social Science Computer Review* 26: 359–368. https://doi.org/10.1177/0894439307309671.

Edgar, J. (2013). Self-administered cognitive interviewing. 68th Annual Conference of the American Association for Public Opinion Research, Boston, MA.

Fitzgerald, R., Widdop, S., Gray, M., and Collins, D. (2011). Identifying sources of error in cross-national questionnaires: application of an error source typology to cognitive interview data. *Journal of Official Statistics* 27: 569–599.

Heath, A., Fisher, S., and Smith, S. (2005). The globalization of public opinion research. *Annual Review of Political Science* 8: 297–333. https://doi.org/10.1146/annurev.polisci.8.090203.103000.

Holland, J.L. and Christian, L.M. (2009). The influence of topic interest and interactive probing on responses to open-ended questions in web surveys. *Social Science Computer Review* 27: 196–212. https://doi.org/10.1177/0894439308327481.

Johnson, T.P. (1998). Approaches to equivalence in cross-cultural and cross-national survey research. In: *ZUMA-Nachrichten Spezial 3: Cross-Cultural Survey Equivalence* (ed. J. Harkness), 1–40. Mannheim: ZUMA.

Kaczmirek, L., Meitinger, K., Behr, D. (2017). Higher data quality in web probing with EvalAnswer: a tool for identifying and reducing nonresponse in openended [sic!] questions. GESIS Papers 2017/01. http://nbn-resolving.de/urn:nbn:de:0168-ssoar-51100-0.

Lee, J. (2014). Conducting cognitive interviews in cross-national settings. *Assessment* 21: 227–240. https://doi.org/10.1177/1073191112436671.

Linde, J. and Ekman, J. (2003). Satisfaction with democracy: a note on a frequently used indicator in comparative politics. *European Journal of Political Research* 42: 391–408. https://doi.org/10.1111/1475-6765.00089.

Meitinger, K. (2017). Necessary but insufficient: why measurement invariance tests need online probing as a complementary tool. *Public Opinion Quarterly* 81: 447–472. https://doi.org/10.1093/poq/nfx009.

Meitinger, K. and Behr, D. (2016). Comparing cognitive interviewing and online probing: do they find similar results? *Field Methods* 28: 363–380. https://doi.org/10.1177/1525822X15625866.

Meitinger, K., Braun, M., and Behr, D. (2018). Sequence matters in online probing: the impact of the order of probes on response quality, motivation of respondents, and answer content. *Survey Research Methods* 12: 103–120. https://doi.org/10.18148/srm/2018.v12i2.7219.

Meredith, W. (1993). Measurement invariance, factor analysis and factorial invariance. *Psychometrika* 58: 525–543.

Metzler, A., Kunz, T., and Fuchs, M. (2015). The use and positioning of clarification features in web surveys. *Psihologija* 48: 379–408. https://doi.org/10.2298/PSI1504379M.

Miller, K. and Willis, G.B. (2016). Cognitive models of answering processes. In: *The SAGE Handbook of Survey Methodology* (eds. C. Wolf, D. Joye, T.W. Smith and Y. Fu), 210–217. London: Sage.

Miller, K., and Maitland, A. (2010). A mixed-method approach for measurement construction for cross-national studies. Joint Statistical Meetings, Vancouver, CA.

Miller, K., Fitzgerald, R., Padilla, J.-L. et al. (2011). Design and analysis of cognitive interviews for comparative multinational testing. *Field Methods* 23: 379–396. https://doi.org/10.1177/1525822X11414802.

Mockovak, W., and Kaplan, R. (2015). Comparing results from telephone reinterview with unmoderated, online cognitive interviewing. 70th Annual Conference of the American Association for Public Opinion Research (AAPOR), Boston, MA

Murphy, J., Keating, M., and Edgar, J. (2013). Crowdsourcing in the cognitive interviewing process. 2013 Federal Committee on Statistical Methodology (FCSM) Research Conference.

Oudejans, M. and Christian, L.M. (2010). Using interactive features to motivate and probe responses to open-ended questions. In: *Social and Behavioral Research and the Internet: Advances in Applied Methods and Research Strategies* (eds. M. Das, P. Ester and L. Kaczmirek), 304–332. London, New York: Routledge.

Pan, Y., Landreth, A., Park, H. et al. (2010). Cognitive interviewing in non-English languages: a cross-cultural perspective. In: *Survey Methods in Multinational, Multiregional, and Multicultural Contexts* (eds. J.A. Harkness, M. Braun, B. Edwards, et al.), 91–113. Hoboken, NJ: Wiley Blackwell.

Prüfer, P. and Rexroth, M. (2005). Kognitive Interviews. *ZUMA How-to-Reihe*, 15. https://www.gesis.org/fileadmin/upload/forschung/publikationen/gesis_reihen/howto/How_to15PP_MR.pdf.

Reja, U., Lozar Manfreda, K., Hlebec, V., and Vehovar, V. (2003). Open-ended vs. close-ended questions in web questionnaires. *Developments in Applied Statistics*, Metodološki zvezki 19: 160–175.

Schonlau, M. and Couper, M.P. (2016). Semi-automated categorization of open-ended questions. *Survey Research Methods* 10: 143–152. https://doi.org/10.18148/srm/2016.v10i2.6213.

Schuman, H. (1966). The random probe: a technique for evaluating the validity of closed questions. *American Sociological Review* 31: 218–222.

Smith, T.W. (2010). The globalization of survey research. In: *Survey Methods in Multinational, Multiregional, and Multicultural Contexts* (eds. J.A. Harkness, M. Braun, B. Edwards, et al.), 477–487. Hoboken, NJ: Wiley Blackwell.

Smyth, J.D., Dillman, D.A., Christian, L.M., and McBride, M. (2009). Open-ended questions in web surveys: can increasing the size of answer boxes and providing extra verbal instructions improve response quality? *Public Opinion Quarterly* 73: 325–337. https://doi.org/10.1093/poq/nfp029.

Steenkamp, J. and Baumgartner, H. (1998). Assessing measurement invariance in cross-national consumer research. *Journal of Consumer Research* 25: 78–107. https://doi.org/10.1086/209528.

Thrasher, J.F., Quah, A.C., Dominick, G. et al. (2011). Using cognitive interviewing and behavioral coding to determine measurement equivalence across linguistic and cultural groups: an example from the International Tobacco Control Policy

Evaluation Project. *Field Methods* 23: 439–460. https://doi.org/10.1177/1525822X11418176.

Van de Vijver, F.J.R. (2013 November). Contributions of internationalization to psychology: toward a global and inclusive discipline. *American Psychologist*: 761–770. https://doi.org/10.1037/a0033762.

Van de Vijver, F.J.R. and Chasiotis, A. (2010). Making methods meet: mixed designs in cross-cultural research. In: *Survey Methods in Multinational, Multiregional, and Multicultural Contexts* (eds. J.A. Harkness, M. Braun, B. Edwards, et al.), 455–473. Hoboken, NJ: Wiley.

Van de Vijver, F.J.R. and Leung, K. (2011). Equivalence and bias: a review of concepts, models, and data analytic procedures. In: *Cross-Cultural Research Methods in Psychology* (eds. F.J.R. van de Vijver and D. Matsumoto), 17–45. Cambridge: Cambridge University Press.

Vandenberg, R. and Lance, C. (2000). A review and synthesis of the measurement invariance literature: suggestions, practices, and recommendations for organizational research. *Organizational Research Methods* 3: 4–70. https://doi.org/10.1177/109442810031002.

Willis, G.B. (2005). *Cognitive Interviewing: A Tool for Improving Questionnaire Design*. Thousand Oaks, CA: Sage.

Willis, G.B. (2015). Research synthesis. The practice of cross-cultural cognitive interviewing. *Public Opinion Quarterly* 79: 359–395. https://doi.org/10.1093/poq/nfu092.

Willis, G.B. and Zahnd, E. (2007). Questionnaire design from a cross-cultural perspective: an empirical investigation of Koreans and non-Koreans. *Journal of Health Care for the Poor and Underserved* 18 (6): 197–217. https://doi.org/10.1353/hpu.2007.0118.

Ziegler, M., Kemper, C.J., and Lenzner, T. (2015). The issue of fuzzy concepts in test construction and possible remedies. *European Journal of Psychological Assessment* 31: 1–4. https://doi.org/10.1027/1015-5759/a000255.

Zuell, C., Menold, N., and Körber, S. (2015). The influence of the answer box size on item nonresponse to open-ended questions in a web survey. *Social Science Computer Review* 33: 115–122. https://doi.org/10.1177/0894439314528091.

Züll (Zuell), C. (2016). Open-ended questions. GESIS Survey Guidelines. Mannheim, Germany: GESIS – Leibniz Institute for the Social Sciences. doi: https://doi.org/10.15465/gesis-sg_en_002.

22

Measuring Disability Equality in Europe: Design and Development of the European Health and Social Integration Survey Questionnaire

Amanda Wilmot

Westat, Rockville, MD, USA

22.1 Introduction

There is general agreement in the literature that designing cross-national surveys is a complex undertaking (Fitzgerald and Zavala-Rojas, Chapter 20 of this book; Harkness et al. 2010a; Mohler 2007; Lynn et al. 2006; Hakim 2000). In addition to the challenges that usually face designers of any type of survey, the designers of cross-national surveys must also aim to achieve cross-national equivalence at all stages of the survey process (Smith 2015). In particular, in cross-national surveys, different languages, cultures, and social structures can hinder the achievement of equivalence related to the performance of the survey questionnaire (Smith 2003).

The development of standards and guidance for cross-national survey design has somewhat lagged behind that for conventional mono-national surveys, which includes industry-wide guidelines provided by National Statistical Institutes, survey methodology book publications, numerous articles in survey methods journals, and even guidelines produced by survey software companies. Although the landscape is changing, most notably with the publication of books such as Survey Methods in Multinational, Multiregional, and Multicultural Contexts (Harkness et al. 2010b) and a more recent publication from the 2016 Survey Methods in Multinational, Multiregional, and Multicultural Contexts (3MC) conference (Johnson et al. 2019), guidelines provided by the European Social Survey[1] and GESIS[2] and the revision in 2016 of the University of Michigan Cross-Cultural guidelines,[3] there is still a paucity of guidance specifically relating to cross-national questionnaire design, development, and

1 http://www.europeansocialsurvey.org.
2 http://www.gesis.org/fileadmin/upload/forschung/publikationen/zeitschriften/zuma_nachrichten_spezial/znspezial11.pdf.
3 http://ccsg.isr.umich.edu.

Advances in Questionnaire Design, Development, Evaluation and Testing, First Edition.
Edited by Paul C. Beatty, Debbie Collins, Lyn Kaye, Jose-Luis Padilla, Gordon B. Willis, and Amanda Wilmot.

testing (Harkness et al. 2010c). This is perhaps surprising since, as Fitzgerald (2015) eloquently explains, "the questionnaire is at the heart of the social survey and the ability of the instrument to measure with validity and reliability is critical." Such guidance that does exist often relates to cross-cultural rather than cross-national studies conducted in a small number of different languages within one country, and advocating the simultaneous development and testing of the different language versions of the survey questionnaire in order to achieve measurement equivalence. This approach may not always be practical in cross-national studies where a large number of countries are involved, such as across the European Union (EU), because of the difficulties of administering such an approach on a large scale, the volume of information produced, and cost. The lack of guidance in cross-national studies may be due to a lack of documentation of the cross-national questionnaire design and development process (Fitzgerald 2015), which in turn makes it difficult to learn from those who have taken on this task.

This chapter contributes to the important area of cross-national comparative instrument design and development by discussing the approach and methods used during the development and implementation of the European Health and Social Integration Survey (EHSIS[4]), which measured the situation of people with disabilities in European society. The chapter discusses the need for comparable data describing the situation of people with disabilities in Europe, the challenges of operationalizing a classification of disability, as well as the process of translation, testing, and refining the survey questionnaire in a cross-national setting.

22.2 Background

With reference to the United Nations Convention on the Rights of People with Disabilities, to which the EU is a signatory,[5] the European Union Disability Action Plan (EU DAP) 2004–2010[6] provided a framework for ensuring that disability issues were considered in all relevant EU policies. Specifically, the second-phase implementation of EU DAP (2006–7) called for statistical data to be produced on the societal integration of people with disabilities in the EU. Through the use of a common survey data collection instrument, cross-national comparable data were to be produced that could also be aggregated at the EU level (Agafitei 2010).

Prior to the implementation of EHSIS in 2012, European data from cross-national surveys on the situation of people with disabilities came from: (i) the European Survey of Income and Living Conditions (EUSILC); (ii) the

4 Formerly known as the European Disability and Social Integration Module (EDSIM).
5 http://ec.europa.eu/social/main.jsp?catId=1137.
6 http://eur-lex.europa.eu/LexUriServ/LexUriServ.do?uri=COM:2010:0636:FIN:en:PDF

European Health Interview Survey (EHIS); and, (iii) forthcoming data from the 2011 European Union Labour Force Survey (EU LFS) ad hoc module on the employment of disabled people (Agafitei 2012). However, these surveys imposed some constraints on the measurement of the integration of people with disabilities and cross-national comparability including: the way in which disability was defined; the limited number of questions about disability that could be asked with respect to the length of the questionnaire; the use of proxy data, which might not provide an accurate measure of disability; the requirement for longitudinal stability in relation to the questioning; the use of different modes of survey administration, which could result in mode effects (Meltzer 2009); and the potential for survey context effects that could affect responses to questions about disability (Baumberg et al. 2015; Bajekal et al. 2004). The EHSIS was conceived in order to attempt to overcome these constraints. Funded by the European Commission (EC), its implementation was overseen by Eurostat, the statistical authority of the European Union.[7]

Described as a "paradigm shift" in the way health and disability are understood and measured (Kostanjsek 2011), the EHSIS approach was different from that used in other disability surveys at the time, which focused on the medical model, defining disability with reference to what is "wrong" with the person and their inability to function (Meltzer et al. 2010). The intention of EHSIS was to provide a measure of disability based on the biopsychosocial model (Meltzer 2011a), introduced by the International Classification of Functioning, Disability, and Health[8] (World Health Organization 2001). The biopsychosocial model approach takes a view of disability as always being an interaction between an individual and the context in which they live – therefore considering both health and social factors when describing disability (Meltzer et al. 2010).

To facilitate cross-national comparison using one data collection instrument, the EHSIS was designed to provide harmonized inputs, in the form of a detailed source questionnaire and accompanying instructions for face-to-face interviewer administration. The English-language source questionnaire provided specified question wording, instructions, order, and routing to be used, which had been tested across a number of European countries.

Although this chapter focuses on the survey questionnaire, it is important to note that in addition to harmonizing the questionnaire cross-nationally, an attempt was made to functionally harmonize other variants that may lead to survey error. In this regard, Eurostat provided a technical specification for each participating country to follow during main-stage implementation of EHSIS. This technical specification included: the data collection field-work period; target population; use of sampling frames; sample design (including

7 http://ec.europa.eu/eurostat/about/overview.

8 The ICF is a classification of health and health-related domains. As the functioning and disability of an individual occurs in a context, ICF also includes a list of environmental factors: http://www.who.int/classifications/icf/en.

minimum sample size); mode of administration[9]; translation protocol; testing requirements; interviewer recruitment and training requirements; respondent-to-interviewer ratio; data processing; validation and weighting rules; and the way in which the data were to be delivered. Guided by a Eurostat quality-reporting template, each in-country contractor was required to provide an end-of-survey quality report describing their implementation process and methods, such that a post-survey cross-national comparability evaluation could be made.

In 2012–13, the main-stage survey was launched and administered across Europe in 26 Member States as well as Norway and Iceland.

22.3 Questionnaire Design

The design of the initial source questionnaire was led by Professor Howard Meltzer from the University of Leicester in the UK. At this stage, Member States were able to provide feedback on the design through the Core Group (HIS[10]) and Technical Group (HIS) meetings, and at the Eurostat meeting on Disability Statistics and the Washington Group Meeting (Dublin, Ireland 2007[11]).

22.3.1 The EHSIS Measure of Disability

The starting point for the EHSIS measure of disability was from the perspective of social integration. The ICF provided the theoretical framework, but this classification needed to be operationalized. The classification needed to be reduced to a manageable size for survey administration, and the language used – which tended to be academic in nature and jargonized – simplified, with the use of more commonly used descriptors (Meltzer 2007).

The ICF contains a great many different domains and domain items. One of the questionnaire design challenges was deciding on the level of detail required. Because of the need to produce a survey questionnaire of reasonable length, plus the need for questions that applied to everyone across Europe, a decision was made to cover the variety of social circumstances rather than collect detailed information on a narrower range (Meltzer 2008). Therefore, 10 life domains were selected pertaining to the two main themes of "environmental factors" and "participation" (Meltzer 2011b). The 10 life domains selected were: mobility, transport, accessibility to buildings, education and

9 It should be noted that although the EHSIS questionnaire was originally designed to be interviewer-administered face to face, the tender specifications for the main-stage survey, although strongly recommending this mode of data collection, also allowed for telephone or web-based data collection.

10 Health surveys, including disability and morbidity statistics.

11 https://www.cdc.gov/nchs/washington_group/wg_meeting7.htm.

training, employment, the internet, social contact and support, leisure pursuits, economic life, and the attitudes and behavior of others. These life domains were chosen to reflect specific statistical EC policy needs, rather than other domains that could have been chosen, for example, to reflect a human rights agenda (Meltzer 2008). Within the two main themes of environment and participation, the concepts on which the chosen life domains would be applied to a survey questionnaire were mapped from the relevant ICF chapters and a description produced to provide a framework from which the survey questions could be drafted. This mapping is shown in Table 22.1 along with the refined concept summary descriptions made after wave-two testing of the data collection instrument (described in Section 22.4.2).

Each section of the questionnaire reflected one of the 10 life domains and established whether respondents had the desire to participate in certain aspects of life, and whether there were any barriers to doing so. Respondents could select as many barrier items as they felt contributed to their lack of social integration within each life domain. Where respondents indicated that a barrier to participation was due to a long-standing health condition or difficulties with basic activities, then a measure of the biopsychosocial model of disability was established.

In summary, the EHSIS considered that people with disabilities are those who face barriers to participation in any of the 10 life domain activities, associated inter alia with a health problem or basic activity limitation. A level of severity of disability in relation to social integration was established by asking two follow-up questions about whether the lack of specialized equipment or personal help was what prevented the barrier from being overcome. The approach was kept consistent throughout the questionnaire (Wilmot and Meltzer 2010). An example of the questioning from the section on education and training is given in Figure 22.1.[12] The main barrier question is asked, and respondents selecting codes 4 or 5, indicating that the barriers faced are associated with a health problem or basic activity limitation, are routed to the follow-up questions determining severity.

22.3.2 Other Measures of Disability in EHSIS

In addition to the EHSIS measure of disability described, the survey questionnaire was supplemented with other prespecified measures of disability used in the cross-national European surveys mentioned in Section 22.2. These prespecified measures were: the three Minimum European Health Module (MEHM[13]) questions on general health (including the Global Activity

12 EDSIM questionnaire and interviewer instructions: https://circabc.europa.eu/w/browse/f420b9de-972b-42ca-b929-a585df7c7b43.

13 http://ec.europa.eu/eurostat/statistics-explained/index.php/Glossary:Minimum_European_Health_Module_(MEHM.

Table 22.1 Conceptual measure of disability.

ICF chapter	Section	Variable description	Section concept summary
d450-d469 WALKING AND MOVING	Mobility	*Barriers* to mobility	To assess whether the person is able to leave their home and go where they want to go, as often as they want to and at a time convenient to the person. The reason for lack of mobility is most likely caused by an interaction of personal factors (such as health) and environmental characteristics (difficult terrain, lack of transport, etc.).
d470-d489 MOVING AROUND USING TRANSPORTATION *e1 PRODUCTS AND TECHNOLOGY* *e5 SERVICES, SYSTEMS, AND POLICIES*	Transport	*Barriers* to using private vehicle Barriers to using other forms of transport	To assess whether the person is able to use various modes of transport as much as they would like to and reasons preventing them from doing so, including health, personal factors (such as desire to use transport) and environmental characteristics (lack of available, suitable or affordable transport, etc.). Includes being a driver and a passenger.
e5 SERVICES, SYSTEMS, AND POLICIES	Accessibility to buildings	*Barriers* to accessing buildings	To establish the barriers associated with accessibility to buildings in terms of entry and exit, getting around a building, and using the facilities. When a problem has been established, the contribution of personal factors and poor architectural features are elucidated.
d810-d839 EDUCATION	Education and Training	*Barriers* to education and training	Social integration is not just about the consequences of past problems in accessing appropriate educational courses or looking at past educational achievement. This section concerns current access to appropriate and desired vocational training and higher education.
d840-d859 WORK AND EMPLOYMENT	Employment	*Barriers* to employment	Social integration is not just about employment status, but about gaining access to appropriate and desired employment. People with a health condition or difficulties with basic activities are more likely to achieve lower employment outcomes. Employer accommodations over working practices can facilitate employment. Hence this section examines barriers to work, and identifies the reasons for this disadvantage.

d350-d369 CONVERSATION AN D USE OF COMMUNICATION DEVICES AND TECHNIQUES *e1 PRODUCTS AND TECHNOLOGY* *e5 SERVICES, SYSTEMS, AND POLICIES*	The internet	*Barriers* to using the internet	The internet can enhance opportunities for obtaining knowledge and widening the scope of communication for all people. This section is concerned with whether people use the internet, and if not, what prevents them from using it including health and personal factors, such as interest or technical ability, and environmental factors, such as cost, suitable, or affordable equipment.
d710-d729 GENERAL INTERPERSONAL INTERACTIONS d730-D779 PARTICULAR INTERPERSONAL RELATIONSHIPS *e3 SUPPORT AND RELATIONSHIPS*	Social contact and support	*Barriers* to speaking with people feel close to	To establish the degree to which people have social contact and support, whether they feel they want more contact, and, if so, what is preventing them from having it. In looking at barriers to increased social contact and support, both personal and environmental factors are examined.
d9 COMMUNITY, SOCIAL, AND CIVIC LIFE	Leisure pursuits	*Barriers* to pursuing hobbies or interests Barriers to attending cultural events	People's desire for involvement in leisure and cultural activities varies. (i) They may be very busy with work and family and do not wish to or have time to do other things. (ii) They may be quiet or shy people who prefer individual to group pursuits. (iii) They may live in such remote communities that regular participation in leisure and cultural activities is not practicable. Therefore participation is measured in a relative sense; relative to the person's desired level of participation. The issue is not so much on the range and frequency of people's participation but the barriers to their desired level of participation.
d860-d879 ECONOMIC LIFE	Economic life	*Barriers* to paying for the essential things in life	To ascertain whether the person has enough money for usual living expenses and if not what the barriers to economic self-sufficiency are – examining the interaction of personal and environmental factors.
e4 ATTITUDES	Attitudes and behavior of others	*Reasons* for feeling treated unfairly	People are asked whether they feel they are subject to discriminatory practices. The question is relevant to everybody, not just those with a health problem or a visible activity limitation. Finally, we look at the sources of negative treatment just for those with a health problem or an activity limitation.

Source: Eurostat (Compiled from Meltzer (2008) and Wilmot and Meltzer (2010)).

EdPrv
APPLIES IF: Not currently studying
DK/REF allowed SHOWCARD

[*] **Is there anything which prevents you from studying for a qualification at the moment? Please use this card as guide and choose all that apply.**
CODE ALL THAT APPLY
1) Financial reasons (lack of money, can't afford it)
2) Too busy (with work, family, other responsibilities
3) Lack of knowledge or information (about what is available)
4) A longstanding health condition, illness, or disease
5) Longstanding difficulties with basic activities (such as seeing, hearing, concentrating, moving around)
6) Difficulties getting on a course
7) Difficulties getting to learning facility
8) Difficulties accessing or using buildings
9) Attitude of employer or teacher
10) Lack of self-confidence or attitudes of other people
11) Other reasons
12) Don't want to study for a qualification
13) No, nothing prevents me from studying for a qualification at the moment.

EdPrv1
APPLIES IF EdPrv = 04 or 05 *'health condition or difficulties with basic activities' DK/REF allowed*
[*] **May I just check, does the lack of special aids or equipment prevent you from studying (for a qualification) at the moment?**
1) Yes
2) No

EdPrv2
APPLIES IF EdPrv = 04 or 05 *'health condition or difficulties with basic activities' DK/REF allowed*
[*] **(May I just check,) does the lack of personal help or assistance prevent you from studying (for a qualification) at the moment?**
1) Yes
2) No

Figure 22.1 Education and training questioning.

Limitations Indicator [GALI]); questions from the EU LFS 2011 Ad Hoc Module on the employment of disabled people that presented a list of health problems and basic activity limitations – seeing, hearing, walking, etc.[14]; questions on activities of daily living (ADL) – feeding, dressing, bathing, etc.; and instrumental activities of daily living (IADL) – preparing meals, shopping, light housework, etc. Therefore, alternative definitions of disability could be derived and compared with the EHSIS definitions (Meltzer et al. 2010).

14 EU LFS 2011 Ad Hoc Module: https://circabc.europa.eu/w/browse/ec58cf67-ed60-4011-a749-4fc397621304.

22.3.3 Core Social Variables

The EHSIS classificatory questions were also prespecified and included the standard core variables for European social surveys.[15] These core social variables were specified according to harmonized outputs (rather than harmonized inputs) where question concepts and definitions, along with the statistical outputs required, are specified, but question wording is not. Support for mapping these outputs to questions during survey implementation was provided by each participating country's national coordinator – usually an employee of the National Statistical Institute.

22.4 Questionnaire Development and Testing

Eurostat grants provided the funding for questionnaire development and testing. The EHSIS source questionnaire was initially developed in English and subsequently translated into the languages of the other countries involved. An iterative and cumulative approach was taken to the development and testing. This included an initial wave of testing and review in three countries and a second wave of testing and review in 10 countries. Prior to fielding, all 28 countries involved in the main-stage survey also conducted a third wave of testing.

Quantitative piloting and qualitative cognitive testing methods were used. Using quantitative piloting, problems with the questionnaire could be identified from interviewer debriefing (e.g. respondents' comments on the questions, questions that respondents are unable to answer, or where respondents provide inappropriate answers) or from data analysis (e.g. response options are not comprehensive, or item non-response is high). Qualitative cognitive testing could identify problems by providing an understanding of the question and answer process from the respondent's perspective, based on a four-stage response process model of (i) comprehension, (ii) retrieval from memory, (iii) decision processes, and (iv) response mapping (Tourangeau 1984). Both methods provided a perspective on data quality and contributed in different ways to the review process.

22.4.1 Wave-One Testing

The first wave of testing took place in three Member States – United Kingdom, Italy, and Lithuania – between 2007 and 2008. Testing took place in the source language, as well as in two non-English-speaking countries representing

15 Core variables: updated 2011 guidelines. eurostat/f/11/dss/01/2.3en, meeting of the European Directors of Social Statistics, Luxembourg, 21–22 September 2011, https://circabc.europa.eu/sd/a/69497b9d-ec9e-4b59-ad7b-1ab709c57d99/CORE%20VARIABLES%20UPDATED%20GUIDELINES%20May%202011.pdf.

different European perspectives. A project advisory group was established to oversee the work and provide appropriate subject matter advice. The group comprised members from Belgium, Denmark, the Netherlands, and the European Disability Forum.[16] The advisory group reported directly to Eurostat and the Directorate-General for Employment, Social Affairs and Inclusion (DG EMPL),[17] who also gave oversight and input into the questionnaire revision process.

As the administration of the EHSIS measure of disability in a survey context could be considered "exploratory," the objectives of this first wave of testing were to examine the question wording, formatting, and conceptual understanding of the survey, and to focus more on the functioning of the questionnaire per se rather than to examine specific cross-national issues at this stage. Hence the focus was also more on using qualitative methods in the form cognitive testing – a method that could provide insight into how potential survey respondents were able to conceptualize and answer the survey questions. In 2007, the English source questionnaire was cognitively tested in the UK, providing an initial test in the source language. Thirty cognitive interviews were conducted with adults,[18] including those self-identifying as disabled and non-disabled. This was supported by a small-scale, interviewer-administered quantitative study in the UK with 70 respondents and included an interviewer in-person debriefing. Following a standard Eurostat team-based translation protocol, commonly considered best practice in order to minimize any translation issues that might affect the questioning (European Social Survey Translation Guidelines 2018), the questionnaire and related testing protocols were translated, prior to simultaneous cognitive testing in Italy and Lithuania, involving 30 participants in each country. During the cognitive testing, standard probing techniques were used based on the response process model, involving both pre-scripted general probes and unscripted expansive probing techniques to explore participants' understanding in more detail (Sebastiani et al. 2010; Meltzer 2007). The questions were revised after each test, taking account of the results presented in the respective reports (Meltzer 2008).

22.4.1.1 Review of Wave-One Findings

The main findings from the wave-one testing were that in general: people did indeed modify their behavior, desires, and expectations to take account of their limitations; asking about why people did not participate worked well; a range of barriers to participation emerged, forming the basis for the design of the barrier-response options; and specific questions worked better than general questions. The main question-specific issues to emerge related to the accessibility questions: the issue was not getting into the building but rather using

16 http://www.edf-feph.org.

17 http://ec.europa.eu/social/main.jsp?catId=1137&langId=en.

18 In Europe, the age of adulthood can vary between 15 and 16.

the facilities in the building; people answered the mobility questions thinking about their mobility facilitators; and questions on interpersonal relationships were difficult to operationalize (Meltzer 2007).

The subsequent cognitive tests in Italy and Lithuania provided additional information on which the questionnaire could be revised and refined: for example, the questions on ability to read, write, and count were interpreted in a variety of ways and were subsequently dropped from the questionnaire (Sebastiani et al. 2010).

22.4.2 Wave-Two Testing

A second wave of testing was carried out in 10 Member States: Bulgaria, the Czech Republic, Estonia, Finland, Greece, Hungary, Latvia, Malta, Slovakia, and Spain. These countries were selected to provide a mix of European localities. The second wave of testing took place between 2009 and 2010. The work was overseen and informed by the Eurostat Task Force on Streamlining Disability Statistics. Dependent on available resources, the questionnaire was cognitively tested in eight European countries and quantitatively piloted in seven (see Table 22.2).

At this stage the main objective was still to make recommendations for possible improvements to the source questionnaire. The focus was expanded to include any difficulties in relation to respondent or interviewer burden, the flow of the interview, routing issues, and order effects. At wave two, country-specific issues were also highlighted related to any cultural or linguistic differences that meant the source questionnaire would not function as anticipated. Eurostat provided a standard template to ensure a structured reporting approach from each country, section by section. Reported findings were based on both qualitative and quantitative testing, as applicable. Findings

Table 22.2 Wave-two testing countries.

Cognitive test	Pilot test
Czech Republic	Bulgaria
Estonia	Czech Republic
Finland	Estonia
Greece	Greece
Hungary	Hungary
Latvia	Latvia
Slovak Republic	Malta
Spain	

Source: Eurostat (Wilmot and Meltzer (2010)).

from the wave-two testing were augmented by recommendations from members of the disability task force; a report from the Nordic group (Ramm and Otnes 2008); and comments from DG EMPL.

22.4.2.1 Wave-Two Methodology Review

Once all of the individual country testing reports were completed, the results were systematically evaluated and reviewed cross-nationally, question by question. A recommendation for change made on the basis of the findings from one country may in fact cause a problem for another that was not there originally. In such cases, a judgment was required based on the findings across all of the testing. However, implementation of the methodologies varied cross-nationally, with those conducting the testing varying in their level of experience, particularly in relation to conducting cognitive testing.

The first stage then in this wave-two review process was to assess the quality and robustness of the testing data collected on which the findings presented were based. An evaluation of the methodologies employed was carried out using specified quality criteria. A "process quality approach" (Haselden and White 2001) was taken to the evaluation at this stage. Quality indicators were devised that covered the areas of sampling, data collection, and analysis for both the qualitative and quantitative testing – see Table 22.3.

All of the indicators contributed an important element to the overall assessment of quality. More weight was given to the findings from those countries adopting a more robust design. For example, the number of cognitive interviews conducted was considered alongside other indicators such as the distribution of characteristics of cognitive interview participants and recruitment methods used. A detailed description of the testing methodologies employed across the various countries taking part can be found in Wilmot and Meltzer (2010).

During the cognitive testing, different countries adopted different cognitive interviewing techniques, such as concurrent or retrospective probing, respondent think-aloud, or a combination. Some countries chose to supplement the one-to-one cognitive interviews with other pretesting methods such as focus groups or observational behavior coding. No assumptions about data quality were made based on these differences.

22.4.2.2 Review of Wave-Two Findings

The second stage of the review involved a synthesis of the findings reported by each country. This involved collating and summarizing all of the data relating to each section of the questionnaire and then examining all suggestions, comments, and recommendations for each question within each section. The outcomes were supplemented with suggestions from the groups advising the project and focused on: the feasibility of asking the questions in a survey setting, question sequence, wording and vocabulary used in each question, and

Table 22.3 Qualitative and quantitative quality indicators.

Qualitative cognitive testing	Quantitative pilot testing
Sample	*Sample*
Achieved sample size	Sample type (random/quota)
Selection criteria	Frame used
Severity of problem/disability	Population
Reasonable spread of respondents by selection criteria or key variables achieved	Selection/stratification criteria
Reasonable spread of urban/rural locations	Reasonable spread a of selection criteria achieved
Reasonable spread of disabilities and health problems achieved (physical/sensory/mental)	Reasonable spread of urban/rural locations
Recruitment method	Response rate (where appropriate)
	Achieved sample size
	Number disabled or with health problem: no disability or health problem
	Reasonable spread of disabilities and health problems achieved (physical/sensory/mental)
	Recruitment method
Data collection	*Data collection*
Length of interview	Length of interview
Interviewer experience and training	Interviewer experience and training
Mode	Mode
Location of interview	Location of interview
Whether interviews were recorded	All questions tested
Whether all questions were tested	Technique
Method	
Probing technique used	
Analysis	*Analysis*
Method	Method
EHSIS translation and testing methodology report form used	EDSIM translation and testing methodology report form used

Source: Eurostat (Wilmot and Meltzer (2010)).

comprehensiveness and completeness of any explanatory notes. Thus, a more structured and coherent data set was produced from which the cross-national review and recommendations for further change or adaption to the source questionnaire could be made (Wilmot and Meltzer 2010).

The underlying principle adhered to when making recommendations at this stage of the testing was that, where possible, the basic approach, structure, and design of the source questionnaire should remain unchanged unless reported problems affected data quality or comparability. Changes were made to the source questionnaire rather than allowing for specific country adaptions to be made at this stage, since the questionnaire had thus far been tested in less than half of the countries that would eventually be administering it.

Decisions about whether recommendations for changes to the source questionnaire were required were informed by the nature of the problem identified and the extent to which it was reported by different countries. Problems that became apparent from the testing can be divided into those which: (i) were cross-national in nature, (ii) were the result of translation errors or issues, and (iii) demonstrated linguistic or cultural differences between countries.

Those that were *cross-national,* in that similar problems were raised by different countries, indicated that problems remained with the design of the original source questionnaire and needed to be addressed. In these circumstances, less weight was given to the initial methodological evaluation using the quality indicators mentioned previously. For example, half of countries taking part in the testing reported respondent problems with understanding the question about "accessing the internet for any reason." The word "*access*" was sometimes misconstrued in the context of using the internet, for example, as referring to having the equipment to access the internet or having the knowledge to use the internet. Six countries queried the definition of "public transport"; and eight asked for clarification of the concept "leaving home" in the section on mobility. Perhaps not surprisingly, the term "pub," used in the section on leisure activities, was also queried and subsequently changed to "bar." Starting the survey with the section about internet access appeared to confuse respondents who could not initially grasp the relevance to the survey topic, while the concept of learning in a group environment was not conveyed sufficiently as evidenced by the fact that activities such as reading a book were included in the responses. However, the majority of cross-national issues related to the variability, or lack of internal harmonization, of the barrier questions where respondents were asked what prevented them from doing something in relation to each of the 10 life domains (see Section 22.3.1). In response to this problem, the format, question stem, and response categories of the barrier questions were standardized as far as possible throughout the questionnaire. This included the use of showcards at these questions since it was apparent that respondents did not always consider the response options as conceptualized by the survey designers. Another problem identified was that the concepts on which some of the sections were based were too broad (for example, the inclusion of both formal and informal education). In response, we reviewed the concepts and purpose of all the sections, resulting in further refinement and clearer summaries.

There were some errors made during *translation,* although fewer than anticipated, probably due to the fact that most countries adopted the Eurostat translation protocol. However, in Greece, for example, "learning opportunities" was translated into "learning program," which may have indicated more formal learning or training than the question intended. A Finnish translation of this question reordered the wording, which also possibly changed the meaning. In addition to particular words or phrases that were difficult to translate and maintain equivalence, in some countries, the question structure did not always translate well, meaning that questions were more cumbersome for interviewers to read out loud in other languages.

There were issues relating to particular words, phrases, and concepts being used in questions or the way in which measures had been designed, that did not translate at all, indicating specific *linguistic or cultural differences*. Where these differences might affect data comparability, the source questionnaire was amended or more detailed interviewer instructions provided. For example, it was said that there was no direct Finnish equivalent to the English concept of "making ends meet," and the term "usual expenses" was not understood in a consistent way by respondents. This question was subsequently changed to: "How easy or difficult is it for you to pay for the essential things in life?" with the addition of an introduction that gave examples "…such as food, clothing, medicine, housing, and transport." Although less likely to impact the functional comparability of the questionnaire, the terms "opera" and "theatre," used in the leisure section, were said to be the same in Estonian and Russian.

The leisure section of the questionnaire was completely revised and simplified after wave-two testing with the battery of leisure activities being replaced by one barrier question referring to cultural events in general, with some examples that included the theater but not the opera. In the Czech Republic, the meaning of the response choices "very often" and "quite often" was considered identical by those respondents who took part in the cognitive testing. These vague quantifying scales were subsequently eliminated from the questionnaire altogether. Internet use among older respondents was considered "sensitive" in countries where older people were unfamiliar with the term, but not in countries where the use of the internet was commonplace among all ages. Additional interviewer instruction was provided with a definition of the term "internet." Some countries queried why those who were economically inactive were being asked the employment section. An explanatory introduction was therefore added: "There are lots of reasons why people choose not to look for paid work." In Latvia, the word "discrimination" was said to be more easily understood by respondents than "unfair treatment." A decision was made to provide an explanation in the introduction: "Discrimination occurs when people are treated unfairly because they are seen as being different from others." Indeed, the title of the survey, which had originally included the word "disability" was said to have negative connotations in some Eastern European countries. This meant that those without impairments were reluctant to take part in the survey at all, thus leading to a decision to change the survey title (Wilmot and Meltzer 2010).

22.4.3 Wave-Three Testing

Wave-three testing involved all 28 countries. Prior to main-stage fieldwork, all countries conducted a pilot study and interviewer debrief. All but five countries also carried out pre-field testing in the form of cognitive interviewing. In over half of the participating countries, this was the first time testing had been carried out. However, there was no provision for any further review of the source questionnaire questions or concepts. Therefore, the purpose at this stage was to test and support the questionnaire translation, where words or phrases may have needed country-specific adaptation if they did not reflect the meaning of the original questions. It is worth noting that contractors conducting the main-stage in-country data collection were not necessarily the same organizations that had been involved in conducting the testing in the previous waves.

The revised English source questionnaire provided the specified question wording, question order, and routing to be used, along with supporting interviewer instructions. The same team-based translation protocol that was used during previous testing was applied prior to main-stage fielding. Countries sharing the same language could organize a coordinated translation process to obtain the same questionnaire, or a version of it, taking into account linguistic and cultural differences. For example, the Austrian questionnaire was based on the German one, while making adaptions to include the Austrian demographic standards to the EU core social variables. The Hungarian questionnaire was also used in Slovakia with minor adaptions made to suit the Slovakian context. Table 22.4 lists the languages into which the questionnaire and interviewer instructions were translated.

Each country was required to complete a translation report form detailing where problems in translation occurred and their proposed solution. Those who had designed the source questionnaire reviewed these forms and gave approval for any proposed in-country amendments, where appropriate. Only two countries deviated from the translation protocol without explanation, using a back-translation process. In general, only minor changes were reported when it was necessary to ensure that the questions were conceptually equivalent. For example, Finland reported that one of the answer options to the barriers to education question was ambiguous when translated directly into Finnish and, therefore, additional instructions were added to the existing interviewer instructions.

22.5 Survey Implementation

All of the countries taking part were contracted to Eurostat and asked to abide by the technical specifications mentioned earlier in relation to the survey implementation, and to administer the questionnaire as specified, so that cross-national comparisons could be made. To support the administration

Table 22.4 Languages used to conduct the survey by country taking part.

Country	Language(s)	Country	Language(s)
Austria	German	Latvia	Latvian; Russian
Belgium	French; Dutch	Lithuania	Lithuanian
Bulgaria	Bulgarian	Luxembourg	French; Luxembourgish; German
Cyprus	Greek	Malta	Maltese; English
Czech Republic	Czech	Netherlands	Dutch
Denmark	Danish	Norway	Norwegian
Estonia	Estonian; Russian	Poland	Polish
Finland	Finish; Swedish	Portugal	Portuguese
France	French	Romania	Romanian
Germany	German	Slovakia	Slovak; Hungarian
Greece	Greek	Slovenia	Slovenian
Hungary	Hungarian	Spain	Spanish; regional official languages (Catalan, Valenciano, Euskera, and Gallego); English
Iceland	Icelandic	Sweden	Swedish
Italy	Italian	United Kingdom	English

Source: Eurostat (Wilmot et al. (2014)).

of the questionnaire, throughout data collection, a list of frequently asked questions (FAQ) was maintained and periodically issued to all participating countries. This worked well in keeping all participating countries informed. Queries addressed in the FAQ related to clarification of some concepts or specific coding queries, as well as data-transmission queries. Other queries related to the standard survey notation used in the source documentation, as different countries and organizations used different standards; and the administration of the EU core social variables, which were the jurisdiction of the in-country national coordinators. Some countries requested permission for proxy data collection, which, in general, was not allowed as the questions were measuring individuals' desires and experiences, and would most likely have affected the accuracy of the data collected.[19] There were some requests for specific country adaptions, to add or change categories outside of the remit of European legislation e.g. at the discrimination question, which were also declined.

In reporting the findings, reference is now made to Wilmot et al. (2014). Following main-stage data collection, all countries were required to report

19 Proxy interviews were only permissible in exceptional cases when respondents were severely impaired, but had the mental capacity to provide consent.

any changes or adaptions made that had not been authorized. The Norwegian contractor reported collecting gross annual income rather than net monthly income according to the specifications of the EU Core Social Variables. All of the survey contractors followed the sequence of the questionnaire except in the UK, where information required to calculate region and household composition was moved closer to the end of the questionnaire. This was because the contractor felt that collecting such information near the start of the interview could have affected response rates. Although the survey was designed for face-to-face administration, some countries used alternative modes (i.e. telephone or internet). This decision was based on cost and the appropriateness of a particular mode in a given country. For example, in some countries such as Denmark, internet coverage is greater than in other countries, and respondents are typically more comfortable with internet use. This had implications for the design of the questionnaire, such as how to administer interviewer-coded responses at the discrimination section when the questionnaire was self-administered, or the repetitive nature of reading out lists over the telephone, which could impact response rates.

In general, individual countries made only minor adaptions to the question wording to account for the mode of administration. As mentioned by one of the contractors, in a computer-assisted telephone interview (CATI) survey, where telephone numbers are randomly generated, it is not possible to send out an advance letter; therefore, the interviewer introductions were adapted accordingly. When the barrier questions were administered via CATI and response options were individually prompted, as opposed to the use of multi-answer selection from a showcard, some contractors chose not to read out the last category "No, nothing prevents me from ....." since if the respondent had not answered "yes" to any of the barriers during the individual prompt, including the "other category," then it could be assumed that the option "No, nothing prevents me…" applied. It should be noted that when response lists were individually prompted, the overall length of the interview increased. In general, adaptions made were designed to minimize mode effects and maintain cross-national comparability.

However, some of the adaptions may have affected the validity of cross-national findings. Denmark, for example, appeared to have added a filter question in the employment section asking "Do you want a paid job?" It is unclear how this was administered and if respondents had the opportunity to hear all of the barriers that might have applied to them before answering. The Slovenian contractor added a category to the list of barriers provided to the question about access to private vehicles: "I do not have anyone to take me." In this regard, the Slovenian data may not be comparable to other countries. Interestingly, Italy reported that because it was no longer compulsory to report marital status on official documents such as identity cards, respondents were more reluctant to provide marital status information in surveys.

22.6 Lessons Learned

This chapter has described the design and development of the EHSIS data collection instrument administered in 28 European countries during 2012–13. In order to attempt to ensure equivalence, the survey questionnaire was tested across countries in three waves, using both qualitative cognitive interview and quantitative pilot pretesting techniques. A cumulative testing approach was taken – at each of three waves of testing, the number of countries involved increased. At waves one and two, the English source language questionnaire – concepts, question wording, formatting, and supporting documentation – was revised, based on the pretesting findings. Prior to fielding the main survey, findings from a third wave of pretesting allowed for country-specific adaptations to take account of any linguistic or cultural issues identified at this stage that could impact on question meaning, but that did not fundamentally change the survey concepts. Throughout the process, subject matter experts and survey methodologists oversaw the design and development of the instrument. Quality controls were in place to ensure that findings from the testing were reliable and that administration of the questionnaire was consistent across countries. Post-survey, a final quality report was provided by participating countries, which included a performance assessment of the questionnaire and reporting of any unauthorized questionnaire amendments, to inform any cross-national data comparison. The rest of this section discusses the key features of the design and successful implementation of the EHSIS data collection instrument, as well as lessons learned along the way, specifically the importance of:

i) Sufficient development time
ii) Simplicity with respect to question concepts and wording
iii) Objectivity and transparent decision-making
iv) A cumulative approach to testing the survey instrument
v) Standardization and harmonization
vi) Translation protocols and pretesting guidelines
vii) Open lines of communication with participating counties throughout data collection
viii) Quality assessment procedures throughout the process

A considerable amount of time was required for the design and subsequent testing of the instrument and main-stage fielding. Indeed, the conceptual approach taken to the measurement of disability from the perspective of societal integration took a lifetime to realize, and the possibility of the application of the concept in the form of a survey began years prior to data collection.

To distill a complex classification into a survey questionnaire that can provide comparable data across countries required clarity of the underlying concepts on which the survey questions were based, along with simple question wording and constructs that were standardized throughout the data collection instrument.

The importance of a steering group with respect to the questionnaire design should not be underestimated. The EHSIS steering group provided subject matter expertise, along with the cultural insight required to support the questionnaire design. It scrutinized the questionnaire design process and in particular the logic and evidence used to make decisions. This external scrutiny afforded a layer of objectivity to the design process. However, to realize the benefits of the steering group there needs to be what Lynn (2003) refers to as effective communication and determination from all parties to work together toward a common, shared goal. Certainly, it was my experience that the commitment on the part of all of those involved in steering the development of EHSIS questionnaire was a strong contributing factor to its success.

The cumulative approach to the testing of the survey questionnaire for EHSIS during the development process worked well. The testing started in the country in which the source questionnaire was initially developed, and built on the findings by testing in different languages and cultures in ever-increasing waves. This cumulative process was essential for enabling the EHSIS questionnaire designers to acquire the knowledge and understanding needed to design an instrument that could provide equivalent validity across countries, and the necessary time to do it. The source questionnaire did not require testing in all participating countries during the development phase. My conclusion here is pragmatic to some degree. The spread of European countries taking part in the initial waves of testing produced a source questionnaire for use by all, which required relatively minor in-country adaption for linguistic or cultural equivalence at main-stage implementation.

Although not all of the countries taking part in the initial waves of testing adopted the same testing techniques or had the same level of experience, particularly in regard to cognitive interviewing, the quality process approach taken to assessing the methodology used and subsequently the findings to inform the cross-national instrument review allowed the questionnaire designers to make judgments on whether or how to amend the source questionnaire. In hindsight, guidelines on how to conduct the cognitive testing would have been useful. Certainly such guidelines are now provided by the Washington Group on Disability Statistics, for example (Miller 2019).

Standardization throughout the questionnaire was extremely important and was one of the key messages arising from the EHSIS wave-two testing. This was particularly true with regard to the important barrier questions. Standardizing these questions – the wording of the question stem and answer options, and the presentation of answer options to respondents using showcards – helped to improve respondent comprehension and reduce cognitive burden and the time it took to administer the survey.

Based on the experience of implementing the EU core social variables, which were output harmonized and unfamiliar to some of the contractors administering the main-stage survey, I would support our use of harmonized inputs for the EHSIS module and recommend their use throughout, where question wording, response categories, and routing are specified.

Bearing in mind that EHSIS was originally designed to be interviewer-administered, I would further recommend that the separate interviewer instructions, provided along with the source questionnaire, should be considered part of the questionnaire design process, and I would argue that they are a critical part of the input harmonization process. In this regard, I would also advocate that these interviewer instructions should be written by those tasked with designing the questionnaire, as they were for EHSIS, to help convey clearly the survey concepts and method of administration.

It would be remiss not to mention the importance of the translation process as integral to the success of the questionnaire to operate as a comparative instrument (Harkness 2003). I have not focused on the translation process in this chapter because it is well documented in the literature, and Eurostat guidelines follow best practice. For the most part, the translation protocol was followed and translation reporting forms from each country vetted as the survey development progressed. Furthermore, very few translation errors were reported. As mentioned earlier, at wave three, the primary purpose of the pretesting was to test and support the questionnaire translation and not the conceptual basis of the questionnaire, and some additional translation issues were identified during wave-three testing. This leads me to consider the advantage of using a different organization to carry out the initial testing than that implementing the main-stage survey within the same country, as some of these additional translation issues were identified by the different organizations within the same language administration.

Although the EHSIS source questionnaire had been designed for face-to-face interviewer administration, at final implementation, allowances (mostly pertaining to cost) were made for countries to select a different mode or multiple modes of administration. Although ideally the survey would have been administered in a single mode across countries, my recommendation in this regard is that questionnaire designers provide instructions on how the questionnaire should be administered in different modes as part of the questionnaire design phase. Of course, there is only so much the questionnaire designer can do to minimize potential mode effects. For example, even where an attempt is made to minimize recency effects by prompting individual response items during telephone administration, there may be effects on the cross-comparability of the data collected with higher levels of reporting to those individually prompted items compared with multiple response selection using a showcard (Nicolaas et al. 2015) – not forgetting the impact of extended interview length and respondent burden when questions are administered this way, which could result in respondent satisficing (Holbrook et al. 2003). The possibility of positivity bias should also be considered when comparing data collected through interviewer administration with that collected using self-administration (Bowling 2005).

So often it is the case that those who designed the survey instrument are not involved at the survey implementation stage. One of the advantages for EHSIS was that the questionnaire designers, with their understanding of how the questionnaire concepts and question wording had developed cross-nationally, and having written the interviewer instructions, were available at implementation to field queries in relation to both the content and the administration of the survey questionnaire. The use of the FAQs document, issued periodically to communicate decision-making in this regard, proved effective for ensuring that all participating countries were aware of the rationale for decisions made and were able to make any necessary amendments to the instrument in unison as fieldwork progressed.

A risk for cross-national surveys involving so many "moving parts" and "actors" is that some decisions on questionnaire wording or formatting are made locally by individual countries rather than centrally, and that deviations from agreed protocols are not reported until after survey completion. The quality reports provided by each participating country post survey allowed for a final comparative review of the questionnaire fielded in each country that informed data analysis.

22.7 Final Reflections

Certainly the design and development of a cross-national comparable survey instrument is a complex undertaking and there are many methodological challenges to overcome. The EHSIS illustrates the value of effort spent on improving the conceptual clarity and multiple waves of testing in a number of countries to refine the source and translated questionnaires. In a large-scale, cross-national enterprise such as the EHSIS it is not always possible to test questions in every country in every wave, but a considered cumulative cross-national testing approach can provide a practicable and effective solution. The EHSIS also highlights the importance of input harmonization with respect to providing a source questionnaire but also the importance of simple and standardized survey questions and response options. Communicating the decision-making that underpins the final questionnaire and administration protocols to the in-country teams can help them understand the measurement aims and objectives of the survey. This can reduce any further customization of the questionnaire during main-stage fielding. Systematically collecting information from each national team at the end of the survey about any modifications or deviations they made from the issued protocol helps to assess the impact on data comparability during analysis.

Despite the methodological challenges in undertaking cross-national research, the potential rewards are great. In the words of the German contractor conducting the main-stage EHSIS: "The strongest point about the EHSIS is its set-up and existence: it deserves a positive assessment that the EHSIS attempts to close the obvious data and information gaps…in context of the international monitoring requirements of the UN Convention on the Rights of Persons with Disabilities and the European Disability Strategy."

Acknowledgments

I would like to acknowledge and thank everybody who contributed to the development of this survey from conceptual inception to main-stage administration. In particular, those at the European Commission who sponsored and oversaw the research, especially Lucian Agafitei; the Task Force for Streamlining Disability Statistics and others involved in steering the project; staff at Sogeti Luxembourg who supported the cross-national administration of the survey; and staff at RTI International who contributed to the final survey quality assessment report on behalf of Eurostat. I am also grateful to the individual survey contractors who contributed to the development of the survey questionnaire and implemented the data collection across Europe, and the members of the public who gave their time and cooperation. In particular, however, I would like to acknowledge the life-time work of my mentor and advocate Professor Howard Meltzer from the University of Leicester, whose wish it was to see Europe measure disability from the perspective of the barriers to social integration as reported by the respondent. Sadly, Professor Meltzer passed away at the beginning of 2013, but it was my great privilege to be able to help support the survey through its implementation across Europe.

Disclaimer

Any opinions expressed in this chapter are those of the author and do not represent the European Commission's official position.

References

Agafitei, L. (2010). Current statistical and data developments relevant to measuring disability equality in Europe. ANED Annual Meeting. http://www.disability-europe.net/content/aned/media/Agafitei_Lucian_Estat%20presentation%20ANED%202010.pdf.

Agafitei, L. (2012). Recent developments on disability statistics in the European Union. 10th meeting of the Washington Group on Disability Statistics. http://www.cdc.gov/nchs/ppt/citygroup/meeting10/WG10_Session1_3_Agafitei.pdf.
Bajekal, M, Harries, T., Breman, R. et al. (2004). Review of disability estimates and definitions. Department of Work and Pensions In-house Report 128. http://www.eurohex.eu/bibliography/pdf/Bajekal_reportDWP_2004-0697451521/Bajekal_reportDWP_2004.pdf.
Baumberg, B., Jones, M., and Wass, V. (2015). Disability prevalence and disability-related employment gaps in the UK 1998e2012: different trends in different surveys? *Social Science & Medicine* 141: 72–81.
Bowling, A. (2005). Mode of questionnaire administration can have serious effects on data quality. *Journal of Public Health* 27 (3): 281–291.
European Social Survey (2018). *ESS Round 9 Translation Guidelines*. London: ESS ERIC Headquarters. Retrieved from: https://www.europeansocialsurvey.org/docs/round9/methods/ESS9_translation_guidelines.pdf.
Fitzgerald, R. (2015). Sailing in unchartered waters: Structuring and documenting cross-national questionnaire design. GESIS-Papers 2015|05. http://www.gesis.org/fileadmin/upload/forschung/publikationen/gesis_reihen/gesis_papers/GESIS-Papers_2015-05.pdf.
Hakim, C. (2000). *Research Design: Successful Designs for Social and Economic Research*, 2e. London: Routledge.
Harkness, J.A. (2003). Questionnaire translation. In: *Cross-Cultural Survey Methods* (eds. J.A. Harkness, F. van de Vijver and P.P. Mohler), 35–56. Hoboken, NJ: Wiley.
Harkness, J.A., Braun, M., Edwards, B. et al. (2010a). Comparative survey methodology. In: *Survey Methods in Multinational, Multiregional, and Multicultural Contexts* (eds. J.A. Harkness, M. Braun, B. Edwards, et al.), 3–16. Hoboken, NJ: Wiley.
Harkness, J.A., Braun, M., Edwards, B. et al. (2010b). *Survey Methods in Multinational, Multiregional, and Multicultural Contexts*. Hoboken, NJ: Wiley.
Harkness, J.A., Edwards, B., Hansen, S. et al. (2010c). Designing questionnaires for multipopulation research. In: *Survey Methods in Multinational, Multiregional, and Multicultural Contexts* (eds. J.A. Harkness, M. Braun, B. Edwards, et al.), 33–57. Hoboken, NJ: Wiley.
Haselden, L. and White, A. (2001). Developing new quality indicators in social surveys. Proceedings of the 2001 Statistics Canada Symposium: Achieving data quality in a statistical agency: a methodological perspective.
Holbrook, A.L., Green, M.C., and Krosnick, J.A. (2003). Telephone versus face-to-face interviewing of national probability samples with long questionnaires comparisons of respondent satisficing and social desirability response bias. *Public Opinion Quarterly* 67: 79–125.
Kostanjsek, N. (2011). Use of the international classification of functioning, disability and health (ICF) as a conceptual framework and common language

for disability statistics and health information systems. *BMC Public Health* 11 (Suppl. 4): S3. Retrieved from https://doi.org/10.1186/1471-2458-11-S4-S3.

Johnson, T.P., Pennell, B., Stoop, I.A.L., and Dorer, B. (2019). *Advances in Comparative Survey Methods: Multinational, Multiregional, and Multicultural Contexts (3MC).* NJL Wiley: Hoboken.

Lynn, P. (2003). Developing quality standards for cross-national research: five approaches. *International Journal of Social Research Methodology* 6: 323–336.

Lynn, P., Japec, L., and Lyberg, L. (2006). What's so special about crossnational surveys? In: *Conducting Cross-National and Cross-Cultural Surveys: Papers from the 2005 Meeting of the International Workshop on Comparative Survey Design and Implementation (CSDI)* (ed. J.A. Harkness), 7–20. http://www.gesis.org/fileadmin/upload/forschung/publikationen/zeitschriften/zuma_nachrichten_spezial/znspezial12.pdf.

Meltzer, H. (2007). Module on social integration of disabled people. Technical Group (HIS) meeting, Luxembourg. https://circabc.europa.eu/webdav/CircaBC/ESTAT/healthtf/Library/technicalsgroupshis/2007_20-21062007/Tech-HIS-2007-5.5%20EDSIM%20project.pdf.

Meltzer, H. (2008). Report on the development of a survey module on disability and social integration. https://circabc.europa.eu/faces/jsp/extension/wai/navigation/container.jsp.

Meltzer, H. (2009). Report from the EU-US Seminar on Employment of Persons with Disabilities. Brussels, 5-6 November 2009. Organised jointly by the European Commission Unit for Integration of People with Disabilities and the U.S. Office of Disability Employment Policy in the Department of Labor.

Meltzer, H. (2011a). Opportunities for statistical indicator development relevant to measuring disability equality in Europe. Academic Network of European Disability (ANED) experts Conference, Brussels. http://www.disability-europe.net/content/aned/media/Powerpoint%20Meltzer_Howard_Presentation.pdf.

Meltzer, H. (2011b) The structure and content of the European Health and Social Integration Survey (EHSIS). Washington Group meeting, Bermuda. https://www.cdc.gov/nchs/ppt/citygroup/meeting11/wg11_session7_2_meltzer.pdf.

Meltzer, H., Wilmot, A., and Demarest, S. (2010). Complimentarity of definitions and methods of measuring disability across European surveys. European Commission. https://circabc.europa.eu/webdav/CircaBC/ESTAT/disabilitystatistics/Library/disability_statistics/meetings/meeting_28-29_2010/Agenda%20item%204.1%20-%20Complimentarity%20paper.pdf.

Miller, K. (2019). Cognitive interviewing methodology to examine survey question comparability. In T.P. Johnson, B. Pennell, I.A.L. Stoop, & B. Dorer (Eds) *Advances in Comparative Survey Methods: Multinational, Multiregional, and Multicultural Contexts (3MC).* (pp 203-225). Wiley.

Mohler, O. (2007). What is being learned from the ESS? In: *Measuring Attitudes Cross-Nationally: Lessons from the European Social Survey* (eds. R. Jowell, C. Roberts, R. Fotzgerald and G. Eva), 79–94. London: Sage Publications.

Nicolaas, G., Campanelli, P., Hope, S. et al. (2015). Revisiting "yes/no" versus "check all that apply": results from a mixed modes experiment. *Survey Research Methods* 9 (3): 189–204.

Ramm, J. and Otnes, B. (2008). The European Module on Disability and Social Integration EDSIM – Nordic project. Statistics Norway Publication 2009/02. https://www.ssb.no/a/english/publikasjoner/pdf/doc_200902_en/doc_200902_en.pdf.

Sebastiani G., Tinto A., Battisti A. & De Palma E. (2010). Cognitive interviewing as a tool for improving data quality in surveys: Experiences in Istat. Paper presented at the 45th Scientific Meeting of the Italian Statistical Society (SIS), Padova, Italy.

Smith, T.W. (2003). Developing comparable questions in cross-national surveys. In: *Cross-Cultural Survey Methods* (eds. J.A. Harkness, F.J.R. Van de Vijver and P.P. Mohler), 69–92. Wiley.

Smith, T.W. (2015). Resources for conduction cross-national survey research. *Public Opinion Quarterly Special Issue: Cross-Cultural Issues in Survey Methodology* 79.

Tourangeau, R. (1984). Cognitive science and survey methods. In: *Cognitive Aspects of Survey Methodology: Building a Bridge between Disciplines* (eds. T. Jabine, M. Straf, J. Tanur and R. Tourangeau), 73–100. Washington, DC: National Academies Press.

Wilmot, A. and Meltzer, H. (2010). Evaluation and review of results from the testing of the European Disability and Social Integration Module (EDSIM). https://circabc.europa.eu/sd/a/c39cab66-3ba6-4d0e-b027-336752948afc/EDSIM%20Final%20report%20for%20Circa.pdf.

Wilmot, A. Peytcheva E,& Swicegood J. (2014) *EU comparative quality and technical report: European Health and Social Integration Survey*. On behalf of Eurostat (restricted access).

World Health Organization. (2001). International classification of functioning, disability and health. https://apps.who.int/iris/bitstream/handle/10665/42407/9241545429.pdf;jsessionid=29E3BE7E5E658E4D7B3E780C6692BE74?sequence=1.

Part V

Extensions and Applications

23

Regression-Based Response Probing for Assessing the Validity of Survey Questions

Patrick Sturgis[1]*, Ian Brunton-Smith*[2]*, and Jonathan Jackson*[1]

[1] *Department of Methodology, London School of Economics and Political Science, London, UK*
[2] *Department of Sociology, University of Surrey, Guildford, UK*

23.1 Introduction

Over the past three decades there has been a marked upturn in interest in and application of formalized methods for the evaluation and testing of survey questions (Willis 2005). Given the centrality of question meaning and interpretation to the accuracy of survey estimates, it is perhaps surprising that the advent of this "pretesting revolution" was so long in gestation. Survey practitioners have, since the early days of survey research, been aware of marginal differences arising from variation in question wording and order (Rugg 1941). Yet, in contrast to the field's treatment of sampling error, the thorny issue of assessing and improving measurement of the validity and reliability of survey questions was for a considerable time left in a state of comparative neglect.

The first International Conference on Questionnaire Design, Development, Evaluation, and Testing in 2002, as well as subsequent meetings and publications (Presser et al. 2004; Saris and Gallhofer 2014; Willis 2005), have been instrumental in advancing the quality and range of pretesting methods available to practitioners. But it would be hubristic to claim that these developments have resolved some of the longstanding limitations of pretesting methodologies. Think for instance of the tension between small samples yielding "rich" data on the one hand, and large-scale field experiments that offer little insight into respondents' cognitive processes on the other.

In this chapter we set out an approach to question testing that combines the strengths of intensive small-sample qualitative approaches with the inferential power of large-scale field trials and experimental manipulations (Creswell and Plano Clark 2007). We call this *regression-based response probing*. We begin with a brief overview of some of the strengths and weaknesses of existing qualitative and quantitative approaches to assessing question validity. We then

Advances in Questionnaire Design, Development, Evaluation and Testing, First Edition.
Edited by Paul C. Beatty, Debbie Collins, Lyn Kaye, Jose-Luis Padilla, Gordon B. Willis, and Amanda Wilmot.

describe the approach we propose to combine the qualitative emphasis on mapping the cognitive frames elicited by different question wordings with the ability to make quantitative inferences about their frequency in the broader population. We illustrate the method with exemplar implementations. The first example focuses on the measurement of generalized trust (Uslaner 2002), drawing on existing published research described in Sturgis and Smith (2010). The second example focuses on the measurement of fear of crime (Farrall et al. 2009) and presents new data and analysis.

23.2 Cognitive Methods for Assessing Question Validity

The historical under-emphasis on non-sampling errors in survey methodology was partly, no doubt, a function of the generally more difficult nature of the problem-what Bob Groves has termed "the tyranny of the easily measurable" (Groves 1991). While the mathematics underpinning complex variance estimation may appear daunting to the non-specialist, the problem is at least theoretically quite tractable. Assessing the validity of questions tapping often vaguely defined psychological states, in contrast, poses significantly greater theoretical and empirical challenges. Another key factor in the skewed nature of the field's focus, however, was the fragmentary and rather atheoretical nature of early work on what came to be loosely known as "context effects." Initial treatments of order and wording effects were limited to empirical demonstrations that seemingly irrelevant aspects of the interview could sometimes have substantial effects on marginal distributions (Belson 1981). Yet, no genuinely sustained or collective effort was made to meld these disparate findings into a coherent theoretical narrative; we knew that wording and context mattered, but not why (Schuman and Presser 1981).

This position began to change with the growing influence of cognitive psychology on survey methodology from the late 1970s. Under the banner of Cognitive Aspects of Survey Methodology (CASM), survey methodologists conceptualized the survey-response process as a series of cognitive stages, each of which was prone to inaccuracies arising out of limitations of human memory, motivation, and processing capacity (Jabine et al. 1984). By drawing on models of attention, memory, comprehension, and judgment, context effects were recast as special cases of more general theories of cognition. This, in turn, led to the development of genuinely explanatory and predictive accounts of survey measurement errors, most notably in the work of Tourangeau, Sudman, and Schwarz (Schwarz and Sudman 1995; Tourangeau et al. 2000), Saris (Saris and Gallhofer 2014), and Krosnick (Krosnick et al. 1996).

The influence of social cognition was not, however, limited to the development of a more robust theoretical foundation to our understanding of the response process. Along with theoretical models, cognitive psychologists also imported their favored methodologies for the delineation of cognitive processes – the "think-aloud" and "protocol" analyses (Ericsson and Simon 1980; Loftus 1984). Applied to draft questionnaires, these "cognitive interview" techniques quickly proved to be a cost-effective means of problem identification. By the mid-2000s, the existence of "cognitive laboratories" in fieldwork agencies had become the rule rather than the exception in the industry (Beatty and Willis 2007).

Albeit covering a wide range of practices, variants of the cognitive interview technique – concurrent, retrospective, probing, and so on (Willis 2005) – soon flourished. Courses, conferences, and books began to appear, to meet the growing demand for research, development, and training in its implementation. Yet despite their popularity and widespread use among survey practitioners, much is still not known about the objective merits of this family of question-testing strategies and techniques (Presser 1984). Based, as they are, on small, usually nonrandom samples, it can be difficult to evaluate the generality of any problems identified, or to determine the effects of changes implemented in the light of findings from cognitive interviews on survey error. Rules of thumb and inductive strategies of "conceptual saturation" have generally operated for determining appropriate sample sizes (Beatty and Willis 2007).

Attempts to quantify sample size requirements in more systematic ways have found that standard assumptions and practices may be inadequate for identifying the full range of potential problems (Conrad and Blair 2009). It has been noted that, for a field of endeavor devoted to accurate quantitative inference, a focus on qualitative evidence for question evaluation is somewhat paradoxical (Tourangeau et al. 2000). And even for the set of potential problem that cognitive interviews identify, even if incomplete, it is not clear whether the amended questions provide more accurate data than the original versions. The small number of large-scale field evaluations conducted to date have been mixed in their conclusions regarding reductions in the mean squared error of estimates (Blair and Presser 1993; Willis et al. 1999). As Willis notes, there remains the very real possibility "that none of them are of help in detecting and correcting question flaws" (Willis 2004, p. 26).

In contrast to the widespread adoption of cognitive interview techniques, the method of experimental field trials for evaluating survey questions has been rather less frequently used. The standard survey experiment randomly assigns respondents to conditions that manipulate aspects of question wording, order, and response format (see Krosnick 2011, for a review). If the experiment is designed and conducted appropriately, differences between experimental conditions can be ascribed to the manipulated variable(s), yielding powerful and generalizable inferences (Tourangeau 2004). The strength of the method is that it incorporates "double randomization," that is, in the selection of the sample and in the allocation of the treatment.

Without doubt, a key factor in the differential uptake of pretest methods relates to cost, with field experiments costing considerably more than small-scale cognitive tests, while generally yielding less-detailed information about potential question problems. But as Presser et al. (2004) note, little is presently known about the *cost-effectiveness* of the different pretesting techniques. The higher total cost of an experimental approach may represent better value for money if it yields robust insights about question problems and solutions. Concomitantly, low-cost, small-sample methods may be evaluated less favorably if they fail to yield genuine improvements in the accuracy of population estimates. And, the rapid growth of web surveys is substantially reducing the cost differential between large-sample field experiments and more intensive, qualitative approaches (Callegaro et al. 2014).

The robustness of the inferences yielded by survey experiments – combining the internal validity of experimental design with the external validity of sample surveys – has been key to understanding how respondents make sense of survey questions (Krosnick 2011). As early as 1941, Benson noted that "increasingly accurate measurement of public opinion has come about largely because of experimentation which has resulted in improved techniques" (Benson 1941, p. 79). Since these early forays into the split-ballot design, number of studies using this type of design have deepened our understanding of the response process and, therefore, of how to write questionnaires (cf. Schuman and Presser 1981; Sudman et al. 1996; Tourangeau et al. 2000).

Be that as it may, there are also well-documented limitations of survey experiments for assessing the quality and accuracy of questionnaire items, beyond their generally high cost. Most importantly in this regard is the rather "black box" nature of the evidence they provide. That is to say, while a survey experiment is capable of providing clean causal inference regarding the effect of the manipulated variable(s) on the outcome (and also of generalizing this causal effect to the broader population) it will often say little about the underlying mechanism, nor which of the implemented treatments should be preferred with regard to the accuracy of estimates. This is particularly so for attitudinal and other psychological variables, where no external criterion can be used as a gauge of accuracy.

For example, imagine we were to randomly assign respondents in a survey to receive one of two alternate versions of an attitude question (where the manipulation might be to include or exclude some verbal qualifier) and, upon analyzing the results, discovered that one version produces significantly higher scores on the response scale than the other. We could be confident that the difference was caused by the manipulation and that if we were to repeat the experiment on a different random sample from the same population, we would find a similar result. However, we would know little or nothing about why scores were higher on one version than the other, nor which version produces the more accurate data. There are of course strategies that can be used to assess the validity of alternate measures, such as criterion and construct validity tests (Krosnick 2011), although these have limitations of their own and require alternative measures of the same, or theoretically related, constructs to be available.

23.3 Regression-Based Response Probing

The approach we describe here seeks to ally the strengths of the richer qualitative data produced in cognitive interviews with the inferential power of large-sample field experiments. Respondents are randomly assigned to treatment conditions that receive alternative versions of a question intended to measure an attitudinal concept in the usual manner. The alternative versions could range from the relatively minor, such as where a question is placed in the questionnaire, to the more substantial, such as different question versions tapping the same underlying concept. Once respondents have answered the questions they were assigned to in their treatment group, they are all administered a verbal probe, which asks them to describe, in their own words, what came to mind when they were answering the question. The exact wording of the probe can of course be tailored to each specific context. In the examples presented here, the probes are retrospective, though there is no reason why concurrent probes could not be used instead.

The verbatim responses to the probes are recorded by the interviewer. Again, in both examples described here, the interviewers transcribed the answers by hand, though they could also be audio-recorded to increase the detail and accuracy of recorded answers (see Sturgis and Luff 2015). The transcribed verbatim responses are then coded to a frame designed to capture the full range of potential cognitive frames elicited. By *cognitive frames* we mean the preexisting systems of thoughts and associated behaviors that organize categories of information in memory (DiMaggio 1997). As is usual in the development of coding frames, there is likely to be some need for iteration and expansion of the initially developed codes as they are applied to the full set of responses. Depending on the specific context and the nature of the concept the questions are intended to measure, the coding frame may need to be adapted or aggregated in order to reflect the conceptual content.

Once the coding has been completed, analysis comprises both descriptive and multivariate components. The distributions of the codes are compared across conditions, as well as across the answer categories of the attitude questions. These assessments enable an evaluation of whether and how cognitive frames differ across question variants and how the observed distributions compare to a priori theoretical expectations. For example, the researcher can use differences in the distribution of codes to assess which of two alternative questions more closely corresponds to the definition of the concept he or she wishes to measure (Ternor et al. 2011).

A limitation of using the codes in this way is that they could simply be reconstructions "after the event," rather than accurate descriptions of what the respondent was thinking as they answered the question (Wilson et al. 1996). These reconstructions could themselves be related to characteristics of respondents, such as their age, sex, gender, and social class. Considering the univariate distributions of the codes could therefore be affected by differences across demographic groups in how frames are elicited and reconstructed. The second stage of analysis therefore assesses the conditional relationship between the attitude questions and the codes by estimating a multivariate regression, with the attitude variable as the outcome and the codes and respondent characteristics as predictors. The model has the following form:

$$logit(\pi_i) = log\left(\frac{\pi_i}{1 - \pi_i}\right) = \mathrm{B}^T X_i \tag{23.1}$$

where the attitudinal outcome, y, is binary and π_i represents the probability that $y_i = 1$ conditional on a vector of observed covariates X_i where $\mathrm{B} = \{\beta_0, \beta_1, \ldots, \beta_k\}$ is a vector of regression parameters and X_i is a vector of covariates, which include the verbatim code dummies and the individual level controls. We specify a binary logistic model here for simplicity of exposition, though the approach can be extended to an ordinal logit or probit model for multi-category outcomes without loss of generality. The key parameters in Eq. (23.1) are the beta coefficients for the code dummies that show how the probability of selecting different answer categories is related to the presence of different cognitive frames during the response process. Additionally, it is straightforward to include interactions between the codes and individual-level covariates. Significant coefficients on these interaction terms indicate that presence of the cognitive frame has a stronger association with the attitude response for the demographic group in question.

It is important to note that the approach we describe here is not intended to be a means of identifying problems such as ambiguities, technical jargon, inconsistencies, and so on in draft questions. Rather, the objective of the method is to elicit the cognitive frames induced in respondents' minds when they read alternative versions of an attitude question. The researcher uses the observed frequency of different cognitive frames across question variants as a means of assessing whether respondents are answering the questions in a way that corresponds to the conceptual definition of the attitude. They also use the (conditional) association between the propensity for a particular cognitive frame to be mentioned and the response selected on the attitude question as a way of better understanding the factors influencing selection of particular response alternatives. The approach is what has been referred to as a form of "cognitive validity" assessment (Karabenick et al. 2007); validity is assessed by how well the thought processes induced in respondents' minds align with the intentions of the question designer.

23.4 Example 1: Generalized Trust

The recent prominence of Robert Putnam's wide-ranging account of America's civic decline has resulted in a proliferation of national and international survey datasets containing items intended to tap the generalized trust dimension of social capital (Uslaner 2002). Usually, the question included in surveys is an item taken from Rosenberg's "faith in people" scale, which requires respondents to choose between the binary alternatives "most people can be trusted" and "you can't be too careful in dealing with people" (Rosenberg 1956). In addition to this generalized trust question (GTQ), an increasingly common approach to measuring generalized trust in surveys requires respondents to state how much they trust people in their "neighborhood" or "local area."

Although the "Trust in Neighbors" (TiN) item is treated as though it taps the same underlying construct as the GTQ, it tends to produce higher estimates of trust in the same populations. For instance, the Social Capital Community Benchmark survey found 47% of respondents reporting that "most people can be trusted," while 83% of respondents to the same survey said that they trust people in their neighborhood either "some" (sic) or "a lot" (Putnam 2007). A key objective of questions tapping generalized trust is that they should measure an individual's level of trust in "people in general" rather than in specific, known individuals such as colleagues, family, and friends.

Sturgis and Smith (2010) used regression-based response probing to assess the validity of the two different trust questions. Respondents in a face-to-face interview survey in the UK were randomly assigned to receive either the GTQ ($n = 481$) or the TiN ($n = 507$). They were then asked, "Who came to mind when you were thinking about 'most people'/'people in your local area'?" Interviewers recorded the verbatim responses, and these were subsequently coded by a team of trained coders to a frame, resulting in 29 distinct categories of response. These initial codes are very literal in nature, describing the range of answers obtained, with no attempt made to group individual responses according to a set of common thematic criteria. Next, the 29 first-stage codes were aggregated into a subset of 5 higher-order codes based on the inferred relationship to the respondent of the person or people mentioned.

The first higher-order code subsumed all verbatim responses in which respondents made reference to a person or persons who would be known to them personally, such as family, friends, neighbors, and colleagues. The second comprised all responses that made reference to abstract aggregations or categories such as "anyone," "people in general," "foreigners," and "no-one in particular" that would not be reducible, in any straightforward manner, to a person or persons with whom the respondent would be personally acquainted. The third code contained responses that referred to people living in the respondent's town, village, or local area, while code four combined responses that directly referenced professional and occupational groups such as policemen, doctors, tradesmen, and politicians. The fifth higher-order code was a residual category, comprising responses that did not mention specific persons or groups but made reference to generic reasons for trusting, or not trusting other people (e.g. "trusting is naive").

A descriptive analysis demonstrated three important points: (i) that both questions elicited high rates of responses that indicated that respondents were thinking about known individuals; and (ii) that this tendency was greater for the TiN question than for the GTQ; but (iii) respondents who reported that people they knew came to mind reported considerably higher rates of trust compared to those who thought of people in general. These effects were confirmed using multivariate regressions; controlling for respondent age, sex, social class, marital status, self-reported health, and education, the odds of selecting a "trusting" response alternative was approximately five times greater if they reported having thought about someone they knew, compared to having thought about people in general.

These results enabled conclusions regarding the validity of the alternative measures of generalized trust based on the cognitive frames that they induced in respondents. For instance, they demonstrated that, counter to theoretical expectation, a substantial minority of respondents think about people who are known to them rather than people in general when answering the GTQ. Because inferences were based on large representative samples and between-subject randomization to treatment groups, the researchers were able to be confident that the findings could be generalized to the broader population and were not influenced by within-subject practice effects (Krosnick 2011).

23.5 Example 2: Fear of Crime

Fear of crime has long been an important topic of research within criminology and related disciplines (for reviews see Farrall et al. 2009; Lorenc et al. 2012). Yet, despite its centrality in the field, debate continues on the foundational issue of how best to define and measure the construct. A good deal of criticism has focused on what some criminologists call the classic single indicator: "How safe do you feel walking alone in your neighborhood after dark?" (response alternatives range from "very safe" to "very unsafe"). For example, Garofalo and Laub (1978) pointed out that this question (i) does not mention crime, (ii) fails to provide a specific geographical reference, and (iii) invokes a hypothetical situation. They argued that the measure likely taps into broader symbolic issues as well as, or perhaps even more than, the "fear" of becoming a victim of crime. The Figgie

Report (Figgie 1980) extended this point by differentiating between "formless" fear (measured by perceptions of safety and abstract threats to one's security) and "concrete" fear (concern about becoming a victim of personal crimes).

Concerns about the conceptual validity of standard measures of fear of crime led to the development of new specifications that: mentioned a particular emotion (most often "fear" or "worry"), differentiated between certain specific types of crimes (e.g. being burgled while away from home, being physically attacked by a stranger in the street), and asked about the intensity or frequency of emotional experience. These indicators deliberately focus less on generalized anxieties about crime and more on concrete concerns about the risk of criminal victimization (Ferraro 1995).

More recently, researchers have combined different types of measures to capture different emotional experiences (Gray et al. 2008; Jackson and Gray 2010). Farrall et al. (2009), for instance, distinguished between an episodic pattern of worrying about risk (an event-based set of experiences) and a more diffuse anxiety about crime and connected issues (a generalized sense of risk and concern about what crime represents in society). They found that people who worried frequently about victimization tended to live in high-crime areas, had extensive experience of victimization, and were especially concerned about local neighborhood breakdown. But for those who lived in more affluent areas, had less experience of crime, and were less concerned about local incivilities, "fear" manifested as a more diffuse anxiety. This generalized attitude was, these authors contend, more akin to an awareness and management of risk and does not neatly map onto the frequency of everyday worry. The outcome of these conceptual debates and empirical investigations has been a diverse range of indicators purporting to measure the same underlying construct and providing widely different estimates of levels of fear of crime when applied to the same populations.

We use this debate around the conceptualization and measurement of fear of crime to motivate our second example of the regression-based response probing methodology. We explore the cognitive frames elicited by three different questions that have been used to measure fear of crime: the intensity of worry, perceptions of safety, and frequency of worry. Our approach is exploratory; we use the content and distribution of elicited cognitive frames to understand what the different questions are measuring and how different cognitive frames reported by respondents are related to expressed levels of fear across questions.

23.6 Data

Data were collected as part of the TNS BMRB Face-to-Face Omnibus survey. The survey has a stratified multistage design with a first stage sample of postcode sectors randomly selected from the Post Office Address file and respondents selected purposively within each sector to match population marginals on age, sex, housing tenure, and working status. The design is not random but achieves a broad geographic coverage and matches the general population closely on these and other characteristics. Fieldwork was conducted during November 2010, with an achieved sample size of 2069. Respondents were randomly allocated to one of three conditions. In condition 1, respondents were administered the generalized worry question:

Thinking about all types of crime in general, how worried are you about becoming a victim of crime?

1. Very worried
2. Fairly worried
3. Not very worried
4. Not at all worried

In condition 2, respondents were administered the "safe in the dark" question:

How safe do you feel walking alone in this area after dark? Would you say you feel...

1. Very safe
2. Fairly safe
3. Not very safe
4. Or not safe at all?

In condition 3, respondents were administered the "episodic worry" question:

During the last 12 months have you ever felt worried about becoming a victim of a crime? Yes/No

Table 23.1 presents the response distributions for the three questions, revealing clear differences in the level of fear expressed, albeit that this is not a straightforward comparison given the substantial differences in wording and response scales. Nonetheless, considering the response alternatives that indicate some level of cognitive/emotional concern about crime, we find 18% said that they felt worried at least once in the past year (episodic fear), a somewhat higher figure of 26% reported not feeling safe walking in their neighborhood in the dark, and approximately three quarters (72%) said they are either very or fairly worried about being a victim of crime.

These estimates are in line with those reported in previous studies in the UK, although the generalized worry estimate is somewhat higher than has been reported elsewhere (Gray et al. 2008). This is likely to be because the questions used in the current study ask about crime in general, while previous studies have named specific crime categories. Taken at face value, then, the generalized worry question indicates the highest level of fear of crime by a considerable margin.

Immediately following the fear of crime questions, respondents in all three conditions were administered the following question:

Table 23.1 Marginal distributions for the three fear-of-crime questions.

Safe in the dark (n = 660)		Episodic worry (n = 699)		Generalized worry (n = 704)	
Very unsafe	9% (57)	Felt worried	18% (129)	Very worried	30% (207)
Not very safe	17% (111)	Not felt worried	82% (575)	Fairly worried	42% (291)
Fairly safe	39% (259)			Not very worried	25% (174)
Very safe	35% (234)			Not at all worried	4% (27)

Now, thinking about your last response, please tell me what came to mind when you were thinking about your answer? There are no right or wrong answers. Please tell me everything that comes into mind in thinking about how you feel.

Responses were recorded verbatim by the interviewers using tablet computers. Interviewers were instructed to probe all respondents with the words "anything else?" until no further reports were forthcoming. The verbatim responses were then coded by trained coders at TNS BMRB. This first stage of coding produced 45 distinct categories of response. As with the generalized trust study, these initial codes are very literal in nature: their function was to describe, with as little subjective judgment as possible, the range of answers obtained. Next, the 45 first-stage codes were aggregated into the following 8 higher-order thematic codes developed by the researchers: 1. no worry/feel safe 2. generalized concerns about crime 3. features of the neighborhood 4. specific crime types 5. individual characteristics/actions 6. been a victim/know victim 7. media reporting 8. nothing came to mind.

The first higher-order code, reported by 39% of respondents, subsumed all verbatim responses in which respondents made reference to feeling safe and not being worried in their neighborhood. The second comprised responses that made reference to abstract or generalized concern about crime and was mentioned by 13% of respondents. The third code, mentioned by 6% of all respondents, aggregates verbatim responses that referred to features of the local environment, such as it being a "neighborhood watch" area, or there being poor street lighting. Under code 4, verbatim responses that made reference to specific types of crime and disorderly behavior were combined ("there is a lot of burglary"; "gangs hanging round on street corners"); 18% of respondents provided a response falling under this heading. Code 5, mentioned by 15% of respondents, was used to aggregate references to individual characteristics of the respondent, such as being old or immobile, or of being able to "take care of myself." Codes 6 and 7 combined, respectively, mentions of having been (or knowing someone who has been) a victim of crime (mentioned by 3%) and references to crime reporting in the media (mentioned by 1%), respectively; while code 8 was for respondents who reported that nothing had come to mind, or refused to provide a response to the question (21% of respondents).

The content of these thematic codes is broadly in line with what would be expected from the existing theoretical and empirical literature on the fear of crime. They comprise a mix of both concrete and diffuse cognitive frames relating to crime, disorder, and the state of society. Concrete frames focus on crime experience and specific features in the individual's neighborhood that symbolize or are associated with crime and offending. In contrast, diffuse frames capture more generalized concerns and anxieties about what crime represents as a social, political, and cultural issue (Gray et al. 2011). It is notable that the most frequent codes by some margin are the two that indicate respondents were either not thinking about crime in formulating their responses, or were unable to report what had come to mind.

Figure 23.1 presents the distribution of each higher-order code by the fear-of-crime question that respondents were administered. While the pattern is similar across items, it is clear that the three questions evoked somewhat different cognitive frames. For instance, 21% of respondents who were administered the "safe in the dark" indicator mentioned generalized concerns, compared to 10% of those administered the "generalized worry" measure, and 8% of those who answered the "episodic worry" item. Similarly, 10% of respondents fielded the "safe in the dark" question mentioned features of their neighborhood, compared to 4% of respondents who were fielded the "generalized worry" measure, and 3% of respondents who answered the "episodic worry" item. There was also a significant difference ($p < 0.05$) in the rates mentioning feelings of safety/lack of worry, with 42% and 40% of those answering the "episodic worry" and "safe in the dark" questions, respectively, mentioning this type of frame, compared to 35% for those administered the "generalized worry" item.

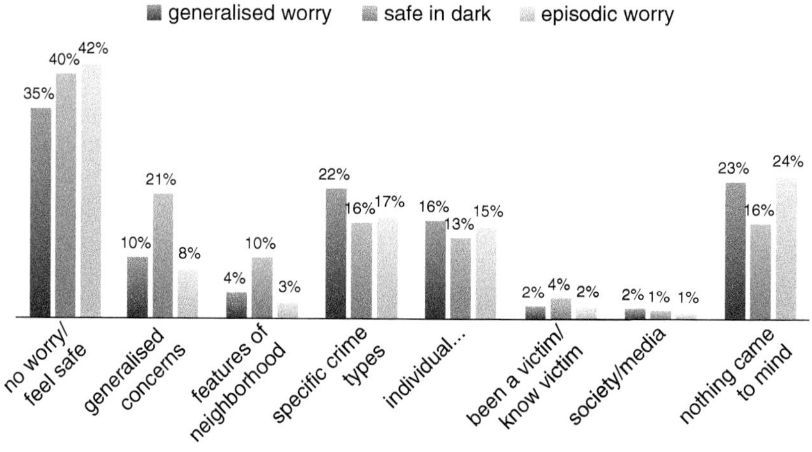

Figure 23.1 Who came to mind across the three fear-of-crime questions.

For the specific crime types, individuals characteristics, been or known a victim, and society/media codes, the rates were statistically indistinguishable across the three different questions. Nonetheless, there is clear evidence here that the questions elicit systematically different frames of reference in respondents' minds. Because respondents were allocated to question conditions randomly, we can be confident that these differences in reported cognitions are a result of differences in the questions, rather than in sample composition or some other aspect of the research design. That these are comparatively large samples also means that we can be confident the estimates generalize to the broader population. In the next section, we use regression models to assess whether these differences might be consequential for expressed levels of fear.

Table 23.2 presents the results of three logit regression models[1] that include the response probe codes as predictors with controls for age, sex, social class, and marital status. The codes for "been/know a victim" and "society/media" are excluded from these models due to their containing an insufficient number of observations. For all three items, the same basic pattern is apparent for the response probe codes, though with some variance in the magnitude of the coefficients.

Three of the cognitive frame codes are positively associated with fear of crime on all three questions, indicating that mentioning any of these frames resulted in higher levels of expressed fear – generalized concerns, features of the neighborhood, and specific crime types – although for the "episodic worry" item, the coefficient on the "features of neighborhood" code is not statistically significant. Respondents whose probed answers expressed feelings of safety/not worrying about crime, unsurprisingly, also reported significantly lower levels of fear of crime across all three items.

The distribution of response codes is informative about the differences in expressed levels of fear across the three items. It is notable, for instance, that the code "specific crime types" is strongly predictive of lower levels of expressed fear on all three questions. This cognitive frame was mentioned most often by respondents in the condition administered the "generalized worry" item, which, as we noted earlier in Table 23.1, elicited the highest level of expressed fear. The evidence from the response probes therefore suggests that the "generalized worry" question invokes systematically different cognitive frames compared to the other two questions and that these frames, in turn, result in higher levels of expressed fear of crime.

---

1 A binary logit model was specified for the "episodic worry" item and ordered logit models for the "generalized worry" and "safe in the dark" items.

Table 23.2 Regressions of fear-of-crime questions on verbatim codes and covariates.

	Safe in the dark		Episodic worry		Generalized worry	
Covariates	**Logit**	**s.e.**	**Logit**	**s.e.**	**Logit**	**s.e.**
	Age (ref = 75+)					
16–24	−0.378	0.377	0.900	0.637	0.432	0.372
25–34	−0.298	0.344	0.284	0.545	0.711[a)]	0.356
35–44	−0.498	0.333	0.325	0.547	0.306	0.334
45–54	−0.816	0.337	0.466	0.559	0.434	0.330
55–64	0.007	0.325	0.509	0.538	0.644	0.343
64–75	−0.159	0.336	−0.731	0.646	0.728	0.358
Sex (male = 1)	−1.087[a)]	0.163	−0.143	0.248	−0.352[a)]	0.151
	Social class (ref = D/E)					
A/B	−0.290	0.226	−0.110	0.364	−0.061	0.212
C1/C2	−0.173	0.193	−0.074	0.32	0.202	0.189
	Marital status (ref = divorced/widow)					
Married	−0.296	0.243	0.088	0.412	0.329	0.242
Single	−0.226	0.304	0.013	0.524	0.231	0.296
	Verbatim code (ref = ind. characteristics/actions)					
no worry/feel safe	−1.590[a)]	0.238	−2.875[a)]	0.462	−1.340[a)]	0.210
Generalized concerns	0.782[a)]	0.238	1.444[a)]	0.373	1.216[a)]	0.272
Features of neighborhood	0.814[a)]	0.259	1.165	0.646	1.201[a)]	0.362
Specific crime types	0.978[a)]	0.252	1.311[a)]	0.313	0.759[a)]	0.217
Nothing came to mind	−0.570	0.277	−1.629[a)]	0.405	−0.169	0.226
Nagelkerke R^2	0.400		0.489		0.263	

a) $p < 0.05$; coefficients in the table are logits.

23.7 Discussion

The use of verbal response probes has a long tradition in survey research as a means of assessing the content validity of questionnaire items (Schwarz and Sudman 1995). This broad class of procedures generally require respondents to vocalize their thought processes as, or after, they read and formulate responses to questions. The recorded and transcribed responses can then be used to identify problems in question wording and to assess whether the cognitive and affective responses elicited by the question align with the concept the researcher wishes to measure. In general these procedures have mostly (though not exclusively) been used in small-group settings where the verbatim response data is analyzed using qualitative methodologies that do not seek to make statistical inferences to the population from which respondents were drawn.

In this chapter, we have described a new method – regression-based response probing – that combines the strengths of intensive small-sample qualitative approaches with the inferential power of large-scale field trials and experimental manipulations. The basic approach is to collect response probe data from a large random sample of respondents, which is then coded to a frame that captures the conceptual and thematic content of the responses. The content and distribution of the elicited codes can then be assessed relative to theoretical and conceptual definitions of the construct the questions are intended to measure. This is an assessment of what Karabenick et al. (2007) have referred to as "cognitive validity": the extent to which the thought processes induced in respondents' minds align with the intentions of the question designer. In this sense, the method differs from similar strategies that seek to identify problems and ambiguities in question wording; rather, the approach we have described and illustrated aims to determine whether questions align with theoretical definitions.

The verbatim response codes are then included as dummy variable predictors in a regression model, where the question that was used to elicit the think-aloud responses is specified as the outcome. Individual-level demographic controls are included, so that the estimated coefficients of the model can be used to interpret whether, and to what extent, the different cognitive frames identified in the think-aloud data align with responses to the question. The technique can be extended to include split-ballot designs, where variants of the target question are randomized across groups to enable an assessment of how question content and response scale options may influence the cognitive frames elicited. We illustrated the procedure using two example concepts: generalized trust and fear of crime.

For the generalized trust example, we demonstrated that, although both questions elicited substantial numbers of verbatim responses that referred to abstract categories of "people in general," a substantial minority mentioned having thought of specific individuals who were known to them. And, although this tendency was greater for the question that asked about TiN, it was also apparent for the "generalized trust" question, counter to what theoretical treatments of the construct suggest should be the case (Uslaner 2002). In the regression analysis, we showed that reported levels of trust were substantially higher when respondents mentioned people who were known to them in their verbatim responses. In this way, we were able to account, at least in part, for the higher levels of trust that are commonly found using the TiN item compared to the GTQ.

Similar results were presented for the fear of crime example. All three questions elicited similar thematic content in the verbatim responses, and this broadly reflected theoretical and conceptual treatments of the construct in the theoretical literature. There were, however, systematic differences in the rates of reporting on all but three of the eight thematic codes; and, because of the randomization of respondents to experimental conditions, we were able to conclude that these differences in the distribution of responses were the result of divergent frames invoked in respondents' minds by the questions. Differences in the rates of reporting of cognitive frames across questions also suggested a reason why the "generalized worry" question tends to produce considerably higher levels of reported fear, compared to the other two questions. The "generalized worry" question produced the highest rates of reporting of cognitive frames that were themselves strongly predictive of reporting higher levels of fear of crime.

An obvious limitation of the method we have proposed here is cost; both the examples presented in this chapter have used large samples and face-to-face interviewing, a design that would be substantially beyond the budgets available to most survey designers for question testing. However, while a large sample size is a requirement if population inference is required, as is the case here, face-to-face interviewing is not a necessary part of the procedure. As online interviewing becomes increasingly prevalent in survey research, the opportunity to implement this type of design at low costs and quick turnaround speeds becomes more feasible (Behr et al. 2014; Meitinger and Behr 2016). Of course, implementation of this method using web interviewing would pose new challenges relating to the quality and completeness of the verbatim responses, as responses would need to be typed into text boxes by respondents rather than recorded by interviewers. However, as web capabilities continue to develop apace, the possibility of audio-recording verbatim responses (Sturgis and Luff 2015) as part of an online self-interview becomes more feasible. These, we contend, are potentially fruitful avenues of research for extending this and similar methods of question testing and evaluation in the future.

References

Beatty, P.C. and Willis, G.B. (2007). Research synthesis: the practice of cognitive interviewing. *Public Opinion Quarterly* 71 (2): 287–311. https://doi.org/10.1093/poq/nfm006.

Behr, D., Braun, M., Kaczmirek, L., and Bandilla, W. (2014). Item comparability in cross-national surveys: results from asking probing questions in cross-national web surveys about attitudes towards civil disobedience. *Quality & Quantity* 48: 127–148.

Belson, W.A. (1981). *The Design and Understanding of Survey Questions*. London: Gower.

Benson, L. (1941). Studies in secret-ballot technique. *Public Opinion Quarterly* 5 (1): 79–82.

Blair, J. and Presser, S. (1993). Survey procedures for conducting interviews to pretest questionnaires: a review of theory and practice. American Statistical Association, Survey Methods Section, Alexandria, VA.

Callegaro, M., Baker, R.P., Bethlehem, J. et al. (2014). *Online Panel Research: A Data Quality Perspective.* Wiley.

Conrad, F.G. and Blair, J. (2009). Sources of error in cognitive interviews. *Public Opinion Quarterly* 73 (1): 32–55. https://doi.org/10.1093/poq/nfp013.

Creswell, J.W. and Plano Clark, V.L. (2007). *Designing and Conducting Mixed Methods Research.* Thousand Oaks: Sage.

DiMaggio, P. (1997). Culture and cognition. *Annual Review of Sociology* 23: 263.

Ericsson, K.A. and Simon, H.A. (1980). Verbal reports as data. *Psychological Review* 87: 215–251.

Farrall, S., Jackson, J., and Gray, E. (2009). *Social Order and the Fear of Crime in Contemporary Times.* Oxford: Oxford University Press.

Ferraro, K.F. (1995). *Fear of Crime: Interpreting Victimization Risk.* New York: SUNY Press.

Figgie, H.E. (1980). *The Figgie Report on Fear of Crime: America Afraid. Part 1: The General Public.* Ohio: Willoughby.

Garofalo, J. and Laub, J. (1978). The fear of crime: broadening our perspective. *Victimology* 3: 242–253.

Gray, E., Jackson, J., and Farrall, S. (2008). Reassessing the fear of crime. *European Journal of Criminology* 5 (3): 363–380.

Gray, E., Jackson, J., and Farrall, S. (2011). Feelings and functions in the fear of crime: applying a new approach to victimisation insecurity. *British Journal of Criminology* 51 (1): 75–94.

Groves, R. (1991). Measurement error across the disciplines. In: *Measurement Errors in Surveys* (eds. P.P. Biemer, L.E. Lyberg, N.A. Mathiowetz and S. Sudman), 1–29. New York: Wiley.

Jabine, T., Straf, M., Tanur, J., and Tourangeau, R. (1984). *Cognitive Aspects of Survey Methodology: Building a Bridge between Disciplines.* Washington D.C.: National Academy of Sciences.

Jackson, J. and Gray, E. (2010). Functional fear and public insecurities about crime. *British Journal of Criminology* 50 (1): 1–21.

Karabenick, S.A., Woolley, M.E., Friedel, J.M. et al. (2007). Cognitive processing of self-report items in educational research: do they think what we mean? *Educational Psychologist* 42 (3): 139–151.

Krosnick, J.A. (2011). Experiments for evaluating survey questions. In: *Question Evaluation Methods* (eds. K. Miller, J. Madans, G. Willis and A. Maitland), 215–239. New York: Wiley.

Krosnick, J.A., Narayan, S.S., and Smith, W.R. (1996). Satisficing in surveys: initial evidence. In: *Advances in Survey Research* (eds. M.T. Braverman and J.K. Slater), 29–44. San Francisco: Jossey-Bass.

Loftus, E. (1984). Protocol analysis of responses to survey recall questions. In: *Methods for Testing and Evaluating Survey Questions* (eds. T. Jabine, M. Straf, J. Tanur and R. Tourangeau), 61–64. Washington DC: National Academy Press.

Lorenc, T., Clayton, S., Neary, D. et al. (2012). Crime, fear of crime, environment, and mental health and wellbeing: mapping review of theories and causal pathways. *Health & place* 18 (4): 757–765.

Meitinger, K. and Behr, D. (2016). Comparing cognitive interviewing and online probing: do they find similar results? *Field Methods* https://doi.org/10.1177/1525822X15625866.

Presser, S. (1984). Is inaccuracy on factual survey items item specific or respondent specific. *Public Opinion Quarterly* 48: 344–355.

Presser, S., Rothgeb, J., Couper, M. et al. (2004). *Methods for Testing and Evaluating Survey Questions*. NJ: Wiley.

Putnam, R.D. (2007). *E Pluribus Unum*: diversity and community in the twenty-first century The 2006 Johan Skytte prize lecture. *Scandinavian Political Studies* 30 (2): 137–174.

Rosenberg, M. (1956). Misanthropy and political ideology. *American Sociological Review* 21: 690–695.

Rugg, D. (1941). Experiments in question wording II. *Public Opinion Quarterly* 5 (1): 91–92.

Saris, W. and Gallhofer, I. (2014). *Design, Evaluation, and Analysis of Questionnaires for Survey Research*. Wiley.

Schuman, H. and Presser, S. (1981). *Questions and Answers in Attitude Surveys: Experiments on Question Form, Wording and Context*. New York: Academic Press.

Schwarz, N. and Sudman, S. (1995). *Answering Questions: Methodology for Determining Cognitive and Communicative Processes in Survey Research*. San Francisco: Jossey-Bass.

Sturgis, P. and Luff, R. (2015). Audio-recording of open-ended survey questions: a solution to the problem of interviewer transcription? In: *Survey Measurements: Techniques, Data Quality and Sources of Error* (ed. U. Engel), 42–57. Routledge.

Sturgis, P. and Smith, P. (2010). Assessing the validity of generalized trust questions: what kind of trust are we measuring? *International Journal of Public Opinion Research* 22 (1): 74–92.

Sudman, S., Bradburn, N.M., and Schwarz, N. (1996). *Thinking about Answers: The Application of Cognitive Processes to Survey Methodology*. San Francisco, CA: Jossey Bass.

Ternor, J.M., Miller, M.K., and Gipson, K.G. (2011). Utilization of a think-aloud protocol to cognitively validate a survey instrument identifying social capital resources of engineering undergraduates. 118th ASEE Annual Conference and Exposition, Vancouver, BC; Canada.

Tourangeau, R. (2004). Experimental design considerations for testing and evaluating questionnaires. In: *Methods for Testing and Evaluating Survey*

Questionnaires (eds. S. Presser, J. Rothgeb, M. Couper, et al.), 209–224. NJ: Wiley.

Tourangeau, R., Rips, L., and Rasinski, K. (2000). *The Psychology of Survey Response*. New York: Cambridge Press.

Uslaner, E.M. (2002). *The Moral Foundations of Trust*. Cambridge: Cambridge University Press.

Willis, G. (2004). Cognitive interviewing revisited: a useful technique, in theory? In: *Methods for Testing and Evaluating Survey Questionnaires* (eds. S. Presser Jennifer M. Rothgeb Mick P. Couper et al.). Hoboken, NJ: Wiley.

Willis, G., DeMaio, T., and Harris-Kojetin, B. (1999). Is the bandwagon headed to the methodological promised land? evaluation of the validity of cognitive interviewing techniques. In: *Cognition and Survey Research* (eds. M. Sirken, D. Herrmann, S. Schechter, et al.), 133–155. New York: Wiley.

Willis, G.B. (2005). *Cognitive Interviewing: A Tool for Improving Questionnaire Design*. Thousand Oaks, CA: Sage Publications.

Wilson, T.D., LaFleur, S.J., and Anderson, D.E. (1996). The validity and consequences of verbal reports about attitudes. In: *Answering Questions: Methodology for Determining Cognitive and Communicative Processes in Survey Research* (eds. N. Schwarz and S. Sudman), 91–114. San Francisco: Jossey-Bass.

24

The Interplay Between Survey Research and Psychometrics, with a Focus on Validity Theory

Bruno D. Zumbo[1] *and José-Luis Padilla*[2]

[1] *Measurement, Evaluation, and Research Methodology Program, and The Institute of Applied Mathematics, University of British Columbia, Vancouver, Canada*
[2] *Faculty of Psychology, University of Granada, Granada, Spain*

24.1 Introduction

The central thesis of this chapter is that there is much to be gained by (re-)initiating a conversation and interplay between survey research and psychometrics, with a particular focus on developments in validity theory over the last three decades. As psychometricians, we are struck by the growing gap between survey research and psychometrics with respect to validity theory.

Two points about the growing gap between the two disciplines are noteworthy. First, although the total survey error paradigm (TSE, Groves and Lyberg 2010) is a very useful integrated theoretical framework that helps survey researchers understand and set a metric for survey errors, it characterizes validity in terms of multiple validities: construct/theoretical validity, and criterion validity (which in turn is divided in to predictive validity and concurrent validity), with no discussion of relevant theoretical developments in contemporary validity theory from psychometrics, toward an unified and contextualized conception of validity. Evidence of the current lack of interest in psychometric validity theory can be found in reference books for survey researchers such as *Handbook of Survey Methodology for the Social Sciences* (Gideon 2012), in which there is no mention of the work of validity theorists in psychometrics post-1970, such as that by Samuel Messick, Susan Embretson, Mike Kane, Denny Borsboom, or Bruno Zumbo. Conversely, psychometric journals, conferences, reference books, etc., do not mention the TSE paradigm, pretesting methods, fitness for use perspective on survey quality, response processes, etc., or works on survey data quality by authors like Paul Biemer, Stanley Presser, and Roger Tourangeau. Therefore, a case can be made that since at least 1980, somewhat independent lines of thought and practice have evolved in survey research, and psychometric measurement validity theory.

Advances in Questionnaire Design, Development, Evaluation and Testing, First Edition.
Edited by Paul C. Beatty, Debbie Collins, Lyn Kaye, Jose-Luis Padilla, Gordon B. Willis, and Amanda Wilmot.

The difficulty in bridging the gap between both fields was to some extent predicted by Groves (1989) when describing how survey research is the "meeting point" for researchers and professionals with different training and backgrounds like statisticians, sociologist, psychologist, etc., to which we can add "psychometricians." Groves (1989) thought that innovations could come from such a "*mélange*" but at the cost of problems of communication and disagreement about the different components of survey quality. Padilla, Benitez, and van de Vijver (Padilla et al. 2018, in press) point out that the different notion of validity is a good example of such disagreements.

There are at least three exceptions to this trend of separate roads taken by survey research and psychometric views of validity theory. These are examples of "meeting points," bridges, or Groves' *mélange* of the two disciplines. A first example of bridges between the two disciplines is the line of research among survey researchers on multitrait-multimethod (MTMM) and structural equation modeling validation approaches (e.g. Saris 1990; Saris and Andrews 1991; Saris et al. 2004; Scherpenzeel and Saris 1997). It is important to note, however, that even though the work of Saris and his colleagues uses current multivariate statistical theories and latent variable models, the MTMM approach to examining construct validity was developed by Campbell and Fiske (1959), significantly predating contemporary psychometric validity theories. A second exception to the separate roads is the use of item response theory (IRT), nonlinear item factor analysis, and latent class mixture models to investigate the internal structure of surveys (e.g. Reeve 2011; Reeve et al. 2011; Sawatzky et al. 2009, 2012). Evidence for internal structure of surveys and questionnaires is the most common form of validity evidence reported; however, when reported on its own it is not sufficient evidence for claims of measurement validity (Kane 1992, 2006, 2013; Messick 1989; Zumbo and Chan 2014). Hubley et al. (2017) recently proposed a novel IRT approach that extends beyond the internal structure of surveys to investigate Messick's (1989, 1995) sense of substantive validity, which focuses on evidence about the process of responding (i.e. how and why people respond to test items and survey questions) as a central tenet to contemporary validation. A third exception to the trend of separate roads is the investigation of cognitive interviewing, think-aloud methods, and response styles as evidence toward Messick's substantive validity (e.g. Benítez et al. 2016; Padilla and Benítez 2014; Zumbo and Hubley 2017). Research on topics like response processes to survey questions by cognitive interviewing has a rich history in survey research, as topics are investigated for their own right to inform survey question design and, more recently, construct validity of survey questions (Miller et al. 2014), but until recently have not been integrated with and discussed in terms of contemporary psychometric validity theory (e.g. Padilla and Leighton 2017).

The chapter is organized into two main sections. First, with somewhat of a historical lens, we provide a brief overview of contemporary validity theory and validation practices in psychometrics, highlighting contrasts with survey research and methodology. Second, we describe two approaches to measurement validity: the ecological model of responding to survey questions, and the argument-based approach to validation, that aim to bridge psychometrics and survey research.

24.2 An Over-the-Shoulder Look Back at Validity Theory and Validation Practices with an Eye toward Describing Contemporary Validity Theories

The era of what we call "contemporary validity theory" was signaled by the work of Lee Cronbach, Samuel Messick, and others starting in the 1970s. A detailed analysis of the history of validity theory in psychometrics is beyond this chapter; interested readers should consult Cronbach (1988), Messick (1989), Hubley and Zumbo (1996), Jonson and Plake (1998), and Shear and Zumbo (2014).

24.2.1 Distinguishing Validity Theory and Validation Methods

Before dealing with the main milestones in the historical development, it is important to note that if one wants to advance the theorizing and practice of measurement, one needs to articulate what they mean by "validity" to go hand-in-hand with the process of validation (Zumbo 2007). As has been noted several times in the validity theory literature (e.g. Messick 1989; Shear and Zumbo 2014; Zumbo 1998, 2007, 2009), when explicit definitions of validity are not provided, the discipline has tended to conflate validity *theory* and validation *methods*. It is therefore important to distinguish them to avoid overly focusing on methods and techniques for data analysis in the absence of a conceptual foundation. For example, although it is not always recognized in the survey research literature, the MTMM approach as originally conceptualized by Campbell and Fiske (1959) is a validation method that is tied to Cronbach and Meehl's (1955) notion of construct validity theory. Likewise, the validation methods of cognitive interviews are loosely founded on the notion of validity involving an explanation for the responding and description of the response process. To be clear, in this latter example, as Zumbo (2007) notes, this theory of validity involves the explanation for the variation in responses to survey questions or test items, and the validation method is the cognitive interview.

24.2.2 Broad Consensus of the Contemporary Concept of Validity

A broad consensus about the contemporary concept of "validity" in psychometrics comes from the legacy of Samuel J. Messick (1931–1998). Messick (1989) stated "validity is an integrated evaluative judgment of the degree to which empirical evidence and theoretical rationales support the adequacy and appropriateness of inferences and actions based on test scores or other modes of assessment" (p. 13). Messick's view is so influential that the latest edition of the *Standards for Educational and Psychological Testing* published by the American Educational Research Association (AERA), the American Psychological Association (APA), and the National Council of Measurement in Education (NCME) states, "validity refers to the degree to which evidence and theory support the interpretations of test scores for proposed uses of tests" (AERA, APA, and NCME 2014, p. 11). Gómez-Benito, Sireci, Padilla, Hidalgo, and Benitez (Gómez-Benito et al. 2018) summarize the arguments associated with the current consensus about validity in psychometrics: (i) tests must be evaluated with respect to a particular purpose; (ii) what needs to be validated are the inferences derived from test scores, not the test itself; (iii) evaluating inferences made from test scores involves several different types of qualitative and quantitative evidence; and (iv) evaluating the validity of inferences derived from test scores is not a one-time event; it is a continuous process. Readers only have to change "test" with "survey" or "survey question," and "test scores" with "survey estimates" to discern our main point: survey research can benefit from psychometric validity theory as an approach to survey data quality, and inferential quality, from a broader and promising perspective.

24.2.3 Eight Conceptualizations of Validity

The following eight conceptualizations trace the historical development of psychometric theory, and may be compared with the commonplace view of validity in survey research. Placing validity and validation within a historical context shows also how these continue to evolve. Table 24.1, based on Shear and Zumbo (2014), lists historical periods for concepts of validity, and corresponding methods of validation. It is important to note that the periods have some overlapping time points and are therefore not necessarily disjoint. Although Shear and Zumbo (2014) originally discussed these theories and methods in terms of tests, measures, and assessments, for our purposes we recast them in the language of survey research and questionnaires and, where possible, relate current survey-validation practices to the appropriate historical period. An important point to take from Table 24.1 is that in our opinion, validity theory as taught and discussed in survey research focuses mainly on the first four of the seven developmental periods.

24.2.3.1 The Era of the Early 1900s to Cronbach and Meehl

As seen in Table 24.1, the earliest view of validity was that a questionnaire/survey is valid if it measures what it is supposed to. This view of validity does not imply a method of validation, and hence is of little practical use to

Table 24.1 Developmental periods for concepts of validity and methods of validation.

Period when concept of validity was introduced and developed	Concept of validity
Early 1900s	A questionnaire/survey is valid if it measures what it is supposed to. *Validation method*: No single implied method.
1920s–1930s	Validity is about establishing whether a questionnaire/survey is a good predictive device or shorthand (criterion validity). *Validation method*: Correlation with a criterion.
1930s–late 1960s	There are multiple "types" of validity. *Validation method*: Depends upon the type of validity.
1950s–1960s	Validity is about evaluating the logical empiricist influenced "nomological network" in "construct validity" (Cronbach and Meehl 1955). *Validation method*: Empirically establishing the nomological network (e.g. multitrait multimethod [MTMM], experiments).
1970s–late 1990s	Validity refers to the degree to which evidence and theory support the interpretations of the questionnaire/survey data entailed by proposed uses of the questionnaire/survey (Cronbach 1971; Messick 1989, 1995). *Validation method*: Multiple sources of evidence used to provide a sound scientific basis for score interpretation.
1980s–present	Construct validity is a universal and interactive system of evidence; emphasizing construct representation and nomothetic span (Embretson 1983, 2007). *Validation method*: Formal cognitive modeling of survey/questionnaire responses and correlational techniques, among others.
2000–present	I. A test is valid for measuring an attribute if and only if the attribute exists and variations in the attribute causally produce variations in the responses to the survey or questionnaire; (Borsboom et al. 2004, 2009). *Validation method*: Formal cognitive modeling, among others. II. Validity is focused on survey or questionnaire development and internal characteristics, such as content representation (Lissitz and Samuelsen, 2007). *Validation method*: Content validation and reliability analysis methods. III. Validity is having a contextualized and pragmatic explanation for variation in the survey or questionnaire data (questions or composite variables) (Zumbo 2007, 2009). *Validation method*: Developing and testing explanatory models; multilevel cognitive and statistical modeling, among others.

survey researchers. This early view of validity was elaborated in the 1920s, in what we describe as the second period in the history of validity, within a behaviorist tradition whose earliest forms can be seen in Hull (1928). In this second period, questionnaires/surveys were seen as serving one of two purposes: a predictive purpose, or serving as a type of "shorthand" approach. In terms of the former, questionnaires/surveys were viewed as a predictive device for assessing future behavior. A contemporary example of this early view of validity would be its use in political predictive polling. In terms of the latter, a questionnaire/survey serves a shorthand function, in that the survey questions provide a brief and simple way of measuring a current characteristic or complex behavior. A parallel example of this can be also seen in attitude or opinion surveys. In this context, validity is defined in purely operational terms and implies a particular approach to validation, albeit a relatively narrow one: correlating questionnaire or survey data with a criterion that tends to involve either future behaviors or present characteristics of the survey respondent.

As can be seen in Table 24.1, the third and fourth periods in the historical development of validity theory saw the advent of a multitude of "validities" – for example, face, predictive, postdictive, criterion, and structural – with a particular focus on content validity, in which survey designers ask experts if the items (or questions) tap the concept/construct of interest. This time period includes two decades prior to the publication of the seminal work by Cronbach and Meehl (1955). This era of multiple validities resulted in practices that equated validation methods with validity theory, resulting in ritualistic validation practices, absent the purposeful guidance of a validity definition. This view also led to the case where one had many "validities" to choose from, and showing evidence for even one of these was sometimes sufficient for claiming your survey or questionnaire was "valid" (see, for example, Hubley and Zumbo 1996; Messick 1989; Zumbo and Chan 2014). As we observed earlier in this chapter, the TSE survey quality framework (e.g. Groves and Lyberg 2010) which is dominant in survey research to model survey quality, reflects the notions of a multitude of "validities" in psychometrics, or what psychometricians could interpret as a certain "*décalage*" in the survey research view of validity theory.

24.2.3.2 The Nascent Era of Cronbach and Meehl

The view from the 1920s onward that a questionnaire or survey was valid for anything with which it correlates led, in good part, to Cronbach and Meehl's (1955) development of the notion of a "nomological network" and "construct validity" in the fourth developmental period in Table 24.1. This development was important because it signaled that questionnaire or survey items changed from just being "predictive devices" to being "signs" of an underlying unobserved attribute or characteristic of the survey respondent. This time period reflected not only a change in validity theory, but also a concomitant change in the conceptualization and role of personality variables, attitudes, and opinions in the psycho-social and health sciences more generally. As such, survey questions started to be seen as indicators, and responses as manifestations and signals of unobserved constructs or characteristics. In this view, validation implied empirically establishing the nomological network. In its initial description by Cronbach and Meehl (1955), construct validity was intended to provide guidance for evaluating test score interpretations when no adequate future behavioral criterion or content definition was available.

Using the philosophical and scientific principles of logical empiricism (Zumbo 2009, 2017), Cronbach and Meehl (1955) outlined an approach to articulating and testing a proposed law-like network of relations among variables and theories (a nomological network), of which responses to survey items were one observable result. They, however, did not provide an operational empirical strategy or statistical technique for validation, but rather a broad philosophical orientation. Subsequently, Campbell and Fiske (1959) introduced the MTMM method as an empirical method to support Cronbach and Meehl's conceptualization of validity. Therefore, it should be noted that although contemporary MTMM validation methods in survey research do not test the "nomological network," per se, these MTMM methods were developed in light of Cronbach and Meehl's view of construct validity, to aid in establishing empirically testable patterns in data to support claims of construct validity. Likewise, this was an era of early factor analysis practice, and that psychometric technique was also used to help understand the structure of the construct.

24.2.3.3 Contemporary Validity Theory, Cronbach, and the Emergence of Messick

Turning now to what we describe as contemporary validity theories, starting in the fifth developmental period in Table 24.1, Cronbach (1971) and later Messick (1989, 1995) refer to validity as the degree to which evidence and theory support the interpretations of questionnaire and survey data entailed by proposed uses of the questionnaires or surveys. As such, it is the interpretations of questionnaire or survey data required by proposed uses that are evaluated, and not the survey questionnaire or survey itself.

By focusing on interpretations of survey data ("survey estimates") from the perspective of users of survey data, contemporary validity theory in psychometrics mirrors the concept of "fitness for use" in survey research. Biemer (2010) conceptualized fitness for use, first developed by Juran and Gryna (1980), as the recognition that producers and users of survey data assess survey quality from different perspectives. Whereas the traditional survey research perspective places a high priority on "data quality" – meaning "accuracy" of key estimates – the "fitness for use" view instead places a high priority on attributes such as "timeliness," "accessibility," "usability of the data," and "relevance of the questionnaire content to their research objective." The concept of fitness for use therefore includes "accuracy" plus all attributes that survey data users are looking for in survey estimates. Similarly, the psychometric definition of validity by Messick (1989) leads validation efforts to support the "adequacy" and "appropriateness" of inferences and actions based on test scores.

Unlike earlier views of validity, which just involved a regression or correlation analysis as evidence, the process of validation starting in the fifth developmental period involves accumulating more diverse evidence to provide a sound scientific basis for the proposed score interpretations. That is, in contrast to earlier developmental periods, three widely accepted guiding tenets emerged that highlight this period: (i) numerous sources of evidence can contribute to a judgment of validity, rather than just a psychometric validity coefficient (e.g. correlation or a regression coefficient); (ii) validity is a matter of degree rather than all or none judgment of "valid/invalid"; and, (iii) as for the "fitness for use" concept in the survey literature, one validates particular uses and interpretations of survey estimates, rather than the questionnaire itself, and as such uses depend on the objectives of the user of the survey estimates ("relevance for their research objectives"), as opposed to being something "static" or "fixed" within the questionnaire as survey producers traditionally and often understand validity. Therefore, for example, a survey instrument itself, such as the Center for Epidemiology Scale of Depression (CESD) (Radloff 1977; Radloff and Locke 1986), is not viewed as either completely valid or invalid. Rather, the inferences one makes from the CESD are validated, hence highlighting population and context dependence of the survey use. This stands in contrast to a common implication of earlier views of validity, which hold that when a survey measure such as the CESD is validated, it can be used indefinitely and without concern for the context or population of respondents (Gelin and Zumbo 2003).

Although Messick's (1989, 1995) definition of validity, described earlier, took root as the dominant view in the field in the fifth, sixth, and seventh developmental periods in Table 24.1, it does not entail a single approach to validation. Messick (1989) describes multiple sources of validity evidence: content, internal structure, relationships with other variables, response process, and the consideration of the consequences of the use of test scores or survey estimates. An important part of Messick's view of validity, which was particularly influential in the last two developmental periods in Table 24.1, is what he called the "substantive" aspect of construct validity. The substantive aspect of construct validity tends to focus on response processes when trying to explain test scores and, similarly, "survey question responses" (Zumbo and Hubley 2017) that bridge to survey pretesting-based methods like cognitive interviews, and think-aloud methods, when they are seen as "construct validation methods" (e.g. Beatty and Willis 2007; Miller et al. 2014).

A much-contested and under-explored idea in Messick's work is the matter of consequences as they related to validity (Hubley and Zumbo 2011; Zumbo and Chan 2014). As an example of the importance of consequences in survey design and survey validation, Stalans (2012) describes how framing effects and context-dependent factors such as cultural issues, varying economic situations, etc., may limit the validity of inferences from survey data to "specific times, cultures, groups, and may also suggest methodological biases that challenge whether the questions are actually measuring what they intend to measure, especially for sensitive topics. For example, questions that implicitly or explicitly address racial relations and beliefs are very sensitive and have become more sensitive across time" (p. 76) – thereby degrading validity. There are many survey projects for which intended or unintended consequences (Hubley and Zumbo 2011) can affect survey data quality. For example, the debate on the extent to which surveys of gender-based violence can affect women re-experiencing victimization through the interview process, or whether men should be included within the target population of such surveys, refers to values and political arguments that a survey researcher cannot ignore when planning and conducting surveys on controversial topics like gender violence.

24.2.3.4 Concepts of Validity Arising in the Last Two Decades

Since the year 2000, the final developmental period in Table 24.1, two approaches have taken hold in validity theory. First, Lissitz and Samuelson (2007) responded to what they saw as Messick's very broad and unmanageable view of validity, and advocated a return to focus on content representation. As such, validation evidence should hinge on content validity evidence, which involves elaborate methods to ask experts if the items tap the construct of interest. Most importantly, this type of validity evidence need not involve survey response data, but rather, expert opinion on item content. A related method in contemporary survey research language could be the extent to what specification errors are avoided. Second, in a series of essays, Zumbo describes his view of "validity" as the explanation of the variation in survey or questionnaire response data, and "validation" as the process of developing and testing the explanation (Stone and Zumbo 2016; Zumbo 2007, 2009, 2015, 2017; Zumbo and Hubley 2016, 2017; Zumbo et al. 2015, 2017).

For example, one may develop and empirically test complex statistical models that account for the variation in survey question responses or test scores from respondents (see, for example, Zumbo et al. 2017; Zumbo and Hubley 2017). Although it has been shown to be aligned with earlier views by Cronbach and Meehl, and Messick, what is unique to Zumbo's approach is that it advocates a contextualized pragmatic explanation (Stone and Zumbo 2016; Zumbo 2009). Zumbo's approach is particularly well suited for questionnaire design, because it is adapted for the validation of multilevel constructs commonly found in survey studies in which the aim of the survey is inferences about aggregate samples or population rather than individual survey respondents (Forer and Zumbo 2011; Zumbo and Forer 2011). In addition, in line with Messick's view, he advocated bringing matters of consequences to the foreground, because they are largely ignored in validation practice (Hubley and Zumbo 2011; Zumbo and Chan 2014; Zumbo and Hubley 2016).

Validity theory is ultimately rooted in an individual differences tradition in which the validity evidence supports inferences made about the test and survey respondents and about differences among test and survey respondents. In a multilevel context, one uses individual survey responses to compute indices at various levels of aggregates of survey respondents who are embedded in a complex ecologically rich system, moving from individual respondents to aggregates of individual respondents. The focus on multilevel construct validation involves the aggregation of individual survey responses, yet inferences are made at a different (higher) level in the system. That is, survey data are collected at the individual respondent level, but inferences are made at a higher aggregate level (e.g. classroom, neighborhood) that may carry with them secondary dimensions that may contaminate or confound the inference. This situation represents a form of inferential fallacy that may occur when inferring across levels of aggregation, because, as Zumbo and Forer (2011) state, the validation needs to be at the same level as the inferences. For example, one may compare schools or neighborhoods based on survey response data, and therefore one needs to consider the school or neighborhood as the level of validation. Validation at the individual survey respondent level for a neighborhood safety survey is therefore not sufficient evidence, when comparisons are only being made at the aggregate neighborhood level. In this case, validation requires complex multilevel statistical methods to support the data aggregation process of to establish a "neighborhood safety index" that is computed from individual survey response data (see, for example, Forer and Zumbo 2011; Zumbo et al. 2017).

24.3 An Approach to Validity that Bridges Psychometrics and Survey Design

Having described the variety of validity theories and validation practices in Table 24.1, we next turn to two contemporary approaches to validation that will help us build the bridge between psychometrics and survey research, and in turn make judgments about inference quality from survey data: (i) an ecological model of responding to survey questions, and (ii) an argument-based approach to validation. Due to space limitations, we will only briefly describe these approaches to validation and provide references for the interested reader.

24.3.1 Ecological Model of Responding to Survey Questions

Figure 24.1 is a depiction of Zumbo et al.'s (2015) ecological model of validation. It is important to note that the ecological model depicted is only one example of such a model, as others could be developed based on the particular

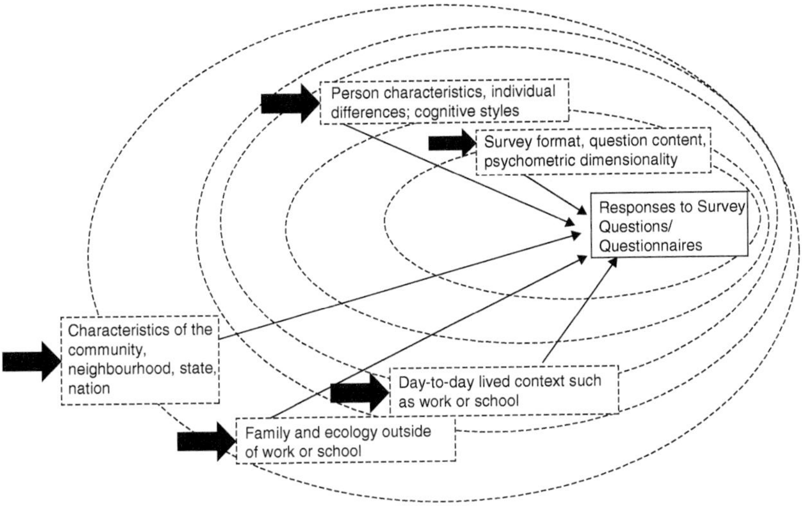

Figure 24.1 An example ecological model for responding to survey questions/questionnaires.

survey context and frame – see Chen and Zumbo (2017) and Zumbo et al. (2017) for more examples. Zumbo et al.'s ecological model has five concentric ovals (each marked with a bold arrow to highlight them) representing: (i) survey format, question content, and psychometric dimensionality; (ii) person characteristics and typical individual differences variables, such as cognition of the survey respondent; (iii) day-to-day lived context such as work or school; (iv) family and ecology outside of work or school; and finally, (v) characteristics of the community, neighborhood, state, and nation. Conventional validation practices have focused on the first oval, with some modest attempts at the second as sources for explanation for survey responding. As such, the ecological model implies that validity depends on the values of these other factors, and is dynamic rather than static.

As an example, Chen and Zumbo (2017) investigated the validity of an international questionnaire on reading attitudes among 15-year old students in 71 nations and jurisdictions. Using an ecological framework of questionnaire item responding similar to that in Figure 24.1, they specified and tested several multilevel regression models designed to account for variation in the gender differences in response to an attitude questionnaire. Their multilevel statistical models investigated the role of reading attitudes as well as national level indicators such as the nations' scores on the United Nations' Human Development Index, which was used as an index of the potential that people can do the things they want to do in their life, as well as the United Nation's Gender Inequality Index, which was used to measure gender disparity. From an ecological modeling point of view, the authors' focus in this study was to provide a method for connecting potential sources of gender differences in attitudes toward reading, to contextual factors outside of test settings – with a particular interest in the potential of using national indices to explain the variability of gender differences across nations.

This description illustrates how a statistical model that is shaped by this ecological framework, or more generally a validation program of research, is applied to explain the variation in responding to survey questions/questionnaires. This, in essence, reflects our thinking about validity and validation. The ecological methods work to establish and support the explanation for item responding – i.e. validity itself – so that validity is the contextualized explanation of the questionnaire item or scale responses via the variables offered in the ecological model, whereas the process of validation involves the myriad methods of psychometric and statistical modeling (Zumbo 2007). Zumbo's approach to validation seeks the enabling conditions via the ecological model, through which a claim about a respondent, as obtained from the item responses on a questionnaire, is supported and makes sense (Stone and Zumbo 2016; Zumbo 2007, 2009).

The ecological point of view is tied closely with the notion of *in vivo* views of psychometrics and validation (Zumbo 2015). Psychometric models are typically considered from a conventional individual differences view of human characteristics, and hence the notion of "ecology" is irrelevant. In first introducing an ecological model, Zumbo et al. (2015) suggest a rhetorical move from how the environment affects the person to a type of interactionist model in which the survey respondent is situated within that environment (context: survey modes, education, social location, etc.). An ecological model of responding to survey questions or questionnaires allows the researcher to focus on sociological, structural, community, and contextual variables (e.g. administration modes), as well as motivational and cognitive factors. There could be more layers in the model, and these layers could include latent classes in a mixture statistical model; and they could, at least in theory, interact, and result in mediated or moderated effects (Zumbo and Gelin 2005).

For example, Zumbo et al. (2015) in their study of an educational survey that contains a reading attitude questionnaire fit a complex statistical model for each item of a survey of reading abilities. This model, the choice and interpretation of the explanatory variables, and the statistical model-building strategy within the validation efforts, were shaped by the ecological framework: looking for explanatory variables in different layers of the model for each individual questionnaire item as the dependent (outcome) variable in the model. The authors investigated many explanatory variables framed by the ecological model, such as the survey item features such as questionnaire item content, survey respondent characteristics such as gender, and whether they like to read fiction or nonfiction, as well as school and teacher or classroom variables, and family and ecology outside of the school such as an index of economic and social status, amount of time outside of school spent reading for enjoyment, parental education levels, and home educational resources.

The ecological approach is connected to Messick's (1989, 1995) view of "substantive validity" and Zumbo's (2007, 2009) view of validity being the explanation of the response variability and response processes (Padilla and Benitez 2017). From our perspective, the question of why and how people respond to survey questions at the core of the ecological approach is closely tied to the interpretative approach in survey research of cognitive interviewing (Miller et al. 2014).

24.3.2 Argument-Based Approaches: How We Bring It All Together

Cronbach (1988), Kane (1992, 2006, 2013), Shepard (1993) and others advocate the building of "validity arguments" as a way to frame or focus validation efforts and clarify intended interpretations of test scores, and, for our purpose in this chapter, of survey data and estimates. The main of objective of this approach to validation is to allocate validation efforts and gauge their progress (Kane 2006, p. 23). As Kane notes, "to validate an interpretation or use of measurements is to evaluate the rationale, or argument, for the proposed conclusions and decisions … Ultimately, the need for validation derives from the scientific and social requirement that public claims and decisions be justified" (p. 17).

Translating Kane's proposal into "survey" language, Kane conceptualizes validation as involving an interpretive argument and a validity argument. The interpretative argument is meant to provide a clear statement of the inferences and assumptions inherent in the proposed interpretations and uses of survey data and estimates. These inferences and assumptions are to be evaluated in a series of analyses and empirical studies. As Kane states, an interpretive argument specifies the proposed interpretations and uses of survey data by laying out a chain or network of inferences and assumptions leading from the survey estimates to the conclusions and decisions based on them. Second, the validity argument provides an evaluation of the interpretive argument's coherence, and of the plausibility of its inferences and assumptions. Kane states that while the interpretations are evaluated in terms of their coherence and plausibility, decisions are evaluated in terms of their outcomes, or consequences.

Castillo-Díaz and Padilla (2012) developed a validation study that was guided by an argument-based approach. They intended to obtain validity evidence to support an interpretative argument for "family support" measures provided by a scale within a national health survey. Some scale items asked if "important decisions are made by all of you together in your home," or if the survey respondent was "satisfied with the time she/he spends with their family." Stakeholder users of the survey data were concerned about if survey respondents maintained a constant meaning of the "family" concept across scale items, given that the scale was included in a household survey. Castillo-Díaz and Padilla

(2012) developed an interpretative argument and a set of assumptions to guide the validation research and respond to the user concern. Evidence from cognitive interviewing pointed out that survey respondents changed the people they included in their "family concept," depending on item content. Therefore, the assumptions were not supported by cognitive interviewing evidence, undermining the validity of the intended interpretation of the "family support" measures.

In conclusion, an argument-based approach to validation can be helpful in reconciling survey producers' and users' views of survey quality. The aim of the validation studies should be to test assumptions that support the interpretative argument "agreed," which may necessitate both qualitative and quantitative evidence to build mixed-method validity arguments.

24.4 Closing Remarks

The aim of this chapter is to explore the interplay between recent developments in psychometric theory (e.g. differing views of validity theory), and questionnaire and survey research approach to draw conclusions about data quality and inferential quality – all with an eye toward bridging concepts and validation practices. It is important to reiterate that the fundamental purpose of validation methods is to help in establishing the degree of validity of the inferences made from the data obtained on a survey. The critical element of degree of validity is highlighted in nearly all current views of validity in psychometrics (see, for example, Hubley and Zumbo 1996; Zumbo 2007). That is, validity is not an all-or-none decision but rather a matter of degree. Furthermore, validation is an ongoing process and is not simply something done once and involving only a correlation with a "gold standard." The notion of validating "inferences," specifically, is critical, because what is being validated are the inferences made from the measurement, and not the measurement tool itself. Therefore, one can have a degree of evidential strength for validity; and, in fact, as we describe, there is a continuum of validity statements that can be articulated in the process of validation.

We argue that psychometrics and survey research have approached validity matters quite differently with some overlap in purposes and intentions. Survey research has tended to place data quality (e.g. estimates of error variance and sampling variability) at the forefront, whereas contemporary psychometric validity theory, by and large, focuses on characteristics of the inferences that are drawn about phenomena such as attitudes, self-reports, or opinions by human agents and the actions that result from these inferences.

Two reasons can explain the gap between psychometrics and survey research with respect to validity: (i) there has been no clear distinction between validity *theory* and validation *methods* in survey research literature; and (ii) validation efforts in survey research have been mainly in the hands of survey producers, with some disconnection of the interests and needs of survey data users. In psychometrics, the first of these limitations was addressed by what is referred to as *contemporary validity theory*, which has enhanced the role of users of test measures in validation: an important development, given the importance and impact on people's lives of test-based decisions in areas of education, health, labor, etc. The "fitness for use" perspective on survey quality offers a chance to bring closer survey research and psychometric views of validity, even though the concept is not typically discussed or framed as "validation" or "validity evidence," per se.

Contemporary validity theory has led to a renewed dissection of what forms of evidence support valid inferences and has brought the focus of investigations back to the survey respondents and a focused lens on what is involved in responding to surveys and questionnaires. When dealing with matters of reliability and validity, we are, in essence, dealing with matters of making inferences from responses to survey and questionnaire items. In other words, data on reliability and validity gathered in the process of measurement aids the survey producer and user of survey data in judging the appropriateness and limitations of their inferences from questionnaire or survey data. Specifically, in terms of how psychometrics speaks to survey research, reliability (error of measurement) is a question of data quality whereas validity is a question of inferential quality from the survey or questionnaire responses. Of course, reliability and validity theory are interconnected research arenas, and quantities derived in the former bound or limit the inferences in the latter, as seen in classical test theory statistics.

From the earlier discussion of the evolution of validity theory, it should be noted that as validity theory enters the late twentieth and early twenty-first centuries, it aims to aid us in the inference from the true score or latent variable score on a survey to the construct of interest. In fact, one of the current themes in validity theory is that construct validity is the totality of validity theory and that its discussion is comprehensive, integrative, and evidence-based. In this sense, construct validity refers to the degree to which inferences can be made legitimately from the observed survey responses to the theoretical constructs about which these data are supposed to contain information. In short, construct validity involves generalizing from survey or questionnaire responses to the concept of our behavioral or social observations and hence, as stated earlier, applies to both self-report subjective indicators and objective indicators. The practice of validation aims to ascertain the extent to which an interpretation of survey responses is conceptually and empirically warranted and should be aimed at making explicit hidden ethical and social values that influence that process (Hubley and Zumbo 2011; Messick 1995; Zumbo and Hubley, 2016).

In the closing section we have highlighted the ecological model and the argument-based approaches to validation, because we believe that survey research would benefit from the ecological viewpoint and from a framework that helps focus survey validation efforts.

References

American Educational Research Association, American Psychological Association, & National Council on Measurement in Education [AERA, APA, & NCME] (2014). *Standards for educational and psychological testing*. Washington, DC: American Educational Research Association.

Beatty, P.C. and Willis, G.B. (2007). Research synthesis: the practice of cognitive interviewing. *Public Opinion Quarterly* 71: 287–311.

Benítez, I., He, J., van de Vijver, F.J.R., and Padilla, J.L. (2016). Linking extreme response styles to response processes: a cross-cultural mixed methods approach. *International Journal of Psychology* 51: 464–473.

Biemer, P.P. (2010). Total survey error: design, implementation and evaluation. *Public Opinion Quarterly* 74: 817–848.

Borsboom, D., Cramer, A.O.J., Kievit, R.A. et al. (2009). The end of construct validity. In: *The concept of validity: Revisions, new directions and applications* (ed. R.W. Lissitz), 135–170. Charlotte: Information Age Publishing.

Borsboom, D., Mellenbergh, G.J., and Van Heerden, J. (2004). The concept of validity. *Psychological Review* 111: 1061–1071.

Campbell, D.T. and Fiske, D.W. (1959). Convergent and discriminant validation by the multitrait-multimethod matrix. *Psychological Bulletin* 56: 81–105.

Castillo-Díaz, M. and Padilla, J.L. (2012). How cognitive interviewing can provide validity evidence of the response processes to scale items. *Social Indicators Research* 114: 963–975.

Chen, M.Y. and Zumbo, B.D. (2017). Ecological framework of item responding as validity evidence: an application of multilevel DIF modeling using PISA data. In: *Understanding and Investigating Response Processes in Validation Research* (eds. B.D. Zumbo and A.M. Hubley), 53–68. New York, NY: Springer.

Cronbach, L.J. (1971). Test validation. In: *Educational measurement*, 2e (ed. R.L. Thorndike), 443–507. Washington, DC: American Council on Education.

Cronbach, L.J. (1988). Five perspectives on validity argument. In: *Test Validity* (eds. H. Wainer and H. Braun), 3–17. Hillsdale, NJ: Erlbaum.

Cronbach, L.J. and Meehl, P. (1955). Construct validity in psychological tests. *Psychological Bulletin* 52: 281–302.

Embretson, S.E. (1983). Construct validity: construct representation versus nomothetic span. *Psychological Bulletin* 93: 179–197.

Embretson, S.E. (2007). Construct validity: a universal validity system or just another test evaluation procedure? *Educational Researcher* 36: 449–455.

Forer, B. and Zumbo, B.D. (2011). Validation of multilevel constructs: validation methods and empirical findings for the EDI. *Social Indicators Research: An*

International Interdisciplinary Journal for Quality of Life Measurement 103: 231–265.

Gelin, M.N. and Zumbo, B.D. (2003). DIF results may change depending on how an item is scored: an illustration with the Center for Epidemiological Studies Depression (CES-D) scale. *Educational and Psychological Measurement* 63: 65–74.

Gideon, L. (ed.) (2012). *Handbook of Survey Methodology for the Social Sciences.* New York, NY: Springer.

Gómez-Benito, J., Sireci, S., Padilla, J.L. et al. (2018). Differential item functioning: beyond validity evidence base on internal structure. *Psicothema* 30 (1): 104–109.

Groves, R.M. (1989). *Survey errors and survey costs.* New York: Wiley.

Groves, R.M. and Lyberg, L. (2010). Total survey error: past, present, and future. *Public Opinion Quarterly* 74: 849–879.

Hubley, A.M., Wu, A.D., Liu, Y., and Zumbo, B.D. (2017). Putting flesh on the psychometric bone: making sense of IRT parameters in non-cognitive measures by investigating the social-cognitive aspects of the items. In: *Understanding and Investigating Response Processes in Validation Research* (eds. B.D. Zumbo and A.M. Hubley), 69–92. New York, NY: Springer.

Hubley, A.M. and Zumbo, B.D. (1996). A dialectic on validity: where we have been and where we are going. *Journal of General Psychology* 123: 207–215.

Hubley, A.M. and Zumbo, B.D. (2011). Validity and the consequences of test interpretation and use. *Social Indicators Research: An International and Interdisciplinary Journal for Quality-of-Life Measurement* 103: 219–230.

Hull, C.L. (1928). *Aptitude Testing.* London: Harrap.

Jonson, J.L. and Plake, B.S. (1998). A historical comparison of validity standards and validity practices. *Educational and Psychological Measurement* 58: 736–753.

Juran, J. and Gryna, F. (1980). *Quality Planning and Analysis.* New York: McGraw-Hill.

Kane, M. (1992). An argument-based approach to validation. *Psychological Bulletin* 112: 527–535.

Kane, M. (2006). Validation. In: *Educational Measurement,* 4e (ed. R.L. Brennan), 17–64. Westport: American Council on Education/Praeger.

Kane, M. (2013). Validating the interpretations and uses of test scores. *Journal of Educational Measurement* 50: 1–73.

Lissitz, R.W. and Samuelsen, K. (2007). A suggested change in terminology and emphasis regarding validity and education. *Educational Researcher* 36: 437–448.

Messick, S. (1989). Validity. In: *Educational Measurement,* 3e (ed. R.L. Linn), 13–103. New York: American Council on Education and Macmillan.

Messick, S. (1995). Validity of psychological assessment: validation of inferences from persons' responses and performances as scientific inquiry into score meaning. *American Psychologist* 50 (9): 741–749. https://doi.org/10.1037/0003-066X.50.9.741.

Miller, K., Chepp, V., Willson, S., and Padilla, J.L. (eds.) (2014). *Cognitive Interviewing Methodology*. Hoboken, NJ: Wiley.

Padilla, J.L. and Benítez, I. (2014). Validity evidence based on response processes. *Psicothema* 26: 136–144.

Padilla, J.L. and Benitez, I. (2017). A rationale for and demonstration of the use of DIF and mixed methods. In: *Understanding and Investigating Response Processes in Validation Research* (eds. B.D. Zumbo and A.M. Hubley), 193–210. New York, NY: Springer.

Padilla, J.L., Benitez, I., and van de Vijver, F.J.R. (in press) (2018). Addressing equivalence and bias in cross-cultural survey research within a mixed methods framework. In: *Advances in Comparative Survey Methods: Multinational, Multiregional and Multicultural Contexts (3MC)* (eds. T.P. Johnson, B.-E. Pennell, I. Stoop and B. Dorer). Hoboken, NJ: Wiley.

Padilla, J.L. and Leighton, J.P. (2017). Cognitive interviewing and think-aloud methods. In: *Understanding and Investigating Response Processes in Validation Research* (eds. B.D. Zumbo and A.M. Hubley), 211–228. New York, NY: Springer.

Radloff, L.S. (1977). The CES-D scale: a self-report depression scale for research in the general population. *Applied Psychological Measurement* (3): 385–401.

Radloff, L.S. and Locke, B.Z. (1986). The community mental health assessment survey and the CES-D scale. In: *Community Surveys of Psychiatric Disorders* (ed. A.E. Slaby) (Series Ed.), 177–189. New Brunswick, NJ: Rutgers University Press.

Reeve, B.B. (2011). Applying item response theory (IRT) for questionnaire evaluation. In: *Question Evaluation Methods: Contributing to the Science of Data Quality* (eds. J. Madans, K. Miller, A. Maitland and G. Willis), 105–123. Hoboken, NJ: Wiley.

Reeve, B.B., Willis, G., Shariff-Marco, S.N. et al. (2011). Comparing cognitive interviewing and psychometric methods to evaluate a racial/ethnic discrimination scale. *Field Methods* 23: 397–419.

Saris, W.E. (1990). Models or evaluation of measurement instruments. In: *Evaluation of Measurement Instruments by Meta-Analysis of Multitrait-Multimethod Studies* (eds. W.E. Saris and A. van Meurs), 52–80. Amsterdam: North Holland.

Saris, W.E. and Andrews, F.M. (1991). Evaluation of measurement instruments using a structural modeling approach. In: *Measurement Errors in Surveys* (eds. P.P. Biemer, R.M. Groves, L.E. Lyberg, et al.), 575–598. New York: Wiley.

Saris, W.E., Satorra, A., and Coenders, G. (2004). A new approach to evaluating quality of measurement instruments: the split-ballot MTMM design. *Sociological Methodology* 34: 311–347.

Sawatzky, R.G., Ratner, P.A., Johnson, J.L. et al. (2009). Sample heterogeneity and the measurement structure of the multidimensional Student's life satisfaction scale. *Social Indicators Research: International Interdisciplinary Journal for Quality of Life Measurement* 94: 273–296.

Sawatzky, R., Ratner, P.A., Kopec, J.A., and Zumbo, B.D. (2012). Latent variable mixture models: a promising approach for the validation of patient reported outcomes. *Quality of Life Research* 21: 637–650.

Scherpenzeel, A.C. and Saris, W.E. (1997). The validity and reliability of survey questions: a meta-analysis of MTMM studies. *Sociological Methods & Research* 25: 341–383.

Shear, B.R. and Zumbo, B.D. (2014). What counts as evidence: a review of validity studies in educational and psychological measurement. In: *Validity and Validation in Social, Behavioral, and Health Sciences* (eds. B.D. Zumbo and E.K.H. Chan), 91–111. New York: Springer.

Shepard, L. (1993). Evaluating test validity. In: *Review of Research in Education* (ed. L. Darling-Hammond), 405–450. Washington, DC: American Educational Research Association.

Stalans, L.J. (2012). Frames, framing effects, and survey responses. In: *Handbook of Survey Methodology for the Social Sciences* (ed. L. Gideon). New York, NY: Springer.

Stone, J. and Zumbo, B.D. (2016). Validity as a pragmatist project: a global concern with local application. In: *Trends in Language Assessment Research and Practice* (eds. V. Aryadoust and J. Fox), 555–573. Newcastle: Cambridge Scholars Publishing.

Zumbo, B.D. (ed.) (1998). *Validity Theory and the Methods Used in Validation: Perspectives from the Social and Behavioral Sciences.* Netherlands: Kluwer Academic Press. [Special issue of the journal *Social Indicators Research*: An International and Interdisciplinary Journal for Quality-of-Life Measurement, Volume 45, No. 1–3].

Zumbo, B.D. (2007). Validity: foundational issues and statistical methodology. In: *Handbook of Statistics*, vol. 26: *Psychometrics* (eds. C.R. Rao and S. Sinharay), 45–79. The Netherlands: Elsevier Science B.V.

Zumbo, B.D. (2009). Validity as contextualized and pragmatic explanation, and its implications for validation practice. In: *The Concept of Validity: Revisions, New Directions and Applications* (ed. R.W. Lissitz), 65–82. Charlotte, NC: IAP - Information Age Publishing, Inc.

Zumbo, B.D. (2015, November). Consequences, side effects and the ecology of testing: Keys to considering assessment 'in vivo'. Keynote address, annual meeting of the Association for Educational Assessment - Europe (AEA-Europe), Glasgow, Scotland. http://brunozumbo.com/aea-europe2015.

Zumbo, B.D. (2017). Trending away from routine procedures, towards an ecologically informed 'in vivo' view of validation practices. *Measurement: Interdisciplinary Research and Perspectives* 15 (3–4): 137–139.

Zumbo, B.D. and Chan, E.K.H. (eds.) (2014). *Validity and Validation in Social, Behavioral, and Health Sciences*. New York: Springer.

Zumbo, B.D. and Forer, B. (2011). Testing and measurement from a multilevel view: psychometrics and validation. In: *High Stakes Testing in Education – Science and Practice in K-12 Settings* (eds. J.A. Bovaird, K.F. Geisinger and C.W. Buckendahl), 177–190. Washington, D.C.: American Psychological Association Press.

Zumbo, B.D. and Gelin, M.N. (2005). A matter of test bias in educational policy research: Bringing the context into picture by investigating sociological/community moderated (or mediated) test and item bias. *Journal of Educational Research and Policy Studies* 5: 1–23.

Zumbo, B.D. and Hubley, A.M. (2016). Bringing consequences and side effects of testing and assessment to the foreground. *Assessment in Education: Principles, Policy & Practice* 23: 299–303.

Zumbo, B.D. and Hubley, A.M. (eds.) (2017). *Understanding and Investigating Response Processes in Validation Research*. New York, NY: Springer.

Zumbo, B.D., Liu, Y., Wu, A.D. et al. (2017). National and international educational achievement testing: a case of multi-level validation framed by the ecological model of item responding. In: *Understanding and Investigating Response Processes in Validation Research* (eds. B.D. Zumbo and A.M. Hubley), 341–362. New York, NY: Springer.

Zumbo, B.D., Liu, Y., Wu, A.D. et al. (2015). A methodology for Zumbo's third generation DIF analyses and the ecology of item responding. *Language Assessment Quarterly* 12: 136–151.

25

Quality-Driven Approaches for Managing Complex Cognitive Testing Projects

Martha Stapleton[1], *Darby Steiger*[1], *and Mary C. Davis*[2]

[1] *Instrument Design, Evaluation, and Analysis (IDEA) Services, Westat, Rockville, MD, USA*
[2] *Office of Survey and Census Analytics, U.S. Census Bureau, Washington, DC, USA*

25.1 Introduction

Cognitive interviewing is a well-established method for pretesting survey questions, and its use is required for many national statistical agencies, including those in the United States (U.S. Office of Management and Budget (OMB) 2016). While these policies describe the circumstances that require cognitive testing, and its purpose and basic elements, they provide little, if any, guidance on how to ensure quality during implementation. An examination of the pretesting methods literature reveals that it too contains little discussion of what quality means in the context of cognitive testing, and what steps researchers can take to ensure quality. In response to that gap, this chapter describes a set of management approaches developed to minimize the risks to quality on four large-scale cognitive testing projects conducted for the U.S. Census Bureau (Census Bureau) between 2014 and 2017. The projects involved hundreds of interviews, dozens of team members, and multiple research objectives, interview guides, languages, locations, and recruiting requirements. Their size and complexity made them vulnerable to misconstrued instructions, unmet research objectives, lost data, procedural mishaps, or other unintentional errors.

The chapter starts by providing more information about the four cognitive testing projects and their features that precipitated the development of a set of quality management approaches. It reflects on the philosophical, ontological, and epistemological debates in the qualitative research methods literature about what quality means and how it might be applied to the use of qualitative methods during survey development and pretesting. It describes 13 quality management approaches and discusses the need for further research.

Advances in Questionnaire Design, Development, Evaluation and Testing, First Edition.
Edited by Paul C. Beatty, Debbie Collins, Lyn Kaye, Jose-Luis Padilla, Gordon B. Willis, and Amanda Wilmot.

25.2 Characteristics of the Four Cognitive Testing Projects

Surveys conducted by the Census Bureau provide data that are invaluable to federal, state, and local governments, researchers, and businesses. Given the importance of the data and the need to ensure continuity with previous data collections, changes to these surveys are only made after rigorous testing. It was in the context of such rigor, together with high visibility, that we developed management approaches for the four cognitive testing projects summarized here:

Content Test (Stapleton and Steiger 2015; Steiger et al. 2015). The American Community Survey (ACS) continuously collects demographic, economic, housing, and social data from households. In preparation for the 2016 American Community Survey Content Test, we conducted cognitive testing of new or revised ACS items that addressed 11 different topics, including Hispanic origin and race; computer and internet use; health insurance; cohabitation; industry and occupation; and retirement income. A team of 25 interviewers conducted 420 interviews across three rounds, in 8 locations and 2 languages. The research design included 48 recruitment targets, 43 versions of the interview guides (based on content, mode of data collection, and question version), and 98 analysis codes.

Burden Test (Robins et al. 2016; Steiger et al. 2017a; Steiger et al. 2017b). It takes an average of 40 minutes per respondent to respond to the 72 questions included on the ACS for each household member.[1] Respondents find some of these questions sensitive, personal, or difficult to answer, and do not always understand why information needs to be collected on some topics. To help reduce this burden, the Census Bureau examined the questions for likely sources of difficulty, sensitivity, and burden to determine potential revisions and engaged a broader set of federal data users to develop recommendations for question modifications. We conducted three rounds of cognitive testing to evaluate the modifications to ACS items that collect data on a variety of topics, including year of naturalization and year of entry; year the housing structure was built; access to telephone service; computer and internet use; work address; weeks worked; and the income series. Across the 3 rounds of testing, 13 interviewers conducted 144 interviews in 4 locations and 2 languages. The research design included 63 recruitment targets, 13 interview guides (based on content, mode, and version), and 119 analysis codes.

School Surveys. The National Teacher and Principal Survey is the primary source of teacher, administrator, and school staff perspectives on public

---

1 https://www.federalregister.gov/documents/2017/12/12/2017-26726/proposed-information-collection-comment-request-the-american-community-survey.

and private K-12 schools. Prior to implementation of the 2017–18 survey cycle, we conducted cognitive testing of new modules being added to three of the four survey questionnaires on the subjects of teacher and principal evaluation, professional development, engagement, and instructional time. A team of 12 cognitive interviewers conducted 183 interviews across 5 rounds in 7 locations. The research design included 12 recruitment targets and 14 interview guides (based on version and changes made from previous rounds).

Victim Survey (Martinez et al. 2017). The purpose of this cognitive research was to test new and revised questions for the 2016 National Crime Victimization Survey (NCVS) Supplemental Victimization Survey instrument. This supplement collects data on stalking victimization with questions designed to measure the prevalence, characteristics, and consequences of nonfatal stalking. The pretesting placed emphasis on the design and performance of the supplement's screener questions, which are used to identify victims of stalking so that they can be asked more detailed follow-up questions. We conducted 60 cognitive interviews with a team of 9 interviewers across 5 rounds of testing. We tested the survey questions with adults and teens who, during recruitment, self-identified as a stalking victim or a non-victim, to measure the performance of the 15 screener items.

25.3 Identifying Detailed, Quality-Driven Management Approaches for Qualitative Research

Cognitive interviewing is a qualitative pretesting method that uses semi-structured, in-depth interview techniques to gain a deeper understanding of participant reactions to and interpretations of survey items and materials; it uses purposive sampling and recruitment methods to identify and obtain cooperation from participants; and it generates rich, descriptive data that require specialized analysis techniques to understand and interpret (OMB 2016). Prior to the Content Test, much of our cognitive testing experience was with small-scale projects, usually involving a handful of respondents, one interview guide, a few screening characteristics, and an interviewing team of two or three that was also responsible for the analysis and report-writing. We were easily able to adapt many of the tools and techniques we had used for successfully conducting the smaller-scale projects with high quality to our four complex cognitive testing projects. Because of the additional risks to quality inherent to these projects' large scale and complexity, we also sought to learn from others' experience managing similar efforts.

There are very few published examples of management tools and techniques that facilitate quality in large and complex cognitive testing projects in particular or, more broadly, in any method of qualitative research. None of the descriptions of management approaches provides sufficient detail for replication, and there is considerable variation in how the approaches are designed to facilitate quality. Some tie quality to specific aspects of the project, such as the ability to meet the project requirements (Sha et al. 2012; Sha and Childs 2014) or the accuracy of the data (Peytcheva et al. 2013; Sha et al. 2010). Others do not mention quality at all (Hunt et al. 2011) or mention it only briefly (Laditka et al. 2009). None positioned their management tools and techniques within a broader, explicit quality framework, and this led us to review the wider qualitative literature on quality, to identify principles we could apply to the development of management approaches for large-scale, complex cognitive testing projects.

25.4 Identifying Principles for Developing Quality-Driven Management Approaches

Qualitative researchers have long held "divergent conceptions of the requirements of rigorous enquiry" in the field, including whether certain kinds of qualitative work such as in-depth interviews can even be claimed as scientific (Hammersley 2007, p. 297). This is not surprising, given that there are myriad different ways to think about and conduct qualitative research (Ormston et al. 2014; Creswell 2007), reflecting the "numerous approaches, paradigms, schools, and movements … that vary in terms of the ontological, epistemological, and methodological assumptions on which they are based" (Spencer et al. 2003, p. 4). These divergent views extend to the concept of quality, what it means, and how to achieve it in the context of qualitative research. As just one example related to one aspect of quality, constructivists reject the very notion of validity because it implies there is a "real world" and some objective criteria by which to measure any representation of it (Maxwell 2012, p. 127).

At the same time, qualitative researchers appear to agree that quality matters (Seale 1999; Creswell 2007; Hammersley 2007; Maxwell 2012), as evidenced by the proliferation of quality checklists, criteria, and frameworks, such as the Quality Framework (Spencer et al. 2003), the Consolidated Criteria for Reporting Qualitative Research (Tong et al. 2007), and the Total Quality Framework (Roller and Lavrakas 2015). (See also Santiago-Delefosse et al. 2016 and Bryman et al. 2008). Like Ormston et al. (2014), we aimed for the cognitive testing projects described in this chapter to be "well-designed [and] well-conducted to generate well-founded and trustworthy evidence" (p. 23). The challenge was to identify specific and practical approaches for facilitating that outcome in an applied survey research setting and for a sponsor operating in a high-profile environment.

A common thread in the literature, regardless of the qualitative research paradigm, is frequent reference to transparency and consistency. Transparency is the concept of providing enough documentation to allow others to evaluate the work, regardless of the criteria, and it facilitates reproducibility, a hallmark of good science. Consistency is the concept of ensuring agreed-upon implementation of procedures across staff and stages of the research. We used these two concepts to guide development of detailed management approaches for maximizing quality on our complex cognitive testing projects.

25.5 Applying the Concepts of Transparency and Consistency

Despite broad acceptance of the need for transparency, there is debate about exactly how transparent the (qualitative) research process can be. Roller and Lavrakas (2015) assert that "qualitative researchers should reveal rich details of all phases of the research to enable the user or reader of the research report(s) to apply or transfer the design features to other contexts" (p. 10). Boeije adds that detailed and accurate documentation ("methodological transparency") is important in facilitating "virtual replication," which allows for assessment of the researcher's choices and the possibility for others to replicate the research design (2009, p. 174). However, Hammersley (2007) questions whether full transparency is achievable given the complexity and messiness of the research process. We recognize this challenge but take the view that transparency is important and that researchers should strive to document as much of the design and decision-making process as possible.

The role of consistency in qualitative research, particularly as it relates to reliability, is also contested. Boeije argues that "standardization of data collection methods," "specific procedures," and "a well-trained interviewer" all serve to increase the "reliability of observations" (2009, p. 169). Roller and Lavrakas also advocate following the same "correct" procedures throughout all of data collection (2015, p. 33). Others question whether the adoption of quantitative conceptualizations of quality, such as reliability, are appropriate evaluative criteria for judging the quality of qualitative research (see, for example, Leininger 1994). We share the view of Miles and Huberman (1984) and Cobb and Hagemaster (1987) that "qualitative researchers must make their work understandable to the 'dominant culture' in the scientific community" (Cobb and Hagemaster 1987, p. 139) so as to demonstrate the credibility of cognitive testing findings and their value in helping to improve survey measurement.

In thinking specifically about cognitive testing, Fitzgerald et al. (2011) argue for the importance of consistency, particularly when testing in more than one language. They adopted quality control measures that included the sharing of research goals with the interviewing team, developing standard interview guides, training interviewers, monitoring interviewer quality, and providing interviewers with feedback on their earlier interviews to address any deviations from the agreed protocol or any bad interviewing habits. However, there is debate about how far cognitive interviews can and should be standardized. For example, Miller et al. (2011) struggled with introducing standardization (consistency) into cognitive interviewing, which they feared would mitigate the method's strength of allowing interviewers to follow up on emergent themes. While standardization of the cognitive interview may risk some of the benefits of reflexivity, we consider some standardization important. As Conrad and Blair's (2009) finding that different kinds of verbal probing yielded significantly different results indicate, whatever interview technique is chosen for a study, it should be consistently reinforced or the findings will be difficult to interpret and may not be useable at all.

25.6 The 13 Quality-Driven Management Approaches

The remainder of this chapter presents practical approaches to managing cognitive testing on a large and complex scale in a way that facilitates consistency and transparency; provides enough rich detail about the approaches that other researchers can easily adapt them to their own projects, regardless of size or complexity; and allows reviewers to form a judgment about the credibility of the findings.

The approaches are organized by three management domains that qualitative researchers generally agree are important and contribute to consistency and transparency – training, communication and documentation, and monitoring. The examples included in the descriptions of each of the management approaches are drawn from our work on the four cognitive testing projects summarized earlier in this chapter.

25.6.1 Training (Management Approaches 1–5)

Project-specific training is essential to any complex qualitative data collection effort. Not only are training sessions the most obvious way to communicate project decisions and approaches to the greatest number of project participants, but they are also the premier opportunity to ensure a consistent and thorough understanding of all data collection materials and procedures, thereby enhancing quality. Roller and Lavrakas (2015) note that in-depth training contributes to consistency, and Tucker (1997) advocates training as a way of injecting consistency into the administration of experimental conditions in the cognitive lab. By training interviewers, recruiters, analysts, and report writers, project leadership ensures all team members have a full understanding of the procedures for each of those aspects of the project. One way to do this is to establish a clear link between the research objectives and the recruitment, interviewing, analysis, and reporting procedures.

25.6.1.1 Management Approach #1: Provide Study-Specific Training to Recruiters

A challenge on the Content Test was imposing consistency on recruitment procedures across recruiting facilities, each of which had its own processes for helping potential respondents understand the screening questions, and for tracking and reporting progress. We needed to ensure that (i) all respondents were exposed to the same stimuli at screening; (ii) any clarification help they might request while answering the questions would not prime them for the survey questions and cognitive probes they would receive during their interviews; (iii) the complex eligibility criteria were accurately balanced across all recruitment locations; and (iv) strict procedures for the protection of personally identifiable information (PII) were followed. To address these potential risks, we developed specific project requirements for administering the screener, for transmitting recruiting and interview data, and for the assignment of unique case identifiers. We then delivered a detailed training to the recruiting teams at all eight facilities. The agenda included the study background and objectives; question-by-question review of the 40 screener items; discussion of how to handle respondent questions or unusual situations; procedures for handling PII; and administrative issues specific to the project. We implemented similar training sessions on the other three projects as well, to facilitate accuracy of the screening results, which ultimately contributed to higher-quality interview data by ensuring survey questions were tested with the target population.

25.6.1.2 Management Approach #2: Go Beyond the Interview Guide to Cover All Aspects of the Interview Process

The role of cognitive interviewer encompasses much more than simply administering the interview guide, especially on large, complex projects. Exhibit 25.1 displays a list of the topics covered in the Content Test interviewer training.

Exhibit 25.1 Minimum List of Training Topics to Cover

- Project background/previous related research
- Project research objectives
- Overview of project approach
- Review of project timeline and team members
- Review of interview guide formatting conventions
- Review of interview techniques
- Review of interview guide questions
- Practice administering interview guide
- Organization of interview materials and folders
- Data storage and security practices
- Other interviewer administrative responsibilities
- Guidelines for writing interview summaries

As one example, interviewers working on the Content Test were trained in how questionnaires should be labeled and returned to the office to comply with the Census Bureau's data security procedures. Interviewers were also provided with written instructions and given the opportunity to discuss the procedures and rules as part of the face-to-face training. These procedures formed an important part of the testing protocol and were essential to the success of the test.

25.6.1.3 Management Approach #3: Allow Multiple and Varied Types of Practice Interviews

All four projects involved multiple versions of the interview guide, some with complex implementation instructions. Interviewer errors can impinge on data quality if the instructions are not implemented consistently, or if the manner in which they are administered causes confusion on the part of the respondent. For example, on the Burden Test, in early practice interviews, some interviewers misunderstood how to follow the flow of the interview guide, incorrectly repeating the question being tested. Even in a detailed and exhaustive training session, interviewers may not be able to absorb everything, making post-training practice interviews all the more important for preventing errors in live interviewing.

On the Content Test, Burden Test, and School Surveys projects, we required interviewers to participate in three different practice activities prior to launch. During training, we conducted interactive (or round robin) interviews, where the trainer played the role of the respondent and the trainees took turns administering the survey questions and probes. We also allowed time during training for interviewers to pair up and act out pre-scripted scenarios, taking turns playing the role of interviewer (administering the instruments as trained) or respondent (reading from the scenario script). Project leaders monitored the paired practice sessions and shared observations with the entire group. After training, interviewers were instructed to take up to an hour to practice administering part or all of any interview guides with which they felt they needed greater familiarization before starting fieldwork. What was most valuable about each of these practice activities was the opportunities they provided for interviewers to ask questions and seek clarification on particular aspects of the protocols, and for the project leaders to observe interviewer practice. Issues were shared and discussed with the group so that everyone came away with a common understanding of how to implement the interview guides.

On the Content Test and for the School Surveys, we built into the early part of data collection the opportunity for retraining by listening to each interviewer's first interview, either live or via audio recording, and providing additional feedback as needed. We highly recommend building into the schedule some time after training to finalize the interview guides, so that additional improvements can be incorporated based on unanticipated issues that arise during practice interviews or suggestions trainees may have.

Building in time for multiple and varied types of practice interviews can facilitate consistency and accuracy in data collection by surfacing unanticipated issues and ensuring deep familiarity with the interview guide by all team members.

25.6.1.4 Management Approach #4: Conduct Formal Training for Writing Interview Summaries

For these large-scale cognitive testing projects, we used interview summaries, rather than transcripts, as the source of data for analysis. Interview summaries are written by listening to the recording of the interview and summarizing what transpired for each item that was tested, including the respondent's answer to the survey question and their responses to each of the cognitive probes. The summary is an early step in analysis because the writer must make informed decisions about what to retain and what to omit from the recording. The interview summary should be a representation of what actually happened in the interview, rather than the summary writer's opinion about how well a question performed against its measurement aims, the implications of what the respondent said or did, or recommendations about observed problems.

As such, these summaries need to be accurate and complete (capturing everything that was discussed), and include judicious selection of verbatim quotes. In the case of the four projects, summaries were completed by a large team of analysts. Analysts used templates formatted for ease of uploading into analysis software. We heeded the advice of d'Ardenne and Collins (2015) to train writers of cognitive interview summaries in the level of detail required and to be clear on the testing objectives. On the Content Test, summary writers participated in a special training module in which the template designer walked them through the structure of the template and reviewed key guidelines for writing summaries, including those outlined in Exhibit 25.2.

Exhibit 25.2 Guidelines for Summary Writers (Content Test)

- Do not alter headings in templates.
- Fill in all fields, even if a probe was skipped, to distinguish what was skipped during the interview versus what was accidentally skipped during summary writing.
- Allow 2 hours of writing for each hour of interviewing.
- You must listen to the recording as you type the summary.
- Record all observations made by the interviewer both in the recording and in written notes.
- Support observations with direct quotes, and clearly distinguish what **actually** happened from your **interpretation** of what happened.

- Note where you do not understand the respondent's comment or reaction.
- Summaries must be clear, specific, and complete.

Trainers also reviewed the project's research objectives with the summary writers, and went over detailed instructions for completing the summary so that all writers did so consistently. This included the wording to use when an interviewer skipped a probe; punctuation rules; and file-naming conventions. A small group of senior team members also reviewed the first three summaries of each writer, providing feedback and retraining as needed, to ensure all writers were following the principles outlined in the training and that they were incorporating earlier feedback into subsequent summaries. Spending time providing detailed instructions and feedback ensured consistency across all the summaries, and a transparent evidence trail supporting each recommendation regarding the wording of the tested questions.

25.6.1.5 Management Approach #5: To Ensure Consistency Among Multiple Report Authors, Set Clear Expectations

Fernald and Duclos stress that when writing collaboratively it is important to "establish guidelines, roles, and expectations early in the process to avoid misunderstandings later" (Fernald and Duclos 2005, p. 363). Besides misunderstandings, using a team of multiple authors may result in different writing styles that can be time-consuming to reconcile in later editing stages. On the Content Test, we found ourselves spending a substantial amount of time working individually with each author to ensure consistency across their report sections. For subsequent reports on that project and for the Burden Test, we convened meetings with all authors before writing began to explain the expectations in detail, including providing good and bad examples to show how to use quotes to support interpretation. We also used reports that had previously been approved by the sponsor as examples of the required level of detail and presentation style. While we still met individually with writers who needed additional guidance, far fewer meetings were required. We found that setting and documenting expectations also made the analysis process more transparent.

25.6.2 Communication and Documentation (Management Approaches 6–10)

The Project Management Institute (PMI) has identified communication as "one of the single biggest reasons for project success or failure" (PMI 2008). On any large-scale project, managing all the possible communication channels to ensure the team is working effectively and in unison toward the same goal is the biggest challenge (Sha et al. 2012).

Members of the HANDOVER Research Collaborative built a "community of practice" designed specifically to ensure consistency in their research process (conducting focus groups and semi-structured interviews) by communicating both formally and informally through weekly or more-frequent meetings in person or by phone, email exchanges, memos, and other documents (Johnson et al. 2012). Documentation across every phase of the project also provides a consistent way to communicate decisions, procedures, and processes. Tucker (1997) asserts that well-documented procedures are a requirement for scientific studies, although difficult to do in cognitive testing because of the somewhat unstructured interview technique it uses. The approaches described here assume that detailed documentation is essential for both supporting the claim of high quality and allowing for replicability.

On our complex cognitive testing projects, it is imperative that all team members understand the research objectives and implement research-based decisions consistently. Our proposed approaches in the communication and documentation domain focus specifically on how communication should stem from and be centered around the research objectives.

25.6.2.1 Management Approach #6: Methodically Document and Communicate Research-Based Decisions

As a large-scale project unfolds, dozens if not hundreds of decisions are made about how to carry out each discrete project task in a way that fulfills the research objectives. To minimize the risk of any given decision not being implemented fully, consistently, or in the manner intended, we sought to thoroughly brief team members on the decisions that were relevant to their work and the research-based reasons for them. We were also careful to document decisions and outcomes of unanticipated situations, so that any subsequent instances could be handled in the same way.

Providing the means to ensure all project staff carry out their work according to a consistent understanding of the research objectives is especially crucial when the project team is large and may be scattered across several locations, increasing the number of communication channels and hence the possibilities for misunderstandings or lost communication. On the Content Test and Burden Test projects, we began by meeting with the project sponsor and stakeholders for a thorough review of the research objectives to ensure we as the project leaders understood and interpreted them as intended. The project sponsor provided an explanation of each objective and engaged in a detailed question-and-answer session. We posted the research objectives to a shared location where all team members could easily access them and refer to them frequently throughout the life of the project.

When dilemmas arose, we made it a habit to ask, “How can the research objectives help us address this situation?” and when contemplating possible solutions, “How does each of the possible solutions serve the research objectives?” For example, the Content Test research design called for the recruitment of college students living in dormitories to test items with those in certain types of group quarters who have different jobs during the school year than they have in the summertime. Because the recruitment phase coincided with the end of the academic year, local universities were unresponsive to our requests for permission to recruit on their campuses. In reassessing the recruitment criteria, the team determined that interviews with residents of other types of group living quarters would still satisfy the research objectives to evaluate items with those who have different types of jobs throughout the year. Similarly, when we had difficulty recruiting Afro-Latinos as respondents, a reassessment determined that collecting information from a (non Afro-Latino) respondent on a household member who was Afro-Latino would provide the information needed to meet the research objectives for the tested items.

We also incorporated the research objectives into as many project documents as relevant, including project meeting minutes, the recruitment plan, interview guides, and training manuals. For both the Content Test and the Burden Test, we included in the minutes from our weekly meetings a section titled "Action Items" and one titled "Actions From Previous Meetings." As shown in Exhibit 25.3, the action items were organized into a table with space to record who was responsible, the date assigned, and notes explaining how and when the item was resolved along with the rationale for the resolution. Whenever applicable, that rationale was tied to the research objectives.

Exhibit 25.3 Action Items From Previous Weekly Meetings (Burden Test)

Item	Item description	Assigned to	Initiation date	Status/next steps
1	Provide questionnaires	Sponsor	3/18/16	Delivered on 3/28/16
2	Provide table of recruitment requirements and soft targets	Project staff	3/18/16	Delivered on 3/22/16; details included in recruitment plan
3	Provide recommendation for third location (or no third location)	Project staff	3/18/16	Decision on 3/30/16 to conduct local interviews only; recruitment targets can be met locally (see details in minutes)

These communication practices contributed to quality by mitigating the risk of inconsistent implementation of, for example, sampling decisions (e.g. choosing a strategy that fails to target the population identified by the research objectives) or interview techniques (e.g. following up on respondent comments that fall outside the research objectives). They also facilitated transparency by providing a historical record of what decisions were implemented and how. As well, we documented when implementation went awry and how those situations were addressed, as learning points for subsequent projects.

25.6.2.2 Management Approach #7: Develop a Sampling and Recruitment Plan

One of the highest priorities on any cognitive testing projects is ensuring the right respondents are recruited. For any groups, and particularly those with low incidence rates in the population, it is imperative to not only have multiple recruiting strategies available, but also establish ahead of time the threshold for deploying them – for example, if database recruiting falls short of goals after the first full week of recruitment efforts. Exhibit 25.4 shows the basic elements of such a plan.

Exhibit 25.4 Basic Elements of a Sampling and Recruitment Plan

- Description of target groups and quotas
- Mapping of target groups and interview locations to research goals
- Interview locations, including facility names and addresses and other logistical information
- Interview dates for each location
- Method for tracking interview progress
- Risk mitigation strategies in case recruitment is lagging
- Recruitment outreach strategies
- Screener questionnaire
- Recruitment advertising language
- Methods for tracking and communicating progress

Risk mitigation procedures are a critical feature of these plans. Our primary strategy for the Content Test was to recruit using the databases of professional focus group facilities. The recruitment plan also outlined additional strategies, including Craigslist advertising, snowballing, intercept recruiting (approaching passersby at, for example, a subway station or shopping mall, and screening them for eligibility), canvassing, and informal social networking. Because these strategies were specified in advance, we were able to quickly and efficiently deploy them when certain target groups were proving difficult to find. Developing a detailed sampling and recruitment plan before recruiting begins requires the team to operationalize decisions made at the study design phase and provides a guide for ensuring decisions are made consistently during recruitment. Comprehensive documentation also facilitates transparency at the reporting stage.

25.6.2.3 Management Approach #8: Develop Clear Formatting Conventions and Handling Procedures for Interview Materials

The interview guide operationalizes the research objectives and provides the means by which the data are gathered systematically and rigorously. Our cognitive testing projects all involved complex interview guides and myriad materials for interviewers to manage during the interview. In some cases interviewers had as many as half a dozen or more different documents to transition between in the course of the interview. While this situation is not unique to large-scale qualitative research, it can quickly interfere with consistency of implementation across the data collection team if procedures for handling multiple documents are not carefully developed. Aside from written instructions, we used color, spacing, and font size to differentiate critical interview questions from lower-priority questions, interviewer instructions, and any other information interviewers needed to remember. For example, instructions for interviewers were printed in red capitalized font; research questions appeared in green font; and probing questions were written in red lowercase font. We also mapped the interview questions to the research objectives (Roller and Lavrakas 2015) and, whenever feasible, stated the research aims within the interview guide (d'Ardenne 2015).[2] Exhibit 25.5 displays some of the features we incorporated into the Content Test and School Surveys interview guides.

Exhibit 25.5 Content Test Interview Guide Formatting – Research Questions, Survey Questions, Cognitive Probes

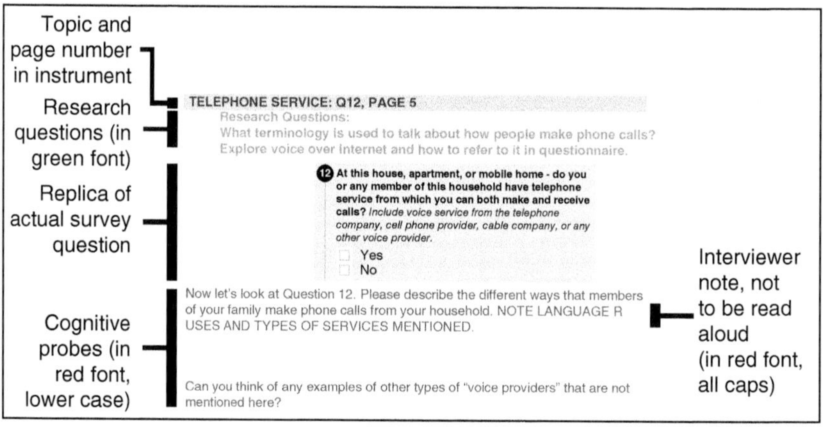

For the Content Test, Burden Test, and School Surveys, we printed the documents in such a way as to facilitate easy transition within and among them. For

2 This is an especially helpful technique in cognitive testing, when many of the probes are quite general (e.g. "Tell me more about your answer").

example, the Burden Test used hard-copy screen shots of the web survey questions we were testing. Interviewers had to manually implement skip patterns to determine the next page to hand to the respondent. We inserted detailed instructions in the interview guide (see Exhibit 25.6) and did several dry runs of the procedures so we could advise interviewers in training how to navigate the materials (also see Management Approach #3).

Exhibit 25.6 Burden Test Interview Guide Formatting – Skip Pattern Instructions

Instructions for using the table below

NEXT, ADMINISTER PERSON 1 ITEMS BY HANDING Packet 1, THEN ASKING REMAINDER OF QUESTIONS ACCORDING TO P1 TABLE BELOW.

Based on answers to items in roster and questionnaire, interviewers hand specific packets of items to respondent

P1 TABLE OF INSTRUCTIONS

P1 Questions	Answer	Roster Q4a Year	Instruction
Q7	Born in US		Go to P2 Selection Question on Page 5 of Protocol
	Born outside US		Use Packet 2
Q8	Us parents, US territory, or non-citizen	1984 or earlier	Use Packet 4a
		1985–1996	Use Packet 4b
		1997–2004	Use Packet 4c
	Citizen by naturalization	1984 or earlier	Use Packet 3a
		1985–1996	Use Packet 3b
		1997–2004	Use Packet 3c

Other strategies include the use of "skip sheets" to help interviewers systematically keep track of answers to previous questions that will be needed for subsequent routing or text fills; a table of contents for lengthy interview guides; and flow charts to provide a bird's-eye view of complex routing. For the Burden Test, we piloted the cognitive interview guide to time the interview and to uncover any unanticipated issues with administering the test questions and cognitive probes, understanding and applying the instructions, or moving between multiple documents correctly.

These strategies aim to ensure that data collection remains focused on the research goals with goals integrated into the interview guide. By adopting consistent formatting conventions to make interview guides and materials clear and easy to administer, we aimed to reduce the risk of interviewers making errors in how the test was carried out.

25.6.2.4 Management Approach #9: Identify and Manage Lines of Communication Among Offsite Recruiters, Interviewers, and Project Leaders

Frequent communication between offsite staff and project leaders is important when data collection is decentralized. When completing a large number of interviews in a short amount of time, project leaders need to be alerted immediately to the number of no-shows and completed interviews, as well as any issues that arise during interview administration. For some of our projects, interviewers were at offsite facilities collecting data for a day or two, and the number of completed interviews affected the recruiting strategy at subsequent locations. On all four projects, interviewers were required to report on the outcome of an appointment right away to facilitate progress monitoring.

We also wanted to be able to apply any early lessons learned from recruiting the first few respondents to later recruitment efforts. The interview teams communicated regularly with project leaders to share lessons learned, and project leaders facilitated constant feedback to and from recruiters. In the Content Test, one of the three interviewers per site was designated as the site lead and was given responsibility for communicating with project leads. This person provided a report of interview results at the end of every interviewing day. Project leaders also created and distributed to the entire team a contact list of names and cell phone numbers of all data collection staff along with contact information for each of the facilities. The list further identified which project staff to notify about what kinds of issues (e.g. recruiting, interview materials, respondent issues). These communication strategies facilitated consistent implementation across teams, locations, and facilities.

25.6.2.5 Management Approach #10: Develop a Detailed Analysis Plan to Improve the Quality of Reporting

Qualitative analysis often involves organizing respondents' narratives, statements, or words into meaningful thematic groupings (Collins 2015; Willis 2015; Ritchie et al. 2014; Krueger and Casey 2014). In the case of complex projects, this may involve hundreds or even thousands of pages of text that need to be organized. Working without an analysis plan risks a haphazard approach to organizing the data reduction, failing to address all the research questions, or overlooking important findings.

In the Burden Test, we developed an analysis plan prior to the launch of data collection for each item being tested. The plan stated each research objective, articulated the qualitative analysis software queries that would be run on the coded data to support that objective, and described the approach to analyzing and writing about the results of the query. An example of this approach is shown (for one research objective) in Exhibit 25.7.

Exhibit 25.7 Sample Item Analysis Guide (Burden Test)

Research Question: Overall, how easy or difficult is it for respondents to answer the questions about where they live and where they work?

Query the variables assessing burden for residential address and place of work address.

Report the number of respondents who found providing an answer for their residential address easy, and how many found it difficult; and the number who reported providing a work address easy, and how many found it difficult.

Assess the reasons respondents gave for why providing one or the other was perceived to be difficult.

The analysis plan accommodated evolving theories as the analysis progressed and documented all analytic decisions.

Having a systematic analysis plan that is motivated by research goals added clarity, efficiency, and structure to the reporting process. It also ensured consistency across analysts in addressing the research goals, and provided a transparent description of how the analysis was conducted for those who wish to reproduce the effort.

25.6.3 Monitoring (Management Approaches 11–13)

An essential component of consistency is monitoring to ensure the work is carried out as documented and communicated to the team, via training or other means (Tucker 1997; Roller and Lavrakas 2015).

Control over the data is especially important on large-scale qualitative research efforts, where the volume of cases and the data generated by them can quickly become a chaotic jumble. While discussion of data processing and transformation as "data management" is common in the literature (Collins 2015; Marshall and Rossman 2006; Fernald and Duclos 2005; Hunt et al. 2011), we focus on monitoring as a means of tracking cases and data, and improving the efficiency of coding and analysis.

25.6.3.1 Management Approach #11: Centralize the Selection of Respondents

For some of our projects, recruitment happened across multiple locations, and target quotas needed to be aggregated across all sites. We realized respondent selection would have to be centralized to prevent targets from being missed or over-sampled. This approach required recruiters to submit their screening data on a daily basis and for selections to be quickly communicated back to them. On the Content Test, the screener questionnaire was programmed as an online survey that was administered by recruiting teams across multiple locations. Screener data were downloaded on a daily basis, and recruiters were notified of selections within one day in order to quickly re-contact eligible respondents and schedule interviews. Monitoring target quotas across locations, languages, and other recruitment criteria ensured all materials were tested with target groups.

25.6.3.2 Management Approach #12: Use Technology to Monitor and Adapt Recruitment Strategies

Willis (2005) calls recruitment the "500 pound gorilla" that determines the feasibility of the research effort. In large scale studies, this "gorilla" can easily become bloated by multiple sampling approaches; complex algorithms for tracking target quotas; extensive screening questionnaires; training and managing large teams of recruiters; and the need to screen, track, schedule, and interview hundreds of potentially eligible respondents. The recruitment phase of a complex cognitive testing project can be characterized by an almost overwhelming flow of information, including screener data, case selections and assignments across interview type, interview schedules, interviewer/moderator assignments, response status, and status of interviewer summaries.

For the Content Test, Burden Test, and School Surveys, we created several customized monitoring tools in Excel. Exhibit 25.8 outlines the elements of each.

Exhibit 25.8 Key Features of Recruitment Monitoring Tools

Screener Tracker

✓ Screener responses
✓ Demographics
✓ Determination of eligibility
✓ Allocation to target group(s)
✓ Assigned version, mode, language, location of interview
✓ Date and time of scheduled interview
✓ Interviewer name
✓ Status of interview (complete vs. no-show/canceled)

Progress Tracker

✓ Counts of completed interviews within each target group
✓ Counts of completed interviews within each version, mode, language, or location
✓ Quotas for target group, version, mode, language, location

Completion Tracker

✓ Counts of cases selected, screened, and completed within each target group
✓ Number of completes by version, mode, language, location
✓ Status of interview summaries for each case

The first was a Screener Tracker that was automatically populated on a daily basis with complete cases from the online screener. We made daily selections of eligible participants from this spreadsheet, allowing us to monitor how they fit into one or more of the many target groups. The second was a Progress Tracker, which monitored the number of completed interviews against specific quotas within each of the target groups, as well as by interview mode and survey version. Once data collection began, the Progress Tracker was updated on a daily basis and directly informed decisions about which additional cases to select from the Screener Tracker. Finally, the Completion Tracker provided an easily digestible snapshot of recruiting progress. It updated automatically via links to the other tracking tools, and displayed the number of cases that had been selected, scheduled, and completed within each location, by each target group. The Completion Tracker contained no PII, so it could be easily shared with the research sponsor via email.

Vigilant monitoring using these tools was critical in helping us meet recruitment goals and ensure consistent implementation of procedures across the dispersed teams of recruiters.

25.6.3.3 Management Approach #13: Track in Detail the Progression of Transformation from Raw to Analyzable Data

On our complex cognitive testing projects, raw data were produced in multiple formats (e.g. recordings, survey responses, interview notes) and arrived at the head office from multiple sources (e.g. large interviewing teams, several locations, more than one data collection organization), increasing the risk of losing data. Project leaders of the Content Test, Burden Test, Victim Survey, and School Surveys all used Excel to track the progress of interview summary production, review, export into the qualitative software, and dissemination to offsite analysts, among other things. Exhibit 25.9 shows a selection of the column labels these projects used, representing case information and dates of key steps in the project's process.

Exhibit 25.9 Data Tracker Column Labels

- Case ID
- Interview Date
- Interviewer
- Recording Uploaded to Drive
- Interview Summary Received
- Interview Summary Reviewed
- Summary Complete or Needs Edit
- Summary Ready for Upload to Software
- Date Uploaded to Software
- Recording Delivered to Sponsor
- Summary Delivered to Sponsor

Such careful tracking takes on added importance in the context of aggressive project timelines. Content Test interviews were still being conducted at the same time that interview summaries were being completed, making it even more critical to know exactly where each case was in the process at any given moment. This level of monitoring provided a transparent record of how cases and data were handled from start to finish on each project.

25.7 Discussion and Conclusion

The quality-driven management approaches described in this chapter are a response to the risks to quality that can arise in large-scale, complex cognitive testing projects. We conceptualized quality as consistency of implementation and transparency in reporting methods that aim to produce robust, credible results and recommendations. Several recurring themes emerged from the 13 practices we describe (listed in Table 25.1). Among them were the need for communication across project team members and sponsors and throughout the entire project; the importance of placing the project research objectives at the center of all decisions; identification of well-defined and detailed procedures at the start of the project, and careful documentation of changes as the project evolves; and the use of technology, tools, and materials to facilitate procedures, reinforce communications, and enhance monitoring.

We view these management approaches as a starting point: we will continue to refine and test them, and work to develop additional approaches. One

Table 25.1 Quality-driven management approaches for large-scale cognitive testing projects.

Training	
1	Provide study-specific training to recruiters.
2	Go beyond the interview guide to cover all aspects of the interview process.
3	Allow multiple and varied types of practice interviews.
4	Conduct formal training for writing interview summaries.
5	To ensure consistency among multiple report authors, set clear expectations.
Communication and documentation	
6	Methodically document and communicate research-based decisions.
7	Develop a sampling and recruitment plan.
8	Develop clear formatting conventions and handling procedures for interview materials.
9	Identify and manage lines of communication among offsite recruiters, interviewers, and project leaders.
10	Develop a detailed analysis plan to improve the quality of reporting.
Monitoring	
11	Centralize the selection of respondents.
12	Use technology to monitor and adapt recruitment strategies.
13	Track in detail the progression of transformation from raw to analyzable data.

area of focus might be management approaches for identifying, selecting, and evaluating qualitative interviewers and recruiters. Sha and her colleagues felt it important for the efficiency of the recruitment effort that recruiters have experience with cold calling and be familiar with the local area in which they were making outreach efforts (Sha et al. 2010; Sha et al. 2012). While interviewer personality characteristics (e.g. friendly, curious) and possession of good listening skills are important for ensuring consistency, as noted by Gray (2015) and Willis (2005), more formal qualifications and experience are also likely to have an impact on data quality. Establishing minimum criteria for the type of training received (e.g. specific instruction in cognitive interview techniques) and years of experience (or number of interviews conducted) when selecting cognitive interviewers may enhance consistency during data collection.

The management approaches discussed in this chapter were designed for and applied to a few large-scale and complex cognitive testing projects conducted within a larger survey production process. The applicability of these approaches to managing other kinds of qualitative data collection that is part of questionnaire development and testing, such as focus groups or in-depth interviews, and to cross-cultural and cross-national qualitative pretesting, requires further consideration. For example, for projects that conduct data collection in multiple languages, interviewer training could be conducted in English, but paired practice interviews, both during training and afterward, could be conducted in the other languages and feedback provided by a project leader who is fluent in that language. The Survey Research Center's guidelines for best practices in cross-cultural surveys (2016) emphasize the importance of quality standards and include a thorough discussion of overall study management tools and techniques. They call for questionnaire pretesting as a requirement yet lack detail about quality criteria or how to implement them.

We are also eager to learn from those in other organizations and industries who use qualitative methods to develop, test, and evaluate survey instruments. Our hope is to establish a dialogue about managing for quality in this environment, particularly as it relates to consistency and transparency. There is also a question about the utility of quality-driven management approaches, such as those described in this chapter, in improving the quality of qualitative pretesting methods, and how this might be assessed. We encourage discussion and research on this issue. Whether adding to or evaluating existing proposed approaches for managing qualitative research projects, we welcome efforts to come to agreement about what the best approaches are and implement them consistently across projects, researchers, organizations, and industries.

References

Boeije, H. (2009). *Analysis in Qualitative Research*. Thousand Oaks, CA: Sage.

Bryman, A., Becker, S., and Sempik, J. (2008). Quality criteria for quantitative, qualitative and mixed methods research: a view from social policy. *International Journal of Social Research Methodology: Theory and Practice* 11 (4): 261–276.

Cobb, A.J. and Hagemaster, J.N. (1987). Ten criteria for evaluating qualitative research proposals. *Journal of Nursing Education* 26 (4): 138–143.

Collins, D. (ed.) (2015). *Cognitive Interviewing Practice*. Thousand Oaks, CA: Sage.

Conrad, F. and Blair, J. (2009). Sources of error in cognitive interviews. *Public Opinion Quarterly* 73 (1): 32–55.

Creswell, J.W. (2007). *Qualitative Inquiry and Research Design: Choosing Among Five Approaches*. Thousand Oaks, CA: Sage.

d'Ardenne, J. (2015). Developing interview protocols. In: *Cognitive Interviewing Practice* (ed. D. Collins), 101–125. London, England: Sage.

d'Ardenne, J. and Collins, D. (2015). Data management. In: *Cognitive Interviewing Practice* (ed. D. Collins), 142–161. London, England: Sage.

Fernald, D.H. and Duclos, C.W. (2005). Enhance your team-based qualitative research. *The Annals of Family Medicine* 3 (4): 360–364.

Fitzgerald, R., Widdop, S., Gray, M., and Collins, D. (2011). Identifying sources of error in cross-national questionnaires: application of an error source typology to cognitive interview data. *Journal of Official Statistics* 27 (4): 569–599.

Gray, M. (2015). Conducting cognitive interviews. In: *Cognitive Interviewing Practice* (ed. D. Collins), 126–114. London, England: Sage.

Hammersley, M. (2007). The issue of quality in qualitative research. *International Journal of Research and Method in Education* 30 (3): 287–305.

Hunt, G., Moloney, M., and Fazio, A. (2011). Embarking on large-scale qualitative research: reaping the benefits of mixed methods in studying youth, clubs and drugs. *Nordic Studies on Alcohol and Drugs* 28 (5–6): 433–452.

Johnson, J.K., Barach, P., and Vernooij-Dassen, M. (2012). Conducting a multicentre and multinational qualitative study on patient transitions. *BMJ Quality and Safety* 21 (Suppl 1): i22–i28.

Krueger, R.A. and Casey, M.A. (2014). *Focus Groups: A Practical Guide for Applied Research*. Thousand Oaks, CA: Sage.

Laditka, S.B., Corwin, S.J., Laditka, J.N. et al. (2009). Methods and management of the healthy brain study: a large multisite qualitative research project. *The Gerontologist* 49 (Suppl1): S18–S22.

Leininger, P. (1994). Evaluation criteria and critique of qualitative research studies. In: *Critical Issues in Qualitative Research Methods* (ed. J.M. Morse). Thousand Oaks, CA: Sage.

Marshall, C. and Rossman, G.B. (2006). *Designing Qualitative Research*, 4e. Thousand Oaks, CA: Sage.

Martinez, M., Henderson, A., Luck, J., and Davis, M.C. (2017). Cognitive pretesting of the national crime victimization survey supplemental victimization survey. In: *Research and Methodology Directorate, Center for Survey Measurement Study Series (Survey Methodology #2017-03).* Washington, DC: U.S. Census Bureau. Retrieved from http://www.census.gov/srd/papers/pdf/rsm2017-03.pdf.
Maxwell, J.A. (2012). *A realist Approach for Qualitative Research.* Thousand Oaks, CA: Sage.
Miles, M.B. and Huberman, A.M. (1984). Drawing valid meaning from qualitative data: toward a shared craft. *Educational Researcher* 13: 20–30.
Miller, K., Fitzgerald, R., Padilla, J. et al. (2011). Design and analysis of cognitive interviews for comparative multinational testing. *Field Methods* 23 (4): 379–396.
Ormston, R., Spencer, L., Barnard, M., and Snape, D. (2014). The foundations of qualitative research. In: *Qualitative Research Practice: A Guide for Social Science Students and Researchers,* 2e (eds. J. Ritchie, J. Lewis, C.M. Nicholls and R. Ormston), 1–25. Los Angeles, CA: Sage.
Peytcheva, E., Sha, M.M., Gerber, E. et al. (2013). Qualitative interviewing with suspected duplicates and cognitive testing of the targeted coverage follow-up (TCFU) interview. In: *Center for Survey Measurement Study Series (Survey Methodology #2013-09).* Washington, DC: U.S. Census Bureau.
Project Management Institute (PMI) (2008). *A Guide to the Project Management Body of Knowledge (PMBOK GUIDE),* 4e. Newton Square, PA: Author.
Ritchie, J., Lewis, J., Nicholls, C.M., and Ormston, R. (eds.) (2014). *Qualitative Research Practice: A Guide for Social Science Students and Researchers,* 2e. Los Angeles, CA: Sage.
Robins, C., Steiger, D., Folz, J. et al. (2016). 2016 American Community Survey respondent burden testing: Final briefing report. 2016 American Community Survey Research and Evaluation Report (Memorandum Series #ACS16-RER-16). U.S. Census Bureau. https://www.census.gov/content/dam/Census/library/working-papers/2016/acs/2016_Westat_01.pdf.
Roller, M.R. and Lavrakas, P.J. (2015). *Applied Qualitative Research Design: A Total Quality Framework Approach.* New York, NY: Guilford Press.
Santiago-Delefosse, M., Gavin, A., Bruchez, C. et al. (2016). Quality of qualitative research in the health sciences: analysis of the common criteria present in 58 assessment guidelines by expert users. *Social Science and Medicine* 148: 142–151.
Seale, C. (1999). Quality in qualitative research. *Qualitative Inquiry* 5 (4): 465–478.
Sha, M. and Childs, J.H. (2014). Applying a project management approach to qualitative survey research projects. *Survey Practice* 7 (4).
Sha, M.M., McAvinchey, G., Rodriguez, S., and Carter, G.R. III, (2010). Respondent recruitment, interviewing, and training: lessons learned from a

Spanish language cognitive interviewing project. In: *Joint Statistical Meeting Proceedings, 65th Annual Conference of the American Association for Public Opinion Research* (ed. American Statistical Association (AMSTAT)), 6372–6381. Alexandria, VA: AMSTAT.

Sha, M., Kenward, K., Feldman, J., and Heimel, S. (2012). Managing quality on a large qualitative research study with complex respondent recruitment criteria. In: *Joint Statistical Meeting Proceedings, 67th Annual Conference of the American Association for Public Opinion Research* (ed. American Statistical Association (AMSTAT)), 5477–5491. Alexandria, VA: AMSTAT.

Spencer, L., Ritchie, J., Lewis, J., and Dillon, L. (2003). *Quality in Qualitative Evaluation: a Framework for Assessing Research Evidence* (for the Cabinet Office). London, England: Government Chief Social Researcher's Office.

Stapleton, M., and Steiger, D. (2015). Cognitive testing of the 2016 American Community Survey content test items: Summary report for round 1 and round 2 interviews. 2016 American Community Survey Research and Evaluation Report (Memorandum Series #ACS16-RER-21). U.S. Census Bureau. https://www.census.gov/content/dam/Census/library/working-papers/2016/acs/2016_Westat_05.pdf.

Steiger, D., Anderson, J., Folz, J. et al. (2015). Cognitive testing of the 2016 American Community Survey content test items: Briefing report for round 3 interviews. 2016 American Community Survey Research and Evaluation Report (Memorandum Series #ACS16-RER-20). U.S. Census Bureau. https://www.census.gov/content/dam/Census/library/working-papers/2016/acs/2016_Westat_04.pdf.

Steiger, D., Robins, C., and Stapleton, M. (2017a). *OY1 American Community Survey Respondent Burden Testing – Spanish: Year of Naturalization, Year of Entry, Weeks Worked, Income Items* (for U.S. Census Bureau). Rockville, MD: Westat.

Steiger, D., Robins, C., and Stapleton, M. (2017b). *OY1 American Community Survey Respondent Burden Testing: Weeks Worked and Income Series* (for U.S. Census Bureau). Rockville, MD: Westat.

Survey Research Center (2016). *Guidelines for Best Practice in Cross-Cultural Surveys*. Ann Arbor, MI: Survey Research Center, Institute for Social Research, University of Michigan. Retrieved 0224/2018 from http://www.ccsg.isr.umich.edu.

Tong, A., Sainsbury, P., and Craig, J. (2007). Consolidated criteria for reporting qualitative research (COREQ): a 32-item checklist for interviews and focus groups. *International Journal for Quality in Health Care* 19 (6): 349.

Tucker, C. (1997). Methodological issues surrounding the application of cognitive psychology in survey research. *Bulletin de Méthodologie Sociologue* 11: 67–92.

U.S. Office of Management and Budget (OMB). (2016). Statistical policy directive no 2. Addendum: standards and guidelines for cognitive interviews. https://obamawhitehouse.archives.gov/sites/default/files/omb/inforeg/directive2/final_addendum_to_stat_policy_dir_2.pdf.

Willis, G. (2005). *Cognitive Interviewing: A Tool for Improving Questionnaire Design*. Thousand Oaks, CA: Sage.

Willis, G. (2015). *Analysis of the Cognitive Interview in Questionnaire Design*. New York, NY: Oxford University Press.

26

Using Iterative, Small-Scale Quantitative and Qualitative Studies: A Review of 15 Years of Research to Redesign a Major US Federal Government Survey

Joanne Pascale

Center for Behavioral Science Methods, US Census Bureau, Suitland, MD, USA

26.1 Introduction

Many surveys used by governments, research organizations, and policymakers around the world are conducted on a regular basis and produce trend data. While there is value in holding the survey methodology constant over time so that any change in the survey estimates can be attributed to real change over time in the phenomena being studied – and not to a change in the methodology – modifying the methodology is sometimes inevitable. One reason for this is real-world change. For example, in the United States the "AFDC" program (Aid to Families with Dependent Children) was introduced in 1935 to provide cash welfare payments to qualified families. More than 60 years later, the program was revamped under welfare reform, which eliminated the term AFDC from the national vocabulary and replaced it with "TANF" (Temporary Assistance to Needy Families) (Office of the Assistant Secretary for Planning and Evaluation 2009). Any US survey aiming to measure income, well-being, poverty, and related topics had to be adjusted to accommodate this change.

Another factor that motivates the producers of large, ongoing surveys to modify the methodology is robust evidence of chronic, persistent measurement error. Sometimes this error is a product of a kind of "inverse cultural lag," where gradual changes in society are not reflected in the tools used to measure key social indicators. For example, the survey used to produce official statistics on labor force participation in the United States was redesigned in the mid-1990s to accommodate broad societal shifts over the prior three decades, such as the growth in service-sector jobs relative to factory jobs and the more prominent role of women in the workforce. "These changes raised issues which were not being fully addressed with the old questionnaire" (Cohany et al. 1994).

Advances in Questionnaire Design, Development, Evaluation and Testing, First Edition.
Edited by Paul C. Beatty, Debbie Collins, Lyn Kaye, Jose-Luis Padilla, Gordon B. Willis, and Amanda Wilmot.

A third rationale for a major questionnaire redesign is unexplained variation in estimates across surveys that aim to measure the same concept. Topics such as household income, labor force participation, disabilities, and health insurance, to name just a few, are measured in multiple, disparate surveys using different questionnaire methodologies and the surveys produce varying estimates. Often little is known about which aspect of the methodology is driving the differences in estimates or, perhaps more important, which estimate is the most accurate. Wide variation in health insurance estimates across surveys in the United States was the driver of a major research effort to redesign the health insurance module in the Current Population Survey (CPS). This particular survey was selected because it produces the most widely cited and used estimates of health coverage in the United States (Blewett and Davern 2006). This chapter documents the process by which the module was redesigned and covers the rationale and highlights of tests and evidence. There are three main objectives. First is to provide researchers who are grappling with the possibility of modifying questions in large, ongoing surveys with a case study on the general approach used to identify the source and extent of measurement error, and to develop and implement improvements. Second is to provide some perspective on cost and timeline for such an undertaking. Third is to discuss the implications of findings on the survey design features that were found to be driving measurement error in this particular module for surveys beyond the CPS health insurance questions.

In the United States, health insurance is obtained through a patchwork of private and public sources, and more than 9% of the population was still uninsured in 2015 despite passage of President Obama's health reform initiative (the Affordable Care Act [ACA]), aka ObamaCare (Barnett and Vornovitsky 2016). Interest in measuring both the number of uninsured and estimating the source of coverage among the insured has been intense both before and after passage of the ACA. Researchers turn to survey data on health insurance to evaluate the cost and effectiveness of public health insurance programs, the dynamics of employer benefits packages, differentials in health outcomes between public and private enrollees, and a host of other reasons. However, since the mid-1980s, wide variation in health insurance estimates across surveys has been well-documented in the literature (Swartz 1986), and this has been a constant source of consternation for researchers.

More than a half-dozen major national surveys include a questionnaire module that asks about health insurance, and estimates of the uninsured across those surveys ranged from about 8–18% prior to the ACA (Bhandari 2004; Lewis et al. 1998; Pascale 2001a; Rosenbach and Lewis 1998). This wide range in estimates and the relevance of the topic prompted a research effort begun in 1999. A comprehensive literature review in 1998 titled "Counting the Uninsured: A Review of the Literature" (Lewis et al. 1998) provided a useful roadmap by summing up the major design differences and estimates across surveys on health insurance at the time. The overarching objective of the research effort was first to examine differences in methodologies across surveys and identify problematic design features. The focus was on the questionnaire (as opposed to other methodological features such as sampling and weighting) because prior research had identified subtle differences in the questionnaire as being a major driver of variation in the estimates (Swartz 1986). Next, for practical purposes, was to select a particular survey as the initial target for developing and testing improvements; and, for reasons cited earlier, the CPS was chosen. It was assumed that evidence of problematic features that cut across surveys could inform interpretation of estimates from other surveys and help craft a research agenda for developing improvements to those questionnaires in the future.

Previous research summarizes the 15-year history of the actual tests and results of the CPS redesign in detail (Pascale 2016). The purpose of this chapter is to discuss the CPS redesign cross-survey investigation and redesign process in the broader context of survey redesign in general. The CPS redesign began as a small, independent research project – not as a "top down" plan with a long-range analysis plan, budget, schedule, etc. It turned out to be 10 years of tests isolating and examining individual questionnaire design features that had been identified in the literature as candidates for contributing to measurement error. These tests were both quantitative (split-ballot field tests) and qualitative (primarily cognitive testing), and they were iterative: results fed into and built upon one another. From these tests, a prototype questionnaire was developed that combined all the revised and improved individual features into a single, redesigned health insurance module. This was pretested and then implemented in a large-scale field test in 2010. That test demonstrated proof of concept; a second large-scale test was conducted in 2013, and the results showed measurable improvements. The redesign was fully implemented in production data collection in 2014. What follows is first a summary of the issues in measuring health insurance, and then a brief history of the key milestones of the CPS redesign. The chapter concludes with a summary and thoughts for applications to other surveys and other topic areas.

26.2 Measurement Issues in Health Insurance

The range of estimates of the uninsured prompted many researchers and agencies to inventory the various surveys, their key design features, and resulting estimates, and there are multiple taxonomies of these issues (Davern 2009; Office of the Assistant Secretary for Planning and Evaluation 2005; State Health Access Data Assistance Center 2013). A review of these taxonomies and related literature indicated that three particular features of questionnaires that measure health insurance were the prime candidates that could be driving differences in the estimates: the reference period, household- versus person-level design, and the general versus specific questions on coverage type. Each is defined and discussed next.

26.2.1 Reference Period

The reference period is "the time frame for which survey respondents are asked to report activities or experiences of interest" (Lavrakas 2008). Some health insurance surveys ask whether respondents had coverage on the day of the interview, rendering a point-in-time estimate of coverage. Other surveys, including the CPS, ask whether individuals had coverage "at any time" during the past calendar year and render an estimate of coverage which, technically speaking, could be anywhere from one day to the entire calendar year. Other surveys use a reference period somewhere in-between and ask about coverage at any time since a given date up to the interview date – usually four to six months. These are often panel surveys, and multiple waves of data can be concatenated to produce estimates of coverage on the day of the interview, or throughout a range of months or even years, depending on the reference period and number of waves.

In comparing estimates across these surveys, it is important to separate measurement error from definitional differences. An estimate of the insured at a given point in time actually *should* be different from an estimate of those insured *at any point* throughout some longer time span, such as the calendar year. However, many of the taxonomies noted demonstrate that CPS estimates of the uninsured throughout the year are roughly in-line with other surveys' estimates of the uninsured at a point in time (State Health Access Data Center 2013). This has led to widespread speculation that the CPS is missing reports of past coverage (Bhandari 2004; Congressional Budget Office 2003). Indeed, a comparative study of the CPS and the Community Tracking Survey (CTS), which used a very similar questionnaire except for the reference period (the former being calendar year, the latter being point in time), found no significant difference in estimates of the uninsured (Rosenbach and Lewis 1998; Pascale 2001a). In sum, research suggests that the phrase "At any time during [past year] were you covered by…" was not adequately eliciting reports of past coverage, and the challenge for the redesign was to figure out an alternative approach to asking about retrospective coverage.

26.2.2 Household- vs. Person-Level Design

A second problematic feature was the specificity with which household members were asked about. Some surveys use a person-level design, where questions ask about each person by name ("Is NAME covered by…"). Other surveys use a household-level design ("Is anyone in the household covered by…"), and if the answer is yes, then a roster of all household members' names is presented and a follow-up question asks who is covered. Note this is distinct from the self versus proxy design feature. For cost reasons, the CPS and many other surveys rely on a single household respondent to answer questions for themselves, and to proxy for all other household members. It was not in the realm of possibilities to modify the CPS to be strictly self-response, but this research examined self versus proxy response within the context of the person- versus household-level design. In single-person households, by default, all answers are self-response. But in multi-person households, the questions are self-response for the household respondent and proxy response for all others.

In terms of background literature, in surveys where a single household respondent answers questions on behalf of all household members, there is some evidence that the household-level design risks the respondent failing to report coverage for some members, particularly in larger or complex households (Blumberg et al. 2004; Hess et al. 2001). This is likely because the survey question itself does not "name names" and instead uses the generic "anyone in the household." Depending on household size and composition (e.g. spouses and children versus distant and non-relatives), it could be a demanding cognitive task to call to mind all household members at once and consider the type of coverage they have. At the same time, other evidence suggests that the person-level design risks respondent fatigue and associated underreporting (Blumberg et al. 2004; Pascale 2001a). This is likely due to redundant questioning, particularly for a topic like health insurance, where multiple household members are often covered by the same type of coverage (e.g. Medicaid) or even by the same exact health insurance plan. For example, in a six-person household where a single parent and five children are covered by the same plan through the parent's job, a series of questions on each coverage type (job-based, Medicaid, etc.) would be asked of the parent, and then the identical series would be repeated for each child. These questions would be asked about each child even if the parent reported in their own series of questions that their job-based plan covered themselves and all five children.

In summary, each design has pros and cons, and the literature is mixed on how the final estimates are affected across all plan types and across households of various sizes and complexity. The challenge for the redesign was how to harness the effectiveness of providing specific names of household members, but avoid the tedium of repeating the entire series for each person.

26.2.3 General vs. Specific Questions on Coverage Type

Most health insurance modules aim to categorize sample members into one of the following types of insurance:

1. Employer-sponsored insurance (ESI)
2. Directly purchased (which now includes coverage obtained on the marketplace)

3. Medicare (primarily for those 65 years old and older)
4. Medicaid/CHIP/state-specific program (primarily for low income)
5. Military coverage
6. Other/unspecified

To meet that goal, surveys generally ask a battery of yes/no questions about each of these types of coverage, along with a few less-common types of coverage (e.g. Indian Health Service, state-sponsored programs) that are then collapsed into one of the six categories. This has come to be called the "laundry list" approach, and the individual questions can be quite detailed and complex. For example, in the pre-redesigned CPS, the first question in the health insurance module asked about employer-sponsored coverage as follows:

At any time in [YEAR], was anyone in this household covered by a health insurance plan provided through their current or former employer or union?
PROBE: Military health insurance will be covered later in another question.

On the face of it, this is a complex question with several compound concepts – *any time* last year, *anyone* in the household (discussed more shortly), and coverage through a current OR former employer OR union. Furthermore, it is not necessarily clear to respondents *whose* employer or union the coverage refers to. Finally, the probe instructs respondents to exclude military coverage, which may not be mutually exclusive of ESI (e.g. for active military who are covered by government military plan).

Perhaps not surprisingly, a host of problems with the laundry-list approach have been identified in the literature (Beatty and Schechter 1998; Loomis 2000; Roman et al. 2002; Pascale 2009a; Pascale 2008; Schaeffer and Presser 2003; Willson 2005). These studies were mostly qualitative and generally found that respondents had difficulty figuring out which category best fit their coverage. For example, dependents on a spouse's or parent's job-based plan were sometimes reluctant to report their coverage as ESI because the coverage was not through *their* job but that of the spouse or parent. Furthermore, individual questions on plan type were often too detailed and complex for respondents to grasp with confidence, which proved particularly problematic when respondents were answering for other household members. For example, in many cases, respondents knew another household member was covered, but could not be confident of the source of coverage (Loomis 2000; Roman et al. 2002; Willson 2005). All these factors led respondents to misreport one plan type as another, report the same plan twice, or fail to report the plan altogether (Loomis 2000; Pascale 2008). While it is very difficult to measure the net effects of these types of reporting errors, two quantitative studies involving linkages between survey data and administrative records found that public coverage was generally under-reported and private coverage was over-reported (Davern et al. 2008; Nelson et al. 2003). Findings suggested that the questions needed to be simplified so that no one item contained too many dimensions, too much ambiguity, and/or too much specificity, but that the series as a whole still captured the necessary detail.

26.3 Methods and Results

Because there were numerous tests conducted over such a long time span, the methods and results have been consolidated into two main phases. Phase I consists of the decade of tests that isolated and investigated individual survey design features and led to the prototype redesign. During this first phase, the problematic features of the questionnaire were intertwined and overlapping, and thus the research path was not linear. Rather than a strict chronological presentation of test and results, thematic highlights are presented. Phase II was more linear and consists of the pretest and the two large-scale field tests used to examine and refine the prototype.

26.3.1 Phase I: Developing a Prototype Redesign

Table 26.1 summarizes the data collection efforts, which were a series of large split-ballot tests, interleaved with cognitive testing. These tests and results are discussed in more detail shortly. (Full reports for each test can be found in the references.)

26.3.1.1 Testing Reference Period and Household- vs. Person-Level Design Features

The Lewis et al. literature review in 1998 set the stage for the first field test (known as QDERS 1999), which isolated the reference period and the household/person-level design features in a two-by-two design. Half the sample received the household-level questions and half received the person-level questions, and within each of these panels, half received the current reference period design and half received the calendar-year design. See Table 26.2. Each cell contained about 320 household-level completed interviews representing about 800 person records for analysis. The test was implemented in the spring of 1999 through the Questionnaire Design Experimental Research Survey (QDERS). This was a research vehicle initiated by the Center for Survey Methods Research at the Census Bureau for the sole purpose of methods testing (Rothgeb 2007). The main findings were: among the person-level questionnaires, the overall percentage of the uninsured was lower, as expected, in the calendar year version than in the current version – 6.9% versus 10.3%. Again, the calendar year version is estimating those covered *at any point* during the entire previous calendar year, not just those covered at the *current* point in time. Thus, one would expect the calendar year version to produce more reports of coverage than the current version, and generate a lower rate of

Table 26.1 Data collection vehicles of the CPS redesign: phase I.

Collection	Type of test	Sample frame	Sample size	Questionnaire feature tested
QDERS 1999 (April–May)	Split-ballot field test	RDD	• 1291 households • 3228 people	• Status quo CPS structure • Household vs. person-level design • Calendar year vs. point-in-time reference period
QDERS 2000 (August–September)	Split-ballot field test	RDD	• 1862 households • 4794 people	• Status quo CPS structure • Sequence of public and private plans manipulated
QDERS 2003 (August)	Split-ballot field test	RDD	• 1919 households • 4805 people	• Status quo CPS structure • Sequence of Medicare and Medicaid questions manipulated
July–September, 2003	Cognitive testing	DC area residents	• 4 interviewers • 20 respondents	• Partial redesign structure: a) Hybrid household/person level b) General-to-specific questions on coverage type • "Last 12 months" reference period
QDERS 2004 (April–June)	Split-ballot field test	RDD	• 4344 households • 10 929 people	• Status quo CPS structure vs. Redesign with a) Hybrid household/person-level b) General-to-specific questions on coverage type • Calendar year vs. point-in-time reference period
September–November, 2004	Cognitive testing	DC area residents	• 6 interviewers • 27 respondents	• Status quo CPS structure • Probing focused on: a) What months of coverage b) Which household members c) Questions on private versus public coverage
June, 2008	Cognitive testing	DC area Residents	• 8 interviewers • 36 respondents	• Full Redesign structure a) Hybrid household/person level b) General-to-specific questions on coverage type c) Integrated point-in-time/calendar year reference period

Table 26.2 QDERS 1999 test: general question wording, cell sizes, and uninsured rate.

	Person- vs. household-level	
Reference period	**Person-level**	**Household-level**
Point in time	*FIRST PERSON:* 1. Are you covered by [plan type X]? 2. Are you covered by [plan type Y]? Etc. for each plan type *NEXT PERSON:* 3. Is [NAME] covered by [plan type X]? 4. Is [NAME] covered by [plan type Y]? Etc. for each plan type	1. Is anyone in this household covered by [plan type X]? 2. [if yes] Who is covered? 3. Is anyone in this household covered by [plan type Y]? 4. [if yes] Who is covered? Etc. for each plan type
	Uninsured = 10.3% *n* = 324 households; 818 people	Uninsured = 12.0% *n* = 332 households; 869 people
Calendar year	*FIRST PERSON:* 1. At any time during [YEAR] were you covered by [plan type X]? 2. At any time during [YEAR] were you covered by [plan type Y]? Etc. for each plan type *NEXT PERSON:* 3. At any time during [YEAR] was [NAME] covered by [plan type X]? 4. At any time during [YEAR] was [NAME] covered by [plan type Y]? Etc. for each plan type	1. At any time during [YEAR] was anyone in this household covered by [plan type X]? 2. [if yes] Who was covered? 3. At any time during [YEAR] was anyone in this household covered by [plan type Y]? 4. [if yes] Who was covered? Etc. for each plan type
	Uninsured = 6.9% *n* = 307 households; 765 people	Uninsured = 12.0% *n* = 316 households; 776 people

Source: 1999 Questionnaire Design Experimental Research Survey (Pascale 2001a).

uninsured. So results seemed reasonable; as the reference period moves from date-of-interview to a full calendar year, sample members' chances of being insured at some point during the year go up, resulting in a calendar year rate of uninsured that is 3.4 percentage points lower ($p = 0.018$) than the rate of uninsured at a point in time.

In the household-level versions of the questionnaires, somewhat surprisingly, there was no difference in the rate of uninsured across calendar year and current treatments; the uninsured rate was 12.0% in both versions. Furthermore, regardless of reference period, the uninsured rate was higher in the household than the person-level design (Pascale 2001a). These results provided some evidence that there is a measurable value to "naming names" to generate reports of coverage, rather than relying on the generic "anyone in the household." Results on the reference period were less straightforward; the "at any time during [YEAR]" phrase appeared to be effective under the person-level but not the household-level design. As noted in the summary paper on the CPS redesign: "This suggested the reference-period wording – 'at any time during [YEAR]' – was effective at eliciting reports of past coverage when respondents were asked to think about only one person at a time, but not when they were asked to think about 'anyone in this household'. In short, there appeared to be some kind of cognitive overtaxing going on within the household/calendar-year design in the CPS." (Pascale 2016).

In this test, the two design features (reference period and the household/person-level structure) were tested simultaneously and thus results are intertwined. First the reference period results of this test will be discussed in the context of other related literature, followed by a similar discussion of the household/person-level design feature.

26.3.1.2 Redesign of the Reference Period Design Feature

The QDERS 1999 test provided strong evidence that the calendar year reference period was problematic and prompted cognitive testing of the traditional CPS question series in 2005. Much of the probing was focused on the phrase "at any time during [YEAR]" and aimed to determine which specific months respondents had in mind, and why. Results showed there were three main patterns of response. Some respondents reported thinking of the correct set of 12 months. Others paid no attention to the reference period and said they were thinking of their current coverage, while a third group was thinking of the time frame that defined the current spell of the insurance (e.g. eligibility due to a pregnancy and young child) (Pascale 2008). Quantitative studies on Medicaid revealed higher underreporting of coverage in the more distant past, and also found that accuracy of past coverage improved if respondents were currently covered (Pascale et al. 2009; Research Project to Understand the Medicaid Undercount 2008). These and other findings led to the development of an alternative strategy integrating questions on coverage at both a point in time and the past calendar year.

A review of relevant general survey methods literature on memory and recall suggested there was some advantage to providing multiple time frames to enhance the accuracy of retrospective reports. Indeed, other national and state surveys included questions on both current and calendar year coverage. A question on current coverage is less taxing and less prone to any recall error than a question on coverage over a given time span. Thus, very generally, the alternative series began by asking about current coverage, and using it as an anchor for reporting retrospective coverage. Specifically, if the respondent did have coverage at the time of the interview (in March of any given year, to be

1) Do you now* have coverage through [plan type X]?
   - Yes → 2
   - No → 3
2) Did that coverage start before or after January 1, [LAST YEAR]?
   - Before January 1, [LAST YEAR] → Was it continuous? If yes, coverage from January 1 through interview data is inferred
   - On or after January 1, [LAST YEAR] → 4
3) DID you have coverage at any time during [LAST YEAR]?
   - Yes → what months?
   - No → verify uninsured status
4) In what month did that coverage start? → 5
5) What other months between January 1 [LAST YEAR] and now were you covered [by plan type X]?

*March of this year

Figure 26.1 CPS redesign routine integrating questions on current and past coverage.

consistent with CPS data collection procedures), a follow-up question asked whether that coverage started before or after January 1 of the prior calendar year. If before, a question asked whether the coverage was continuous. If so, no more questions about time frame were asked and it was inferred that the coverage lasted from at least January 1 of the prior year through the date of the interview (roughly a 15-month time span). If the coverage began after January 1 of the prior year, a question was asked to determine the month the current spell began, and then ascertain whether there were any other spells of that same coverage type at any time between January 1 of the prior year up to the beginning of the current spell. By identifying start and end dates of coverage spells, asking if coverage was continuous within a spell, and probing at the month-level when coverage was reported for only part of the reference period, the questionnaire rendered month-level data for each person and each type of coverage. The basic routine of the alternative design is shown in Figure 26.1. In sum, while the primary goal of changes related to the reference period was to improve reporting of retrospective coverage, the particular changes in the service of that goal also rendered more data: a measure of point-in-time coverage and coverage at any time throughout the past year, and person-level monthly data on each type of coverage.

Cognitive testing of this basic routine was conducted in 2008 (Pascale 2009b). Minor issues were identified and refined during the test rounds, but no fatal flaws with this basic routine were detected.

26.3.1.3 Redesign of the Household- vs. Person-Level Design Feature

Results from the 1999 test in Table 26.2 provided evidence that asking about each household member by name was important for eliciting reports of coverage. However, a deeper look indicated this was not the case across all household sizes and types of coverage. Medicaid provided a good opportunity to

Table 26.3 QDERS 1999 test: estimates of Medicaid coverage.

	Overall			4+ Person households			Children < 18		
Reference period	Person	HH	Person-HH	Person	HH	Person-HH	Person	HH	Person-HH
Point in time (PIT)	5.4%	4.8%	0.6	4.0%	4.9%	−0.9	9.6%	8.5%	1.1
Calendar year (CAL)	6.1%	7.7%	−1.6	6.2%	11.5%	−5.3*	11.6%	18.0%	−6.4*
CAL-PIT	0.7	2.9*	n/a	2.2	6.6*	n/a	2.0	9.5*	n/a

* $P < 0.05$.
Source: 1999 Questionnaire Design Experimental Research Survey (Pascale 2001b).

explore this more. Prior to the ACA, Medicaid was a joint federal-state program designed primarily for low-income women and children. While age cut-offs for children's eligibility varied across states, in many households if one child was covered, all were covered. Table 26.3 presents results from the QDERS 1999 test for Medicaid reporting overall, among large households (those with four or more members), and for children under 18. Overall there is relatively little difference between the household- and person-level designs, in either the point in time reference period design (0.6) or the calendar year design (−1.6). In large households, however, there is a substantial difference in reported Medicaid in the calendar year design, and it is in the opposite direction from the uninsured estimates in Table 26.2. In the middle panel of Table 26.3, results show that in large households, the household-level design yields much higher reporting of Medicaid than the person-level design. The far-right panel of Table 26.3 shows the same pattern, and results are even more pronounced for reporting of Medicaid for children under 18. These results suggest that once the focus is on large households and children, the household-level design generates higher reporting of Medicaid. This could be due to the person-level design becoming particularly redundant in larger households, and respondent fatigue setting in, resulting in under-reporting.

As a follow-up to this test, cognitive testing in the fall of 2004 sought to learn which household members respondents had in mind with regard to the phrase "anyone in the household." Results showed that respondents had difficulty with the phrase, especially in relatively large, complex, or nontraditional households. Furthermore, some respondents had difficulty reporting for other household members because they had only limited knowledge of the person's coverage; they may have known the person was covered but did not know the source of coverage. In sum, the problems seemed to stem from a combination of the sheer number of household members for whom a respondent was asked to report, and the respondent's familiarity with the details of coverage for those people.

These problems did not appear to be directly associated with the "closeness" of the relationship between the respondent and the household member for whom he or she was reporting. Errors were observed among adult children and their parents, and among live-in unmarried partners and children. The issue seemed to have more to do with whether the respondent shared the same type of coverage with the people about whom he/she was reporting. For example, few errors or issues were observed for respondents who were policyholders reporting on their own policies and individuals covered under those policies (Pascale 2008). A later record-check study that linked Medicaid records to CPS survey data supported these findings. Respondents who were, themselves, covered by Medicaid were more likely to accurately report Medicaid for other household members than respondents who were not, themselves, also covered by Medicaid (Pascale et al. 2009). This suggests that the issue may not be tied to being the policyholder per se, but being covered by the same general *type* of coverage as other household members.

These findings all pointed toward the development of a hybrid person-household level design, where the benefits of asking about each household member by name would be retained, but the efficiencies of the household-level approach would be built in to reduce chances of respondent fatigue. This was accomplished by a question that first established whether a household member had any coverage and, if so, the type and months of coverage. Then, rather than repeating this same battery of questions for each person, as in the strict (and often redundant) person-level series, the hybrid routine determined whether other household members shared that coverage, as shown in Figure 26.2.

Up to this point, there was some precedent in other surveys of identifying a policyholder and then asking who else in the household was a dependent on that same policy. But this general routine had not been extended to public plans. Though public plans are not administered using the same policyholder/dependent structure, in many cases, multiple household members are covered by the same type of plan. For example, elderly couples are usually both covered by Medicare, and in low-income households often all or most members are covered by Medicaid. Cognitive testing of this basic routine was first conducted in 2003. Results demonstrated the viability of extending the "who else was covered" routine to public plans: "In most cases when there were multiple people in a household covered by the same plan type, respondents reported the household members as a 'unit'." (Pascale 2003).

26.3.1.4 Redesign of the General vs. Specific Questions on Type of Coverage

Turning now to the issue of the general versus specific questions on coverage type, as noted earlier, multiple studies suggested that very specific questions were problematic, and that reporting difficulty was exacerbated when answering questions for other household members. In addition to that literature, there was speculation that real-world change was muddying the waters. In the 1990s,

Person 1 Series (any coverage; if so, type and months of coverage); if Person 1 was reported covered by plan type X → 1

1) Is anyone else in this household also covered by [plan type X]?
   - Yes → 2
   - No → 4
2) Who?
   - Person 2
   - Person 3
   - Etc.

   → 3
3) Was/Were [Person(s) 2, 3, etc.] covered by plan type X the same months as Person 1? [if not, capture months each person was covered].
4) Now I'd like to ask you about Person 2. Other than [plan type X], is Person 2 now covered by any other type of health coverage?
   - Yes → [capture type and months of coverage for Person 2, then ask if others also covered]
   - No → 5
5) Now I'd like to ask you about Person 3. Other than [plan type X], is Person 3 now covered by any type of health coverage?
   - Yes → [capture type and months of coverage for Person 2, then ask if others also covered]
   - No →

Figure 26.2 CPS redesign routine on hybrid person- and household-level questions.

there was a general shift in which many Medicaid enrollees were moved into managed care plans. This prompted some policy researchers to speculate that these public enrollees would misreport their coverage as private. There was also emerging literature that respondents confused Medicaid and Medicare plans in particular (Loomis 2000). These findings prompted two different split-ballot surveys, executed through the QDERS vehicle, both designed to explore context/sequencing effects within the laundry-list approach to asking about coverage type. The first, in 2000, manipulated the sequencing of questions on private and public coverage type. The second, in 2003, manipulated the sequencing of Medicaid and Medicare. This study also manipulated definitions of these programs – embedding them into the questions themselves versus providing them on the help screen to be read at the interviewer's discretion.

In the 2000 study, half the sample was asked the CPS status quo (the "control" version – private coverage first, followed by public), and for half the sample the public plans were asked about first (the "test" version). The overall sample was subdivided into two groups: those where the household included at least one person eligible for Medicare (as defined by reporting a disability or being age 65 or older), and those households without anyone eligible for Medicare. The questionnaires in the former subset did not include a question about Medicare in the battery of questions on plan type, while for the latter group there was a Medicare question (see Table 26.4). In terms of sequencing effects, the

Table 26.4 QDERS 2000 test: estimates by coverage type.

Coverage type	Non-Medicare sample (n = 3971)			Medicare sample (*n* = 823)		
	Control (ESI 1st)	Test (Mcaid 1st)	C − T	Control (ESI 1st)	Test (Mcare 1st)	C − T
Private						
ESI	70.1	72.9	2.8*	47.9	41.8	6.1*
Direct purchase	5.7	4.6	1.2	22.6	19.7	2.9
Someone outside the household	4.5	2.5	2.0*	2.7	2.0	0.7
Public						
Medicare	n/a	n/a	n/a	68.4	61.5	6.9*
Medicaid	10.9	9.8	1.1	8.0	5.1	2.9*
Military	3.5	3.2	0.3	2.7	5.8	−2.8*
OTHER	3.4	2.8	0.6	10.9	8.3	2.6
Uninsured	11.4	10.9	0.5	6.6	10.1	−3.5*

* $P < 0.05$.

hypothesis was that when a given coverage type was asked early in the sequence it would garner higher reporting than when it was sequenced later in the series. Results in Table 26.4 show this was not the case among the non-Medicare sample. For example, in the control version, ESI was asked first in the series and resulted in an estimate of 70.1%, compared to 72.9% in the test version, where ESI was asked after questions on Medicaid and military coverage. Similarly, in the test version for the non-Medicare sample, Medicaid was asked about first and produced an estimate of 9.8%, compared to 10.9% when Medicaid was asked about later in the series. Among the Medicare sample, with the exception of military coverage, the control version garnered higher reporting across coverage types than the test version. For Medicare in particular, which was asked about first in the series in the test version, the control estimate was almost seven percentage points higher than the test version (Pascale 2001c).

Though counterintuitive, it appears that when a question was asked earlier in the series, it actually suffered more under-reporting than when it was placed later in the series. This had downstream effects on the uninsured rate because the first coverage type in sequence for both samples was the most prevalent. That is, for the non-Medicare subgroup, ESI is the most prevalent coverage type: it was under-reported by 2.8 percentage points when it was asked first (in the control) and the uninsured rate was 0.5 percentage points higher in the control than the test version. Similarly, for the Medicare subgroup, Medicare is the most prevalent coverage type, and it was under-reported by 6.9 percentage points when it was asked first (in the test version), and the uninsured rate was 3.5 percentage points higher in the test than the control version. One possible explanation for under-reporting of the first plan in the sequence is the complexity and specificity of the structure of the series. Respondents are provided with only a very minimal introduction to the series, and then they are asked a detailed set of questions that may not map very well on to their own way of thinking about health insurance coverage. Given the structure of these items, respondents may actually have an easier time answering questions about specific coverage types when those questions are placed later in the sequence, once the gist of the series has become clear.

The QDERS 2003 study examined the sequencing of Medicaid and Medicare within the status quo CPS design. The sequence of questions on coverage type, as shown in Table 26.4 for the Medicare sample/control version, was the control version in QDERS 2003; and the test version was this same battery but the order of the Medicare and Medicaid items was switched, so that Medicaid was asked first. Results showed no main effects for estimates of Medicaid or Medicare; there were no significant differences in coverage estimates regardless of sequence. However, for Medicare reporting, there was an effect among low-income households, where household members were more likely to be covered by Medicaid than the overall sample. Among these households, when Medicare was asked first, Medicaid estimates were unaffected but Medicare reporting was higher – by just over 6 percentage points (24.8% versus 18.6%). There was also significantly more reporting of both Medicare and Medicaid in low-income households; both programs were reported for 6.3% of household members when Medicare was asked first, but only 3.8% when Medicaid was asked first ($p = 0.04$). Findings suggest that when respondents are reporting Medicaid for household members, if they get the two programs confused they may mistakenly say "yes" to Medicare when it is asked first because it sounds vaguely familiar, and then when the true Medicaid question is asked they say "yes" again since they recognize the program name as being more precise. In contrast, when Medicaid is asked first, they may correctly report Medicaid; then, when the Medicare question is asked, they do not also say "yes" to Medicare since they feel more confident that they have correctly reported Medicaid already (Pascale 2004).

Results from these split-ballot tests were not entirely straightforward and, at times, were counterintuitive. But the findings generally corroborated qualitative evidence that respondents have difficulty with questions on very specific coverage types. However, most surveys seek to go beyond insured/uninsured, or even public/private, and categorize respondents into specific types of coverage, such as ESI, Medicaid, and Medicare. This prompted the development of an alternative approach to the question series where the first question just asked a basic yes/no question on insured/uninsured. Some surveys do begin with this kind of global question, but they follow it with a single question listing all the various specific types of coverage. In a departure from that approach, the redesign alternative structure was to follow up the yes/no question on coverage with a question asking about general source of the coverage (job, government, or other). Later questions tailored to each general source would then be asked to tease out the detail needed to place household members in the specific coverage type. This basic structure was first examined through cognitive testing in 2003 (Pascale 2003) and then in a split-ballot test in 2004 (Pascale 2007) and another cognitive test in 2008 (Pascale 2009b). Several refinements were made along the way based on the testing, and the general series for the prototype is shown in Figure 26.3.

26.3.2 Phase II: Evaluating the Prototype

Table 26.5 summarizes the pretest of the initial prototype questionnaire, the qualitative research to adapt the prototype for health reform, and the two large field tests.

26.3.2.1 Pretest and First Large-Scale Field Test

A small pretest of the prototype redesign was conducted with just over 50 cases via telephone from one of the Census Bureau call centers in March 2009. The purpose was to assess the length of the module and to identify any fatal flaws in a production data collection setting prior to a larger, more costly field study. Results showed that the general flow worked well for a range of households with a wide array of health coverage situations. The general routine was also found to work well from the interviewers' perspective, and the respondent debriefing showed that respondents felt the questions accurately captured the health coverage situation for their household (Pascale 2009c). Because the pretest revealed no outstanding issues with regard to the wording or flow of the prototype redesign, plans proceeded toward a large-scale split-ballot field test comparing the status quo and redesigned CPS in March 2010. For that test – the Survey of Health Insurance and Program Participation (SHIPP) – a sample was drawn from two sources: a random digit dial (RDD) frame and Medicare enrollment files. The purpose of the latter sample was to "seed" the sample with individuals known to be enrolled in a given coverage type in order to examine whether reporting was more accurate in the status quo or the redesigned CPS. Ideally, enrollment files from other, more relevant/prevalent coverage types (such as Medicaid) would have been used, but those records were not available. The Medicare files were available, and in order to make them more relevant, the sample was stratified to over-sample those under 65 and those who enrolled within the past 15 months of the sample draw.

SHIPP results showed that among the RDD sample, there were very few significant differences between estimates from the two different survey

1) Do you have any type of health plan or health coverage?
   - Yes → 2
   - No → Follow up series on coverage types typically under-reported
2) Is it provided through a job, the government, or some other way?
   - Job → 5
   - Government → 3
   - Other → 8
3) Is that coverage related to a JOB with the government?
   - Yes → 5
   - No → 4
4) What type of government plan is it – Medicare, Medicaid, Medical Assistance or S-CHIP, military or Veterans' Administration coverage, or something else?
   - Medicare → questions on months covered
   - Medicaid, Medical Assistance or S-CHIP → questions on months covered
   - Military or Veterans' Administration care → questions on months covered
   - Other → 7
5) Is that plan related to military service in any way?
   - Yes → 6
   - No → 9
6) Which plan are you covered by? Is it TRICARE, CHAMPVA, Veterans Administration care, military health care, or something else?
   - TRICARE
   - TRICARE for Life
   - CHAMPVA
   - VA
   - Military health care
   - Other (specify)

→ 9

- Is it a government assistance-type plan?
  - Yes → questions on months covered
  - No → 8
- How is that coverage provided? Is it through a parent or spouse, direct purchase from the insurance company, a union or business association, a school, or some other way?
  - parent or spouse
  - direct purchase from the insurance company
  - union or business association
  - school
  - some other way?
- Who is the policyholder? [include "Someone outside household"] → 10
- Is that coverage provided through their job, direct purchase from the insurance company, or some other way?
  - Job
  - Direct purchase
  - Other way

Figure 26.3 Prototype of CPS redesign for 2009 pretest.

Table 26.5 Data collection vehicles of the CPS redesign: phase II.

Collection	Type of test	Sample frame	Sample size	Questionnaire feature tested
Prototype pretest, March, 2009	Pretest	RDD and convenience sample	• 5 interviewers • 54 interviews (47 RDD; 7 convenience)	Prototype redesign: • Integrated point-in-time/calendar year reference period • Hybrid household/person-level design • General-to-specific questions on coverage type
SHIPP 2010 (March)	Split-ballot field test	RDD and Medicare enrollees	• 5 376 households • 12 743 people	Prototype redesign: • Integrated point-in-time/calendar year reference period • Hybrid household/person-level design • General-to-specific questions on coverage type
ACA research: September, 2011–September, 2012	Expert consultation, focus groups, cognitive testing	Enrollees in Massachusetts marketplace; Medicaid	• 6 conference calls with 12 experts • 4 focus groups; 39 participants • 4 rounds of cognitive testing; 72 participants	Expert consultation: • Understand mechanics of marketplace • Determine marketplace eligibility in order to identify subgroups for focus groups and cognitive testing Focus groups and cognitive testing: • Pathways to enrollment; process and experiences • Terms enrollees use to describe coverage type
CPS content test 2013 (March)	Field test	CPS production sample	• 13 228 people (control) • 16 401 (test)	Status quo CPS design vs. redesign under production data collection conditions

designs – across plan types or within subgroups. For the Medicare sample, the overall uninsured rate was lower in the redesign than in the status quo, and this was driven primarily by higher reporting of public coverage in the redesign (Pascale 2011). To examine accuracy, the Medicare portion of the SHIPP survey data was matched back to the Medicare records from which that sample was drawn originally, and indicators of enrollment from the two data sources were compared. Results were favorable to the redesign in terms of both under-reporting (i.e. among those who have coverage according to the records, how many reported it) and over-reporting (i.e. among those who reported coverage, how many could be validated in the records to actually have that coverage). Results on over-reporting were particularly robust. The false-positive (over-reporting) error rate in the status quo CPS was 6.9% and in the redesign the error rate was 2.0% – a 4.9 percentage-point difference that was statistically significant (Resnick 2013).

Given the literature that suggests the CPS misses reports of past coverage, it was hoped that the redesign would garner higher reports of past coverage than the status quo CPS. Results were flat in this regard, but at least the redesign did not lose ground on this measure. Coverage bias due to the landline-only sample could explain the lack of differences in the estimates (Blumberg and Luke 2011). Furthermore, the modified structure of questions on retrospective coverage rendered person/plan-level data for each month from January of the prior year up to the interview date. An examination of patterns of monthly churning on and off coverage showed the redesigned CPS had face validity. For example, there was no evidence of monthly transitions peaking in any particular month with the exception of January, which is to be expected given the "open season" opportunities to change coverage at the turn of the calendar year (Pascale 2011). These and other factors raised in a separate analysis of SHIPP (Boudreaux et al. 2013) warranted a follow-up test with a sample more representative of the CPS.

26.3.2.2 Adaptation in Anticipation of Health Reform

By chance, the day the SHIPP test was launched in the field (March 23, 2010), President Obama signed the ACA into law. Key provisions of the law – namely the introduction of the marketplace and the expansion of public coverage – were set to go into effect in January 2014. This set the stage for adapting the newly redesigned CPS health coverage module with questions accommodating health reform. The only opportunity to explore the experiences of applicants and enrollees in the marketplace, and how they might respond to questions about this type of coverage in a survey, was in Massachusetts, which had passed state-level legislation much like the ACA in 2006. Thus in 2011, a three-part research project was launched. The first phase was an expert consultation in which a dozen research and policy experts were called on to provide a basic understanding of the mechanics of the marketplace in terms of eligibility rules, application, and enrollment procedures. The second stage was a set of four focus groups with marketplace enrollees in Massachusetts in order to flesh out terminology and prototype questions for asking about this coverage type within the context of the redesigned CPS. The third stage was cognitive interviews (four rounds with a total of 74 respondents) with both marketplace enrollees and Medicaid enrollees, given the fine distinction between the two programs in terms of eligibility and enrollment procedures. Results from the focus groups and cognitive testing suggested that the basic structure of the CPS redesign could be maintained largely intact in order to categorize respondents into the conventional sources of coverage (ESI, Medicaid, etc.). To accommodate health reform and produce output variables that could help categorize coverage post-reform, the Massachusetts testing suggested that three new items could be inserted into the existing redesigned module, asking whether the coverage was obtained on the marketplace, whether there was a monthly premium, and whether the premium was subsidized (Pascale et al. 2013).

26.3.2.3 Second Large-Scale Field Test

These three new items were embedded within the CPS redesign in preparation for a final large-scale field test under production CPS conditions. This test – called the CPS Content Test – was launched in March 2013. The control panel was derived from a subsample of production CPS cases that were administered the status quo CPS. The test panel was derived from respondents who had been in-sample for production CPS during the past and had cycled out. Results showed that the redesigned CPS yielded higher reporting of past calendar year coverage than the status quo design, and that the redesign had reduced presumed underreporting of past year coverage (Pascale et al. 2015). Further analysis found that both the status quo and the redesigned CPS did equally well in single-person households and when the respondent had to report only for him/herself and/or his/her child. But in larger households (with two to four members), and in situations where the household respondent was more socially distant from the person whose coverage they were reporting (i.e. where the relationship was something other than a self-report or a report for one's own child), the differences between the status quo and redesign CPS were especially pronounced. In other words, where the redesigned CPS demonstrated significant, measurable improvements was in households with two to four members, and with proxy reports for individuals who were more socially distant from the respondent. To estimate how many additional people were reported as insured under the redesigned CPS solely due to its different impact on these two subgroups, a rough calculation was conducted. Results showed a total of about four and a half million individuals reported as uninsured in the status quo CPS would have been reported as insured in the new CPS. Taking this a step further, the insured rate under the redesign came to 89.9%, compared to the insured rate of 88.0% under the status quo CPS, representing a 1.9 percentage-point difference (Pascale 2016). This difference is in line with findings from the 2013 content test, which showed a difference of 1.4 percentage points in the uninsured rate between the old and new CPS. It is also in line with a comparison of the 2013 and 2014 production estimates, which employed the old and new CPS, respectively. While there were many caveats to that comparison, a 1.6 percentage-point difference was approximated to be attributable to the change in questionnaire design (Pascale et al. 2015).

26.4 Discussion

While often the problem with an existing survey instrument is obvious, the fix is usually not so obvious. Furthermore, the cost of a redesign (apart from the research and development) is a break in the time series. The goal of this research effort was to examine and reduce measurement error in a very systematic way. First, all relevant evidence – from the wider literature and from purpose-built tests – was constantly brought to bear, and past evidence was reevaluated in light of new findings. Second, survey design features that appeared most likely to be contributing to measurement error were experimentally isolated for quantitative test/control comparisons. Third, qualitative cognitive testing was conducted to probe the reasons for observed differences in the quantitative experiments. This process enabled a thorough understanding of not just the problems but the magnitude of their effects. It also allowed for an empirical demonstration of improvements. Finally, conducting a large-scale quantitative test under production data collection conditions mitigated the risk of instrumentation and programming bugs and glitches that could compromise the redesign if it proved to be warranted to adopt it for production data collection. The final result was a redesigned CPS that reduced measurement error in the form of an uninsured rate that is more in line with other major surveys. The modified approach to asking about past coverage also yielded additional data in the form of month-person-plan level variables on coverage. These data will enable analysis on month-to-month churn – both on and off coverage, and transitions from one type of coverage to another – which was not possible with the data rendered under the old design.

This research effort was long and protracted and for the first decade was still an independent research project versus an official "redesign" effort. There were also gaps in the timeline due to staffing and resource constraints. This research shows that empirically demonstrating a reduction in measurement error is often elusive and can take years. In the context of other more official redesign efforts, an exact figure on how many years hinges on when one marks the starting point – the first recommendation by an agency or outside technical review board, the first draft of a redesign strategy, the first test implemented, etc. Documented reports of the CPS labor force module indicate about 10 years (mid-1980s to 1994) (Cohany et al. 1994; Martin and Polivka 1995); and for a major redesign of the National Crime Victimization Survey (NCVS), the timeline was closer to 20 years (mid-1970s to 1992) (Kindermann et al. 1997). The Bureau of Labor Statistics initiated a redesign of the Consumer Expenditure Survey in 2009 and, pending budget, implementation is expected in 2023 (Edgar et al. 2013). From that perspective, in terms of timeline, the CPS redesign was on par with other, more official efforts.

Cost estimates are a different matter. It is, theoretically, possible to produce cost figures on each of the tests conducted as part of the CPS redesign. However, the usefulness of the figures is dubious due to multiple variables. For example, for several of the tests there were in-kind contributions from other areas within the Census Bureau; in some cases, the instrument programming was contracted out and in others it was conducted in-house; and for some tests, additional unrelated experiments were included that brought in other sources of funding. Nevertheless, Tables 26.1 and 26.5 provide very basic parameters on the nature and scope of the tests (e.g. sample size, number of interviewers), that could be used to derive very rough cost estimates. It is hoped that these tables, combined with the timeline, provide some context and guidance on the scope, costs, staffing, and timeline for other redesign efforts. For example, the US Department of Health and Human Services (HHS) recently completed an assessment of its survey portfolio and has begun a program to bring its survey methodologies into closer alignment with each other (Office of the Assistant Secretary for Planning and Evaluation 2016). To the extent that this initiative may result in an effort to standardize particular questionnaire modules (e.g. on tobacco use, use of health services) across surveys, the experience with the CPS health insurance module could provide a useful point of reference.

Finally, with regard to implications for other topic areas and surveys, while many of the tests discussed here were purpose-built for health insurance measurement in the CPS, some of the survey design features that were targeted for improvement were first identified as problematic from research on other topic areas. For example, research on the 12-month reference period in a series of questions on Food Stamps found a lack of attentiveness to the reference period (Hess and Singer 1995), and related literature showed that respondents tended to underreport receipt of benefits from the more distant past (Lynch 2006; Resnick et al. 2004; Ringel and Klerman 2005). This is also consistent with findings from the survey methods literature more generally (Schaeffer and Presser 2003). Hence, coming full circle, for retrospective questions in general there is some evidence to support shifting away from phrasing like, "At any time in the [past months/year] did you [receive/experience X]?" An alternative that begins with a question anchoring respondents in the present ("Do you now receive X?") and then establishes dates of past coverage with follow-up questions (in this case by identifying start dates of spells) holds promise to reduce underreporting of retrospective events generally. Indeed, the Survey of Income and

Program Participation (SIPP) recently moved to this approach across a number of topic areas including employment status, income, and benefits receipt (Fields 2017).

With regard to the household- versus person-level design, given that many surveys rely on a single household respondent to report on complex issues for all members of the household, there appears to be something of a blind spot in the literature. There is, of course, a healthy literature on proxy reporting in general, but much of it centers on the relationship to the "householder," which the US Census Bureau defines as "a person (or one of the people) in whose name the housing unit is owned or rented" (Mathiowetz 2010; Moore 1988). There seems to be a dearth of research on the particular relationship between the household respondent and the person on whom they are reporting. The 2013 CPS split-ballot test (Pascale et al. 2015) and the 2009 Medicaid record-check study (Pascale et al. 2009) discussed here provided compelling evidence that the "social distance" between the two is tied to data quality. Recent research that examined questions on year of naturalization (Grieco and Armstrong 2014) and sexual orientation and gender identity (Ellis et al. 2017; Holzberg et al. 2017) support these findings. Thus, this issue – examining the relationship between respondent and the person on whom they are reporting in relation to the nature of the questions being asked (e.g. in terms of their sensitivity, complexity, and difficulty) – appears relatively unexplored and rich with potential implications. It could prove particularly useful given the trend toward an increase in complex households (Schwede et al. 2005).

The last survey design feature – asking a general yes/no question on any coverage and then drilling down to the specific type of coverage with follow-up questions – is a subset of "decomposition" or breaking up a complex concept into its smaller parts. Schaeffer and Dykema (2011) note common implementations are "Decomposing a general category into a set of more specific categories…" and "…when a researcher divides a global behavioral frequency question into two or more mutually exclusive, less cognitively taxing subquestions…and provides cues along a relevant dimension, such as time, place, person…" Decomposition of complex questions has been demonstrated to result in higher data quality in several contexts (Loftus et al. 1990; Redline 2013; Fowler 2004). The advantages of decomposition, of course, hinge on the nature and complexity of the topic itself (Beatty 2010). Another factor is how the reporting task relates to the subject (i.e. the person being asked about), and how that person relates to the respondent. In other words, there is an intersection between the complexity of the question and who it is being asked about. A very complex question (e.g. annual earnings after taxes and other deductions such as retirement savings) may be easy for a respondent to answer for him/herself, but much more difficult to answer for multiple household members who are socially distant. Likewise, a simple question (such as sex of other household members) may be easy for a respondent regardless of household size and relationships among members.

One thing is certain, however: many topics asked about in surveys – such as income receipt (from a wide array of sources such as an interest-earning checking account, stocks, alimony, and rental property), enrollment in pensions and retirement plans, receipt of public benefits, and employment status – pose a dual challenge of being arcane and also potentially sensitive and personal. While options for reducing the perceived sensitivity of questions may be limited, the general-to-specific questioning approach could address the arcane part of the problem, particularly to the extent it is intertwined with social distance. For example, if asked about whether household members own various types of retirement plans (401 K, annuities, pensions), a household respondent may be quite knowledgeable about the specifics for their spouse. But, if asked about a non-relative/renter, the respondent may know only that the renter has a job and that they have some kind of retirement plan. The approach of asking a general yes/no question, then, has the potential to capture at least the "top-line" knowledge respondents do have across household members (regardless of social distance), and still allow for capture of more details, should the respondent happen to know those details, even for unrelated/distant individuals.

26.5 Final Reflections

This research began as a single experiment designed to chip away at a measurement conundrum that cut across several surveys. Results from that one experiment spawned several other tests, the sum total of which helped to identify and improve survey design features that are commonly employed in multiple surveys across a range of topic areas. This chapter documents the high-level methods and results of each test in the interest of providing some perspective for other redesign endeavors. Large-scale redesign efforts, with detailed long-range test plans, schedules, and budgets, are often saddled by such a high price tag that the research does not get out of the starting gate. Another pitfall of large-scale efforts is that it is not possible to predict the outcome from one test to another, so mapping out a realistic schedule and budget is based on so many assumptions that the research agenda can sometimes crumble under its own weight. This research project demonstrated that a small-scale test with a specific objective that is linked to a larger methodological problem can be a much more cost-effective and efficient approach to a major redesign. It also allows for many nuances of implementation (e.g. technical instrument specifications, training materials, wording of interviewer instructions) to be refined and perfected along the way so that by the time the product is final, there are no surprises in the production data collection. This small-scale, iterative approach is not incompatible with a long-range plan; it merely suggests the long-range plan should be nimble and flexible enough to accommodate the findings from each test in order to properly design the next test.

References

Barnett, J.C. and Vornovitsky, M.S. (2016). Health insurance coverage in the United States: 2015. Current Population Reports, P60-257(RV). Washington, DC: U.S. Government Printing Office.

Beatty, P. (2010). Considerations regarding the use of global survey questions. Consumer Expenditures Survey Methods Workshop, Hyattsville, MD.

Beatty, P. and Schechter, S.. (1998). Questionnaire evaluation and testing in support of the Behavioral Risk Factor Surveillance System (BRFSS), 1992–98. Office of Research and Methodology, NCHS, working paper series, 26: 12–17.

Bhandari, S. (2004). People with health insurance: A comparison of estimates from two surveys. Survey of Income and Program Participation (SIPP) working paper no. 243. Washington, D.C.: U.S. Census Bureau. https://www.census.gov/sipp/workpapr/wp243.pdf.

Blewett, L.A. and Davern, M.E. (2006). Meeting the need for state-level estimates of health insurance coverage: use of state and Federal Survey Data. *Health Services Research* 41: 946–975. https://doi.org/10.1111/j.1475-6773.2006.00543.x.

Blumberg, S.J. and Luke, J.V. (2011). Wireless substitution: early release of estimates from the National Health Interview Survey, July–December 2010. National Center for Health Statistics. http://www.cdc.gov/nchs/nhis.htm.

Blumberg, S.J., Osborn, L., Luke, J.V. et al. (2004). Estimating the prevalence of uninsured children: an evaluation of data from the National Survey of children with special health care needs, 2001. *Vital Health Statistics* 2 (136). Hyattsville, MD: National Center for Health Statistics.

Boudreaux, M.H., Fried, B., Turner, J., and Call, K.T. (2013). SHADAC Analysis of the survey of health insurance and program participation. State Health Assistance Data Center. http://www.shadac.org/files/shadac/publications/SHIPP_final_report.pdf.

Cohany, S., Polivka, A., and Rothgeb, J. (1994). Revisions in the current population survey effective January 1994. *Employment and Earnings* 41 (2): 13–37.

Congressional Budget Office. (2003). How many people lack health insurance and for how long? CBO Report. The Congress of the United States. https://www.cbo.gov/publication/14426.

Davern, M. (2009). Unstable ground: Comparing income, poverty & health insurance estimates from major national surveys. Academy Health Annual Research Meeting, Chicago, IL.

Davern, M., Call, K.T., Ziegenfuss, J. et al. (2008). Validating health insurance coverage survey estimates: a comparison of self-reported coverage and administrative data records. *Public Opinion Quarterly* 72: 241–259.

Edgar, J., Nelson, D.V., Paszkiewicz, L., and Safir, A. (2013). The Gemini Project to redesign the consumer expenditure survey: Redesign proposal. Bureau of Labor Statistics. http://www.bls.gov/cex/ce_gemini_redesign.pdf.

Ellis, R., Virgile, M., Holzberg, J., et al. (2017). Assessing the feasibility of asking about sexual orientation and gender identity in the current population survey: Results from cognitive interviews. Report presented to the Bureau of Labor Statistics. https://www.bls.gov/osmr/pdf/cps_sogi_cognitive_interview_report.pdf.

Fields, J. (2017). Introduction to, and re-engineering of, the Survey of Income and Program Participation. Johns Hopkins Bloomberg School of Public Health Population, Family and Reproductive Health Noon Seminar.

Fowler, F.J. (2004). The case for more Split-sample experiments in developing survey instruments. In: *Methods for Testing and Evaluating Survey Questionnaires* (eds. S. Presser, J. Rothgeb, M. Couper, et al.), 173–188. Hoboken, NJ: Wiley.

Grieco, E.M. and Armstrong, D.M. (2014). Assessing the 'year of naturalization' data in the American community survey: Characteristics of naturalized foreign born who report – and don't report – the year they obtained citizenship. Applied Demography Conference, San Antonio, TX.

Hess, J., Moore, J., Pascale, J. et al. (2001). The effects of person-level vs. household-level questionnaire design on survey estimates and data quality. *Public Opinion Quarterly* 65: 574–584.

Hess, J. and Singer, E. (1995). The role of respondent debriefing questions in questionnaire development. In: *Proceedings of the Section on Survey Research Methods, August 13–17, 1995*, 1075–1080. Washington, DC: American Statistical Association. Available at: http://www.amstat.org/sections/srms/proceedings/papers/1995_187.pdf (accessed 5 February 2016).

Holzberg, J., Ellis, R., Virgile, M., et al (2017). Assessing the feasibility of asking about gender identity in the current population survey: Results from focus groups with members of the transgender population. Report presented to the Bureau of Labor Statistics. https://www.bls.gov/osmr/pdf/cps_sogi_focus_group_report.pdf.

Kindermann, C., Lynch, J., and Cantor, D. (1997). Effects of the redesign on victimization estimates. NCJ-164381. National Crime Victimization Survey. Washington, D.C.: U.S. Department of Justice, Office of Justice Programs. Bureau of Justice Statistics.

Lavrakas, P.J. (2008). *Encyclopedia of Survey Research Methods*. Thousand Oaks, CA: SAGE Publications Ltd https://doi.org/10.4135/9781412963947.

Lewis, K., Ellwood, M., and Czajka, J. (1998). *Counting the Uninsured: A Review of the Literature*. Washington, D.C.: Mathematica Policy Research.

Loftus, E.F., Klinger, M.R., Smith, K.D., and Fiedler, J. (1990). A tale of two questions: benefits of asking more than one question. *Public Opinion Quarterly* 54: 330–345.

Loomis, L. (2000). *Report on Cognitive Interview Research Results for Questions on Welfare Reform Benefits and Government Health Insurance for the March 2001 Income Supplement to the CPS*. Washington, DC: Center for Survey Methods Research, Statistical Research Division, U.S. Census Bureau.

Lynch, V. (2006). *Causes of Error in Survey Reports About Who in the Household Gets Welfare. Unpublished*. College Park, Md.: Joint Program in Survey Methodology.

Martin, E. and Polivka, A.E. (1995). Diagnostics for redesigning survey questionnaires: measuring work in the current population survey. *Public Opinion Quarterly* 59 (4) (Winter 1995)): 547–567. https://doi.org/10.1086/269493.

Mathiowetz, N. (2010). Self and proxy reporting in the consumer expenditure survey program. Consumer Expenditures Methods Workshop, Bureau of Labor Statistics. https://www.bls.gov/cex/methwrkshp_pap_mathiowetz.pdf.

Moore, J.C. (1988). Miscellanea, self/proxy response status and survey response quality, a review of the literature. *Journal of Official Statistics* 4 (2): 155–172.

Nelson, D.E., Powell-Griner, E., Town, M., and Kovar, M.G. (2003). A comparison of National Estimates from the National Health Interview Survey and the behavioral risk factor surveillance system. *American Journal of Public Health* 93: 1335–1341.

Office of the Assistant Secretary for Planning and Evaluation (ASPE), Health and Human Services.(2005). Understanding estimates of the uninsured: putting the differences in context. https://aspe.hhs.gov/basic-report/understanding-estimates-uninsured-putting-differences-context-0.

Office of the Assistant Secretary for Planning and Evaluation (ASPE), U.S. Department of Health and Human Services. (2009). Aid to Families with Dependent Children (AFDC) and Temporary Assistance for Needy Families (TANF) – Overview. https://aspe.hhs.gov/aid-families-dependent-children-afdc-and-temporary-assistance-needy-families-tanf-overview-0.

Office of the Assistant Secretary for Planning and Evaluation (ASPE), U.S. Department of Health and Human Services. (2016). Improving data for decision making: HHS data collection strategies for a transformed health system. https://aspe.hhs.gov/improving-data-decision-making-hhs-data-collection-strategies-transformed-health-system.

Pascale, J. (2001a). Methodological issues in measuring the uninsured. In: *Proceedings of the Seventh Health Survey Research Methods Conference, Williamsburg*. Centers for Disease Control and Prevention http://www.cdc.gov/nchs/data/hsrmc/hsrmc_7th_proceedings_1999.pdf.

Pascale, J. (2001b). The role of questionnaire design in medicaid estimates: Results from an experiment. Seminar presented at the Washington Statistical Society. http://washstat.org/seminars/sem2001.html#010321.
Pascale, J. (2001c). Measuring private and public health coverage: results from a split-ballot experiment on order effects. Annual Meetings of the American Association for Public Opinion Research, Proceedings of the Section on Survey Research Methods, American Statistical Association.
Pascale, J. (2003). Questionnaire Design Experimental Research Survey (QDERS) 2004 cognitive testing results on health insurance questions. Center for Survey Methods research report.
Pascale, J. (2004). Medicaid and Medicare reporting in surveys: an experiment on order effects and program definitions. In: *Proceedings of the American Association for Public Opinion Research*, 4976–4983. Available at: http://www.amstat.org/sections/SRMS/Proceedings. American Statistical Association, May 13–16, 2004 (accessed 2 April 2015).
Pascale, J. (2007). Questionnaire Design Experimental Research Survey (QDERS) 2004: Results overview. Center for Survey Methods research report.
Pascale, J. (2008). Measurement error in health insurance reporting. *Inquiry* 45 (4): 422–437. doi: https://doi.org/10.5034/inquiryjrnl_45.04.422 Available at: . http://inq.sagepub.com/content/45/4/422.full.pdfþhtml (accessed 4 February 2015).
Pascale, J. 2009a. Health insurance measurement: a synthesis of cognitive testing results. Questionnaire Evaluation Standards (QUEST) meeting, Bergen, Norway.
Pascale, J. (2009b). Cognitive testing results of experimental questions on integrated current and calendar year coverage. Center for Survey Methods research report.
Pascale, J. (2009c). Findings from a pretest of a new approach to measuring health insurance in the current population survey. Federal Committee on Statistical Methodology Research Conference.
https://nces.ed.gov/FCSM/2009research.asp.
Pascale, J. (2011). Findings from a split-ballot experiment on a new approach to measuring health insurance in the current population survey. Center for Survey Methods research report.
Pascale, J. (2016). Modernizing a major Federal Government Survey: a review of the redesign of the current population survey health insurance questions. *Journal of Official Statistics* 32 (2, 2016): 461–486. https://doi.org/10.1515/JOS-2016-0024.
Pascale, J., Boudreaux, M., and King, R. (2015). Understanding the new current population survey health insurance questions. *Health Services Research* 51: 240–261. https://doi.org/10.1111/1475-6773.12312.
Pascale, J., Rodean, J., Leeman, J. et al. (2013). Preparing to measure health coverage in Federal Surveys Post-Reform: lessons from Massachusetts. *Inquiry:*

The Journal of Health Care Organization, Provision, and Financing 50: 106–123. https://doi.org/10.1177/0046958013513679.

Pascale, J., Roemer, M.I., and Resnick, D.M. (2009). Medicaid underreporting in the CPS: results from a record check study. *Public Opinion Quarterly* 73: 497–520. Available at http://inq.sagepub.com/content/45/4/422.full.pdfþ html. (accessed 2 April 2015).

Redline, C. (2013). Clarifying categorical concepts in a web survey. *Topics in Survey Measurement and Public Opinion.* Special issue, Public Opinion Quarterly 77: 89–105.

Research Project to Understand the Medicaid Undercount. (2008). Phase II research results: Examining discrepancies between the national Medicaid Statistical Information System (MSIS) and the Current Population Survey (CPS) Annual Social and Economic Supplement (ASEC). https://www.census.gov/did/www/snacc/docs/SNACC_Phase_II_Full_Report.pdf.

Resnick, D. (2013). Microsimulation support for tax, transfer & health insurance policy analysis summary. U.S. Department of Commerce, U.S. Census Bureau, and U.S. Department of Health and Human Services.

Resnick, D., Love, S., Taeuber, C., and Staveley, J.M. (2004). Analysis of ACS Food Stamp program participation underestimate. Joint Statistical Meeting, Toronto, Canada.

Ringel, J.S. and Klerman, J.A. (2005). Today or last year? How do interviewees answer the CPS health insurance questions? RAND Labor and Population working paper series WR-288. Santa Monica, Calif.: RAND.

Roman, A.M., Hauser, A., and Lischko, A. (2002). Measurement of the insured population: The Massachusetts experience. Annual Meetings of the American Association for Public Opinion Research, St. Pete's Beach, Fla.

Rosenbach, M. and Lewis, K. (1998). *Estimates of Health Insurance Coverage in the Community Tracking Study and the Current Population Survey.* Washington, D.C.: Mathematica Policy Research, Document No. PR98-54.

Rothgeb, J. (2007). A valuable vehicle for question testing in a field environment: the U.S. Census Bureau's questionnaire design experimental research survey. Center for Survey Methods research report, Survey Methodology #2007–17. https://www.census.gov/srd/papers/pdf/rsm2007-17.pdf.

Schaeffer, N.C. and Dykema, J. (2011). Questions for surveys: current trends and future directions. *Public Opinion Quarterly* 75 (5): 909–961. https://doi.org/10.1093/poq/nfr048.

Schaeffer, N.C. and Presser, S. (2003). The science of asking questions. *Annual Review of Sociology* 29: 65–88. https://doi.org/10.1146/annurev.soc.29.110702.110112.

Schwede, L.K., Blumberg, R.L., and Chan, A.Y. (eds.) (2005). *Complex Ethnic Households in America.* Editors: Rowman & Littlefield.

State Health Access Data Assistance Center (SHADAC) (2013). *Comparing Federal Government Surveys that Count the Uninsured*. State Health Access Data Center.

Swartz, K. (1986). Interpreting the estimates from four National Surveys of the number of people without health insurance. *Journal of Economic and Social Measurement* 14: 233–242.

Willson, S. (2005). Cognitive interviewing evaluation of the national immunization survey insurance module: Results of fieldwork and laboratory interviews. Hyattsville, MD: National Center for Health Statistics.

27

Contrasting Stylized Questions of Sleep with Diary Measures from the American Time Use Survey

Robin L. Kaplan, Brandon Kopp, and Polly Phipps

Office of Survey Methods Research, Bureau of Labor Statistics, Washington, DC, USA

27.1 Introduction

Researchers, government agencies, and health institutes have become increasingly interested in collecting data on how people spend their time, as time use can have important economic, health, and policy implications. Two common methods of collecting time-use data are time diaries and stylized questions, both of which can be interviewer or self-administered. Time diaries involve prompted recall, where respondents report on all of their activities for a specified period of time, such as the previous 24 hours. Stylized questions ask respondents to report the amount of time they spend on different activities on an *average*, *typical*, or *usual* day or week (e.g. "How many hours do you work on a typical day?").

Although both methods collect data on time use, diary measures are typically considered more valid and reliable than stylized measures (Juster et al. 2003; Kan and Pudney 2008), as they focus on a set reference period and are less prone to recall and estimation bias. However, diary methods are more expensive to administer and burdensome for respondents to complete compared to stylized measures (Schulz and Grunow 2012). Diary and stylized measures also produce different time use estimates across a variety of activities (Kan and Pudney 2008). For instance, researchers have observed a "gap" in estimates of self-reported paid work hours using stylized questions and diary measures, with estimates from stylized questions exceeding diary estimates by an average of 3.52 hours per week (Lin 2012). A similar pattern is found for hours spent on household chores, where stylized estimates exceed diary estimates by about 0.79 hours for women and 1.96 hours for men (Kan 2008). Similar gaps occur with a wide range of other activities, including religious service attendance (Brenner 2011), exercise (Adams et al. 2005), and sleep (Miller et al. 2015) – the topic of this chapter.

Advances in Questionnaire Design, Development, Evaluation and Testing, First Edition.
Edited by Paul C. Beatty, Debbie Collins, Lyn Kaye, Jose-Luis Padilla, Gordon B. Willis, and Amanda Wilmot.

27.2 The Sleep Gap

Unlike other activities, for sleep, diary measures tend to exceed stylized measures. For instance, the American Time Use Survey (ATUS), which measures all activities (including sleep) a respondent did on the previous day, found in 2014 that Americans aged 18 years and over slept 8.7 hours per night on average. In contrast, other national US surveys (e.g. the National Health Interview Survey [NHIS]) use stylized questions to collect data on how many hours per night people sleep, such as, "On average, how many hours of sleep do you get in a 24-hour period?" Surveys using stylized questions consistently find that people report sleeping between 6.9 and 7.1 hours per night (Ford et al. 2015). This constitutes roughly a 1.7 hour gap between diary (ATUS) and stylized measures of sleep duration. Despite these surveys being nationally representative with similar sampling methodologies, they produce different estimates of sleep duration. These differences may therefore be partly due to the way questions about sleep are asked.

27.2.1 Diary Measures of Sleep in the American Time Use Survey

The ATUS measures how Americans allocate their time in a one-day time frame using a sample of approximately 26 000 people each year (Phipps and Vernon 2009). One individual from each sample household (aged 15 years or older) is selected to respond, and he or she participates in a computer-assisted telephone interview (CATI). The interviewer asks the respondent about what he or she did over a 24-hour period from 4:00 a.m. on the diary day until 4:00 a.m. on the interview day. Each activity is recorded along with either the duration or the start and stop times for the activity. For survey estimates, the total duration of time that people spent doing various activities is calculated.

In addition to the ATUS, other time diary surveys have shown that Americans 18 years and older report sleeping an average of 7.7 hours per night (Hale 2005) and 8.1 hours per night (Biddle and Hamermesh 1990). Statistics Canada's General Social Survey, which uses a similar methodology to the ATUS, found that adults over 15 years old reported mean sleep durations between 8.0 and 8.3 hours (Hurst 2008). Across multiple studies, time-use methodologies tend to yield sleep estimates that are eight hours or longer in duration.

27.2.2 Stylized Measures of Sleep

Unlike diary measures, stylized questions ask respondents directly about the amount of sleep they get in a typical, usual, or average day. Stylized questions can also differ in terms of the time frame they ask about, ranging from the previous day, to a typical day, or a typical week. For example, the National Center for Health Statistics (NCHS) collects sleep-duration data via stylized questions. Two of its surveys, the NHIS and the Behavioral Risk Factor Surveillance System (BRFSS) are both conducted with samples of adults aged

18 years or over. Both surveys ask respondents, "On average, how many hours of sleep do you get in a 24-hour period?" Responses are recorded as integer values (i.e. decimal or fractional reports are rounded to the nearest whole hour). Respondents in the 2014 NHIS and BRFSS reported an average of 7.1 hours of sleep per 24-hour period; the median amount of sleep respondents reported was 7.0 hours for both surveys.[1] Thus, national surveys using stylized sleep questions produce lower sleep duration estimates than those of the ATUS and other diary measures.

27.2.3 Response Processes in Self-Reported Sleep

When respondents answer survey questions, they often follow a four-step process where they try to understand (comprehend) the survey question, retrieve the relevant information to arrive at an answer, make a judgment (e.g. calculate an average), and finally give a response (Tourangeau 1984). Measurement error can occur at any stage of the survey response process. Here, we hypothesize how diary and stylized measures may affect reports of sleep at each stage of the response process and contribute to the sleep gap.

Comprehension. How respondents define "sleep" may affect how they report on their sleep. Some respondents may include naps, resting with their eyes closed, dozing off, or trying to fall asleep, while others may not. Some may interpret stylized questions as asking only about continuous episodes of nighttime sleep, excluding naps or times they were awake at night (Canfield et al. 2003). Since diaries ask respondents to report on a full day of activities, they are more likely to capture daytime naps. Thus, diary measures may capture additional sleep that a stylized question might not. Survey context may also play a role – a survey about health may cause respondents to interpret the term "sleep" differently than a general survey about time use.

Recall. Both time diaries and stylized questions may be prone to recall error. Time diaries rely on respondents' ability to recall the activities they did the previous day. It may be difficult for some respondents to recall the precise times they fell asleep and woke up. In contrast, stylized questions require respondents to search their memories for a representative set of days that reflect their typical sleep pattern, adjusting for weekends, holidays, or other events that may have affected their usual sleep (Kan and Pudney 2008).

Judgment. Both diary and stylized sleep measures may be prone to measurement error if respondents cannot directly recall information that would help them report on sleep, and estimate it instead. For time diaries, respondents may rely on their typical routine to infer what time they must have fallen asleep and woken up. In contrast, stylized questions require respondents to make a judgment about the typical or average amount they slept during that period

1 There is no equivalent question for weekend nights, which typically show longer sleep durations, so this number is likely a low estimate of respondents' total sleep.

(Kan and Pudney 2008). This judgment requires respondents to use an estimation strategy (e.g. rate retrieval, rate and adjustment, averaging; Conrad et al. 1998). Estimation strategies are prone to systematic biases, such as rounding or calculation errors.

Response. The last stage is reporting a response, which can be prone to filtering and social-desirability concerns. Respondents may also edit their answers differently depending on the survey topic (e.g. Couper et al. 2007). Diary measures of sleep are usually collected within the context of time-use activities, whereas stylized questions ask directly about sleep, which may affect respondents' answers (Schwarz et al. 1991), for example, by encouraging them to report on norms and beliefs about the appropriate amount one should sleep rather than their actual behaviors (Bonke 2005). This difference in question context may contribute to observed differences in sleep estimates. Furthermore, it has been suggested that stylized questions may be more prone to errors arising from respondents' editing or rounding their answers than diary measures (Kan and Pudney 2008).

Aside from the survey response process, another potential contributor to the sleep gap is the way in which a survey defines and measures sleep duration. ATUS includes napping, falling asleep, and sleeplessness in its sleep estimate. Other surveys using stylized questions tend to define sleep as the longest continuous episode of sleep (Silva et al. 2007). It takes the average American about 20 minutes to fall asleep (Silva et al. 2007), but if respondents report falling asleep within 30 minutes of going to bed, the ATUS records that time as sleep, which may inflate ATUS sleep estimates. In addition, NCHS records sleep duration as integer values (i.e. decimal or fractional reports are rounded to the nearest whole hour), which may affect the distribution of responses.

27.3 The Present Research

This research drew on a range of questionnaire evaluation methods (i.e. behavior coding, qualitative interviews, a quantitative survey, and a validation study) to assess possible reasons for the discrepancy between diary and stylized sleep measures. We explored the cognitive processes involved in answering diary and stylized sleep questions to identify possible sources of measurement error in both measures. We used a sequential mixed research approach (Creswell and Creswell 2017) to garner unique insights from each method, to build off the previous research findings, and to contribute in different ways to our understanding of the sleep gap and measurement error.

Study 1 involved behavior coding of ATUS interview transcripts to investigate issues related to interviewer and respondent behaviors that may affect the ATUS sleep estimates. Behavior coding allows researchers to identify concepts and tasks that respondents and interviewers may struggle with and detect practices that could be associated with survey measurement error (e.g. Van der

Zouwen and Smit 2004; Dykema et al. 1997). Drawing on the findings from the behavior coding research, in Study 2 we carried out cognitive interviews to uncover more about respondents' cognitive processes when answering diary and stylized questions about sleep. Cognitive interviews provide an in-depth understanding of a respondent's thought processes and reactions to a question, and can reveal the content validity of a question (i.e. does the question measure what it is intended to measure) and what possible sources of error may underlie the question (Willis 2005). Findings from the cognitive interviews were used to generate a set of hypotheses for Study 3, which involved a quantitative approach. We tested hypotheses about diary and stylized measures, survey context, and providing definitions of sleep, using a large online sample to make statistical comparisons across experimental groups. Finally, in Study 4, we carried out a validation study comparing self-reported sleep (diary and stylized measures) against sensor data obtained from devices worn by study participants for one week that tracked their activity level, including their sleep.

In the following sections, we will describe each study and its findings, the methods used, and how each method revealed different potential sources of measurement error associated with diary and stylized sleep measures that may help explain the sleep gap.

27.4 Study 1: Behavior Coding

Study 1 used behavior coding, a questionnaire evaluation method in which researchers systematically code interviewer and respondent interactions during the survey interview, either "live" during the interview or from an audio recording of the interaction (e.g. Van der Zouwen and Smit 2004; Dykema et al. 1997). We started with behavior coding of ATUS interview transcripts to gain a deeper understanding of issues related to interviewer and respondent behaviors that may affect the ATUS sleep estimates.

Sample and demographics. A total of 104 ATUS interviews conducted by 36 interviewers during 2008 were audio-recorded with respondent consent. Interviewers recorded up to three consecutive interviews each. The demographics of the study sample were compared to the full 2008 ATUS sample, and no significant differences were found for sex, age, race, and other demographic variables (p-values > 0.05). Thus, while the interviews were not randomly selected, the study sample can be viewed as demographically representative of the full ATUS sample (Denton et al. 2012).

Coding scheme development. A team of survey methodologists developed the coding scheme, coded the interview transcripts, and analyzed the data. The main unit of analysis was a *sleep episode*, defined as a full conversation around sleep (e.g. the first and last conversational turn related to sleep). Within a sleep episode, several items were coded: the type of sleep (sleep or nap); mentions of sleep time, which consisted of reported wake times, sleep times, or sleep duration; and whether the respondent used a recall strategy (e.g. alarm clock) or qualifier when providing time (e.g. about, around). We also coded interviewer behaviors (e.g. use of scripted versus unscripted probes).

For reliability purposes, 12 transcripts were double coded, or 11.5% of the total transcripts. For quantitative variables (e.g. sleep duration), percent agreement[2] was calculated, and ranged from 0.85 to 1.00. For categorical variables, kappas were calculated and resulted in moderate to almost perfect agreement, ranging from 0.45[3]–1.0.

27.4.1 Behavior Coding Results

Sleep episodes. Respondents reported an average of 2.2 episodes of sleep (*SD* = 0.57, range = 1–4) during the 24-hour period covered by the diary. These sleep episodes were most often found to occur at the beginning and end of the diary day. Respondents often reported the time they woke up and fell asleep on the diary day and the time they woke up on the day of the interview.

Sleep duration. On average, respondents reported sleeping 8.53 hours (*SD* = 1.98; range = 2.42–15.48 hours) per 24-hour period. Just over a fifth (22.0%) of respondents reported at least one nap. Of the respondents who reported taking at least one nap, the total time spent napping was 1.35 hours on average (*SD* = 0.80 hours). These sleep estimates were comparable to the ATUS published sleep estimates.

Interviewer/Respondent interactions. We analyzed the number of conversational turns from when a respondent first mentioned sleep until the final sleep-related turn. The number of turns gives an indication of how complicated the process of recording sleep can be. In an average interview, 21 interviewer-respondent interactions were needed to record sleep, amounting to 11.9% of the interactions in the interview. An average of 6.7 turns was needed to record the time of waking or falling asleep. Interestingly, the mean number of turns required to record the time that respondents fell asleep was much higher (*M* = 11; range = 1 52 turns) than the number of turns to record wake times (*M* = 3.9, range = 1–13), suggesting greater cognitive task complexity in reporting falling asleep times versus waking up times.

Wake-time probes. At the beginning of the diary day, if a respondent reports that he or she was sleeping at 4:00 a.m., interviewers are trained to ask a

2 Percent agreement was used because it is a measure of inter-rater reliability between two coders using quantitative variables (see McHugh 2012). Pearson *r* correlations were similar and ranged from 0.89–1.00.

3 Only one category, respondent qualifications about their sleep, had a kappa of 0.45, or moderate agreement. The remaining kappas were all 0.85 or above. Qualifications may have been more difficult to code due to ambiguity in the language respondents used to describe their activities and that it was sometimes unclear if the qualification, or multiple qualifications within a sleep episode, referred to sleep or other pre-sleep activities.

non-leading question, such as "What time did you wake up?" Often, however, interviewers will rephrase this question in a potentially leading way. For example, an interviewer might ask "What time did you get up?" which could be interpreted as asking about the time they physically left their bed rather than when they woke up. Interviewers used leading question wording 68.9% of the time. The most common leading question was "What time did you get up?" with "What time did you wake up yesterday morning?" or "What time did you wake up this morning?" also being commonly asked. The latter is problematic, because it suggests to respondents that the interviewer is not interested in capturing times awake during the night.

"Went to bed" probes. Another place where the use of language may increase measurement error is the transition between wakefulness and sleep. Near bed time, respondents may use phrases such as "I went to bed." This could mean when they went to sleep, but it could also mean when they laid in bed awake, e.g. watching TV, reading, or trying to fall asleep. This time should not be recorded as sleep. When respondents say they "went to bed," interviewers are trained to use the following scripted probes:

1. What time did you fall asleep?
2. Did you groom, read, watch television or something else before you fell asleep?

Respondents used the phrase "went to bed" in a total of 76 (or 73.0%) of interviews, with interviewers probing respondents about this in 72 (or 95.0%) of instances. Of the probes used by interviewers, 42.0% were scripted and 53.0% unscripted. The most common unscripted probe was some variation of "Did you go to sleep immediately?"

Respondent qualification of answers. We coded whether respondents used qualifiers (e.g. about, around, or maybe) when reporting the time they woke up or fell asleep. The use of qualifiers gives an indication of whether the respondent was confident in their answer. Respondents averaged 1.8 qualifiers per interview across the average of 3.1 reports of falling asleep or waking. Overall, of the 56.9% of the time respondents who reported a transition between sleep and wakefulness, they did so with some amount of reservation. Respondents were slightly more likely to qualify their responses when reporting the time they fell asleep (60.6% of the time) than times they woke up (49.3% of the time), indicating they may have had more difficulty recalling or estimating their sleep versus wake times.

Recall strategies. To better understand how respondents formulate their response of when they woke up or went to sleep, any explicitly mentioned response strategy was coded. That is, if a respondent said "I always get up at 8 a.m.," their response strategy would be coded as "typical routine." If the respondent said "My alarm went off at 8 a.m.," their response strategy would be coded as "alarm clock." In 83 cases (25.1% of the total 331 sleep mentions

Table 27.1 Frequency of respondents' strategies when reporting sleep and wake times.

Response strategy	Uses	%
Alarm	33	40.0
TV	32	38.6
Viewed clock	7	8.4
Direct recall	0	0.0
Guess	0	0.0
Typical routine	9	10.8
Other	2	2.4

across all interviews), respondents mentioned some type of response strategy. Table 27.1 shows the frequency of those strategies.

The most common response strategies were reports of an alarm clock going off (usually to recall a wake time) or specific mentions of a TV show, e.g. "The news was on so it must have been 11 p.m.," (usually to recall a sleep time).

Summary. Behavior coding provided insight into how sleep is reported and recorded in ATUS interviews. We found that the interactions, especially those surrounding when a person falls asleep, are often complex, indicating that respondents may have difficulty recalling or estimating their sleep time. Interviewers often used unscripted or leading probes when requesting this additional information. This could lead to underreporting of both time spent awake during the night and time spent lying awake in bed before physically getting up, which might inflate ATUS sleep estimates and contribute to the sleep gap.

While the behavior coding study was useful to understand interviewer-respondent interactions in the ATUS, we did not have insight into how respondents answer questions about sleep. To expand on the knowledge garnered by the coded interactions, we brought participants to the lab to conduct cognitive interviews.

27.5 Study 2: Cognitive Interviews

Cognitive interviews provide an in-depth understanding of a respondent's thought processes and reactions to a question. Cognitive interviews can uncover the content validity of a question (i.e. does the question measure what it is intended to measure) and what possible sources of error may underlie the question (Willis 2005). To follow-up on the results of the behavior coding, we conducted cognitive interviews to gain insight into how respondents report on their sleep. We explored differences between diary and stylized measures at each stage of the response process (comprehension, retrieval, judgment, and reporting) with the aim of identifying possible sources of measurement error that may contribute to the sleep gap.

The cognitive interviews for this study were conducted by Bureau of Labor Statistics (BLS) researchers using the ATUS interview protocol and lasted approximately one hour each. Participants were asked to complete an abbreviated ATUS daily-recall interview about the prior 24-hour period (from 4:00–4:00 a.m.) and answer a set of stylized questions about their sleep, as follows:

1. *Diary measure.* The total number of hours participants reported sleeping in the previous 24-hour period (the abbreviated ATUS interview measure).
2. *General stylized measure.* "How many hours do you sleep at night on an average weekday?"
3. *Last week stylized measure.* "Thinking about the past week, on average, how many hours did you sleep each night?"

Interviews took place Tuesday–Friday, meaning all the cognitive ATUS interviews covered a weekday.[4] The ATUS interview focused on times when participants were likely to have woken up and gone to sleep (i.e. 4:00 a.m. to 10:00 a.m. and 7:00 p.m. to 3:59 a.m.). Participants were also asked about any naps taken between 10:00 a.m. and 7:00 p.m.[5] The order of the ATUS interview and stylized questions was randomized. Following the administration of the survey questions, participants answered retrospective probes aimed at understanding how they arrived at their answers to each of the three questions.

Participants. We recruited 29 participants (11 male, 18 female) from the Washington, DC metro area. The mean age was 46.0 ($SD = 14.1$), with a range of 21–69 years old. Nine participants had a high school diploma or equivalent, six had some college, seven had a college degree (associate's/bachelor's), and five had an advanced degree (master's/doctorate).

27.5.1 Cognitive Interview Results

Comprehension. First participants were asked to describe what the word "sleep" meant to them, what activities they included as part of sleep, and whether these activities were included in their answers. Responses varied from narrow definitions (e.g. being fully unconscious) to broader ones (e.g. dozing off, trying to fall asleep). As seen in Table 27.2, participants were fairly evenly divided between those using a narrow or broad definition of sleep, with those having a broad definition reporting sleeping approximately one hour more on average across each of the three measures than those with a narrow definition of sleep.

---

4 People tend to get more sleep on weekends, so we limited the study to weekdays only (Ford et al. 2015).

5 Results did not change significantly whether including or excluding naps in the ATUS sleep measures.

Table 27.2 Mean sleep duration across varying participant definitions of sleep.

	Diary	General stylized	Last week stylized
Narrow sleep definition (n = 15)	7.25	6.90	5.92
Broad sleep definition (n = 13)	8.18	7.15	6.93

Diary recall. While most participants could confidently recall what time they woke up in the ATUS interview because they followed a structured schedule and set an alarm for the same time each morning, the majority could not directly recall what time they fell asleep. Many used the TV program they were watching that evening to infer what time they must have fallen asleep. Some looked at the clock or guessed, and only two knew because they have a regular, scheduled bedtime.

Stylized recall/estimation. Participants were asked to describe how they answered the last week stylized sleep question. Responses fell into the following categories:

- *Recalled directly.* Participants reported having a structured schedule and knew what time they fell asleep and woke up (e.g. "I went to bed at about the same time every night. I knew I went to bed at the same time; I have the same schedule").
- *Rate retrieval.* Participants recalled the typical number of hours they sleep in a night, and used that as the average (e.g. "My usual hours of sleep are between 11 and 6 for the work week").
- *Rate and adjustment.* Participants recalled the typical number of hours they slept in the past week, and then made adjustments for events that happened that week (e.g. "Since it was a long weekend, that came to my mind. Thought of the average and then subtracted a bit because yesterday was busier than usual").
- *Calculation.* Participants summed the number of hours slept each night that week, and then took the average (e.g. "Tried to apply a median. Some nights that were shorter, some were greater. 4 + 3 + … and came up with actual average").
- *Estimate/Guess.* Participants could not recall how much they slept in the past week, so they estimated or guessed (e.g. "I took a guess. I went by the activities I was doing last week").

Table 27.3 shows the frequency of participants using the various recall and estimation techniques to answer the stylized last week question. Of the 29 participants interviewed, 10 reported that they could recall directly how much sleep they got last week, either because they had a very structured schedule

Table 27.3 Frequency of recall or estimation strategies used to answer the stylized last week question.

Strategy	Frequency
Recalled directly	10
Rate retrieval	9
Rate and adjustment	4
Calculation	4
Estimate/Guess	2

or had looked at the clock. The remainder used some other strategy to arrive at their answer. Nine participants reported using a rate-retrieval strategy, where they estimated they slept about seven hours per night. Participants who used a rate and adjustment or calculation strategy reported lower sleep estimates, around 5.5 hours of sleep. It is possible these participants adjusted their estimates downward too much due to calculation errors (Edgar 2009).

Reporting. We asked participants about possible social-desirability concerns in reporting on sleep. For instance, we asked if they believed there is an appropriate number of hours that people should sleep in one night, and the minimum and maximum number of hours a person should sleep per night. Of the 29 participants, 21 indicated they believed there is an appropriate number of hours people should sleep in one night, while 8 indicated that it depends on the individual. On average, participants reported that 7.5 hours was the appropriate amount of sleep, with a range between 6.0 and 9.0 hours.

Of the 21 participants who indicated they believed there is an appropriate number of hours people should sleep in one night, all of them reported sleeping less than that amount. However, sleep estimates only deviated by an average of 25 minutes between participants' self-reported appropriate sleep duration and diary-reported sleep duration. Deviations were higher for the stylized estimates, approximately 1.45 hours less than their self-reported appropriate sleep duration and their general and last week stylized-reported sleep duration.

Participants reported that oversleeping would be more embarrassing than undersleeping. Common examples of reasons for embarrassment at oversleeping included "looking lazy" or "being teased for sleeping too much." Participants were also asked whether they would be more embarrassed at over- versus undersleeping if the survey was about employment or health. Of the 29 participants, 17 thought it would be embarrassing to admit sleeping too much in a survey about employment and jobs, and 13 thought it would be embarrassing to admit sleeping too little in a survey about health. Thus, survey context may differentially impact social desirability concerns. Sleeping too much in the context of a survey about employment may appear "lazy," whereas undersleeping in the context of a survey about health may appear as though a respondent does not get adequate sleep.

Summary. Through the cognitive interviews, we uncovered factors at each stage of the survey response process that may affect reports of sleep duration across diary and stylized measures and contribute to the sleep gap. We also found that, like the national surveys, participants reported getting more sleep in the diary than stylized measure. These insights allowed us to generate hypotheses about what factors might contribute to differences between diary and stylized sleep reports and the measurement error associated with each of them. For instance, perhaps providing respondents with a standardized definition of what counts as "sleep" would bring diary and stylized sleep estimates closer together. A survey about health versus jobs may cause people to report getting more or less sleep, respectively. However, the qualitative nature of the cognitive interviews limited the generalizability of the findings and the ability to make statistical comparisons. To determine whether these factors affect reports of sleep in diary and stylized measures, we conducted a larger-scale, quantitative study.

27.6 Study 3: Quantitative Study

A large-scale online experiment collecting quantitative data was designed to make statistical comparisons of reported sleep duration across diary and stylized measures and test hypotheses generated from Studies 1 and 2. First, we wanted to determine whether we would replicate the sleep gap in our online sample. Based on the results of the cognitive interview study, we also wanted to determine whether providing a definition of sleep affects reported sleep duration. Finally, we wanted to assess context effects by comparing sleep duration estimates across participants who thought the survey was about jobs versus health.

Method. Participants were recruited using Amazon.com's Mechanical Turk (MTurk) platform, an online crowdsourcing website where research participants receive small incentives for completing surveys or other tasks (Buhrmester et al. 2011). Although MTurk samples are not representative of the US population, MTurk yields samples that are large and more demographically diverse than those obtained in other convenience samples, such as college students or local participants brought into cognitive laboratories (Casey et al. 2017; Stewart et al. 2017; Edgar et al. 2016). Participants were routed to a web survey, where they completed a modified version of the ATUS daily recall diary interview and answered a set of stylized questions about their activities (e.g. working, physical exercise, sleep). Again, participants completed the survey

Table 27.4 Experimental design for the quantitative study.

Between-subjects factors		Within-subjects factors
Survey framing	**Activity definitions provided**	**Type of measure (order randomized)**
Framing of "Survey about Jobs and Employment"	• Sleep definition • No definition	• Diary then stylized • Stylized then diary
Framing of "Survey about Health and Wellness"	• Sleep definition • No definition	• Diary then stylized • Stylized then diary
Framing of "Survey about Time Use" (control)	• Sleep definition • No definition	• Diary then stylized • Stylized then diary

considering weekdays (both the diary day and previous 24-hour period were weekdays).

Participants and design. A total of 1233 participants living in the United States (54% female, with an average age of 36.34) complete the survey. A total of 62.1% were employed full time, 22.0% were employed part-time, 9.4% were unemployed, 3.5% were students, and 3.0% were retired; and the average household size was 2.63. Demographics did not vary by condition ($ps > 0.08$). Participants were randomly assigned to one of three different survey framing conditions, in which they were told the survey was about health, jobs, or general time use. They were then randomly assigned definitions of terms (including sleep[6]) or received no definitions. Finally, the order in which participants completed the diary and stylized questions in the survey was randomized. This yielded a $3 \times 2 \times 2$ mixed-model design with 2 between-subjects factors (framing and definitions) and 1 within-subjects factor (question type). See Table 27.4 for an illustration of the different survey variants.

Participants completed an online version of the ATUS time diary in which they entered all of their activities from the prior 24-hour period (from 4:00 a.m.–4:00 a.m.). They selected activities from a subset of the 12 most commonly reported activities in the ATUS (e.g. working, commuting) from a drop-down menu, and indicated the start and stop time of each activity, entering up to a maximum of 20 activities. They also answered a set of stylized questions about their activities throughout the previous week. Embedded within these questions were the same stylized sleep measures used in Study 2.

6 The definition read, "By sleep, we mean the number of hours you actually spend sleeping. This may be different from the number of hours you spend in your bed, time you spend preparing to go to sleep, or resting with your eyes closed but not actually asleep. Please include any times you were sleeping during the day (or napping)." This was embedded with definitions of other common activities, such as work and exercise.

27.6.1 Quantitative Study Results

We calculated the total time participants reported sleeping in the prior 24-hour period to create a time diary estimate. We then compared participants' self-reported diary versus stylized hours of sleep. We found a sleep gap where the diary measure (M = 7.95 hours; SD = 1.76) exceeded stylized reports (M = 7.27 hours; SD = 1.39) of sleep. As seen in Figure 27.1, we found the participants' stylized sleep reports peaked at rounded numbers, such as six and seven, and dropped quickly after eight hours, indicating potential measurement error due to rounding.

To examine the effects and interactions of framing, definitions, and question order, we conducted a 3 (framing type – jobs vs. health vs. time use) × 2 (definitions – provided a definition vs. no definition) × 2 (question type order – diary first vs. stylized first) mixed-model ANOVA, where the dependent variable was mean sleep duration as measured by the diary and past week stylized questions. Table 27.5 shows the results of this analysis.

As seen in Table 27.5, we did not find a significant three-way interaction between definition, question order, and framing on sleep question type. We did find a significant interaction between question type and definition where participants who read the definitions reported more similar hours of sleep across the diary (M = 7.87, SD = 1.74) and stylized (M = 7.33, SD = 1.52) measures versus those who did not read definitions for the diary (M = 8.00, SD = 1.77) and stylized (M = 7.22, SD = 1.26) measures.

An interaction between question type and order was found, where participants who completed the diary first (M = 8.39 hours; SD = 1.82) reported more sleep in the time diary than participants who completed the stylized questions first (M = 7.49 hours; SD = 1.57); see Figure 27.2. In contrast, participants who

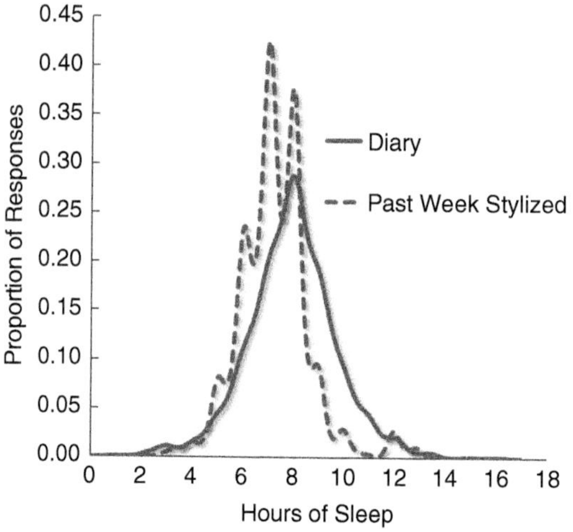

Figure 27.1 Response distribution of sleep reports across diary and past week stylized questions.

Table 27.5 Results of mixed-model ANOVA on sleep duration by question type (framing type definition) and order (dependent variable = mean hours of sleep).

	df	F value	p-Value	η^2_p
Question type	1	200.50	$p < 0.01$	0.14
Question type × definition	1	5.84	$p = 0.01$	0.01
Question type × order	1	29.49	$p < 0.01$	0.02
Question type × framing	2	1.55	$p = 0.21$	0.00
Question type × definition × order	1	0.16	$p = 0.69$	0.00
Question type × definition × framing	2	0.77	$p = 0.46$	0.00
Question type × order × framing	2	0.26	$p = 0.77$	0.00
Question type × definition × order × framing	2	1.46	$p = 0.23$	0.00
Residuals	1221			

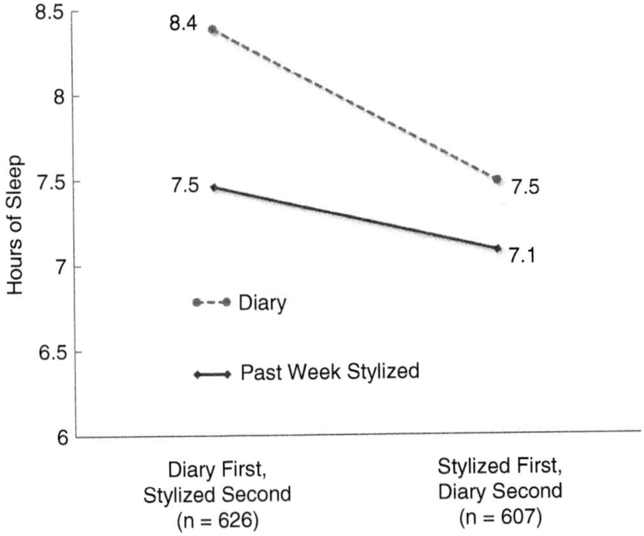

Figure 27.2 Mean hours of sleep reported as a function of whether participants answered the diary versus stylized questions first.

completed the stylized questions first ($M = 7.08$ hours; $SD = 1.40$) reported less sleep for the stylized question than those who completed the diary questions first ($M = 7.47$ hours; $SD = 1.37$).

We also conducted a Fisher r-to-z transformation to test the magnitude of the correlation between the two measures. We found the correlation between the diary and stylized measures was significantly greater ($z = -4.38$, $p < 0.001$) when participants answered the stylized questions first ($r = 0.54$), versus completing the diary first, ($r = 0.34$). Thus, answering the stylized question first seemed to pull the diary and stylized sleep measures closer together. Finally, we found a main effect of question type where the diary sleep measure ($M = 7.95$; $SD = 1.76$) exceeded the stylized sleep measure ($M = 7.28$; $SD = 1.40$), replicating the sleep gap. No other significant main effects or interactions were found.

Summary. In the quantitative study, we found a sleep gap, where diary sleep estimates exceeded stylized sleep estimates. We also found that certain features of diary and stylized sleep questions (definitions, question order) may have a larger impact than others (survey context) on a web survey, as the framing of the survey (health, jobs, or time use) did not affect participants' reports of sleep. Context effects may have been minimized because the survey was self-administered and anonymous, which reduces social-desirability concerns (Kreuter et al. 2008). Also consistent with the previous studies, providing a definition of sleep brought diary and stylized estimates slightly closer together, pointing to potential measurement error due to comprehension of the term "sleep."

We found that when stylized questions preceded diary questions, the gap between diary and stylized estimates was much smaller. This may indicate that participants alter their response strategies (e.g. recall or estimation) when answering both diary and stylized questions within the same survey. Although we did not anticipate this question-order effect, stylized questions focus attention on a particular topic (in this case sleep), and this focus on sleep may have affected answers to subsequent questions (Schulz and Grunow 2012). In contrast, when the time diary came first, the total amount of sleep reported was masked and embedded within reports of many other activities. This is consistent with the literature on order effects, showing that when specific questions (e.g. stylized questions) precede general, broader questions (e.g. diary questions), respondents anchor their answers to the more specific question that came first (Schwarz et al. 1991).

Our research so far has relied solely on self-reported sleep duration and could not assess whether diary or stylized sleep measures were more accurate. In our final study, we aimed to assess the accuracy of diary and stylized sleep measures for individuals by comparing them to sleep recorded by a sensor.

27.7 Study 4: Validation Study

The use of sensor data to conduct behavioral research has become more common over the past few years, providing an alternative, objective measure that is free from the measurement error associated with self-reports (Evenson et al. 2015; Wright et al. 2017). As these technologies have emerged, researchers have become interested in whether they can be used to validate survey questions (e.g. Downs et al. 2014). We tracked respondents' activities (including sleep) to compare their self-reported sleep to objectively measured sleep via a wearable device (the Fitbit Charge). We selected the Fitbit Charge because it automatically records sleep without any user action and does not display sleep data on its interface, so participants cannot view their sleep data. The Fitbit Charge has sensors that automatically measure the angle and movement of the device. The device then interprets these measurements as physical actions (e.g. walking or running). The device also measures the absence of movement, or detects only subtle movements, which is interpreted as sleep. Wearable devices are light and unobtrusive so they can be worn most of the time and during sleep. It has recently been suggested that the popularity of such devices may facilitate conducting larger-scale studies that compare self-reported sleep to objectively measured sleep (e.g. Miller et al. 2015). However, the validity and reliability of these devices varies, sometimes overestimating or underestimating different activities (Evenson et al. 2015). For sleep measurement, these devices are considered to be fairly reliable in healthy adults without sleep disorders (e.g. Kang et al. 2017; Lee et al. 2017; Cook et al. 2017).

Method. Participants were interviewed twice about one week apart. At Visit 1, they answered general questions about their typical routine and were instructed to wear the Fitbit Charge at all times over the next week, except while showering or bathing.[7] At Visit 2, participants completed the same abbreviated ATUS diary interview used in Study 2. We also asked participants a set of stylized sleep questions (the same ones used in Studies 2 and 3). Afterward, we compared the total sleep duration from the diary, stylized questions, and Fitbit measures, asking participants targeted probe questions to understand the differences (if any) between the measures. Among other data, the Fitbit records the number of hours slept per day and periods of wakefulness or restlessness during the night.

Participants. We recruited 44 participants in the Washington, DC metro area. Only participants who wore the Fitbit the day before the second visit were included in analyses to ensure sleep comparisons were possible across the diary, stylized questions, and Fitbit-recorded sleep durations.[8] A total of 35 participants (13 female with an average age of 44.58 years) complied with these instructions and were included in our final analyses.

27.7.1 Validation Study Results

Table 27.6 shows the mean sleep duration across each of the sleep measures. The Fitbit-estimate of sleep fell in between the diary and stylized questions estimates.

We conducted a within-subjects ANOVA contrasting each of the sleep measures to assess whether participants' sleep duration estimates differed by

7 When removing the device for bathing/showering, respondents were asked not to clasp it shut so as to avoid the device registering sleep.

8 Two participants lost the device during the week, six could not return for their second interview, and one participant's device fell off during the night before their scheduled interview.

Table 27.6 Mean hours of sleep across diary, stylized, and Fitbit measures ($N = 35$).

Measure	Mean and SD hours of sleep
Diary	7.26 (1.79)
General stylized	6.62 (1.14)
Last week stylized	6.56 (1.14)
Fitbit (prior 24 hours)	7.11 (1.64)
Fitbit (over past week)	6.88 (1.36)

Table 27.7 Intra-class correlation coefficients (ICCs) and confidence intervals (CIs) across sleep-duration measures.

Measures	ICC and CI	*p*-Value	Agreement
Diary and stylized (weekdays)	0.49 (0.19–0.71)	$p = 0.002$	Fair
Diary and Fitbit-recorded sleep (prior 24 hours)	0.76 (0.60–0.88)	$p < 0.001$	Excellent
Stylized and Fitbit-recorded sleep (average over week)	0.40 (0.08–0.64)	$p = 0.01$	Fair
Stylized and Fitbit-recorded sleep (average over weekdays)	0.62 (0.35–0.79)	$p < 0.001$	Good
Stylized and Fitbit-recorded sleep (average over weekend)	0.30 (0.05–0.58)	$p > 0.05$, n.s.	Poor

sleep measure. We found the measures differed significantly, $F(4, 136) = 3.48$, $p = 0.10$, $\eta^2_p = 0.09$. Post-hoc pairwise comparisons showed that the diary and Fitbit-recorded sleep from the previous 24-hour period exceeded both stylized measures, (ps < 0.05). No other differences were found.

We also assessed how well each measure agreed with one another using intra-class correlation coefficients (ICCs).[9] Table 27.7 shows the ICCs, confidence intervals, significance level, and agreement between each of the measures.

The diary and stylized measures had fair agreement with one another, consistent with prior research (e.g. Schulz and Grunow 2012). The diary and Fitbit-recorded sleep from the previous 24-hour period yielded the best

9 ICCs were calculated because this statistic provides a measure of how well related variables (e.g. sleep measures from the same participant) that measure the same construct agree with one another (see Kang et al. 2017). Results using a Spearman correlation coefficient were similar to those obtained using the ICCs.

agreement among all of the measures, falling in the excellent agreement range, consistent with literature showing that diary measures may be a more reliable measure of time use (Juster et al. 2003; Kan and Pudney 2008). Overall, the stylized and Fitbit-recorded sleep over the week had fair agreement, with good agreement on weekdays and poor agreement on weekends. This is consistent with our findings from the cognitive interviews, suggesting that people may be better at estimating their sleep and wake times on weekdays, when they tend to follow a more structured schedule, versus weekends where schedules are less structured and estimating sleep duration may be more difficult. Similar to Study 3, we again observed rounding in the stylized measure, with participants providing responses of five, six, or seven hours of sleep.

During respondent debriefing, we found that in some instances the Fitbit-recorded data aided participants' recall of their wake and sleep times, (e.g. recalling they hit the snooze button and got some extra sleep, or woke up a little earlier than normal on that day). Participants could generally recall waking up during the night once or twice, but the Fitbit tended to show many awakenings during the night that participants could not recall. Thus, the Fitbit may have its own set of measurement error where the absence of movement does not always correspond to sleep and movement does not always correspond to being awake (Wright et al. 2017). For example, the Fitbit may have overestimated the amount of wakefulness experienced during the night (e.g. recording tossing and turning as time awake), underestimating the total sleep duration for each night. In other cases, the Fitbit may have overestimated sleep, recording periods of time lying still watching television or reading as naps.

Summary. We found evidence that sleep duration recorded via sensor data may fall somewhere in between diary and stylized sleep estimates. Diary measures tended to agree more with the sensor data, consistent with prior research showing that diary measures may be more reliable and valid than stylized measures. However, sensor data are also prone to measurement and user error, and the Fitbit-recorded sleep may not always accurately reflect actual sleep.

27.8 General Discussion

This research investigated the gap between diary and stylized sleep measures and potential sources of measurement error associated with them. We used different questionnaire evaluation methods (behavior coding, cognitive interviews, quantitative research, and a validation study) to address our research questions. In Table 27.8, we summarize the main findings from each method, the sources of measurement error each uncovered, and the pros and cons associated with each method.

We found a sleep gap across Studies 2–4, where diary sleep measures led participants to report more sleep than stylized measures. Each method revealed

Table 27.8 Summary of the main findings, sources of measurement, and pros and cons of each method.

	Sleep gap observed?	Comprehension	Recall	Judgment	Reporting	Sources of measurement error	Pros of method	Cons of method
Behavior coding (Study 1)	N/A	N/A	Respondents used alarms or TV programs to help recall wake and sleep times	More conversational turns and qualifications occurred in reporting sleep time than wake times	N/A	• Interactions surrounding sleep times are complex, may be imprecise • Interviewers use unscripted, leading probes that may affect ATUS sleep estimates	• Use of production interviews • Code numerous features of ATUS interview • Identify problematic concepts, tasks, and practices	• Little insight into respondents' cognitive processes • No comparison to stylized sleep questions
Cognitive interviews (Study 2)	Yes	Broad definitions of sleep were associated with reporting more sleep than narrow definitions of sleep	Respondents used alarms or TV programs to help recall wake and sleep times	Respondents used rate retrieval, rate and adjustment, calculation, and guessing for stylized questions	Survey context (employment versus health) may affect self-reports of sleep	• -Recall and estimation bias may be present in reporting sleep times • -Survey context may push sleep estimates up or down based on social desirability	• Rich understanding of cognitive processes surrounding sleep questions • Hypothesis generation	• Small, non-representative sample • Cannot make statistical inferences and comparisons
Quantitative study (Study 3)	Yes	Providing a definition of sleep narrowed the sleep gap	N/A	Rounding observed in stylized sleep estimates	No survey framing or context effect observed (employment versus health) on self-reports of sleep	Definitions of sleep and question order affected self-reports of sleep, but not survey framing	• Collected large amount of data in short timeframe • Large sample allowed for statistical comparisons	Non-probability sample; cannot make generalizations to the U.S. population
Validation study (Study 4)	Yes	N/A	Sensor data aided recall for sleep and wake times	Rounding observed in stylized sleep estimates	N/A	• Sensor data agreed more with diary sleep estimates • Wearable devices have their own set of measurement and user error to consider	Objective sleep measure does not rely on self-report data	• Sensor-recorded data also prone to measurement error • User error with wearable device

sources of measurement error that may have caused diary measures to exceed stylized measures, helping to explain reasons for the sleep gap.

Study 1 used behavior coding to identify issues in the ATUS interviews that may lead to measurement error. We found that interviewer and respondent interactions surrounding sleep are often complex. Respondents had difficulty recalling or estimating their sleep and wake times. Interviewers often used leading questions and unscripted probes that may encourage respondents to believe they should define sleep as a continuous episode, potentially inflating ATUS sleep estimates. One strength of behavior coding is the use of actual production interviews to identify potential issues and the ability to investigate and code numerous features of the diary interview, including questions, probes, and answers, as well as unscripted conversation that may contribute to error in the ATUS. A limitation is that it did not provide direct insight into how respondents arrived at their answers to these questions, and we had no direct comparison to stylized questions. This led us to conduct cognitive interviews.

Study 2 involved conducting cognitive interviews with the aim of understanding how respondents report on sleep at each stage of the response process. Building on our knowledge from the behavior-coding research, we found that definitions of sleep, recall of sleep times, and social-desirability biases were areas where measurement error is likely to occur. The cognitive interviews provided a rich understanding of these issues and allowed us to generate hypotheses about what factors might contribute to differences between diary and stylized sleep reports. One downside was the small, geographically limited sample that could not be used to make statistical comparisons. This led us to conduct a larger-scale quantitative study.

In Study 3, we designed a quantitative experimental study to compare participants' diary and stylized sleep estimates, where we also found a sleep gap in which diary measures exceeded stylized measures of sleep duration. Drawing on the results of the previous two studies, we found that how respondents define sleep affected their answers for the diary and stylized last week measures, but not the general stylized measure. It may be that people rely on the typical amount of sleep they get overall (e.g. seven hours) when reporting in general versus considering activities that took place over the week, an area for future investigation. We also found that when stylized questions preceded diary questions, the sleep gap narrowed, indicating that participants may have anchored their answers to the stylized estimate. Context effects were less apparent in the quantitative study – perhaps due to mode – being an online, anonymous survey rather than an interviewer-administered survey (e.g. Kreuter et al. 2008).

One benefit of using online crowdsourcing panels such as MTurk is that it enabled us to collect a large amount of data in a short timeframe (Edgar et al. 2016). We were able to obtain a larger, more geographically diverse sample from participants around the country than would be possible to obtain in traditional laboratory studies, such as cognitive interviews (Casey et al. 2017;

Stewart et al. 2017). Although MTurk samples are not representative of the general population, they are useful for experimental purposes and as a research tool since we were interested in assessing internal validity rather than representativeness of any particular population. As such, crowdsourced panels are not a replacement for probability samples, and the results should be interpreted with caution. Also, these findings could not tell us whether diary or stylized sleep estimates were more accurate, which led us to our validation study.

Finally, in Study 4, we conducted a validation study that compared sleep duration across diary, stylized, and sensor data. We found evidence that an objective measure of sleep may fall somewhere in between diary and stylized measures, but the sensor data agreed more with diary than stylized self-reports. Viewing the sensor data also helped participants recall their sleep and wake times in some instances; however, it was not without its own set of measurement and user error. Depending on the researcher's goals, such devices could be a useful tool to assess sources of question measurement error for surveys where participants are asked to recall their activities or time use.

27.9 Implications and Future Directions

These results have broad implications for researchers interested in measuring time use. Researchers should be aware that diary and stylized questions might yield different results and understand the sources of measurement error associated with both measures. Future research should explore additional reasons for the gap between diary and stylized measures beyond just the response process. This might include sampling, context effects, and interviewer effects in the administration of both diary and stylized questions, data collection procedures, and how the survey organization defines and calculates time spent on activities. Researchers might also analyze other activities that show a gap in diary and stylized measures, such as work or exercise. As wearable devices improve, researchers may want to capitalize on these new technologies to investigate potential sources of survey measurement error. In future research, we also recommend using a multimethod approach (refer chapter 5 by d'Ardenne and Collins), as each method can capture unique sources of measurement error and can build off the preceding results and insights. We believe this approach will be highly useful to researchers designing, evaluating, testing, or validating survey questions.

References

Adams, S.A., Matthews, C.E., Ebbeling, C.B. et al. (2005). The effect of social desirability and social approval on self-reports of physical activity. *American Journal of Epidemiology* 161 (4): 389–398.

Biddle, J.E. and Hamermesh, D.S. (1990). Sleep and the allocation of time. *Journal of Political Economy* 98: 922–943.

Bonke, J. (2005). Paid work and unpaid work: diary information versus questionnaire information. *Social Indicators Research* 70 (3): 349–368.

Brenner, P.S. (2011). Identity importance and the overreporting of religious service attendance: multiple imputation of religious attendance using the American Time Use Study and the General Social Survey. *Journal for the Scientific Study of Religion* 50 (1): 103–115.

Buhrmester, M., Kwang, T., and Gosling, S.D. (2011). Amazon's Mechanical Turk: a new source of inexpensive, yet high-quality, data? *Perspectives on Psychological Science* 6: 3–5.

Canfield, B., Miller, K., Beatty, P., et al. (2003). Adult questions on the Health Interview Survey – Results of cognitive testing. Internal NCHS peport.

Casey, L., Chandler, J., Levine, A.S., et al. (2017). Demographic characteristics of a large sample of us workers. https://osf.io/preprints/psyarxiv/8352x.

Conrad, F.G., Brown, N.R., and Cashman, E.R. (1998). Strategies for estimating behavioural frequency in survey interviews. *Memory* 6 (4): 339–366.

Cook, J.D., Prairie, M.L., and Plante, D.T. (2017). Utility of the Fitbit flex to evaluate sleep in major depressive disorder: a comparison against polysomnography and wrist-worn actigraphy. *Journal of Affective Disorders* 217: 299–305.

Couper, M.P., Conrad, F.G., and Tourangeau, R. (2007). Visual context effects in web surveys. *Public Opinion Quarterly* 71 (4): 623–634.

Creswell, J.W. and Creswell, J.D. (2017). *Research Design: Qualitative, Quantitative, and Mixed Methods Approaches*. Sage Publications.

D'Ardenne, J. and Collins, D. (2020). Combining multiple question evaluation methods – what does it mean when the data appear to conflict? In: *Advances in Questionnaire Design, Development, Evaluation, and Testing* (eds. P. Beatty, D. Collins, L. Kaye, et al.). Hoboken, NJ: Wiley.

Denton, S., Edgar, J., Fricker, S., and Phipps, P. (2012). Exploring conversational interviewing in the American Time Use Study: Behavior coding study report. Internal BLS report.

Downs, A., Van Hoomissen, J., Lafrenz, A., and Julka, D.L. (2014). Accelerometer-measured versus self-reported physical activity in college students: implications for research and practice. *Journal of American College Health* 62 (3): 204–212.

Dykema, J., Lepkowski, J.M., and Blixt, S. (1997). The effect of interviewer and respondent behavior on data quality: analysis of interaction coding in a validation study. In: *Survey Measurement and Process Quality* (eds. L. Lyberg et al.), 287–310. https://doi.org/10.1002/9781118490013.ch12.

Edgar, J. (2009). What does "usual" usually mean? American Association for Public Opinion Research.

Edgar, J., Murphy, J., and Keating, M. (2016). Comparing traditional and crowdsourcing methods for pretesting survey questions. *SAGE Open* 6 (4) https://doi.org/10.1177/2158244016671770.

Evenson, K.R., Goto, M.M., and Furberg, R.D. (2015). Systematic review of the validity and reliability of consumer-wearable activity trackers. *International Journal of Behavioral Nutrition and Physical Activity* 12 (1): 1.

Ford, E.S., Cunningham, T.J., and Croft, J.B. (2015). Trends in self-reported sleep duration among US adults from 1985 to 2012. *Sleep* 38 (5): 829–832.

Hale, L. (2005). Who has time to sleep? *Journal of Public Health* 27: 205–211.

Hurst, M. (2008). Who gets any sleep these days? Sleep patterns of Canadians. *Canadian Social Trends* 85: 39–45. Statistics Canada Catalogue no. 11-008-XWE. Retrieved from: http://www.statcan.gc.ca/pub/11-008-x/2008001/article/10553-eng.htm.

Juster, F.T., Ono, H., and Stafford, F.P. (2003). An assessment of alternative measures of time use. *Sociological Methodology* 33 (1): 19–54.

Kan, M.Y. (2008). Measuring housework participation: the gap between "stylised" questionnaire estimates and diary-based estimates. *Social Indicators Research* 86 (3): 381–400.

Kan, M.Y. and Pudney, S. (2008). Measurement error in stylized and diary data on time use. *Sociological Methodology* 38 (1): 101–132.

Kang, S.G., Kang, J.M., Ko, K.P. et al. (2017). Validity of a commercial wearable sleep tracker in adult insomnia disorder patients and good sleepers. *Journal of Psychosomatic Research* 97: 38–44.

Kreuter, F., Presser, S., and Tourangeau, R. (2008). Social desirability bias in CATI, IVR, and web surveys: the effects of mode and question sensitivity. *Public Opinion Quarterly* 72 (5): 847–865.

Lee, H.A., Lee, H.J., Moon, J.H. et al. (2017). Comparison of wearable activity tracker with actigraphy for sleep evaluation and circadian rest-activity rhythm measurement in healthy young adults. *Psychiatry Investigation* 14 (2): 179–185.

Lin, K.H. (2012). Revisiting the gap between stylized and diary estimates of market work time. *Social Science Research* 41 (2): 380–391.

McHugh, M.L. (2012). Interrater reliability: the kappa statistic. *Biochemia Medica* 22 (3): 276–282.

Miller, C.B., Gordon, C.J., Toubia, L. et al. (2015). Agreement between simple questions about sleep duration and sleep diaries in a large online survey. *Sleep Health* 1 (2): 133–137.

Phipps, P.A. and Vernon, M.K. (2009). Twenty-four hours: an overview of the recall diary method and data quality in the American Time Use Survey. In: *Calendar and Time Diary Methods in Life Course Research* (eds. R.F. Belli, F.P. Stafford and D.F. Alwin), 109–128. Thousand Oaks, CA: Sage.

Schulz, F. and Grunow, D. (2012). Comparing diary and survey estimates on time use. *European Sociological Review* 28 (5): 622–632.

Schwarz, N., Strack, F., and Mai, H.P. (1991). Assimilation and contrast effects in part-whole question sequences: a conversational logic analysis. *Public Opinion Quarterly* 55 (1): 3–23.

Silva, G.E., Goodwin, J.L., Sherrill, D.L. et al. (2007). Relationship between reported and measured sleep times: the sleep heart health study (SHHS). *Journal of Clinical Sleep Medicine* 3: 622–630.

Stewart, N., Chandler, J., and Paolacci, G. (2017). Crowdsourcing samples in cognitive science. *Trends in Cognitive Sciences* 21 (10): 736–748.

Tourangeau, R. (1984). Cognitive sciences and survey methods. In: *Cognitive Aspects of Survey Methodology: Building a Bridge Between Disciplines* (eds. T. Jabine et al.), 73–100. Washington, DC: National Academy Press.

Van der Zouwen, J. and Smit, J.H. (2004). Evaluating survey questions by analyzing patterns of behavior codes and question–answer sequences: a diagnostic approach. In: *Methods for Testing and Evaluating Survey Questionnaires* (eds. R.M. Groves et al.), 109–130.

Willis, G.B. (2005). *Cognitive Interviewing: A Tool for Improving Questionnaire Design*. Sage Publications https://doi.org/10.1002/0471654728.ch6.

Wright, S.P., Brown, T.S.H., Collier, S.R., and Sandberg, K. (2017). How consumer physical activity monitors could transform human physiology research. *American Journal of Physiology-Regulatory, Integrative and Comparative Physiology* 312 (3): 358–367.

28

Questionnaire Design Issues in Mail Surveys of All Adults in a Household

Douglas Williams, J. Michael Brick, W. Sherman Edwards, and Pamela Giambo

Westat, Rockville, MD, USA

28.1 Introduction

Mail surveys have experienced a renaissance in the last decade in the United States (Link et al. 2008), due largely to the development of a sampling frame of addresses that is nearly complete (Iannacchione 2011). The commercial availability of this frame coincided with a rapid decline in response rates for random digit dial telephone surveys, fueling renewed interest in mail surveys. A large body of research has established methods for mail surveys (e.g. Dillman 1978) and telephone surveys (e.g. Lepkowski et al. 2008), but the changing environment required rethinking the relative merits of these modes of data collection.

Messer and Dillman (2011) showed that mail surveys can achieve high response rates and provide good demographic representation of the general population. Montaquila et al. (2013) summarized a set of experiments showing that well-conducted mail surveys can have substantially higher response rates and lower costs compared to telephone surveys. Brick et al. (2016) also found mail surveys can achieve better response rates and response quality compared to telephone surveys. These results suggest that mail surveys can be an attractive alternative to telephone surveys. However, the new sampling frame of addresses does not contain detailed and complete data on the members of the household, and this presents difficulties for surveying a specific person in a self-administered mode where there is no interviewer present to control the selection process.

In this situation, researchers have two design options. They can attempt to sample a person from within the household to respond (as is done in telephone surveys), or they may allow one person to report data for all persons or adults in the household. Even in telephone surveys, the accuracy of within-household sampling is relatively poor due to respondents not following the interviewer's sampling instructions (Lavrakas et al. 2000); and in self-administered surveys,

Advances in Questionnaire Design, Development, Evaluation and Testing, First Edition.
Edited by Paul C. Beatty, Debbie Collins, Lyn Kaye, Jose-Luis Padilla, Gordon B. Willis, and Amanda Wilmot.

respondents fail to select the appropriate person more often (Olson et al. 2014). When one person responds for all the members of the household, sampling within household is not an issue. However, in this situation, concerns do arise about the accuracy of the data that the household respondent reports for the other household members. Questionnaire design can play an important role in supporting accurate reporting for other household members, but little research has been published on this topic.

Most US federal government surveys of all adults or persons in a household, such as the Current Population Survey, the Consumer Expenditure Survey, and the National Survey on Drug Use and Health are interviewer-administered. The US Census Long Form and its successor, the American Community Survey (ACS), are notable surveys with a mail component that collect data on all household members. Since the introduction of the mail survey in the 1960 Census, the questionnaires for these surveys have always begun by rostering all the members and then collecting data for each member. Most of the extensive research conducted for the Long Form and the ACS has focused on obtaining higher response rates, more timely returns, and improved measurement (US Census Bureau 2014). However, the basic structure of rostering followed by items for each person has remained consistent and has drawn less attention.

In this chapter, we explore these issues within the context of a survey collecting data on criminal victimization. We begin, in the next section, by providing some background on the survey we used for our study and then describe the key challenges we faced in developing a self-administered version of the survey.

28.2 Background

The Bureau of Justice Statistics (BJS) sponsors the National Crime Victimization Survey (NCVS), which produces national estimates on crime victimization, including incidents not reported to the police. The NCVS uses an area probability sample with a rotating panel design (Rand 2007). The US Bureau of the Census collects the data for the BJS. The NCVS uses a combination of face-to-face and telephone interviewing modes to interview all household members who are 12 years old and older.

Although the NCVS provides extremely valuable information at the national level, criminal victimization estimates, including incidents not reported to the police, are very much in demand at lower levels of geography (Groves and Cork 2008). The current design and cost make it infeasible to extend the NCVS design to produce estimates for local areas such as large cities and police jurisdictions. A mail survey approach is an attractive, lower-cost alternative for producing local area estimates because an address-based sample, selected from files derived from United States Postal Service delivery service, can finely target geographic areas such as cities and police jurisdictions.

Collecting data on all adults in the household is also a compelling approach for the NCVS as crime victimization is a rare event. A survey of all adults in the household more than doubles the person-level sample size, giving much more precise estimates than would be achieved by sampling one adult from the household. This all-adult approach also avoids the difficulties of sampling a specific adult from within the household in a self-administered context.

A test of a self-administered mail survey was therefore undertaken to assess its feasibility for producing estimates of victimization at the local level for comparisons across local areas, and for tracking change in the estimates over time. We refer to the mail survey as the National Crime Victimization Survey-Companion Survey (NCVS-CS)[1] to distinguish it from the core NCVS conducted by the Census Bureau.

The NCVS-CS contained items included in the core NCVS, adapted for self-administration. One important modification was that the NCVS-CS reference period for reporting victimization (how far back the respondent is asked to recall incidents) was 12 months rather than the 6-month period used in the core NCVS. The 12-month recall period was chosen because victimization events are rare. Having a longer reference period increases the chance of capturing eligible victimization events.

28.3 The NCVS and Mail Survey Design Challenges

We faced two key questionnaire design issues when designing the NCVS-CS as a household questionnaire for mail administration.

- How to structure the questionnaire to collect data from all adults in the household (especially when victimization may occur infrequently)
- How to order the content of the questionnaire to enhance response rates and assist complete reporting of data

We addressed these issues in the field test by developing two different questionnaires and using an experimental design to support the analysis of the results. The experimental design is described in Section 28.4. First, we describe the two different questionnaire designs tested.

28.3.1 The Person-Level Survey (PLS) and Incident-Level Survey (ILS) Designs

The ACS approach, used on many self-administered household surveys, involves the respondent listing all household members and then answering a

1 The survey instrument was labeled as the American Crime Survey, and other reports from the survey are calling the survey the American Victimization Survey. We use NCVS-CS here.

series of questions about each adult in the household in turn. In the NCVS-CS, the ACS approach was modified: the roster was dropped, and instead a series of screening questions was asked with regard to each adult in the household, to identify those who had experienced victimization during the reference period. We call this a *person-level survey* (PLS) approach because it focuses on one person at a time.

An alternative questionnaire design is to focus on the victimization incidents that adults in the household may have experienced rather than focusing on each adult separately. In the NCVS-CS, a household roster was completed prior to victimization screening questions. After the victimization screening was completed, details on the specific incidents were captured and linked to the adult(s) who experienced them. We call this the *incident-level survey* (ILS) approach because it focuses on incidents rather than persons. The ILS mimics the interviewer-administered questionnaire design of the core NCVS. The design was inspired by research that showed that asking screening questions about rare victimization incidents first can provide cues that help respondents to remember such events and elicit more complete reporting of the incidents (Lehnen and Skogan 1981).

The differences in navigational structures between these two survey approaches are significant and affect the survey content. By developing two survey versions – one based on the ILS approach and one on the PLS approach – we aimed to examine the effects of these different questionnaire designs on key outcomes. Figure 28.1 shows the flow of each survey version.

The ILS and PLS structures also required different approaches to other aspects of the questionnaire design. In the ILS version, full details are collected on the two most recent incidents per type of crime, along with the date of each incident. In the PLS, instead of details about incidents, the incident characteristics are collected for each type of crime along with the date of the most recent incident. The date of the most recent incident determines the eligibility of that victimization type (by confirming that an incident occurred during the 12-month reference period). Thus, the recall burden associated with recalling victimization characteristics, especially dates, is less for the PLS than it is for the ILS.

Another difference is that the PLS starts with a household section, to collect data about household-level victimizations (property crime), and then proceeds to questions about each adult sequentially to determine if that adult experienced any of several personal victimization types (see Figure 28.2). The ILS, on the other hand, begins with personal victimizations, as is done in the core NCVS. The sequence of asking household and then personal items in the PLS was necessary because it would have been difficult for respondents to skip forward in the questionnaire to a household section at the back.

A further difference between the two versions is that the ILS requests the names of adults in the household to link adults to victimizations. These names

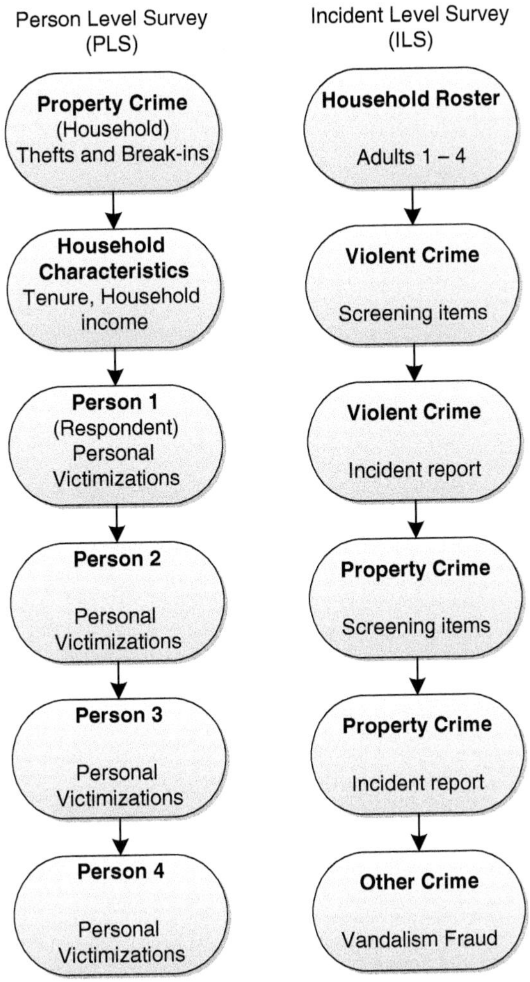

Figure 28.1 Flow of items in the ILS and PLS questionnaires.

are not needed in the PLS because the victimizations are reported at the person level, as shown in Figure 28.2. In the ILS, the respondent uses the first name and person number (defined in the instrument by the position in the roster) to link the personal victimizations to the appropriate household member. Figure 28.3 shows how the rostering began in the ILS (this example is for the first adult). Demographic data such as age, sex, race/ethnicity, and education were collected for each adult listed in the roster. Figure 28.4 shows how the rostered adult was linked to a personal victimization using the adult's name and person number from the roster.

We assumed that the different navigational structures and complexities (such as the need to link adults in the ILS) would have important effects

Questions about You (Adult 1)

You are Adult 1. Please answer questions 31 to 80 for yourself (Adult 1)

Physical Attacks

31. In the last 12 months, has anyone physically attacked you?

☐ Yes → GO TO 32
☐ No → GO TO 41

Figure 28.2 First item in PLS about crimes for Adult 1 in the household.

(YOU) Adult 1

▸ **Starting with you, complete each column for each person age 18 or older living in this household. You will be Adult 1.**

This information you provide will help you with some later questions.

1. What is your first name? For later questions this is Adult number 1.

[ ]

First Name

Figure 28.3 Question wording for rostering first adult in the household for ILS.

6. Who did this happen to? Write in the adult number of the person(s) this happened to from pages 3 and 4. Then write in that person's first name.

Later questions will refer to this person or these person as the "victim."

Adult #	***First Name*** (Refer to Adults listed on pages 3 and 4.)

Figure 28.4 Question wording for ILS linking adult to victimization.

on victimization prevalence, unit response rates, completeness of reporting (especially in households with more than one adult), and item missing rates. In particular, the earlier research in the NCVS on the use of screening items suggested the two survey versions might differ in their effects on estimates of the number of adults who experienced victimizations.

28.3.2 Ordering Content to Enhance Response Rates

Our second research interest was to examine methods to improve response rates by manipulating the questionnaire content. We were especially interested in households that had not experienced any victimization in the last year because they might not view a survey about crime as being relevant to them. The NCVS-CS questionnaire included items about community safety and police interactions that were not specific to victimizations, but were intended to provide context for the analysis of the victimizations. We refer to this set of items as the community and policing questions (CPQs). These items were designed to be relevant to all households. These were new questions developed in conjunction with BJS to target attitudes and perceptions that may correlate with victimization. To address our research question we tested the effect of placing the CPQ items at either the beginning or the end of the survey.

Respondent-friendly design principles (e.g. Dillman et al. 2009 and Tourangeau et al. 2000) suggest that questionnaire designs that engage respondents should increase response rates and lower item-missing data rates. Edwards et al. (2002) reviewed the evidence on the effect of specific design features in mail surveys, but they could not draw conclusions about the placement of general-interest content (like the CPQ items) because only one study had assessed this feature.

We also suspected that the placement of the CPQ items could interact with other features of the questionnaires. Shapiro (1987) found respondents reported more incidents when they were asked a lengthy set of attitudinal items similar to the CPQs before asking about victimization incidents. We thought that effect might arise in the NCVS-CS if the CPQs preceded the victimization questions. Similarly, Williams et al. (2016) theorized that early requests for personally identifying information (such as the respondent's name) may discourage response, but if a set of general-interest items like the CPQ items are asked first, respondents may be reluctant to discard the work they have already done and be more likely to complete the survey. Their study indeed found increases in unit response when content relevant to all households was placed before a request for personal information. However, this study showed that respondents were also more likely to simply skip the sensitive request (increasing item nonresponse). However, the study by Williams et al. (2016) was done in the context of asking for names of children, which can be particularly sensitive, and it is not clear if asking for names of adults would have the same effect.

The CPQ items consist of nine questions on perceptions of the safety of the respondent's community and perceptions of police interactions. The CPQ items were:

- On the whole, how much of the time is the community where you live safe?
- Is there any place within a mile of your home where you would be afraid to walk alone at night?
- How often does fear of crime prevent you from doing things you would like to do?
- When you leave your home, how often do you think about it being broken into or vandalized while you're away?
- In the last 3 years, do you believe your community has become safer, stayed the same, or become less safe?
- Overall, how much of the time is the place where you work safe?
- While living at this address, have you ever contacted the local police department for assistance?
- If so, how satisfied were you with the police response?
- How would you rate the job the local police department is doing in your community?

To examine the effect of the placement of the CPQs, we developed two forms – Form A with the CPQ items in the beginning of the instrument and Form B with the CPQs near the end of the instrument. This was done for both the ILS and PLS versions of the instrument. Thus, four questionnaires were constructed – ILS Form A, ILS Form B, PLS Form A, and PLS Form B.

28.4 Field Test Methods and Design

The survey forms were administered in a field test of the NCVS-CS with the overall goal of testing the feasibility of using a mail survey to collect data on victimization at local levels. For the field test, an address-based sample was selected in each of the 40 largest core-based statistical areas (CBSAs) in the United States. These metropolitan areas were selected to test the potential for the mail survey to produce valid comparisons across jurisdictions or local areas. Since victimization is a rare event, it was important to have a large sample. The initial sample size for the field test was 229 251 addresses. The protocol for the survey mailings followed the mail data collection procedures outlined by Dillman et al. (2009). This protocol included the initial survey mailing, a reminder contact, and two subsequent nonresponse follow-up mailings.

Key objectives of the questionnaire experiments in the field test were to determine the effects of (i) the structure for collecting data on all adults from a

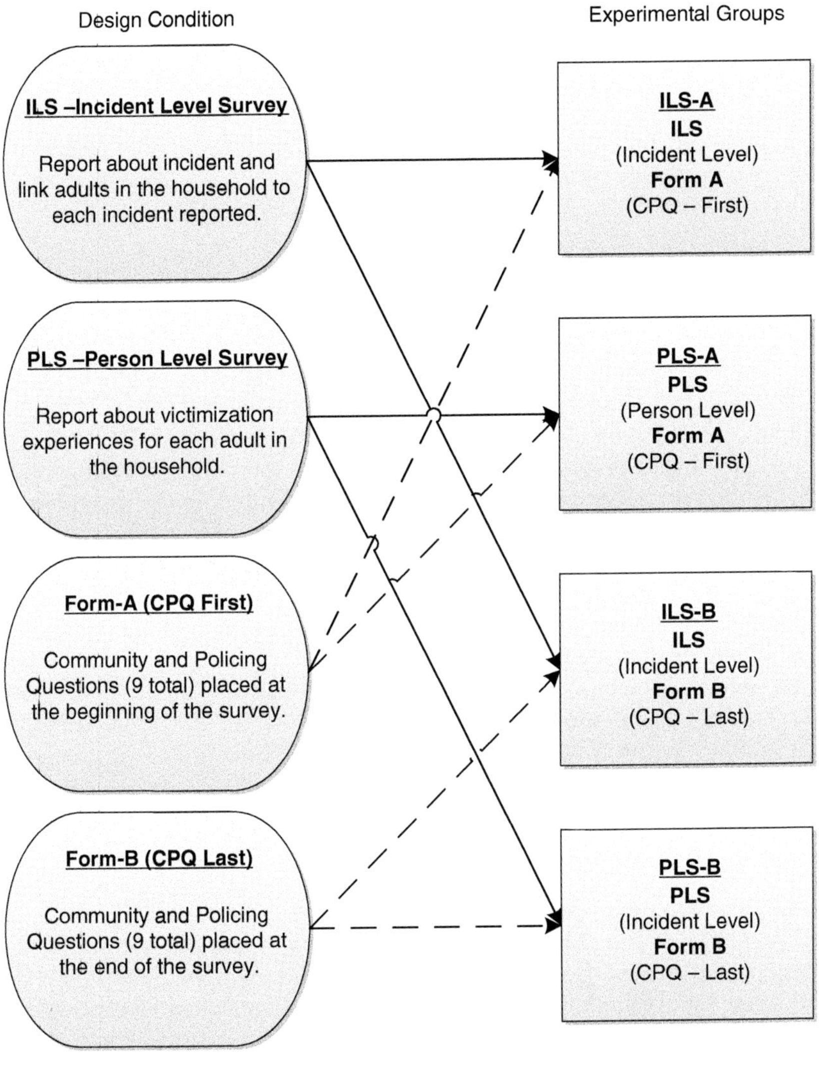

Figure 28.5 Mapping of survey design treatments to experimental groups.

household respondent (ILS or PLS version) and (ii) the placement of CPQ items of general interest (Form A or Form B). As a result, we constructed four questionnaires as shown in Figure 28.5.

To compare the versions and forms efficiently, the field test used a blocked randomized design. The blocks, the units in which randomization was performed, were the CBSAs. Within each block, the four version-by-form treatments (ILS-A, ILS-B, PLS-A, and PLS-B) were randomly assigned. Each CBSA thus had an equal sample size for the four treatments. In the three oversampled cities, the sample was allocated by jurisdiction within the CBSA, and those units were the blocks. This design was chosen because the CBSAs are representative of local areas where the new mail survey approach would be used. The goal was to evaluate how the mail methodology performs in each of the CBSAs rather than in aggregates across the 40 CBSAs.

28.5 Outcome Measures

As discussed, the version (ILS/PLS) and the form (A and B) treatments could affect response rates and reporting. Thus, the same outcome measures were used to evaluate the ILS and PLS and the Form A and Form B results. Specifically, the key outcome measures were: victimization prevalence, unit response rates, completeness of reporting in households with more than one adult, and item-missing rates. The specifics of the outcome measures are described next.

28.5.1 Response Rates

We present findings for response rates (unit nonresponse) for each combination of survey version and form. While differential response rates by the survey version were not expected, we did hypothesize (H_1) that placing the CPQs first in the questionnaire would give higher unit response rates than placing them at the end. This hypothesis was based on the theory that having relevant items of interest to all households early in the instrument would increase response rates.

28.5.2 Item Nonresponse

We examine item nonresponse for two key survey questions: adult household member identification (first name, initials, or nickname) and victimization date. As noted earlier, the name of adult household members is essential in the ILS since it allows adults to be associated with a reported victimization. If the name is not reported, it may not be possible to confirm which adults in the household were victimized, so accurate victimization rates cannot be computed. We could only examine item-nonresponse for adult names for the ILS because name is not collected or needed in the PLS due to the structure of that instrument. We hypothesized (H_2) that ILS-A, with its early placement of CPQs, would decrease unit nonresponse but could increase item nonresponse for adult names. Due to the importance of names in linking to victimizations in the ILS, even a low item nonresponse rate could be problematic.

The victimization date is used to determine eligibility – whether a victimization incident occurred in the reference period. If the victimization date is not provided, then the eligibility of the incident cannot be determined. The PLS collects only the most recent victimization date, while the ILS collects dates for all victimizations over the prior 12 months. Thus, direct comparisons between questionnaire versions are not possible. However, the date for the most recent incident can be compared for both versions. We did not expect differences in item nonresponse for dates between survey versions, but we did expect differences by form within version. Our hypothesis (H_3) was that placing the CPQs first would prompt respondents to report less salient victimizations (an effect identified by Shapiro 1987) because they are harder to recall without the CPQ prompting. This would therefore result in higher item nonresponse for dates – a detail too specific for a low saliency event. In contrast, our hypothesis assumed that without having the CPQs early in the instrument, some respondents would fail to recall less salient victimizations, especially those that happened longer ago, resulting in lower estimates of victimization.

28.5.3 Completeness of Reporting

Completeness of reporting is defined as reporting of a key outcome crime characteristic, touched by crime (TBC), where an adult (household) is TBC if the adult (household) experienced a victimization in the reference period. Since victimization is often underreported, higher reports of TBC are generally assumed to be more accurate (Biderman et al. 1986).

We expected that one adult in the household would report the victimization experiences of all adults in the same household. We hypothesized (H_4) that the structure of the ILS would make this a more cognitively challenging task and lead to lower completeness. The rationale is that in the ILS the respondent must first recall all victimizations within a specific type (e.g. violent victimizations), determine their temporal order to decide which incidents should be reported, and then report the specific household member who was victimized in the incident. In comparison, the PLS requires the respondent to think about each adult separately, report if that adult experienced a specific victimization type, and only report the most recent date an incident happened.

28.5.4 Correlations in Victimization Rates with Core NCVS

If a survey version (either the ILS or the PLS) is highly correlated with victimization rate estimates from the core NCVS, this would help demonstrate external validity. We did not have a prior hypothesis about one version being more highly correlated with the core NCVS than another, but did hypothesize (H_5) that Form A might be more highly correlated than Form B due to previous research which showed that items like the CPQs at the start of the survey tended to increase victimization reporting (Shapiro 1987).

Table 28.1 Number of households by response status and response rates (AAPOR RR1), by questionnaire version (ILS/PLS) and form type (A/B).

Survey version		Overall	Form A CPQ first	Form B CPQ last
ILS	Respondent	46 508	23 683	22 825
	Non-respondent	60 132	29 539	30 593
	Ineligible	8 098	4 147	3 951
	Response rate, %	43.6	44.5	42.7
PLS	Respondent	47 097	23 518	23 579
	Non-respondent	59 447	29 718	29 729
	Ineligible	8 193	4 133	4 060
	Response rate, %	44.2	44.2	44.2

28.6 Findings

28.6.1 Response Rates

The response rate across all versions and form conditions was 47.1% using AAPOR RR3 (AAPOR 2016). This rate speaks to the willingness of households to respond to a mail survey about crime and victimization. Even though this response rate is lower than the face-to-face core NCVS household response rate of 82% (Truman and Morgan 2016), it is much higher than the unit response rate of 11.9% for a similar effort using the telephone (Edwards et al. 2012). For the remainder of our analysis, we use AAPOR RR1 rather than AAPOR RR3. RR1 assumes all nonresponding addresses are eligible, while RR3 allows for some of the non-respondents to be ineligible. The RR1 rate, which is the minimum rate that could be computed, allows for greater consistency when comparing rates over versions and forms. The overall AAPOR RR1 was 43.9%.

Table 28.1 shows the dispositions and unit response rates for each version and form. The response rate for the ILS is nearly equal to that of the PLS (the ILS rate of 43.6% is 0.6 of a percentage point lower than the PLS rate of 44.2%). While the 0.6 percentage point difference is statistically significant ($z = -2.75$, $P = 0.006$), the magnitude of the difference is not substantive. Overall, we found that both versions have approximately equal unit response rates.

Looking at response rates by the placement of the CPQs (forms A and B); the findings are mixed with respect to our hypothesis (H_1) that placing the CPQs early in the instrument would increase unit response rates. For the PLS, the unit response rates are the same for forms A and B (44.2%), which is inconsistent with H_1. However, for the ILS, the response rate is significantly higher for Form A than Form B – nearly 2 percentage points ($z = 5.83$, $P = 0.003$) – which is

Table 28.2 Item nonresponse rates in the ILS for adult's first name by form (A and B) and person number.

	Person 1	Person 2	Person 3	Person 4
Form A	4.5% (23 587)	4.6% (16 655)	4.6% (4 287)	4.1% (1 464)
Form B	3.1% (22 796)	3.3% (16 228)	3.5% (4 341)	4.2% (1 516)
Difference	1.4%*	1.3%*	1.1%*	−0.1%

* p-value < 0.01; sample sizes shown in ().

consistent with H_1. Table 28.1 shows that PLS-A, PLS-B, and ILS-A all have almost the same response rates, while the ILS-B has a low response rate.

We also examined the sample composition for the four treatments. Household characteristics, such as whether the home was owned or rented, and person characteristics such as age and sex, were computed for each combination of version and form. The composition of these characteristics across the treatments was very similar (not shown), and no evidence of nonresponse bias was found.

28.6.2 Item Nonresponse

28.6.2.1 Request for Names (ILS)

We begin by examining item nonresponse rates for the adult name in the ILS across the two forms. Table 28.2 shows that placing CPQ items first (Form A) has higher item nonresponse than Form B for each person (except Person 4). The differences are relatively small, just over 1 percentage point for Persons 1–3, but statistically significant. For Person 4, the rates are almost equal and not statistically different.

These item nonresponse findings, when combined with the earlier unit nonresponse findings, support our hypothesis (H_2) that ILS-A, with its early placement of CPQs, decreases unit nonresponse and increases item nonresponse for adult names compared with ILS-B where the CPQs are placed later in the questionnaire.

28.6.2.2 Victimization Date

Next, we examine item nonresponse for the date of the reported victimization. We begin by giving results for the ILS and PLS separately in Tables 28.3 and 28.4 because of the different data collection structures of the ILS and PLS. Recall, the ILS collects multiple dates for victimizations that occurred over the past 12 months, while the PLS asks about victimization experiences (within the same recall period) for different victimization types – collecting a date only for the most recent event.

Table 28.3 Percent of reported victimizations with missing date for the ILS, by victimization type (violent and property) and form.

	Both forms	Form A (CPQ first)	Form B (CPQ last)
Violent crime			
Number 1	6.0% (1741)	6.2% (965)	5.7% (776)
Number 2	14.2% (247)	12.4% (137)	16.4% (110)
Property crime			
Number 1	5.1% (5747)	5.6%** (3042)	4.4% (2705)
Number 2	10.6% (839)	12.3%** (494)	8.1% (345)
Number 3	19.4% (175)	19.4%* (98)	40.3% (77)

* p-value < 0.01; ** p-value < 0.05; sample size in ().

Table 28.4 Percent of reported victimizations with missing date for the PLS, by victimization type (property and violent categories), form, and person number.

	Both forms	Form A	Form B
Property crimes			
Break-in	9.1% (2125)	9.8% (1250)	8.2% (875)
Theft	4.8% (2472)	4.9% (1320)	4.6% (1170)
Violent crimes			
Attack			
Person 1	12.5% (639)	14.6%† (342)	10.1% (297)
Person 2	15.6% (262)	13.1% (145)	18.8% (117)
Person 3	11.6% (129)	10.7% (84)	13.3% (45)
Threat			
Person 1	6.9% (1598)	8.4%* (906)	5.1% (692)
Person 2	8.4% (525)	10.3% (300)	5.8% (225)
Person 3	7.9% (165)	9.4% (96)	5.8% (69)

* p-value < 0.01; † p-value < 0.10; sample size in ().

Table 28.3 shows the proportion of dates that are missing for the ILS by form. In the ILS, collection of victimization reports occurs in reverse chronological order, i.e. it starts with the most recent. Victimizations beyond the first incident reported should be more distant temporally and more difficult to recall. As expected for both violent and property type victimizations, item nonresponse increases for the older (higher sequential numbered victimization) incidents. For violent crimes the differences are not significant. For the most recent violent crime there is nominally more item-nonresponse when CPQ items are placed first, but for the second most recent violent crime the pattern is reversed. For property crime, which is a more common event, the differences are significant, but again inconsistent. For the two most recent property crimes, item nonresponse is higher when the CPQ items are first, but for the third-most-recent, item-nonresponse is dramatically higher when the CPQ items are last. The number of victimizations of each type drops substantially after the most recent report, and this instability contributes to the inconsistent pattern between form types.

The results in Table 28.3 for the first incident by type of crime lends some support to our hypothesis (H_3) that the early placement of CPQs (consisting of general-interest items) helps provide additional framing that orients the respondent's thoughts to crime and helps them to recall less salient victimizations. However, the inconsistent results for the more temporally distant incidents do not support this hypothesis.

Table 28.4 gives the item nonresponse rates for the victimization dates in the PLS. The PLS collects information for up to four adults, but we only show data for three adults since very few responding households had four adults. Unlike the ILS, only the date of the most recent victimization is collected in the PLS, but this is done by type of incident (e.g. a date is collected for property crimes separately for the most recent break-in and for the most recent theft). The same approach is taken for each type of violent crime. The differences in item nonresponse by form type for the PLS are almost all insignificant, with the only exception being violent victimizations for Person 1; the difference in item nonresponse for threats is significant and for attacks the difference in item nonresponse is marginally significant. The instrument asks the respondent to complete the form as Person 1, and thus they would be the only household member exposed to the survey content (especially the CPQs). The results for Person 1 are in the expected direction supporting our hypothesis (H_3) that the additional framing from early placement of the CPQs increases item nonresponse for victimization date. This finding may be due an increase in the recall of less salient incidents.

We next assess aggregate differences between the ILS and PLS. For the ILS, we aggregate across victimization types including all incidents reported – up to two reports of violent crime, and three reports for property crime as shown in Table 28.3. For the PLS, we aggregate across victimization types including the types listed in Table 28.4 for up to the first three persons reported. In Form A, the item-nonresponse rate for the ILS was 6.9% versus 8.6% for the PLS, and this difference is statistically significant ($p < 0.01$). In Form B, the differences were not significant with 6.0% for the ILS and 6.8% for the PLS. It is important to note that in this comparison, the PLS incidents are all the most recent but the ILS includes incidents beyond just the most recent – where the date is more difficult to recall. These findings suggest that early placement of the CPQs,

Table 28.5 Unweighted percentage of adults who were touched by violent crime, by version (ILS/PLS) and person number.

	Adult 1		Adult 2		Adult 3		Adult 4	
	ILS	PLS	ILS	PLS	ILS	PLS	ILS	PLS
Violent Incident, %	1.2	1.2	0.7	0.8	1.2	1.0	1.0	1.2

providing general-interest questions that orient the respondent's thoughts to crime incidents, increased the recall of less-salient victimizations in the PLS version more than in the ILS version.

28.6.3 Completeness of Reporting

As noted earlier, we refer to differences in the percentage of adults TBC as an indicator of the completeness of reporting. It is also important to note that the victimization rates typically reported in the core NCVS estimates are victimization rates defined as the total number of victimization incidents divided by the population size. The NCVS-CS does not collect the same level of detail as the core NCVS and cannot estimate victimization rates equivalently.

Table 28.5 shows the percentage of adults who were TBC, for violent crime, by version and person number. The percentage who are TBC with a violent incident during the year is very small regardless of the version or person number. The results observed in Table 28.5 suggest that both survey versions perform equivalently when examining exposure to violent victimizations. We had hypothesized (H_4) that the ILS would be more cognitively difficult and that might lead to lower reporting of victimizations, especially for persons other than person 1 (the adult responding for the household). The data do not support this hypothesis.

To confirm the results in Table 28.5, we used logistic regression to model TBC for violent crime (the dependent binary outcome variable, where it is equal to 1 if they experience a violent crime and 0 otherwise) with predictor variables that could affect TBC reporting. The predictor variables were: the person number (roster number in the ILS), gender, age (3 categories – 18–29 years, 30–54 years, and 55 and older), number of adults in household (1, more than one), and questionnaire treatment (version and form ILS-A, ILS-B, PLS-A, and PLS-B).

Our main goal is to assess the effect of the questionnaire treatments on potential differences in the completeness of violent TBC reporting. Interactions of the questionnaire treatment with person number are most relevant in this analysis as it would indicate differences across questionnaire treatment in victimization reporting for other adults in the household. Analysis (not shown) did not find any substantial interactions between questionnaire treatment and person number. Table 28.6 shows the results of the final model. As expected,

Table 28.6 Estimated mixed logistic regression coefficients for predicting violent touched by crime.

Variable	Estimate	*p*
Intercept	−4.361	<0.0001
Treatment (ILS-A)	−0.060	0.1803
Treatment (ILS-B)	−0.116	0.0162
Treatment (PLS-A)	0.133	0.0021
Reference (PLS version B)		
Person number (1)	0.272	<0.0001
Person number (2)	−0.182	0.0008
Person number (3)	0.087	0.1721
Reference (Person 4)		
Gender	0.074	0.0012
Reference (female)		
Age (18–29)	0.647	<0.0001
Age (30–54)	0.160	<0.0001
Reference (55+)		
Number of adults (1 adult)	0.450	<0.0001
Reference (>1 adult)		
Treatment (ILS-A)*Age (18–29)	***−0.147***	***0.0105***
Treatment (ILS-A)*Age (30–54)	0.025	0.6582
Treatment (ILS-B)*Age (18–29)	0.087	0.124
Treatment (ILS-B)*Age (30–54)	−0.054	0.3537
Treatment (PLS-A)*Age (18–29)	0.090	0.0909
Treatment (PLS-A)*Age (30–54)	−0.070	0.1952
Reference (Treat PLS B/Age 55+)		
Treatment (ILS-A)*Number of adults (1)	0.033	0.4645
Treatment (ILS-B)*Number of adults (1)	***−0.101***	***0.0345***
Treatment (PLS-A)*Number of adults (1)	−0.011	0.7927
Reference (treat PLS-B/number of adults (>1))		

several main effects are closely associated with experiencing victimizations, but these are not relevant in our analysis of the treatments effects. All of the interactions with the treatments are small, and most are not statistically significant. These findings are consistent with those in Table 28.5 that there are no substantial differences in reporting by questionnaire treatment.

Table 28.7 Definition of TBC estimates from the NCVS-CS.

Variable	Description
Household level	
TBC-Property1	Households touched by property crime, excludes attempts
TBC-Property2	Households touched by property crime, includes attempts
TBC-Vehicle theft	Households touched by motor vehicle theft
TBC-H violent1	Households touched by violent crime, excluding threats
TBC-H violent2	Households touched by violent crime, including threats
Person level	
TBC-P violent1	Persons touched by violent crime, excluding threats
TBC-P violent2	Persons touched by violent crime, including threats
TBC-P serious violent	Persons touched by serious violent crime

28.6.3.1 Correlations of NCVS-CS and Core NCVS TBC Estimates

A key reason for testing the questionnaire versions and forms was to identify if any of the approaches would produce "better" estimates of victimization at the local level (including incidents not reported to police) to track trends over time. A goal of the low-cost alternative NCVS-CS is to be able to produce estimates that are correlated to the core NCVS estimates. This result would lend support to the idea that the NCVS-CS items measure the same constructs as the core NCVS. We examine estimates of TBC from the NCVS-CS and from the core NCVS for each of the CBSAs by version and form for this purpose.

Table 28.7 defines the TBC estimates from the NCVS-CS used in this analysis. Both household and person-level estimates of violent crime are computed, where the household measure shows if any adult in the household reported a violent crime. For these estimates, the NCVS-CS estimates were fully weighted and adjusted to population totals for the individual CBSAs.

The core NCVS estimates were computed at the Census Bureau using the weights on the core public use file. Because the sample size for the core NCVS was small at the CBSA level, we combined data for three years (2013–2015) to produce the core NCVS estimates. This added stability to the core NCVS estimates. The TBC definitions used for the core NCVS were similar to those used for the NCVS-CS.

The differences in the data collection schemes for the core NCVS and the NCVS-CS discussed earlier are very important for this analysis. The core NCVS asks respondents about victimizations in the last six months but the NCVS-CS asks about the last 12 months. Two successive interviews would

Table 28.8 Pearson correlation coefficients for NCVS and NCVS-CS CBSA-level summary statistics.

NCVS-core	NCVS-CS	ILS-both	ILS A	ILS B	PLS-both	PLS A	PLS B
TBC-Property	TBC-Property1	0.64***	0.67***	0.52***	0.65***	0.67***	0.56***
TBC-Property	TBC-Property2	0.67***	0.68***	0.56***	0.62***	0.60***	0.58***
TBC-Vehicle theft	TBC-Vehicle theft	0.34*	0.34*	0.18	0.59***	0.71***	0.26
TBC-H violent	TBC-H violent1	0.54***	0.40*	0.44**	0.47**	0.33*	0.24
TBC-H violent	TBC-H violent2	0.54***	0.43**	0.45**	0.45**	0.26	0.41**
TBC-P violent	TBC-P violent1	0.45**	0.14	0.48**	0.50***	0.39*	0.29
TBC-P violent	TBC-P violent2	0.46**	0.17	0.48**	0.49**	0.31*	0.47**
TBC-P serious violent	TBC-P serious violent	0.47**	0.14	0.50***	0.51***	0.44**	0.30

* p-value < 0.05; ** p-value < 0.01;*** p-value < 0.001.

need to be linked to calculate statistics about percentage TBC in the last 12 months from the core NCVS; however, the rotating panel design and the fact that NCVS does not follow households that move means that there would be substantial missing data for these types of estimates from the core. Instead, TBC estimates for the core NCVS were calculated using victimizations reported for the 6-month interview recall period, and we assume that the percentage TBC in the last 6 months would be highly correlated with the percentage TBC in the last 12 months.

Table 28.8 shows the Pearson correlation coefficients estimated from the core NCVS TBC estimates[2] for each CBSA and the estimates calculated for ILS (both forms), ILS-A, ILS-B, PLS (both forms), PLS-A, and PLS-B.

The correlations between the core NCVS and the NCVS-CS estimates are positive and highly statistically significant for both property and violent crime. This finding demonstrates that the NCVS-CS measures the same victimization constructs as the NCVS. With respect to our hypothesis (H_5) that Form A might be more correlated than Form B, there is limited support. For both the ILS and the PLS, the correlations with core NCVS estimates for property crime are higher for Form A than for Form B. The correlations for violent crime at the person level are higher for ILS Form B than for ILS Form A. The correlations for motor vehicle theft are somewhat erratic, probably due to the very low rates of this type of crime – in several CBSAs the NCVS-CS estimates were zero.

2 Since the core NCVS is collected by the Census Bureau for BJS, restrictions exist on the release of estimates computed from these data. The Disclosure Review Board at the Census Bureau reviewed and approved the release of the correlation estimates presented here.

Table 28.9 Summary results for hypotheses.

	Hypothesis	Result
H_1	CPQ first = higher unit response across survey versions	Supported for ILS version, but no support for PLS version. Content a mediating factor.
H_2	CPQ first = higher item nonresponse for adult names (ILS only)	Supported confirming previous research.
H_3	CPQ first = higher item nonresponse for incident date; while CPQ last = lower TBC estimates	Partially supported; location within the questionnaire a mediating factor.
H_4	ILS would have lower correlations with core NCVS compared to PLS version due to difficult structure of ILS	No support for the hypothesis.
H_5	CPQ first would have lower correlations with core NCVS across both survey versions	Partially supported; location within the questionnaire a mediating factor.

28.7 Summary

We began with five hypotheses about the effects the two survey versions and two form types would have on unit response, item nonresponse, and correlations with the core NCVS. Table 28.9 provides a high-level summary of the results we observe for each. Overall, we observed some level of support for all, with the exception of one: H_4. The null finding speaks to the robustness of the two different approaches. However, observations for our other hypothesis suggest there are a number of trade-offs to consider between these designs.

28.8 Discussion

Overall, the findings for the two questionnaire versions and form treatments were encouraging in terms of overall response and ability to estimate TBC. Neither version was clearly better than the other, and our findings suggest researchers should consider their research objectives to determine the optimal questionnaire design. We next discuss our findings and suggest areas for further research.

28.8.1 Version Treatment

The results of this study showed relatively small differences by survey version, and the evidence suggests both versions are capturing victimization experiences adequately, including incidents not reported to the police. Given this result, conclusions on a preferable approach should take into account factors other than estimates of victimization alone.

We expected the major questionnaire design issue with the ILS would be the difficulty respondents would have linking the adults to victimization incidents (H_4). Despite our concerns, this did not have a large effect on estimates of victimization. A possible explanation for this result is that the task was not as difficult for respondents as we hypothesized. Victimization is generally rare, but can be a very salient event, particularly when it involves family members. The extent to which recall is affected by personal relationships is an area we could not explore. However, we would expect this to be an important area for further research.

While the task of linking adults to victimizations in the ILS did not affect prevalence estimates, this requirement does add burden, requiring the respondent to pay more attention to the response task. Additionally, the ILS requires respondents to directly report the name of adults who have been victimized, which can be perceived as intrusive. These are both disadvantages for the ILS.

If the researcher's goal is to simply estimate exposure to incidents (in this case victimization incidents, but these results could also extend to other measures, such as periods of unemployment, or chronic health conditions), then the PLS may be the preferred approach for the reasons just noted. However, the ILS has the advantage of giving more detail on the properties of victimization incidents and may therefore be a better option when a greater level of detail is important to meet research objectives.

28.8.2 Form Treatment

The comparison of form treatments did show important differences with respect to the placement of the CPQs and the way these items impacted the results was often influenced by the different structure of the questionnaire (either ILS or PLS).

We hypothesized that placing the CPQ items early in the questionnaire would have a positive effect on unit response (H_1), but this result was only true for the ILS. However, this finding is generally consistent with the findings of Williams et al. (2016). They found that having supplemental items (similar in function to our CPQ items) to engage the respondent at the start of the questionnaire resulted in higher unit response rates and concluded that the initial questions improved the perceived relevance of the self-administered questionnaire.

For the ILS-B questionnaire, the CPQs appear at the end of the questionnaire and the first items in the survey are not about victimization. Rather, the first items ask for personal information about each adult member of the household. This content may not match respondent expectations of a survey about victimization, resulting in lower response rates. In contrast, having the CPQs at the beginning of the instrument in the ILS-A matches respondent expectations that the survey contains questions about victimization, increasing response rates.

While we did not see the same result for the PLS, this may be because the PLS-B does not require a roster (like the ILS) and instead starts with household-level questions about victimization. In other words, both the PLS-A and the PLS-B begin with items directly relevant to a crime victimization survey, and this may explain the lack of a difference in the PLS unit response rates by form.

We also hypothesized (H_2 *and* H_3), that the difference in forms would lead to differences in item nonresponse rates for the names of adults (in the ILS) and for incident dates (in both the ILS and PLS). As expected, item nonresponse was generally high for the request of names in the ILS-A (H_2), despite the higher unit response rates. This finding is also consistent with Williams et al. (2016). Although the respondent, for the ILS-A, may feel that a question asking for personally identifying information (i.e. names) is intrusive, they are more likely to continue with the survey when the request appears later in the instrument and skip the name request. Further research to identify strategies that would encourage respondents to supply names instead of skipping the item, may help to further improve the performance of the ILS-A.

However, our hypothesis that Form A would also lead to higher item nonresponse for incident dates (H_3) was only partially confirmed, and our results revealed a complex pattern between forms, which does not lend itself to a simple explanation. For the ILS-A there was no difference in item nonresponse for the dates of violent crimes, but there were significant increases for the most recent property crimes. The pattern was reversed for the PLS-A where item nonresponse for violent crimes was significantly higher for the adult completing the questionnaire, but not different for property (household level) crimes.

A similar complex pattern was evident when comparing the differences in correlations between the NCVS-CS and core NCVS TBC estimates (H_5). For the ILS, Form A had higher correlations for property crime, but lower correlations for violent crime. For the PLS, the correlation for violent crime was higher for Form A (when incidents only involving threats are excluded) and property crime was nearly the same for both forms.

To some extent, these findings may be explained by the inability to link adults when respondents skipped the name item, thereby reducing the number of eligible violent crimes compared to property crimes, which do not require the same linkage. However, the order of the questionnaire content clearly has some influence. In Form A, the CPQ items come first in the questionnaire and victimization items come later. However, as described earlier, violent crime precedes property crime in the ILS, while in the PLS property crime precedes violent crime for each adult. Results show that both the item nonresponse for dates and the correlations with the core NCVS are higher for items that appear later in the questionnaire. It is unclear why the CPQ placement differentially affected results depending on items' position within the questionnaire. To better understand this, additional research would be useful.

We hypothesized that the early placement of the CPQs would help orient the respondent and provide a contextual framing that would help them to recall victimization incidents. However, we did not observe differences in TBC rates. This may be because respondents are stimulated to recall less-salient incidents that are too difficult to place temporally. Without a date, the eligibility of the incident cannot be verified and the incident is therefore excluded from the data.

Although the early placement of the CPQs did not improve recall of this key component of victimization (the incident date), some imprecision in the recall is acceptable for the NCVS-CS as long as there is confidence that the incident occurred within the reference period. Results suggest that respondents may be able to recall an incident, or the additional framing provided by the CPQs may help them to remember that an incident occurred during the reference period, but that specifying a specific month within that period may be too difficult. Offering categorical time spans (e.g. one to three months ago) may help to signal to respondents that some imprecision in recall is acceptable. Respondents also use external events or time periods to anchor incidents. For example, "it happened in the summer," or "it happened after I changed jobs." Designing items to prompt these kind of recall strategies may help respondents to date events and could increase the number of eligible incidents recalled. In combination with the use of CPQs early in the questionnaire, such techniques could lead to improvements in reports of crime victimization. This is an area for future research.

28.9 Conclusion

We first conclude with a few comments about the overall feasibility of the NCVS-CS as a self-administered household survey because it is a major departure from the approach used in the core NCVS. The field test results discussed indicate that the NCVS-CS is viable as a low-cost alternative approach to collecting data on victimization experiences of households; it appears especially valuable for making local area estimates. The unit response rates for this mail survey were close to 50%, which is considerably higher than what most other low-cost alternatives can achieve. The NCVS-CS also had relatively low item nonresponse with respect to key data items such as date of victimization. Another important finding is that the key estimates of being TBC from the NCVS-CS were highly correlated to the corresponding estimates from the core NCVS, showing the NCVS-CS estimates provide a reasonable measure of victimization. The correlations were high for both property crime and personal or violent crime. Overall, the findings of the field test were very encouraging.

Turning to a summary of our discussion of the two questionnaire versions, we conclude that both the ILS and PLS performed well for estimating TBC. However, when comparing our results between form types, we found that early placement of topically relevant questions (the CPQs) was only effective in a specific context. Early placement of these items improved victimization reports for the ILS specifically, which would otherwise have begun with items that were not directly related to the topic of the survey (a roster of household members). For questionnaires that are structured in this way, it may improve response to include items of general interest at the start of the questionnaire. However, when the initial content of the questionnaire is already relevant to the survey topic, inserting general-interest items is unlikely to increase unit response.

Overall, the findings described in this chapter show that it is viable to collect crime victimization data via a self-administered household questionnaire. Although the increased item nonresponse for adult names in the ILS, and dates in both forms, are problematic, strategies that improve the respondent's willingness to supply a name and their ability to date incidents (such as those described earlier) should improve data quality and increase the correlations between NCVS and the new self-administered NCVS-CS.

References

Biderman, A.D., Cantor, D., Lynch, J.P., and Martin, E. (1986). *Final Report of Research and Development for the Redesign of the National Crime Survey*. Washington, DC: Bureau of Social Science Research.

Brick, J.M., Andrews, W.R., and Mathiowetz, N.A. (2016). Single-phase mail survey design for rare population subgroups. *Field Methods* 28: 381–395.

Dillman, D.A. (1978). *Mail and Telephone Surveys: The Total Design Method*. New York: Wiley.

Dillman, D.A., Smyth, J.D., and Christian, L.M. (2009). *Internet, Mail, and Mixed-Mode Surveys. The Tailored Design Method*, 3e. Hoboken, NJ, USA: Wiley.

Edwards, P., Roberts, I., Clarke, M. et al. (2002). Increasing response rates to postal questionnaires: systematic review. *BMJ* 324 (7347): 1183.

Edwards, W.S., Brick, J.M., and Lohr, S. L. (2012). Designing a low(er)-cost companion to the National Crime Victimization Survey. Federal Committee on Statistical Methodology conference.

Groves, R.M. and Cork, D.L. (eds.) (2008). *Surveying Victims: Options for Conducting the National Crime Victimization Survey*. National Academies Press.

Iannacchione, V.G. (2011). The changing role of address-based sampling in survey research. *Public Opinion Quarterly* 75 (3): 556–575.

Lavrakas, P.J., Stasny, E.A., and Harpuder, B. (2000). A further investigation of the last birthday respondent selection method and within-unit coverage error. In: *Proceedings the American Statistical Association, Survey Research Methods Section*, 890–895.

Lehnen, R.G. and Skogan, W.G. (eds.) (1981). *The National Crime Survey: Working Papers. Volume I: Current and Historical Perspectives*. U.S. Department of Justice, Bureau of Justice Statistics https://www.ncjrs.gov/pdffiles1/nij/75374.pdf.

Lepkowski, J.M., Tucker, N.C., Brick, J.M. et al. (2008). *Advances in Telephone Survey Methodology*. New York: Wiley.

Link, M.W., Battaglia, M.P., Frankel, M.R. et al. (2008). A comparison of address-based sampling (ABS) versus random-digit dialing (RDD) for general population surveys. *Public Opinion Quarterly* 72 (1): 6–27.

Messer, B.L. and Dillman, D.A. (2011). Surveying the general public over the internet using address-based sampling and mail contact procedures. *Public Opinion Quarterly* 75 (3): 429–457.

Montaquila, J.M., Brick, J.M., Williams, D. et al. (2013). A study of two-phase mail survey data collection methods. *Journal of Survey Statistics and Methodology* 1 (1): 66–87.

Olson, K., Stange, M., and Smyth, J. (2014). Assessing within-household selection methods in household mail surveys. *Public Opinion Quarterly* 78 (3): 656–678. nfu022.

Rand, M.R. (2007). The National Crime Victimization Survey at 34: looking back and looking ahead. *Crime Prevention Studies* 22: 145.

Shapiro, G.M. (1987). Interviewer-respondent bias resulting from adding supplemental questions. *Journal of Official Statistics* 3 (2): 155–168.

The American Association for Public Opinion Research (2016). *Standard Definitions: Final Dispositions of Case Codes and Outcome Rates for Surveys*, 9e. AAPOR.

Tourangeau, R., Rips, L.J., and Rasinski, K. (2000). *The Psychology of Survey Response*. Cambridge University Press.

Truman, J.L. and Morgan, R.E. (2016). Criminal victimization, 2015. *Criminal Victimization Series*. NCJ 250180. Bureau of Justice Statistics. http://www.bjs.gov/index.cfm?iid=5804&ty=pbdetail.

US Census Bureau. (2014). American Community Survey design and methodology. https://www.census.gov/programs-surveys/acs/methodology/design-and-methodology.html.

Williams, D., Michael Brick, J., Montaquila, J., and Han, D. (2016). Effects of screening questionnaires on response in a two-phase postal survey. *International Journal of Social Research Methodology* 19 (1): 51–67.

29

Planning Your Multimethod Questionnaire Testing Bento Box: Complementary Methods for a Well-Balanced Test

Jaki S. McCarthy

National Agricultural Statistics Service, US Department of Agriculture, Washington, DC, USA

29.1 Introduction

Increasingly, survey practitioners are using multiple methods for testing and evaluating survey questionnaires, combining results to inform the design (OMB 2016; Pascale 2016; Willimack 2013, pp. 253–302; Yan et al. 2012; Madans et al. 2011; Campanelli 2008, pp. 176–200; Kaplowitz et al. 2004; Presser et al. 2004). Similar to multimethod approaches to research (Creswell 2014), multiple testing methods are used to complement each other and expand on the information obtained in each to provide a more robust questionnaire evaluation (Yan et al. 2012; Demaio and Landreth 2004; Willis et al. 1999).

Qualitative methods, such as focus groups and cognitive interview testing, are often combined with quantitative methods in later field tests (which may include split-ballot experiments), each providing information to inform, develop, and refine survey questionnaires (Pascale 2016; Murphy et al. 2015; Persson et al. 2015; Campanelli 2008). Schaeffer and Dykema (2004, pp. 475–502) for example, included focus groups and cognitive interviews along with quantitative behavior coding of survey interviews as part of their multimethod approach. Others have combined early expert review by survey methodologists with questionnaire design expertise with traditional cognitive interviews and other methods (Murphy et al. 2015).

Although much of the focus has been on using multiple methods to test household survey questionnaires, multimethod approaches have to a lesser degree been applied to establishment surveys (Tuttle et al. 2010; McCarthy and Buysse 2010; Bavdaz 2009; Giesen and Hak 2005; Phipps et al. 1995). For testing of establishment survey questionnaires, the involvement of subject matter experts, both in questionnaire development and revision, may be more common due to the often technical terminology or concepts included. Tuttle

Advances in Questionnaire Design, Development, Evaluation and Testing, First Edition.
Edited by Paul C. Beatty, Debbie Collins, Lyn Kaye, Jose-Luis Padilla, Gordon B. Willis, and Amanda Wilmot.

et al. (2010) used multiple methods to test questions about foreign direct investment used by the US Bureau of Economic Analysis. Survey methodologists initially interviewed subject matter experts to supplement their expertise before conducting cognitive interviews. Similarly, Giesen and Hak (2005) explicitly included a "stakeholder check" to gather input from subject matter experts in their multimethod testing for the Dutch Annual Business Inquiry. This was conducted between the survey methodologists' expert review of the questionnaire and subsequent testing, again due to the complexity of the concepts involved.

The US Department of Agriculture's National Agricultural Statistics Service (NASS) conducts the US Census of Agriculture (COA) every five years. Due to the complexity and scope of the COA, a multimethod approach was also adopted in testing the COA questionnaire. More specific detail on testing for the COA and results have been reported elsewhere (McCarthy and Buysse 2010; McCarthy et al. 2018; McCarthy 2016; Moore et al. 2016; Ott et al. 2016) with the idea that the Japanese bento box was an apt metaphor for this approach. The objective of this chapter is go further to describe the strengths and weaknesses of common testing methods and, more importantly, how and why others can combine them to create a better questionnaire evaluation than any individual test. We think this approach of building a "bento box" can be leveraged by other survey methodologists to improve questionnaire testing efforts. Survey organizations can make the most of combining multiple tests when they plan for this from the start.

A traditional Japanese bento box is a meal constructed according to the principles of balance – different elements of colors, flavors, and cooking methods provide an optimum eating experience (Sekiguchi 2009). While any ingredients may be included, if multiple ingredients and complementary elements are used, the resulting meal will be nutritious and well balanced. Like a bento box, the specific elements of a multimethod questionnaire test can be customized to each situation. The combination will provide information in different forms that can be tailored to the "needs" of the particular survey. This chapter presents a discussion of common testing methods with several examples from the 2017 COA to illustrate how the bento box approach can be applied.

NASS conducts the COA every five years, mailing a 24-page questionnaire to approximately three million known or potential agricultural establishments. Agricultural operations are defined as those which raise and sell (or normally would sell) $1000 or more of agricultural products in the reference year. Information about crop and livestock production and inventory, production practices, farm economics, and operator demographics are collected and published for all counties in the United States. Respondents can complete the self-administered paper or online questionnaire. Initial non-respondents receive additional questionnaire mailings before being contacted by telephone. Limited in-person nonresponse follow up is also used for operations based on their size, location, or other characteristics. Data quality is extremely important due to the wide-ranging and extensive uses of the resulting data. A considerable amount of resources are devoted to data collection and processing; any improvements that can be made to the questionnaire also have the potential to save resources. Therefore, questionnaire testing is an integral part of the COA cycle. Prior to each COA, NASS forms a team of survey methodologists with questionnaire design expertise, subject matter experts, and survey operations staff to review, revise, and oversee the entire testing process of the COA questionnaires. The methods in our COA testing bento box and discussed in this chapter are:

1. Evaluation of historical data
2. Expert review
3. Cognitive interviews
4. Field testing
5. Follow-up interviews

Each method has its own strengths and weaknesses and complements the others by providing different kinds of information. The methods can be used sequentially, or in parallel, with results used together to revise questionnaires. Each of our testing methods is briefly described, with special consideration given for establishment survey testing. Examples of their strength in combination within the context of the COA are provided.

29.2 A Questionnaire Testing Bento Box

29.2.1 Evaluation of Historical Data

Like rice in the bento box, analysis of existing survey data can provide a good base to any multimethod test of an ongoing survey program. A review of existing survey data is a useful starting point in identifying problems likely to be found in later survey cycles. Since the COA is conducted every five years, previous data are readily available. Items with high imputation or edit rates are potential candidates for improvement and help to narrow the focus of subsequent testing by highlighting in advance areas of potential concern. The number of edits, in both absolute and relative terms, is important. For establishment surveys, which often include distinct subpopulations or widely varied establishments, low-frequency items with high error rates may also merit special consideration. To address this, NASS ranks all COA items both by the number and rate of edits and imputations for each section of the questionnaire. This information is then used to identify particularly problematic questions within each section for further team review. To supplement the historical data analysis, NASS also reviews the number of respondent phone calls for assistance completing various sections of the forms in the previous COA. The areas where respondents requested help most often are tallied and also considered to help prioritize questionnaire items for testing. In addition, paradata from previous survey cycles, i.e. information about survey interviews, such as interview lengths, item response times, items that respondents change in online forms, etc. may also be useful in questionnaire evaluation (Kreuter 2013), although that is not commonly evaluated in NASS.

The key strengths of evaluation of historical data are that it provides objective quantitative evidence of potential problems, can fully represent the population of interest including rare items and small subpopulations, and reflects the use of operational data collection procedures. For an establishment survey like the COA, many diverse types of farming and ranching operations are represented, and it is near impossible to include all less-common types of operations (for example, mushrooms, lentils, ducks, or mink) in smaller-scale qualitative evaluations. Thus problems in small subpopulations that may be difficult to find in small-scale testing may be uncovered in evaluations of historical survey data. However, while reviews of items like historical edit and imputation rates can identify problems, they give little direct information about the reason for those problems. Therefore, this type of quantitative evaluation is best supplemented with qualitative methods, such as cognitive interviews or expert reviews, which can provide greater insight into the reasons for observed errors.

29.2.2 Expert Review

Another testing method included in our bento box is expert review. Like wasabi in the bento box, a little can go a long way. Expert review typically involves collecting qualitative feedback from individuals or small groups of experts about the functioning of the questionnaire and potential problems. NASS uses several types of experts in this role.

29.2.2.1 Questionnaire Design Experts

Questionnaire design experts are often asked to critique questionnaires and may be the most commonly consulted type of experts (Willis et al. 1999; Presser and Blair 1994). Although formal questionnaire appraisal systems (QASs) have been developed by survey methodologists to guide the expert review process, both manual, e.g. the Questionnaire Appraisal System (Lessler and Forsyth 1996; Willis and Lessler 1999), and automated, e.g. the Question Understanding Aid (QUAID) system (Graesser et al. 2006), the questionnaire design expert review is typically a qualitative review based on the individual expert's experience and training in questionnaire design. This type of evaluation can provide valuable information at little expense. NASS does not use questionnaire design experts in one-time stand-alone reviews, but rather they are part of the questionnaire evaluation and testing team providing suggestions and recommendations throughout the questionnaire development and testing process. Questionnaire design experts typically take a general view of the questionnaire and potential problems based on principles such as question structure, problems with similar questions from other surveys, and the impact of format and layout of questions. They may also use knowledge of the performance of similar existing questions. This can lead to them identifying more (and different) potential problems than other techniques such as cognitive interviews (Demaio and Landreth 2004; Presser and Blair 1994).

29.2.2.2 Subject Matter Experts

Subject matter experts may also provide valuable input into the questionnaire evaluation process with specialized background knowledge relevant to the questionnaire. This may be particularly important in establishment surveys where knowledge of specific technical topic areas can be critical for questionnaire development (Willimack 2013; Giesen and Hak 2005; Ramirez 2002). Subject matter experts may have better insight as to whether terminology is understood by respondents and captures intended concepts, or whether information is likely to match records used by respondents to report. This allows them to provide insightful critiques of proposed questions. For example, NASS questionnaire design experts may have limited knowledge of the appropriate units for agricultural production for specific crops (i.e. bales, pounds, hundredweight, tons, etc.), typical land leasing arrangements, or terminology used for government program participation. Subject matter experts provide this type of information. During questionnaire evaluation, NASS relies on internal agency staff to contribute subject matter expertise to the expert review of survey questionnaires. These staff draw on their knowledge of the topic area and nuances of questions that may impact statistics or potential inconsistencies with other agency products. In addition, internal subject matter experts at NASS have experience with earlier data collections and existing areas of concern for data quality, which can inform their review.

Expert opinion from outside the organization can also be useful, bringing a perspective more likely to suggest fundamental changes to survey concepts. External experts are often data users, so they have a unique perspective and may understand the shortcomings of existing data. This may include data that do not appear to capture the intended concepts (for example, data inconsistent with other known information) or needed data that are not available given existing questionnaires. At NASS, concerns were raised by data users (both USDA staff involved with farm programs and non-government industry associations) that the number of women, beginning, and younger operators shown in previous COAs understated their role. In response to this, NASS commissioned a panel of external subject matter and questionnaire design experts to address these concerns (Ridolfo et al. 2016). The panel met several times and provided recommendations for 2017, such as adding questions and making revisions to the terminology used. For example, they recommended that NASS collect information on those involved in any decisions for the operation, rather than just those making day-to-day decisions, as well as information about specific kinds of decisions. They also recommended collecting detailed information for up to four individuals rather than three as in the past, so couples operating jointly could be identified. These types of changes were unlikely to be initiated from within the survey organization more focused on continuing existing data series.

Because external experts may suggest more fundamental changes to concepts, their input is best used early in survey development to allow sufficient time to incorporate and test their recommendations. In addition, external experts may have little knowledge of operational constraints associated with proposed changes. For this reason, their input is best paired with internal survey staff input and can be tested in subsequent field tests.

29.2.2.3 Survey Operations Experts

Another often-overlooked resource are experts in survey operations. In particular, internal staff involved in prior data collections may be able to provide unique insights into the questionnaire performance from their experience with data collection and data processing. For example, survey field directors can identify questions that were particularly problematic for interviewer administration, or required additional interviewer training. Those involved in processing may identify high-edit-rate items and also provide insight into how data were corrected and possible ways to reduce editing in the future.

After all major data collections, NASS consults field office staff as a key group of internal experts, formally soliciting feedback on all aspects of the survey operations, materials, and procedures. Field staff are asked to provide comments and suggestions for improvements for future data collections. Comments were solicited after the 2012 COA and formed part of the initial review in preparation for the 2017 COA.

A noteworthy limitation in each group of experts is typically narrow consideration of information only in their area of expertise. Indeed, while useful, Olson (2010) found that reviews by questionnaire design experts tend to provide idiosyncratic and inconsistent results. Subject matter experts are necessarily limited to the domain of their expertise while operations experts will focus on survey production issues. Thus at NASS we recommend that expert review should not be used in isolation when resources for additional evaluation are available. Like wasabi, it may be powerful, but does not make for a very good meal on its own. However, experts can provide quick and inexpensive insight into potential reasons for errors, suggest items to examine more closely, and provide input for new or modified questions that can be tested in cognitive interviews or field tests.

29.2.3 Cognitive Interviews

Cognitive interviews are staples of survey questionnaire development (Willis 2015; Beatty and Willis 2007; Willis 2004). Like the ubiquitous California roll in a bento box, cognitive interviews are found in many questionnaire tests. The objective of cognitive interviewing is to understand how respondents interpret and answer survey questions and to identify the source of problems to inform the questionnaire design. Respondents are asked to describe their reporting processes either during or after completing questionnaires. Additional follow-up probes are also used to understand respondents' question interpretation and reporting process. In establishment surveys, information such as who the appropriate respondent is, respondents' use of records to report, any establishment restrictions on releasing data, the interpretation of technical terms, and whether establishment records match survey questions can also be examined in cognitive interviews (Willimack and Nichols 2010). For establishment surveys, cognitive interviews may be more difficult to conduct, as recruiting is often more challenging. Some of the establishment interviewee recruiting challenges include identifying and contacting appropriate respondents within an establishment and gaining cooperation from respondents who must take time from business activities and may have concerns about releasing proprietary business data. In addition, interviewers often have to travel to establishment sites where respondents and their records are located rather than interviewing centrally in a cognitive lab.

Cognitive interviews typically use small convenience samples compared to field tests and production data collection, so it may be difficult to generalize to the broader population. However, cognitive interviews can provide rich insight into the question response process beyond that which can be gleaned from quantitative data alone. Cognitive interviews are most productive as a testing method when they are conducted in iterative rounds following a test-modification-retest approach (Willis 2016; Collins 2015). Since cognitive interviews typically result in specific changes to questionnaires, an iterative approach allows recommended changes from initial rounds to be tested to verify that they have improved the questionnaire.

Several iterative rounds of cognitive interviews were conducted prior to the 2017 COA. An initial round of eight interviews was conducted, specifically focused on questions and changes prompted by the NASS commissioned expert panel on women and beginning farmers discussed in Section 29.2.2.2. Following this, 37 additional cognitive interviews were completed, testing subsets of the form and targeting specific types of commodities. Recommendations from these interviews were incorporated into the forms used in the field test described later in this chapter.

Due to delays in scheduling the cognitive interviews, revised forms were tested in another round of 71 cognitive interviews in parallel (rather than prior) to the field test. Subsets of these interviews focused on types of operations that had not been included in the initial interviews, alternative formats to some of the sections, and a short form. A final round of 20 interviews was conducted to evaluate alternative formats for preprinting commodity names and codes within the questionnaire and instruction booklet. Recommendations from each round of cognitive interviews were used to revise the forms for subsequent cognitive testing and inclusion in the final 2017 questionnaires.

In each round, different types of operations were targeted and selected in multiple states to include a diversity of operation types and sizes and to account for regional variation in production of commodities. Because of the lengthy nature of the COA questionnaire, NASS conducted cognitive interviews targeting subsets of the form. Results from the earlier review of historical data and expert reviews were used to identify sections of the questionnaire to focus on and issues to probe respondents on. In addition, NASS focused on new items being added to the questionnaire.

One of the limitations of a using a qualitative method such as cognitive interviewing is that it is not possible to quantify the extent of problems or their impact on survey estimates. This makes it a natural partner to more quantitative assessments such as the historical data review or larger-scale field tests. In addition, it may be difficult to recruit respondents with rare items or unique characteristics. For example, in testing questions about land, we would want to include respondents who own land, who rent land to others, and who rent land from others. Ideally we would also want to include those renting land for cash, for shares of the crop, on an animal unit month basis, on an exchange basis, for free, or under other types of rental arrangements. Including all of these types of respondents is difficult if not impossible in small-scale testing.

One important characteristic that often impacts the way in which questions are answered in the COA is establishment size. Large corporate establishments may answer quite differently than small operations selling from a farm stand. Including a variety of establishment sizes in the cognitive testing is therefore also an important consideration when testing a questionnaire like that used in the COA. Other considerations for the COA were the type of commodities produced. Operations may answer questions quite differently depending on whether they are growing corn, tomatoes, peaches, shellfish, cows, emus, or cut flowers. Likewise, farms may also differ based on their geographic location, and farms in remote rural locations may be difficult to recruit for in-person interviews.

When interpreting data from cognitive interviews, it is important to bear in mind that cognitive interview samples are often unrepresentative convenience samples and often do not report in conditions replicating the survey itself. COA respondents provide their information via a self-administered form completed at the respondent's convenience. In contrast, our cognitive interviews feature a trained interviewer observing respondents, which may influence the amount of effort expended by the respondent or their normal reporting process. Thus, we consider our cognitive interview results alongside information from methods more closely resembling survey conditions.

29.2.4 Field Testing

Quantitative field tests are carried out as a matter of course for the COA using revised forms based on the expert reviews and initial cognitive testing. The field test is often the largest compartment of the testing bento box – the entrée requiring the most planning, assembly, and resources. A field test typically refers to a larger scale "dry run" of the survey using materials and procedures similar to those in the production survey, typically conducted once questionnaires are near final. For a program as large as the COA, NASS conducts one or two field tests with samples of several thousand to tens of thousands. As in other establishment survey populations, NASS has extensive auxiliary information about its establishments on the sample frame, either from previous census and surveys or from outside sources. This allows the sample to be tailored to ensure coverage of the sections of the questionnaire that are being tested. For example, for the 2017 COA we selected records to ensure representation of different types of operations (specific types of livestock, crops, production practices, etc.).

An effective way to test questionnaires in field tests is to administer alternative versions of questionnaires to randomly assigned split samples of respondents (Fowler 2004; Tourangeau 2004). In the COA, alternative versions and orderings of several sections of the questionnaire suggested in the cognitive interview results were tested this way, allowing a direct comparison of data. COA field test sample sizes were determined on the basis of the number of split samples required and a desired minimum number of specific types and sizes of operations.

A field test provides data collected with procedures similar to the operational survey and should more closely resemble typical survey data. Although it may or may not be designed to make population estimates, a field test with a large sample provides quantitative measures for evaluation. The size of the sample allows for identification of issues that might arise in rare circumstances or for only a small percentage of respondents. Indeed, a field test may be the only opportunity to evaluate in advance how questionnaires will perform under operational survey conditions prior to the survey proper. If near-operational procedures are used, the field test is also an opportunity to test survey processing. If several possible changes have been suggested by early testing, split samples in a field test comparing alternative versions can be used to select the optimum version for adoption. For other items it may be too late in the survey development to make changes to the questionnaire but it may still be possible to flag items to develop item edits or target for extra review during data processing.

In 2016, a COA field test was conducted with sample of approximately 30 000 selected agricultural operations (Ott et al. 2016). The field test for the COA mimicked operational procedures, to the extent possible. The field test included alternative versions of the questionnaire mailed to randomized subsets of respondents. The alternative versions differed in format, layout, and the order in which sections of the questionnaire were presented. Subsamples were also included to test a new short form. This included subsamples selected using the criteria defining the operations eligible for the short form. Data from the alternative questionnaire versions were then compared on several metrics including unit response rates and data quality measures such as item edit and imputation rates.

Much of the necessary fixed overhead costs for the survey will be required for preliminary field tests, since a field test will require all of the infrastructure necessary for the main-stage survey. Just as some bento boxes will include fresh organic salmon, while others feature tofu, whether or not you can afford a field test (and of what size) will depend on your available resources. While we conducted a large field test for the COA, smaller field tests or those eliminating some of the more typical survey overhead are still of significant benefit to survey organizations. For example, NASS also conducted a smaller COA field test of a few thousand restricted to an evaluation of the online version of the questionnaire. Results from that test were used to improve the online questionnaire.

For establishment surveys, the burden for establishments' participation can be an obstacle, since certain subsets of establishments often may be selected in many survey samples. Survey organizations may be hesitant to jeopardize participation of establishments critical to other surveys for field tests. For example, NASS was conducting a separate organic production survey during our field test. All operations in that survey were excluded from the field test so as not to adversely impact response in the organic survey.

In addition, as with all questionnaire testing, time must be available to evaluate the results of the tests and act on any findings prior to the survey proper. Because of the larger size of most field tests and complexity of processing the data, the analysis of results may take longer than smaller tests, such as cognitive interviews.

29.2.5 Follow-Up Interviews

Another technique in our testing bento box is the follow-up interview, sometimes known as respondent debriefing interviews. Like using pickled ginger in your bento box to cleanse the palate, multimethod testing can use these interviews strategically after consumption of production or field test interviews. Survey organizations have used response analysis surveys to draw random samples of survey respondents for recontact (Goldenberg 1994, pp. 1357–1362; Phipps et al. 1995). Qualitative follow-up interviews can also be targeted to specific cases and provide a valuable method for assessing data quality post production. For the COA such interviews are targeted to outliers, individuals who have provided specific types of data, or suspected problems identified during data analysis. In either case, follow-up interviews can be conducted with selected respondents to verify or obtain further explanation directly from them.

The limitations of these interviews are similar to cognitive interviews: that is, the limited number usually possible and difficulty generalizing to the wider population. Furthermore, if used following a field test, results are available late in the survey development process, and thus there may be limited time or opportunity to make major questionnaire changes based on the findings.

29.3 Examples from the Census of Agriculture Questionnaire Testing Bento Box

All five of the bento box elements described were employed during testing for the 2017 COA questionnaire. In this section, several examples are presented to illustrate how different combinations of the five testing methods help strengthen the evaluation process. While the specific results of our testing are of secondary interest, questionnaire designers should be able to see why combining methods gave us more confidence in our results. Simply put, the sum of a multimethod test is greater than its parts.

29.3.1 Example 1: Historical Data Review Supplemented with Cognitive Interview Results

To inform the design of the 2017 COA, we reviewed historical data to evaluate changes made to the COA questionnaire in the past, by comparing data from 2007 with data from the 2012 COA. For example, in 2007, data were collected on acres owned, rented from others, and rented to others. Respondents were instructed to subtract the acreage rented to others from the total acreage owned and rented from others to arrive at the total acres operated. On the facing page, they were asked to report the total acres operated by land use and instructed to add these items to again arrive at their total acres operated – see Figures 29.1 and 29.2. Equivalent data was required for both total acres variables and was edited for consistency or imputed when not reported.

Data for these items from 2007 showed that total acres operated often did not equal the sum of the sub-acreages, or that the two total acres operated amounts were not equal (see Table 29.1). These sections had been redesigned for the 2012 COA based on recommendations from NASS survey methodologists and limited cognitive interviewing. The 2012 redesign first walked respondents

		None	Number of Acres
1. All land owned.	0043	☐	
2. All land rented or leased **from others,** including land worked by you on shares, used rent free, in exchange for services, payment of taxes, etc. Include Federal, State, and railroad land leased on a per-acre basis. Exclude land (i.e. private, Federal, State, railroad, etc.) used on a per-head or animal unit month (AUM) basis under a grazing permit. . . .	0044	☐	
3. All land rented or leased **to others,** including land worked on shares by others and land subleased.	0045	☐	
4. **TOTAL ACRES** in this operation for this census - Add items 1 and 2, then subtract item 3. If the entry is zero, please refer to the enclosed Instruction Sheet, section 1. These acres are referred to as **THIS OPERATION** for the remainder of this report. ➡		0046	

Figure 29.1 2007 COA acreage questions.

SECTION 2 LAND

Report how the acres reported in SECTION 1, item 4 were used in 2007. Include land in CRP, WRP, and other State and Federal programs. **Exclude land rented to others.** Report land only once, in the first item that applies. For example: Land that was both pastured and had a crop harvested should be reported only in cropland harvested (item 1a).

		None	Number of Acres
1. Cropland - Exclude cropland pasture.			
a. Cropland harvested - Include all land from which crops were harvested or hay was cut, all land in orchards, citrus groves, vineyards, berries, and nursery and greenhouse crops, Christmas trees, and short rotation woody crops.	0787	☐	
b. Cropland on which all crops failed or were abandoned - Exclude land in orchards and vineyards.	0790	☐	
c. Cropland in cultivated summer fallow.	0791	☐	
d. Cropland idle or used for cover crops or soil-improvement but not harvested and not pastured or grazed.	1062	☐	
2. Pasture			
a. Permanent pasture and rangeland - Exclude cropland pasture.	0796	☐	
b. Woodland pastured.	0794	☐	
c. Cropland used **only** for pasture or grazing - Include rotation pasture and grazing land that could have been used for crops without additional improvements.	0788	☐	
3. Woodland not pastured - Include woodlots, timber tracts, and sugarbush. .	0795	☐	
4. All other land - Include land in farmsteads, buildings, livestock facilities, ponds, roads, wasteland, etc.	0797	☐	
5. **TOTAL ACRES** - Add the acres reported in items 1 through 4 above. Should be the same acres as those reported in SECTION 1, Item 4. . . .	0798	☐	

Figure 29.2 2007 COA land section.

through the steps of the mathematical calculation required to arrive at the total acres operated, and more clearly specified the subtypes of land to be included or excluded. A check question was incorporated to prompt respondents to explicitly verify that the two total acres amounts were the same – see Figures 29.3 and 29.4.

Table 29.1 Number and percentage of missing values for land and acreage sections for 2007 and 2012 COA.

	2007 Census of Agriculture		2012 Census of Agriculture	
	Count	**%**	**Count**	**%**
Either total acres missing	174 642	13.8	89 654	8.5
2 total acres figures *not* equal	197 949	15.6	77 034	7.4
Total acres correct	894 893	70.6	882 837	84.1
Total	1 267 484	100.0	1 049 525	100.0

$X^2 = (2, N = 2\,317\,009) = 60\,612.48, p < 0.0001.$

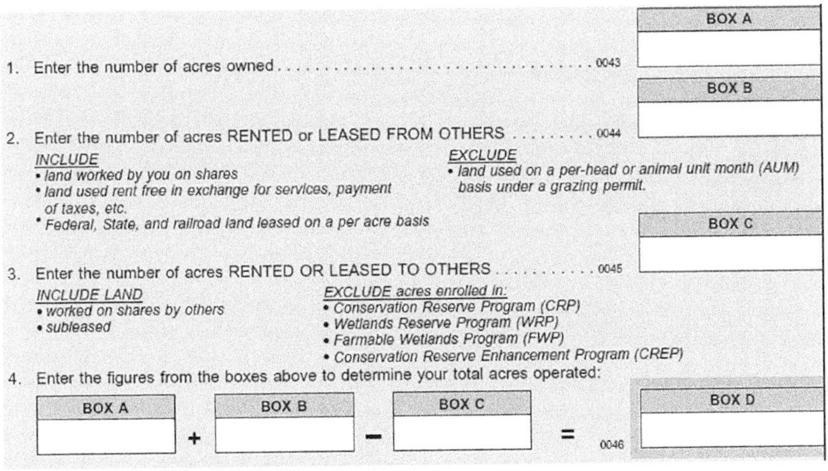
BOX A

1. Enter the number of acres owned 0043

BOX B

2. Enter the number of acres RENTED or LEASED FROM OTHERS 0044

INCLUDE
- land worked by you on shares
- land used rent free in exchange for services, payment of taxes, etc.
- Federal, State, and railroad land leased on a per acre basis

EXCLUDE
- land used on a per-head or animal unit month (AUM) basis under a grazing permit.

BOX C

3. Enter the number of acres RENTED OR LEASED TO OTHERS 0045

INCLUDE LAND
- worked on shares by others
- subleased

EXCLUDE acres enrolled in:
- Conservation Reserve Program (CRP)
- Wetlands Reserve Program (WRP)
- Farmable Wetlands Program (FWP)
- Conservation Reserve Enhancement Program (CREP)

4. Enter the figures from the boxes above to determine your total acres operated:

BOX A + BOX B − BOX C = 0046 BOX D

Figure 29.3 2012 COA acreage questions.

Unedited data as reported by respondents in 2007 was compared to unedited data reported in 2012 and showed that in 2012, using the revised format, there were fewer cases with the total acres missing and a greater number of cases where the two total acre numbers were equal in the 2012 COA, as shown in Table 29.1.

In addition to the error rates, information from calls made to a toll-free telephone helpline also provided evidence that the changes made in 2012 had improved the quality of the data collected. During the 2007 COA, there were 99 993 calls from respondents asking for help completing their questionnaires. In 2012, 107 675 calls for help were received, but the percent of respondents requesting help on the land-use section of the form declined (Table 29.2).

In addition to the questionnaire changes, there were other changes between the two censuses that could have contributed to changes in the data, such as

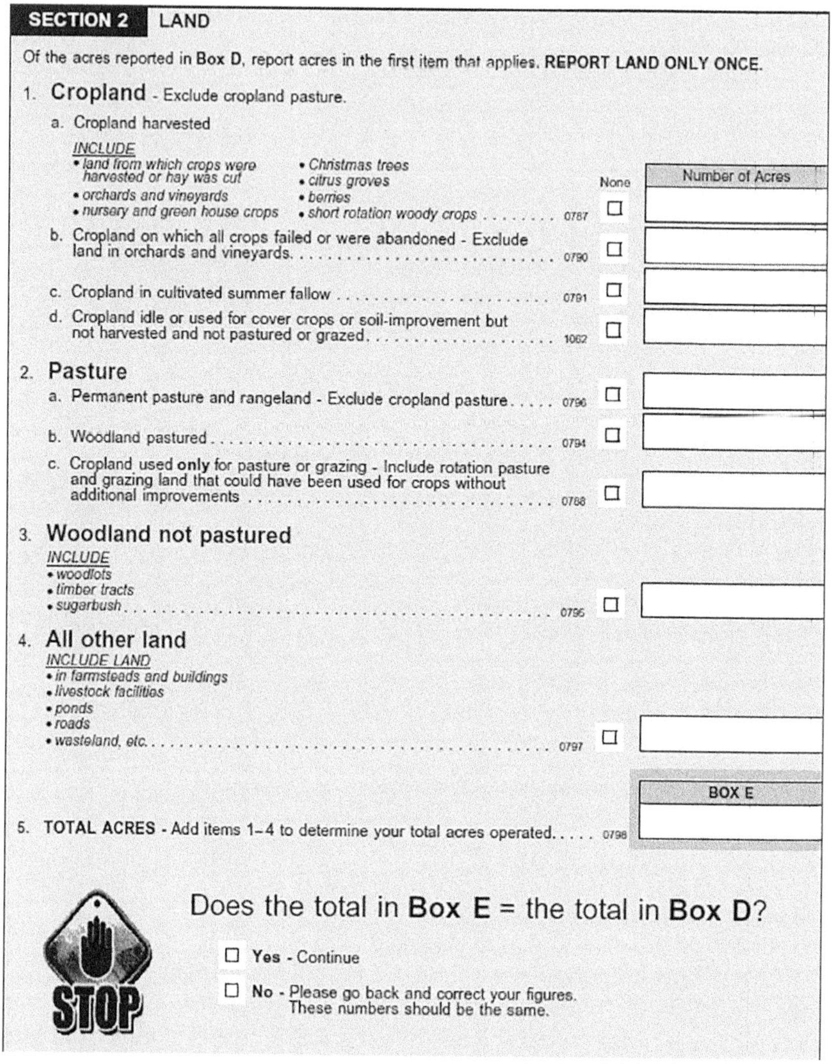

SECTION 2 LAND

Of the acres reported in **Box D**, report acres in the first item that applies. **REPORT LAND ONLY ONCE.**

1. **Cropland** - Exclude cropland pasture.

   a. Cropland harvested

   INCLUDE
   - *land from which crops were harvested or hay was cut*
   - *orchards and vineyards*
   - *nursery and green house crops*
   - *Christmas trees*
   - *citrus groves*
   - *berries*
   - *short rotation woody crops* . . . . 0787 ☐ None | Number of Acres

   b. Cropland on which all crops failed or were abandoned - Exclude land in orchards and vineyards. . . . . 0790 ☐

   c. Cropland in cultivated summer fallow . . . . 0791 ☐

   d. Cropland idle or used for cover crops or soil-improvement but not harvested and not pastured or grazed. . . . . 1062 ☐

2. **Pasture**

   a. Permanent pasture and rangeland - Exclude cropland pasture. . . . . 0796 ☐

   b. Woodland pastured . . . . 0794 ☐

   c. Cropland used **only** for pasture or grazing - Include rotation pasture and grazing land that could have been used for crops without additional improvements . . . . 0788 ☐

3. **Woodland not pastured**

   INCLUDE
   - *woodlots*
   - *timber tracts*
   - *sugarbush*. . . . . 0795 ☐

4. **All other land**

   INCLUDE LAND
   - *in farmsteads and buildings*
   - *livestock facilities*
   - *ponds*
   - *roads*
   - *wasteland, etc.*. . . . . 0797 ☐

5. **TOTAL ACRES** - Add items 1–4 to determine your total acres operated. . . . . 0798 BOX E

STOP

Does the total in **Box E** = the total in **Box D**?

☐ **Yes** - Continue

☐ **No** - Please go back and correct your figures. These numbers should be the same.

Figure 29.4 2012 COA land use section.

different survey staff and changes to the pre-census screening procedures. Changes to the members of the population and farming practices were also possible during that time. Cognitive interviews for the 2017 COA with questions formatted using the 2012 format were conducted after the data review. Respondents in the cognitive interviews were observed answering the new check question and going back and correcting the acreage reported. Together these three sources of information were used to recommend retaining the

Table 29.2 Calls to the help line by section for the 2007 and 2012 COA.

Section[a]	2007 COA (*n*)	%	2012 COA (*n*)	%
Land	23 780	23.78	14 340	13.32
Crops	4 927	4.93	2 920	2.71
Livestock	8 205	8.21	4 750	4.41
Production contracts	543	0.54	437	0.41
Economic data	7 299	7.30	3 498	3.25
Operator characteristics	7 027	7.03	4 296	3.99
Address label	1 265	1.27	1 754	1.63
Conclusion	4 025	4.03	2 254	2.09
Web form issues	679	0.68	4 778	4.44
Other	70 778	70.78	83 780	77.81

$X^2 \pm (9, N = 251\,425) = 10\,425.83, p < 0.0001$.
a) Multiple sections could be indicated in a single call.

2012 format (even though this required more space on the questionnaire). The historical data comparisons between the censuses showed improvements in data quality over time, but cognitive interviews helped verify that these improvements were at least in part due to the questionnaire changes.

29.3.2 Example 2 – Expert Reviews, and Field Test Evaluations Informed by Cognitive Interview Results

In several sections of the COA form, information about acreage, production, and value of sales for different types of commodities is collected in tables with each commodity listed on a separate line. In previous COAs, the most common crops or livestock were prelisted within the tables, with a list of additional items (and their associated codes) listed on the same page beneath the table. An example of this format for field crops (Version A) is shown in Figure 29.5. Because new content was requested for the 2017 COA but the number of pages available in the form was limited, alternatives to this format that would require less space in the form were drafted. Alternative formats were tested for the field crops, fruit, vegetable, nursery, aquaculture, other livestock, and production contract sections.

One alternative format removed the prelisted commodities from the table and provided a list of commodities and codes below the table. Figure 29.6 shows this format for field crops (Version B). The number of commodities listed on the page was limited by the available space, with the most common and their codes listed (the full list appeared in the separate instruction sheet). Another version was drafted with no commodities prelisted in the table or within the

SECTION 6 FIELD CROPS

1. Were any field crops, such as corn, soybeans, wheat, etc., harvested from this operation in 2012?
   INCLUDE
   • *your landlord's share and crops grown under contract*
   EXCLUDE
   • *crops grown on land rented to others*

   1011 1 ☐ **Yes** - Complete this section 3 ☐ **No** - Go to SECTION 7

2. Report quantity harvested in the unit specified with the crop name. For those crops not printed in the following table, enter the field crop name and code from the list below for any other field crop harvested in 2012. Report gross value of agricultural products sold from this operation in 2012. Include the value of your landlord's share, marketing charges, taxes, hauling, etc. Exclude value of items produced under production contracts.

Field Crop	Code	Acres Harvested: Acres	Acres Harvested: Tenths	Total Quantity Harvested		Acres Irrigated: Acres	Acres Irrigated: Tenths	Value of Sales (Dollars)
Tobacco - all types (tenth acres)	0094				Lbs.			$ .00

Field Crop	Code	Acres Harvested	Total Quantity Harvested		Acres Irrigated	Value of Sales (Dollars)
Barley for grain or seed	0070			Bu.		$ .00
Corn for grain or seed	0067			Bu.		$ .00
Corn for silage or greenchop	0070			Tons		$ .00
Dry edible beans - kidney, black, etc. - Exclude Limas	0554			Cwt.		$ .00
Oats for grain or seed	0076			Bu.		$ .00
Popcorn - pounds shelled	0662			Lbs.		$ .00
Rye for grain or seed - Exclude ryegrass	0686			Bu.		$ .00
Sorghum for grain or seed - Include milo	0082			Bu.		$ .00
Sorghum for silage or greenchop - Report Sorghum-Sudan crosses in Section 7	0085			Tons		$ .00
Soybeans for beans	0088			Bu.		$ .00
Wheat, Spring for grain or seed, other than Durum	0728			Bu.		$ .00
Wheat, Winter for grain or seed harvested in 2012	0572			Bu.		$ .00
						$ .00
						$ .00
						$ .00
						$ .00

If more space is needed, use a separate sheet of paper.

FIELD CROPS	CODE
Alfalfa hay - Report in SECTION 7.	
Alfalfa seed (pounds)	0542
Birdsfoot trefoil seed (pounds)	0566
Bromegrass seed (pounds)	0569
Buckwheat (bushels)	0575
Camelina (pounds)	0608
Canola, edible (pounds)	0614
Clover, red clover seed (pounds)	0671
Cotton, Upland (bales) - Include cottonseed in value of sales only	0581
Emmer and spelt (bushels)	0599
Fescue seed (pounds)	0602
Flaxseed (bushels)	0605
Hay - Report in SECTION 7.	
Herbs, dried (pounds)	0620
Kentucky bluegrass seed (pounds)	0629
Lespedeza seed (pounds)	0638
Mint, peppermint (pounds of oil)	0047
Mint, spearmint (pounds of oil)	0050
Mint, tea leaves (pounds)	0767
Miscanthus (tons)	0641
Orchardgrass seed (pounds)	0653
Peas, dry edible (hundredweight)	0659
Potatoes - Report in SECTION 10.	
Proso millet for grain or seed (bushels)	0665
Ryegrass seed (pounds)	0689
Sorghum for syrup (gallons)	0704
Sorghum-Sudan crosses - Report in SECTION 7.	
Sudangrass seed (pounds)	0713
Sugarbeets for seed (pounds)	0716
Sugarbeets for sugar (tons)	0719
Sunflower seed, non-oil variety (pounds)	0776
Sunflower seed, oil variety (pounds)	0773
Sweet potatoes - Report in SECTION 10	
Switchgrass (tons)	0647
Timothy seed (pounds)	0746
Triticale for grain (bushels)	0749
Vetch seed (pounds)	0755
Wheatgrass seed (pounds)	0758
Wild rice (hundredweight)	0764
Other field crop, specify above	0752

Figure 29.5 Version A – example of preprinted items in a commodity table with additional items listed on the page (prior format).

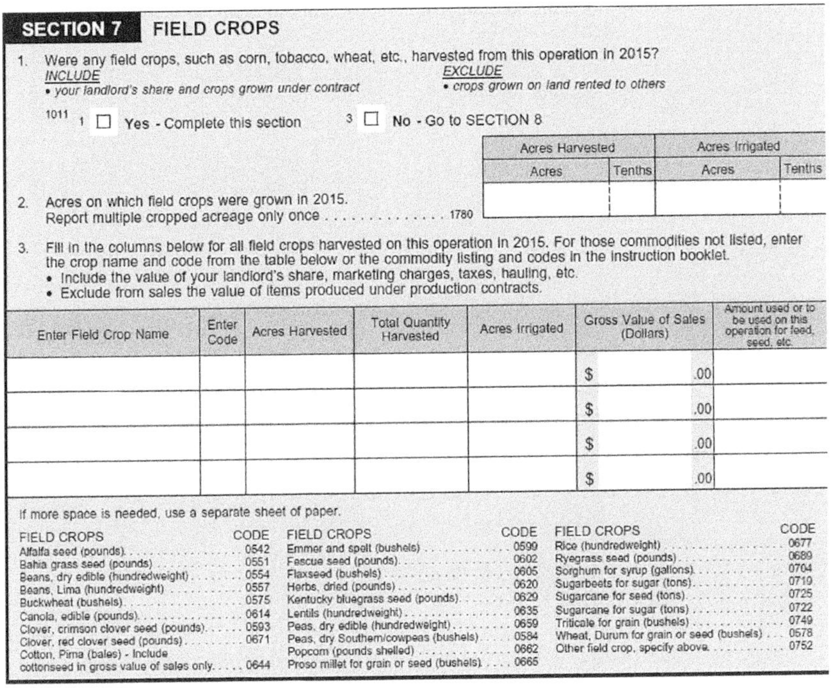

SECTION 7 FIELD CROPS

1. Were any field crops, such as corn, tobacco, wheat, etc., harvested from this operation in 2015?

<u>INCLUDE</u>
- *your landlord's share and crops grown under contract*

<u>EXCLUDE</u>
- *crops grown on land rented to others*

1011 1 ☐ **Yes** - Complete this section 3 ☐ **No** - Go to SECTION 8

	Acres Harvested		Acres Irrigated	
	Acres	Tenths	Acres	Tenths
2. Acres on which field crops were grown in 2015. Report multiple cropped acreage only once 1780				

3. Fill in the columns below for all field crops harvested on this operation in 2015. For those commodities not listed, enter the crop name and code from the table below or the commodity listing and codes in the instruction booklet.
   - Include the value of your landlord's share, marketing charges, taxes, hauling, etc.
   - Exclude from sales the value of items produced under production contracts.

Enter Field Crop Name	Enter Code	Acres Harvested	Total Quantity Harvested	Acres Irrigated	Gross Value of Sales (Dollars)	Amount used or to be used on this operation for feed, seed, etc.
					$ .00	
					$ .00	
					$ .00	
					$ .00	

If more space is needed, use a separate sheet of paper.

FIELD CROPS	CODE	FIELD CROPS	CODE	FIELD CROPS	CODE
Alfalfa seed (pounds)	0542	Emmer and spelt (bushels)	0599	Rice (hundredweight)	0677
Bahia grass seed (pounds)	0551	Fescue seed (pounds)	0602	Ryegrass seed (pounds)	0689
Beans, dry edible (hundredweight)	0554	Flaxseed (bushels)	0605	Sorghum for syrup (gallons)	0704
Beans, Lima (hundredweight)	0557	Herbs, dried (pounds)	0620	Sugarbeets for sugar (tons)	0719
Buckwheat (bushels)	0575	Kentucky bluegrass seed (pounds)	0629	Sugarcane for seed (tons)	0725
Canola, edible (pounds)	0614	Lentils (hundredweight)	0635	Sugarcane for sugar (tons)	0722
Clover, crimson clover seed (pounds)	0593	Peas, dry edible (hundredweight)	0659	Triticale for grain (bushels)	0749
Clover, red clover seed (pounds)	0671	Peas, dry Southern/cowpeas (bushels)	0584	Wheat, Durum for grain or seed (bushels)	0578
Cotton, Pima (bales) - Include cottonseed in gross value of sales only	0644	Popcorn (pounds shelled)	0662	Other field crop, specify above	0752
		Proso millet for grain or seed (bushels)	0665		

Figure 29.6 Version B – example of prelisted commodities removed from table with commodities and codes listed below the table.

questionnaire as shown in Figure 29.7, and the complete commodity list and codes were only provided in the separate instruction sheet (Version C). The complete list of valid commodity codes was provided in a separate instruction sheet for all versions (A, B, and C). Survey methodologists reviewed the forms and recommended that commodities should be prelisted on the forms to reduce respondent burden, act as memory prompts, and show respondents the needed level of specificity of crops (for example, to show "corn for grain" and "corn for silage" separately instead of reporting "corn"). But whether listings within the table were necessary and how much measurement error might be introduced with other formats was unknown.

The new versions of the questionnaire without commodities prelisted within the tables (as in versions B and C) were included in initial rounds of cognitive interviews. None of the respondents where commodities were not listed on the questionnaire referred to the separate instruction sheet. This included all respondents with no commodities listed on the questionnaire (Version C) and Version B respondents whose commodities were not included in the printed list due to space constraints. Without referencing the separate instruction sheet, multiple problems with reporting occurred including uncodable responses

SECTION 7 FIELD CROPS

1. Were any field crops, such as corn, tobacco, wheat, etc., harvested from this operation in 2015?
   INCLUDE
   • *your landlord's share and crops grown under contract*
   EXCLUDE
   • *crops grown on land rented to others*

   1011 1 ☐ Yes - Complete this section   3 ☐ No - Go to SECTION 8

	Acres Harvested		Acres Irrigated	
	Acres	Tenths	Acres	Tenths
2. Acres on which field crops were grown in 2015. Report multiple cropped acreage only once 1780				

3. Fill in the columns below for all field crops harvested on this operation in 2015. **Refer to the commodity listing and codes in the instruction booklet to fill in the crop name and code.**
   • Include the value of your landlord's share, marketing charges, taxes, hauling, etc.
   • Exclude from sales the value of items produced under production contracts.

Enter Field Crop Name	Enter Code	Acres Harvested	Total Quantity Harvested	Acres Irrigated	Gross Value of Sales (Dollars)	Amount used or to be used on this operation for feed, seed, etc.
					$ .00	
					$ .00	
					$ .00	
					$ .00	

If more space is needed, use a separate sheet of paper.

Figure 29.7 Version C – example of alternate version without commodities prelisted in the table or on the page. Commodity list and codes only in a separate instruction sheet.

and entering information for commodities in the wrong sections (i.e. entering aquaculture production in the field crops section). The results confirmed the experts' recommendations to list commodities within the questionnaire.

The later final round of cognitive interviews included Version A with commodities preprinted within the table and listed on the questionnaire to explicitly evaluate this format against those tested in the earlier cognitive interviews. Respondents were able to easily report their information for the commodities that were either prelisted in the table or listed below it. But the primary problems with Version A of the questionnaire were for less-common commodities grown that were not listed on the page. In this case respondents were instructed to refer to the separate instruction sheet for the appropriate item name and code (although they typically did not). As with Versions B and C, when cognitive interview respondents had commodities not listed on the questionnaire page, some did not report them, a problem that could not be identified from data review. Other respondents simply reported their commodities as "other" rather than with the appropriate code. In addition, many did not have enough empty lines in the table to report all of their commodities, so omitted some, combined them into one line, or wrote them in the margin. Each of these actions would require an analyst review and in many cases a loss of data if they could not be coded to the correct commodity. Cognitive interviews indicated that the format which required all respondents to use a separate instruction sheet (Version C) negatively impacted data quality most (potentially increasing underreporting or editing). The write-in space limitations of Version A also prompted errors.

The cognitive interviews therefore supported adoption of Version B but could not provide any measures of the errors.

Subsequent to the cognitive interviews, a field test with a sample of approximately 30 000 agricultural operations was conducted. In the field test, versions of the form with each of the three alternative commodity section formats were mailed to randomized split samples of respondents using procedures similar to the 2017 COA. Comparisons of the alternative formats (in multiple types of commodity sections) showed that respondents who received a form with commodities listed (Versions A and B) were significantly more likely to report at least one item in the section than those receiving the version without commodities listed (Version C). This suggested that data may be underreported in Version C. However, for respondents reporting any commodities in a given section, significantly more individual commodities were reported when they were not preprinted within the table (Versions B and C) and more open write-in lines were available. This finding again suggests potential underreporting, in this case for Version A. Differences between the comparison groups for the versions of the field crops section were statistically different at the $p = 0.05$ level. Due to the small number of respondents reporting commodities in the other sections (vegetables, berries, aquaculture, etc.), the differences for these comparisons were not statistically significant but were consistent and in the same direction.

Another measure from the field test was the rate of items that could not be coded. If respondents report something in the section that cannot be coded or must be interpreted before it can be coded, it will be summarized in a general category of "other" crops or livestock. These "other" entries included those not valid for that commodity section, unknown crops, illegible entries, entries that do not correspond to valid commodities, and other misreporting. In the COA, these entries would have to be reviewed individually by an analyst. If they are able to be assigned to existing codes, measurement error might not be impacted. For example, if a respondent writes in "S. Beans" this may be referred for analyst review. If an analyst correctly sets this to the code for "soybeans" no measurement error has been introduced, although resources for the analyst review have been expended. However, given the volume of records in the COA, "other" crop and livestock entries that have to be reviewed should be minimized to contain staff time and costs.

In the field test, we compared the number of "other" items reported for the different formats. The overall number of respondents reporting "other" was low. However, respondents in a group with no commodity codes listed (Version C) reported something in the "other" category significantly more often than comparable groups of respondents who had commodity codes listed on the form (Version B). These relatively rare occurrences (under 10%) could not be adequately explored in small cognitive interview studies, where something that infrequent might be missed or seen in only a single interview.

Based on these results, the recommendation was made to remove the crops preprinted in the tables and retain the commodity and code listings on the form near the table where they are needed (i.e. Version B). While this recommendation was informed by the initial cognitive interview results, the empirical evidence from the field test strengthened this recommendation (see Ott et al. 2016 for additional details).

29.3.3 Example 3 – Expert Review, Cognitive Interviews, and Field Test Results Combined with Qualitative Follow-Up Interviews

In addition to the traditional COA form, NASS also developed a short form for a subset of operations. The objective of this form was to reduce costs and respondent burden by eliminating unnecessary content. For example, if a respondent is known to have historically grown only corn and soybeans, questions about fruit, aquaculture, poultry, and other items can be removed from the form. NASS had not been satisfied with short forms used in prior censuses and so engaged external experts in order to generate new ideas for the design of the form. In addition, using external experts allowed testing to begin before resources for testing were available within the agency (Moore et al. 2016). The short form was designed by removing several pages of the long form that were not relevant to targeted respondents identified using information from the sample frame. Operations known to have specific commodities in the removed sections were ineligible to receive the short form. For example, a full page of questions asking for detail regarding the acreage and production of fruits was replaced by a question asking only if the operation had grown any fruit in the previous calendar year. This was intended to verify the expectation that the respondent did not have any of the specified commodities. If respondents reported having them, they could be re-contacted to collect the detailed information for those commodities. Several other commodity pages were also removed from the form this way.

A subset of 40 cognitive interviews (from the main round of 71 cognitive interviews) tested the short form. Respondents did not appear to have problems answering the questions, verifying that they did not have any of the commodities removed from the short form. For example, they were easily able to report that they did not produce any aquaculture, berries, vegetables, etc.

The short form was also included in the larger field test described earlier to provide information on how it would work using regular survey data collection procedures. Respondents were selected from the proposed short-form population and randomly assigned to receive either a short or long form. In addition, randomly assigned split samples of respondents from the long-form population were also mailed either a short or long form. Important results from this test included a measure of how often respondents targeted for the short form reported having items that we did not expect. For the 13 items that would potentially prompt follow-up contacts, respondents reported unexpected items from less than 1% for some items to over 6% for respondents reporting production of vegetables. With an anticipated mailing of 400 000 short forms in the 2017 COA, the number of re-contacts needed for all of the removed content was deemed unacceptably high.

From the field test data, it was impossible to tell whether reports of these commodities were legitimate, and we were left with several questions about the accuracy of reported commodities and potential underreporting. In order to learn more about these respondents, a small set of qualitative follow-up interviews were conducted. Short-form respondents who reported having unexpected commodities were re-contacted and asked to provide more detail about them. A few reported data that simply did not match our auxiliary list frame control data for the record. However, many of the follow-up respondents who reported vegetable production (the most problematic question) stated that they had small home gardens for their own use, and *not* commercial vegetable production. On the long form, instructions to exclude home-use gardens is provided in the section collecting vegetable data, but this instruction did not appear on the short form. From these few targeted interviews it was clear that instructions were necessary to help respondents to interpret the question in the way we intended. Based on the field test data and the follow-up interviews, the questionnaire was modified to include those additional instructions and also to ask for the acreage or number of each commodity instead of a simple Yes/No question.

Results from the short-form subsample that was mailed the long form also showed that some respondents had reported commodities they were not expected to have. For most, the amounts reported were minimal, but a handful of respondents reported large amounts (for example, thousands of hogs raised). This led to a recommendation to collect acreages or number on the short form, rather than just an indication of the presence of the commodity. The expert review injected new ideas into the design of the form and allowed some early testing. Cognitive interviews showed the feasibility of the short form, but did not allow examination of rarely occurring problems. The field test was needed to understand the magnitude of potential misreporting and directly compare alternative forms. Finally, qualitative interviews could be strategically targeted to understand why respondents reported in unexpected ways. Together the multiple testing methods provided a strong case to revise the questionnaire.

29.4 Conclusion

For any large and complex data collection, a mixed-methods bento box approach to questionnaire testing can be more effective than any single method. At NASS, using evidence from multiple testing methods provided us with confidence that we had correctly identified problems and their source and scope, providing stronger justification for proposed changes to the 2017 COA questionnaire. Although using multiple testing methods is more costly, for such a large and complex data collection as the COA, the dividends, in better quality data and savings in staff time for data editing, processing, and analysis, more than offset that cost. The examples discussed illustrate how information from multiple methods of questionnaire evaluation was combined to provide a fuller picture of potential data quality problems. Reviews of historical data from past surveys lay the foundation for subsequent testing by identifying areas of the questionnaire that may have been problematic for respondents. Expert reviews from questionnaire design specialists, subject matter specialists, and operational survey managers provided insight into potential solutions for additional testing. Cognitive interviews provided rich qualitative information on how respondents interpret questions and instructions, their reporting strategies, and navigation through the questionnaire, which could not be obtained from data or expert review. A large-scale field test using operational survey procedures embedded split-sample experiments, and allowed direct comparisons of reported data, item nonresponse rates, imputation, and edit rates to provide a measure of the size of potential errors and how they differed between questionnaire versions. It also allowed us to examine low-frequency problems. Finally, follow-up qualitative interviews were able to further explore unexpected data. Other methods used for evaluating survey questionnaires such as focus groups, behavior coding, latent class modeling, and usability testing (OMB 2016) could also be included in your bento box, but the five described here are the ones most commonly used by NASS. Depending on the needs of the survey, NASS may add these additional elements to our bento box in the future. For example, usability testing has become more important as online reporting increases.

For a balanced and nutritious meal, a bowl of rice is no match for a well-constructed bento box. For questionnaire testing, the multiple-method evaluation elements of our bento box, taken together, provide a more satisfying meal than any of the evaluation methods consumed alone.

References

Bavdaz, M. (2009). Conducting research on the response process in business surveys. *Statistical Journal of the International Association for Official Statistics* 26: 1–14.

Beatty, P. and Willis, G. (2007). Research synthesis: the practice of cognitive interviewing. *Public Opinion Quarterly* 27: 287–311.

Campanelli, P. (2008). Testing questionnaires. In: *International Handbook of Survey Methodology* (eds. E. De Leeuw, J. Hox and D. Dillman). NY: Lawrence Erlbaum Associates.

Collins, D. (ed.) (2015). *Cognitive Interviewing Practice*. London, UK: SAGE Publications, Inc.
Creswell, J.W. (2014). *A Concise Introduction to Mixed Methods Research*. Thousand Oaks, CA: SAGE Publications, Inc.
Demaio, T. and Landreth, A. (2004). Do different cognitive interview techniques produce different results? In: *Methods for Testing and Evaluating Survey Questionnaires* (eds. S. Presser, J.M. Rothgeb, M.P. Couper, et al.), 89–108. NY: Wiley.
Fowler, F. (2004). The case for more split-sample experiments in developing survey instruments. In: *Methods for Testing and Evaluating Survey Questionnaires* (eds. S. Presser, J.M. Rothgeb, M.P. Couper, et al.), 173–188. NY: Wiley.
Giesen, D. and Hak, T. (2005). Revising the structural business survey: from a multi-method evaluation to design. Federal Committee on Statistical Methodology Research Conference, Washington, DC.
Goldenberg, K. (1994). Answering questions, questioning answers: evaluating data quality in an establishment survey. In: *Proceedings of the ASA Section on Survey Research Methods*. Alexandria, VA: American Statistical Association.
Graesser, A., Cai, Z., Louwerse, M., and Daniel, F. (2006). Question understanding aid (QUAID): a web facility that tests question comprehensibility. *Public Opinion Quarterly* 70: 3–22.
Kaplowitz, M., Lupi, F., and Hoehn, J. (2004). Multiple methods for developing and evaluating a stated-choice questionnaire to value wetlands. In: *Methods for Testing and Evaluating Survey Questionnaires* (eds. S. Presser, J.M. Rothgeb, M.P. Couper, et al.), 503–524. NY: Wiley.
Kreuter, F. (ed.) (2013). *Improving Surveys with Paradata: Analytic Uses of Process Information*, vol. 581. NY: Wiley.
Lessler, J. and Forsyth, B. (1996). A coding system for appraising questionnaires. In: *Answering Questions* (eds. N. Schwarz and S. Sudman), 259–292. San Francisco: Jossey-Bass Publishers.
Madans, J., Miller, K., Maitland, A., and Willis, G. (2011). *Question Evaluation Methods: Contributing to the Science of Data Quality*. NY: Wiley.
McCarthy, J. (2016). Combining multiple questionnaire testing methods: the bento box approach in the 2017 census of agriculture. In: *Proceedings of the Fifth International Conference of Establishment Surveys*, June 20–23, 2016. Geneva, Switzerland: American Statistical Association.
McCarthy, J. and Buysse, D. (2010). Bento box questionnaire testing: multi-method questionnaire testing for the 2012 Census of Agriculture. American Association for Public Opinion Research Annual Conference.
McCarthy, J., Ott, K., Ridolfo, H. et al. (2018). Combining multiple methods in establishment questionnaire testing: the 2017 census of agriculture testing bento box. *Journal of Official Statistics* 34 (2): 341–364.
Moore, D., Ott, K., and Gertseva, A. (2016). Developing and evaluating a short form: results and recommendations from tests of a form designed to reduce

questionnaire length. In: *Proceedings of the Fifth International Conference of Establishment Surveys*, June 20–23, 2016. Geneva, Switzerland: American Statistical Association.

Murphy, J., Mayclin, D., Richards, A., and Roe, D. (2015). A Multi-method Approach to Survey Pretesting. Federal Committee on Statistical Methodology Research Conference, Washington, DC.

Office of Management and Budget (2016). Statistical Policy Working Paper 47: Evaluating Survey Questions: An Inventory of Methods. Washington, DC.

Olson, K. (2010). An examination of questionnaire evaluation by expert reviewers. *Field Methods* 22 (4): 295–318.

Ott, K., McGovern, P., and Sirkis, R. (2016). Using analysis of field test results to evaluate questionnaire performance. In: *Proceedings of the Fifth International Conference of Establishment Surveys*, June 20–23, 2016. Geneva, Switzerland: American Statistical Association.

Pascale, J. (2016). Modernizing a major federal government survey: a review of the redesign of the current population survey health insurance questions. *Journal of Official Statistics* 32: 461–486.

Persson, A., Björnram, A., Elvers, E., and Erikson, J. (2015). A strategy to test questionnaires at a national statistical office. *Statistical Journal of the IAOS* 31 (2): 297–304.

Phipps, P., Butani, S., and Chun, Y. (1995). Research on establishment survey questionnaire design. *Journal of Business and Economic Statistics* 7: 337–346.

Presser, S. and Blair, J. (1994). Survey pretesting: do different methods produce different results? *Sociological Methodology* 24: 73–104.

Presser, S., Rothgeb, J., Couper, M. et al. (eds.) (2004). *Methods for Testing and Evaluating Survey Questionnaires*. New York: Wiley.

Ramirez, C. (2002). Strategies for subject matter expert review in questionnaire design. International Conference on Questionnaire Development, Evaluation, and Testing Methods, Charleston, SC.

Ridolfo, H., Harris, V., McCarthy, J. et al. (2016). Developing and testing new survey questions: the example of new question on the role of women and new/beginning farm operators. In: *Proceedings of the Fifth International Conference of Establishment Surveys*, June 20–23, 2016. Geneva, Switzerland: American Statistical Association.

Sekiguchi, R. (2009). The Power of Five. Savory Japan. http://www.savoryjapan.com/learn/culture/power.of.five.html.

Schaeffer, N.C. and Dykema, J. (2004). A multiple-method approach to improving the clarity of closely related concepts. In: *Methods for Testing and Evaluating Survey Questionnaires* (eds. S. Presser, J.M. Rothgeb, M.P. Couper, et al.). NY: Wiley.

Tourangeau, R. (2004). Experimental design considerations for testing and evaluating questionnaires. In: *Methods for Testing and Evaluating Survey*

Questionnaires (eds. S. Presser, J.M. Rothgeb, M.P. Couper, et al.), 209–224. NY: Wiley.

Tuttle, A.D., Morrison, R.L., and Willimack, D.K. (2010). From start to pilot: a multi-method approach to the comprehensive redesign of an economic survey questionnaire. *Journal of Official Statistics* 26: 87–103.

Willimack, D. (2013). Methods for the development, testing, and evaluation of data collection instruments. In: *Designing and Conducting Business Surveys* (eds. G. Snijkers, G. Haraldsen, J. Jones and D. Willimack). NY: Wiley.

Willimack, D. and Nichols, E. (2010). A hybrid response process model for business surveys. *Journal of Official Statistics* 26: 3–24.

Willis, G.B. (2004). *Cognitive Interviewing: A Tool for Improving Questionnaire Design*. SAGE Publications, Inc.

Willis, G.B. (2015). *Analysis of the Cognitive Interview in Questionnaire Design*. Oxford University Press.

Willis, G.B. (2016). "Questionnaire Pretesting). *The SAGE Handbook of Survey Methodology*. SAGE Publications, Inc.

Willis, G.B. and Lessler, J.T. (1999). *Question Appraisal System, QAS-99*. Rockville, MD: Research Triangle Institute.

Willis, G.B., Schechter, S., and Whitaker, K. (1999). A comparison of cognitive interviewing, expert review, and behavior coding: What do they tell us? Annual Meeting of the American Statistical Association, Baltimore, MD.

Yan, T., Kreuter, F., and Tourangeau, R. (2012). Evaluating survey questions: a comparison of methods. *Journal of Official Statistics* 28: 503–529.

30

Flexible Pretesting on a Tight Budget: Using Multiple Dependent Methods to Maximize Effort-Return Trade-Offs

Matt Jans[1], Jody L. Herman[2], Joseph Viana[3*], David Grant[4*], Royce Park[5], Bianca D.M. Wilson[2], Jane Tom[6*], Nicole Lordi[7], and Sue Holtby[7]*

[1] ICF, Rockville, MD, USA
[2] The Williams Institute, UCLA School of Law, Los Angeles, CA, USA
[3] Los Angeles County Department of Public Health, Los Angeles, CA, USA
[4] RAND Corporation, Santa Monica, CA, USA
[5] UCLA Center for Health Policy Research, Los Angeles, CA, USA
[6] Independent scholar, USA
[7] Public Health Institute, Oakland, CA

30.1 Introduction

Questionnaire pretesting is a complex endeavor with many decision points including choosing a pretesting method, implementing it, interpreting results, and applying them to individual questions and the questionnaire overall. Hypothetically, an optimal pretest should take only the amount of time and effort required to gain actionable insights, without expending additional effort. Any additional effort expended should gather additional insights that further clarify question problems and aid in making revisions. This effort/return trade-off is a difficult optimization problem for methodologists to solve at the beginning of a pretest. Problems inherent in a question's grammatical or conceptual structure may not be immediately obvious, and specific cognitive or interaction problems, and other difficulties interviewers have asking questions may not be known until observed during administration. Thus, the survey methodologist risks one of two suboptimal outcomes in every pretest design. If they design a pretest optimized for one type of problem (e.g. interviewer administration difficulty), they may miss other problems. If they design an in-depth test, committing significant hours to multiple methods and phases of pretesting, they may not have enough time to complete the test before results are needed. Further, including multiple pretest methods can lead to conflicting

* The majority of this work was completed while the author was affiliated with the UCLA Center for Health Policy Research.

Advances in Questionnaire Design, Development, Evaluation and Testing, First Edition.
Edited by Paul C. Beatty, Debbie Collins, Lyn Kaye, Jose-Luis Padilla, Gordon B. Willis, and Amanda Wilmot.

findings and "results overload," which can stymie question revisions. However, at the end of the pretest, the methodologist needs to have enough data and findings to either know how to make the tested questions better, or attest to their quality.

This chapter focuses on pretesting for interviewer-administered modes, and telephone surveys more specifically. It illustrates the use of multiple methods (randomized experimental design, qualitative observation, and coding from audio recordings) in a single pretest, and describes how the specific combination and order of methods can be a flexible component of the design. It also demonstrates a meta-method, *dependent pretesting,* in which pretesting methods are modified on-the-fly based on initial presenting results.

30.1.1 Pretesting Assumptions: Choosing the "Best" Method

Professional judgment plays a large role in every stage of pretesting, beginning with selecting the method (or methods) to use, and ending with applying pretest results to question revisions. That judgment includes several implicit assumptions about the match between pretesting goals and chosen methods. (i) The first assumption of any pretest is that the chosen method identifies actual problems that will arise in live data collection with real respondents. If not (i.e. if results are artifacts of the testing process or random noise), the methodologist cannot confidently make question revisions. (ii) The methodologist assumes that the pretest will produce actionable results, whether the action involves retaining the question unmodified, modifying the question, or rejecting the question altogether. (iii) When methodologists choose a pretesting method for a specific pretesting goal, they assume that there is not a simpler method that can identify the same problems for less effort. In other words, the methodologist judges that they have optimized the cost/quality (i.e. effort/return) trade-off of the pretest. This is accomplished by aiming to have enough of the right kind of "data" (e.g. observations, experimental results, expert input) to make sound decisions, but no more of any one kind of data than the minimum needed.

A brief review of available pretesting methods provides some context for understanding this conundrum. Focus groups, which are relatively low-effort, are helpful for exploring question constructs even before writing questions, and obtaining feedback from multiple people at a time. However, they provide little insight at the later stages of question development, because it is difficult to test candidate questions with this method. Once questions are written, expert reviews can be a quick method for identifying the most obvious problems, and can be made transparent and standard by using a tool like the Question Appraisal System (Willis and Lessler 1999; Dean et al. 2007). However, expert reviews generally suffer from reviewer subjectivity. More in-depth methods, such as cognitive interviewing and behavior coding, tend to require significantly more effort than the previously mentioned techniques, and the effort/return trade-offs can vary widely depending on specifics of the implementation. Behavior and interaction coding carry statistical rigor but require larger sample sizes than qualitative approaches like cognitive interviewing. Behavior coding entails developing a detailed coding scheme and definitions, training coders, and conducting reliability coding or checks. As objectively-observable phenomena are coded more reliably than subjective phenomena, inferences made about a respondent's psychological state from this method can be tenuous. Alternately, cognitive interviewing allows for rich information regarding respondents' thoughts and feelings about survey questions, but employs smaller sample sizes and can be time-consuming and complicated to analyze and summarize. Further, cognitive interviewing is, by definition, conducted "out of context" of the typical survey interview, whereas behavior coding can happen during active data collection.

In addition to methods specific to question pretesting, the methodologist may employ basic scientific and statistical methods to supplement the pretest. For example, a randomized experimental design (e.g. split-ballot experiment) can be used to compare alternative versions of questions. Randomly assigning respondents to question versions can offer insights into the bottom-line performance of each, independent of other characteristics. Randomized experiments are concrete, reproducible, and provide quantitative statistical rigor not available with many pretesting methods. However, inferences about the cognitive processes leading to observed responses can be difficult to make from the resulting proportions and means alone. In addition to analyzing substantive response distributions, "don't know" and refusal responses (i.e. item missing data) can sometimes offer insight into cognitive or interaction problems with questions (e.g. Beatty and Herrmann 2002). For example, respondents will be less likely to answer a question that they do not understand, or that is too personal, than one that is simple and perceived as benign. Analysis of paradata can further supplement analysis of response distributions and provide outcomes that are interpretable, either as practicality checks or as indicators of psychological or interaction problems with the question. The time required for the interviewer to read and respondent to answer (i.e. question-answer duration) is an example of such a metric (Bargh and Chartrand 2000; Draisma and Dijkstra 2004; Olson and Smyth 2015). As a practicality check, it can help answer the question "How much additional time will be added to the interview?" As an indicator of potential interaction problems, longer durations indicate problems with reading the question, answering the question, or both.

The following case study describes the use of a flexible *dependent pretesting* approach in a real-world question pilot that was part of a live random digit dial phone survey. The case study provides insights into the following questions:

1) How can multiple pretesting methods be combined into a single pretest protocol?

2) How are decisions made to add, remove, or modify methods during the pretest?
3) What lessons were learned that could apply to other pretesting situations?

30.2 Evolution of a Dependent Pretesting Approach for Gender Identity Measurement

The pretest assessed gender identity (GI) (i.e. transgender status) measurement within the California Health Interview Survey (CHIS, www.chis.ucla.edu). The test began with four gender identity measures that had been used in other surveys and proposed as candidate measures for CHIS by collaborators at The Williams Institute (https://williamsinstitute.law.ucla.edu). These measures each represented a common or best practice in GI measurement for general population surveys (GenIUSS Group 2014). This section discusses how the pretest method evolved and describes the dependent pretesting decision-making process. Section 30.3 summarizes the pretest results.

30.2.1 The Measurement Challenge: Complexities in Measuring Gender Identity

Measuring GI and transgender identification in the general public poses several interesting survey measurement challenges. First, sexual and gender minority communities have very descriptive terminology for self-identification (e.g. transgender, gender queer, non-binary, demi-gender, gender-fluid). While this nuanced lexicon may be essential for a high-quality measure in some survey contexts (e.g. a survey of a transgender population specifically or an LGBT youth center), general population survey measures must include terms and concepts that all respondents either already understand or can quickly learn. Respondents will likely only endorse identities during a survey question that they have consciously identified with outside of the survey. No doubt, transgender and gender-nonconforming people have thought about their GI extensively before being asked about it during a survey interview. Comparatively, it seems uncontroversial to assert that most people are cisgender (i.e. have a GI that matches their sex assigned at birth), and that most cisgender people have never given much thought to their GI separate from their biological sex. That is, they accept without question the sex they were assigned at birth as their GI. A good general population measure of GI must be answerable by even the most gender-nonreflective, cisgender respondent.

Second, in addition to being understandable and answerable by the vast cisgender population, the terminology used to measure GI must be precise and meaningful to the transgender and gender-nonconforming population.

Like every survey question, GI measures must provide mutually exclusive and exhaustive response options, and enough options so that every respondent has a way to answer. Given the potential sensitivity and complexity of GI measurement, item nonresponse is a risk, and providing adequate, meaningful response options helps abate that risk.

Third, the questions must be specific and meaningful enough to transgender respondents to avoid false negatives (i.e. transgender people who do not identify themselves as such in the survey) and false positives (i.e. cisgender respondents mistakenly identifying as transgender). Accuracy is as important as answering, particularly for low-frequency characteristics like transgender.

There are two competing approaches to measuring GI in contemporary survey practice. One involves a single survey question (a *one-step* measure) that defines GI and transgender status and asks respondents if they identify with that definition. The other uses a two-question series (a *two-step* measure) that asks about the respondent's sex at birth and current GI. The Behavioral Risk Factor Surveillance System (BRFSS), among other surveys, employs the one-step measure (Centers for Disease Control and Prevention 2017), while leading best practices strongly recommend the two-step measure when possible (e.g. GenIUSS Group 2014; Jans et al. 2018). Although two questions may seem more burdensome than one, the single-question measure has a few challenging characteristics. (i) It simply takes longer for the interviewer to read and requires more respondent attention. (ii) The respondent must understand the definition of transgender presented to them, which may include unfamiliar terms and concepts. (iii) The respondent must judge how their own GI (which may be completely subconscious for many cisgender respondents) maps onto the definition. (iv) Response can be difficult for gender-nonconforming respondents for whom the specific transgender definition provided does not match their identity. Comparatively, the two-question measure is shorter in overall words, uses common terms that most respondents will know, presents a simple gender classification for cisgender respondents, and does not require that gender-nonconforming respondents identify with a particular transgender definition.

The four measures tested in this pretest included two one-step measures, each with a different definition of transgender identity, and two two-step measures (shown next).

One-Step Version 1

"Some people describe themselves as transgender when they experience a different gender identity from their sex at birth. For example, a person born into a male body, but who feels female or lives as a woman. Do you consider yourself to be transgender?"

- *YES*[1]
- *NO*
- *DON'T KNOW*
- *REFUSED*

Description: One-Step Version 1 was used by Conron et al. (2012) with data collected from the Massachusetts BRFSS, and its language has influenced the current national BRFSS sexual orientation and gender identity (SOGI) question module, which the CDC has employed since 2014.[2]

One-Step Version 2

"Sex is what a person is born. Gender is how a person feels. When a person's sex and gender do not match, they might think of themselves as transgender. Are you transgender?"

- *YES*
- *NO*
- *DON'T KNOW*
- *REFUSED*

Description: This question is a simplified version of Conron et al.'s (2012) MA BRFSS question, adapted by the Gay, Lesbian, and Straight Education Network (GLSTEN).[3] It retains the same concepts as One-Step Version 1 but uses simpler sentences and grammar.

Two-Step Version 1

Q1 "Earlier in the survey I asked you your gender. What sex were you assigned at birth, on your original birth certificate?"

- *MALE*
- *FEMALE*
- *DON'T KNOW*
- *REFUSED*

---

1 Text in "*quoted italics*" was read to respondents, and ALL CAPS ITALICS were not read out loud).

2 "Do you consider yourself to be transgender?" Interviewer note: if asked about definition of transgender: "Some people describe themselves as transgender when they experience a different gender identity from their sex at birth. For example, a person born into a male body, but who feels female or lives as a woman would be transgender. Some transgender people change their physical appearance so that it matches their internal gender identity. Some transgender people take hormones, and some have surgery. A transgender person may be of any sexual orientation – straight, gay, lesbian, or bisexual." Interviewer note: If asked about definition of gender nonconforming: "Some people think of themselves as gender nonconforming when they do not identify only as a man or only as a woman." https://www.cdc.gov/brfss/questionnaires/index.htm.

3 See discussion in (GenIUSS Group 2014).

Q2 "Do you currently identify (describe yourself[4]) as male, female, or transgender?"

- o *MALE*
- o *FEMALE*
- o *TRANSGENDER*
- o *NONE OF THESE ("What is your current gender identity?")*
- o *DON'T KNOW*
- o *REFUSED*

Description: This two-step measure is recommended by The William's Institute's GenIUSS Group as a best practice for general population surveys because it uses simple language and concepts understandable by all respondents, and lowers the burden of identifying as transgender or gender nonconforming by not requiring explicit identification with a specific label.

Two-Step Version 2

Q1 "Earlier in the survey I asked you your gender. What sex were you assigned at birth, on your original birth certificate?"

- o *MALE*
- o *FEMALE*
- o *DON'T KNOW*
- o *REFUSED*

Q2 "Do you currently identify (describe yourself) as male, female, transgender, are you not sure yet, or do you not know what this question means?"

- o *MALE*
- o *FEMALE*
- o *TRANSGENDER*
- o *NOT SURE YET*
- o *DON'T KNOW WHAT THE QUESTION MEANS*
- o *DON'T KNOW*
- o *REFUSED*

Description: This modified two-step version is identical to the first one with the addition of "are you not sure yet, or do you not know what this question means?" The qualifying phrase came from a version administered to teens. Given the novelty of asking transgender questions in a

4 The question entered the pilot test with the term "identify." During an initial listening session, we discussed whether "describe yourself" would be a better phrase, particularly for transgender people who simply think of themselves as male or female, and made that change in the remaining pilot cases.

general population survey, this qualifier was kept in case respondents were uncomfortable admitting that they did not understand the question.

30.2.2 The Pretesting Challenge and Methods Selection

Given the measurement challenges of asking gender identity of a primarily cisgender population, interaction difficulties were expected. Thus, the initial pretesting plan included three arms: group listening sessions, interaction coding (done through solo listening and coding), and analysis of substantive responses and item nonresponse. The group listening sessions provided a chance for all team members to hear the questions implemented with real respondents, and discuss initial modification ideas. Behavior and interaction coding, conducted via solo listening, was used to provide a more systematic and statistical view of problems observed during group listening, and to identify additional problems not detected in the group sessions. Analysis of responses offered a way to see the bottom-line performance of each question version. This included transgender and gender-nonconforming identification rates and item nonresponse rates, with the latter serving as indicators of sensitivity and difficulty (see Grant et al. 2015 for additional details about the pilot).

30.2.3 Pretest Implementation

30.2.3.1 Sample and Experimental Design

The pretest was implemented during the 2014 fall and winter production interviewing for CHIS. CHIS is a general population telephone survey of the noninstitutionalized household population of California, conducted using a dual-frame random digit dialing (RDD) sample.[5] During the pretest, 2828 respondents were randomly assigned to one of the four GI measures, which were included in the CHIS instrument within a section on sexual orientation, sexual behavior, and sexual health.

30.2.3.2 Group Listening Sessions vs. Solo Listening and Behavior/Interaction Coding

The group listening sessions were essentially meetings of the pretesting team, during which team members reviewed 66 digital recordings of pretest questions. The sessions lasted between 60 and 90 minutes each. English and Spanish interviews were both monitored. Key problems and observations were summarized in meeting minutes circulated to team members following the listening sessions. Contributions to the minutes from all listeners were requested but not

5 http://healthpolicy.ucla.edu/chis/design/Pages/methodology.aspx.

mandatory, and participants took their notes in whatever format was convenient for them. The listening sessions were an integral part of the overall pretest, and directly informed the coding effort's evolution.

Behavior and interaction coding occurred in separate sessions (also 60–90 minutes each) where staff listened to additional pilot test recordings and applied codes to interviewer and respondent behaviors in a standardized, Excel-based form. All coders worked with English interviews, but only one of the three coders worked with Spanish interviews as well. The form evolved significantly from the detailed behavior and interaction coding form shown in Figures 30.1a,b to the simpler duration-focused form shown in Figure 30.2. The original form was designed to record information about the question and interviewer behaviors at the top (Figure 30.1a) and respondent behaviors and paralanguage at the bottom (Figure 30.1b). The form asked coders to assess several interviewer and respondent behaviors that are assessed in traditional behavior coding (Cannell et al. 1975; Fowler and Cannell 1996; Ongena and Dikstra 2006), and paralanguage found to be related to respondent difficulty (Conrad et al. 2008; Ehlen et al. 2007).

However, after a few sessions, and bolstered by the group listening, the coded dimensions were reduced significantly because of the lack of observed behavior (Figure 30.2). The revised form removed most of the codes in the original version, leaving only the question version (known from the recording), the respondent's sex and age (used as identifiers for the recordings), and the coder's assessment of the sex of the interviewer. Question duration was calculated automatically in Excel once the coder entered the GI measure's start and end time (recordings included questions other than the GI measures). The revised form also retained a field for notes or interesting behaviors, although these notes were not used in any substantive way in the question revisions. Two coders used this form to make detailed measurements of question durations on 220 interviews (165 English and 55 Spanish). The transition point between behavior coding and duration coding, and how that decision was made, are discussed more in Section 30.3.

30.2.3.3 Quantitative Measures Used

For our revised approach to question evaluation, three quantitative measures were used to evaluate item performance: question-answer duration, item nonresponse rates, and transgender identification. Question-answer duration was used both as a measure of the practical impact of the question on the larger CHIS survey (i.e. how much time the question would add) and as a measure of interviewer and respondent difficulty. It included the entire reading of the question and completion of an answer (or nonresponse) from the respondent. Because the primary goal was overall impact on the interview and dynamic, the full administration time was more appropriate than response latency

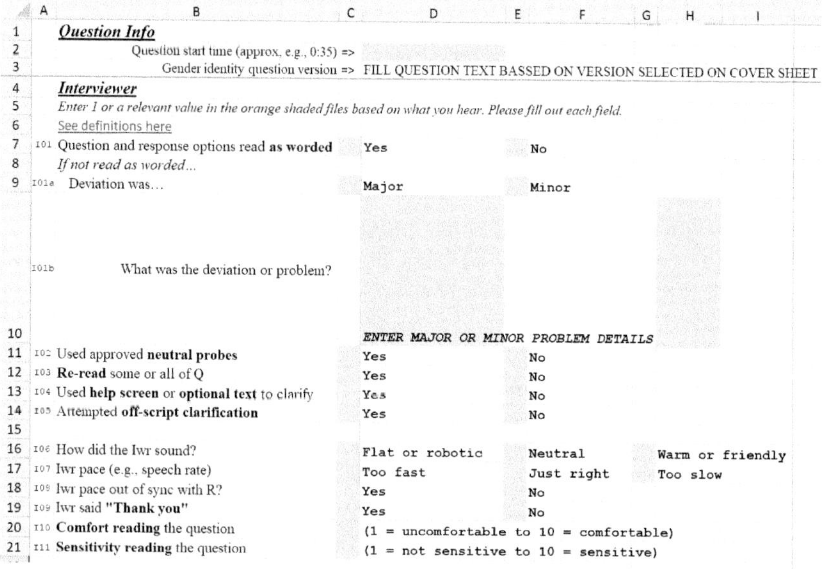

	A	B	C/D	E/F	G/H/I
1		***Question Info***			
2		Question start time (approx, e.g., 0:35) =>			
3		Gender identity question version =>	FILL QUESTION TEXT BASSED ON VERSION SELECTED ON COVER SHEET		
4		***Interviewer***			
5		*Enter 1 or a relevant value in the orange shaded files based on what you hear. Please fill out each field.*			
6		See definitions here			
7	I01	Question and response options read **as worded**	Yes	No	
8		*If not read as worded...*			
9	I01a	Deviation was...	Major	Minor	
10	I01b	What was the deviation or problem?	***ENTER MAJOR OR MINOR PROBLEM DETAILS***		
11	I02	Used approved **neutral probes**	Yes	No	
12	I03	**Re-read** some or all of Q	Yes	No	
13	I04	Used **help screen** or **optional text** to clarify	Yes	No	
14	I05	Attempted **off-script clarification**	Yes	No	
15					
16	I06	How did the Iwr sound?	Flat or robotic	Neutral	Warm or friendly
17	I07	Iwr pace (e.g., speech rate)	Too fast	Just right	Too slow
18	I08	Iwr pace out of sync with R?	Yes	No	
19	I09	Iwr said **"Thank you"**	Yes	No	
20	I10	**Comfort reading** the question	(1 = uncomfortable to 10 = comfortable)		
21	I11	**Sensitivity reading** the question	(1 = not sensitive to 10 = sensitive)		

(a)

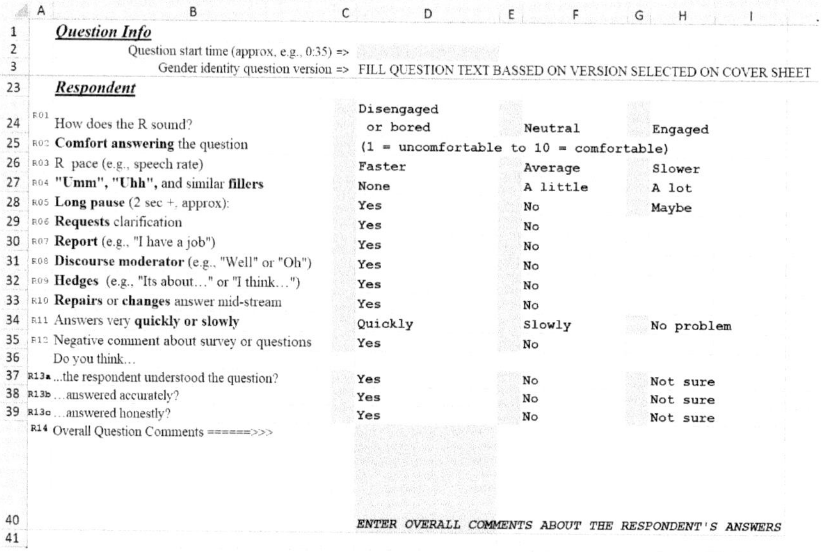

	A	B	C/D	E/F	G/H/I
1		***Question Info***			
2		Question start time (approx, e.g., 0:35) =>			
3		Gender identity question version =>	FILL QUESTION TEXT BASSED ON VERSION SELECTED ON COVER SHEET		
23		***Respondent***			
24	R01	How does the R sound?	Disengaged or bored	Neutral	Engaged
25	R02	**Comfort answering** the question	(1 = uncomfortable to 10 = comfortable)		
26	R03	R pace (e.g., speech rate)	Faster	Average	Slower
27	R04	**"Umm", "Uhh"**, and similar **fillers**	None	A little	A lot
28	R05	**Long pause** (2 sec +, approx):	Yes	No	Maybe
29	R06	**Requests** clarification	Yes	No	
30	R07	**Report** (e.g., "I have a job")	Yes	No	
31	R08	**Discourse moderator** (e.g., "Well" or "Oh")	Yes	No	
32	R09	**Hedges** (e.g., "Its about..." or "I think...")	Yes	No	
33	R10	**Repairs** or **changes** answer mid-stream	Yes	No	
34	R11	Answers very **quickly or slowly**	Quickly	Slowly	No problem
35	R12	Negative comment about survey or questions	Yes	No	
36		Do you think...			
37	R13a	...the respondent understood the question?	Yes	No	Not sure
38	R13b	...answered accurately?	Yes	No	Not sure
39	R13c	...answered honestly?	Yes	No	Not sure
40	R14	Overall Question Comments ======>>>	***ENTER OVERALL COMMENTS ABOUT THE RESPONDENT'S ANSWERS***		
41					

(b)

Figure 30.1 (a) Interviewer variables section (top half) of initial interaction coding sheet; (b) respondent variables section (bottom half) of initial interaction coding sheet.

K21

	A	B	C	D	E	F	G	H
1	Type	Rsex	Rage	Isex	Total length	Start of GI	Duration of GI	Notes/behaviors
2	"Some people consider themselves...for example...Do you consider yourself to be transgender?"							
3	1sv1	*M/F*	*AGE*	*SEX*	*M:SS*	*M:SS*	*M:SS*	*CASE NOTES RECORDED HERE*
4	...							
5							*AVERAGE WITHIN 1SV1 CALCULATED HERE*	
6	"Sex is what a person is born...When a person's sex and gender do not match...Are you transgender?"							
7	1sv2	*M/F*	*AGE*	*SEX*	*M:SS*	*M:SS*	*M:SS*	*CASE NOTES RECORDED HERE*
8	...							
9							*AVERAGE WITHIN 1SV2 CALCULATED HERE*	
10	"What sex were you assigned at birth...Do you currently describe yourself as male, female, or transgender?"							
11	2sv1	*M/F*	*AGE*	*SEX*	*M:SS*	*M:SS*	*M:SS*	*CASE NOTES RECORDED HERE*
12	...							
13							*AVERAGE WITHIN 2SV1 CALCULATED HERE*	
14	"What sex were you assigned at birth...Do you currently identify as male, female, transgender, **are you not sure yet, or do you not know what this question means?"**							
15	2sv2	*M/F*	*AGE*	*SEX*	*M:SS*	*M:SS*	*M:SS*	*CASE NOTES RECORDED HERE*
16	...							
17							*AVERAGE WITHIN 2SV2 CALCULATED HERE*	
18								

Figure 30.2 Simplified coding form focusing on question-answer duration.

measures used by others (e.g. Bassili and Fletcher 1991; Couper 2000; Draisma and Dijkstra 2004; Olson and Smyth 2015; Yan and Tourangeau 2008).

Item nonresponse may also indicate cognitive or interaction difficulties (Beatty and Herrmann 2002; Jans 2010) and thus was used as an indicator of problematic questions. Further, item nonresponse has a practical interpretation, as the amount of usable data obtained for later analyses has implications for data users. Transgender identification rate was used to ensure that the questions were obtaining frequencies similar to other studies, usually between 0.1% and 0.5% (Conron et al. 2012; Gates 2011; Reed et al. 2009; Olyslager and Conway 2007).

30.3 Analyzing and Synthesizing Results

30.3.1 Qualitative Assessment of Question Performance

In both solo and group sessions, the absence of problem behaviors and nonparadigmatic sequences was striking. The vast majority of respondents expressed no confusion and had no problem answering. One respondent verbally expressed her belief that the questions were motivated by "liberal bias," and another expressed some difficulty answering due to a genetic condition with which she was diagnosed at birth. However, both respondents were able to answer the questions. Interviewer problems reading the questions were rare as well. Among the 26 cases monitored using the original in-depth coding form, the most frequent and severe interviewer deviations were incidental stammers (eight cases). In one case, the interviewer dropped the word "of" in the phrase "...might think *of* themselves...." In two instances, interviewers began to stumble over the word "transgender," but slowed down their reading and recovered the question delivery. There were fewer than five other slight stumbles but details were not recorded. However, no instance of interviewer misreading seemed to affect question understanding.

One clear benefit of the group listening sessions was having multiple people observe and discuss the same interviewer-respondent interactions in real time. This allowed the team to move quickly from insight to pretest design changes. For example, the group observed that the two-step measures were read and answered much more quickly than either one-step measure. This qualitative observation led the team to seek quantitative support in question-answer duration. Thus, the coding effort changed focus to this measure exclusively.

30.3.2 Question-Answer Duration

Men and women were statistically similar in question-answer duration, $t(218) = 0.057$, $p = 0.95$, and there was no relationship between age and duration, $r = -0.006$, $p = 0.94$. Across GI measures, English questions ($M = 16$ seconds, $SD = 0.42$) were quicker to administer than Spanish questions ($M = 22$ seconds, $SD = 0.1.25$), $t(218) = -5.75$, $p < 0.0001$. Figure 30.3a shows question-answer duration by GI measure for English-speaking respondents, and Figure 30.3b shows duration by measure for Spanish-speaking respondents. The means and percentiles are plotted with, and statistical tests include outliers, but outliers are not displayed for disclosure protection. Among English interviews, GI measures differed significantly in question-answer duration, $F(3, 216) = 5.75$, $p < 0.0001$, and Spanish interviews showed marginally significant differences in duration between measures, $F(3, 51) = 2.39$, $p = 0.079$. Post-hoc tests with Šidák adjustment for multiple testing showed that Two-Step Version 1 was the fastest in Spanish interviews and tied for fastest in English interviews. In English it was about 1.5 seconds faster than the next fastest measure (One-Step Version 2), but that difference was not significant ($p = 0.8$). It was significantly faster than One-Step Version 1 ($p = 0.03$) and Two-Step Version 2 ($p = 0.001$). One-Step Version 2 was marginally significantly faster than One-Step ($p = 0.05$). The two one-step questions were not different from each other in duration, nor were One-Step Version 1 and Two-Step Version 2.

In Spanish, Two-Step Version 1 was six seconds faster than the next fastest question (One-Step Version 2). The only marginally significant difference was between One-Step Version 2 and Two-Step Version 1. The small number of significant differences is likely due to small sample size. Overall, despite limited statistical information and a small sample size, these results were combined with the qualitative listening experiences to conclude that the Two-Step Version 1 was the quickest to administer.

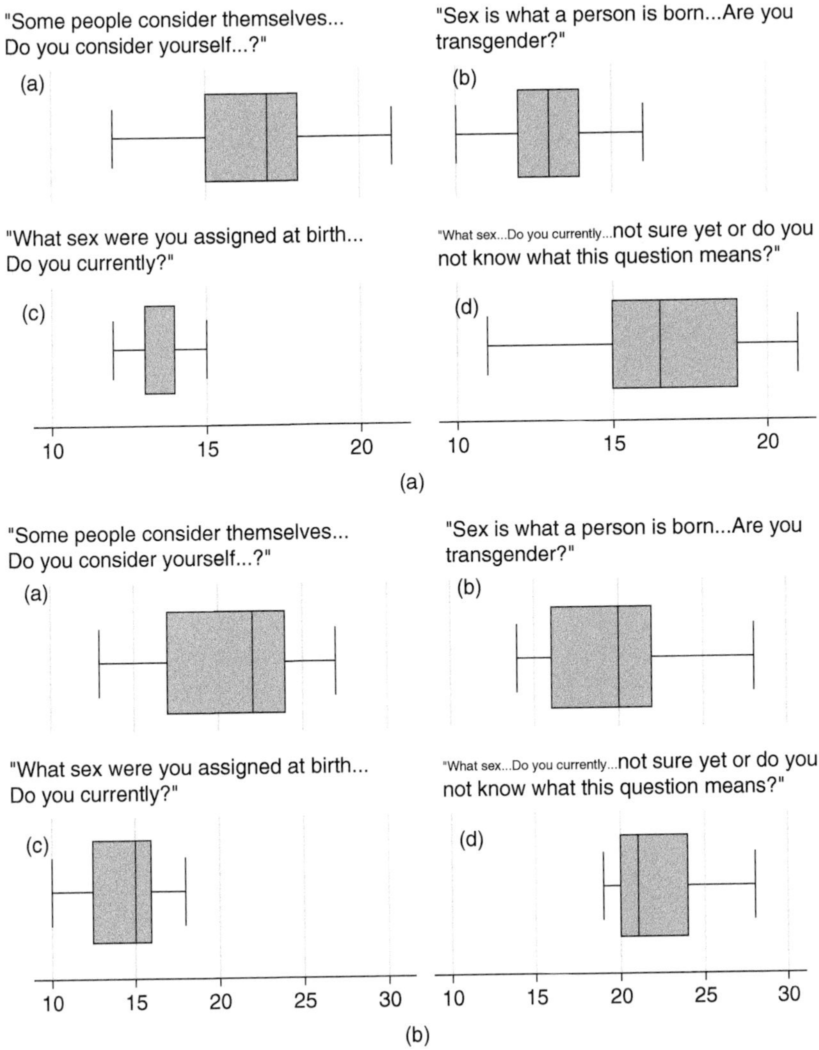

Figure 30.3 (a) English interview gender identity question duration in seconds (outliers removed, $n = 165$); (b) Spanish interview gender identity question duration in seconds (outliers removed, $n = 55$).

30.3.3 Transgender Identification Rates and Item Nonresponse

Figure 30.4 shows transgender identification and item nonresponse (i.e. not ascertained) rates for the four measures. Ultimately, each of the measures obtained the same transgender identification rate 0.3–0.4% (likelihood ratio

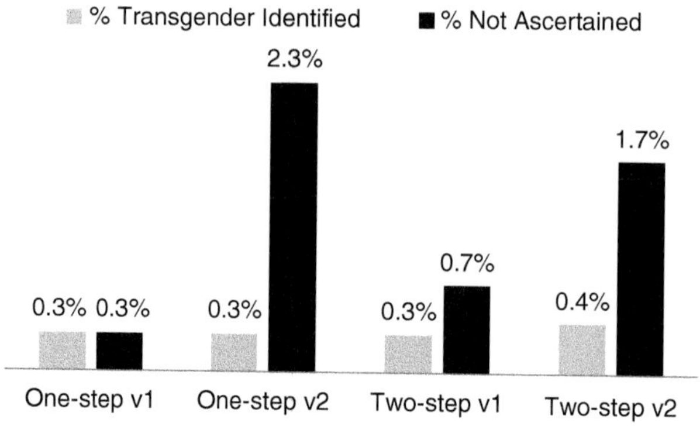

Figure 30.4 Transgender identification and item nonresponse rates across pretest versions (unweighted).

$\chi^2 = 0.29$, $p = 0.96$).[6] However, measure version and item nonresponse rate[7] were significantly associated (likelihood ratio $\chi^2 = 15.38$, $p = 0.0015$). The two measures with the lowest item nonresponse rates (One-Step Version 1 and Two-Step Version 1) were not significantly different from each other, and neither were the two measures with the highest item nonresponse rates (One-Step Version 2 and Two-Step Version 2). Yet, both "Version 2s" obtained significantly more item nonresponse than "Version 1s" (likelihood ratio $\chi^2 = 13.32$, $p = 0.0003$). For One-Step Version 2, the higher item nonresponse is difficult to understand because the question is much simpler than One-Step Version 1, which includes a long definition of transgender. Perhaps the shorter definition made it harder for respondents to understand what was being asked. In Two-Step Version 2, the high item nonresponse is almost certainly due to offering an explicit nonresponse (i.e. "…do you not know or do you not know what this question means?").

Closely related to item nonresponse is break-off. If respondents found the GI items extraordinarily difficult or objectionable in content, they might hang up during or immediately after the questions. Despite a few respondents expressing political opposition or personal distaste for the questions, no respondent hung up during the questions or immediately after them. Even the most hostile respondents either answered the question or refused to answer and continued with the interview.

6 Identification rates are calculated out of substantive responses, i.e. excluding don't know, refusal, and other not ascertained.

7 For two-step measures, the item nonresponse rates reflect an aggregation across all component questions in each measure.

30.3.4 Synthesizing Across Evidence Sources

While every scientifically-oriented pretester wants solid data to support their decisions, having too much data or conflicting evidence can be just as problematic as having none. In this case study, evidence for and against each GI measure came in various forms and at different time points. The group listening sessions were the first piece of evidence, followed by item nonresponse and transgender identity rates, and finally question-answer duration. Table 30.1 shows how the pretesting team weighed the various pieces of evidence and adapted its decision

Table 30.1 Decision tracking for gender identity question choice during pretest.

Step	Evidence source	Evidence summary	Tentative decision	Decision modification
1	Group listening	*Two-Step Version 1* sounds more natural and appears to go more quickly for both interviews and respondents	*Two-Step Version 1*	N/A
2	Item nonresponse	Highest rates on *One-Step Version 2* (2.3%), and *two-Step version2* (1.7%) Both *One-Step Version 1* (0.3%) and *Two-Step Version 1* (0.7%) are very low	*One-Step Version 1* or *Two-Step Version 1*	Consider both *One-Step Version 1* and *Two-Step Version 1*
3	Transgender identification rates	No significant differences, although *Two-Step Version 2 slightly higher* (0.4% v. 0.3% for all other versions)	No guidance from this evidence. All versions equal.	None
4	Question-answer duration	Tie for quickest between *Two-Step Version 1* and *One-Step Version 2*, but distributions show tighter range around *two-Step version one.* With larger sample size, this would probably be the quickest.	*Two-Step Version 1*	*Two-Step Version 1*

over the course of the pretest. The process was essentially a "cognitive Bayesian" approach by which initial decisions were modified as more data became available.

30.4 Discussion

This chapter presented a real-world question pretest where multiple methods were used in an integrated and dependent manner, such that the pretesting protocol changed as the pretest progressed. The pretest began by simply listening to recorded interviews and developing plans to behavior code. When behavior coding was found to be unproductive, it was abandoned to focus on question-answer duration as a metric of both the overall time required to administer the question, and the combined interviewer and respondent difficulty administering it. During the decision process, GI and item nonresponse rates were used as measures of question yield, with the latter also being a measure of question difficulty and sensitivity.

In this case study, dependent pretesting produced results that the original pretest plan could have easily missed, specifically the duration finding. While duration was essentially being recorded during the short behavior coding phase (i.e. question start time was recorded, and each digital recording ended at the end of the GI measure), the findings we observed could have easily been lost in crowd of other outcomes if we had chosen to continue behavior coding and analyze those data. The mid-test refocus helped bring this finding to the fore. Also of note is that using additional quantitative measures (duration, item nonresponse, and transgender identification) did not lead to a different decision than what it seems the team would have made from listening alone. While the quantitative evidence certainly supported the decision, it could also be seen as superfluous in this case. Overall, however, this case study demonstrates how multiple evaluation metrics and evidence sources can be used to asses question fit, and how methodologists can adapt pretesting methods to specific contexts. The authors encourage other pretesters to try similar adaptive methods and decisions processes to see if they arrive at similar conclusions.

30.4.1 Lessons Learned and Guidance for Future Pretests

30.4.1.1 Lesson 1: Combining Methods (e.g. Randomized Experimentation, Group Listening, Behavior Coding, and Response Distribution Analysis) Creates a Robust Pretest

The strength of this pretest was its multimethod approach. At its core was a randomized experiment, which allowed comparison of response distributions and other metrics across question versions. Unstructured group listening was a helpful addition to this, allowing all team members to hear how the questions were working. The group listening was a notable aspect, in that it was relatively unstructured. There was no coding, and listeners just reflected on what they heard, while taking whatever level of notes was helpful for them. Through this listening process, the team also determined which quantitative metrics to capture in coding sessions. Although traditional behavior coding was initially a core component of the pretest, it became evident that it was unnecessary due to the extremely low rates of problem behaviors and other indicators of confusion from both interviewers and respondents. If behavior coding been used exclusively, little actionable information would have been gained from the effort. Without the transgender identification rates and item nonresponse rates, it would have been difficult to see the eventual product of the measures. Ultimately, question-answer durations and response distributions were the quantitative metrics that informed the pretest conclusions most.

30.4.1.2 Lesson 2: Pretesting Objectives Should Be Balanced Against Resources and Project Goals, and Be Flexible Enough to Change Midstream

The pretest was designed to be extremely cost-efficient, meaning completed with low-effort and quickly. The goal of the coding component was to provide quantitative support for the observations from the group listening sessions and to identify problematic behaviors not heard in those sessions. However, because there was very little problematic verbal behavior, paralanguage, or other discourse markers to code, the team had to quickly reduce the elements coded. It would have been more efficient to reduce the coding scope earlier and skip traditional behavior coding altogether. This initial inefficiency arose because the group listening and behavior coding began at the same time. Practitioners are strongly encouraged to do significant qualitative listening before developing and applying a behavior coding protocol, designing forms, and other time-consuming tasks. This pretest started big and was simplified as it developed. We would recommend starting small and expanding. We recommend listening to 10–20 interactions first, as a group, to hear the most obvious and common issues, and design coding protocols around those.

30.4.1.3 Lesson 3: More Detailed Coding, or More in-Depth Pretesting Methods, Do Not Necessarily Lead to Different or Better Design Decisions

It is striking that the pretest's design conclusions were essentially arrived at from the group listening alone. While impossible to know this in advance, it suggests that just listening is sometimes sufficient to make a sound pretesting decision. While it can be tempting to choose a specific pretest method a priori, this can lead to expending effort that is not needed and may not provide any additional insight. At worst, it could provide misleading evidence. Practitioners are, again, encouraged to be judicious in the pretesting method selection and to conduct sufficient listening before putting effort into other methods.

30.4.1.4 Lesson 4: Truly Multimethod Pretesting Is Not as Complex as It May Sound, but Be Prepared to Use Expert Judgment to Resolve Differences

This pretest successfully combined quantitative (duration measures and response distributions) and qualitative information (group listening observations) to decide which GI measure to carry forward in CHIS. This decision would have been much less grounded had any piece of the pretesting evidence been missing (even though the same decisions may have resulted without the quantitative evidence). The qualitative component of this pretest was not particularly sophisticated but was sufficient for the pretest goals. There was one situation in which qualitative and quantitative results were incongruent. Looking at response distributions alone suggested that One-Step Version 1 was the best measure because it had the lowest item nonresponse rates and a transgender identification rate equal to the other versions. However, the qualitative group listening, bolstered by the duration results, pushed the decision toward Two-Step Version 1.

30.4.1.5 Lesson 5: The Importance of a Strong Questionnaire Development Team

This pretest was a highly collaborative effort between CHIS and Williams Institute staff at UCLA, with strong support from Westat. The Williams Institute has published extensive guidelines on best practices for measuring GI, and interested readers should review those in detail (e.g. GenIUSS Group 2014). Pretests should always include close collaborations between topic experts, survey methodologists, and survey practitioners, facilitated with techniques such as group listening sessions. Many hands certainly make for lighter work.

Acknowledgments

The authors thank Sherman Edwards, Susan Fraser, and Denise Buckley from Westat for scheduling and running the listening sessions, and for their commitment to CHIS data quality. Gary Gates were also instrumental in developing and funding of this project. Finally, we sincerely thank the Arcus Foundation, The Bohnett Foundation, Ford Foundation, The Gil Foundation, and Mr. Weston Milliken for funding this pilot test. A complete list of CHIS funders can be found at http://healthpolicy.ucla.edu/chis/about/Pages/funds.aspx.

References

Bargh, J.A. and Chartrand, T.L. (2000). The mind in the middle. In: *Handbook of Research Methods in Social and Personality Psychology* (eds. H.T. Reis and C.M. Judd), 253–279. Cambridge, UK: Cambridge University Press.

Bassili, J.N. and Fletcher, J.F. (1991). Response-time measurement in survey research: A method for CATI and a new look at nonattitudes. *Public Opinion Quarterly* 55 (3): 331–346.

Beatty, P. and Herrmann, D. (2002). To answer or not to answer: Decision processes related to survey item nonresponse. In: *Survey Nonresponse* (eds. R.M. Groves, D.A. Dillman, J.L. Eltinge and R.J.A. Little), 71–85. Hoboken, NJ: Wiley.

Cannell, C.F., Lawson, S.A., and Hausser, D.L. (1975). *A Technique for Evaluating Interviewer Performance: A Manual for Coding and Analyzing Interviewer Behavior from Tape Recordings of Household Interviews.* Survey Research Center, Institute for Social Research, University of Michigan.

Centers for Disease Control and Prevention. (2017). BRFSS Questionnaires. Centers for Disease Control and Prevention. https://www.cdc.gov/brfss/questionnaires/index.htm.

Conrad, F.G., Schober, M., and Dijkstra, W. (2008). Cues of communication difficulty in telephone interviews. In: *Advances in Telephone Survey Methodology* (eds. J.M. Lepkowski, C. Tucker, J.M. Brick, et al.), 212–230. New York: Wiley.

Conron, K.J., Scott, G., Stowell, G.S., and Landers, S.J. (2012). Transgender health in Massachusetts: Results from a household probability sample of adults. *American Journal of Public Health* 102 (1): 118–122.

Couper, M.P. (2000). Usability evaluation of computer-assisted survey instruments. *Social Science Computer Review* 18 (4): 384–396.

Dean, E., Caspar, R., McAvinchey, G. et al. (2007). Developing a low-cost technique for parallel cross-cultural instrument development: The question appraisal system (QAS-04). *International Journal of Social Research Methodology* 10 (3): 227–241. https://doi.org/10.1080/13645570701401032.

Draisma, S. and Dijkstra, W. (2004). Response latency and (para) linguistic expressions as indicators of response error. In: *Methods for Testing and Evaluating Survey Questionnaires* (eds. S. Presser, J.M. Rothgeb, M.P. Couper, et al.), 131–148. Hoboken, N.J: Wiley.

Ehlen, P., Schober, M.F., and Conrad, F.G. (2007). Modeling speech disfluency to predict conceptual misalignment in speech survey interfaces. *Discourse Processes* 44 (3): 245–265.

Fowler, F.J. and Cannell, C.F. (1996). Using behavioral coding to identify cognitive problems with survey questions. In: *Answering Questions: Methodology for Determining Cognitive and Communicative Processes in Survey Research* (eds. N. Schwarz and S. Sudman), 15–36. San Francisco, CA: Jossey-Bass.

Gates, G.J. (2011). *How Many People Are Lesbian, Gay, Bisexual, and Transgender?* Los Angeles: The Williams Institute, The UCLA School of Law.

GenIUSS Group. (2014). Best practices for asking questions to identify transgender and other gender minority respondents on population-based surveys. http://williamsinstitute.law.ucla.edu/research/census-lgbt-demographics-studies/geniuss-report-sept-2014.

Grant, D., Jans, M., Park, R. et al. (2015). Putting the "T" in LBGT: A transgender question pilot test in the California Health Interview Survey. In: *Proceedings of the Survey Research Methods Section*. American Statistical Association (AAPOR) http://www.asasrms.org/Proceedings/y2015/files/234234.pdf (Accessed 6/3/19).

Jans, M. (2010). Verbal paradata and survey error: Respondent speech, voice, and question-answering behavior can predict income item nonresponse. http://deepblue.lib.umich.edu/handle/2027.42/75932.

Jans, M., Wilson, B.D.M., and Herman, J.L. (2018). Measuring Aspects of Sexuality and Gender: A Sexual Human Rights Challenge for Science and Official Statistics. *CHANCE* 31 (1): 12–20. http://dx.doi.org/10.1080/09332480.2018.1438704.

Olson, K. and Smyth, J.D. (2015). The effect of CATI questions, respondents, and interviewers on response time. *Journal of Survey Statistics and Methodology* 3 (3): 361–396. https://doi.org/10.1093/jssam/smv021.

Olyslager, F. and Conway, L. (2007). On the calculation of the prevalence of transsexualism. In: *World Professional Association for Transgender Health 20th International Symposium, Chicago, Illinois*, vol. 22, 2010. http://citeseerx.ist.psu.edu/viewdoc/download?doi=10.1.1.692.8704&rep=rep1&type=pdf

Ongena, Y. and Dijkstra, W. (2006). Methods of behavior coding of survey interviews. *Journal of Official Statistics* 22 (3): 419–451.

Reed, B., Rhodes, S., Schofield, P. et al. (2009). Gender variance in the UK: Prevalence, incidence, growth and geographic distribution. Report of the Gender Identity Research and Education Society. https://www.gires.org.uk/wp-content/uploads/2014/10/GenderVarianceUK-report.pdf

Willis, G.B., and Lessler, J.T. (1999). Question appraisal system QAS-99. Rockville, MD: Research Triangle Institute. https://www.researchgate.net/profile/Gordon_Willis/publication/267938670_Question_Appraisal_System_QAS-99_By/links/54b7b26a0cf2e68eb2803f6a.pdf (Accessed 8/14/19).

Yan, T. and Tourangeau, R. (2008). Fast times and easy questions: The effects of age, experience and question complexity on web survey response times. *Applied Cognitive Psychology* 22 (1): 51–68. http://doi.org/10.1002/acp.1331.

Index

Advances in Questionnaire Design, Development, Evaluation and Testing, First Edition.
Edited by Paul C. Beatty, Debbie Collins, Lyn Kaye, Jose-Luis Padilla, Gordon B. Willis, and Amanda Wilmot.

q